# Essential Statistics: Exploring the World through Data

Robert Gould

**University of California, Los Angeles**

Colleen Ryan

**California Lutheran University**

**PEARSON**

Boston   Columbus   Indianapolis   New York   San Francisco   Upper
Amsterdam   Cape Town   Dubai   London   Madrid   Milan   Munich   Pa
Delhi   Mexico City   Sao Paulo   Sydney   Hong Kong   Seoul   Singap

D1410033

Editor in Chief: Deirdre Lynch
Acquisitions Editor: Marianne Stepanian
Sponsoring Editor: Christina Lepre
Executive Market Development Manager: Dona Kenly
Executive Director of Development: Carol Trueheart
Editorial Assistant: Sonia Ashraf
Senior Managing Editor: Karen Wernholm
Associate Managing Editor: Tamela Ambush
Senior Production Supervisor: Peggy McMahon
Associate Director of Design, USHE EMSS/HSC/EDU: Andrea Nix
Art Director/Cover Designer: Beth Paquin
Marketing Manager: Erin Lane
Marketing Assistant: Kathleen DeChavez
Senior Author Support/Technology Specialist: Joe Vetere
Executive Manager, Course Production: Peter Silvia
Procurement Manager: Evelyn Beaton
Procurement Specialist: Debbie Rossi
Media Procurement Specialist: Ginny Michaud
Design, Production Coordination, Technical Illustrations, and Composition: Cenveo Publisher
Services/Nesbitt Graphics, Inc.
Cover photo: Boardwalk along the salt marsh, Kouchibouguac National Park, New Brunswick,
Canada by Darwin Wiggett/All Canada Photos/Getty Images

Credits and acknowledgments borrowed from other sources and reproduced, with permission, in this
text appear on page C-1.

Portions of information contained in this publication/book are printed with permission of Minitab Inc.
All such material remains the exclusive property and copyright of Minitab Inc. All rights reserved.

Microsoft® and Windows® are registered trademarks of the Microsoft Corporation in the U.S.A. and
other countries. Screen shots and icons reprinted with permission from the Microsoft Corporation.
This book is not sponsored or endorsed by or affiliated with the Microsoft Corporation.

Many of the designations used by manufacturers and sellers to distinguish their products are claimed
as trademarks. Where those designations appear in this book, and Pearson was aware of a trademark
claim, the designations have been printed in initial caps or all caps.

The student edition of this book has been cataloged by the Library of Congress as follows:
**Library of Congress Cataloging-in-Publication Data**
Gould, Robert, 1965-
   Essential statistics / Robert Gould, Colleen N. Ryan. — 1st ed.
     p. cm.
   Includes bibliographical references.
   ISBN 978-0-321-83698-4
1. Mathematical statistics. 2. Statistics. I. Ryan, Colleen N.
(Colleen Nooter), 1939- II. Title.
QA276.12.G6867 2014
519.5—dc23

                                    2012010221

1 2 3 4 5 6 7 8 9 10—CRK—15 14 13 12

**PEARSON**

ISBN-10: 0-321-83698-7
ISBN-13: 978-0-321-83698-4

# *Dedication*

To my parents and family, my friends, and my colleagues who are also friends. Without their patience and support, this book would not have been possible.

—Rob

To my teachers and students, and to my family who have helped me in so many different ways.

—Colleen

# About the Authors

## Robert Gould

**Robert L. Gould** (Ph.D., University of California, San Diego) is a leader in the statistics education community. He has served as chair of the American Statistical Association's Committee on Teacher Enhancement, has served as chair of the ASA's Statistics Education Section, and served on a panel of co-authors for the *Guidelines for Assessment in Instruction on Statistics Education (GAISE) College Report*. As the associate director of professional development for CAUSE (Consortium for the Advancement of Undergraduate Statistics Education), Rob has worked closely with the American Mathematical Association of Two-Year Colleges (AMATYC) to provide traveling workshops and summer institutes in statistics. For over ten years, he has served as Vice-Chair of Undergraduate Studies at the UCLA Department of Statistics, and he is director of the UCLA Center for the Teaching of Statistics. In 2009, Rob was elected president of the Southern California Chapter of the American Statistical Association. Rob was selected 2012 ASA Fellow for innovative, far-reaching contributions to statistics education.

In his free time, Rob plays the cello, runs, and is an ardent reader of fiction.

## Colleen Ryan

**Colleen N. Ryan** has taught statistics, chemistry, and physics to diverse community college students for decades. She taught at Oxnard College from 1975 to 2006, where she earned the Teacher of the Year Award. Colleen currently teaches statistics part-time at California Lutheran University. She often designs her own lab activities. Her passion is to discover new ways to make statistical theory practical, easy to understand, and sometimes even fun.

Colleen earned a B.A. in physics from Wellesley College, an M.A.T. in physics from Harvard University, and an M.A. in chemistry from Wellesley College. Her first exposure to statistics was with Frederick Mosteller at Harvard.

In her spare time, Colleen sings with the Oaks Camerata, has been an avid skier, and enjoys time with her family.

# Contents

# Preface

## About This Book

This is a book about data. Data are at the heart of the everyday life of modern citizens, and in particular modern students. Google, Facebook, smartphones, and highways are just a few examples of everyday objects that routinely collect, organize, and transmit data. Students who can think about and reason with data will have an advantage over those who cannot. To become "data literate," students need to develop quantitative literacy, understand how to handle computational technology, understand the role of variability in interpreting data, and learn to communicate clearly. This book teaches these skills.

We realize that students usually do not take this course voluntarily or with great enthusiasm. But we intend to convert them to enthusiasts by showing them that data are everywhere and that everyone can profit both intellectually and financially (and, dare we add, emotionally?) by achieving an understanding of basic statistical concepts. We take special care to accommodate students who have not had a mathematics course in a while and students who may not feel confident in their mathematical abilities. Wherever possible, we build on students' intuition in order to demonstrate that statistics is often just formalized common sense. However, we are sensitive to the fact that students' intuitions can lead to misconceptions, and we proceed carefully to correct common mistakes.

Students might not take another formal course in statistics, but this course is not the end of their relationship with data. Indeed, students will use the concepts and critical thinking skills they learn in this class throughout their lives. The number of careers that require statistical reasoning is growing rapidly—from English majors using digitized texts to analyze works of literature in a larger context, to community activists using statistics to understand patterns of community change. In this class, students will learn how they can use their new-found data skills to expand their world.

## Approach

Our text is concept-based, as opposed to method-based. We do teach useful statistical methods, but we emphasize that applying the method is secondary to understanding the concept.

In the real world, computers do most of the heavy lifting for statisticians. We therefore adopt an approach which frees the instructor from having to teach tedious procedures and leaves more time for teaching deeper understanding of concepts. Accordingly, we present formulas as an aid to understanding the concepts rather than as the focus of study.

We believe students need to learn how to

- determine which statistical procedures are appropriate,

- instruct the software to carry out the procedures, and

- interpret the output.

We understand that students will probably see only one type of statistical software in class. But we believe it is useful for students to compare output from several different sources, so in some examples we ask them to read output from two or more software packages.

One of the authors (Rob Gould) served on a panel of co-authors for the collegiate version of the American Statistical Association-endorsed *Guidelines for Assessment and Instruction in Statistics Education* (*GAISE*). We firmly believe in its main goals and have adopted them in the preparation of this book.

- We emphasize understanding over rote performance of procedures.
- We use real data whenever possible.
- We encourage the use of technology both to develop conceptual understanding and to analyze data.
- We believe strongly that students learn by doing. For this reason, the homework problems offer students both practice in basic procedures and challenges to build conceptual understanding.

## Coverage

The first few chapters of this book are concept-driven and cover exploratory data analysis and inferential statistics—fundamental concepts that every introductory statistics student should learn. The last part of the book builds on that strong conceptual foundation and is more methods-based.

Our ordering of topics is guided by the Cycle of Data.

*Chapters 1–4: Exploratory Data Analysis.* The first four chapters cover data collection and summary. Chapter 1 introduces the important topic of data collection and compares and contrasts observational studies with controlled experiments. This chapter also teaches students how to handle raw data so that the data can be uploaded to their statistical software. Chapters 2 and 3 discuss graphical and numerical summaries of single variables based on samples. We emphasize that the purpose is not just to produce a graph or a number but, instead, to explain what those graphs and numbers say about the world. Chapter 4 introduces simple linear regression and presents it as a technique for providing graphical and numerical summaries of relationships between two numerical variables.

We feel strongly that introducing regression early in the text is beneficial in building student understanding of the applicability of statistics to real-world scenarios. After completing the chapters covering data collection and summary, students have acquired the skills and sophistication they need to describe two-variable associations and to generate informal hypotheses. Two-variable associations provide a rich context for class discussion and allow the course to move from fabricated problems (because one-variable analyses are relatively rare in the real world) to real problems that appear frequently in everyday life.

*Chapters 5–8: Inference.* These chapters teach the fundamental concepts of statistical inference. The main idea is that our data mirror the real world, but imperfectly; although our estimates are uncertain, under the right conditions we can quantify our uncertainty. Verifying that these conditions exist and understanding what happens if they are not satisfied are important themes of these chapters.

*Chapters 9–10: Methods.* Here we return to the themes covered earlier in the text and present them in a new context by introducing additional statistical methods, such as estimating population means and analyzing categorical variables. We also provide guidance (in Section 10.3) for reading scientific literature, to offer students the experience of critically examining real scientific papers.

## Organization

Our preferred order of progressing through the text is reflected in the Contents, but there are some alternative pathways as well.

*10-week Quarter.* The first eight chapters would provide a full, one-quarter course in introductory statistics. If time remains, cover Sections 9.1 and 9.2 as well, so that students can solidify their understanding of confidence intervals and hypothesis tests by revisiting the topic with a new parameter.

*Proportions First.* Ask two statisticians, and you will get three opinions on whether it is best to teach means or proportions first. We have come down on the side of proportions for a variety of reasons. Proportions are much easier to find in popular news media (particularly around election time), so they can more readily be tied to students' everyday lives. Also, the mathematics and statistical theory is simpler; because there's no need to provide a separate estimate for the population standard deviation, inference is based on the Normal distribution, and no further approximations (that is, the *t*-distribution) are required. Hence, we can quickly get to the heart of the matter with fewer technical diversions.

The basic problem here is how to quantify the uncertainty involved in estimating a parameter and how to quantify the probability of making incorrect decisions when posing hypotheses. We cover these ideas in detail in the context of proportions. Students can then more easily learn how these same concepts are applied in the new context of means (and any other parameter they may need to estimate).

*Means First.* Conversely, many people feel that there is time for only one parameter and that this parameter should be the mean. For this alternative presentation, cover Chapters 6, 7, and 9, in that order. On this path, students learn about survey sampling and the terminology of inference (population vs. sample, parameter vs. statistic) and then tackle inference for the mean, including hypothesis testing.

To minimize the coverage of proportions, you might choose to cover Sections 6.1 and 6.2 of Chapter 6, Section 7.1 (which covers in detail the language and framework of statistical inference), and then Chapter 9. Chapters 7 and 8 develop the concepts of statistical inference more slowly than in Chapter 9, but essentially, Chapter 9 develops the same ideas in the context of the mean.

If you present Chapter 9 before Chapters 7 and 8, we recommend that you devote roughly twice as much time to Chapter 9 as you have devoted to previous chapters, because many challenging ideas are explored in this chapter. If you have already covered Chapters 7 and 8 thoroughly, Chapter 9 can be covered more quickly.

# Features

We've incorporated into this book variety of features to aid student learning and to facilitate the use of this text in any classroom.

## Integrating Technology

Modern statistics is inseparable from technology. We have worked to make this textbook accessible for any classroom, regardless of the level of in-class exposure to technology, while still remaining true to the demands of the analysis. We know that students sometimes do not have access to technology when doing homework, so many exercises provide output from software and ask students to interpret and critically evaluate that given output.

Using technology is important because it enables students to handle real data, and real data sets are often large and messy. The following features are designed to guide students.

- **TechTips** outline steps for performing calculations using TI-83/84® graphing calculators, Excel®, Minitab®, and StatCrunch®. We do not want students to get stuck because they don't know how to reproduce the results we show in the book, so whenever a new method or procedure is introduced, an icon, Tech, refers students to the TechTips section at the end of the chapter. Each set of TechTips contains at least one mini-example, so that students are not only learning to use the technology but also practicing data analysis and reinforcing ideas discussed in the text.

- **Check Your Tech** examples help students understand that statistical calculations done by technology do not happen in a vacuum and assure them that they can get the same numerical values by hand. Although we place a higher value on interpreting results and verifying conditions required to apply statistical models, the numerical values are important, too.

- The companion **CD-ROM** includes Technology Tutorial Videos for Excel, Minitab, and the TI83/84, as well as all data sets used in the exposition and exercises. These data are also available at http://www.pearsonhighered.com/mathstatsresources/.

## Guiding Students

- Each chapter opens with a **Theme**. Beginners have difficulty seeing the forest for the trees, so we use a theme to give an overview of the chapter content.

- Each chapter begins by posing a real-world **Case Study**. At the end of the chapter, we show how techniques covered in the chapter helped solve the problem presented in the Case Study.

- **Margin Notes** draw attention to details that enhance student learning and reading comprehension.

  **Caution** notes provide warnings about common mistakes or misconceptions.

  **Looking Back** reminders refer students to earlier coverage of a topic.

  **Details** clarify or expand on a concept.

- **Key Points** highlight essential concepts to draw special attention to them. Understanding these concepts is essential for progress.

- **Snapshots** break down key statistical concepts introduced in the chapter, quickly summarizing each concept or procedure and indicating when and how it should be used.

- An abundance of worked-out **examples** model solutions to real-world problems relevant to students' lives. Each example is tied to an end-of-chapter exercise so that students can practice solving a similar problem and test their understanding. Within the exercise sets, the icon TRY indicates which problems are tied to worked-out examples in that chapter, and the numbers of those examples are indicated.

- The **Chapter Review** that concludes each chapter provides a list of important new terms, student learning objectives, a summary of the concepts and methods discussed, and sources for data, articles, and graphics referred to in the chapter.

## Active Learning

- For each chapter we've included an activity, **Exploring Statistics**, that students are intended to do in class as a group. We have used these activities ourselves, and we have found that they greatly increase student understanding and keep students engaged in class.

- All exercises are located at the end of the chapter. **Section Exercises** are designed to begin with a few basic problems that strengthen recall and assess basic knowledge, followed by mid-level exercises that ask more complex, open-ended questions. **Chapter Review Exercises** provide a comprehensive review of material covered throughout the chapter.

  The exercises emphasize good statistical practice by requiring students to verify conditions, make suitable use of graphics, find numerical values, and interpret their findings in writing. All exercises are paired so that students can check their work on the odd-numbered exercise and then tackle the corresponding even-numbered exercise.

Challenging exercises, identified with an asterisk (*), ask open-ended questions and sometimes require students to perform a complete statistical analysis. For exercises marked with a ✳, accompanying data sets are available on the companion CD-ROM included with the book. The answers to all odd-numbered exercises appear in the back of the book.

- Most chapters include select exercises, marked with a $g$ within the exercise set, to indicate that problem-solving help is available in the **Guided Exercises** section. If students need support while doing homework, they can turn to the Guided Exercises to see a step-by-step approach to solving the problem.

# Acknowledgments

We are grateful for the attention and energy that a large number of people devoted to making this a better book. We extend our gratitude to Elaine Newman (Sonoma State University), Fred J. Rispoli (Dowling College), Delray Schultz (Millersville University), and Catalina Yang (Oxnard Community College), who checked the accuracy of this text and its many exercises. Thanks also to David Chelton, our developmental editor, to Carol Merrigan, who handled production, to Peggy McMahon, senior production project manager, and to Connie Day, our copyeditor. Many thanks to John Norbutas for his technical advice and help with the TechTips. We thank Deirdre Lynch, editor-in-chief, for signing us up and sticking with us, and we are grateful to Dona Kenly for her market development efforts. Special thanks to our acquisitions editor, Marianne Stepanian, whose patience and good advice made it all possible, and to our sponsoring editor, Christina Lepre, who kept us on task and never steered us wrong.

We would also like to extend our sincere thanks for the suggestions and contributions made by the following reviewers, class testers, and focus group attendees.

Arun Agarwal, *Grambling State University*

Anne Albert, *University of Findlay*

Michael Allen, *Glendale Community College*

Eugene Allevato, *Woodbury University*

Dr. Jerry Allison, *Trident Technical College*

Polly Amstutz, *University of Nebraska*

Patricia Anderson, *Southern Adventist University*

MaryAnne Anthony-Smith, *Santa Ana College*

David C. Ashley, *Florida State College at Jacksonville*

Diana Asmus, *Greenville Technical College*

Kathy Autrey, *Northwestern State University of Louisiana*

Wayne Barber, *Chemeketa Community College*

Roxane Barrows, *Hocking College*

Jennifer Beineke, *Western New England College*

Diane Benner, *Harrisburg Area Community College*

Norma Biscula, *University of Maine, Augusta*

KB Boomer, *Bucknell University*

David Bosworth, *Hutchinson Community College*

Diana Boyette, *Seminole Community College*

Elizabeth Paulus Brown, *Waukesha County Technical College*

Leslie Buck, *Suffolk Community College*

RB Campbell, *University of Northern Iowa*

Stephanie Campbell, *Mineral Area College*

Anne Cannon, *Cornell College*

Carolyn Chapel, *Western Technical College*

Christine Cole, *Moorpark College*

Linda Brant Collins, *University of Chicago*

James A. Condor, *Manatee Community College*

Carolyn Cuff, *Westminster College*

Phyllis Curtiss, *Grand Valley State University*

Monica Dabos, *University of California, Santa Barbara*

Greg Davis, *University of Wisconsin, Green Bay*

Bob Denton, *Orange Coast College*

Julie DePree, *University of New Mexico–Valencia*

Jill DeWitt, *Baker Community College of Muskegon*

Paul Drelles, *West Shore Community College*

Rob Eby, *Blinn College*

Nancy Eschen, *Florida Community College at Jacksonville*

Karen Estes, *St. Petersburg College*

Mariah Evans, *University of Nevada, Reno*

Harshini Fernando, *Purdue University North Central*

Stephanie Fitchett, *University of Northern Colorado*

Elaine B. Fitt, *Bucks County Community College*

Michael Flesch, *Metropolitan Community College*

Melinda Fox, *Ivy Tech Community College, Fairbanks*

Joshua Francis, *Defiance College*

Michael Frankel, *Kennesaw State University*

Heather Gamber, *Lone Star College*

Debbie Garrison, *Valencia Community College, East Campus*

Kim Gilbert, *University of Georgia*

Stephen Gold, *Cypress College*

Nick Gomersall, *Luther College*

Ken Grace, *Anoka Ramsey Community College*

Larry Green, *Lake Tahoe Community College*

Jeffrey Grell, *Baltimore City Community College*

Albert Groccia, *Valencia Community College, Osceola Campus*

David Gurney, *Southeastern Louisiana University*

Chris Hakenkamp, *University of Maryland, College Park*

Melodie Hallet, *San Diego State University*

Donnie Hallstone, *Green River Community College*

Cecil Hallum, *Sam Houston State University*

Josephine Hamer, *Western Connecticut State University*

Mark Harbison, *Sacramento City College*

Beverly J. Hartter, *Oklahoma Wesleyan University*

Laura Heath, *Palm Beach State College*

Greg Henderson, *Hillsborough Community College*

Susan Herring, *Sonoma State University*

Carla Hill, *Marist College*

Michael Huber, *Muhlenberg College*

Kelly Jackson, *Camden County College*

Bridgette Jacob, *Onondaga Community College*

Chun Jin, *Central Connecticut State University*

Maryann Justinger, Ed.D., *Erie Community College*

Joseph Karnowski, *Norwalk Community College*

Susitha Karunaratne, *Purdue University North Central*

Mohammed Kazemi, *University of North Carolina–Charlotte*

Robert Keller, *Loras College*

Omar Keshk, *Ohio State University*

Raja Khoury, *Collin County Community College*

Brianna Killian, *Daytona State College*

Yoon G. Kim, *Humboldt State University*

Greg Knofczynski, *Armstrong Atlantic University*

Erica Kwiatkowski-Egizio, *Joliet Junior College*

Sister Jean A. *Lanahan, OP, Molloy College*

Katie Larkin, *Lake Tahoe Community College*

Michael LaValle, *Rochester Community College*

Deann Leoni, *Edmonds Community College*

Lenore Lerer, *Bergen Community College*

Quan Li, *Texas A&M University*

Doug Mace, *Kirtland Community College*

Walter H. Mackey, *Owens Community College*

Keith McCoy, *Wilbur Wright College*

Elaine McDonald-Newman, *Sonoma State University*

William McGregor, *Rockland Community College*

Bill Meisel, *Florida State College at Jacksonville*

Wendy Miao, *El Camino College*

Ashod Minasian, *El Camino College*

Megan Mocko, *University of Florida*

Sumona Mondal, *Clarkson University*

Kathy Mowers, *Owensboro Community and Technical College*

Mary Moyinhan, *Cape Cod Community College*

Junalyn Navarra-Madsen, *Texas Woman's University*

Stacey O. Nicholls, *Anne Arundel Community College*

Helen Noble, *San Diego State University*

Lyn Noble, *Florida State College at Jacksonville*

Keith Oberlander, *Pasadena City College*

Pamela Omer, *Western New England College*

Nabendu Pal, *University of Louisiana at Lafayette*

Irene Palacios, *Grossmont College*

Adam Pennell, *Greensboro College*

Joseph Pick, *Palm Beach State College*

Philip Pickering, *Genesee Community College*

Robin Powell, *Greenville Technical College*

Nicholas Pritchard, *Coastal Carolina University*

Linda Quinn, *Cleveland State University*

William Radulovich, *Florida State College at Jacksonville*

Fred J. Rispoli, *Dowling College*

Nancy Rivers, *Wake Technical Community College*

Corlis Robe, *East Tennesee State University*

Thomas Roe, *South Dakota State University*

Dan Rowe, *Heartland Community College*

Carol Saltsgaver, *University of Illinois–Springfield*

Radha Sankaran, *Passaic County Community College*

Delray Schultz, *Millersville University*

Jenny Shook, *Pennsylvania State University*

Danya Smithers, *Northeast State Technical Community College*

Larry Southard, *Florida Gulf Coast University*

Dianna J. Spence, *North Georgia College & State University*

René Sporer, *Diablo Valley College*

Jeganathan Sriskandarajah, *Madison Area Technical College–Traux*

David Stewart, *Community College of Baltimore County–Cantonsville*

Linda Strauss, *Penn State University*

John Stroyls, *Georgia Southwestern State University*

Joseph Sukta, *Moraine Valley Community College*

Lori Thomas, *Midland College*

Malissa Trent, *Northeast State Technical Community College*

Ruth Trygstad, *Salt Lake Community College*

Gail Tudor, *Husson University*

Manuel T. Uy, *College of Alameda*

Lewis Van Brackle, *Kennesaw State University*

Mahbobeh Vezvaei, *Kent State University*

Joseph Villalobos, *El Camino College*

Barbara Wainwright, *Sailsbury University*

Henry Wakhungu, *Indiana University*

Dottie Walton, *Cuyahoga Community College*

Jen-ting Wang, *SUNY, Oneonta*
Jane West, *Trident Technical College*
Michelle White, *Terra Community College*
Bonnie-Lou Wicklund, *Mount Wachusett Community College*
Sandra Williams, *Front Range Community College*
Rebecca Wong, *West Valley College*
Alan Worley, *South Plains College*

Jane-Marie Wright, *Suffolk Community College*
Haishen Yao, *CUNY, Queensborough Community College*
Lynda Zenati, *Robert Morris Community College*
Yan Zheng-Araujo, *Springfield Community Technical College*
Cathleen Zucco-Teveloff, *Rider University*
Mark A. Zuiker, *Minnesota State University, Mankato*

# Supplements

## Student Resources

*Student Solutions Manual*, by Robert Keller (Loras College), provides detailed, worked-out solutions to all odd-numbered text exercises. (ISBN-13: 978-0-321-83825-4; ISBN-10: 0-321-83825-4).

*Study Cards for Statistics Software* This series of study cards, available for Excel®, Minitab®, JMP®, SPSS®, R, StatCrunch®, and the TI-83/84 graphing calculators, provides students with easy, step-by-step guides to the most common statistics software. Visit www.myPearsonStore .com for more information.

## Instructor Resources

*Instructor's Edition* contains answers to all text exercises, as well as a set of Instructor Resource Pages that offer chapter-by-chapter teaching suggestions and commentary. (ISBN-13: 978-0-321-83821-6; ISBN-10: 0-321-83821-1)

*Instructor's Solutions Manual*, by Robert Keller (Loras College), contains worked-out solutions to all the text exercises. (ISBN-13: 978-0-32183839-1; ISBN-10: 0-321-83839-4)

*Online Test Bank* (download only), by Jill DeWitt (Baker College of Muskegon), includes three sets of tests for each chapter. The Test Bank is available for download at www .pearsonhighered.com/irc. (ISBN-13: 978-0-321-75639-8; ISBN-10: 0-321-75639-8)

*Instructor Podcasts* are brief audio podcasts from author Robert Gould that focus on the key points of each chapter, helping both new and experienced instructors prepare for class. Available in MyStatLab or at www.pearsonhighered .com/irc. (ISBN-13: 978-0-321-82423-3; ISBN-10: 0-321-82423-7)

*PowerPoint® Lecture Slides,* by Larry Green (Lake Tahoe Community College) provide an outline to use in a lecture setting, presenting definitions, figures, Guided Exercises, Case Studies, and Snapshots from the text. These slides are available within MyStatLab or at www.pearsonhighered .com/irc. (ISBN-13: 978-0-321-88172-4; ISBN-10: 0-321-88172-9)

*Active Learning Questions* Prepared in PowerPoint®, these questions are intended for use with classroom response systems. Several multiple-choice questions are available for each chapter of the book, enabling instructors to assess mastery of material quickly in class. The Active Learning Questions are available to download from within MyStatLab™ or at www .pearsonhighered.com/irc.

## Technology Resources

On the Companion CD-ROM

- **Data sets** from the book are provided in multiple formats.
- **Technology Tutorial Videos.** These brief video clips walk students through common statistical procedures for Minitab, Excel, and the TI-83/84 graphing calculator. They are also available within MyStatLab.

### MyStatLab™ Online Course (access code required)

MyStatLab is a course management systems that delivers **proven results** in helping individual students succeed.

- MyStatLab can be successfully implemented in any environment–lab-based, hybrid, fully online, traditional–and demonstrates the quantifiable difference that integrated usage has on student retention, subsequent success, and overall achievement.

- MyStatLab's comprehensive online gradebook automatically tracks students' results on tests, quizzes, homework, and in the study plan. Instructors can use the gradebook to provide positive feedback or intervene if students have trouble. Gradebook data can be easily exported to a variety of spreadsheet programs, such as Microsoft Excel.

MyStatLab provides **engaging experiences** that personalize, stimulate, and measure learning for each student. In addition to the resources below, each course includes a full interactive online version of the accompanying textbook.

- **Tutorial Exercises with Multimedia Learning Aids:** The homework and practice exercises in MyStatLab align with the exercises in the textbook, and they regenerate algorithmically to give students unlimited opportunity for practice and mastery. Exercises offer immediate helpful feedback, guided solutions, sample problems, animations, videos, and eText clips for extra help at point-of-use.

- **StatTalk Videos:** *24 Conceptual Videos to Help You Actually Understand Statistics*. Fun-loving statistician Andrew Vickers takes to the streets of Brooklyn, NY, to demonstrate important statistical concepts through interesting stories and real-life events. These fun and engaging videos will help students actually understand statistical concepts. They are available with an instructor's user guide and assessment questions.

- **Getting Ready for Statistics:** A library of questions now appears within each MyStatLab course to offer the developmental math topics students need for the course. These can be assigned as a prerequisite to other assignments, if desired.

- **Conceptual Question Library:** In addition to algorithmically regenerated questions that are aligned with your textbook, there is a library of 1,000 Conceptual Questions available in the assessment manager that require students to apply their statistical understanding.

- **StatCrunch:** MyStatLab integrates the web-based statistical software, StatCrunch, within the online assessment platform so that students can easily analyze data sets from exercises and the text. In addition, MyStatLab includes access to **www.StatCrunch.com,** a web site where users can access more than 14,000 shared data sets, conduct online surveys, perform complex analyses using the powerful statistical software, and generate compelling reports.

- **Statistical Software Support:** Knowing that students often use external statistical software, we make it easy to copy our data sets, both from the ebook and the MyStatLab questions, into software such as StatCrunch, Minitab, Excel, and more. Students have access to a variety of support tools—Technology Instruction Videos, Technology Study Cards, and Manuals for select titles—to learn how to effectively use statistical software.

- **Expert Tutoring:** Although many students describe the whole of MyStatLab as "like having your own personal tutor," students also have access to live tutoring from Pearson. Qualified statistics instructors provide tutoring sessions for students via MyStatLab.

And, MyStatLab comes from a **trusted partner** with educational expertise and an eye on the future.

Knowing that you are using a Pearson product means knowing that you are using quality content. That means that our eTexts are accurate, that our assessment tools work, and that our questions are error-free. And whether you are just getting started with MyStatLab, or have a question along the way, we're here to help you learn about our technologies and how to incorporate them into your course.

To learn more about how MyStatLab combines proven learning applications with powerful assessment, visit **www.mystatlab.com** or contact your Pearson representative.

## MyStatLab™Plus

MyLabsPlus combines effective teaching and learning materials from MyMathLab® and MyStatLab™ with convenient management tools and a dedicated services team.

Designed to support growing math and statistics programs, it includes additional features, such as

- **Batch Enrollment:** Your school can create the login name and password for every student and instructor, so everyone can be ready to start class on the first day. Automation of this process is also possible through integration with your school's Student Information System.

- **Login from Your Campus Portal:** You and your students can link directly from your campus portal into your MyLabsPlus courses. A Pearson service team works with your institution to create a single sign-on experience for instructors and students.

- **Advanced Reporting:** MyLabsPlus's advanced reporting enables instructors to review and analyze students' strengths and weaknesses by tracking their performance on tests, assignments, and tutorials. Administrators can review grades and assignments across all courses on your MyLabsPlus campus for a broad overview of program performance.

- **24/7 Support:** Students and instructors receive 24/7 support, 365 days a year, by phone, email, or online chat.

MyLabsPlus is available to qualified adopters. For more information, visit our website at **www.mylabsplus.com** or contact your Pearson representative.

## MathXL® for Statistics Online Course (access code required)

MathXL® is the homework and assessment engine that runs MyStatLab. (MyStatLab is MathXL plus a learning management system.) With MathXL for Statistics, instructors can:

- Create, edit, and assign online homework and tests using algorithmically generated exercises correlated at the objective level to the textbook.

- Create and assign their own online exercises and import TestGen tests for added flexibility.

- Maintain records of all student work, tracked in MathXL's online gradebook.

With MathXL for Statistics, students can.

- Take chapter tests in MathXL and receive personalized study plans and/or personalized homework assignments based on their test results.

- Use the study plan and/or the homework to link directly to tutorial exercises for the objectives they need to study.

- Students can also access supplemental animations and video clips directly from selected exercises.

- Knowing that students often use external statistical software, we make it easy to copy our data sets, both from the ebook and the MyStatLab questions, into software like StatCrunch, Minitab, Excel and more.

MathXL for Statistics is available to qualified adopters. For more information, visit our website at **www.mathxl .com,** or contact your Pearson representative.

## StatCrunch™

Powerful web-based statistical software that allows users to perform complex analyses, share data sets, and generate compelling reports of their data. The vibrant online community offers more than 14,500 data sets for students to analyze.

- **Collect**. Users can upload their own data to StatCrunch or search a large library of publicly shared data sets, spanning almost any topic of interest. Also, an online survey tool allows users to quickly collect data via web-based surveys.

- **Crunch**. A full range of numerical and graphical methods allow users to analyze and gain insights from any data set. Interactive graphics help users understand statistical concepts, and are available for export to enrich reports with visual representations of data.

- **Communicate**. Reporting options help users create a wide variety of visually-appealing representations of their data.

Full access to StatCrunch is available with a MyStatLab kit, and StatCrunch is available by itself to qualified adopters. **StatCrunch Mobile** is now available to access from your mobile device. For more information, visit our website at www.statcrunch .com or contact your Pearson representative.

**TestGen**® (www.pearsoned.com/testgen) enables instructors to build, edit, print, and administer tests using a computerized bank of questions developed to cover all the student learning objectives of the text. TestGen is algorithmically based, and instructors can create multiple but equivalent versions of the same question or test with the click of a button. Instructors can also modify Test Bank questions or add new questions. The software and Test Bank are available for download from Pearson Education's online catalog.

**The Student Edition of MINITAB** is a condensed version of the Professional release of MINITAB statistical software. It offers the full range of statistical methods and graphical capabilities, along with worksheets that can include up to 10,000 data points. Individual copies of the software can be bundled with the text (ISBN-13: 978-0-321-11313-9; ISBN-10: 0-321-11313-6) (CD only)

**JMP Student Edition** is an easy-to-use, streamlined version of JMP desktop statistical discovery software from SAS Institute, Inc. and is available for bundling with the text. (ISBN-13: 978-0-321-89164-8; ISBN-10: 0-321-89164-3)

**XLSTAT™ for Pearson.** Used by leading businesses and universities, XLSTAT is an Excel® add-in that offers a wide variety of functions to enhance the analytical capabilities of Microsoft Excel, making it the ideal tool for your everyday data analysis and statistics requirements. XLSTAT is compatible with all Excel versions (except Mac 2008) and is available for bundling with the text. (ISBN-13: 978-0-321-75932-0; ISBN-10: 0-321-75932-X)

# Index of Applications

# Essential Statistics: Exploring the World through Data

# 1 Introduction to Data

## THEME

Statistics is the science of data, so we must learn the types of data we will encounter and the methods for collecting data. The method used to collect data is very important because it determines the types of conclusions we can reach and, as you'll learn in later chapters, the types of analyses we can do. By organizing the data we've collected, we can often spot patterns that are not otherwise obvious.

This book will teach you to examine data to better understand the world around you. If you know how to sift data to find patterns, can communicate the results clearly, and understand whether you can generalize your results to other groups and contexts, you will be able to make better decisions, offer more convincing arguments, and learn things you did not know before. Data are everywhere these days, and making effective use of them is such a crucial task that one prominent economist has proclaimed statistics one of the most important professions of the decade (*McKinsley Quarterly* 2009).

The use of statistics to make decisions and convince others to take action is not new. Some statisticians date the current practice of statistics back to the mid-nineteenth century. One famous example occurred in 1854, when the British were fighting the Russians in the brutal Crimean War. A British newspaper had criticized the military medical facilities, and a young but well-connected nurse, Florence Nightingale, was appointed to study the situation and, if possible, improve it.

Nightingale carefully recorded the numbers of deaths, the causes of the deaths, and the times and dates of the deaths. She organized these data graphically, and these graphs enabled her to see a very important pattern: A large percentage of deaths were due to contagious disease, and many deaths could be prevented by improving sanitary conditions. Within six months, Nightingale had reduced the death rate by half. Eventually she convinced Parliament and military authorities to completely reorganize the medical care they provided. Accordingly, she is credited with inventing modern hospital management.

In modern times, we have equally important questions to answer. Do cell phones cause brain tumors? Are alcoholic drinks healthy in moderation? Which diet works best for losing weight? What percentage of the public is concerned about job security? **Statistics**—the science (and art!) of collecting and analyzing observations to learn about ourselves, our surroundings, and our universe—helps answer questions such as these.

Data are the building blocks of statistics. This chapter introduces some of the basic types of data and explains how we collect them, store them, and organize them. These ideas and skills will provide a basic foundation for your study of the rest of the book.

## CASE STUDY

### Deadly Cell Phones?

In September 2002, Dr. Christopher Newman, a resident of Maryland, sued Motorola, Verizon, and other wireless carriers, accusing them of causing a cancerous brain tumor behind his right ear. As evidence, his lawyers cited a study by Dr. Lennart Hardell. Hardell had studied a large number of people with brain tumors and had found that a greater percentage of them used cell phones than of those who did not have brain tumors (CNN 2002; Brody 2002).

Speculation that cell phones might cause brain cancer began as early as 1993, when (as CNN reports) the interview show *Larry King Live* featured a man who claimed that his wife died because of cancer caused by her heavy cell phone use. However, more recent studies have contradicted Hardell's results, as well as earlier reports about the health risks of heavy cell phone use.

The judge in Dr. Newman's trial was asked to determine whether Hardell's study was compelling enough to support allowing the trial to proceed. Part of this determination involved evaluating the method that Hardell used to collect data. If you were the judge, how would you rule? You will learn the judge's ruling at the end of the chapter. You will also see how the methods used to collect data about important *cause-and-effect relationships*—such as that which Dr. Newman alleged to exist between cell phone use and brain cancer—can affect the conclusions we can draw.

# What Are Data?

The study of statistics rests on two major concepts: variation and data. **Variation** is the more fundamental of these concepts. To illustrate this idea, draw a circle on a piece of paper. Now draw another one, and try to make it look just the same. Now another. Are all three exactly the same? We bet they're not. They might be slightly different sizes, for instance, or slightly different versions of round. This is an example of variation. How can you reduce this variation? Maybe you can get a coin and outline the coin. Try this three times. Does variation still appear? Probably it does, even if you need a magnifying glass to see, say, slight variations in the thickness of the penciled line.

**Data** are observations that you or someone else records. The drawings in Figure 1.1 are data that record our attempts to draw three circles that look the same. Analyzing pictorial data such as these is not easy, so we often try to turn such observations into numbers. How would you measure whether these three circles were the same? Perhaps you would compare diameters or circumferences, or somehow try to measure how and where these circles depart from being perfect circles. Whatever technique you chose, these measurements could also be considered data.

Data are more than just numbers, though. David Moore, a well-known statistician, defined data as "numbers in context." By this he meant that data consist not only of the numbers we record, but also of the story behind the numbers. For example,

$$10.00, 9.88, 9.81, 9.81, 9.75, 9.69, 9.5, 9.44, 9.31$$

are just numbers. But in fact these numbers represent "Weight in pounds of the ten heaviest babies in a sample of babies born in North Carolina in 2004." Now these numbers have a context and have been elevated to data. See how much more interesting data are than numbers?

These data were collected by the state of North Carolina in part to help researchers understand the factors that contribute to low-weight and premature births. If doctors

### Details

**Data Are What Data Is**

If you want to be "old school" grammatically correct, then the word *data* is plural. So we say "data *are*" and not "data *is*." The singular form is *datum*. However, this usage is changing over time, and some dictionaries now say that *data* can be used as both a singular and a plural noun.

▶ **FIGURE 1.1 (a)** Three circles drawn by hand. **(b)** Three circles drawn using a coin. It is clear that the circles drawn by hand show more variability than the circles drawn with the aid of a coin.

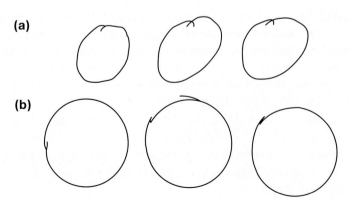

(a)

(b)

understand the causes of premature births, they can work to prevent this—perhaps by helping expectant mothers change their behavior, perhaps by medical intervention, and perhaps by a combination of both.

 **KEY POINT**  Data are "numbers in context."

The news media regularly collect data through surveys to assess what the public thinks about an issue or a politician. Every few months, approval ratings are taken to see how people think the U.S. president and other politicians are doing. Politicians may say that they don't believe in polls (particularly if they don't like the outcome), but these polls often shape the behavior of the government. This is partly because, when they are done right, polls are an accurate gauge of what the public is thinking.

You collect data. You save pictures of events or people you want to remember. These pictures tell a story that, you hope, others can reconstruct when they flip through your picture book. Perhaps you save records of your spending behavior so that you'll know how much money you have left in your account. Maybe you digitize your music collection on your computer; if so, you have data that can tell us much about your musical tastes. These data are not numbers, but often we can use numbers to describe these things. For example, a computer can tell you how many Rock and how many Classical tracks are on your mp3 player, as well as how long each track lasts.

People collect data about you. Maybe you have a "preferred customer" card for your local market or bookstore. These cards enable the market to track your purchases so that it can offer you coupons that you might be interested in—and thus encourage you to remain a loyal customer. Your school also keeps data on you. Administrators know how many classes you took, when you took them, and what grades you received. Even your car keeps data about you: how many miles you've driven the car and maybe the number of miles since your last fill-up of gas. (Some rental cars record even more data, such as the speed you drove the car at every moment.)

You can see that in our society, data are everywhere. That is one reason why it's important to understand how to work with and analyze data.

In this book you will study the science of data. You will learn (a little) about how data are stored and coded, and how you can organize data to see patterns that otherwise might be invisible. Most important, you'll see how data can sometimes be used to make generalizations about large groups of people or things.

## SECTION 1.2

# Classifying and Storing Data

The first step to understanding data is to understand the different types of data you will encounter. As you've seen, data are numbers in context. But that's only part of the story; data are also recorded observations. Your photo from your vacation to Carhenge in Nebraska is data (Figure 1.2). The ultraviolet images streaming from the Earth Observer Satellite system are data (see Figure 1.3 on the next page). These are just two examples of data that are not numbers. Statisticians work hard to help us analyze complex data, such as images and sound files, just as easily as we study numbers. Most of the methods involve recoding the data into numbers. For example, your photos can be digitized in a scanner, converted into a very large set of numbers, and then analyzed. You might have a digital camera that gives you feedback about the quality of a photo you've taken. If so, your camera is not only collecting data but also analyzing it!

▲ **FIGURE 1.2** A photo of Carhenge, Nebraska.

Almost always, our data sets will consist of characteristics of people or things (such as gender and weight). These characteristics are called **variables**. Variables are not "unknowns" like those you studied in algebra. We call these characteristics variables because they have variability: The values of the variable can be different from person to person.

> **KEY POINT**   Variables in statistics are different from variables in algebra. In statistics, variables record characteristics of people or things.

When we work with data, they are grouped into a collection, which we call either a **data set** or a **sample**. The word *sample* is important, because it implies that the data we see are just one part of a bigger picture. This "bigger picture" is called a **population**. Think of a population as the Data Set of Everything—it is the data set that contains all of the information about everyone or everything with respect to whatever variable we are studying. Quite often, the population is really what we want to learn about, and we learn about it by studying the data in our sample. However, many times it is enough just to understand and describe the sample. For example, you might collect data from students in your class simply because you want to know about the students in your class, and not because you wish to use this information to learn about all students at your school. Sometimes, data sets are so large they effectively *are* the population, as you'll see in the data from North Carolina that follow.

## Context Is Key

The context is the most important aspect of data, although it is frequently overlooked. Table 1.1 shows a few lines from the data set of births in 2004 in North Carolina (Holcomb 2006).

To understand these data, we need to ask and answer some questions: What are the objects of interest? What variables were measured? How were they measured? What are the units of measurement? Who collected the data? How did they collect the data? Where were the data collected? Why were the data collected?

Many, but not all, of these questions can be answered for these data by reading the information provided on the website that hosts the data. Other times we are not so lucky and must rely on very flimsy supporting documentation. If you collect the data yourself, you should be careful to record this extra supporting information. Or, if you get a chance to talk with the people who collected the data, then you should ask them these questions.

- *What are the objects of interest?* By "objects of interest" we mean who or what was measured. In this case, the object of interest is a baby. Each line in the table represents a newborn baby born in North Carolina in 2004. If we were to see the whole table, we would see a record of every baby born in 2004 in North Carolina.

- *What variables were measured?* For each baby, the state records the weight, the gender, and whether the mother smoked.

- *How were the variables measured?* Unknown. Presumably, most measurements on the baby were taken from a medical care-giver at the time of the birth, but we don't know how or when information about the mother was collected.

- *What are the units of measurement?* Units of measurement are important. The same variable can have different units of measurement. For example, weight could be measured in pounds, in ounces, or in kilograms. For Table 1.1,

  Weight: reported in pounds
  Gender: reported as M for boys and F for girls.
  Smoke: reported as a 1 if the mother smoked during the pregnancy, as a 0 if she did not.

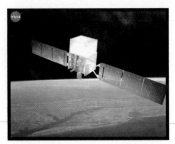

▲ **FIGURE 1.3** Satellites in NASA's Earth Observing Mission record ultraviolet reflections and transmit these data back to earth. Such data are used to construct images of our planet. Earth Observer (http://eos.gsfc.nasa.gov/).

> ### Details
>
> **More Grammar**
> We're using the word *sample* as a noun—it is an object, a collection of data that we study. Later we'll also use the word *sample* as a verb—that is, to describe an action. For example, we'll sample ice cream cones to measure their weight.

| Weight | Gender | Smoke |
|--------|--------|-------|
| 7.69 | F | 0 |
| 0.88 | M | 1 |
| 6.00 | F | 0 |
| 7.19 | F | 0 |
| 8.06 | F | 0 |
| 7.94 | F | 0 |

▲ **TABLE 1.1** Birth Data from North Carolina in 2004

- *Who collected the data?* The government of the state of North Carolina.

- *How did they collect the data?* Data were recorded for *all* births that occurred in hospitals in North Carolina. Later in the chapter you'll see that data can be collected by drawing a random sample of subjects, or by assigning subjects to receive different treatments, as well as through other methods. The exact method used for Table 1.1 is not clear, but the data were probably compiled from publicly available medical records and from reports by the physicians and caregivers.

- *Where were the data collected?* The location that the data were collected often gives us information about who (or what) the study is about. These data were collected in North Carolina and consist of babies born in that state. We should therefore be very wary about generalizing our findings to other states or other countries.

- *Why were the data collected?* Sometimes, data are collected to learn about a larger population. At other times, the goals are limited to learning more about the sample itself. In this case the data consist of all births in North Carolina, and it is most likely that researchers wanted to learn how the health of infants was related to the smoking habits of mothers within this sample.

**KEY POINT** The first time you see a data set, ask yourself these questions:
- What are the objects of interest?
- What variables were measured?
- How were the variables measured?
- What are the units of measurement?
- Who collected the data?
- How did they collect the data
- Where were the data collected?
- Why did they collect the data?

## Two Types of Variables

The variables you'll find in your data set come in two basic types, which can themselves be broken into smaller divisions, as we'll discuss later.

**Numerical variables** describe quantities of the objects of interest. The values will be numbers. The weight of an infant is an example of a numerical variable.

**Categorical variables** describe qualities of the objects of interest. These values will be categories. The gender of an infant is an example of a categorical variable. The possible values are the categories "male" and "female." Eye color of an infant is another example; the categories might be brown, blue, black, and so on. You can often identify categorical variables because their values are *usually* words, phrases, or letters. (We say "usually" because we sometimes use numbers to represent a word or phrase. Stay tuned.)

> **Details**
>
> **Quantitative and Qualitative Data**
> Some statisticians use the word *quantitative* to refer to numerical variables (think "quantity") and *qualitative* to refer to categorical variables (think "quality"). We prefer *numerical* and *categorical*. Both sets of terms are commonly used, and you should be prepared to hear and see both.

## EXAMPLE 1 Crash-Test Results

The data in Table 1.2 on the next page are an excerpt from crash-test dummy studies in which cars are crashed into a wall at 35 miles per hour. Each row of the data set represents the observed characteristics of a single car. This is a small sample of the database, which is available from the National Transportation Safety Administration. The *head injury* variable reflects the risk to the passengers' heads. The higher the number, the greater the risk.

▶ **TABLE 1.2** Crash-Test Results for Cars

| Make | Model | Doors | Weight | Head Injury |
|------|-------|-------|--------|-------------|
| Acura | Integra | 2 | 2350 | 599 |
| Chevrolet | Camaro | 2 | 3070 | 733 |
| Chevrolet | S-10 Blazer 4X4 | 2 | 3518 | 834 |
| Ford | Escort | 2 | 2280 | 551 |
| Ford | Taurus | 4 | 2390 | 480 |
| Hyundai | Excel | 4 | 2200 | 757 |
| Mazda | 626 | 4 | 2590 | 846 |
| Volkswagen | Passat | 4 | 2990 | 1182 |
| Toyota | Tercel | 4 | 2120 | 1138 |

> **Details**
>
> **Categorical Doors**
> Some people might consider *doors* a categorical variable, because nearly all cars have either 2 doors or 4 doors, and for many people, the number of doors designates a certain type of car (small or larger). There's nothing wrong with that.

**QUESTION**   For each variable, state whether it is numerical or categorical.

**SOLUTION**   The variables *make* and *model* are categorical. Their values are descriptive names. The units of *doors* are, quite simply, the number of doors. The units of *weight* are pounds. The variables *doors* and *weight* are numerical because their values are measured quantities. The units for *head injury* are unclear; head injury is measured using some scale that the researchers developed.

**TRY THIS!**   Exercise 1.1

> **Caution**
>
> **Don't Just Look for Numbers!**
> You can't always tell whether a variable is categorical simply by looking at the data table. You must also consider what the variable represents. Sometimes, researchers code categorical variables with numerical values.

## Coding Categorical Data with Numbers

Sometimes categorical variables are "disguised" as numerical. The smoke variable in the North Carolina data set (Table 1.1) has numbers for its values (0 and 1), but in fact those numbers simply indicate whether or not the mother smoked. Mothers were asked, "Did you smoke?" and if they answered "Yes," the researchers coded this categorical response with a 1. If they answered "No," the response was coded as a 0. These particular numbers represent categories. *Smoke* is a categorical variable.

Coding is used to help both humans and computers understand what the values of a variable represent. For example, a human would understand that a "yes" under the "Smoke" column would mean that the person was a smoker, but to the computer, the word "yes" is just a string of symbols. If instead, we follow a convention where a 1 means "yes" and a 0 means "no," then a human understands that the 1's represent smokers, and a computer can easily add the values together to determine, for example, how many smokers are in the sample.

This approach for coding categorical variables is quite common and useful. If a categorical variable has only two categories, as do *gender* and *smoke*, then it is almost always helpful to code the values with 0 and 1. To help readers know what a "1" means, rename the variable with one its category names. A "1" then means the person belongs to that category, and a 0 means the person belongs to the other category. For example, instead of calling a variable *gender*, we rename it *female*. And then if the baby is a boy we code a 0, and if it's a girl we code a 1.

Sometimes your computer does the coding for you without your needing to know anything about it. So even if you see the words *female* and *male* on your computer, the computer has probably coded these with values of 0 and 1 (or vice versa).

## Storing Your Data

The format in which you record and store your data is very important. Computer programs will require particular formats, and by following a consistent convention, you can be confident that you'll better remember the qualities of your own data set if you need to revisit it months or even years later. Data are often stored in a spreadsheet-like format in which each row represents the object (or person) of interest. Each column represents a variable. In Table 1.3, each row represents a baby. The columns are variables: *Weight*, *Female*, and *Smoke*. All the previous tables use this format, which is sometimes referred to as the **stacked data** format.

When you collect your own data, the stacked format is almost always the best way to record and store your data. One reason is that it allows you to easily record several different variables for each subject. Another reason is that it is the format that most software packages will assume you are using for most analyses. (The exceptions are TI-83/84 and Excel.)

Some technologies, such as the TI calculators, require, or at least accommodate, data stored in a different format, called **unstacked data**. Unstacked data tables are also common in some books and media publications. In this format, each column represents a variable from a different group. For example, one column could represent men's heights, and another column could represent women's heights. The data set, then, is a single variable (*height*) broken into two groups. The groups are determined by a categorical variable. Table 1.4 shows an example of unstacked data, and Figure 1.4 shows the same data in TI-83/84 input format.

By way of contrast, Table 1.5 shows the same data in stacked format.

The great disadvantage of the unstacked format is that it can store only two variables at a time: the variable of interest (for example, height), and a categorical variable that tells us which group the observation belongs in (for example, gender). However, most of the time, we record many variables for each observation. For example, we record a baby's weight, gender, and whether or not the mother smoked. The stacked format enables us to display as many variables as we wish.

## EXAMPLE 2 Downhill Skiers

The accompanying table shows downhill skiing times (seconds) for the top five men and the top five women finalists in the 2010 Winter Olympics in Vancouver. The data are in unstacked form.

(Note that the women raced a different course than the men.)

| Men | Women |
|---|---|
| 114.31 | 104.19 |
| 114.38 | 104.75 |
| 114.4 | 105.65 |
| 114.52 | 105.68 |
| 114.64 | 106.17 |

QUESTION   Create a new data table that displays the data in stacked form.

SOLUTION   At first it appears that only one numerical variable, time, is displayed. In fact, there is also a categorical variable, gender, which is used to distinguish the two groups. To stack the data, we identify the gender that is associated with each time.

| Weight | Female | Smoke |
|---|---|---|
| 7.69 | 1 | 0 |
| 0.88 | 0 | 1 |
| 6.00 | 1 | 0 |
| 7.19 | 1 | 0 |
| 8.06 | 1 | 0 |
| 7.94 | 1 | 0 |

▲ **TABLE 1.3** Data for Newborns from Table 1.1 Using Coding

| Men's Heights | Women's Heights |
|---|---|
| 70 | 59 |
| 68 | 70 |
| 71 | 61 |
|  | 62 |

▲ **TABLE 1.4** Data by Groups (Unstacked)

▲ **FIGURE 1.4** TI-83/84 data input screen (unstacked data).

| Height | Gender |
|---|---|
| 70 | male |
| 68 | male |
| 71 | male |
| 59 | female |
| 70 | female |
| 61 | female |
| 62 | female |

▲ **TABLE 1.5** The Same Data as in Table 1.4, Shown Here in Stacked Format

| Time (seconds) | Gender |
|---|---|
| 114.31 | M |
| 114.38 | M |
| 114.4 | M |
| 114.52 | M |
| 114.64 | M |
| 104.19 | F |
| 104.75 | F |
| 105.65 | F |
| 105.68 | F |
| 106.17 | F |

**TRY THIS!** Exercise 1.5

# Organizing Categorical Data

Once we have a data set, we next need to organize and display the data in a way that helps us see patterns. This task of organization and display is not easy and we discuss it throughout the entire book. In this section we introduce the topic for the first time, in the context of categorical variables.

With categorical variables, we are usually concerned with knowing how often a particular category occurs in our sample. We then (usually) want to compare how often a category occurs for one group versus another (liberal/conservative, man/woman). To do these comparisons, you need to understand how to calculate percentages and other rates.

A common method for summarizing two potentially related categorical variables is to use a two-way table. **Two-way tables** show how many times each combination of categories occurs. For example, Table 1.6 is a two-way table from the Youth Behavior Risk Survey that shows gender and whether or not the respondent always (or almost always) wears a seat belt when riding in or driving a car. The actual Youth Behavior Risk Survey has over 10,000 respondents, but we are practicing on a small sample from this much larger data set.

The table tells us that 2 people were male and did not always wear a seat belt. Three people were female and did not always wear seat belts. These counts are also called frequencies. A **frequency** is simply the number of times a value is observed in a data set.

Some books and publications present two-way tables as though they displayed the original data collected by the investigators. However, two-way tables do not consist of "raw" data but, rather, are summaries of data sets. For example, the data set that produced Table 1.6 is shown in Table 1.7.

To summarize this table, we simply count how many of the males (a 1 in the Male column) also do not always wear seat belts (a 1 in the Not Always column). We then count how many both are male and always wear seat belts (a 1 in the Male column, a 0 in the Not Always column); how many both are female and don't always wear seat

| | Male | Female |
|---|---|---|
| Not Always | 2 | 3 |
| Always | 3 | 7 |

▲ **TABLE 1.6** This two-way table shows counts for 15 youths who responded to a survey about wearing seat belts.

belts (a 0 in the Male column, a 1 in the Not Always column); and, finally, how many both are female and always wear a seat belt.

　　Example 3 illustrates that summarizing the data in a two-way table can make it easy to compare groups.

## EXAMPLE 3 Percentages of Seat Belt Wearers

The 2008 Youth Behavior Risk Survey is a national study that asks American youths about potentially risky behaviors. We show the two-way summary again. All of the people in the table were between 14 and 17 years old. The participants were asked whether they wear a seat belt while driving or riding in a car. The people who said always or almost always were put in the Always group. The people who said sometimes or rarely were put in the Not Always group.

|  | **Male** | **Female** |
|---|---|---|
| Not Always | 2 | 3 |
| Always | 3 | 7 |

**QUESTIONS**

a. How many men are in this sample? How many women? How many people do not always wear seat belts? How many always wear seat belts?

b. What percent of the sample are men? What percent are women? What percent don't always wear seat belts? What percent always wear seat belts?

c. Are the men in the sample more likely than the women in the sample to take the risk of not wearing a seat belt?

**SOLUTIONS**

a. We can count the men by adding the first column: 2 + 3 = 5 men. Adding the second column gives us the number of women: 3 + 7 = 10.

　We get the number who do not always wear seat belts by adding the first row: 2 + 3 = 5 people don't always wear seat belts. Adding the second row gives us the number who always wear seat belts: 3 + 7 = 10.

b. This question asks us to convert the numbers we found in part (a) to percentages. To do this, we divide the numbers by 15, because there were 15 people in the sample. To convert to percentages, we multiply this proportion by 100%.

　The proportion of men is $5/15 = 0.333$. The percentage is $0.333 \times 100\% = 33.3\%$. The proportion of women must be $100\% - 33.3\% = 66.7\%$ ($10/15 \times 100\% = 66.7\%$).

　The proportion who do not always wear seat belts is $5/15 = 0.333$, or 33.3%. The proportion who always wear seat belts is $100\% - 33.3\% = 66.7\%$.

c. You might be tempted to answer this question by counting the number of males who don't always wear seat belts (2 people) and comparing that to the number of females who don't always wear seat belts (3 people). However, this is not a fair comparison because there are more females than males in the sample. Instead, we should look at the percentage of those who don't always wear seat belts in each group. This question should be reworded as follows:

　Is the percentage of males who don't always wear seat belts greater than the percentage of females who don't always wear seat belts?

| Male | Not Always |
|---|---|
| 1 | 1 |
| 1 | 1 |
| 1 | 0 |
| 1 | 0 |
| 1 | 0 |
| 0 | 1 |
| 0 | 1 |
| 0 | 1 |
| 0 | 0 |
| 0 | 0 |
| 0 | 0 |
| 0 | 0 |
| 0 | 0 |
| 0 | 0 |
| 0 | 0 |

▲ **TABLE 1.7** This data set is equivalent to the two-way summary shown in Table 1.6. We highlighted in red those who did not always wear a seat belt (the risk takers).

Because 2 out of 5 males don't always wear seat belts, the percent of males who don't always wear seat belts is $(2/5) \times 100\% = 40\%$.

Because 3 out of 10 females don't always wear seat belts, the percent of females who don't always wear seat belts is $(3/10) \times 100\% = 30\%$.

In fact, females in this sample are less likely than males to engage in this risky behavior. Among all U.S. youth, it is estimated that about 28% of males do not always wear their seat belt, compared to 23% of females.

**TRY THIS!** Exercise 1.7

The calculations in Example 3 took us from frequencies to percentages. Sometimes, we want to go in the other direction. If you know the total number of people in a group, and are given the percentage that meets some qualification, you can figure out *how many* people in the group meet that qualification.

## EXAMPLE 4 Numbers of Seat Belt Wearers

A statistics class has 300 students, and they are asked if they always ride or drive with a seat belt.

**QUESTIONS**

a. Suppose that 30% of the students do not always wear a seat belt. How many students is this?

b. Suppose we know that in another class, 20% of the students do not always wear seat belts, and this is equal to 43 students. How many students are in the class?

**SOLUTIONS**

a. We need to find 30% of 300. When working with percentages, first convert the percentage to its decimal equivalent:

$$30\% \text{ of } 300 = 0.30 \times 300 = 90$$

Therefore, 90 students don't always wear seat belts.

b. The question tells us that 20% of some unknown larger number (call it $y$) must be equal to 43.

$$0.20y = 43$$

Divide both sides by 0.20 and you get

$$y = 215$$

There are 215 total students in the class, and 43 of them don't always wear seat belts.

**TRY THIS!** Exercise 1.9

Sometimes, you may come across data summaries that are missing crucial information. Suppose we wanted to know which team sports are the most dangerous to play. Table 1.8 shows the number of sports-related injuries that were treated in U.S. emergency rooms in 2004 (National Safety Council 2004). (Note that this table is not the table of original data but is, instead, a summary of the original data.)

Wow! It's a dangerous world out there. Which would you conclude is the most dangerous sport? Which is the least dangerous?

| Sport | Injuries |
|-------|----------|
| Baseball | 178,668 |
| Basketball | 615,546 |
| Bowling | 21,133 |
| Football | 387,948 |
| Ice hockey | 16,435 |
| Soccer | 173,519 |
| Softball | 125,875 |
| Tennis | 19,633 |
| Volleyball | 59,225 |

▲ **TABLE 1.8** Summary of Counts of Sports Injuries

Did you answer that basketball was the most dangerous sport? It does have the most injuries (615,546)—in fact, almost twice as many as in football (387,948). Ice hockey is known for its violence (you've heard the old joke, "I went to a fight and suddenly a hockey match broke out"), but here, it seems to have caused relatively few injuries and looks safe.

The problem with comparing the numbers of injuries for these sports is that the sports have different numbers of participants. Injuries might be more common in basketball simply because more people play basketball. Also, there might be relatively few injuries in ice hockey merely because fewer people play. One important component is missing in Table 1.8, and the lack of this component makes our analysis impossible.

Table 1.9 includes the component missing from Table 1.8: the number of participants in each sport. We can't directly compare the number of injuries from sport to sport, because the numbers of members of the various groups are not the same. This improved table shows us the total membership of each group.

| Sport | Injuries | Participants |
|---|---|---|
| Baseball | 178,668 | 15,600,000 |
| Basketball | 615,546 | 28,900,000 |
| Bowling | 21,133 | 43,900,000 |
| Football | 387,948 | 17,700,000 |
| Ice hockey | 16,435 | 2,100,000 |
| Soccer | 173,519 | 14,500,000 |
| Softball | 125,875 | 13,600,000 |
| Tennis | 19,633 | 11,000,000 |
| Volleyball | 59,225 | 11,500,000 |

◀ **TABLE 1.9** Summary of Counts of Sports Injuries and Numbers of Participants

Which sport is the most dangerous? We now have the information we need to answer this question. Specifically, we can find the percentage of participants injured in each sport. For example, what percent of basketball players were injured? There were 28,900,000 participants and 615,546 were injured, so the percent injured is ($615,546/28,900,000$) $\times$ 100% = 2.1299%.

Sometimes, with percentages as small as this, we understand the numbers more easily if we report not a percentage, but "number of events per 1000 objects" or maybe even "per 10,000 objects". We call such numbers **rates**. To get the injury rate per 1000 people, instead of multiplying ($614,546/28,900,000$) by 100 we multiply by 1000: ($615,546/28,900,000$) $\times$ 1000 = 21.299 injuries per 1000 people.

These results are shown in Table 1.10.

| Sport | Injuries | Participants | Injury/Part | Per Thousand |
|---|---|---|---|---|
| Baseball | 178,668 | 15,600,000 | 0.01145 | 11.45 |
| Basketball | 615,546 | 28,900,000 | 0.02130 | 21.30 |
| Bowling | 21,133 | 43,900,000 | 0.00048 | 0.48 |
| Football | 387,948 | 17,700,000 | 0.02192 | 21.92 |
| Ice hockey | 16,435 | 2,100,000 | 0.00783 | 7.83 |
| Soccer | 173,519 | 14,500,000 | 0.01197 | 11.97 |
| Softball | 125,875 | 13,600,000 | 0.00926 | 9.26 |
| Tennis | 19,633 | 11,000,000 | 0.00178 | 1.78 |
| Volleyball | 59,225 | 11,500,000 | 0.00515 | 5.15 |

◀ **TABLE 1.10** Summary of Rates of Sports Injuries

We see now that football is the most dangerous sport: 21.92 players are injured out of every 1000 players. Basketball is slightly less risky, with 21.30 injuries per 1000 players.

### EXAMPLE 5 Comparing Rates of Stolen Cars

Which model of car has the greatest risk of being stolen? The Highway Loss Data Institute reports that the Cadillac Escalade is the most stolen car; 13.2 Escalades are reported stolen out of every 1000 Escalades on the road. By way of contrast, the Ford Taurus is the least stolen; 0.3 Taurus is reported stolen for every 1000 Tauruses on the road (Insurance Institute for Highway Safety 2006).

QUESTION Why does the Highway Loss Data Institute report theft rates rather than the number of each type of car stolen?

SOLUTION We need to take into account the fact that some cars are more popular than others. Suppose there were only 10 Escalades stolen all year long, but 350 Tauruses were stolen. It might seem that Tauruses are at a higher risk of being stolen. However, Escalades are expensive compared to Tauruses, so there are probably many more Tauruses on the road available to be stolen. By looking at the *theft rate*, we adjust the total number of cars *of that kind* on the road.

TRY THIS! Exercise 1.19

> **KEY POINT** In order for us to compare groups, the groups need to be similar. When the data consist of counts, then percentages or other rates are often better for comparisons because they take into account possible differences among the sizes of the groups.

## SECTION 1.4

# Collecting Data to Understand Causality

Often, the most important questions in science, business, and everyday life are questions about **causality**. These are usually phrased in the form of "what if" questions. What if I take this medicine; will I get better? What if I change my Facebook profile; will my profile get more hits?

Questions about causality are often in the news. The *Los Angeles Times* reported that many people believe a drink called peanut milk can cure gum disease and slow baldness (Glionna 2006). The BBC News (2010) reported that "Happiness wards off heart disease." Statements such as these are everywhere we turn these days. How do we know whether to believe these claims?

The methods we use to collect data determine the types of conclusions we can make. Only one method of data collection is suitable for making conclusions about causal relationships, but as you'll see, that doesn't stop people from making such conclusions anyway. In this section we talk about three methods commonly used to collect data for questions about causality: anecdotes, observational studies, and controlled experiments.

Most questions about causality can be understood in terms of two variables: the **treatment variable** and the **outcome variable**. (The outcome variable is also sometimes called the **response variable**, because it responds to changes in the treatment.)

We are essentially asking whether the treatment variable causes changes in the outcome variable. For example, the treatment variable might record whether or not a person drinks Peanut Milk, and the outcome variable records whether or not that person's gum disease improved. Or the treatment variable might record whether or not a person is generally happy, and the outcome variable might record whether or not that person suffered from heart disease in a ten-year period.

People who receive the treatment of interest (or have the characteristic of interest) are said to be in the **treatment group**. Those who do not receive that treatment (or do not have that characteristic) are in the **control group**, which is also called the **comparison group.**

## Anecdotes

Peanut milk is a drink invented by Jack Chang, an entrepreneur in San Francisco, California. He noticed that after he drank peanut milk for a few months, he stopped losing hair and his gum disease went away. According to the *Los Angeles Times* (Glionna 2006), another regular drinker of peanut milk says that the beverage caused his cancer to go into remission. Others have reported that drinking the beverage has reduced the severity of their colds, has helped them sleep, and has helped them wake up.

This is exciting stuff! Peanut milk could very well be something we should all be drinking. But can peanut milk really solve such a wide variety of problems? On the face of it, it seems that there's evidence that peanut milk has cured people of illness. The *Los Angeles Times* reports the names of people who claim that it has. However, the truth is that this is simply not enough evidence on the basis of which to judge whether the beverage is helpful, harmful, or without any effect at all.

These testimonials are examples of anecdotes. An **anecdote** is essentially a story that someone tells about her or his own (or a friend's or relative's) experience. Anecdotes are an important type of evidence in criminal justice because eyewitness testimony can carry a great deal of weight in a criminal investigation. However, for answering questions about groups of people with great variability or diversity, anecdotes are essentially worthless.

The primary reason why anecdotes are not useful for reaching conclusions about cause-and-effect relationships is that the most interesting things that we study have so much variety that a single report can't capture the variety of experience. For example, have you ever bought something because a friend recommended it, only to find that after a few weeks it fell apart? If the object was expensive, such as a car, you might have been angry at your friend for recommending such a bad product. But how do you know whose experience was more typical, yours or your friend's? Perhaps the car is in fact a very reliable model, and you just got a lemon.

A very important question to ask when someone claims that a product makes some kind of change is to ask, "Compared to what?" Here the claim is that drinking peanut milk will make you healthier. The question to ask is "Healthier compared to what?" Compared to people who don't drink peanut milk? Compared to people who take medicine for their particular ailment? To answer these questions, we need to examine the health of these other groups of people who do not drink peanut milk.

Anecdotes do not give us a comparison group. We might know that a group of people believe that peanut milk made them feel better, but we don't know how the milk drinkers' experiences compare to those of people who did not drink peanut milk.

 When someone makes a claim about causality, ask, "Compared to what?"

Another reason for not trusting anecdotal evidence is a psychological phenomenon called the placebo effect. People often react to the idea of a treatment, rather than to the treatment itself. A **placebo** is a harmless pill (or sham procedure) that a patient believes is actually an effective treatment. Often, the patient taking the pill feels better, even though the pill actually has no effect whatsoever. In fact, a survey of U.S. physicians published in the *British Medical Journal* found that up to half of physicians prescribe sugar pills—placebos—to manage chronic pain. This psychological wish fulfillment—we feel better because we think we *should* be feeling better—is called the **placebo effect**.

## Observational Studies

The identifying mark of an **observational study** is that the subjects in the study are put into the treatment group or the control group either by their own actions or by the decision of someone else who is not involved in the research study. For example, if we wished to study the effects on health of smoking cigarettes (as many researchers have), then our treatment group would consist of people who had chosen to smoke, and the control group would consist of those who had chosen not to smoke.

Observational studies compare the outcome variable in the treatment group with the outcome variable in the control group. Thus, if many more people are cured of gum disease in the group that drinks peanut milk (treatment) than in the group that does not (control), then we would say that drinking peanut milk is associated with improvement in gum disease; that is, there is an **association** between the two variables. If the group of happy people tend to have less heart disease than the not-happy people, we would say that happiness is associated with improved heart health.

Note that we do not conclude that peanut milk *caused* the improvement in gum disease. In order for us to draw this conclusion, the treatment group and the control group must be very similar in every way except that one group gets the treatment and the other doesn't. For example, if we knew that the group of people who started drinking peanut milk and the group that did not drink peanut milk were alike in every way—both groups have the same overall health, are roughly the same ages, include the same mix of genders and races, education levels, and so on—then if the peanut milk group is healthier after a year, we would be fairly confident in concluding that peanut milk is the reason for their better health.

Unfortunately, in observational studies this goal of having very similar groups is *extremely* difficult to achieve. *Some* characteristic is nearly always different in one group than in the other. This means that the groups may experience different outcomes because of this different characteristic, not because of the treatment. A difference between the two groups that can explain why the outcomes were different is called a **confounding variable**.

For example, early observational studies on the effects of smoking found that a greater percent of smokers than of nonsmokers had lung cancer. However, some scientists argued that genetics was a confounding variable (Fisher 1959). They maintained that the smokers differed genetically from the nonsmokers. This genetic difference made some people more likely to smoke and also more susceptible to lung cancer.

This was a convincing argument for many years. It not only proposed a specific difference between the groups (genetics) but also explained how that difference might come about (genetics makes some people smoke more, perhaps because it tastes better to them or because they have addictive personalities). And the argument also explained why this difference might affect the outcome (the same genetics cause lung cancer). Therefore, the skeptics said, genetics—and not smoking—might be the cause of lung cancer.

Later studies established that the skeptics were wrong about genetics. Some studies compared pairs of identical twins in which one twin smoked and the other did not. These pairs had the same genetic makeup, and still a higher percentage of the smoking twins had cancer than of the nonsmoking twins. Because the treatment and control groups had the same genetics, genetics could not explain why the groups had different cancer rates. When we compare groups in which we force one of the variables to be the same, we say that we are *controlling for* that variable. In these twin studies, the researchers controlled for genetics by comparing people with the same genetic makeup (Kaprio and Koskenvuo 1989).

A drawback of observational studies is that we can never know whether there exists a confounding variable. We can search very hard for it, but just because we don't find a confounding variable does not mean it isn't there. For this reason, we can never make cause-and-effect conclusions from observational studies.

 **KEY POINT** We can never draw cause-and-effect conclusions from observational studies because of potential confounding variables. A single observational study can conclude only that there is an *association* between the treatment variable and the outcome variable.

## EXAMPLE 6 Does Kimchi Cause Gastric Cancer?

In a paper titled "Kimchi and Soybean Pastes Are Risk Factors of Gastric Cancer," researchers noted that gastric cancer rates were ten times higher in Korea and Japan than in the United States. Koreans and Japanese also eat many times more kimchi (a heavily spiced, fermented cabbage dish) than those living in the United States (Nan HongMei et al. 2005).

QUESTION On the basis of this information, can we conclude that kimchi causes gastric cancer? If yes, explain why. If no, state a potential confounding variable.

SOLUTION No, we cannot. The treatment group consists of Japanese and Korean people, and the control group consists of Americans. Many other factors differ between these two groups. A potential confounding variable is diet, which differs between Koreans and Americans in many ways other than in the amount of kimchi they eat. These differences in diet may also affect cancer rates.

TRY THIS! Exercise 1.31

## Controlled Experiments

In order to answer cause-and-effect questions, we need to create a treatment group and a control group that are alike in every way possible, except that one group gets a treatment and the other does not. As you've seen, this cannot be done with observational studies because of confounding variables. In a **controlled experiment**, researchers take control by assigning subjects to the control or treatment group. If this assignment is done correctly, it ensures that the two groups can be nearly alike in every relevant way except whether or not they receive the treatment under investigation.

Well-designed and well-executed controlled experiments are the only means we have for definitively answering questions about causality. However, controlled experiments are difficult to carry out (this is one reason why observational studies are often done instead). Let's look at some of the attributes of a well-designed controlled experiment.

A well-designed controlled experiment has four key features:

- The sample size must be large so that we have opportunities to observe the full range of variability in the humans (or animals or objects) we are studying.

- The subjects of the study must be assigned to the treatment and control groups at random.

- Ideally, the study should be "double-blind," as explained below.

- The study should use a placebo if possible.

These features are all essential in order to ensure that the treatment group and the control group are as similar as possible.

To understand these key design features, imagine that a friend wants to lose weight and has asked your advice about whether he should go on the Atkins diet. (See Dansinger et al. 2005 for an account of a study related to the hypothetical one described here.) The Atkins diet is fairly severe: Eating bread in any form is essentially forbidden. Does this diet work? For a control group, we might compare the Atkins diet to a more traditional diet, such as that advocated by Weight Watchers. The Weight Watchers diet restricts calories and is generally considered a reliable way to lose weight. How do we know whether the Atkins diet (the treatment) works, compared to the Weight Watchers diet (the control)?

## Sample Size

A good controlled experiment designed to determine whether the Atkins diet works should have a large number of people participate in the study. People react to changes in their diet in a variety of ways, and the effects of a diet can vary greatly from person to person. To observe the full range of variability, you therefore need a large number of people.

The question of exactly how many people are required is difficult to answer, and most medical studies hire statisticians to determine the number of participants required. In general, the more the better, and you should be more critical of studies with very few participants.

## Random Assignment

The next step is to assign people to the treatment and control groups such that the two groups are similar in every way possible. As we saw when we discussed observational studies, letting the participants themselves decide doesn't work, because people who choose the more severe Atkins diet over the more traditional Weight Watchers might differ in other important ways (such as level of motivation) that affect the outcome. Instead, a good controlled experiment uses **random assignment**. One way of doing this is to flip a coin. Heads means the participant goes into the treatment group, tails means the control group (or the other way around—as long as the researchers are consistent). In practice, the randomizing might instead be done with a computer or even with the random number generator on a calculator, but the idea is always the same: No human determines group assignment. Rather, assignment is left to chance.

If both groups have a large enough number of people, random assignment will "balance" the groups. The variation in weights, the mix of ages and genders, and the mix of most variables will be similar in both groups. Note that by "similar" we don't mean exactly the same. We don't expect both groups to have exactly the same percentage of men, for example. Except in rare cases, random variation results in slight differences in the mixes of the groups. But these differences should be small.

If you read about a controlled experiment that does not use random assignment, then there is a very real possibility that the results of the study are invalid. The technical term for what happens with nonrandomized assignment is bias. We say that a study exhibits **bias** when the results are influenced in one particular direction. A researcher who puts the heaviest people in one diet group, for example, is biasing the outcome. It's not always easy or possible to predict what the effects of the bias will be, but the important point is that the bias creates a confounding variable and makes it difficult, or impossible, to determine whether the treatment we're investigating really affects the outcome we're observing.

**KEY POINT** — **Random assignment** (assignment to treatment groups by a randomization procedure) helps balance the groups to minimize bias. This helps make the groups comparable.

## Blinding

So far, we've recruited a large number of people and randomly assigned half to use the Atkins diet and the other half to use the Weight Watchers diet. In principle, these two groups will be very similar. However, there are still two potential differences. First, we might know who is in which group. This means that when we interact with a participant, we might consciously or unconsciously treat that person differently, depending on which group he or she belongs to. For example, if we believe strongly in the Atkins diet, we might give special advice or encouragement to people on the Atkins diet that we don't give to people on the Weight Watchers diet. If so, then we've biased the study.

To prevent this from happening, researchers should be **blind** to assignment. This means that an independent party—someone who does not regularly see the participants and who does not participate in determining the results of the study—handles the assignment to groups. The researchers do not know who is in which group until the study has ended.

Second, we must consider the participants themselves. If they know they are in the treatment group, they might behave differently than if they know they are in the control group. Perhaps they will work harder at losing weight. Or perhaps they will work less hard, because they might have a false sense of confidence in the diet. Why would we have made the Atkins diet our treatment group, they might reason, if we didn't think it was the better diet? To prevent this from happening, the participants should also not know whether they are in the treatment group or the control group.

When neither the researchers nor the participants know whether they are in the treatment or the control group, we say that the study is **double-blind**. The double-blind format helps prevent the bias that can result if one group acts differently from the other because they know they are being treated differently, or because the researchers treat the groups differently or evaluate them differently because of what the researchers expect.

## Placebos

The treatment and control groups might differ in another way. People often react not just to a particular medical treatment, but also to the *idea* that they are getting medical treatment. This means that patients who receive a pill, a vaccine, or some other form of treatment often feel better even when the treatment in truth does absolutely nothing. Interestingly, this placebo effect also works in the other direction: If told that a certain pill might cause side effects (for example, a rash), some patients experience the side effects even though the pill they were given is just a sugar pill.

To neutralize the placebo effect, it is important that the control group receive attention similar to what the treatment group receives, so that both groups feel they are being treated the same by the researchers. In our diet study, both groups are put on a special diet, so no difference between the groups should arise from the placebo effect.

In our hypothetical study about diets, then, a placebo is not necessary, because we are comparing two diets to each other and are not interested in comparing the dieters to a group of subjects who do not go on any diet at all. But if we were studying whether peanut milk improves baldness, we would require the control group to take a placebo drink so that we could rule out any placebo effect and thus perform a valid comparison between treatment and control.

**KEY POINT**   The Gold Standard for Experiments:

*Large sample size.* This ensures that the study captures the full range of variation among the population and allows small differences to be noticed.

*Controlled and randomized.* Random assignment of subjects to treatment or control groups to minimize bias.

*Double-blind.* Neither subjects nor researchers know who is in which group.

*Placebo (if appropriate).* This format controls for possible differences between groups that occur simply because subjects think their treatment is effective.

## EXAMPLE 7 Brain Games

Brain training video games, such as Nintendo's Brain Age, claim to improve basic intelligence skills, such as memory. A study published in the journal *Nature* investigated whether playing such games can actually boost intelligence (Owen et al. 2010). The researchers explain that 11,430 people logged onto a Web page and were randomly assigned to one of three groups. Group 1 completed six training tasks that emphasized "reasoning, planning and problem-solving." Group 2 completed games that emphasized a broader range of cognitive skills. Group 3 was a control group and didn't play any of these games; instead, members were prompted to answer "obscure" questions. At the end of six weeks, the participants were compared on several different measures of thinking skills. The results? The control group did just as well as the treatment groups.

QUESTION Which features of a well-designed controlled experiment does this study have? Which features are missing?

SOLUTION Sample size: The sample size of 11,430 is quite large. Each of the three groups will have about 3800 people.

Randomization: The authors state that patients were randomly assigned to one of the three groups.

Double-blind format: Judging on the basis of this description, there was no double-blind format. It's possible (indeed, it is likely) that the researchers did not know, while analyzing the outcome, to which treatment group individuals had been assigned. But we do not know whether *participants* were aware of the existence of the three different groups and how they differed.

Placebo: The control group participated in a "null" game, in which they simply answered questions. This activity is a type of placebo, because the participants could have thought that this null game was a brain game.

TRY THIS! Exercise 1.33

> **! Caution**
>
> **At Random**
> The concept of randomness is used in two different ways in this section. *Random assignment* is used in a controlled experiment. Subjects are randomly assigned to treatment and control groups in order to achieve a balance between groups. This ensures that the groups are comparable to each other and that the only difference between the groups is whether or not they receive the treatment under investigation. **Random selection** occurs when researchers select subjects from some larger group via a random method. We must employ random selection if we wish to extend our results to a larger group.

## Extending the Results

In both observational studies and controlled experiments, researchers are often interested in knowing whether their findings, which are based on a single collection of people or objects, will extend to the world at large.

The researchers in Example 7 concluded that brain games are not effective, but might it just be that the games weren't effective for those people who decided to participate? Maybe if the researchers tested people in another country, for example, the findings would be different.

It is usually not possible to make generalizations to a larger group of people unless the subjects for the study are representative of the larger group. The only way to collect a sample that is representative is to collect the objects we study at random. We will discuss how to collect a random sample, and why we can then make generalizations about people or objects who were not in the sample, in Chapter 7.

Selecting subjects using a random method is quite common in polls and surveys (which you'll also study in Chapter 7), but it is much less common in other types of studies. Most medical studies, for example, are not conducted on people selected randomly, so even when a cause-and-effect relationship emerges between the treatment and the response, it is impossible to say whether this relationship will hold for a larger (or different) group of people. For this reason, medical researchers often work hard to replicate their findings in diverse groups of people.

## Statistics in the News

When reading in a newspaper or blog about a research study that relies on statistical analysis, you should ask yourself several questions to evaluate how much faith you can put in the conclusions reached in the study.

1. *Is this an observational study or a controlled experiment?*

   If it's an observational study, then you can't conclude that the treatment caused the observed outcome.

2. *If the study is a controlled experiment, was there a large sample size? Was randomization used to assign participants to treatment groups? Was the study double-blind? Was there a placebo?*

   See the relevant section of this chapter for a review of the importance of these attributes.

3. *Was the paper published in a peer-reviewed journal? What is the journal's reputation?*

   "Peer-reviewed" means that each paper published in the journal is rigorously evaluated by at least two anonymous researchers familiar with the field. The best journals are very careful about the quality of the research they report on. They have many checkpoints to make sure that the science is as good as it can be. (But remember, this doesn't mean the science is perfect. If you read a medical journal regularly, you'll see much debate from issue to issue about certain results.) Other journals, by contrast, sometimes allow sloppy research results, and you should be very wary of these journals.

4. *Did the study follow people for a long enough time?*

   Some treatments take a long time to work, and some illnesses take a long time to show themselves. For example, many cost-conscious people like to refill water bottles again and again with tap water. Some fear that drinking from the same plastic bottle again and again might lead to cancer. If this is true, it might take a very long time for a person to get cancer from drinking out of the same bottle day after day. So researchers who wish to determine whether drinking water from the same bottle causes cancer should watch people for a very long time.

Often it is hard to get answers to all of these questions from a newspaper article. Fortunately, the Internet has made it much easier to find the original papers, and your college library or local public library will probably have access to many of the most popular journals.

Even when a controlled experiment is well designed, things can still go wrong. One common way in which medical studies go astray is that people don't always do what their doctor tells them to do. Thus, people randomized to the treatment group might not actually take their treatments. Or people randomized to the Atkins diet might switch to Weight Watchers because they don't like the food on the Atkins diet. A good research paper will report on these difficulties and will be honest about the effect on their conclusions.

## EXAMPLE 8 Does City Living Raise Blood Pressure?

The *New York Times* linked to an article that claims higher blood pressure occurs in people who live in urban areas (www.scienceblog.com 2010).

QUESTION  Is this more likely to be an observational study or a controlled experiment? Why? Does this mean that moving to a more urban area will result in an increase in your blood pressure?

SOLUTION This is most likely an observational study. The treatment variable is whether or not a person lives in an urban area. We are not told that participants were randomly assigned to live in urban or nonurban areas for some period of time, and in fact it is pretty unlikely that such a study could be done. Researchers probably measured blood pressure in people who chose to live in urban or nonurban settings. Because the participants themselves chose where to live, this is an observational study.

Because this is an observational study, we cannot conclude that the *treatment* (living in urban areas) causes the *outcome* (increased blood pressure). This means that simply moving to (or away from) an urban area may not change your blood pressure. One potential confounding variable could be personality, because people who prefer the fast pace of city life might have other personality traits that increase their blood pressure.

TRY THIS! Exercise 1.39

## EXAMPLE 9 Crohn's Disease

Crohn's disease is a bowel disease that causes cramping abdominal pain, fever, and fatigue. A study reported in the *New England Journal of Medicine* (Columbel et al. 2010) tested two medicines for the disease: injections of infliximab (Inflix) and oral azathioprine (Azath). The participants were randomized into three groups. All groups received an injection and a pill (some were placebos, but still a pill and an injection). One group received Inflix injections alone (with placebo pills), one received Azath pills alone (with placebo injections), and one group received both injections and pills. A good outcome was defined as the disease being in remission after 26 weeks. The accompanying table shows the numbers.

|  | Combination | Inflix Alone | Azath Alone |
|---|---|---|---|
| Remission | 96 | 75 | 51 |
| Not in Remission | 73 | 94 | 119 |

QUESTIONS

a. Compare the percentages in remission for the three treatments. Which treatment was the most effective and which the least effective for this sample?

b. Can we conclude that the combination treatment causes a better outcome? Why or why not?

SOLUTIONS

a. For the combination: 96 / 169, or 56.8%, success
   For the Inflix alone: 75 / 169, or 44.4%, success
   For the Azath alone: 51 / 170, or 30%, success

The combination treatment was the most effective for this sample, and Azath alone was the least effective.

b. Yes, we can conclude that the combination of drugs causes a better outcome than the single drugs. The study was placebo-controlled and randomized. The sample size was reasonably large. Blinding was not mentioned, but at least, thanks to the placebos, the patients did not know what treatment they were getting.

**TRY THIS!** Exercise 1.41

## CASE STUDY REVISITED

Do cell phones cause cancer? The evidence presented at the beginning of the chapter seems to suggest that they might. But let's consider the nature of the evidence. The man who appeared on *Larry King Live* said that his wife, who had died of brain cancer, talked on her cell phone a great deal. But this is just an anecdote. We can conclude nothing about the effects of cell phones on brain cancer, because we don't know how many other people who talked the same amount of time on their cell phones did not get brain cancer, and we don't know how many people who *did* get brain cancer did not use cell phones at all.

The Hardell study was an observational study. The two groups compared were a group that had brain cancer and a group that did not. Clearly, the researchers could not assign subjects to one of these two groups, so all they could do was observe that one group had a different brain cancer rate than the other. However, many potential confounders might explain this difference. For example, critics of the study point out that Hardell examined only brain cancer survivors, which means that he excluded a large percentage of those with brain cancer. Survivors are different from others, and whether or not people survived cancer determined whether they were in this study or not and affected their responses. (Clearly, it's easier to determine how much time people spend talking on the phone if they are alive and able to answer questions.) Researchers in Australia actually did a controlled experiment with mice. They found that the brain cancer risk for mice that received cell phone levels of radiation was no different from that for mice that did not receive the radiation.

Ruling that the evidence was insufficient to suggest a causal relationship between cell phone use and brain cancer, the judge threw out the lawsuit.

# EXPLORING STATISTICS
## CLASS ACTIVITY

## Collecting a Table of Different Kinds of Data

**GOALS**

In this activity, you will learn about different types of data and discuss how to summarize data derived from members of the class.

**MATERIALS**

A board and a marker (or a computer spreadsheet).

**ACTIVITY**

As a class, you will suggest possible variables that would help you describe your class to others. You will then collect data and summarize and describe these variables. How would you describe the people in your class as a group? How does your class compare to another class, maybe the class next door? Suggest variables you might collect on individuals that would help you describe your class and make comparisons. (Examples of such variables might include gender, number of credits currently taken, distance the individual lives from campus, and the like.) Be prepared to give your own values for these variables.

Some variables (such as family income, for example) might be too private; inquiring about others might be offensive to some students. You should not feel compelled to provide information that might embarrass you; try to avoid suggesting variables that you think might embarrass others. Table A shows an example of the format for the input

| Gender | Age | Favorite Class | Units | ____ | ____ |
|--------|-----|----------------|-------|------|------|
|        |     |                |       |      |      |

▲ **TABLE A**

**BEFORE THE ACTIVITY**

If someone asked you to describe the students who make up your class, which characteristics would you focus on?

**AFTER THE ACTIVITY**

1. Which variables that were used are numerical? Which are categorical?

2. Write a one-paragraph summary of the data you collected that you think would help someone understand the composition of your class.

## CHAPTER REVIEW

## Key Terms

statistics, *3*
variation, *4*
data, *4*
variables, *6*
data set, *6*
sample, *6*
population, *6*
numerical variable, *7*
categorical variable, *7*

stacked data, *9*
unstacked data, *9*
two-way table, *10*
frequency, *10*
rate, *13*
causality, *14*
treatment variable, *14*
outcome variable
    (or response variable), *14*

treatment group, *15*
control group
    (or comparison group), *15*
anecdotes, *15*
placebo, *15*
placebo effect, *16*
observational study, *16*
association, *16*
confounding variable, *16*

controlled experiment, *17*
random assignment, *18*
bias, *18*
blind, *19*
double-blind, *19*
random selection, *20*

## Learning Objectives

After reading this chapter and doing the assigned homework problems, you should

- Be able to distinguish between numerical and categorical variables and understand methods for coding categorical variables.

- Know how to find and use rates (including percentages) and understand when and why they are more useful than counts for describing and comparing groups.

- Understand when it is possible to infer a cause-and-effect relationship from a research study and when it is not.

- Be able to explain how confounding variables prevent us from inferring causation, and suggest confounding variables that are likely to occur in some situations.

- Be able to distinguish between observational studies and controlled experiments.

## Summary

Statistics is the science (and art) of collecting and analyzing observations (called data) and communicating your discoveries to others. Often, we are interested in learning about a population on the basis of a sample taken from that population. Here are some questions to consider when first examining a data set:

What are the objects of interest?
What variables were measured?
How were they measured?
What are the units of measurement?
Who collected the data?
How did they collect the data?
Where were the data collected?
Why were the data collected?

With categorical variables, we are often concerned with comparing rates or frequencies between groups. A two-way table is sometimes a useful summary. Always be sure that you are making valid comparisons by comparing proportions or percentages of groups, or that you are comparing the appropriate rates.

Many studies are focused on questions of causality: If we make a change to one variable, will we see a change in the other? Anecdotes are not useful for answering such questions. Observational studies can be used to determine whether associations exist between treatment and outcome variables, but because of the possibility of confounding variables, observational studies cannot support conclusions about causality. Controlled experiments, if they are well designed, do allow us to draw conclusions about causality.

A well-designed controlled experiment should have the following attributes:

A large sample size
Random assignment of subjects to a treatment group or a control group
A double-blind format
A placebo

## Sources

BBC News, February 18, 2010. http://news.bbc.co.uk (accessed May 13, 2010).

Britt, C. "U.S. Doctors Prescribe Drugs for Placebo Effects Survey Shows." Bloomberg.com (accessed October 23, 2008).

Brody, J. 2002. Cellphone: A convenience, a hazard or both? *New York Times*, October 1.

cnn.com. September 11, 2002. Cancer study may be used in Motorola suit (accessed September 23, 2002).

Colombel, Sandborn, Reinisch, Mantzaris, Kornbluth, Rachmilewitz, Lichtiger, D'Haens, Diamond, Broussard, Tang, van der Woude, and Rutgeerts. 2010. Infliximab, azathioprine, or combination therapy for Crohn's disease. *New England Journal of Medicine* 362, 1383–1395.

Dansinger, M., J. Gleason, J. Griffith, H. Selker, and E. Schaefer. 2005. Comparison of the Atkins, Ornish, Weight Watchers, and Zone diets for weight loss and heart disease risk reduction: A randomized trial. *JAMA* 293(1), January 5.

Decmick, B. 2006. Koreans' kimchi adulation, with a side of skepticism. *Los Angeles Times*, May 21.

Fisher, R. 1959. *Smoking: The cancer controversy*. Edinburgh, UK: Oliver and Boyd.

Glionna, J. 2006. Word of mouth spreading peanut milk. *Los Angeles Times*, May 17.

Holcomb, J. 2006. http://www2.irss.unc.edu/ncvital/index.html (accessed May 17, 2006).

Insurance Highway Safety Institute. 2006. http://www.iihs.org/news/2006/hldi_news_060706.pdf (accessed July 12, 2006).

Kaprio, J., and M. Koskenvuo. 1989. Twins, smoking and mortality: A 12-year prospective study of smoking-discordant twin pairs. *Social Science and Medicine* 29(9), 1083–1089.

*McKinsley Quarterly.* 2009. Hal Varian on how the Web challenges managers. Business Technology Office, January 2009.

Nan HongMei, Park JinWoo, Song YoungJin, Yun HyoYung, Park JooSeung, Hyun TaiSun, Youn SeiJin, Kim YongDae, Kang JongWon, and Kim Heon. 2005.

Kimchi and soybean pastes are risk factors of gastric cancer. *World Journal of Gastroenterology*, June 2005.

National Safety Council. 2004. *Injury Facts*, 2004 edition. Itasca, Illinois.

Owen, A., et al. 2010. Letter: Putting brain training to the test. *Nature* advance online publication, April 20, www.nature.com (accessed May 15, 2010).

www.scienceblog.com. 2010. Higher blood pressure found in people living in urban areas (accessed May 2010).

## SECTION EXERCISES

## SECTION 1.2

*The data in Table 1A were collected from one of the authors' statistics classes. The column heads give the variable, and each of the other rows represents a student in the class. Refer to this table for Exercises 1.1–1.4, 1.11, and 1.12.*

| Male | Age | Eye Color | Shoe Size | Height (inches) | Weight (pounds) | Number of Siblings | College Units This Term | Handedness |
|------|-----|-----------|-----------|-----------------|-----------------|--------------------|-------------------------|------------|
| 1 | 20 | Brown | 9.5 | 71 | 170 | 1 | 16 | Right |
| 0 | 19 | Blue | 8 | 66 | 135 | 1 | 13 | Right |
| 0 | 42 | Brown | 7.5 | 63 | 130 | 3 | 5 | Right |
| 0 | 19 | Brown | 8.5 | 65 | 150 | 0 | 15 | Left |
| 1 | 21 | Brown | 11 | 70 | 185 | 5 | 19.5 | Right |
| 0 | 20 | Hazel | 5.5 | 60 | 105 | 2 | 11.5 | Right |
| 1 | 21 | Blue | 12 | 76 | 210 | 2 | 9.5 | Right |
| 0 | 21 | Brown | 10 | 70 | 140 | 0 | 8 | Left |
| 0 | 32 | Brown | 8 | 64 | 165 | 1 | 13.5 | Right |
| 1 | 23 | Brown | 7.5 | 63 | 145 | 6 | 12 | Right |
| 0 | 21 | Brown | 6.5 | 61.5 | 110 | 4 | 14 | Right |

▲ **TABLE 1A**

TRY **1.1 (Example 1)** Are the following variables, from Table 1A, numerical or categorical? Explain.

　a. Handedness

　b. Age

**1.2** Are the following variables, from Table 1A, numerical or categorical? Explain.

　a. Shoe size

　b. Eye color

**1.3 Coding** Explain why the variable Male, in Table 1A, is categorical, even though its values are numbers. Often, it does not make sense, or is not even possible, to add the values of a categorical variable. Does it make sense for Male? If so, what does the sum represent?

**1.4 Coding** Students with fewer than 12 units in the current term are considered part-time. Create a new categorical variable that classifies each student in Table 1A as full-time (12 or more units) or part-time. Call this variable Full. Report the values in a column in the same order as those in the table.

TRY **1.5 Brain Size (Example 2)** In 1991, researchers conducted a study on brain size as measured by pixels in a magnetic resonance imagery (MRI) scan. The numbers are in hundreds of thousands of pixels. The data table provides the sizes of the brains and the gender. (Source: www.lib.stat.cmu.edu/DASL)

　a. Is the format of the data set stacked or unstacked?

　b. Explain the coding. What do 1 and 0 represent?

　c. If you answered "stacked" in part a, then unstack the data into two columns labeled Male and Female. If you answered "unstacked," then stack the data into one column; choose an appropriate name for the stacked variable.

| Brain | Male |
|-------|------|
| 9.4 | 1 |
| 9.5 | 0 |
| 9.5 | 1 |
| 9.5 | 1 |
| 9.5 | 0 |
| 9.7 | 1 |
| 9.9 | 0 |

**1.6 Students' Ages** The accompanying table gives ages for some of the students in two statistics classes. One class met at noon and the other at 5 p.m.

a. Is the format for the data set stacked or unstacked?

b. If you answered "stacked," then unstack the data set. If you answered "unstacked," then stack and code the data set.

| 5 p.m. | Noon |
|--------|------|
| 31 | 24 |
| 34 | 18 |
| 46 | 21 |
| 47 | 20 |
| 50 | 20 |

## SECTION 1.3

TRY **1.7 Changing Answers (Example 3)** On a multiple-choice exam, if you think you have made a mistake, is it better to stick with your first choice, or to erase and change your answer? One of the authors instructed her students to change the answer if the second answer they got seemed better than the first (Change!). Her colleague told his students that their first answer was usually the best and they should avoid changing answers (Don't Change!). After both classes had taken several multiple-choice exams, the exams were studied to see whether answers were erased and replaced with different answers. The results are shown in the accompanying table.

a. What is the percentage of answers that were changed for the students who were told not to change their answers?

b. What is the percentage of answers that were changed for the students who were told to change their answers?

| | | Instruction | |
|---|---|---|---|
| | | Don't Change! | Change! |
| **Behavior** | Changed | 189 | 124 |
| | Unchanged | 29,428 | 14,389 |

**1.8 Changing Answers** Read the information in Exercise 1.7. The table below summarizes the outcomes based on the instructions given by the teacher.

a. For the teacher who said Don't Change!, what percentage of the changes were from wrong to right? What percentage of the changes were from right to wrong? Did the results of the changes raise students' grades or lower them?

b. For the teacher who said Change!, what percentage of the changes were from wrong to right? What percentage of the changes were from right to wrong? Did the results of the changes raise students' grades or lower them?

c. On the basis of these data, what advice would you give to a student who asked you whether it's better to stick with your first hunch or to change your answer?

| | Don't Change! | Change! |
|---|---|---|
| Wrong to right | 91 | 86 |
| Right to wrong | 48 | 24 |
| Wrong to wrong | 50 | 14 |

TRY **1.9 Finding and Using Percentages (Example 4)**

a. A statistics class is made up of 15 men and 23 women. What percentage of the class is male?

b. A different class has 234 students, and 64.1% of them are men. How many men are in the class?

c. A different class is made up of 40% women and has 20 women in it. What is the total number of students in the class?

**1.10 Finding and Using Percentages**

a. A hospital employs 346 nurses and 35% of them are male. How many male nurses are there?

b. An engineering firm employs 178 engineers and 112 of them are male. What percentage of these engineers are *female*?

c. A large law firm is made up of 65% male lawyers, or 169 male lawyers. What is the total number of lawyers at the firm?

**1.11 Women** Find the frequency, proportion, and percentage of women in Table 1A on page 26.

**1.12 Right-Handed People** Find the frequency, proportion, and percentage of right-handed people in Table 1A on page 26.

**1.13 Population Prediction** The *2009 World Almanac and Book of Facts* predicted that the United States will have an elderly population (65 and older) of 88,547,000 in the year 2050 and that this will be 20.2% of the population. What is the total predicted U.S. population in 2050?

**1.14 2007 Population** The *2009 World Almanac and Book of Facts* reported that in 2007 there were 12,608,000 people age 16 or older who had a "go outside the home" disability and that this was 5.5% of the U.S. population (of this age group). These are people who cannot go outside the home without help. How large was the total population (of this age group) in 2007?

g **1.15 Living with AIDS** The table gives the number of people living with AIDS in 2007 in the five states with the largest number of cases, as well as the District of Columbia, as reported by the

U.S. Centers for Disease Control and Prevention (CDC). It also shows the population of those regions at that time from the U.S. Census Bureau.

Find the number of people living with AIDS per thousand residents in each region, and rank the six regions from highest rate (rank 1) to lowest rate (rank 6). Compare these rankings (of rates) with the ranks of total number of cases. If you moved to one of these regions and met 50 random people, in which region would you be most likely to meet at least one person living with AIDS? In which of these regions would you be least likely to meet at least one person with AIDS? *See page 33 for guidance.*

| State | AIDS | Population |
|---|---|---|
| New York | 75,253 | 19,297,729 |
| California | 65,582 | 36,553,215 |
| Florida | 48,059 | 18,251,243 |
| Texas | 34,940 | 23,904,380 |
| Pennsylvania | 19,236 | 12,432,792 |
| District of Columbia | 8,895 | 588,292 |

**1.16 Population Density** The accompanying table gives the population of the six U.S. states with the largest populations in 2008 and the area of these states. (Source: www.infoplease.com)

| State | Population | Area (square miles) |
|---|---|---|
| Pennsylvania | 12,448,279 | 44,817 |
| Illinois | 12,901,563 | 55,584 |
| Florida | 18,328,340 | 53,927 |
| New York | 19,490,297 | 47,214 |
| Texas | 24,326,974 | 261,797 |
| California | 36,756,666 | 155,959 |

a. Determine and report the rankings of the population density by dividing each population by the number of square miles to get the population density (in people per square mile). Use rank 1 for the highest density.

b. If you wanted to live in the state (of these six) with the lowest population density, which would you choose?

c. It you wanted to live in the state (of these six) with the highest population density, which would you choose?

**1.17 Marriage Rates** The number of married people in the United States and the total number of adults in the United States (in millions) are provided in the accompanying table for several years.

Find the percentage of people married in each of the given years, and describe the trend over time. (Source: *2009 World Almanac and Book of Facts*)

| Year | Married | Total |
|---|---|---|
| 1990 | 112.6 | 191.8 |
| 1997 | 116.8 | 207.2 |
| 2000 | 120.2 | 213.8 |
| 2007 | 129.9 | 235.8 |

**1.18 Elderly Population** The number of people in the United States age 65 or older and the total number of people living in the United States (in thousands) are reported in the accompanying table for several years. Find what percentage of population is elderly for each year, and describe the trend of these percentages. (Source: *2009 World Almanac and Book of Facts*)

| Year | Elderly | Population |
|---|---|---|
| 1900 | 3,080 | 75,100 |
| 1920 | 4,933 | 105,000 |
| 1940 | 9,019 | 133,000 |
| 1960 | 16,560 | 180,000 |
| 1980 | 25,549 | 226,000 |
| 2000 | 34,992 | 282,000 |

TRY **1.19 Course Enrollment Rates (Example 5)** Two sections of statistics are offered, the first at 8 a.m. and the second at 10 a.m. The 8 a.m. section has 25 women, and the 10 a.m. section has 15 women. A student claims this is evidence that women prefer earlier statistics classes than men do. What information is missing that might contradict this claim?

* **1.20 Pedestrian Fatalities** In 2008, the National Highway Traffic Safety Administration reported that the number of pedestrian fatalities in Miami-Dade County, Florida, was 65 and that the number in Hillsborough County, Florida, was 45. Can we conclude that pedestrians are safer in Hillsborough County? Why or why not?

## SECTION 1.4

*For Exercises 1.21–1.28, indicate whether the study is an observational study or a controlled experiment.*

**1.21** A student watched picnickers with a large cooler of soft drinks to see whether teenagers were less likely than adults to choose diet soft drinks over regular soft drinks.

**1.22** The medical records of people who live near high-power lines are compared with the medical records of people who do not live near high power lines to see whether people living near high-power lines are more likely to have cancer.

**1.23** A researcher wonders whether the order in which people taste beverages influences their preference. Students are told by the researcher to drink from two unmarked cups. The researcher assigns one group of students to drink Coke first and then Pepsi. The researcher assigns the other group to drink Pepsi first and then Coke. The subjects are asked whether they prefer the first or the second beverage.

**1.24** Patients with Alzheimer's disease are randomly divided into two groups. One group is given a new drug, and the other is given a placebo. After six months they are given a memory test to see whether the new drug fights Alzheimer's better than a placebo.

**1.25** A group of boys is randomly divided into two groups. One group watches violent cartoons for one hour, and the other group watches cartoons without violence for one hour. The boys are then observed to see how many violent actions they take in the next two hours, and the two groups are compared.

**1.26** A local public school encourages, but does not require, students to wear uniforms. The principal of the school compares the grade point averages of students at this school who wear uniforms with the GPAs of those who do not wear uniforms to determine whether those wearing uniforms tend to have higher GPAs.

**1.27** A researcher was interested in the effect of exercise on academic performance in elementary school children. She went to the recess area of an elementary school and identified some students who were exercising vigorously and some who were not. The researcher then compared the grades of the exercisers with the grades of those who did not exercise.

**1.28** A researcher was interested in the effect of exercise on memory. She randomly assigned half of a group of students to run up a stairway three times and the other half to rest for an equivalent amount of time. Each student was then asked to memorize a series of random digits. She compared the numbers of digits remembered for the two groups.

**1.29 Effects of Tutoring on Math Grades** A group of educators want to determine how effective tutoring is in raising students' grades in a math class, so they arrange free tutoring for those who want it. Then they compare final exam grades for the group that took advantage of the tutoring and the group that did not. Suppose the group participating in the tutoring tended to receive higher grades on the exam. Does that show that the tutoring worked? If not, explain why not and suggest a confounding variable.

**1.30 Treating Depression** A doctor who believes strongly that antidepressants work better than "talk therapy" tests depressed patients by treating half of them with antidepressants and the other half with talk therapy. After six months the patients are evaluated on a scale of 1 to 5, with 5 indicating the greatest improvement.

a. The doctor is concerned that if his most severely depressed patients do not receive the antidepressants, they will get much worse. He therefore decides that the most severe patients will be assigned to receive the antidepressants. Explain why this will affect his ability to determine which approach works best.

b. What advice would you give the doctor to improve his study?

c. The doctor asks you whether it is acceptable for him to know which treatment each patient receives and to evaluate them himself at the end of the study to rate their improvement. Explain why this practice will affect his ability to determine which approach works best.

d. What improvements to the plan in part c would you recommend?

TRY **1.31 Treating Heart Disease (Example 6)** A study reported by Hannon et al. looked at people treated for heart disease and reported lower death rates for those who received a coronary bypass (CABG) compared to those who received a stent. A stent is a device to keep blood vessels open. The beginning of the abstract is given below.

*Methods* We identified patients with multivessel disease who received drug-eluting stents or underwent CABG in New York State between October 1, 2003, and December 31, 2004, and we compared adverse outcomes (death, death or myocardial infarction, or repeat revascularization) through December 31, 2005, after adjustment for differences in baseline risk factors among the patients.

The adverse outcomes were greater with stenting. Was this an observational study or a controlled experiment? How do you know? Can we say that the use of CABG causes a better success rate? Why or why not? (Source: E. H. Hannan et al., Drug-eluting stents vs. coronary-artery bypass grafting in multivessel coronary disease, *New England Journal of Medicine*, vol. 358: 331–341, #4, January 24, 2008)

**1.32 Effects of Exercise on Aging** In a study reported on HealthNews.com, Dr. Yonas E. Geda and colleagues at Mayo Clinic in Rochester, Minnesota, analyzed data on 1324 individuals without dementia who completed a questionnaire about physical activity between 2006 and 2008 as part of the Mayo Clinic Study of Aging. The participants had an average age of 80 and were classified as having either normal cognition (1126) or mild cognitive impairment (MCI; 198). Those who reported performing moderate exercise such as yoga, aerobics, or brisk walking during midlife were 39% less likely to develop MCI, and moderate exercise later in life was associated with a 32% reduction.

Was this more likely to have been a controlled experiment or an observational study, and how do you know? Can we say that exercise reduces cognitive impairment? Why or why not?

TRY **1.33 Vitamin C Study (Example 7)** A pair of college students have decided to test vitamin C to see whether it prevents colds. They recruit 500 students with a sign-up sheet, containing a numbered list. The first half of those on the sheet (Numbers 1–250) are asked to take 500 mg of vitamin C per day, and the second half are told *not* to use vitamin C. At the end of the school year, participants are asked how many colds they had. How would you improve this study, and why?

**1.34 Weight Loss Study** A group of overweight people are asked to participate in a weight loss program. Participants are allowed to choose whether they want to go on a vegetarian diet or follow a traditional low-calorie diet that includes some meat. Half of the people choose the vegetarian diet, and half choose to be in the control group and continue to eat meat. Suppose that there is greater weight loss in the vegetarian group.

a. Suggest a plausible confounding variable that would prevent us from concluding that the weight loss was due to the lack of meat in the diet. Explain why it is a confounding variable?

b. Explain a better way to do the experiment that is likely to remove the influence of confounding variables.

**1.35 Are Optimists Healthier?** In 2004 "On Health," the *Consumer Reports* medical newsletter, reported a study on the effects of optimism on health. Researchers studied the emotional styles of 334 subjects and then squirted a cold virus in the noses of these subjects. Those who scored high on "energy, happiness, and relaxation" were significantly less likely to develop colds. Is this study evidence that you can reduce your chance of catching colds by training yourself to be more relaxed and happy? Explain your answer.

**1.36 Reducing Migraines** A 2004 article in *The Nutrition Reporter* stated that melatonin supplements were effective in preventing migraine headaches. The article quoted a Brazilian doctor who used melatonin for 3 months to treat 34 patients with a history of two to eight migraine headaches per month. Thirty-two patients completed the study, and 25 patients reported experiencing at least a 50% reduction in headaches. None of the patients had an increase in headaches. What is the major flaw in this research design that might prevent us from concluding that melatonin helps prevent migraine headaches?

**1.37 Flu Vaccine** In the fall of 2004, there was a shortage in flu vaccine in the United States after it was discovered that vaccines from one of the manufacturers were contaminated. The *New England Journal of Medicine* reported on a study that was done to see whether a smaller dose of the vaccine could be used successfully. If that were the case, then a small amount of vaccine could be divided into more flu shots. In this study, the usual amount of vaccine was injected into half the patients, and the other half of the patients had only a small amount of vaccine injected. The response was measured by looking at the production of antibodies (more antibodies generally result in less risk of getting the flu). In the end, the lower dose of vaccine was just as effective as a higher dose for those under 65 years old. What more do we need to know to be able to conclude that the lower dose of vaccine was equally effective at preventing the flu for those under 65? (Source: Beishe et al., Serum antibody responses after intradermal vaccination against influenza, *New England Journal of Medicine*, 2004)

**1.38 Effect of Confederates on Compliance** A study was conducted to see whether participants would ignore a sign that said, "Elevator may stick between floors. Use the stairs." The study was done at a university dorm on the ground floor of a three-level building. Those who used the stairs were said to be compliant, and those who used the elevator were said to be noncompliant. There were three possible situations, two of which involved confederates. A confederate is a person who is secretly working with the experimenter. In the first situation, there was no confederate. In the second situation, there was a compliant confederate (one who used the stairs), and in the third situation, there was a noncompliant confederate (one who used the elevator). The subjects tended to imitate the confederates. What more do you need to know about the study to determine whether the presence or absence of a confederate causes a change in the compliance of subjects? (Source: Wogalter, et al., (1987), reported in Shaffer & Merrens (2001), *Research Stories in Introductory Psychology,* Allyn and Bacon: Boston.)

TRY **1.39 Vitamin C and Allergies (Example 8)** The March 2005 issue of *The Nutrition Reporter* reported on the effects of vitamin C in breast milk on preventing allergies in infants. Researchers analyzed levels of vitamin C in the breast milk of some new mothers who had decided whether to take vitamin C during their pregnancy. The highest levels of vitamin C were associated with a 30 percent lower risk of allergies in the infants. Was this an observational study or a controlled experiment? On the basis of this study, can you conclude that vitamin C lowers the risk of allergies in infants? Why or why not?

**1.40 Childhood Exposure to Tobacco Smoke and Emphysema** A headline on the website e! Science News in December 2009 stated, "Exposure to tobacco smoke in childhood home [is] associated with early emphysema in adulthood."

Is this likely to have been a controlled experiment or an observational study? How do you know? Is it possible to conclude, on the basis of this study, that exposure to tobacco smoke for children causes early emphysema in adulthood?

TRY **1.41 Effects of Light Exposure (Example 9)** A study carried out by Baturin and colleagues looked at the effects of light on female mice. Fifty mice were randomly assigned to a regimen of 12 hours of light and 12 hours of dark (LD), while another fifty mice were assigned to 24 hours of light (LL). Researchers observed the mice for two years, beginning when the mice were two months old. Four of the LD mice and 14 of the LL mice developed tumors. The accompanying table summarizes the data. (Source: Baturin et al., The effect of light regimen and melatonin on the development of spontaneous mammary tumors in mice, *Neuroendocrinology Letters*, 2001)

|  | LD | LL |
|---|---|---|
| Tumors | 4 | 14 |
| No tumors | 46 | 36 |

a. Determine the percentage of mice that developed tumors from each group (LL and LD). Compare them and comment.

b. Was this a controlled experiment or an observational study? How do you know?

c. Can we conclude that light for 24 hours a day causes an increase in tumors in mice? Why or why not?

**1.42 Preschool Attendance and High School Graduation Rates** The Perry preschool project was started in the early 1960s by David Weikart in Ypsilanti, Michigan. In this project, 123 African American children were randomly assigned to one of two groups: One group enrolled in the Perry preschool, and one group did not enroll in preschool. Follow-up studies were done on the children after 27 years and after 40 years. The accompanying table shows whether the students graduated from high school or not. The treatment variable is attendance at preschool (or not) and the response variable is graduation from high school (or not). (Source: Schweinhart, L.J. et al., The High/Scope Perry Preschool Study through age 40. *Journal of Experimental Criminology,* (1)1, April, 2005)

|  | Preschool | No Preschool |
|---|---|---|
| Grad HS | 37 | 29 |
| No Grad | 20 | 35 |

a. What percentage of those who went to preschool graduated from high school?

b. What percentage of those who did not go to preschool graduated from high school?

c. Is the better graduation rate from those who attended preschool or those who did not attend preschool?

d. Can you conclude that attending Perry preschool affects graduation rates? Explain.

## CHAPTER REVIEW EXERCISES

**1.43 Obesity and Marital Status** A 2009 study analyzed data from the National Longitudinal Study of Adolescent Health. Participants were studied into adulthood. Each study participant was categorized as to whether they were obese or not and whether they were dating, cohabiting, or married. Obesity was defined as having a Body Mass Index of 30 or more. (Source: The and Larsen, Entry into romantic partnership is associated with obesity, *Obesity*, vol. 17, no. 7: 1441–1447, 2009)

|  | Dating | Cohabiting | Married |
|---|---|---|---|
| Obese | 81 | 103 | 147 |
| Not Obese | 359 | 326 | 277 |
|  | 440 | 429 | 424 |

a. What percentage of those who were dating were obese?
b. What percentage of those who were cohabiting were obese?
c. What percentage of those who were married were obese?
d. Which group had the highest rate of obesity? Does this imply that marital status causes obesity? Why or why not? If not, can you name a confounding variable?

**1.44 Are Cell Phones More Dangerous Than Cigarettes?** Fox News reported on a study in 2008, headed by Dr. Vini Khurana, that suggested that there is some evidence that using cell phone head sets for 10 years or more doubles the risk of brain cancer. He said that three billion people now use cell phones, but only one billion smoke and that smoking kills about five million each year. He concluded that cell phones are more dangerous than cigarettes, and so he suggested that people should avoid using cell phones. (Source: www.independent.co.uk)

a. Is this study more likely to be a controlled experiment or an observational study? Explain.
b. Do you think this study shows that using cell phones causes brain cancer? Also, does it show that cell phones are more dangerous than cigarettes? Why or why not?

**1.45 Probation** A statistics student conducted a study on young male and female criminals 15 years of age and under who were on probation. The purpose of the study was to see whether there was an association between type of crime and gender. The subjects of the study lived in Ventura County, California. Violent crimes involve physical contact such as hitting or fighting. Nonviolent crimes are vandalism, robbery, or verbal assault. The raw data are shown in the accompanying table; v stands for violent, n for nonviolent, b for boy, and g for girl.

a. Make a two-way table that summarizes the data. Label the columns (across the top) Boy and Girl. Label the rows Violent and Nonviolent.
b. Find the percentage of girls on probation for violent crimes and the percentage of boys on probation for violent crimes, and compare them.

c. Are the boys or the girls more likely to be on probation for violent crimes?

| Gen | Viol? | Gen | Viol? | Gen | Viol? |
|---|---|---|---|---|---|
| b | n | b | n | g | n |
| b | n | b | n | g | n |
| b | n | b | n | g | n |
| b | n | b | n | g | v |
| b | n | b | v | g | v |
| b | n | b | v | g | v |
| b | n | b | v | g | v |
| b | n | b | v | g | v |
| b | n | b | v | g | v |
| b | n | b | v | g | v |
| b | n | b | v | g | v |
| b | n | b | v | g | v |
| b | n | b | v | g | v |
| b | n | b | v |  |  |
| b | n | g | n |  |  |

**1.46 Scorpion Antivenom** A study was done on children (6 months to 18 years of age) who had (nonlethal) scorpion stings. Each child was randomly assigned to receive an experimental antivenom or a placebo. Good results were no symptoms after four hours. Make a summary of the data in the form of a two-way table. Label the columns Antivenom and Placebo. Label the rows Better and Not Better. Compare the percentage better for the antivenom group and the placebo group. (Source: Boyer, Leslie V. et al., Antivenom for critically ill children with neurotoxicity from scorpion stings, *New England Journal of Medicine*, vol. 360: 2090–2098, no. 20, May 14, 2009)

| Antivenom | Better | Antivenom | Better |
|---|---|---|---|
| 1 | 1 | 0 | 0 |
| 1 | 1 | 0 | 0 |
| 1 | 1 | 0 | 0 |
| 1 | 0 | 0 | 1 |
| 1 | 1 | 0 | 0 |
| 1 | 1 | 0 | 0 |
| 1 | 1 | 1 | 1 |
| 0 | 0 |  |  |

**\* 1.47 Activated Charcoal** A man posting on an Internet message board claimed that his wife was sick for more than 6 months with many problems, including nausea, confusion, and night sweats. Doctors were unable to help her. However, when she started taking "activated charcoal," her condition began to improve. The posting on the message board also explained that his wife had possibly ingested herbicides and might have been poisoned by these "toxins."

Describe the design of a controlled experiment to determine whether activated charcoal can cure people suffering from what we will call "toxin disease," assuming that such a disease exists. Assume that you have access to 200 people suffering from this condition. Your description of your experiment should address all of the major features of a controlled experiment.

**\* 1.48 Allergy Experiment** A post on an Internet message board reported that a person suffering from allergies took grapeseed extract and had a remarkable decrease in allergy attacks.

Describe the design of a controlled experiment to determine whether grapeseed extract can relieve the symptoms of allergy sufferers. Assume that you have access to 200 people suffering from allergies. Your description of your experiment should address all of the major features of a controlled experiment.

**1.49 Clinical Trials** In a study of more than 2000 British patients with heart failure, patients were asked whether they would be willing to participate in a clinical trial. After 55 months, those who said they would be willing to participate were about half as likely to have died as those who were not willing to participate. Does this imply that if you have heart failure, you can increase your chance of living by being willing to participate in clinical trials? Why or why not? (Source: Clark et al., Is taking part in clinical trials good for your health? A cohort study, *European Journal of Heart Failure*, vol. 11: 1078–1083, 2009)

**1.50 Recidivism Rate** In May 2010 *Time* magazine published an article about an innovative Norwegian prison, Halden Fengsel, which provides prisoners with dorm-like accommodations, flat-screen televisions, and mini-fridges. The staff does not carry guns. Within two years of release from Halden Fengsel, the rate of recidivism (being imprisoned again) is 20%. In the United States that rate is between 50% and 60%. Is this sufficient evidence to conclude that if the United States changed its prisons to be more like Halden Fengsel, the recidivism rate in the United States would go down? Explain.

**1.51 Death Row and Head Trauma** A study conducted by Lewis et al. in 1986 looked at 14 juveniles awaiting execution. They found that 57% (8 of the 14) had had a serious brain injury. Can we conclude that head trauma causes bad behavior later in life? What primary factor is not present here that should be present in both observational studies and controlled experiments? (Source: Psychiatric, neurological, and psychoeducational characteristics of 15 death row inmates in the United States, *American Journal of Psychiatry*, vol. 143: 838–845. 1986)

**1.52 Brief Exercise and Diabetes** As part of a study, sixteen young men performed high-intensity exercise that totaled only 15 minutes in a two-week period. At the end of two weeks, several (but not all) tests for diabetes, such as an insulin sensitivity test, showed improvement. Do these results indicate that brief, high-intensity exercise causes an improvement in markers for diabetes? What essential component of both controlled experiments and observational studies is missing from this study? (Source: Babraj et al., Extremely short duration high intensity interval training substantially improves insulin action in young healthy males, *BMC Endocrine Disorders*, vol. 9: 3doi: 10.1186/1472-6823-9-3, January 2009)

# gUIDED EXERCISES

**g 1.15 Living with AIDS** The accompanying table gives the numbers of people living with AIDS in 2007 in the five states with the largest numbers of cases, as well as the District of Columbia, as reported by the U.S. Centers for Disease Control and Prevention (CDC). It also shows the population of those regions at that time, from the U.S. Census Bureau.

**Question** Find the number of people living with AIDS per thousand in each region, and rank the six regions from highest rate (rank 1) to lowest rate (rank 6). Compare these rankings (of rates) with the ranks of total number of cases. If you moved to one of these regions and met 50 random people, in which region would you be most likely to meet at least one person living with AIDS? In which of these regions would you be least likely to meet at least one person living with AIDS?

**Step 1** ▶ **Figure out the populations of the remaining regions in thousands, and add them to the table.**

**Step 2** ▶ **For each region, divide the number of people living with AIDS by the population in thousands, and fill in column 6.**

**Step 3** ▶ **Enter the ranks for the rates of AIDS patients per 1000 population, using 1 for the largest value and 6 for the smallest.**

**Step 4** ▶ **Are the ranks for the rates the same as the ranks for the numbers of cases? If not, describe at least one difference.**

**Step 5** ▶ **Finally, if you moved to one of these regions and met 50 random people, in which region would you be most likely to meet at least one person living with AIDS? In which region would you be least likely to meet at least one person living with AIDS?**

| State | AIDS | Rank Cases | Population | Population (thousands) | AIDS per 1000 population | Rank Rate |
|---|---|---|---|---|---|---|
| New York | 75,253 | 1 | 19,297,729 | 19,298 | 3.90 | 2 |
| California | 65,582 | 2 | 36,553,215 | 36,553 | | |
| Florida | 48,059 | 3 | 18,251,243 | | | |
| Texas | 34,940 | 4 | 23,904,380 | | | |
| Pennsylvania | 19,236 | 5 | 12,432,792 | | | |
| District of Columbia | 8,895 | 6 | 588,292 | | | |

# 2 Picturing Variation with Graphs

Any collection of data exhibits variation. The most important tool for organizing this variation is called the distribution of the sample, and visualizing this distribution is the first step in every statistical investigation. We can learn much about a numerical variable by focusing on three components of the distribution: the shape, the center, and the variability, or horizontal spread. Examining a graph of a distribution can lead us to deeper understanding of the situation that produced the data.

One of the major concepts of statistics is that although individual events are hard to predict, large numbers of events usually exhibit predictable patterns. The search for patterns is a key theme in science and business. An important first step in this search is to identify and visualize the key features of your data.

Using graphics to see patterns and identify important trends or features is not new. One of the earliest statistical graphs dates back to 1786, when a Scottish engineer named William Playfair published a paper examining whether there was a relationship between the price of wheat and wages. To help answer this question, Playfair produced a graph (shown in Figure 2.1) that is believed to be the first of its kind. This graph became the prototype of two of the most commonly used tools in statistics: the bar chart and the histogram.

Graphics such as these can be extraordinarily powerful ways of organizing data, detecting patterns and trends, and communicating findings. The graphs that we use have changed somewhat since Playfair's day, but graphics are of fundamental importance to analyzing data. The first step in any statistical analysis

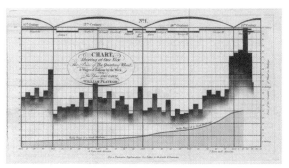

▲ **FIGURE 2.1** Playfair's chart explores a possible relationship between wages and the price of wheat. (Source: Playfair 1786)

is to make a picture of some kind in order to check our intuition against the data. If our intuitions are wrong, it could very well be because the world works differently than we thought. Thus, by making and examining a display of data (as illustrated in this chapter's Case Study), we gain some insight into how the world works.

In Chapter 1 we discussed some of the methods used to collect data. In this chapter we'll cover some of the basic graphics used in analyzing the data we collect. Then, in Chapter 3, we'll comment more precisely on measuring and comparing key features of our data.

## CASE STUDY

## Student-to-Teacher Ratio at Colleges

Are private four-year colleges better than public four-year colleges? That depends on what you mean by "better." One measure of quality that many people find useful (and there are many other ways to measure quality) is the student-to-teacher ratio: the number of students enrolled divided by the number of teachers. For schools with small student-to-teacher ratios, we expect class sizes to be small; students can get extra attention in a small class.

The data in Table 2.1 on the next page were collected from some schools that award four-year degrees. The data are for the 2004–2005 academic year; 85 private colleges and 57 state-supported (public) colleges were sampled. Each ratio was rounded to the nearest whole number for simplicity. For example, the first private

| 85 Private Colleges | | | | | | | | | 57 Public Colleges | | | | | |
|---|---|---|---|---|---|---|---|---|---|---|---|---|---|---|
| 10 | 20 | 24 | 10 | 11 | 12 | 20 | 14 | 9 | 15 | 20 | 22 | 18 | 1 | 23 |
| 16 | 10 | 11 | 13 | 15 | 32 | 14 | 7 | 7 | 20 | 20 | 21 | 20 | 29 | 21 |
| 10 | 13 | 11 | 10 | 10 | 17 | 11 | 11 | 21 | 15 | 25 | 20 | 13 | 23 | 16 |
| 15 | 12 | 15 | 38 | 10 | 14 | 8 | 13 | 11 | 15 | 15 | 16 | 16 | 18 | 17 |
| 16 | 5 | 17 | 14 | 9 | 11 | 10 | 16 | 55 | 20 | 20 | 21 | 15 | 16 | 13 |
| 11 | 13 | 13 | 16 | 9 | 8 | 11 | 16 | | 15 | 16 | 18 | 16 | 21 | 17 |
| 14 | 11 | 16 | 9 | 14 | 32 | 10 | 15 | | 28 | 16 | 18 | 11 | 17 | 17 |
| 12 | 30 | 10 | 10 | 9 | 10 | 12 | 24 | | 12 | 15 | 17 | 20 | 20 | |
| 16 | 10 | 12 | 11 | 6 | 15 | 13 | 11 | | 17 | 9 | 15 | 12 | 22 | |
| 12 | 10 | 11 | 15 | 11 | 8 | 11 | 11 | | 13 | 14 | 17 | 20 | 24 | |

▲ **TABLE 2.1** Ratio of Students to Teachers at Private and Public Colleges
(Source: *2006 World Almanac and Book of Facts*)

college listed has a student-to-teacher ratio of 10, which means that there are about 10 students for every teacher. What differences do you expect between the two groups? What similarities do you anticipate?

It is nearly impossible to compare the two groups without imposing some kind of organization on the data. In this chapter you will see several ways in which we can graphically organize groups of data like these so that we can compare the two types of colleges. At the end of this chapter, you'll see what the right graphical summaries can tell us about how these types of colleges compare.

SECTION 2.1

# Visualizing Variation in Numerical Data

One of the most important conceptual tools in statistics and data analysis is the distribution. The **distribution of a sample** of data is simply a way of organizing the data. Think of the distribution of the sample as a list that records (1) the values that were observed and (2) the frequencies of these values. **Frequency** is another word for *count*, of how many times the value occurred in the collection of data.

 **KEY POINT** The distribution of a sample is one of the central organizational concepts of data analysis. The distribution organizes data by recording all of the values observed in a sample, as well as how many times each value was observed.

Distributions are important because they capture much of the information we need in order to make comparisons between groups, examine data for errors, and learn about real-world processes. Distributions allow us to examine the variation of the data in our sample, and from this variation we can often learn more about the world.

The first step of almost every statistical investigation is to visualize the distribution of the sample. By creating an appropriate graphic, we can see patterns that might otherwise escape our notice.

For example, here are some raw data from the National Collegiate Athletic Association (NCAA), available online. This set of data shows the number of goals scored per game for Division III women soccer players for all games in the 2009 season. (Division III schools are colleges or universities that are not allowed to offer scholarships to athletes.) To make the data set smaller, we show only first-year students.

0.727, 0.737, 0.737, 0.737, 0.750, 0.750, 0.750, 0.762, 0.765, 0.765, 0.789, 0.789, 0.813, 0.833, 0.875, 0.889, 0.909, 0.941, 0.947, 1.053, 1.091, 1.158, 1.294, 1.400

This list includes only the values. A distribution lists the values and also the frequencies. The distribution of this sample is shown in Table 2.2.

It's hard to see patterns when the distribution is presented as a table. A picture makes it easier for us to answer questions such as "What's the typical number of goals per game scored by a player?" and "Is 1.4 goals per game an unusually high number?" Data are also available for male soccer players and for other divisions and classes. A picture would make it easier to compare the numbers of goals for different groups. For example, in a season, do men typically score more goals per game or fewer goals per game than women?

When examining distributions, we use a two-step process:

1. See it.

2. Summarize it.

In this section we explain how to visualize the distribution. In the next section, we discuss the characteristics you should look for to help you summarize it.

All of the methods we use for visualizing distributions are based on the same idea: Make some sort of mark that indicates how many times each value occurred in our data set. In this way, we get a picture of the sample distribution so that we can see at a glance which values occurred and how often.

Two very useful methods for visualizing distributions of numerical variables are dotplots and histograms. Dotplots are simpler; histograms are more commonly used and perhaps more useful.

## Dotplots

In constructing a **dotplot**, we simply put a dot above a number line where each value occurs. We can get a sense of frequency by seeing how high the dots stack up. Figure 2.2 shows a dotplot for the number of goals for first-year female soccer players.

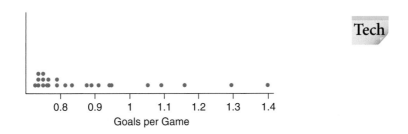

Tech

With this simple picture, we can see more than we could from Table 2.2. We can see from this dotplot that most women scored an average of less than 1 goal per game, but a few scored more. Also, we can see that the woman who scored 1.4 goals per game was exceptional within this group. Not only is 1.4 the largest value in this data set, but it stands apart from most of the others by a large gap.

| Value | Frequency |
|-------|-----------|
| 0.727 | 1 |
| 0.737 | 3 |
| 0.750 | 3 |
| 0.762 | 1 |
| 0.765 | 2 |
| 0.789 | 2 |
| 0.813 | 1 |
| 0.833 | 1 |
| 0.875 | 1 |
| 0.889 | 1 |
| 0.909 | 1 |
| 0.941 | 1 |
| 0.947 | 1 |
| 1.053 | 1 |
| 1.091 | 1 |
| 1.158 | 1 |
| 1.294 | 1 |
| 1.400 | 1 |

▲ **TABLE 2.2** Distribution of the number of goals per game scored by first-year women soccer players in NCAA Division III in 2009.

**Details**

**Making Dotplots**
Dotplots are easy to make with pen and paper, but don't worry too much about recording values to great accuracy. The purpose of a plot like this is to help us see the overall shape of the distribution, not to record details about individual observations.

◀ **FIGURE 2.2** Dotplot of number of goals per game scored by first-year women soccer players in NCAA Division III, 2009. Each dot represents a soccer player. Note that the horizontal axis begins at 0.7.

**SNAPSHOT**　THE DOTPLOT

| | | |
|---|---|---|
| **WHAT IS IT?** | ▶ | A graphical summary. |
| **WHAT DOES IT DO?** | ▶ | Shows a picture of the distribution of a numerical variable. |
| **HOW DOES IT DO IT?** | ▶ | Each observation is represented by a dot on a number line. |
| **HOW IS IT USED?** | ▶ | To see patterns in samples of data with variation. |

## Histograms

While dotplots have one dot for each observation in the data set, **histograms** produce a smoother graphic by grouping observations into intervals, called bins. These groups are formed by dividing the number line into bins of equal width and then counting how many observations fall into each bin. To represent the bins, histograms display vertical bars, where the height of each bar is proportional to the number of observations inside that bin.

For example, with the goals scored in the dotplot in Figure 2.2, we could create a series of bins that go from 0.7 to 0.8, 0.8 to 0.9, 0.9 to 1.0, and so on. Twelve women scored between 0.7 and 0.8 goals per game, so the first bar has a height of 12. The second bin contains four observations and consequently has a height of 4. The finished graph is shown in Figure 2.3. (Note that some statisticians use the word *interval* in place of *bin*. You might even see another word that means the same thing: *class*.) Figure 2.3 shows, among other things, that three women scored between 0.9 and 1.0 goals per game, two scored between 1.0 and 1.1 goals per game, and so on.

▶ **FIGURE 2.3** Histogram of goals per game for female first-year soccer players in NCAA Division III, 2009. The first bar, for example, tells us that 12 players scored between 0.7 and 0.8 goals per game during the season.

**Tech**

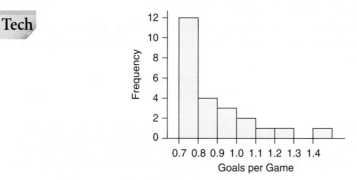

Making a histogram requires paying attention to quite a few details. For example, we need to decide on a rule for what to do if an observation lands exactly on the boundary of two bins. In which bin would we place an observation of 0.8 goals per game? A common rule is to decide always to put "boundary" observations in the bin on the right, but we could just as well decide always to put them in the bin on the left. The important point is to be consistent. The graphs here use the right-hand rule and put boundary values in the bin to the right.

Figure 2.4 shows two more histograms of the same data. Even though they display the same data, they look very different—both from each other and from Figure 2.3. Why?

Changing the width of the bins in a histogram changes its shape. Figure 2.3 has bins with width of 0.1. In contrast, Figure 2.4a has much smaller bins, and Figure 2.4b has wider bins. Note that when we use small bins, we get a spiky histogram. When we use wider bins, the histogram gets less spiky. Using wide bins hides more detail. If you chose very wide bins, you would have no details at all. You would see just one big rectangle!

How large should the bins be? Too small and you see too much detail (as in Figure 2.4a). Too large and you don't see enough (as in Figure 2.4b). Most computer software will automatically make a good choice. Our software package (Minitab, in this case) automatically chose the binwidth of 0.1 that you see in Figure 2.3. Still,

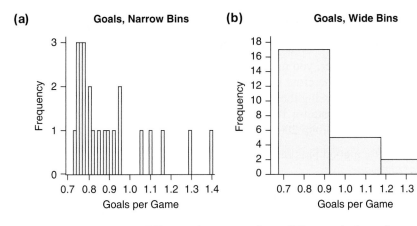

**(a)** Goals, Narrow Bins

**(b)** Goals, Wide Bins

◀ **FIGURE 2.4** Two more histograms of goals scored per game, the same data as in Figure 2.3. **(a)** This histogram has narrow bins and is spiky. **(b)** This histogram has wide bins and offers less detail.

if you can, you should try different sizes to see how different choices change your impression of the distribution of the sample. Fortunately, most statistical software packages make it quite easy to change the bin width.

## Relative Frequency Histograms

A variation on the histogram (and statisticians, of course, love variation) is to change the units of the vertical axis from frequencies to relative frequencies. A **relative frequency** is simply a proportion. So instead of reporting that the first bin had 12 observations in it, we would report that the proportion of observations in the first bin was $12/24 = 0.50$. We divide by 24 because there were a total of 24 observations in the data set. Figure 2.5 is the same as the first histogram shown for the distribution of goals per game (Figure 2.3); however, Figure 2.5 reports relative frequencies, and Figure 2.3 reports frequencies.

Using relative frequencies does not change the shape of the graph; it just communicates different information to the viewer. Rather than answering the question "How many players scored between 0.7 and 0.8 goals per game?" (12 players), it now answers the question "What *proportion* of players scored between 0.7 and 0.8 goals per game?" (0.50).

◀ **FIGURE 2.5** Relative frequency histogram of goals scored per game by first-year women soccer players in NCAA Division III, 2009.

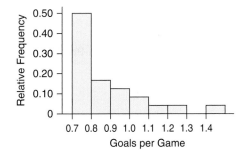

**SNAPSHOT** THE HISTOGRAM

| | |
|---|---|
| **WHAT IS IT?** ▶ | A graphical summary for numerical data. |
| **WHAT DOES IT DO?** ▶ | Shows a picture of the distribution of a numerical variable. |
| **HOW DOES IT DO IT?** ▶ | Observations are grouped into bins, and bars are drawn to show how many observations (or what proportion of observations) lie in each bin. |
| **HOW IS IT USED?** ▶ | By smoothing over details, histograms help our eyes pick up on more important, large-scale patterns. Be aware that making the bins wider hides detail, and making the bins smaller can show too much detail. The vertical axis can display frequency, relative frequency, or percents. |

## EXAMPLE 1 Visualizing Bar Exam Pass Rates at Law Schools

In order to become a lawyer, you must pass your state's bar exam. When you are choosing a law school, it therefore makes great sense to choose one that has a high percentage of graduates passing the bar exam. The Internet Legal Research Group website provides the pass rate for 181 law schools in the United States in 2006 (Internet Research Legal Group 2008).

**QUESTION** Is 80% a good pass rate for a law school?

**SOLUTION** This is a subjective question. An 80% pass rate might be good enough for a prospective student, but another way of looking at this is to determine whether there are many schools that do better than 80%, whether 80% is a typical pass rate, or whether there are very few schools that do so well. Questions such as these can often be answered by considering the distribution of our sample of data, so our first step is to choose an appropriate graphical presentation of the distribution.

Either a dotplot or a histogram would show us the distribution. Figure 2.6a shows a dotplot generated by Minitab. It's somewhat difficult (but not impossible) to read a dotplot with 181 observations; there's too much detail to allow us to answer broad questions such as these. It is easier to use a histogram, as in Figure 2.6b. In this histogram, each bar has a width of 10 percentage points, and the *y*-axis tells us how many law schools had a pass rate within those limits.

▶ **FIGURE 2.6 (a)** A dotplot for the bar-passing rate for 181 law schools. Each dot represents a law school, and the dot's location indicates the bar-passing rate for that school. **(b)** A histogram shows the same data as in part (a), except that the details have been smoothed.

**(a)**

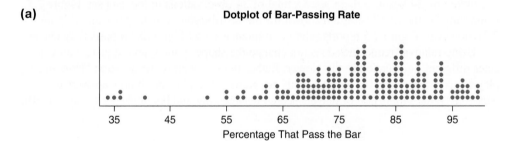

**(b)**

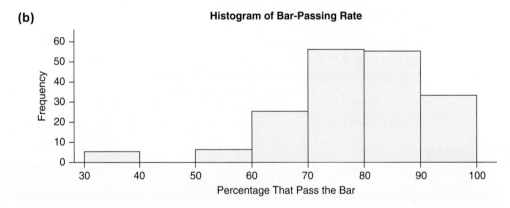

We see that a little more than 50 law schools had a pass rate between 80% and 90%. Another 30 or so had a pass rate over 90%. Adding these together, about 80 schools had a pass rate higher than 80%, and this is $80/181 = 0.4420$ or about 44% of the schools. This tells us that a pass rate of 80% is not all that unusual; about 44% of all law schools do this well or better.

**TRY THIS!** Exercise 2.5

## Stemplots

Stemplots, which are also called stem-and-leaf plots, are useful for visualizing numerical variables when you don't have access to technology and the data set is not large. Stemplots are also useful if you want to be able to easily see the actual values of the data.

To make a **stemplot**, divide each observation into a "stem" and a "leaf." The **leaf** is the last digit in the observation. The **stem** contains all the digits that precede the leaf. For the number 60, the *6* is the stem and the *0* is the leaf. For the number 632, the *63* is the stem and the *2* is the leaf. For the number 65.4, the *65* is the stem and the *4* is the leaf.

A stem-and-leaf plot can help us understand data such as drinking behaviors. Alcohol is a big problem at many colleges and universities. For this reason, a collection of college students who said that they drink alcohol were asked how many alcoholic drinks they had consumed in the last seven days. Their answers were

    1, 1, 1, 1, 1, 2, 2, 2, 3, 3, 3, 3, 3, 4, 5, 5, 5, 6, 6, 6, 8, 10, 10, 15, 17, 20, 25, 30, 30, 40

For one-digit numbers, imagine a 0 at the front. The observation of 1 drink becomes 01, the observation of 2 drinks becomes 02, and so on. Then each observation is just two digits; the first digit is the stem, and the last digit is the leaf. Figure 2.7 shows a stemplot of these data.

If you rotate a stemplot 90 degrees counterclockwise, it looks not too different from a histogram. Unlike histograms, stemplots display the actual values of the data. With a histogram, you know only that the values fall somewhere within an interval.

Stemplots are often organized with the leaves in order from lowest to highest, which makes it easier to locate particular values, as in Figure 2.7. This is not necessary, but it makes the plot easier to use.

From the stemplot, we see that most students drink a moderate amount in a week but that a few drink quite a bit. Forty drinks per week is almost six per day, which qualifies as problem drinking by some physicians' definitions.

Figure 2.8 shows a stemplot of some exam scores. Note the empty stems at 4 and 5, which show that there were no exam grades between 40 and 59. Most of the scores are between 60 and 100, but one student scored very low relative to the rest of the class.

Tech

| Stem | Leaves |
|------|--------|
| 0 | 11111222333334555666 8 |
| 1 | 0057 |
| 2 | 05 |
| 3 | 00 |
| 4 | 0 |

▲ **FIGURE 2.7** A stemplot for alcoholic drinks consumed by college students. Each digit on the right (the leaves) represents a student. Together, the stem and the leaf indicate the number of drinks for one student who is a drinker.

| Stem | Leaves |
|------|--------|
| 3 | 8 |
| 4 | |
| 5 | |
| 6 | 0257 |
| 7 | 00145559 |
| 8 | 0023 |
| 9 | 0025568 |
| 10 | 00 |

▲ **FIGURE 2.8** A stemplot for exam grades. Two students had scores of 100, and no students scored in the 40's or 50's.

**SNAPSHOT**   **THE STEMPLOT**

| | |
|---|---|
| **WHAT IS IT?** ▶ | A graphical summary for numerical data. |
| **WHAT DOES IT DO?** ▶ | Shows a picture of the distribution of a numerical variable. |
| **HOW DOES IT DO IT?** ▶ | Numbers are divided into leaves (the last digit) and stems (the preceding digits). Stems are written in a vertical column, and associated leaves are "attached." |
| **HOW IS IT USED?** ▶ | In very much the same way as a histogram. It is often useful when technology is not available and the data set is not large. |

SECTION 2.2

# Summarizing Important Features of a Numerical Distribution

When examining a distribution, pay attention to describing the shape of the distribution, the **typical value (center)** of the distribution, and the **variability (spread)** in the distribution. The typical value is subjective, but the location of the center of a distribution often gives us an idea which values are typical for this variable. The variability is reflected in the amount of horizontal spread the distribution has.

> **KEY POINT**   When examining distributions of numerical data, pay attention to the shape, center, and horizontal spread.

Figure 2.9 compares distributions for two groups. You've already seen histogram (a)—it's the histogram for the goals scored per game by first-year women soccer players in Division III. Histogram (b) shows the same goals per game for first-year male soccer players in Division III. How do these two distributions compare?

▶ **FIGURE 2.9** Distributions of the goals scored per game for **(a)** first-year women and **(b)** first-year men in Division III soccer in 2009.

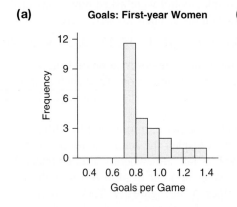

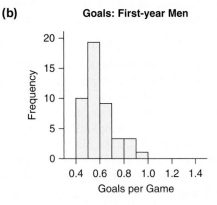

1. *Shape.* Are there any interesting or unusual features about the distributions? Are the shapes very different? (If so, this might be evidence that men play the game differently than women.)

2. *Center.* What is the typical value of each distribution? Is the typical number of goals scored per game different for men than for women?

3. *Spread.* The horizontal spread presents the variation in goals per game for each group. How do the amounts of variation compare? If one group has low variation, it suggests that the soccer skills of the members are pretty much the same. Lots of variability might mean that there is a wider variety of skill levels.

Let's consider these three aspects of a distribution one at a time.

## Shape

You should look for three basic characteristics of a distribution's shape:

1. Is the distribution symmetric or skewed?

2. How many mounds appear? One? Two? None? Many?

3. Are unusually large or small values present?

### Symmetric or Skewed?

A symmetric distribution is one in which the left-hand side of the graph is roughly a mirror image of the right-hand side. The idealized distributions in Figure 2.10 show several possibilities. Figure 2.10a is a **symmetric distribution** with one mound. (Statisticians often describe a distribution with this particular shape as a **bell-shaped distribution**. Bell-shaped distributions play a major role in statistics, as you will see throughout this book.)

▶ **FIGURE 2.10** Sketches of **(a)** a symmetric distribution and **(b)** a right-skewed distribution.

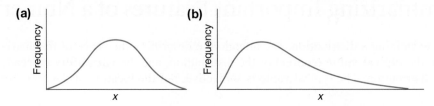

---

**Details**

**Vague Words**
For now, we have left the three ideas *shape*, *center*, and *spread* deliberately vague. In Chapter 3 you'll learn ways of measuring more precisely where the center is and how much spread exists. However, the first task in a data analysis is to examine a distribution to informally evaluate the shape, center, and spread.

Figure 2.10b represents a nonsymmetric distribution with a skewed shape that has one mound. Note that it has a "tail" that extends out to the right (toward larger values). Because the tail goes to the right, we call it a **right-skewed distribution**. This is a typical shape for the distribution of a variable in which most values are relatively small but that also has a few very large values. If the tail goes to the left, it is a **left-skewed distribution**, where most values are relatively large but that also has a few very small values.

Figure 2.11 shows a histogram of 123 college women's heights. How would you describe the shape of this distribution? This is a good real-life example of a symmetric distribution. Note that it is not perfectly symmetric, but you will never see "perfect" in real-life data.

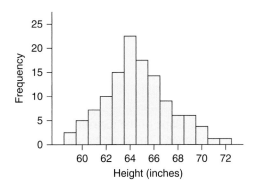

◀ **FIGURE 2.11** Histogram of heights of women. (Source: Brian Joiner in Tufte 1983)

Suppose we asked a sample of people how many hours of TV they watched in a typical week. Would you expect a histogram of these data to be bell-shaped? Probably not. The smallest possible value for this data set would be 0, and most people would probably cluster near a common value. However, a few people probably watch quite a bit more TV than most other people. Figure 2.12 shows the actual histogram. We've added an arrow to emphasize that the tail of this distribution points to the right; this is a right-skewed distribution.

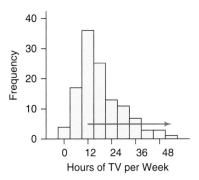

▲ **FIGURE 2.12** This data set on TV hours viewed per week is skewed to the right (Source: Minitab Program)

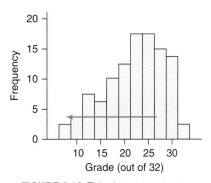

▲ **FIGURE 2.13** This data set on test scores is skewed to the left. (Source: Ryan 2006)

Figure 2.13 shows a left-skewed distribution of test scores. This is the sort of distribution that you should hope your next exam will have. Most people scored pretty high, so the few people who scored low create a tail on the left-hand side. A very difficult test, one in which most people scored very low and only a few did well, would be right-skewed.

Another circumstance in which we often see skewed distributions is when we collect data on people's income. When we graph the distribution of incomes of a large group of people, we quite often see a right-skewed distribution. You can't make less than 0 dollars per year. Most people make a moderate amount of money, but there is no upper limit to how much a person can make, and a few people in any large sample will make a very large amount of money.

Example 2 shows that you can often make an educated guess as to the shape of a distribution even without collecting data.

## EXAMPLE 2 Roller Coaster Endurance

A morning radio show is sponsoring a contest in which contestants compete to win a car. About 40 contestants are put on a roller coaster, and whoever stays on it the longest wins. Suppose we make a histogram of the amount of time the contestants stay on (measured in hours or even days).

(QUESTION) What shape do we expect the histogram to have and why?

(SOLUTION) Probably most people will drop out relatively soon, but a few will last for a very long time. The last two contestants will probably stay for a very long time indeed. Therefore, we would expect the distribution to be right-skewed.

(TRY THIS!) Exercise 2.9

## How Many Mounds?

What do you think would be the shape of the distribution of heights if we included men in our sample as well as women? The distributions of women's heights by themselves and men's heights by themselves are usually symmetric and have one mound. But because we know that men tend to be taller than women, we might expect a histogram that combines men's and women's heights to have two mounds.

The statistical term for a one-mound distribution is a **unimodal distribution**, and a two-mound distribution is called a **bimodal distribution**. Figure 2.14a shows a bimodal distribution. A **multimodal distribution** has more than two modes. The modes do not have to be the same height (in fact, they rarely are). Figure 2.14b is perhaps the more typical bimodal distribution.

▶ **FIGURE 2.14** Idealized bimodal distributions. **(a)** Modes of roughly equal height. **(b)** Modes that differ in height.

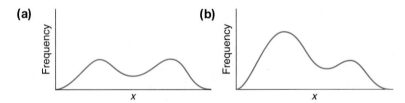

These sketches are idealizations. In real life you won't see distributions this neat. You will have to make a decision about whether a histogram is close enough to be called symmetric or whether it has one mound, two mounds, no mounds, or many mounds. The existence of multiple mounds is sometimes a sign that two very different groups have been combined into a single collection (such as combining men's heights with women's heights). When you see multimodal distributions, you may want to look back at the original data and see whether you can examine the groups separately, if separate groups exist. At the very least, whenever you see more than one mound, you should ask yourself, "Could these data be from different groups?"

## EXAMPLE 3 Two Marathons, Merged

Data were collected on the finishing times for two different marathons. One marathon consisted of a small number of elite runners: the 2004 Olympic Games. The other marathon included a large number of amateur runners: a marathon in Ontario, Canada.

(QUESTION) What shape would you expect the distribution of this sample to have?

(SOLUTION) We expect the shape to be bimodal. The elite runners would tend to have faster finishing times, so we expect one mound on the left for the Olympic runners and another, taller mound (because there were more amateur runners) on the right. Figure 2.15 is a histogram of the data.

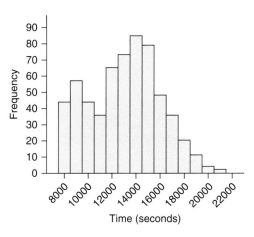

◀ **FIGURE 2.15** Histogram of finishing times for two marathons.

There appears to be one mound centered at about 9,000 seconds (2.5 hours) and another centered at about 14,000 seconds (about 3.9 hours).

**TRY THIS!** Exercise 2.11

When we view a histogram, our understanding of the shape of a distribution is affected by the width of the bins. Figure 2.15 reveals the bimodality of the distribution partly because the width of the bins is such that we see the right level of detail. If we had made the bins too big, we would have got less detail and might not have seen the bimodal structure. Figure 2.16 shows what would happen. Experienced data analysts usually start with the bin width the computer chooses and then also examine histograms made with slightly wider and slightly narrower bins. This lets them see whether any interesting structure emerges.

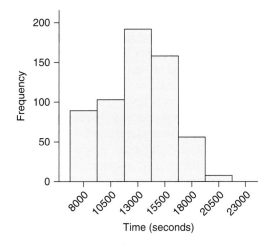

◀ **FIGURE 2.16** Another histogram of the same running times as in Figure 2.15. Here, the bins are wider and "wash out" detail, so the distribution no longer looks bimodal, even though it should.

## Do Extreme Values Occur?

Sometimes when you make a histogram or dotplot, you see extremely large or extremely small observations. When this happens, you should report these values and, if necessary, take action (although *reporting* is often action enough). Extreme values can appear when an error is made in entering data. For example, we asked students in one of our classes to record their weights in pounds. Figure 2.17 shows the distribution. The student who wrote 1200 clearly made a mistake. He or she probably meant to write 120.

Extreme values such as these are called **outliers**. The term *outlier* has no precise definition. Sometimes you may think an observation is an outlier, but another person might disagree, and this is fine. However, if there is no gap between the observation in question and the bulk of the histogram, then the observation probably should not be considered an outlier. Outliers are points that don't fit the pattern of the rest of the data, so if a large percentage of the observations are extreme, it might not be accurate to label them as outliers. After all, if lots of points don't fit a pattern, maybe you aren't seeing the right pattern!

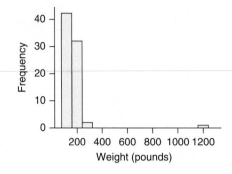

► **FIGURE 2.17** Histogram of weights with an extreme value. (Source: Ryan 2005)

> **KEY POINT** Outliers are values so large or small that they do not fit into the pattern of the distribution. There is no precise definition of the term *outlier*. Outliers can be caused by mistakes in data entry, but genuine outliers are sometimes unusually interesting observations.

Sometimes outliers result from mistakes, and sometimes they do not. The important thing is to make note of them when they appear and to investigate further, if you can. You'll see in Chapter 3 that presence of outliers can affect how we measure center and spread.

## Center

An important question to ask about any data set is "What is the typical value?" The typical value is the one in the center, but we use the word *center* here in a deliberately vague way. We might not all agree on precisely where the center of a graph is. But the idea here is to get a rough impression so that we can make comparisons later. For example, judging on the basis of the histogram shown in Figure 2.9, the center for the women soccer players is about 0.8 goals per game. Thus we could say that the typical first-year woman soccer player scored about 0.8 goals per game in 2009. In contrast, the center of the distribution for the male soccer players is about 0.55. It would seem that the typical male soccer player scores fewer goals per game than the typical female player, perhaps indicating that men's soccer is stronger on defense than women's soccer, at least among Division III first-year players.

If the distribution is bimodal or multimodal, it may not make sense to seek a "typical" value for a data set. If the data set combines two very different groups, then it might be more useful to find separate typical values for each group. What is the typical finishing time of the runners in Figure 2.15? There is no single typical time, because there are two distinct groups of runners. The elite group have their typical time, and the amateurs have a different typical time. However, it *does* make sense to ask about the typical test score for the student scores in Figure 2.13, because there is only one mound and only one group of students.

## EXAMPLE 4 Typical Bar-Passing Rate for Law Schools

Examine the distribution of bar-passing rates for law schools in the United States (see Figure 2.6b, which is repeated here for convenience).

**QUESTION** What is a typical bar-passing rate for a law school?

**SOLUTION** We show Figure 2.6b again. The center is somewhere in the range of 70% to 80%, so we would say that the typical bar-passing rate is between 70% and 80%.

> **! Caution**
>
> **Multimodal Centers**
> Be careful about giving a single typical value for a multimodal or bimodal distribution. These distributions sometimes indicate the existence of different and diverse groups combined into the same data set. It may be better to report a different typical value for each group, if possible.

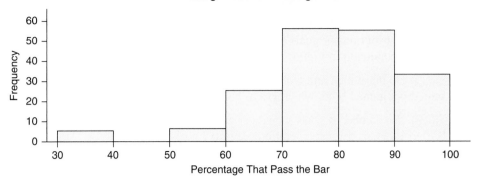

**Histogram of Bar-Passing Rate**

◀ **FIGURE 2.6b** (repeated)
Percentage passing the bar exam
from different law schools.

TRY THIS!  Exercise 2.13

## Variability

The third important feature to consider when examining a distribution is the amount of variation in the data. If all of the values of a numerical variable are the same, then the histogram (or dotplot) will be skinny. On the other hand, if there is great variety, the histogram will be spread out, thus displaying greater variability.

Here's a very simple example. Figure 2.18a shows a family of four people who are all very similar in height. Note that the histogram of these heights (Figure 2.18b) is quite skinny; in fact, it is just a single bar!

**(a)**

**(b)**

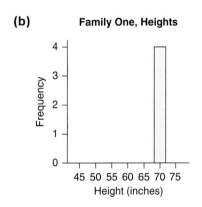

Family One, Heights

▲ **FIGURE 2.18** Family One with a Small Variation in Heights

Figure 2.19a, on the other hand, shows a family that exhibits large variation in height. The histogram for this family (Figure 2.19b) is more spread out.

**(a)**

**(b)**

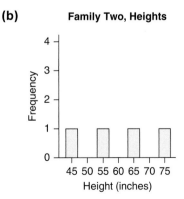

Family Two, Heights

▲ **FIGURE 2.19** Family Two with a Large Variation in Height

## EXAMPLE 5 NCAA Soccer Players

Consider the distributions of goals scored per game for women and men soccer players that we first encountered in Figure 2.9.

**QUESTION** Do these men and women soccer players have different variability in the goals scored per game? If so, which group has the greatest variability?

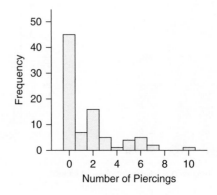

**SOLUTION** The number of goals scored per game for women ranges from about 0.70 to 1.4 goals per game; a difference of 0.7 goals per game. Goals scored per game by men range from 0.40 to 1, a range of 0.6 goals per game. We see that the women have more variability in the number of goals scored per game. One interpretation of this difference in spread is that women's soccer has a wider range of skill levels. This is also supported by the shape of the distribution, which suggests that most women score in the neighborhood of about 0.80 goals per game, a few as many as 1 goal per game, but a very few do quite a bit better. Men, on the other hand, are much more similar to each other, at least in terms of the number of goals scored per game. In Chapter 3 you'll see some more precise ways of measuring the spread of a numerical variable.

**TRY THIS!** Exercise 2.15

## Describing Distributions

When you are asked to describe a distribution or to compare two distributions, your description should include the center of the distribution (What is the typical value?), the spread (How much variability is there?), and the shape. If the shape is bimodal, it might not be appropriate to mention the center, and you should instead identify the approximate location of the mounds. You should also mention any unusual features, such as extreme values.

## ✳ EXAMPLE 6 Body Piercings

How common are body piercings among college students? How many piercings does a student typically have? One statistics professor asked a large class of students to report (anonymously) the number of piercings they possessed.

**QUESTION** Describe the distribution of body piercings for students in a statistics class.

**SOLUTION** The first step is to "see it" by creating an appropriate graphic of the distribution of the sample. Figure 2.20 shows a histogram for these data. To summarize this distribution, we examine the shape, center, and spread and comment on any unusual features.

▶ **FIGURE 2.20** Numbers of piercings of students in a statistics class.

The distribution of piercings is right-skewed, as we might expect, given that many people will have 0, 1, or 2 piercings, but a few people are likely to have more. The typical number of piercings (the center of the distribution) seems to be about 1, although a majority of students have none. The number of piercings ranges from 0 to about 10. An interesting feature of this distribution is that it appears to be multimodal. There are three peaks: at 0, 2, and 6, which are even numbers. This makes sense, because piercings often come in pairs. But why is there no peak at 4? (The authors do not know.) What do you think the shape of the distribution would look like if it included only the men? Only the women?

**TRY THIS!** Exercise 2.17

## SECTION 2.3

# Visualizing Variation in Categorical Variables

When visualizing data, we treat categorical variables in much the same way as numerical variables. We visualize the distribution of the categorical variable by displaying the values (categories) of the variable and the number of times each value occurs.

To illustrate, consider the Statistics Department at UCLA. UCLA offers an introductory statistics course every summer, and it needs to understand what sorts of students are interested in this class. In particular, understanding whether the summer students are mostly first-year students (eager to complete their general education requirements) or seniors (who put off the class as long as they could) can help the department better plan its course offerings.

Table 2.3 shows data from a sample of students in an introductory course offered during the 2009 summer term at UCLA. The "unknown" students are probably not enrolled in any university (adult students taking the course for business reasons or high school students taking the class to get a head start).

*Class* is a categorical variable. Table 2.4 on the next page summarizes the distribution of this variable by showing us all of the values in our sample and the frequency with which each value appears. Note that we added a row for first-year students.

Two commonly used graphs to display the distribution of a sample of categorical data are bar charts and pie charts. Bar charts look, at first glance, very similar to histograms, but they have several important differences, as you will see.

## Bar Charts

A **bar graph** (also called a **bar chart**) shows a bar for each observed category. The height of the bar is proportional to the frequency of that category. Figure 2.21a on the next page shows a bar graph for the UCLA statistics class data. The vertical axis measures frequency. We see that the sample has one graduate student and four juniors. We could also display relative frequency if we wished (Figure 2.21b). The shape does not change; only the numbers on the vertical axis change.

Note that there are no first-year students in the sample. We might expect this of a summer course, because entering students are unlikely to take courses in the summer before they begin college, and students who have completed a year of college are generally no longer first-year students (assuming they have completed enough units).

### Bar Charts vs. Histograms

Bar charts and histograms look a lot alike, but they have some very important differences.

- In a bar chart, it sometimes doesn't matter in which order you place the bars. Quite often, the categories of a categorical variable have no natural order. If they

| Student ID | Class |
|---|---|
| 1 | Senior |
| 2 | Junior |
| 3 | Unknown |
| 4 | Unknown |
| 5 | Senior |
| 6 | Graduate |
| 7 | Senior |
| 8 | Senior |
| 9 | Unknown |
| 10 | Unknown |
| 11 | Sophomore |
| 12 | Junior |
| 13 | Junior |
| 14 | Sophomore |
| 15 | Unknown |
| 16 | Senior |
| 17 | Unknown |
| 18 | Unknown |
| 19 | Sophomore |
| 20 | Junior |

▲ **TABLE 2.3** Identification of Classes for Students in Statistics

| Class | Frequency |
|-------|-----------|
| unknown | 7 |
| First-year student | 0 |
| Sophomore | 3 |
| Junior | 4 |
| Senior | 5 |
| Graduate | 1 |
| **Total** | **20** |

▲ **TABLE 2.4** Summary of Classes for Students in Statistics

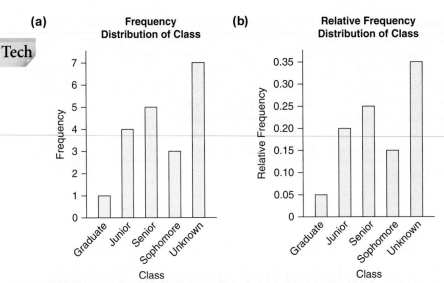

▲ **FIGURE 2.21** **(a)** Bar chart showing numbers of students in each class enrolled in an introductory statistics section. The largest "class" is the group made up of seven unknowns. First-year students are not shown because there were none in the data set. **(b)** The same information as shown part (a), but now with relative frequencies. The unknowns are about 0.35 (35%) of the sample.

do have a natural order, you might want to sort them in that order. For example, in Figure 2.22a we've sorted the categories into a fairly natural order, from "Unknown" to "Graduate." In Figure 2.22b we've sorted them from most frequent to least frequent. Either choice is acceptable.

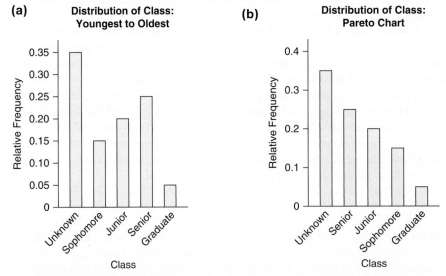

▲ **FIGURE 2.22** **(a)** Bar chart of classes using natural order. **(b)** Pareto chart of the same data. Categories are ordered with the largest frequency on the left and arranged so the frequencies decrease to the right.

Figure 2.22b is interesting because it shows more clearly that the most populated category consists of "Unknown." This result suggests that there might be a large demand for this summer course outside of the university. Bar charts that are sorted from most frequent to least frequent are called **Pareto charts**. (These charts were invented by the Italian economist and sociologist Vilfredo Pareto (1848–1943).) They are often an extremely informative way of organizing a display of categorical data.

- Another difference between histograms and bar charts is that in a bar chart, it doesn't matter how wide or narrow the bars are. The widths of the bars have no meaning.

- A final important difference is that a bar chart has gaps between the bars. This indicates that it is impossible to have observations between the categories. In a histogram, a gap indicates that no values were observed in the interval represented by the gap.

## SNAPSHOT  THE BAR CHART

| | |
|---|---|
| **WHAT IS IT?** ▶ | A graphical summary for categorical data. |
| **WHAT DOES IT DO?** ▶ | Shows a picture of the distribution of a categorical variable. |
| **HOW DOES IT DO IT?** ▶ | Each category is represented by a bar. The height of the bar is proportional to the number of times that category occurs in the data set. |
| **HOW IS IT USED?** ▶ | To see patterns of variation in categorical data. The categories can be presented in order of most frequent to least frequent, or they can be arranged in another meaningful order. |

## Pie Charts

Pie charts are another popular format for displaying relative frequencies of data. A **pie chart** looks, as you would expect, like a pie. The pie is sliced into several pieces, and each piece represents a category of the variable. The area of the piece is proportional to the relative frequency of that category. The largest piece in the pie in Figure 2.23 belongs to the category "Unknown" and takes up about 35% of the total pie.

Some software will label each slice of the pie with the percentage occupied. This isn't always necessary, however, because a primary purpose of the pie chart is to help us judge how frequently categories occur relative to one another. For example, the pie chart in Figure 2.23 shows us that "Unknown" occupies a fairly substantial portion of the whole data set. Also, labeling each slice gets cumbersome and produces cluttered graphs if there are many categories.

Although pie charts are very common (we bet that you've seen them before), they are not commonly used by statisticians or in scientific settings. One reason for this is that the human eye has a difficult time judging how much area is taken up by the wedge-shaped slices of the pie chart. Thus, in Figure 2.23, the "Sophomore" slice looks only slightly smaller than the "Junior" slice. But you can see from the bar chart (Figure 2.22) that they're actually quite different. Pie charts are also extremely difficult to use to compare the distribution of a variable across two different groups (such as comparing males and females as well as which classes the students come from).

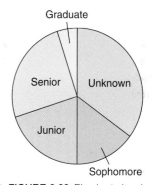

▲ **FIGURE 2.23** Pie chart showing the distribution of the categorical variable *Class* in a statistics course.

## SNAPSHOT  THE PIE CHART

| | |
|---|---|
| **WHAT IS IT?** ▶ | A graphical summary for categorical data. |
| **WHAT DOES IT DO?** ▶ | Shows the proportion of observations that belong to each category. |
| **HOW DOES IT DO IT?** ▶ | Each category is represented by a wedge in the pie. The area of the wedge is proportional to the relative frequency of that category. |
| **HOW IS IT USED?** ▶ | To understand which categories are most frequent and which are least frequent. Sometimes it is useful to label each wedge with the proportion of occurrence. |

# Summarizing Categorical Distributions

The concepts of *shape*, *center*, and *spread* that we used to summarize numerical distributions sometimes don't make sense for categorical distributions, because we can often order the categories any way we please. The center and shape would be different for every ordering of categories. However, we can still talk about typical outcomes and the variability in the sample.

## The Mode

When describing a distribution of a categorical variable, pay attention to which category occurs most often. This value, the one with the tallest bar in the bar chart, can sometimes be considered the "typical" outcome. There might be a tie for first place, and that's okay. It just means there's not as much variability in the sample. (Read on to see what we mean by that.)

The category that occurs most often is called the **mode**. This meaning of the word *mode* is similar to its meaning when we use it with numerical variables. However, one big difference between categorical and numerical variables is that we call a categorical variable bimodal only if two categories are nearly tied for most frequent outcomes. (The two bars don't need to be exactly the same height, but they should be very close.) Similarly, a categorical variable's distribution is multimodal if more than two categories all have roughly the tallest bars. For a numerical variable, the heights of the mounds do not need to be the same height for the distribution to be multimodal.

For an example of a mode, as part of their Internet and American Life Project, the Pew Research Center interviewed 2054 Americans in 2002, 2006, and 2007 to ask them how they used various electronic devices. One question they asked was whether it would be "very hard to give up" their cell phone, or Internet service, or television, or landline phone, or email service, or a wireless email device such as a BlackBerry. Figure 2.24 shows the results in 2002.

▶ **FIGURE 2.24** Percents of respondents who reported, in 2002, that they would find it very hard to give up a cell phone, access to the Internet, television, a landline phone, email service, or a wireless email device in 2002. Note that the percents do not add up to 100% because people could find it difficult to give up more than one of these devices (and some might even claim that every one of them would be hard to live without).

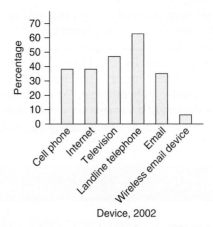

We see that many people (about 60%) would find it very hard to give up their landline telephones. The landline phone is the mode of this distribution. Very few people in 2002 said they could not live without a BlackBerry (wireless email) device, whereas email, a cell phone, and access to the Internet were all considered about equally dispensable.

## EXAMPLE 7 Electronic Devices: What Can't You Live Without?

In 2007, the Pew survey asked a new group of people whether they would find it very hard to give up their electronic devices. The bar chart in Figure 2.25 shows the distribution of responses.

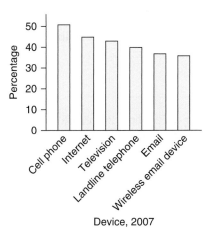

◀ **FIGURE 2.25** Percents of respondents who reported, in 2007, that they would find it very hard to give up a cell phone, access to the Internet, television, a landline phone, email service, or a wireless email device in 2007. Compare to Figure 2.24, where the data for 2002 are given.

**QUESTION** What is the typical response? How would you compare the responses in 2007 to to those in 2002?

**SOLUTION** The mode is cell phone, which 50% of the sample would find hard to give up. Between 2002 and 2007, the mode has changed from landline phone to cell phone.

**TRY THIS!** Exercise 2.31

## Variability

When thinking about the variability of a categorical distribution, it is sometimes useful to think of the word *diversity*. If the distribution has a lot of diversity (many observations in many different categories), then variability is high. If the distribution has only a little diversity (many of the observations fall into the same category), then variability is low.

For example, Figure 2.26 shows bar charts of the ethnic composition of two schools in the Los Angeles City School System. Which school has the greater variability in ethnicity?

School A (Figure 2.26a) has much more diversity. School B (Figure 2.26b) consists almost entirely of a single ethnic group, with very small numbers of the other groups.

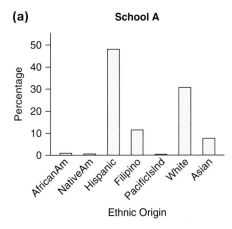

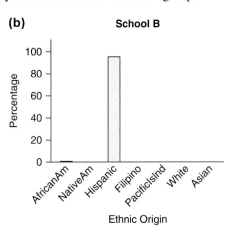

◀ **FIGURE 2.26** Percents of students at two Los Angeles elementary schools who are identified with several ethnic groups. Which school has more ethnic variability?

## EXAMPLE 8 Electronic Devices: Growing Toward Consensus?

Compare the responses to the Pew study in 2002 (Figure 2.24) with the responses in 2007 (Figure 2.25).

**QUESTION** In which year is more variability apparent? Explain.

**SOLUTION** In 2002, the bar chart shows that the landline phone was the clear favorite: there was little variability in the sample. But in 2007, while *cell phone* is, strictly speaking, the most frequent category, the others are close behind. All of these devices seem popular with the respondents. There is much more variability and diversity in choices in 2007 than in 2002.

**TRY THIS!** Exercise 2.33

You might be surprised by the answer to Example 8 because the bars in the 2007 bar chart are nearly all the same height, so it looks like things don't change much from category to category. However, variability is not just about the heights of the bars; variability is measured by how many different categories have responses in them. The most diverse distribution for categorical data would be represented by a bar chart in which all categories had bars of exactly the same height. The least variability would be represented by one in which there was only one bar, and all the other categories were empty.

## Describing Distributions of Categorical Variables

When describing a distribution of categorical data, you should mention the mode (or modes) and say something about the variability. Example 9 illustrates what we mean.

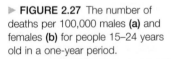 When summarizing graphs of categorical data, report the mode or modes and describe the variability (diversity).

## EXAMPLE 9 Causes of Death

According to some experts, about 51.5% of babies currently born in the United States are male. But among people between 100 and 104 years old, there are four times as many women as men (U.S. Census Bureau, 2000). How does this happen? One possibility is that the percent of boys born has changed over time. Another possibility is presented in the two bar charts in Figure 2.27. These bar charts show the numbers of deaths per 100,000 people in one year, for people aged 15–24 years, for both males and females.

**QUESTION** Compare the distributions depicted in Figure 2.27. Note that the categories are put into the same order on both graphs.

► FIGURE 2.27 The number of deaths per 100,000 males **(a)** and females **(b)** for people 15–24 years old in a one-year period.

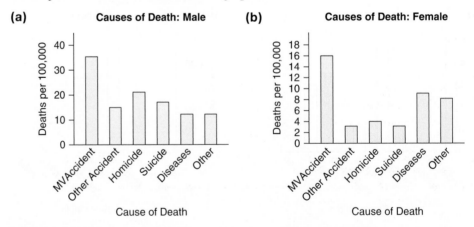

**SOLUTION** First, note that the histograms are not on the same scale, as you can see by comparing the values on the *y*-axes. This presentation is typical of most software and is sometimes desirable, because otherwise, one of the bar charts might have such small bars that we couldn't easily discern differences.

Although motor vehicle accidents are the mode for both groups, males show a consistently high death rate for all other causes of death, whereas females have relatively low death rates in the categories for other accident, homicide, and suicide. In other words, the cause of death for females is less variable than that for males. It is also worth noting that the death rates are higher for males in every category. For example, roughly 16 out of every 100,000 females died in a motor vehicle accident in one year, while roughly 35 out of every 100,000 males died in car accidents in the same year.

TRY THIS!  Exercise 2.37

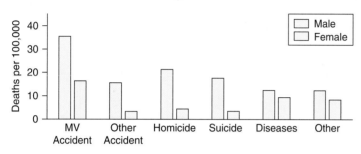

We can also make graphics that help us compare two distributions of categorical variables. When comparing two groups across a categorical variable, it is often useful to put the bars side-by-side, as in Figure 2.28. This graph makes it easier to compare rates of death for each cause. The much higher death rate for males is made clear.

◀ **FIGURE 2.28** Death rates of males and females, graphed side by side.

SECTION 2.5

# Interpreting Graphs

The first step in every investigation of data is to make an appropriate graph. Many analyses of data begin with visualizing the distribution of a variable, and this requires that you know whether the variable is numerical or categorical. When interpreting these graphics, you should pay attention to the center, spread, and shape.

Often, you will come across graphics produced by other people, and these can take extra care to interpret. In this section, we'll warn you about some potential pitfalls when interpreting graphs and show you some unusual visualizations of data.

## Misleading Graphs

A well-designed statistical graphic can help us discover patterns and trends and can communicate these patterns clearly to others. However, our eyes can play tricks on us, and manipulative people can take advantage of this to use graphs to give false impressions.

The most common trick—one that is particularly effective with bar charts—is to change the scale of the vertical axis so that it does not start at the origin (0). Figure 2.29a shows the number of violent crimes per year in the U.S. as reported by the FBI (http://www.fbi.gov). The graphic seems to indicate a dramatic drop in crime in 2009. Notice that the vertical axis starts at about 1,240,000 (1.24 million) crimes.

Because the origin begins at 1.24 million and not at 0, the bars are all shorter than they should be. This creates a much more dramatic apparent decline in crime, since the shortest bar is roughly one-tenth the height of the tallest bar. What does this chart look like if we make the bars the correct height? Figure 2.29b shows the same data, but to the correct scale. It's still clear that there has been a decline, but the decline doesn't look nearly so dramatic now, does it? Why not?

The reason is that when the origin is correctly set to 0, as in Figure 2.29b, it is clear that violent crimes were only slightly lower in 2009 than in 2008. The actual decline from 2008 to 2009 was about 5 percentage points. The largest percentage decline in several years, and certainly good news. But not as dramatically good as the first graph erroneously suggests.

(a)
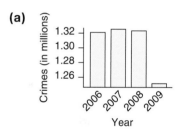

(b)
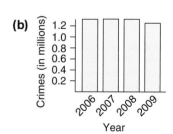

▲ **FIGURE 2.29** (a) This bar chart shows a dramatic decline in the number of violent crimes in 2009. The origin for the vertical axis begins at about 1.24 million, not at 0. (b) This bar chart reports the same data as part (a), but here the vertical axis begins at the origin (0).

Most of the misleading graphics you will run across exploit a similar theme: Our eye tends to compare the relative sizes of objects. Many newspapers and magazines like to use pictures to represent data. For example, the plot in Figure 2.30 attempts to illustrate the number of homes sold for some past years, with the size of each house representing the number of homes sold that year. Such graphics can be very misleading, because the pictures are often not to scale with the actual numbers.

In Figure 2.30 the *heights* of the homes are indeed proportional to the sales numbers, but our eye reacts to the *areas* instead. The smallest house is 69% as tall as the largest house (because 4.9 is 69% of 7.1), but the area of the smallest house is only about 48% of that of the largest house, so our tendency to react to area rather than to height exaggerates the difference.

► **FIGURE 2.30** Deceptive graphs: Image **(a)** represents 7.1 million homes sold in 2005, image **(b)** represents 6.5 million homes sold in 2006, image **(c)** represents 5.8 million homes sold in 2007, and image **(d)** represents 4.9 million homes sold in 2008. (Source: *L.A. Times*, April 30, 2008)

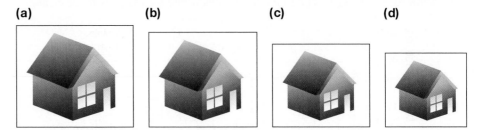

**(a)**   **(b)**   **(c)**   **(d)**

## The Future of Statistical Graphics

The Internet allows for a great variety of graphical displays of data that take us beyond simple visualizations of distributions. Many statisticians, computer scientists, and designers are experimenting with new ways to visualize data. Most exciting is the rise of interactive displays. The *State of the Union Visualization*, for example (http://stateoftheunion .onetwothree.net), makes it possible to compare the content of State of the Union speeches. Every U.S. president delivers a State of the Union address to Congress near the beginning of each year. This interactive graphic enables users to compare words from different speeches and "drill down" to learn details about particular words or speeches.

For example, Figure 2.31 is based on the State of the Union address that President Barack Obama delivered on January 27, 2010. The largest words are the words that appear most frequently in the speech. The words that appear to the left are words that typically appear earlier in the speech, and the words located to the right typically appear later. Thus we see that near the beginning, President Obama talked about the economy, using words such as *wall*, *street*, *recovery*, and *businesses*. Toward the end, there was talk about more general things, and we see *families*, *values*, and *deficit*. The word *Americans* is more

► **FIGURE 2.31** President Obama's 2010 State of the Union speech, visualized as a "word cloud." This array shows us the most commonly used words, approximately where these words occurred in the speech, how often they occurred, and how unusual they are compared to the content of other State of the Union speeches. (Courtesy of Brad Borevitz, http://onetwothree.net. Used with permission.)

or less in the middle, which suggests that it was used frequently throughout the speech. The vertical position of a word indicates how unusual it is compared to the content of other State of the Union addresses. Words that appear near the top, such as *businesses* and *jobs*, are words that are particular to this speech, compared to all other State of the Union addresses. Words near the bottom, such as *kids, talk,* and *cuts,* are words that occur relatively frequently throughout all State of the Union addresses.

Figure 2.31 is packed with information. On the bottom, in red, you see dots that represent the grade level required to read and understand the speech. Each dot represents a different speech, and the speeches are sorted from first (George Washington) to most recent (Barack Obama). Near the bottom right we see that the grade level of the speech is 9.0, or about ninth grade, and that this is considerably lower than it was 100 years ago but consistent with more recent speeches. The white "bars" represent the numbers of words in the State of the Union addresses. This speech contains 7068 words, which is fairly typical of the last few years. The tallest spike represents Jimmy Carter's speech in 1981, which had 33,613 words and a grade level of 15.3.

## CASE STUDY REVISITED

## Student-to-Teacher Ratio at Colleges

How do public colleges compare to private colleges in student-to-teacher ratios? The data were presented at the beginning of the chapter. This list of raw data makes it hard to see patterns, so it is very difficult to compare groups. But because the student-to-teacher ratio is a numerical variable, we can display these distributions as two histograms to enable us to make comparisons between public and private schools.

In Figure 2.32, the student-to-teacher ratio for the private schools is shown in the left histogram, and that for state schools is shown in the right histogram. The distributions are quite different. The typical student-to-teacher ratio for private schools is around 10 or 11 students per teacher, while for state schools it's around 16 or 20. Although both schools have quite a bit of spread in these ratios, there may be a little less spread—a little less variation—in the state schools. The private schools' distribution is right-skewed. The state schools' distribution appears bimodal (at least with these bins), with mounds centered at about 16 and 20 students per teacher. It makes us wonder whether there are two kinds of state colleges.

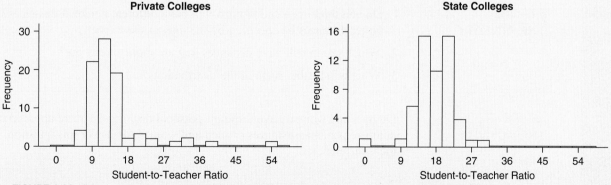

▲ **FIGURE 2.32** Histograms of the student-to-teacher ratio at samples of private colleges and state colleges.

By the way, many of the schools with the smallest student-to-teacher ratios deal with specialties such as health or design. In case you are wondering about the state college with the smallest student-to-teacher ratio in Figure 2.32b, it is the University of Colorado at Denver Health Science Center. At medical facilities like this one, the physicians are often listed as faculty. The private school with the 54-to-1 ratio is Western Governors University at Salt Lake City, Utah. It is an online university offering accredited courses in distance learning.

## Personal Distance

How much personal distance do people require when they're using an automatic teller machine?

| GOALS | MATERIALS |
|---|---|
| In this activity, you will learn to make graphs of sample distributions in order to answer questions about data in comparing two groups of students. | Meter stick (or tape measure). |

**ACTIVITY**

Work in groups of three students. Each group must have a meter stick. The first person stands (preferably in front of a wall) and imagines that she or he is at an ATM getting cash. The second student stands behind the first. The first student tells the second student how far back he or she must stand for the first student to be just barely comfortable, saying, for example, "Move back a little, now move forward just a tiny bit," and so on. When that distance is set, the third student measures the distance between the heel of the first person's right shoe to the toe of the second person's right shoe. That will be called the "personal distance."

For each student in your group, record the gender and personal distance. Your instructor will help you pool your data with the rest of the class.

*Note:* Be respectful of other people's personal space. Do not make physical contact with other students during this activity.

**BEFORE THE ACTIVITY**

1. Do you think men and women will have different personal distances? Will the larger distances be specified by the men or the women?

2. Which group will have distances that are more spread out?

3. What will be the shape of the distributions?

**AFTER THE ACTIVITY**

Do men and women have different personal distances? Create appropriate graphics to compare personal distances of men and women to answer this question. Then describe these differences.

## CHAPTER REVIEW

## Key Terms

distribution of a sample, *36*
frequency, *36*
dotplot, *37*
histogram, *38*
relative frequency, *39*
stemplot (or stem-and-leaf plot), *41*

leaf, *41*
stem, *41*
typical value (center), *41*
variability, *41*
symmetric distribution, *42*
bell-shaped distribution, *42*

right-skewed distribution, *43*
left-skewed distribution, *43*
unimodal distribution, *44*
bimodal distribution, *44*
multimodal distribution, *44*
outlier, *45*

bar graph (bar chart), *49*
Pareto chart, *50*
pie chart, *51*
mode (in categorical variables), *52*

## Learning Objectives

After reading this chapter and doing the assigned homework problems, you should

- Understand that a distribution of a sample of data displays a variable's values and the frequencies (or relative frequencies) of those values.

- Know how to make graphs of distributions of numerical and categorical variables and how to interpret the graphs in context.

- Be able to compare centers and spreads of distributions of samples informally.

## Summary

The first step in any statistical investigation is to make plots of the distributions of the data in your data set. You should identify whether the variables are numerical or categorical so that you can choose an appropriate graphical representation.

If the variable is numerical, you can make a dotplot, histogram, or stemplot. Pay attention to the shape (Is it skewed or symmetric? Is it unimodal or multimodal?), to the center (What is a typical outcome?), and to the spread (How much variability is present?). You should also look for unusual features, such as outliers.

Be aware that many of these terms are deliberately vague. You might think a particular observation is an outlier, but another

person might not agree. That's okay, because the purpose isn't to determine whether such points are "outliers" but to indicate whether further investigation is needed. An outlier might, for example, be caused by a typing error someone made when entering the data.

If you see a bimodal or multimodal distribution, ask yourself whether the data might contain two or more groups combined into the single graph.

If the variable is categorical, you can make a bar chart, a Pareto chart (a bar chart with categories ordered from most frequent to least frequent), or a pie chart. Pay attention to the mode (or modes) and to the variability.

## Sources

Internet Research Legal Group. 2008. http://www.ilrg.com/rankings/law/index.php/1/desc/LSATLow/2006

Minitab 15 Statistical Software (2007). [Computer software]. State College, PA: Minitab, Inc. (www.minitab.com)

National Safety Council. 2002. *Injury Facts*, 2002 edition. Itasca, Illinois.

Playfair, W. 1786. Commercial and political atlas: Representing, by copper-plate charts, the progress of the commerce, revenues, expenditure, and debts of England, during the whole of the eighteenth century. London: Corry. Reprinted in H. Wainer and I. Spence (eds.), *The commercial*

*and political atlas and statistical breviary*, New York: Cambridge University Press, 2005. ISBN 0-521-85554-3. http://www.math.yorku.ca/SCS/Gallery/milestone/refs.html#Playfair:1786)

Tufte, E. 1983. *The visual display of quantitative information*, Graphics Press: Cheshire, Connecticut. 1st ed., p. 141. Photo by Brian Joiner.

*World Almanac and Book of Facts*. 2006. World Almanac.

U.S. Census Bureau. 2000. http://ceic.mt.gov/C2000/SF12000/Pyramid/pptab00.htm

## SECTION EXERCISES

### SECTIONS 2.1 AND 2.2

**2.1 Body Mass Index** The dotplot shows body mass index (BMI) for 134 people according to the National Health and Nutrition Examination Survey (NHANES) in 2010, as reported in *USA Today*.

  a. A BMI of more than 40 is considered morbidly obese. Report the number of morbidly obese shown in the dotplot.

  b. Report the percentage of people who are morbidly obese. Compare this with an estimate from 2005 that 3% of people in the United States at that time were morbidly obese.

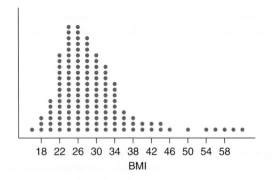

**2.2 Cholesterol Levels** The dotplot shows the cholesterol level of 93 adults from the 2010 NHANES data.

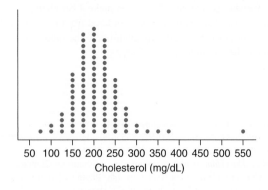

  a. A total cholesterol level of 240 mg/dL (milligrams per deciliter) or more is considered unhealthy. Report the number of people in this group with unhealthy cholesterol levels.

  b. Knowing there are a total of 93 people in this sample, report the percentage of people with unhealthy total cholesterol levels. How does this compare with an estimate from 2010 that 18% of people in the United States had unhealthy cholesterol levels?

**2.3 Ages of CEOs** The histogram shows frequencies for the ages of 25 CEOs listed at Forbes.com. Convert this histogram to one showing relative frequencies by relabeling the vertical axis with the appropriate relative frequencies. You may just report the new labels for the vertical axis because that is the only thing that changes.

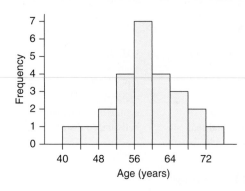

**2.4 Private College Tuition** The histogram shows the distribution of tuition at 85 private colleges, as reported in the *2006 World Almanac and Book of Facts*. Approximately how many colleges charged $30,000 or more in tuition?

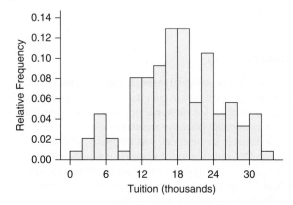

**2.5 Televisions (Example 1)** The histogram shows the distribution of the number of televisions in the homes of 90 community college students.

  a. According to the histogram, about how many homes do not have a television?

  b. How many televisions are in the homes that have the most televisions?

  c. How many homes have three televisions?

  d. How many homes have six or more televisions?

  e. What proportion of homes have six or more televisions?

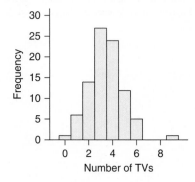

**2.6 Exercise Hours** The histogram shows the distribution of self-reported number of hours of exercise per week for 50 community college students. This graph uses a right-hand rule: If someone exercised for exactly 4 hours, they would be in the third bin, the bin to the right of 4.

a. According to the histogram, there are two possible values for the maximum number of hours of exercise. What are they?

b. How many people exercised 0 or 1 hour (less than 2 hours)?

c. How many people exercised 10 or more hours?

d. What proportion of people exercised 10 or more hours?

**2.7 Rental Prices** Have you ever wondered whether you could afford to move to another city? The dotplot shows rental prices (dollars per month) in three cities in December 2009 for units with at least one bedroom and one bathroom. (Source: Realtor.com)

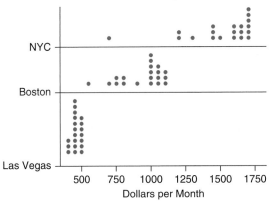

a. Center: Which city typically has the highest rents?

b. Spread: Which city's distribution has the smallest variation?

c. Shape: Which city's distribution is the least skewed?

**2.8 Rental Prices** The dotplot shows rental prices (dollars per month) in three cities in December 2009 for units with at least one bedroom and one bathroom. (Source: Realtor.com)

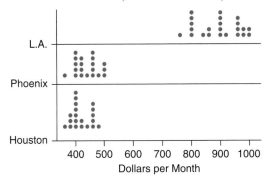

a. Center: Which city typically has the highest rental prices?

b. Which city's distribution has the greatest variation?

c. Describe the shape of the distribution of prices in Houston.

TRY **2.9 CEO Salaries (Example 2)** Predict the shape of the distribution of the salaries of 25 chief executive officers (CEOs).

A typical value is about 50 million per year, but there is an outlier at about 200 million.

**2.10 Cigarettes** A physician asks all of his patients to report the approximate number of cigarettes they smoke in a day. Predict the shape of the distribution of number of cigarettes smoked per day.

TRY **2.11 Armspans (Example 3)** According to the ancient Roman architect Vitruvius, a person's armspan (the distance from fingertip to fingertip with the arms stretched wide) is approximately equal to his or her height. For example, people 5 feet tall tend to have an armspan of 5 feet. Explain, then, why the distribution of armspans for a class containing roughly equal numbers of men and women might be bimodal.

★ **2.12 Tuition** The distribution of in-state annual tuition for all colleges and universities in the United States is bimodal. What is one possible reason for this bimodality?

TRY **2.13 Ages of CEOs (Example 4)** From the histogram in Exercise 2.3, approximately what is a typical age of a CEO in this sample?

**2.14 Private College Tuition** From the histogram shown in Exercise 2.4, approximately what is a typical tuition, in 2005, for a private college?

TRY **2.15 Commute Times (Example 5)** Use the histograms to compare the times spent commuting for community college students who drive to school in a car with the times spent by those who take the bus. Which group typically has the longer commute time? Which group has the more variable commute time?

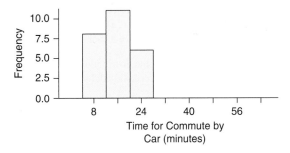

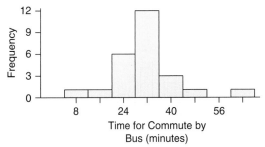

**2.16 Spending on Clothes** The histograms show the distribution of the estimated number of dollars per month spent on clothes for college women (left) and college men (right).

a. Compare and describe the shape of the distributions.

b. Which group tends to spend more?

c. Which group has more variation in its expenses?

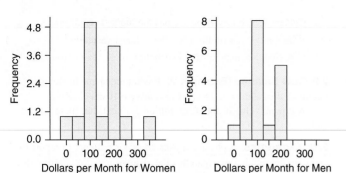

Dollars per Month for Women     Dollars per Month for Men

**TRY** **2.17 Education (Example 6)** In 2008, the General Social Survey (GSS), a national survey conducted nearly every year, reported the number of years of formal education for 2018 people. The histogram shows the distribution of data.

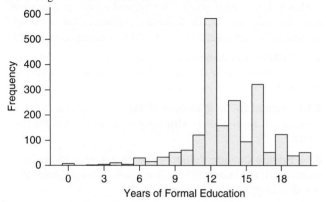

Years of Formal Education

a. Describe and interpret the distribution of years of formal education. Mention any unusual features.

b. Assuming that those with 16 years of education completed a bachelor's degree, estimate how many of the people in this sample got a bachelor's degree or higher.

c. The sample includes 2018 people. What percentage of people in this sample have a bachelor's degree or higher? How does this compare with Wikipedia's estimate that 27% have a bachelor's degree?

**2.18 Siblings** The histogram shows the distribution of the numbers of siblings (brothers and sisters) for 2000 adults surveyed in the 2008 General Social Survey.

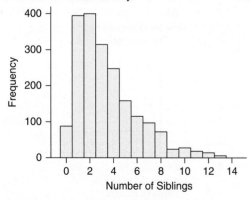

Number of Siblings

a. Describe the shape of the distribution.

b. What is the typical number of siblings, approximately?

c. About how many people in this survey have no siblings?

d. What percentage of the 2000 adults surveyed have no siblings?

**\*2.19 Years of Education** The GSS asked people how many years of education they had and how many years their mothers had.

If people who responded to the survey (respondents) completed high school but had no further education, they reported 12 years of education. If they stopped after a bachelor's degree, they reported 16 years. There were 2018 people who answered the question about their own education and 1780 who answered the question about the education of their mothers. Compare the distribution of years of education of the respondents with the distribution of years of education of their mothers.

**Education of Respondent**

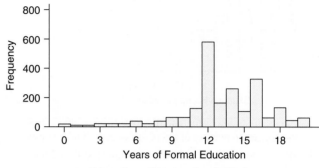

Years of Formal Education

**Education of Respondent's Mother**

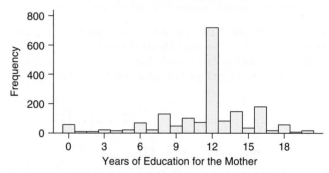

Years of Education for the Mother

**\*2.20 Hours Worked** In the 2008 General Social Survey, 636 male paid employees and 537 female paid employees were asked how many hours they worked in the last week. (Those who said, "I don't know" or "I don't work" were not included in the data set.) Compare the distributions of hours of work for the men and the women.

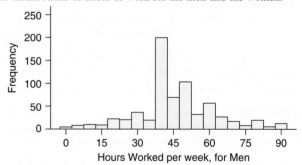

Hours Worked per week, for Men

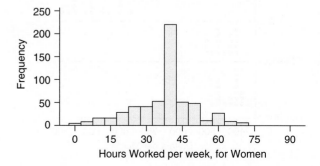

Hours Worked per week, for Women

**2.21 Matching Histograms** Match each of the following histograms to the correct situation.

1. The age of death of a sample of 19 typical women in the U.S.
2. The yearly tuition (according to the 2006 *World Almanac and Book of Facts*) for 142 colleges, 85 of which are private and 57 of which are state-supported.
3. The outcomes of rolling a fair die (with six sides) 5000 times.

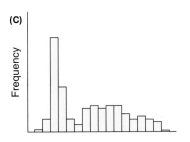

**2.22 Matching Histograms** Match each of the following histograms to the correct situation.

1. Test scores for students on an easy test.
2. The numbers of hours of television watched by a large, typical group of Americans.
3. The heights of a large, typical group of adults in the U.S.

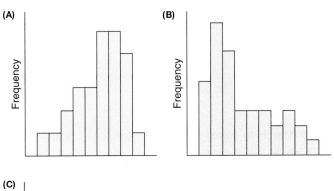

**2.23 Eating Out and Jobs** College student Jacqueline Loya asked students who had full-time jobs and students who had part-time jobs how many times they went out to eat in the last month.

Write a brief comparison of the distributions of the two groups. Include appropriate graphics. *See page 71 for guidance.*

Full-time: 5, 3, 4, 4, 4, 2, 1, 5, 6, 5, 6, 3, 3, 2, 4,
5, 2, 3, 7, 5, 5, 1, 4, 6, 7
Part-time: 1, 1, 5, 1, 4, 2, 2, 3, 3, 2, 3, 2, 4,
2, 1, 2, 3, 2, 1, 3, 3, 2, 4, 2, 1

**2.24 Comparing Weights of Baseball and Soccer Players** College students Edward Lara and Anthony Dugas recorded the self-reported weights (in pounds) of some community college male baseball players and male soccer players. Write a brief comparison of the distributions of weight for the two groups. Include appropriate graphics. For one approach, look at the guidance given for exercise 2.23.

**Baseball Players' Weights**

| | | | | |
|---|---|---|---|---|
| 190 | 205 | 230 | 198 | 195 |
| 200 | 185 | 195 | 180 | 182 |
| 187 | 177 | 169 | 182 | 193 |
| 181 | 207 | 186 | 193 | 190 |
| 192 | 225 | 210 | 200 | 186 |

**Soccer Players' Weights**

| | | | | |
|---|---|---|---|---|
| 165 | 189 | 167 | 173 | 184 |
| 190 | 170 | 190 | 158 | 174 |
| 185 | 182 | 185 | 150 | 190 |
| 187 | 172 | 156 | 172 | 156 |
| 183 | 180 | 168 | 180 | 163 |

**2.25 Textbook Prices** The table shows prices of 50 college textbooks in a community college bookstore, rounded to the nearest dollar. Make an appropriate graph of the distribution of the data, and describe the distribution.

| | | | | | | | | | |
|---|---|---|---|---|---|---|---|---|---|
| 76 | 19 | 83 | 45 | 88 | 70 | 62 | 84 | 85 | 87 |
| 86 | 37 | 88 | 45 | 75 | 83 | 126 | 56 | 30 | 33 |
| 26 | 88 | 30 | 30 | 25 | 89 | 32 | 48 | 66 | 47 |
| 115 | 36 | 30 | 60 | 36 | 140 | 47 | 82 | 138 | 50 |
| 126 | 66 | 45 | 107 | 112 | 12 | 97 | 96 | 78 | 60 |

**2.26 SAT scores** The table shows a random sample of 50 quantitative SAT scores of first-year students admitted at a university. Make an appropriate graph of the distribution of the data, and describe the distribution.

| | | | | | | | | | |
|---|---|---|---|---|---|---|---|---|---|
| 649 | 557 | 734 | 653 | 652 | 538 | 674 | 705 | 729 | 737 |
| 672 | 583 | 729 | 677 | 618 | 662 | 692 | 692 | 672 | 624 |
| 669 | 529 | 609 | 526 | 665 | 724 | 557 | 647 | 719 | 593 |
| 624 | 611 | 490 | 556 | 630 | 602 | 573 | 575 | 665 | 620 |
| 629 | 593 | 665 | 635 | 700 | 665 | 677 | 653 | 796 | 601 |

**2.27 Animal Longevity** The table below exercise 2.28 shows the average lifespan for some mammals in years, according to infoplease.com. Graph these average lifespans and describe the distribution. What is a typical lifespan? Identify the three outliers and report their lifespans. If you were to include humans in this graph, where would the data point be? Humans average about 75 years.

**2.28 Animal Gestation Periods** The accompanying table also shows the gestation period (in days) for some animals. The gestation period is the length of pregnancy. Graph the gestation period and describe the distribution. If there are any outliers, identify the animal(s) and give their gestation periods. If you were to include humans in this graph where would the data point be? The human gestation period is about 266 days.

| Animal | Gestation (days) | Lifespan (years) | Animal | Gestation (days) | Lifespan (years) |
|---|---|---|---|---|---|
| Baboon | 187 | 12 | Hippo | 238 | 41 |
| Bear, grizzly | 225 | 25 | Horse | 330 | 20 |
| | | | Leopard | 98 | 12 |
| Beaver | 105 | 5 | Lion | 100 | 15 |
| Bison | 285 | 15 | Monkey, rhesus | 166 | 15 |
| Camel | 406 | 12 | | | |
| Cat, domestic | 63 | 12 | Moose | 240 | 12 |
| Chimp | 230 | 20 | Pig, domestic | 112 | 10 |
| Cow | 284 | 15 | Puma | 90 | 12 |
| Deer | 201 | 8 | Rhino, black | 450 | 15 |
| Dog, domestic | 61 | 12 | Sea Lion, Cal | 350 | 12 |
| Elephant, African | 660 | 35 | Sheep | 154 | 12 |
| Elephant, Asian | 645 | 40 | Squirrel, gray | 44 | 10 |
| Elk | 250 | 16 | Tiger | 105 | 16 |
| Fox, red | 52 | 7 | Wolf, maned | 63 | 5 |
| Giraffe | 457 | 10 | | | |
| Goat | 151 | 8 | Zebra, Grant's | 365 | 15 |
| Gorilla | 258 | 20 | | | |

**2.29 Matching** Match each description with the correct histogram of the data.

1. Heights of students in a large UCLA statistics class that contains about equal numbers of men and women.
2. Numbers of hours of sleep the previous night in the same large statistics class.
3. Numbers of driving accidents for students in a large university in the U.S.

(A)

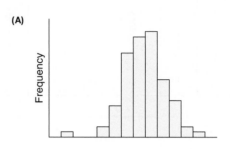

(B)

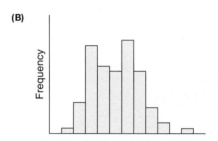

(C)

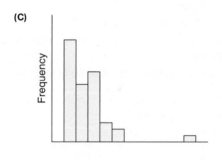

**2.30 Matching** Match each description with the correct histogram of the data.

1. Quantitative SAT scores for 1000 students who have been accepted to a private university.
2. Weights of over 500 adults, about half of whom are men and half of whom are women.
3. Ages of all 39 students in a community college statistics class that is made up of full-time students and meets during the morning.

**(A)**

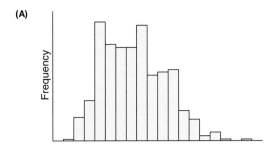

**(B)**

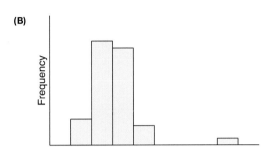

**(C)**

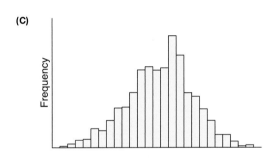

## SECTIONS 2.3 AND 2.4

TRY **2.31 Changing Multiple-Choice Answers When Told *Not* to Do So (Example 7)** One of the authors wanted to determine the effect of changing answers on multiple-choice tests. She studied the tests given by another professor, who had told his students before their exams that if they had doubts about an answer they had written, they would be better off *not changing* their initial answer. The author went through the exams to look for erasures, which indicate that the first choice was changed. In these tests, there is only one correct answer for each question. Do the data support the view that students should not change their initial choice of an answer?

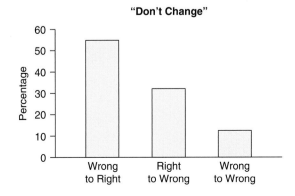

**2.32 Changing Multiple-Choice Answers When Told to Do So** One of the authors wanted to determine the effect of changing answers on multiple-choice tests. She had advised her students that if they had changed their minds about a previous answer, they should replace their first choice with their new choice. By looking for erasures on the exam, she was able to count the number of changed answers that went from wrong to right, from right to wrong, and from wrong to wrong. The results are shown in the bar chart.

a. Do the data support her view that it is better to replace your initial choice with the revised choice?

b. Compare this bar chart with the one in Exercise 2.31. Does changing answers generally tend to lead to higher or to lower grades?

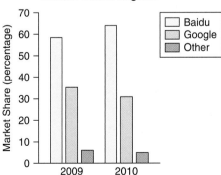

TRY **2.33 Search Engines (Example 8)** The bar chart shows market share for the search engines used in China. (Source: Data from *Wall Street Journal*, April 27, 2010)

a. Report the mode for both time periods. What changes do you see from 2009 to 2010?

b. In which time period, 2009 (fourth quarter) or 2010 (first quarter) was there more variability in the market share for search engines used for computers in China?

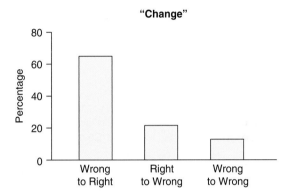

**2.34 Browsers** The bar chart shows the market share of browsers according to w3schools, an online Web development school. Write a sentence or two describing changes in the market share of these browsers from 2007 to 2010. (Source: Data from www.w3schools.com)

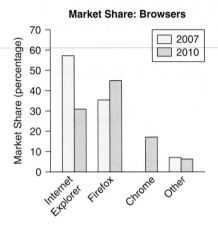

**Market Share: Browsers**

**2.35 Political Party Affiliation: Men** The 2008 General Social Survey (GSS) asked its respondents to report their political party affiliation. The graphs show the results for 926 men.

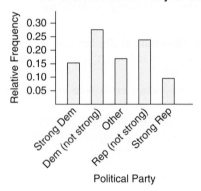

**Bar Chart of Political Party for Men**

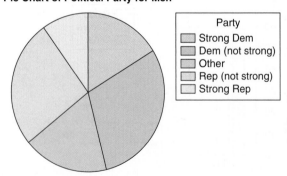

**Pie Chart of Political Party for Men**

a. Which political affiliation has the most men? Is this easier to determine with the bar chart or with the pie chart? Why?

b. Which political affiliation has the second largest number of men?

**2.36 Political Party Affiliation: Women** The 2008 GSS asked its participants to report their political party affiliation. The graphs show the distribution of political party affiliation for 1084 women.

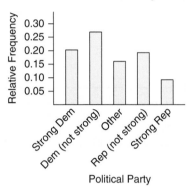

**Bar Chart of Political Party for Women**

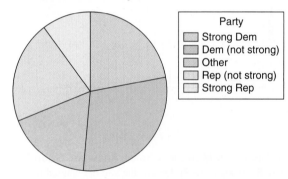

**Pie Chart of Political Party for Women**

a. Which political affiliation has the most women? Is this easier to see from the bar chart or the pie chart?

b. Which political affiliation has the second largest number of women?

c. Some people believe that women tend to lean more towards liberal political positions (like Democrats). By comparing the shape of the distribution of political affiliation for women to that for men (Exercise 2.35), do you see evidence of this? Explain.

TRY **2.37 Age by Year (Example 9)** The bar chart shows the projected percentage of U.S. residents in different age categories by year, according to the 2009 *World Almanac and Book of Facts*.

a. Comment on the predicted changes from 2010 through to 2030. Which age groups are predicted to become larger, which are predicted to become smaller, and which are predicted to stay roughly the same?

b. Comment on the effect this might have on Social Security, a government program that collects money from those currently working and gives it to retired people.

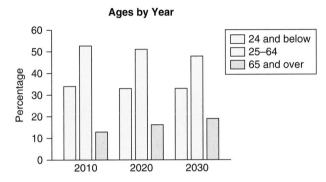

**2.38 Retail Car Sales** With gas prices rising, as they did between 1985 and 2007, you might expect people to move toward buying smaller cars. Compare the types of cars sold in 1985, 2000, and 2007 as shown in the figure. (Source: 2009 *World Almanac and Book of Facts*.)

a. Which type of car sold the most in all three years?

b. What is the trend for small cars? Has a higher or a lower percentage of small cars sold in more recent years?

c. What is the trend for large cars?

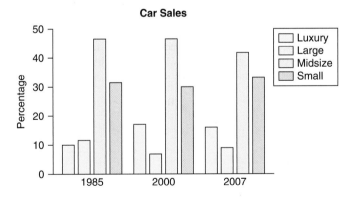

**2.39 Majors** The table gives information on college majors at Wellesley College, a women's college outside of Boston. Sketch an appropriate graph of the distribution, and comment on its important features.

| Major | Percentage |
|---|---|
| Humanities | 30% |
| Social Science | 38% |
| Math and Science | 15% |
| Interdisciplinary | 17% |

**2.40 Adoptions** The table gives information on the top five countries from which U.S. residents adopted children in 2007. Sketch an appropriate graph of the distribution, and comment on its important features. (Source: 2009 *World Almanac and Book of Facts*)

| Country | Number |
|---|---|
| China | 5453 |
| Guatemala | 4728 |
| Russia | 2310 |
| Ethiopia | 1255 |
| South Korea | 939 |

## CHAPTER REVIEW EXERCISES

**2.41 Television** The table shows the first few entries for the number of hours of television viewed per week for some fifth-grade students, stacked and coded, where 1 represents a girl and 0 represents a boy.

What would be appropriate graphs to compare the distributions of hours of TV watched per week for boys and girls if you had all the data? Explain.

| TV | Girl |
|----|------|
| 3  | 1    |
| 8  | 1    |
| 11 | 1    |
| 12 | 1    |
| 7  | 0    |
| 5  | 0    |
| 4  | 0    |

**2.42 Job** The table shows the job categories for some employees at a business. What type of graph(s) would be appropriate to compare the distribution of jobs for men and for women if you had all of the data? Explain.

| Male | Job |
|------|-----|
| 1    | Custodial  |
| 0    | Clerical   |
| 0    | Managerial |
| 0    | Clerical   |
| 1    | Managerial |

**2.43 Hormone Replacement Therapy** The use of the drug Prempro, a combination of two female hormones that many women take after menopause, is called hormone replacement therapy (HRT). In July 2002, a medical article reported the results of a study that was done to determine the effects of Prempro on many diseases. (Source: Women's Writing Group, Risks and benefits of estrogen plus progestin in healthy postmenopausal women, *JAMA*, 2002).

The study was placebo-controlled, randomized, and double-blind. From studies like these, it is possible to make statements about cause and effect. The figure shows comparisons of disease rates in the study.

a. For which diseases was the disease rate higher for those who took HRT? And for which diseases was the rate lower for those who took HRT?

b. Why do you suppose we compare the rate per 10,000 women (per year), rather than just reporting the numbers of women observed who get the disease?

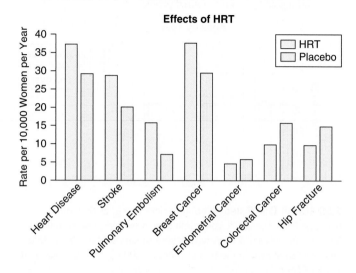

**2.44 ER Visits for Injuries** The graph shows the rates of visits to the ER for injuries by gender and by age. Note that we are concerned with the rate per 100 people of that age and gender in the population. (Source: National Safety Council 2004)

a. Why does the National Safety Council give us rates instead of numbers of visits?

b. For which ages are the males more likely than the females to have an ER visit for an injury? For which ages are the men and women similar? For which ages do the women have more visits?

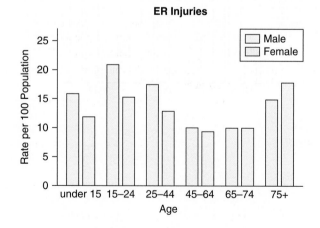

**2.45 Hormone Replacement Therapy Again** The bar chart shows a comparison of breast cancer rates for those who took HRT and those who took a placebo. Explain why the graph is deceptive, and indicate what could be done to make it less so.

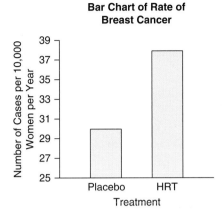

**Bar Chart of Rate of Breast Cancer**

**2.46 Holding Breath** A group of students held their breath as long as possible and recorded the times in seconds. The times went from a low of 25 seconds to a high of 90 seconds, as you can see in the stemplot. Suggest improvements to the histogram below generated by Excel, assuming that what is wanted is a histogram of the data (not a bar chart).

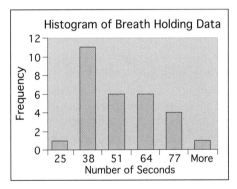

**2.47 Gender**

a. An elementary school class includes 12 girls and 8 boys. Which type of graph would be appropriate to show the distribution of gender in the class: a pie chart or a histogram?

b. In a school district, there are 45 elementary school classrooms, each containing 20 students. Which type of graph would be appropriate to show the distribution of the number of boys in the classrooms: a pie chart or a histogram?

**2.48 Handedness**

a. An elementary school class is made up of 3 children who are left-handed and 17 who are right-handed. The objects of study are the children, and the variable of interest is whether the child is right- or left-handed. Which type of graph would be appropriate to show the distribution of handedness in the class: a pie chart or a histogram?

b. In a school district, there are 45 elementary school classrooms, each containing 20 students. The unit of study is a classroom, and the variable is the number of right-handed students in each class. Which type of graph would be appropriate to show the distribution of the number of right-handed students in the classrooms: a pie chart or a histogram?

**\* 2.49 Global Temperatures** The histograms show the average global temperature per year for two 26-year ranges in degrees Fahrenheit. The range for 1880 to 1905 is on the top, and the range for 1980 to 2005 is on the bottom. Compare the two histograms for the two time periods, and explain what they show. Also estimate the difference between the centers. That is, about how much does the typical global temperature for the 1980–2005 time period differ from that for the 1880–1905 period? (Source: Goddard Institute for Space Studies, NASA, 2009)

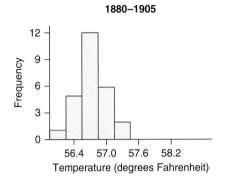

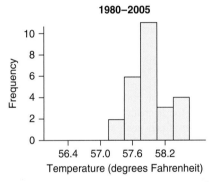

**\* 2.50 Employment after Law School** Accredited law schools were ranked from 1 for the best (Harvard) down to number 181 by the Internet Research Legal Group (ILRG). When you decide on a law school to attend, one of the things you might be interested in is whether, after graduation, you will be able to get a job for which your law degree is required. We split the group of 181 law schools in half, with the top-ranked schools in one group and the lower ranked schools in the other. The histograms show the distribution of the percentages of graduates who, 9 months after graduating, have obtained jobs that require a law degree. Compare the two histograms.

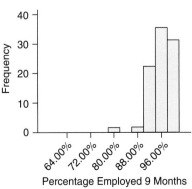

**Top 90 Law Schools**

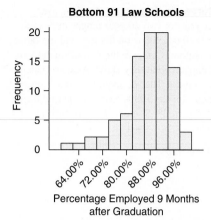

**Bottom 91 Law Schools**

Percentage Employed 9 Months after Graduation

**2.51** Create a dotplot that has at least 10 observations and is right-skewed.

**2.52** Create a dotplot that has at least 10 observations and does not have skew.

**2.53 Home Prices** How do home prices compare in West Los Angeles (WLA) and Midtown (MT) Los Angeles? Using appropriate graphics, write a few sentences comparing the distribution of home prices. The data are sampled from a real estate website, and prices are in hundreds of thousands of dollars. (Thus 10 represents $1,000,000.) The data are on the disk that comes with the book.

**2.54 Traffic Cameras** College students Jeannette Mujica, Ricardo Ceja Zarate, and Jessica Cerda conducted a survey in Oxnard, California, of the number of cars going through a yellow light at intersections with and without traffic cameras that are used to automatically send notification of fines to drivers who run red lights. The cameras were very noticeable to drivers. The amount of traffic was constant throughout the study period (the afternoon commute.) The data record the number of cars that crossed the intersection during a yellow light for each light cycle. A small excerpt of the data is shown in the table; see the disk for all the data. What differences, if any, do you see between intersections with cameras and those without? Use an appropriate graphical summary, and write a comparison of the distributions.

| # Cars | Cam |
|--------|--------|
| 1 | Cam |
| 2 | Cam |
| 1 | No Cam |
| 3 | No Cam |

**2.55 Forbes 100 Companies** The 100 companies with the highest net incomes (for 2008) were obtained from Forbes.com. First predict the shape of a histogram of the data, and explain your prediction. Then make a histogram of the data (on the disk) and describe it. Mention the names of any companies that are outliers.

**2.56 Ideal Weight** Thirty-nine students (26 women and 13 men) reported their ideal weight (in most cases, not their current weight). The tables show the data.

| Women | | | | | |
|------|-----|-----|-----|-----|-----|
| 110 | 115 | 123 | 130 | 105 | 110 |
| 130 | 125 | 120 | 115 | 120 | 120 |
| 120 | 110 | 120 | 150 | 110 | 130 |
| 120 | 118 | 120 | 135 | 130 | 135 |
| 90 | 110 | | | | |

| Men | | | | | |
|------|-----|-----|-----|-----|-----|
| 160 | 130 | 220 | 175 | 190 | 190 |
| 135 | 170 | 165 | 170 | 185 | 155 |
| 160 | | | | | |

a. Explain why the distribution of ideal weights is likely to be bimodal if men and women are both included in the sample.

b. Make a histogram combining the ideal weights of men and women. Use the default histogram provided by your software. Report the bin width and describe the distribution.

c. Vary the number of bins, and print out a second histogram. Report the bin width and describe this histogram. Compare the two histograms.

# GUIDED EXERCISES

g
✳

**2.23 Eating Out and Jobs**   College student Jacqueline Loya asked students who had full-time jobs and students who had part-time jobs how many times they went out to eat in the last month.

Full-time: 5, 3, 4, 4, 4, 2, 1, 5, 6, 5, 6, 3, 3, 2, 4, 5, 2, 3, 7, 5, 5, 1, 4, 6, 7

Part-time: 1, 1, 5, 1, 4, 2, 2, 3, 3, 2, 3, 2, 4, 2, 1, 2, 3, 2, 1, 3, 3, 2, 4, 2, 1

**Question**   Compare the two groups by following the steps below. Include appropriate graphics.

**Step 1 ▶ Create graphs.**
Make dotplots (or histograms) using the same axis with one set of data above the other, as shown in the figure.

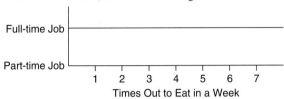

**Step 2 ▶ Examine shape.**
What is the shape of the data for those with part-time jobs? What is the shape of the data for those with full-time jobs?

**Step 3 ▶ Examine center.**
Which group tends to go out to eat more often, those with full-time or those with part-time jobs?

**Step 4 ▶ Examine variation.**
Which group has a wider spread of data?

**Step 5 ▶ Check for outliers.**
Were there any numbers separated from the other numbers as shown in the dotplots? (In other words, were there any gaps in the dotplots?)

**Step 6 ▶ Summarize.**
Finally, in one or more sentences, compare the shape, center, and variation (and mention outliers if there were any).

# TechTips

## General Instructions for All Technology

**EXAMPLE A ▶** Use the following ages to make a histogram:

7, 11, 10, 10, 16, 13, 19, 22, 42

### Columns
Data sets are generally put into columns, not rows. The columns may be called variables, lists, or the like. Figure 2A shows a column of data.

| L1 | L2 | L3 | 1 |
|---|---|---|---|
| **7** | ------ | ------ | |
| 11 | | | |
| 10 | | | |
| 10 | | | |
| 16 | | | |
| 13 | | | |
| 19 | | | |

L1(1)=7

▲ **FIGURE 2A** TI-83/84
Data Screen

### TI-83/84

**Resetting the Calculator (Clearing the Memory)**
If you turn off the TI-83/84, it does not reset the calculator. All the previous data and choices remain. If you are having trouble and want to start over, you can reset the calculator.

1. **2nd Mem** (with the + sign)
2. **7** for **Reset**
3. **1** for **All RAM**
4. **2** for **Reset**

It will say **RAM cleared** if it has been done successfully.

**Entering Data into the Lists**

1. Press **STAT**, and select **EDIT** (by pressing **ENTER** when **EDIT** is highlighted).
2. If you find that there are data already in **L1** (or any list you want to use), you have three options:
   a. Clear the entire list by using the arrow keys to highlight the **L1** label and then pressing **CLEAR** and then **ENTER**. Do not press **DELETE**, because then you will no longer have an **L1**. If you delete **L1**, to get it back you can **Reset** the calculator.

b. Delete the individual entries by highlighting the top data entry, then pressing **DELETE** several times until all the data are erased. (The numbers will scroll up.)

c. Overwrite the existing data. CAUTION: Be sure to **DELETE** data in any cells not overwritten.

3. Type the numbers from the example into **L1** (List1). After typing each number, you may press **ENTER** or use ▼ (the arrow down on the keypad). Double-check your entries before proceeding.

## Histogram

1. Press **2nd**, **STATPLOT** (which is in the upper left corner of the keypad).

2. If more than the first plot is **On**, press **4** (for **PlotsOff**) and **ENTER** to turn them all off.

3. Press **2nd**, **STATPLOT**, and **1** (for **Plot 1**).

4. Turn on **Plot1** by pressing **ENTER** when **On** is flashing; see Figure 2B.

▲ **FIGURE 2B** TI-83/84

5. Use the arrows on the keypad to get to the histogram icon (highlighted in Figure 2B) and press **ENTER** to choose it. Use the down arrow to get to Xlist and press **2nd** and **1** for **L1**. The settings shown in Figure 2B will lead to a histogram of the data in List 1 (L1).

6. Press **GRAPH**, **ZOOM** and **9** (**ZoomStat**) to create the graph.

7. To see the numbers shown in Figure 2C, press **TRACE** (in the top row on the keypad) and move around using the group of four arrows on the keypad to see other numbers.

Figure 2C shows a histogram of the numbers in the example.

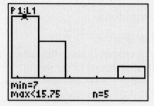

▲ **FIGURE 2C** TI-83/84 Histogram

The TI-83/84 cannot make stemplots, dotplots, or bar charts.

# Downloading Numerical Data from a Computer into a TI-84

Before you can use your computer with your TI-84, you must install (on your computer) the software program, TI Connect™, and a driver for the calculator. You need do this only once. If you have done steps 1–3 and are ready to download the data, start with Step 4.

1. Downloading and saving the TI Connect™ setup program. Insert the CD that came with the TI-84 into the computer disk drive and follow the on screen instructions. This will copy the setup file into the Downloads folder in your hard drive. (If you no longer have the CD, the programs can be downloaded free from the TI website, www.TI.com.)

2. Installing TI Connect™.
Click on the globe in the lower left corner of your desktop and in the **Search** box, type **Downloads**. When you get to the **Downloads**, double click on **TIConnect** to begin the installation.

Follow the on screen instructions. This should also create a TI Connect icon on your desktop screen.

3. Installing the calculator driver on the computer. To do this, connect the TI-84 to the computer using the USB cable that came with the calculator; then follow the on screen instructions.
This completes the installation and the CD may now be removed from the computer

4. Double click the TI Connect icon on your desktop screen. If it doesn't exist, Click **Start**, **Programs**, **TI Tools**, and **TI-Connect**.

5. Click on **TI-Data Editor**.

6. Refer to Figure 2D.
Click on the icon for the white sheet of paper. The arrow points to it. This will give you a white column to use for the data. If you have more than one column of data click on the piece of paper until you have the correct number of columns for your use. Figure 2D shows two columns.

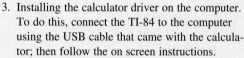
TI DataEditor

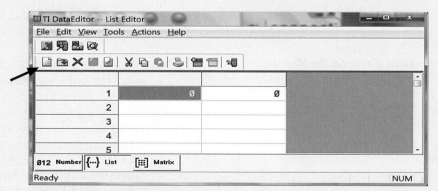

▲ **FIGURE 2D** TI Data Editor

7. Copy a column or more than one column of *numerical* data from your computer. You may use Excel, Minitab, or any spreadsheet for the source of the column(s) of numbers but do not include any labels or words. (If there are any letters or special characters in the column, they will not transfer and may show up as zeroes.) Then click on the 0 in the first cell of the column (see Figure 2D). The cell *must* become colored (blue, as shown in the figure) before pasting. If it is not colored, click out of the column (for example on the gray area to the right of the column) and then click in the first cell of the first column again. When that cell is blue, paste the column(s) of numbers into the TI-84 data editor. Alternatively you can just type numbers in the column on the data editor.

8. While your cursor is in the column you want to name, choose **File** and **Properties** from the **Data Editor**. Then refer to Figure 2E: Check a list number like **L1** as shown (or if you want a name, it cannot be more than 8 characters) and click **OK**. Then go back to any other column and do the same but pick a different name such as **L2**.

9. Connect the TI-84 to the computer with the cable and turn on the TI-84 calculator.

10. Refer to Figure 2D. To paste the column(s) of data click **Actions, Send All Lists**.

11. When you get the **Warning**, click **Replace or Replace All** to overwrite the old data in the lists.

Look in your calculator lists (**STAT**, **EDIT**) to see the data there. If it is not there, check the cable connection, check that the TI-84 is turned on and start again with step 6.

Caution: While the data is transferring you will not be able to use the calculator; it is thinking.

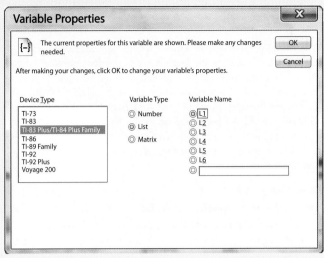

▲ **FIGURE 2E** Variable Properties

## Using Data from the Disk or Website

If uploading data from the disk or website to your calculator, the column headers will have variable names, such as **HT**, instead of list numbers, such as **L1**. If there are two files each with one variable (for example, one with **HT** and one with **WT**), open both of the files in TI-Connect. You can do this by opening one file in TI-Connect and then, while it is open, double-clicking on the other file. You should see both variables in the same TI-Connect screen. Next, double-click on the headers, one at a time, to change them to numbered lists, such as **L1** and **L2**. Then when you click **Actions** and **Send All Lists**, the data will go into the lists you have chosen for headers.

## MINITAB

### Entering the Data

When you open Minitab, you will see a blank spreadsheet for entering data. Type the data from the example into **C1**, column 1. Be sure your first number is put into Row 1, not above Row 1 in the label region. Be sure to enter only numbers. (If you want a label for the column, type it in the label region *above* the numbers.) You may also paste in data from the computer clipboard. Double-check your entries before proceeding.

### All Minitab Graphs

After making the graph, double click on what you want to change, such as labels.

### Histogram

1. Click **Graph > Histogram**
2. Leave the default option **Simple** and click **OK**.
3. Double click **C1** (or the name for the column) and click **OK**. (Another way to get **C1** in the big box is to click **C1** and click **Select**.)
4. After obtaining the histogram, if you want different bins (intervals), double click on the *x*-axis and look for **Binning**.

Figure 2F shows a Minitab histogram of the ages.

### Stemplot
Click **Graph > Stem-and-Leaf**

### Dotplot
Click **Graph > Dotplot**

### Bar Chart
Click **Graph > Bar Chart**

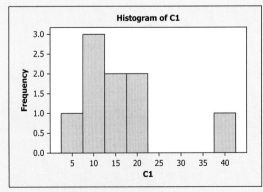

▲ **FIGURE 2F** Minitab Histogram

When you open Excel, you will see a blank spreadsheet for entering data. But first, click the **Data** tab (top of screen, middle); you should see **Data Analysis** just below and on the far right. If you do not see it, you will need to load the Data Analysis Toolpak (instructions below). Now click the **Add-Ins** tab; you should see the XLSTAT icon ◪ just below and on the far left. If you do not see it, you should install XLSTAT (instructions below). You will need both of these add-ins in order to perform all of the statistical operations described in this textbook

### Data Analysis Toolpak

1. Click **File** > **Options** > **Add-Ins**.
2. In **Manage** Box, select **Excel Add-Ins**, and click **Go**.
3. In **Add-ins available** box, check **Analysis Toolpak** and click **OK**.

### XLSTAT

1. Close Excel.
2. Download XLSTAT from www.myPearsonstore.com.
3. Install XLSTAT.
4. Open Excel, click **Add-Ins tab**, click the XLSTAT icon ◪.

You need to install the Data Analysis Toolpak only once. The Data Analysis tab should be available now every time Excel is opened. For XLSTAT, however, step 4 above may need to run each time Excel is opened (if you expect to run XLSTAT routines).

### Entering Data

See Figure 2G. Enter the data from the example into column A with **Ages** in cell A1. Double check your entries before proceeding.

| | A |
|---|---|
| 1 | Ages |
| 2 | 7 |
| 3 | 11 |
| 4 | 10 |
| 5 | 10 |
| 6 | 16 |
| 7 | 13 |
| 8 | 19 |
| 9 | 22 |
| 10 | 42 |

▲ **FIGURE 2G** Excel Data Screen

### Histogram

1. Click **Add-Ins**, **XLSTAT**, **Describing data**, and **Histograms**.
2. When the box under **Data** is activated, drag your cursor over the column containing the data including the label **Ages**.
3. Click **OK** and **Continue**.

Figure 2H shows the histogram.

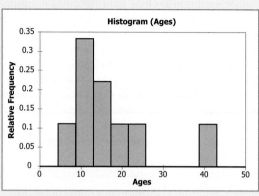

▲ **FIGURE 2H** XLSTAT Histogram

### Dotplot and Stemplot

1. Click **Add-Ins**, **XLSTAT**, **Visualizing data**, and **Univariate plots**.
2. When the box under **Quantitative Data** is activated, drag your cursor over the column containing the data including the label **Ages**.

#### Dotplot

3. Click the **Options, Charts, Charts(1)** and select **Scattergrams** and **Horizontal**.
4. Click **OK** and **Continue**.

#### Stemplot

3. Click **Options, Charts, Charts (1)** and select **Stem-and-leaf plots**.
4. Click **OK** and **Continue**.

### Bar Chart

After typing a summary of your data in table form (including labels), drag your cursor over the table to select it, click **Insert**, **Column** (in the **Charts** group), and select the first option.

For Help: After logging in, click on **Help** or **Resources** and **Watch StatCrunch Video Tutorials on YouTube**.

### Entering Data

1. Click **Open StatCrunch** and you will see a spreadsheet as shown in Figure 2I.
2. Enter the data from the example into the column labeled var1.
3. If you want labels on the columns, click on the variable label, such as **var1**, and backspace to remove the old label and type the new label. Double-check your entries before proceeding.

### Pasting Data

1. If you want to paste data from your computer clipboard, click **Paste** (on the left side of Figure 2I), which will open a screen that says **Load Data from Paste**.
2. Click in the **Input** box and then paste your data.
3. Note that it will use the first line as variable name(s) unless you uncheck it. Scroll down to the very bottom and click **Load Data**.

### Histogram

1. Click **Graphics > Histogram**
2. Under **Select Columns**, click the variable you want a histogram for.
3. Click **Create Graph!**
4. To copy the graph for pasting into a document for submission, click **Options** and **Copy** and then paste it into a document.

Figure 2J shows the StatCrunch histogram of the ages.

### Stemplot

Click **Graphics > Stem-and-leaf**

### Dotplot

Click **Graphics > Dotplot**

### Bar Chart

Click **Graphics > Chart > Columns**

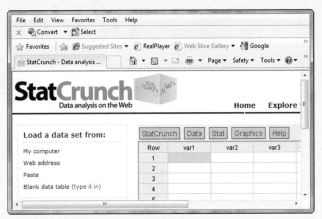

▲ **FIGURE 2I** StatCrunch Data Table

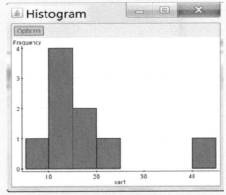

▲ **FIGURE 2J** StatCrunch Histogram

# 3 Numerical Summaries of Center and Variation

# THEME

The complexity of numerical distributions can often be captured with a small number of summaries. In many cases, two numbers—one to measure the typical value and one to measure the variability—are all we need to summarize a set of data and make comparisons between groups. We can use many different ways of measuring what is typical and also of measuring variability; which method is best to use depends on the shape of the distribution.

Google the phrase *average American* and you'll find lots of entertaining facts. For example, the average American lives within 3 miles of a McDonald's, showers 10.4 minutes a day, and prefers smooth peanut butter to chunky. The average American woman is 5'4" tall and weighs 140 pounds, while the average American female fashion model is considerably taller and weighs much less: 5'11" and 117 pounds.

Whether or not these descriptions are correct, they are attempting to describe a typical American. The reason for doing this is to try to understand a little better what Americans are like, or perhaps to compare one group (American women) to another (female fashion models).

These summaries can seem odd, because we all know that people are too complex to be summarized with a single number. Characteristics such as weight, distance from a McDonald's, and length of a shower vary quite a bit from person to person. If we're describing the "typical" American, shouldn't we also describe how much variation exists among Americans? If we're making comparisons between groups, as we do in this chapter's Case Study, how do we do it in a way that takes into account the fact that individuals may vary considerably?

In Chapter 2, we talked about looking at graphs of distributions of data to get an intuitive, informal sense of the typical value (the center) and the amount of variation (the spread). In this chapter, we explore ways of making these intuitive concepts more precise by assigning numbers to them. We will see how this step makes it much easier to compare and interpret sets of data, for both symmetric and skewed distributions. These measures are important tools that we will use throughout the book.

# CASE STUDY

## Living in a Risky World

Our perception of how risky an activity is plays a role in our decision making. If you think that flying is very risky, for example, you will be more willing to put up with a long drive to get where you want to go. A team of UCLA psychologists were interested in understanding how people perceive risk and whether a simple reporting technique was enough to detect differences in perceptions between groups. The researchers asked over 500 subjects to consider various activities and rate them in terms of how risky they thought the activities were. The ratings were on a scale of 0 (no risk) to 100 (greatest possible risk). For example, subjects were asked to assign a value to the following activities: "Use a household appliance" and "Receive a diagnostic X-ray every 6 months." One question of interest to the researchers was whether men and women would assign different risk levels to these activities (Carlstrom et al. 2000).

Figure 3.1a on the next page shows a histogram for the perceived risk of using a household appliance, and Figure 3.1b shows a histogram for the risk of twice-annual X-rays. (Women are represented in the left panels, men in the right.) What differences do you see between the genders? How would you quantify these differences? In this chapter, you'll learn several techniques for answering these questions. And at the end of the chapter, we'll use these techniques to compare perceived risk between men and women.

▶ **FIGURE 3.1** Histograms showing the distributions of perceived
risk by gender, for two activities.
**(a)** Perceived risk of using everyday
household appliances. **(b)** Perceived
risk of receiving a diagnostic X-ray
twice a year. The left panel in each
part is for women and the right
panel for men.

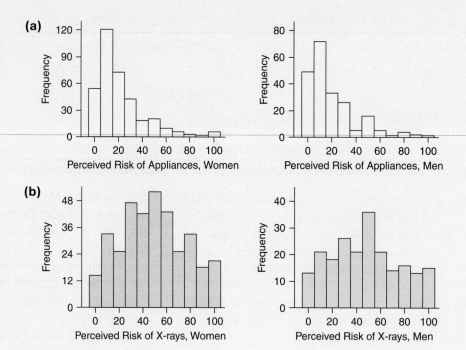

# Summaries for Symmetric Distributions

In Chapter 2 you learned that we can characterize the typical value of a distribution by the center of the distribution, and the variability in that distribution by the horizontal spread. We left these concepts somewhat vague, but now our goal is to quantify them—to measure these concepts with numbers. However, assigning a number to the center and spread of a distribution is not all that straightforward. Statisticians have different ways of thinking about both center and spread, and these different ways of thinking play different roles, depending on the context of the data.

In this chapter, the two different ways in which we will think about the concept of center are (1) center as the balancing point (or center of mass), and (2) center as the halfway point. In this section we introduce the idea of the center as a balancing point, useful for symmetric distributions. Then in Section 3.3, we introduce the idea of the center as the halfway point, useful for skewed distributions. Each of these approaches results in a different measure, and our choice also affects the method we use to measure variability.

## The Center as Balancing Point: The Mean

The most commonly used measure of center is the mean. The **mean** of a collection of data is the arithmetic average. The mean can be thought of as the balancing point of a distribution of data, and when the distribution is symmetric, the mean closely matches our concept of the "typical value."

### Visualizing the Mean

Different groups of statistics students often have different backgrounds and levels of experience. Some instructors collect student data to help them understand whether one class of students might be very different from another. For example, if one class is offered earlier in the morning, or later in the evening, it might attract a different composition of students. An instructor at Peoria Junior College in Illinois collected data from two classes, including the students' ACT scores. (ACT is a standardized national college entrance exam.) Figure 3.2 shows the distribution of self-reported ACT scores for one statistics class.

 **Looking Back**

**Symmetric Distributions**
Recall that symmetric distributions are those for which the left-hand side of the graph of the distribution is roughly a mirror image of the right-hand side.

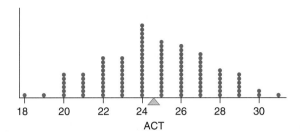

◄ **FIGURE 3.2** The distribution of ACT scores for one class of Statistics students. The mean is indicated here with a fulcrum (triangle). The mean is at the point on the dotplot that would balance if the points were placed on a seesaw. (Source: StatCrunch, Statistical_Data_499)

The balancing point for this distribution is roughly in the middle, because the distribution is fairly symmetric. The mean ACT for these students is calculated to be 24.6. You'll see how to do this calculation shortly. At the moment, rather than worrying about how to get the precise number, note that 24.6 is about the point where the distribution would balance if it were on a seesaw.

When the distribution of the data is more or less symmetric, the balancing point is roughly in the center, as in Figure 3.2. However, when the distribution is not symmetric, as in Figure 3.3, the balancing point is off-center and the average may not match what our intuition tells us is the center.

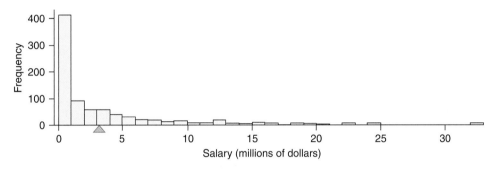

◄ **FIGURE 3.3** Salaries of Professional Baseball Players in 2010

Figure 3.3 shows a very skewed distribution: the salaries of professional baseball players in 2010. Salaries are in millions of dollars. (The highest paid player was Alex Rodriguez at $33 million. The lowest salary was $400,000, which 58 players received. The first bin includes over 400 players who earned $1 million dollars or less.) Where would you place the center of this distribution? Because the distribution is skewed to the right, the balancing point is fairly high; the mean is at 3.2 million dollars. However, when you consider that almost 70% of the players made less than this amount, you might not think that 3.2 million is what the "typical" player made. In other words, in this case the mean might not represent our idea of a typical salary, even for this group of famously well-paid professionals.

## EXAMPLE 1 Math Scores

Figure 3.4 shows the distribution of math achievement scores for 46 countries as determined by the National Assessment of Education Progress. The scores are meant to measure the accomplishment of each country's eighth grade students with respect to mathematical achievement.

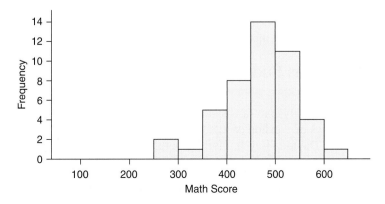

◄ **FIGURE 3.4** Distribution of international math scores for 46 countries. The scores measure the mean level of mathematical achievement for a country's eighth grade students.

QUESTION Based on the histogram, approximately what value do you think is the mean math achievement score?

SOLUTION The mean score is at the point where the histogram would balance if it were placed on a seesaw. The distribution looks roughly symmetric, and the balancing point looks to be somewhere between 400 and 500 points.

CONCLUSION The mean international math achievement score looks to be about 450 points. By the way, the U.S. score was 504, and the highest three scores were from Singapore (highest at 605), South Korea, and Hong Kong. This tells us that the typical score, among the 46 countries, on the international math achievement test was between 400 and 500 points.

TRY THIS! Exercise 3.3

## The Mean in Context

As mentioned, the mean tells us the typical value in a data set with variability. We know that different countries had different math achievement scores, but typically, what is the math score of these countries? When you report the mean of a sample of data, you should not simply report the number but, rather, should report it in the context of the data so that your reader understands what you are measuring: The typical score on the international math test among these countries is between 400 and 500 points.

Knowing the typical value of one group allows us to compare this group to another group. For example, that same instructor at Peoria Junior College recorded ACT scores for a second classroom and found that the mean there was also 24.6. This tells us that these two classrooms were comparable, at least in terms of their mean ACT score. However, one classroom had a slightly younger mean age: 26.9 years compared to 27.2 years.

> KEY POINT — The mean of a collection of data is located at the "balancing point" of a distribution of data. The mean is one representation of the "typical" value of a variable.

## Calculating the Mean for Small Data Sets

The mean is used so often in statistics that it has its own symbol: $\bar{x}$, which is pronounced "x-bar." To calculate the mean, find the (arithmetic) **average** of the numbers; that is, simply add up all the numbers and divide that sum by the number of observations. Formula 3.1 shows you how to calculate the mean, or average.

$$\text{Formula 3.1:} \quad \text{Mean} = \bar{x} = \frac{\sum x}{n}$$

The symbol $\sum$ is the Greek capital sigma, or capital S, which stands for *summation*. The $x$ that comes after $\sum$ represents the value of a single observation. Therefore, $\sum x$ means that you should add all the values. The letter $n$ represents the number of observations. Therefore, this equation tells us to add the values of all the observations and divide that sum by the number of observations.

The mean shown in Formula 3.1 is sometimes called the sample mean in order to make it clear that it is the mean of a collection (or sample) of data.

## EXAMPLE 2 Gas Buddy

According to GasBuddy.com (a website that invites people to submit prices at local gas stations), the prices of 1 gallon of regular gas at the 12 service stations for which data were provided in Daytona Beach, Florida, were as follows on one summer day in 2010:

$2.59, $2.59, $2.62, $2.65, $2.65, $2.65, $2.65, $2.67, $2.67, $2.67, $2.69, $2.69

A dotplot (not shown) indicates that the distribution is fairly symmetric.

**QUESTION** Find the mean price of a gallon of gas for these service stations. Explain what the value of the mean signifies in this context (in other words, interpret the mean).

**SOLUTION** Add the 12 numbers together to get $31.79. We have 12 observations, so we divide $31.79 by 12 to get $2.649.

$$\bar{x} = \frac{2.59 + 2.59 + 2.62 + 2.65 + 2.65 + 2.65 + 2.65 + 2.67 + 2.67 + 2.67 + 2.69 + 2.69}{12}$$

$$= \frac{31.79}{12} = 2.649$$

**CONCLUSION** The typical price of 1 gallon of gas at these gas stations in Daytona Beach, Florida, is $2.65 on this particular day.

**TRY THIS!** Exercise 3.5a

## Calculating the Mean for Larger Data Sets

Formula 3.1 tells you how to compute the mean if you have a small set of numbers that you can easily type into a calculator. But for large data sets, you are better off using a computer or a statistical calculator. That is true for most of the calculations in this book, in fact. For this reason, we will often just display what you would see on your computer or calculator and describe in the TechTips section the exact steps used to get the solution.

For example, the histogram in Figure 3.5 shows the distribution of the amount of particulate matter, or smog, in the air in 333 cities in the United States in 2008, as reported by the Environmental Protection Agency. (The units are micrograms of particles per cubic meter.) When inhaled, these particles can affect the heart and lungs, so you would prefer your city to have a fairly low amount of particulate matter. (The EPA says that levels over 15 micrograms per cubic meter are unsafe.) Looking at the histogram, you can estimate the mean value using the fact that, since the distribution is fairly symmetric, the average will be about in the middle (around 11 micrograms per cubic meter). If you were given the list of 333 values, you could find the mean using Formula 3.1. But you'll find it easier to use the pre-programmed routines of your calculator or software. Example 3 demonstrates how to do this.

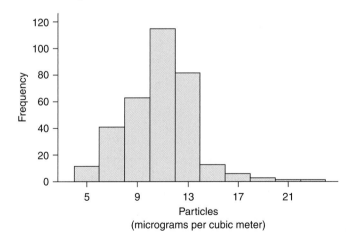

◀ **FIGURE 3.5** Levels of particulate matter for 333 U.S. cities in 2008. Because the distribution is fairly symmetric, the balancing point is roughly in the middle at about 11 micrograms per cubic meter.

## ❋ EXAMPLE 3 Mean Smog Levels

We used four different statistical software programs to find the mean particulate level for the 333 cities. The data were uploaded into StatCrunch, Minitab, Excel, and the TI-83/84 calculator.

Tech

**QUESTION** For each of the computer outputs shown in Figure 3.6, find the mean particulate matter. Interpret the mean.

▶ **FIGURE 3.6a** StatCrunch output.

◯ ◯ ◯                  Column Statistics          Java Applet Window ⚠

[ Options ]

Summary statistics:

| Column | n | Mean | Variance | Std. Dev. | Std. Err. | Median | Range | Min | Max | Q1 | Q3 |
|--------|---|------|----------|-----------|-----------|--------|-------|-----|-----|-----|-----|
| pm25_wtd | 333 | 10.738439 | 6.6927953 | 2.5870438 | 0.14176913 | 10.8 | 19.1 | 4.4 | 23.5 | 9.2 | 12.3 |

▶ **FIGURE 3.6b** Minitab output.

**Minitab Descriptive Statistics: pm25_wtd**

```
Variable     N    N*    Mean  SE Mean   StDev  Minimum      Q1  Median      Q3
pm25_wtd   333   213  10.738    0.142   2.587    4.400   9.200  10.800  12.300

Variable  Maximum
pm25_wtd   23.500
```

▶ **FIGURE 3.6c** TI-83/84 calculator output.

```
1-Var Stats
 x̄=10.73843844
 Σx=3575.9
 Σx²=40621.59
 Sx=2.587043708
 σx=2.583156337
↓n=333
```

```
1-Var Stats
↑n=333
 minX=4.4
 Q₁=9.2
 Med=10.8
 Q₃=12.3
 maxX=23.5
```

▶ **FIGURE 3.6d** Excel output.

| | *Column1* |
|---|---|
| Mean | 10.73843844 |
| Standard Error | 0.141769122 |
| Median | 10.8 |
| Mode | 11.4 |
| Standard Deviation | 2.587043708 |
| Sample Variance | 6.692795145 |
| Kurtosis | 2.494762112 |
| Skewness | 0.542367196 |
| Range | 19.1 |
| Minimum | 4.4 |
| Maximum | 23.5 |
| Sum | 3575.9 |
| Count | 333 |

**SOLUTION** We have already seen, from Figure 3.5, that the distribution is close to symmetric, so the mean is a useful measure of center. With a large data set like this, it makes sense to use a computer to find the mean. Minitab and many other statistical software packages produce a whole slew of statistics with a single command, and your job is to choose the correct value.

**CONCLUSION** The software outputs give us a mean of 10.7 micrograms per cubic meter. We interpret this to mean that the typical level of particulate matter for these cities is 10.7 micrograms per cubic meter. For StatCrunch, Minitab, and Excel the mean is easy to find; it is labeled clearly. For the TI output, the mean is labeled as $\bar{x}$, "x-bar."

**TRY THIS!** Exercise 3.9

In this textbook, pay more attention to how to apply and interpret statistics than to individual formulas. Your calculator or computer will nearly always find the correct values without your having to know the formula. However, you need to tell the calculator *what* to compute, you need to make sure the computation is meaningful, and you need to be able to explain what the result tells you about the data.

## SNAPSHOT    THE MEAN OF A SAMPLE

**WHAT IS IT?** ▶ A numerical summary.

**WHAT DOES IT DO?** ▶ Measures the center of a sample distribution.

**HOW DOES IT DO IT?** ▶ The mean identifies the "balancing point" of the distribution, which is the arithmetic average of the values.

**HOW IS IT USED?** ▶ The mean represents the typical value in a set of data when the distribution is roughly symmetric.

## Measuring Variation with the Standard Deviation

The mean amount of particulate matter in the air, 10.7 micrograms per cubic meter, does not tell the whole story. Just because the mean of the 333 cities is at a safe level (below 15 micrograms per cubic meter) does not imply that any particular city—yours for example—is at a healthful level. Are most cities close to the mean level of 10.7? Or do cities tend to have levels of particulate matter far from 10.7? Values in a data set vary, and this variation is measured informally by the horizontal spread of the distribution of data. A measure of variability, coupled with a measure of center, helps us understand whether most observations are close to the typical value or far from it.

### Visualizing the Standard Deviation

The histograms in Figure 3.7 show daily high temperatures in degrees Fahrenheit recorded over one recent year at two locations in the United States. The histogram shown in Figure 3.7a records data collected in Provo, Utah, a city far from the ocean and at an elevation of 4500 feet.

The histogram shown in Figure 3.7b records data collected in San Francisco, California, which sits on the Pacific coast and is famously chilly in the summer. (Mark Twain is alleged to have said, "The coldest winter I ever spent was a summer in San Francisco.")

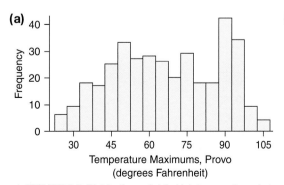
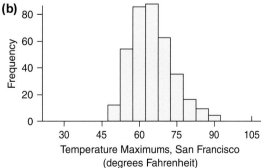

▲ **FIGURE 3.7** Distributions of daily high temperatures in two cities: **(a)** Provo, Utah; **(b)** San Francisco. Both cities have about the same mean temperature, although the variation in temperatures is much greater in Provo than in San Francisco.

The distributions of temperatures in these cities are similar in several ways. Both temperature distributions are fairly symmetric. You can see from the histograms that both cities have about the same mean temperature. For San Francisco the mean daily high temperature was about 65 degrees; for Provo it was about 67 degrees. But note the difference in the spread of temperatures!

One effect of the ocean on coastal climate is to moderate the temperature: The highs are not that high, and the lows not too low. This suggests that a coastal city should have less spread in the distribution of temperatures. How can we measure this spread, which we can see informally from the histograms?

Note that in San Francisco, most days were fairly close to the mean temperature of 65 degrees; rarely was it more than 10 degrees warmer or cooler than 65. (In other words, rarely was it colder than 55 or warmer than 75 degrees.) Provo, on the other hand, had quite a few days that were more than 10 degrees warmer or cooler than average.

The **standard deviation** is a number that measures how far away the typical observation is from the mean. Distributions such as San Francisco's temperatures have smaller standard deviations because more observations are fairly close to the mean. Distributions such as Provo's have larger standard deviations because more observations are farther from the mean.

As you'll soon see, for most distributions, a majority of observations are within one standard deviation of the mean value.

> **KEY POINT**   The standard deviation should be thought of as the typical distance of the observations from their mean.

## EXAMPLE 4 Comparing Standard Deviations from Histograms

Each of the three graphs in Figure 3.8 shows a histogram for a distribution of the same number of observations, and all the distributions have a mean value of about 3.5.

**QUESTION** Which distribution has the largest standard deviation, and why?

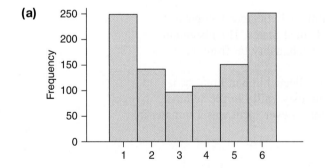

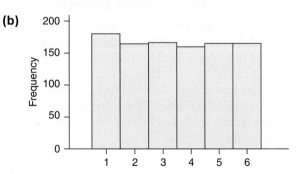

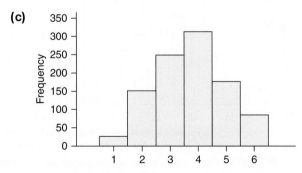

▶ **FIGURE 3.8** All three of these histograms have the same mean, but each has a different standard deviation.

SOLUTION All three groups have the same minimum and maximum values. However, the data shown in Figure 3.8a have the largest standard deviation. Why? The standard deviation measures how widely spread the points are from the mean. Note that the histogram in Figure 3.8a has the greatest number of observations farthest from the mean (at the values of 1 and 6). Figure 3.8c has the smallest standard deviation because so many of the data are near the center, close to the mean, which we can see because the taller bars in the center show us that there are more observations there.

CONCLUSION Figure 3.8a has the largest standard deviation, and Figure 3.8c has the smallest standard deviation.

TRY THIS! Exercise 3.13

## The Standard Deviation in Context

The standard deviation is somewhat more abstract, and harder to understand, than the mean. Fortunately, in a symmetric, unimodal distribution, a handy rule of thumb helps make this measure of spread more comprehensible. In these distributions, the majority of the observations (in fact, about two-thirds of them) are less than one standard deviation from the mean.

For temperatures in San Francisco (see Figure 3.7), the standard deviation is about 8 degrees, and the mean is 65 degrees. This tells us that in San Francisco, on a majority of days, the high temperature is within 8 degrees of the mean temperature of 65 degrees—that is, usually it is no colder than $65 - 8 = 57$ degrees and no warmer than $65 + 8 = 73$ degrees. In Provo, the standard deviation is substantially greater: 21 degrees. On a typical day in Provo, the high temperature is within 21 degrees of mean. Provo has quite a bit more variability in temperature.

## EXAMPLE 5 Standard Deviation of Smog Levels

The mean particulate matter in the 333 cities graphed in Figure 3.5 is 10.7 micrograms per cubic meter, and the standard deviation is 2.6 micrograms per cubic meter.

QUESTION Find the level of particulate matter one standard deviation above the mean and one standard deviation below the mean. Keeping in mind that the EPA says that levels over 15 micrograms per cubic meter are unsafe, what can we conclude about the air quality of most of the cities in this sample?

SOLUTION The typical city has a level of 10.7 micrograms per cubic meter, and because the distribution is unimodal and (roughly) symmetric, most cities have levels within 2.6 micrograms per cubic meter of this value. In other words, most cities have levels of particulate matter between

$$10.7 - 2.6 = 8.1 \text{ micrograms per cubic meter and}$$
$$10.7 + 2.6 = 13.3 \text{ micrograms per cubic meter}$$

CONCLUSION Because the value of 13.3 (one standard deviation above the mean) is lower than 15, most cities are below the safety limit. (The three cities reporting the highest levels of particulate matter were Phoenix, Arizona; Visalia, California; and Hilo, Hawaii. The three cities reporting the lowest levels of particulate matter were Cheyenne, Wyoming; Santa Fe, New Mexico; and Dickinson, North Dakota.)

As you'll soon see, we can say even more about this example. In a few pages you'll learn about the Empirical Rule, which tells us that about 95% of all cities should be within two standard deviations of the mean particulate level.

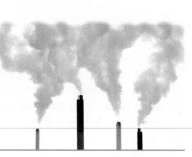

**TRY THIS!** Exercise 3.15

## Calculating the Standard Deviation

The formula for the standard deviation is somewhat more complicated than that for the mean, and a bit more work is necessary to calculate it. A calculator or computer is pretty much required for all but the smallest data sets. Just as the mean of a sample has its own symbol, the standard deviation of a sample of data is represented by the letter $s$.

**Formula 3.2:**    Standard deviation $= s = \sqrt{\dfrac{\sum (x - \bar{x})^2}{n - 1}}$

Think of this formula as a set of instructions. Essentially, the instructions say that we need to first calculate how far away each observation is from the mean. This distance, including the positive or negative sign, $(x - \bar{x})$, is called a **deviation**. We square these deviations so that they are all positive numbers, and then we essentially find the average. (If we had divided by $n$, and not $n - 1$, it would have been the average. You will see in Chapter 9, Exercise 9.81, why we divide by $n - 1$ and not $n$.) Finally, we take the square root, which means that we're working with the same units as the original data, not with squared units.

## EXAMPLE 6 A Gallon of Gas

From the website Gasbuddy.com, we collected the prices of a gallon of regular gas at all 12 reported gas stations in Daytona Beach, Florida, for one day in June 2010.

$2.59, $2.59, $2.62, $2.65, $2.65, $2.65, $2.65, $2.67, $2.67, $2.67, $2.69, $2.69

**QUESTION** Find the standard deviation for the prices. Explain what this value means in the context of the data.

**SOLUTION** We show this result two ways. The first way is by hand, which illustrates how to apply Formula 3.2. The second way uses a statistical calculator.

The first step is to find the mean. We did this earlier in Example 2, using Formula 3.1, which gave us a mean value of $2.65. We substitute this value for $\bar{x}$ in Formula 3.2.

Table 3.1 shows the first two steps. First we find the deviations (in column 2). Next we square each deviation (in column 3). The numbers are sorted so you can more easily compare the differences.

The sum of the squared deviations—the sum of column 3—is 0.0125. Dividing this by 11 (because $n - 1 = 12 - 1 = 11$), we get 0.001136364. The last step is to take the square root of this. The result is our standard deviation:

$$s = \sqrt{0.001136364} = 0.03371$$

| $x$ | $x - \bar{x}$ | $(x - \bar{x})^2$ |
|---|---|---|
| 2.59 | −0.06 | 0.0036 |
| 2.59 | −0.06 | 0.0036 |
| 2.62 | −0.03 | 0.0009 |
| 2.65 | 0 | 0 |
| 2.65 | 0 | 0 |
| 2.65 | 0 | 0 |
| 2.65 | 0 | 0 |
| 2.67 | 0.02 | 0.0004 |
| 2.67 | 0.02 | 0.0004 |
| 2.67 | 0.02 | 0.0004 |
| 2.69 | 0.04 | 0.0016 |
| 2.69 | 0.04 | 0.0016 |

▲ **TABLE 3.1**

To recap:

$$s = \sqrt{\frac{\sum(x - \bar{x})^2}{n - 1}} = \sqrt{\frac{0.0125}{12 - 1}} = \sqrt{\frac{0.0125}{11}} = \sqrt{0.001136364}$$

$$= 0.03371, \text{ or about 3 cents}$$

When doing these calculations, your final result will be more accurate if you do not round any of the intermediate results. For this reason, it is far easier, and more accurate, to use a statistical calculator or statistical software to find the standard deviation. Figure 3.9 shows the standard deviation as $Sx = 0.0336987546$, which we round to about 0.03 dollars (or, if you prefer, 3 cents). Note that the value reported by the calculator in Figure 3.9 is not exactly the same as the figure we obtained by hand. In part, this is because we used an approximate value for the mean in our hand calculations.

Tech

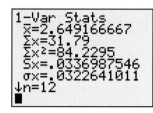

▲ **FIGURE 3.9** The standard deviation is denoted "Sx" in the TI calculator output.

**CONCLUSION** The standard deviation is about 3 cents, or $0.03. Therefore, at most of these gas stations, the price of a gallon of gas is within 3 cents of $2.65.

**TRY THIS!** Exercise 3.17

One reason why we suggest using statistical software rather than the formulas we present is that we nearly always look at data using several different statistics and approaches. We nearly always begin by making a graph of the distribution. Usually the next step is to calculate a measure of the center and then a measure of the spread. It does not make sense to have to enter the data again every time you want to examine them; it is much better to enter them once and use the functions on your calculator (or software).

**SNAPSHOT**  **THE STANDARD DEVIATION OF A SAMPLE**

| | |
|---|---|
| **WHAT IS IT?** ► | A numerical summary. |
| **WHAT DOES IT DO?** ► | Measures the spread of a distribution. |
| **HOW DOES IT DO IT?** ► | It measures the typical distance of the observations from the mean. |
| **HOW IS IT USED?** ► | To measure the amount of variability in a sample when the distribution is fairly symmetric. |

## Variance, a Close Relative of the Standard Deviation

Another way of measuring spread—a way that is closely related to the standard deviation—is the variance. The **variance** is simply the standard deviation squared, and it is represented symbolically by $s^2$.

**Formula 3.3:**   Variance $= s^2 = \dfrac{\sum(x - \bar{x})^2}{n - 1}$

In Example 5, the standard deviation of the concentration of particulate matter in the cities in our sample was 2.6 micrograms per cubic meter. The variance is therefore $2.6 \times 2.6 = 6.76$ micrograms squared per cubic meter squared. The standard deviation in daily high temperatures in Provo is 21 degrees, so the variance is $21 \times 21 = 441$ degrees squared.

For most applications, the standard deviation is preferred over the variance. One reason is that the units for the variance are always squared (degrees squared in the last paragraph), which implies that the units used to measure spread are different from the units used to measure center. The standard deviation, on the other hand, has the same units as the mean.

# What's Unusual? The Empirical Rule and *z*-Scores

Finding the standard deviation and the mean is a useful way to compare different samples and to compare observations from one sample with those in another sample.

## The Empirical Rule

The **Empirical Rule** is a rough guideline, a rule of thumb, that helps us understand how the standard deviation measures variability. This rule says that if the distribution is unimodal and symmetric, then

- Approximately 68% of the observations (roughly two-thirds) will be within one standard deviation of the mean.

- Approximately 95% of the observations will be within two standard deviations of the mean.

- Nearly all the observations will be within three standard deviations of the mean.

When we say that about 68% of the observations are within one standard deviation of the mean, we mean that if we count the observations that are between the mean minus one standard deviation and the mean plus one standard deviation, we will have counted about 68% of the total observations.

The Empirical Rule is illustrated in Figure 3.10 in the context of the data on particulate matter in 333 U.S. cities, introduced in Example 3. Suppose we did not have access to the actual data and knew only that the distribution is unimodal and symmetric, that the mean particulate matter is 10.7 micrograms per cubic meter, and that the standard deviation is 2.6 micrograms per cubic meter. The Empirical Rule predicts that about 68% of the cities will fall between 8.1 micrograms per cubic meter (10.7 − 2.6 = 8.1) and 13.3 micrograms per cubic meter (10.7 + 2.6 = 13.3).

The Empirical Rule predicts that about 95% of the cities will fall within two standard deviations of the mean, which means that about 95% of the cities will be between 5.5 and 15.9 micrograms per cubic meter (10.7 − (2 × 2.6) = 5.5 and 10.7 + (2 × 2.6) = 15.9). Finally, nearly all cities, according to the Empirical Rule, will be between 2.9 and 18.5 micrograms per cubic meter. This is illustrated in Figure 3.10.

▶ **FIGURE 3.10** The Empirical Rule predicts how many observations we will see within one standard deviation of the mean (68%), within two standard deviations of the mean (95%), and within three standard deviations of the mean (almost all).

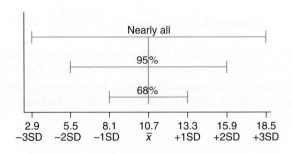

KEY POINT

In a relatively large collection of observations, if the distribution is unimodal and roughly symmetric, then about 68% of the observations are within one standard deviation of the mean; about 95% are within two standard deviations of the mean, and almost all observations are within three standard deviations of the mean. Not all unimodal, symmetric distributions are the same, so your actual outcomes might differ from these values, but the Empirical Rule works well enough in a surprisingly large number of situations.

## EXAMPLE 7 Comparing the Empirical Rule to Actual Smog Levels

Because the distribution of levels of particular matter (PM) in 333 U.S. cities is roughly unimodal and symmetric, the Empirical Rule predicts that about 68% of the cities will have particulate matter levels between 8.1 and 13.3 micrograms per cubic meter, about 95% of the cities will have PM levels between 5.5 and 15.9 micrograms per cubic meter, and nearly all will have PM levels between 2.9 and 18.5 micrograms per cubic meter.

QUESTION   Figure 3.11 shows the actual histograms for the distribution of PM levels for these 333 cities. The location of the mean is indicated, as well as the boundaries for points within one standard deviation of the mean (a), within two standard deviations of the mean (b), and within three standard deviations of the mean (c). Using these figures, compare the actual number of cities that fall within each boundary to the number predicted by the Empirical Rule.

SOLUTION   From Figure 3.11a, by counting the heights of the bars between the two boundaries, we find that there are about $20 + 50 + 52 + 60 + 48 = 230$ cities that actually lie between these two boundaries. (No need to count very precisely; we're after approximate numbers here.) The Empirical Rule predicts that about 68% of

**(a)**
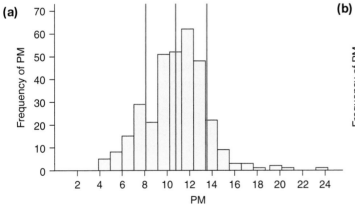

**(b)**
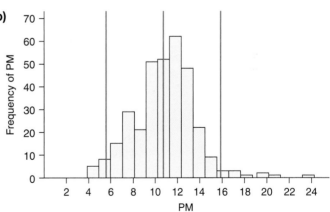

**(c)**
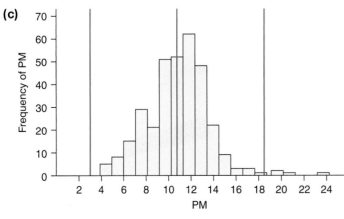

▲ FIGURE 3.11 (a) Boundaries are placed one standard deviation below the mean (8.1) and one standard deviation above the mean (13.3). The middle line in all three figures indicates the mean (10.7). (b) The Empirical Rule predicts that about 95% of observations will fall between these two boundaries (5.5 and 15.9 micrograms per cubic meter). (c) The Empirical Rule predicts that nearly all observations will fall between these two boundaries (2.9 and 18.5 micrograms per cubic meter).

the cities, or $0.68 \times 333 = 226$ cities, will fall between these two boundaries. The Empirical Rule is pretty accurate in this case.

From Figure 3.11b, we count that there are about $15 + 30 + 20 + 50 + 52 + 60 + 48 + 20 + 8 + 2 = 305$ cities within two standard deviations of the mean. The Empirical Rule predicts that about 95% of the cities, or $0.95 \times 333 = 316$ cities, will fall between these two boundaries. So again, the Empirical Rule is not too far off.

Finally, from Figure 3.11c, we clearly see that all but a few cities (about 4 or 5), are within three standard deviations of the mean. We summarize in a table.

CONCLUSION

| Interval | Empirical Rule Prediction | Actual Number |
|---|---|---|
| within 1SD of mean | 226 | 230 |
| within 2SD of mean | 316 | 305 |
| within 3SD of mean | nearly all | all but 5 |

TRY THIS! Exercise 3.21

# EXAMPLE 8 Temperatures in San Francisco

We have seen that the mean daily high temperature in San Francisco is 65 degrees Fahrenheit and that the standard deviation is 8 degrees.

QUESTION Using the Empirical Rule, decide whether it is unusual in San Francisco to have a day when the maximum temperature is colder than 49 degrees.

SOLUTION One way of answering this question is to find out about how many days in the year we would expect the high temperature to be colder than 49 degrees. The Empirical Rule says that about 95% of the days will have temperatures within two standard deviations of the average—that is, within $2 \times 8 = 16$ degrees of 65 degrees. Therefore, on most days the temperature will be between $65 - 16 = 49$ degrees and $65 + 16 = 81$ degrees. Because only (approximately) 5% of the days are outside this range, we know that days either warmer or cooler than this are rare. According to the Empirical Rule, which assumes a roughly symmetric distribution, about half the days outside the range—2.5%—will be colder, and half will be warmer. This shows that about 2.5% of 365 days (roughly 9 or 10 days) should have a maximum temperature colder than 49 degrees.

CONCLUSION A maximum temperature colder than 49 degrees is fairly unusual for San Francisco. According to the Empirical Rule, only about 2.5% of the days should have maximum temperatures below 49 degrees, or about 9 or 10 days out of a 365-day year. Of course, the Empirical Rule is only a guide. In this case, we can compare with the data to see just how often these cooler days occurred.

Your concept of "unusual" might be different from ours. The main idea is that "unusual" is rare, and in selecting two standard deviations, we chose to define temperatures that occur 2.5% of the time or less as rare and therefore unusual. You might very reasonably set a different standard for what you wish to consider unusual.

TRY THIS! Exercise 3.23

## z-Scores: Measuring Distance from Average

The question "How unusual is this?" is perhaps the statistician's favorite question. (It is just as popular as "Compared to what?") Answering this question is complicated because the answer depends on the units of measurement. Eighty-four is a big value if we are measuring a man's height in inches, but it is a small value if we are measuring his weight in pounds. Unless we know the units of measurement and the objects being measured, we can't judge whether a value is big or small.

One way around this problem is to change the units to standard units. **Standard units** measure a value relative to the sample rather than with respect to some absolute measure. A measurement converted to standard units is called a *z*-**score**.

### Visualizing *z*-Scores

Specifically, a standard unit measures how many standard deviations away an observation is from the mean. In other words, it measures a distance, but instead of measuring in feet or miles, it counts the number of standard deviations. A measurement with a *z*-score of 1.0 is one standard deviation above the mean. A measurement with a *z*-score of −1.4 is 1.4 standard deviations below the mean.

Figure 3.12 shows a dotplot of the heights (in inches) of 247 adult men. The average height is 70 inches, and the standard deviation is 3 inches (after rounding). Below the dotplot is a ruler that marks off how far from average each observation is, measured in terms of standard deviations. The average height of 70 inches is 0 because 70 is zero standard deviations away from the mean. The height of 76 is marked as 2 because it is two standard deviations above mean. The height of 67 is marked as −1 because it is one standard deviation *below* mean.

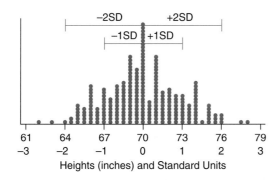

◀ **FIGURE 3.12** A dotplot of heights of 247 men marked with *z*-scores as well as heights in inches.

We would say that a man from this sample who is 73 inches tall has a *z*-score of 1.0 standard unit. A man who is 67 inches tall has a *z*-score of −1.0 standard unit.

### Using *z*-Scores in Context

*z*-Scores allow us to compare observations in one group with those in another, even if the two groups are measured in different units or under different conditions. For instance, some students might choose their math class on the basis of which professor they think is an easier grader. So if one student gets a 65 on an exam in a hard class, how do we compare his score to another student who gets a 75 in an easy class? If we converted to standard units, we would know how far above (or below) the average each test score was, so we could compare these students' performances.

## EXAMPLE 9 Exam Scores

Maria scored 80 out of 100 on her first stats exam in a course and 85 out of 100 on her second stats exam. On the first exam, the mean was 70 and the standard deviation was 10. On the second exam, the mean was 80 and the standard deviation was 5.

QUESTION  On which exam did Maria perform better when compared to the whole class?

SOLUTION     On the first exam, Maria is 10 points above the mean because 80 − 70 is 10. Because the standard deviation is 10 points, she is one standard deviation above the mean. In other words, her z-score for the first exam is 1.0.

On the second exam, she is 5 points above average because 85 − 80 is 5. Because the standard deviation is 5 points, she is one standard deviation above the mean. In other words, her z-score is again 1.0.

CONCLUSION     The second exam was a little easier; on average, students scored higher and there was less variability in the scores. But Maria scored one standard deviation above average on both exams, so she did equally well on both when compared to the whole class.

TRY THIS!     Exercise 3.27

## Calculating the z-Score

It's straightforward to convert to z-scores when the result is a whole number, as in the last few examples. More generally, to convert a value to its z-score, first subtract the mean. Then divide by the standard deviation. This simple recipe is summarized in Formula 3.4.

$$\text{Formula 3.4:} \quad z = \frac{x - \bar{x}}{s}$$

Let's apply this to the data shown in Figure 3.12. What is the z-score of a man who is 75 inches tall? Remember that the mean height is 70 inches and the standard deviation is 3 inches. Formula 3.4 says first to subtract the mean height.

$$75 - 70 = 5 \text{ inches}$$

Next divide by the standard deviation:

$$5/3 = 1.67 \text{ (rounding off to two decimal digits.)}$$

$$z = \frac{x - \bar{x}}{s} = \frac{75 - 70}{3} = \frac{5}{3} = 1.67$$

This person has a z-score of 1.67. In other words, we would say that he is 1.67 standard deviations taller than average.

## EXAMPLE 10 Daily Temperatures

The mean daily high temperature in San Francisco is 65 degrees F, and the standard deviation is 8 degrees. On one day, the high temperature is 49 degrees.

QUESTION     What is this temperature in standard units? Assuming that the Empirical Rule applies, does this seem unusual?

SOLUTION

$$z = \frac{x - \bar{x}}{s} = \frac{49 - 65}{8} = \frac{-16}{8} = -2.00$$

CONCLUSION     This is an unusually cold day. From the Empirical Rule, we know that 95% of z-scores are between −2 and 2 standard units, so it is fairly unusual to have a day as cold as or colder than this one.

TRY THIS!     Exercise 3.29

## SNAPSHOT THE z-SCORE

| | | |
|---|---|---|
| **WHAT IS IT?** | ▶ | A standardized observation. |
| **WHAT DOES IT DO?** | ▶ | Converts a measurement into standard units. |
| **HOW DOES IT DO IT?** | ▶ | By measuring how many standard deviations away a value is from the sample mean. |
| **HOW IS IT USED?** | ▶ | To compare values from different groups, such as two exam scores from different exams, or to compare values measured in different units, such as inches and pounds. |

## SECTION 3.3

# Summaries for Skewed Distributions

As you saw earlier, for a skewed distribution, the center of balance is not a good way of measuring a "typical" value. Another concept of center, which is to think of the center as being the location of the *middle* of a distribution, works better in these situations. You saw one example of this in Figure 3.3, which showed that the mean baseball player salary was quite a bit higher than what a majority of the players actually earned. Figure 3.13 shows another example of a strongly right-skewed distribution. This is the distribution of incomes of 788 New York State residents, drawn from the 2000 U.S. census. The mean income of $27,198 is marked with a triangle. However, note that the mean doesn't seem to match up very closely with what we think of as typical. The mean seems to be too high to be typical. In fact, over half (about 55%) of residents earn less than this mean amount.

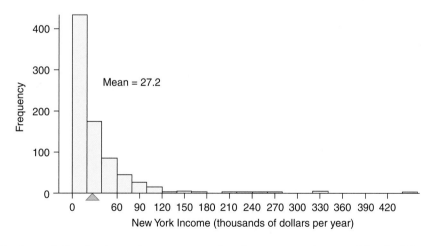

**◀ FIGURE 3.13** The distribution of annual incomes for a collection of New York State residents is right-skewed. Thus the mean is somewhat greater than what many people would consider a typical income. The location of the mean is shown with a triangle.

In skewed distributions, the mean can often be a poor measure of "typical." Instead of using the mean and standard deviation in these cases, we measure the center and spread differently.

## The Center as the Middle: The Median

The median provides an alternative way of determining a typical observation. The **median** of a sample of data is the value that would be right in the middle if you were to sort the data from smallest to largest. The median cuts a distribution down the middle, so about 50% of the observations are below it and about 50% are above it.

## Visualizing the Median

One of the authors found herself at the grocery store, trying to decide whether to buy ham or turkey for sandwiches. Which is more healthful? Figure 3.14 shows a dotplot of the percentage of fat for each of ten types of sliced ham. The vertical line marks the location of the median at 23.5 percent. Note that five observations lie below the median and five lie above it. The median cuts the distribution exactly in half. In Example 12, you'll see how the median percentage of fat in sliced turkey compares to that in sliced ham.

◀ **FIGURE 3.14** A dotplot of percentage fat from ham has a median of 23.5%. This means that half the observations are below 23.5% and half are above it.

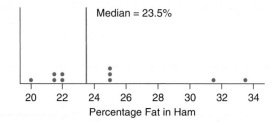

Finding the median for the distribution of incomes of the 778 New York State residents, shown again in Figure 3.15, is slightly more complicated because we have many more observations. The median (shown with the red vertical line) cuts the total area of the histogram in half. The median is at $15,800, and roughly 50% of the observations are below this value and about 50% are above it.

◀ **FIGURE 3.15** The distribution of incomes of New York State residents, with the median indicated by a vertical line. The median has about 50% of the observations above it and about 50% below it.

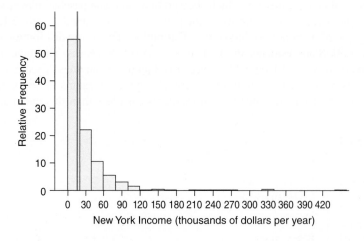

 **KEY POINT**     The median is the value that cuts a distribution in half. The median value represents a "typical" value in a data set.

## The Median in Context

The median is used for the same purpose as the mean: to give us a typical value of a set of data. Knowing the typical value of one group helps us to compare it to another. For example, as we've seen, the typical median income of this sample of New York State residents is $15,800. How does the typical income in New York compare to the typical income in Florida? A representative sample of 1000 Florida residents (where many New Yorkers go to retire) has a median income of $18,000, which is higher than the median income in New York (Figure 3.16).

The typical person in our data set of New York incomes makes less than the typical person in our Florida data set, as measured by the median. Because the median for Florida is $18,000, we know that more than half of Florida residents in the sample make more than the median New York salary of $15,800.

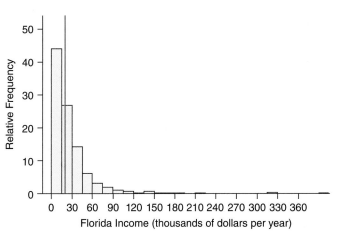

◀ **FIGURE 3.16** Distribution of incomes of a selection of Florida residents (in thousands of dollars). The median income is $18,000.

The median is often reported in news stories when the discussion involves variables with distributions that are skewed. For example, you may hear reports of "median housing costs" and "median salaries."

## Calculating the Median

To calculate the value of the median, follow these steps:

1. Sort the data from smallest to largest.

2. If the set contains an odd number of observed values, the median is the middle observed value.

3. If the set contains an even number of observed values, the median is the average of the two middle observed values. This places the median precisely halfway between the two middle values.

## EXAMPLE **11** Twelve Gas Stations

The prices of a gallon of regular gas at 12 Daytona Beach gas stations in June 2010 (see Example 6) were

$2.65, $2.67, $2.65, $2.65, $2.59, $2.65, $2.69, $2.67, $2.67, $2.59, $2.62, $2.69

QUESTION Find the median price for a gallon of gas and interpret the value.

SOLUTION First we sort the data from smallest to largest.

2.59, 2.59, 2.62, 2.65, 2.65, 2.65, 2.65, 2.67, 2.67, 2.67, 2.69, 2.69

Because the data set contains an even number of observations (12), the median is the average of the two middle observations, the sixth and seventh: 2.65 and 2.65.

2.59, 2.59, 2.62, 2.65, 2.65, 2.65, 2.65, 2.67, 2.67, 2.67, 2.69, 2.69

                         Med

CONCLUSION The median is $2.65, which is the typical price of a gallon of gas at these 12 gas stations. (Note: Because the distribution of gas prices at these stations is roughly symmetric, the mean and median prices are very similar. We'll explore this more in a few pages.)

TRY THIS! Exercise 3.35a

Example 12 demonstrates how to find the median in a sample with an odd number of observations.

## EXAMPLE 12 Sliced Turkey

Figure 3.14 showed that the median percentage of fat from the various brands of sliced ham for sale at a grocery store was 23.5%. How does this compare to the median percentage of fat for the turkey? Here are the percentages of fat for the available brands of sliced turkey:

14, 10, 20, 20, 40, 20, 10, 10, 20, 50, 10

**QUESTION** Find the median percentage of fat and interpret the value.

**SOLUTION** The data are sorted and displayed below. Because we have 11 observations, the median is the middle observation, 20.

▲ **FIGURE 3.17** Some TI-83/84 output for the percentage of fat in the turkey.

**Tech**  10  10  10  10  14  20  20  20  20  40  50

Med

**CONCLUSION** The median for the turkey is 20% fat. Thus the typical percentage of fat for these types of sliced turkey is 20%. This is (slightly) less than that for the typical sliced ham, which has 23.5% fat. Figure 3.17 provides TI-83/84 output that confirms our calculation.

**TRY THIS!** Exercise 3.37a

---

## SNAPSHOT  THE MEDIAN OF A SAMPLE

**WHAT IS IT?**  ▶  A numerical summary.

**WHAT DOES IT DO?**  ▶  Measures the center of a distribution.

**HOW DOES IT DO IT?**  ▶  It is the value that has roughly the same number of observations above it and below it.

**HOW IS IT USED?**  ▶  To measure the typical value in a data set, particularly when the distribution is skewed.

---

## Measuring Variability with the Interquartile Range

The standard deviation measures how spread out observations are with respect to the mean. But if we don't use the mean, then it doesn't make sense to use the standard deviation. When a distribution is skewed and you are using the median to measure the center, an appropriate measure of variation is called the interquartile range. The **interquartile range (IQR)** tells us, roughly, how much space the middle 50% of the data occupy.

### Visualizing the IQR

To find the IQR, we cut the distribution into four parts with roughly equal numbers of observations. The distance taken up by the middle two parts is the interquartile range.

The dotplot in Figure 3.18 shows the distribution of weights for a class of introductory statistics students. The vertical lines slice the distribution into four parts so that each part has about 25% of the observations. The IQR is the distance between the first "slice" (at about 121 pounds) and the third slice (at about 160 pounds). This is an interval of 39 pounds (160 − 121 = 39).

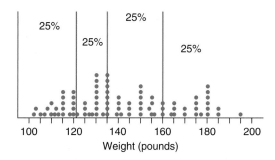

◀ **FIGURE 3.18** The distribution of weights (in pounds) of students in a class is divided into four sections so that each section has roughly 25% of the observations. The IQR is the distance between the outer vertical lines (at 121 pounds and 160 pounds).

Figure 3.19 shows distributions for the same students, but this time the weights are separated by gender. The vertical lines are located so that about 25% of the data are below the leftmost line, and 25% are above the rightmost line. This means that about half of the data lie between these two boundaries. The distance between these boundaries is the IQR. You can see that the IQR for the males, about 38 pounds, is much larger than the IQR for the females, which is about 20 pounds. The females have less variability in their weights.

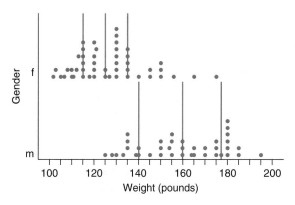

◀ **FIGURE 3.19** The dotplot of Figure 3.18 with weights separated by gender (the women on top). The men have a much larger interquartile range than the women.

The IQR focuses only on the middle 50% of the data. We can change values outside of this range without affecting the IQR. Figure 3.20 shows the men's weights, but this time we've changed one of the men's weights to be very small. The IQR is still the same as in Figure 3.19.

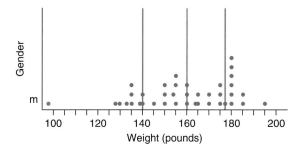

◀ **FIGURE 3.20** The men's weights are given with a fictitious point (in red) below 100. Moving an extremely large (or small) value does not change the interquartile range.

## The Interquartile Range in Context

The IQR for the incomes of the New Yorkers in the data set previously shown in Figure 3.15 is $33,200, as shown in Figure 3.21 at the top of the next page. This tells us that the middle 50% of people in our data set had incomes that varied by as much as $33,200. Compare this to the incomes from Florida, which have an IQR of $26,400. There is less variability among the Floridians; they are more similar (at least in terms of their incomes).

An IQR of $33,200 for New Yorkers seems like a pretty large spread. However, considered in the context of the entire distribution (see Figure 3.21 on the next page), which includes incomes near $0 and as large as nearly $400,000, the IQR looks fairly small. The reason is that lots of people (half of our data set) have incomes in this fairly narrow interval.

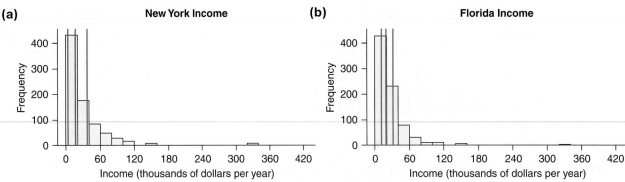

▲ **FIGURE 3.21** **(a)** The distribution of incomes for New Yorkers. **(b)** The distribution of income for Floridians. In both figures, vertical bars are drawn to divide the distribution into areas with about 25% of the observations. The IQR is the distance between the outer vertical lines, and it is wider for the New York incomes.

## Calculating the Interquartile Range

Calculating the interquartile range involves two steps. First, you must determine where to "cut" the distribution. These points are called the **quartiles**, because they cut the distribution into quarters (fourths). The **first quartile (Q1)** has roughly one-fourth, or 25%, of the observations at or below it. The **second quartile (Q2)** has about 50% at or below it; actually, Q2 is just another name for the median. The **third quartile (Q3)** has about 75% of the observations at or below it. The second step is the easiest: To find the interquartile range you simply find the interval between Q3 and Q1—that is, Q3 − Q1.

To find the quartiles:

- First find the median, which is also called the second quartile, Q2. The median cuts the data into two regions.

- The first quartile (Q1) is the median of the lower half of the sorted data. (Do not include the median observation in the lower half if you started with an odd number of observations.)

- The third quartile (Q3) is the median of the upper half of the sorted data. (Again, do not include the median itself if your full set of data has an odd number of observations.)

> **Formula 3.5:**   Interquartile range = Q3 − Q1

## EXAMPLE 13 Heights of Children

A group of eight children have the following heights (in inches):

48.0, 48.0, 53.0, 53.5, 54.0, 60.0, 62.0, and 71.0

They are shown in Figure 3.22.

**QUESTION** Find the interquartile range for the distribution of the children's heights.

**SOLUTION** As before, we first explain how to do the calculations by hand and then show the output of technology.

First we find Q2 (the median). Note that the data are sorted and there are four observed values below the median and four observed values above the median.

| Duncan | Charlie | Grant | Aidan | | Sophia | Seamus | Cathy | Drew |
|--------|---------|-------|-------|--|--------|--------|-------|------|
| 48 | 48 | 53 | 53.5 | | 54 | 60 | 62 | 71 |

Q2
53.75

To find Q1, examine the numbers below the median and find the median of them, as shown.

| Duncan | Charlie | | Grant | Aidan |
|--------|---------|--|-------|-------|
| 48 | 48 | | 53 | 53.5 |

Q1
50.50

**(a)**

**(b)**

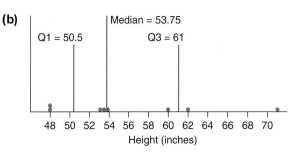

▲ **FIGURE 3.22** **(a)** Eight children sorted by height. **(b)** A dotplot of the heights of the children. The median and quartiles are marked with vertical lines. Note that two dots, or 25% of the data, appear in each of the four regions.

To find Q3, examine the numbers above the median and find the median of them.

| Sophia | Seamus | | Cathy | Drew |
|--------|--------|--|-------|------|
| 54 | 60 | | 62 | 71 |

$$Q3$$
$$61.00$$

Together, these values are

| Duncan | Charlie | Grant | Aidan | Sophia | Seamus | Cathy | Drew |
|--------|---------|-------|-------|--------|--------|-------|------|
| 48 | 48 | 53 | 53.5 | 54 | 60 | 62 | 71 |

$$Q1 \qquad\qquad Q2 \qquad\qquad Q3$$
$$50.50 \qquad\quad 53.75 \qquad\quad 61.00$$

Here's how we calculated the values:

$$Q1 = \frac{48 + 53}{2} = \frac{101}{2} = 50.50 \qquad \text{(Halfway between Charlie and Grant)}$$

$$Q2 = \frac{53.5 + 54}{2} = \frac{107.5}{2} = 53.75 \qquad \text{(Halfway between Aidan and Sophia)}$$

$$Q3 = \frac{60 + 62}{2} = \frac{122}{2} = 61.00 \qquad \text{(Halfway between Seamus and Cathy)}$$

The last step is to subtract:

$$IQR = Q3 - Q1 = 61.00 - 50.50 = 10.50$$

Figure 3.22b shows the location of Q1, Q2, and Q3. Note that 25% of the data (two observations) lie in each of the four regions created by the vertical lines.

Figure 3.23 shows the TI-83/84 output. The TI-83/84 does not calculate the IQR directly; you must subtract Q3 – Q1 yourself. The IQR is 61 − 50.5 = 10.5, which is the same as the IQR done by hand above.

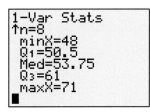

▲ **FIGURE 3.23** Some output of a TI-83/84 for eight children's heights.

**Tech**

CONCLUSION  The interquartile range of the heights of the eight children is 10.5 inches.

TRY THIS!  Exercise 3.39b

## Finding the Range, Another Measure of Variability

Another measure of variability is similar to the IQR but much simpler. The **range** is the distance spanned by the entire data set. It is very simple to calculate: the largest value minus the smallest value.

**Formula 3.6:**   Range = maximum − minimum

For the heights of the eight children (Example 13), the range is $71.0 - 48.0 = 23.0$ inches.

The range is useful for a quick measurement of variability because it's very easy to calculate. However, because it depends on only two observations—the largest and the smallest—it is very sensitive to any peculiarities in the data. For example, if someone makes a mistake when entering the data and enters 710 inches instead of 71 inches, then the range will be very wrong. The IQR, on the other hand, depends on many observations and is therefore more reliable.

## SNAPSHOT   THE INTERQUARTILE RANGE

| | |
|---|---|
| **WHAT IS IT?** ▶ | A numerical summary. |
| **WHAT DOES IT DO?** ▶ | It measures the spread of the distribution of a data set. |
| **HOW DOES IT DO IT?** ▶ | It computes the distance taken up by the middle half of the sorted data. |
| **HOW IS IT USED?** ▶ | To measure the variability in a sample, particularly when the distribution is skewed. |

## SECTION 3.4

# Comparing Measures of Center

Which should you choose, the mean (accompanied by the standard deviation) or the median (with the IQR)? These pairs of measures have different properties, so you need to choose the pair that's best for the data you're considering. Our primary goal is to choose a value that is a good representative of the typical values of the distribution.

### Look at the Shape First

This decision begins with a picture. The shape of the distribution will determine which measures are best for communicating the typical value and the variability in the distribution.

### ✳ EXAMPLE 14 MP3 Song Lengths

One of the authors created a data set of the songs on his mp3 player. He wants to describe the distribution of song lengths.

QUESTION What measures should he use for the center and spread: the mean (250.2 seconds) and standard deviation (152.0 seconds) or the median (226 seconds) and interquartile range (117 seconds)? Interpret the appropriate measures. Refer to the histogram in Figure 3.24.

SOLUTION Before looking at the histogram, you should think about what shape you expect the graph to be. No song can be shorter than 0 seconds. Most songs on the radio are around 4 minutes (240 seconds), with some a little longer and some a little shorter. However, a few songs, particularly classical tracks, are quite long, so we might expect the distribution to be right-skewed. This suggests that the median and IQR are the best measures to use.

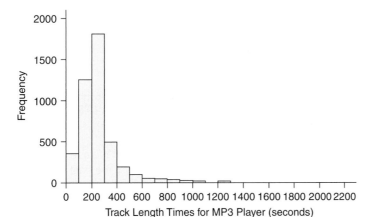

◀ **FIGURE 3.24** The distribution of lengths of songs (in seconds) on the mp3 player of one of the authors.

Figure 3.24 confirms that, as we predicted, the distribution is right-skewed, so the median and interquartile range would be the best measures to use.

**CONCLUSION** The median length is 226 seconds (roughly 3 minutes and 46 seconds), and the interquartile range is 117 seconds (close to 2 minutes). In other words, the typical track on the author's mp3 player is about 4 minutes, but there's quite a bit of variability, with the middle 50% of the tracks differing by about 2 minutes.

**TRY THIS!** Exercise 3.47

Sometimes, you don't have the data themselves, so you can't make a picture. If so, then you must deduce a shape for the distribution that is reasonable and choose the measure of center on the basis of this deduction.

When a distribution is right-skewed, as it is with the mp3 song lengths, the mean is generally larger than the median. You can see this in Figure 3.24; the right tail means the balancing point must be to the right of the median. With the same reasoning, we can see that in a left-skewed distribution, the mean is generally less than the median. In a symmetric distribution, the mean and median are approximately the same.

**KEY POINT** In a symmetric distribution, the mean and median are approximately the same. In a right-skewed distribution, the mean tends to be greater than the median, and in a left-skewed distribution, the mean tends to be less than the median.

## The Effect of Outliers

Even when a distribution is mostly symmetric, the presence of one or more outliers can have a large effect on the mean. Usually, the median is a more representative measure of center when an outlier is present.

The average height of the eight children from Example 13 was 56.2 inches. Imagine that we replace the tallest child (who is 71 inches tall) with retired basketball player Shaquille O'Neal, whose height is 85 inches. Our altered data set is now

48, 48, 53, 53.5, 54, 60, 62, and 85 inches

In order to keep the balance of the data, the mean has to shift higher. The mean of this new data set is 57.9 inches—over 1 inch higher. The median of the new data set, however, is the same: 53.75 inches, as shown in Figure 3.25.

When outliers are present, the median is a good choice for a measure of center. In technical terms, we say that the median is **resistant to outliers**; it is not affected by the size of an outlier and does not change even if a particular outlier is replaced by an even more extreme value.

**Looking Back**

**Outliers**
Recall from Chapter 2 that an outlier is an extremely large or small observation relative to the bulk of the data.

▶ **FIGURE 3.25** The effect of changing the tallest child's height into Shaquille O'Neal's height. Note that the mean (shown with triangles) changes, but the median (the vertical line) stays the same.

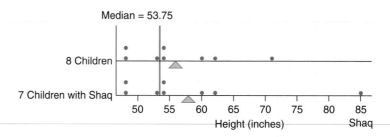

KEY POINT   The median is resistant to outliers; it is not affected by the size of the outliers. It is therefore a good choice for a measure of the center if the data contain outliers and you want to reduce their effect.

## EXAMPLE 15 Fast Food

A (very small) fast-food restaurant has five employees, all of whom work full-time for $7 per hour. Each employee's annual income is about $16,000 per year. The owner, on the other hand, makes $100,000 per year.

QUESTION   Find both the mean and the median. Which would you use to represent the typical income at this business—the mean or the median?

SOLUTION   Figure 3.26 shows a dotplot of the data. The mean income is $30,000, and the median is $16,000. Nearly all of the employees earned less than the mean amount!

▶ **FIGURE 3.26** Dotplot of salaries for five employees and their boss at a fast-food company.

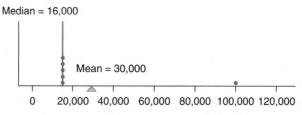

CONCLUSION   Given the choice between mean and median, the median is better at showing the typical income.

Why are the mean and median so different? Because the owner's salary of $100,000 is an outlier.

TRY THIS!   Exercises 3.59

## Many Modes: Summarizing Center and Spread

What should you do if the distribution is bimodal or has several modes? Unfortunately, the answer is "It's complicated."

You learned in Chapter 2 that multiple modes in a graphical display of a distribution sometimes indicate that the data set combines different groups. For example, perhaps a data set containing heights includes both men and women. The distribution could very well be bimodal, because we're combining two groups of people who differ markedly in terms of their heights. In this case, and in many other contexts, it is more useful to separate the groups and report summary measures separately for each group. If we know which observations belong to the men and which to the women, then we can separate the data and compute the mean height for men separately from the mean height for women.

For example, Figure 3.27 shows a histogram of the finishing times of female marathon runners. The most noticeable feature of this distribution is that there appear to be two modes. When confronted with this situation, a natural question to ask is "Are two different groups of runners represented in this data set?"

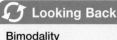 **Looking Back**

**Bimodality**
Recall from Chapter 2 that we said that a mode was a major mound in a graph (such as a histogram) of a single numerical variable. A bimodal distribution has two major mounds.

⚠ **Caution**

**Don't Get Modes from Computer Output**
Avoid using your computer to calculate the modes. For data sets with many numerical observations, many software programs give meaningless values for the modes. For the data shown in Figure 3.27, for example, most software packages would report the location of five different modes, none of them corresponding to the high points on the graph.

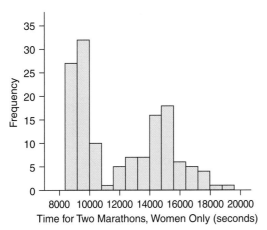

◀ **FIGURE 3.27** Marathon times reported for two groups: amateur and Olympic athletes. Note the two modes. Only women runners were included.

As it turns out, the answer is "yes." Table 3.2 shows the first few lines of the data set. From this, we see that the times belong to runners from two different events. One event was the 2008 Olympics, which includes the best marathoners in the world. The other event was an amateur marathon in Canada (London, Ontario) in 2009. The "L" in Table 3.2 means the event took place in London, Ontario, and the "O" means it took place during the Olympics.

| Time (seconds) | Event |
|---|---|
| 12,981 | L |
| 12,590 | L |
| 13,722 | O |
| 11,905 | O |

▲ **TABLE 3.2** First four lines of data set of 2008 marathon times for women.

Figure 3.28 shows the data separately for each of these events. We could now compute measures for center and spread separately for the Olympic and amateur events. However, you will sometimes find yourself in situations where a bimodal distribution occurs but it *does* make sense to compute a single measure of center. We can't give you advice for what to do in all situations. Our best advice is always to ask, "Does my summary make sense?"

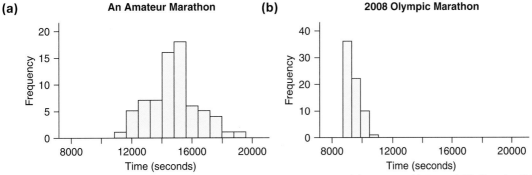

▲ **FIGURE 3.28** Women's times for a marathon, separated into two groups: **(a)** amateur athletes and **(b)** Olympic athletes.

## Comparing Two Groups with Different-Shaped Distributions

Note that the times for the amateur runners in Figure 3.28a are fairly symmetric, which suggests we should use the mean to report the typical finishing time. But the Olympic times are right-skewed, which suggests we should compute the median. Which do you think is the better measure for comparing these two groups, the mean or the median?

When comparing two distributions, you should always use the same measures of center and spread for both distributions. Otherwise, the comparison is not valid.

## EXAMPLE 16 Marathon Times, Revisited

In Figure 3.27 we lumped all of the marathon runners' finishing times into one group. But in fact, our data set had a variable that told us which time belonged to an Olympic runner and which to an amateur runner, so we could separate the data into groups. Figure 3.28 shows the same data, separated by groups.

**! Caution**

**Average**
A guy comes across a statistician who is standing with one foot in a pot of boiling water and one foot frozen in a block of ice. "Isn't that painful?" the man asks. "Maybe so," says the statistician, "but on average it feels just right." Beware of applying the mean to situations where it will not provide a "typical" measure!

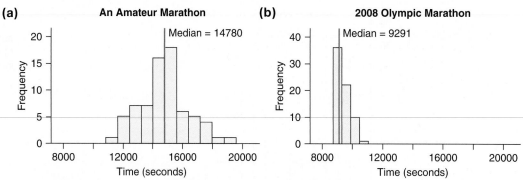

**(a)**

**An Amateur Marathon**

Median = 14780

**(b)**

**2008 Olympic Marathon**

Median = 9291

▲ **FIGURE 3.29** Women's times for a marathon, separated into two groups: **(a)** amateur athletes and **(b)** Olympic athletes. Medians are shown for each group.

QUESTION Typically, which group has the fastest finishing times?

SOLUTION The distribution of Olympic runners is right-skewed, so the median would be the best measure. Although the distribution of amateur runners is relatively symmetric, we report the median because we want to compare the typical running time of the amateurs to the typical running time of the Olympic runners. The median of the women Olympic runners is 9291 seconds (about 2.6 hours), and the median for the amateur women runners is 14,780 seconds (about 4.1 hours). Figure 3.29 shows the location of each group's median running time.

CONCLUSION The typical woman Olympic runner finishes the marathon considerably faster: a median time of 9291 seconds (about 2.6 hours) compared to 14,780 seconds (about 4.1 hours) for the amateur athlete.

TRY THIS! Exercise 3.61

KEY POINT When you are comparing groups, if any one group is strongly skewed or has outliers, it is usually best to compare the medians and interquartile ranges for all groups.

# Using Boxplots for Displaying Summaries

**Boxplots** are a useful graphical tool for visualizing a distribution, especially when comparing two or more groups of data. A boxplot shows us the distribution divided into fourths. The left edge of the box is at the first quartile (Q1) and the right edge is at the third quartile (Q3). Thus the middle 50% of the sorted observations lie inside the box. Therefore, the length of the box is proportional to the IQR.

A vertical line inside the box marks the location of the median. Horizontal lines, called whiskers, extend from the ends of the box to the smallest and largest values, or nearly so. (We'll explain soon.) Thus the entire length of the boxplot spans most, or all, of the range of the data.

Figure 3.30 compares a dotplot (with the quartiles marked with vertical lines) and a boxplot for the price of gas at stations in Daytona, Florida, as discussed in Examples 6 and 11.

Unlike many of the graphics used to visualize data, boxplots are relatively easy to draw by hand, assuming that you've already found the quartiles. Still, most of the time you will use software or a graphing calculator to draw the boxplot. Most software packages produce a variation of the boxplot that helps identify observations that are extremely large or small compared to the bulk of the data.

These extreme observations are called potential outliers. Potential outliers are different from the outliers we discussed in Chapter 2, because sometimes, points that look extreme in a boxplot are not that extreme when shown in a histogram or dotplot. Such points are called *potential* outliers because you should consult a histogram or dotplot of the distribution before deciding whether the observation is too extreme to fit the pattern of the distribution. (Remember, whether or not an observation is an outlier is a subjective decision.)

**Potential outliers** are identified by this rule: They are observations that are a distance of more than 1.5 interquartile ranges below the first quartile (the left edge of a horizontal box) or above the third quartile (the right edge).

To allow us to see these potential outliers, the whiskers are drawn from the edge of each box to the most extreme observation that is not a potential outlier. This implies that before we can draw the whiskers, we must identify any potential outliers.

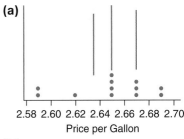

**(a)**

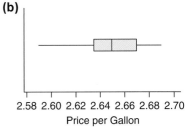

**(b)**

▲ **FIGURE 3.30 (a)** A dotplot with Q1, Q2, and Q3 indicated and **(b)** a boxplot for the price of regular, unleaded gas at stations in Daytona Beach, Florida.

 **KEY POINT**  Whiskers in a boxplot extend to the most extreme values that are not potential outliers. Potential outliers are points that are more than 1.5 IQRs from the edges of the box.

Figure 3.31 is a boxplot of temperatures in San Francisco (see Examples 8 and 10). From the boxplot, we can see that

$$IQR = 70 - 59 = 11$$

$$1.5 \times IQR = 1.5 \times 11 = 16.5$$

$$\text{Right limit} = 70 + 16.5 = 86.5$$

$$\text{Left limit} = 59 - 16.5 = 42.5$$

Any points below 42.5 or above 86.5 would be potential outliers.

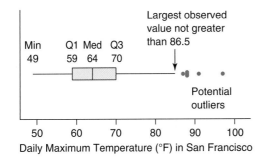

Min 49    Q1 59    Med 64    Q3 70    Largest observed value not greater than 86.5    Potential outliers

Daily Maximum Temperature (°F) in San Francisco

◄ **FIGURE 3.31** Boxplot of maximum daily temperatures in San Francisco.

The whiskers go to the most extreme values that are not potential outliers. On the left side of the box, observations smaller than 42.5 would be potential outliers. However, there are no observations that small. The smallest is 49, so the whisker extends to 49.

On the right, several values in the data set are larger than 86.5. The whisker extends to the largest temperature that is less than (or equal to) 86.5, and the larger values are shown in Figure 3.31 with dots. These represent days that were unusually warm in San Francisco. Figure 3.32 shows a boxplot made with a TI-83/84.

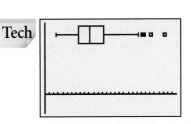

**Tech**

▲ **FIGURE 3.32** TI-83/84 output for a boxplot of San Francisco temperatures.

## Investigating Potential Outliers

What do you do with potential outliers? The first step is always to investigate. A potential outlier might not be an outlier at all. Or a potential outlier might tell an interesting story, or it might be the result of an error in entering data.

Figure 3.33a is a boxplot of the NAEP International math scores for 42 countries (International math scores 2007). One country (South Africa, as it turns out) is flagged as a potential outlier. However, if we examine a histogram, shown in Figure 3.33b, we see that this outlier is really not that extreme. Most people would not consider South Africa to be an outlier in this distribution, because it is not separated from the bulk of the distribution in the histogram.

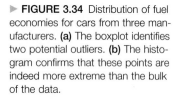

► **FIGURE 3.33 (a)** Distribution of International Math Scores for eighth grade achievement. The boxplot indicates a potential outlier. **(b)** The histogram of the distribution of math scores shows that although South Africa's score of 264 might be the lowest, it is not that much lower than the bulk of the data.

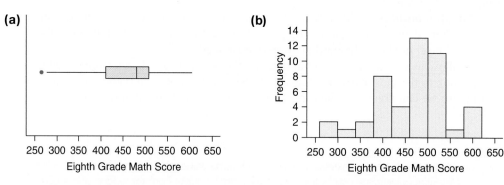

Figure 3.34 shows a boxplot and histogram for the fuel economy (in city driving) of the 2010 model sedans from Ford, Toyota, and GM, in miles per gallon, as listed on their websites. Two potential outliers appear, which are far enough from the bulk of the distribution as shown in the histogram that many people would consider them real outliers. These outliers turn out to be hybrid cars: the Ford Fusion and the Toyota Prius. These hybrids run on both electricity and gasoline, so they have much better fuel economy.

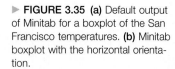

► **FIGURE 3.34** Distribution of fuel economies for cars from three manufacturers. **(a)** The boxplot identifies two potential outliers. **(b)** The histogram confirms that these points are indeed more extreme than the bulk of the data.

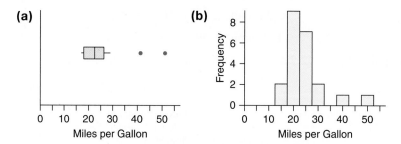

## Horizontal or Vertical?

Boxplots do not have to be horizontal. Many software packages (such as Minitab) provide you with the option of making vertical boxplots. (See Figure 3.35a). Which direction you choose is not important. Try both to see which is more readable.

► **FIGURE 3.35 (a)** Default output of Minitab for a boxplot of the San Francisco temperatures. **(b)** Minitab boxplot with the horizontal orientation.

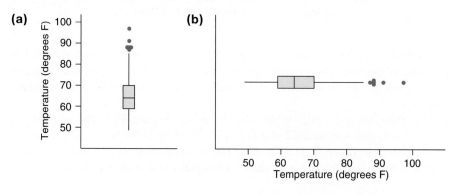

## Using Boxplots to Compare Distributions

Boxplots are often a very effective way of comparing two or more distributions. How do temperatures in Provo compare with those in San Francisco? Figure 3.36 shows boxplots of daily maximum temperatures for San Francisco and Provo. At a glance, we can see how these two distributions differ and how they are similar. Both cities have similar typical temperatures (the median temperatures are about the same). Both distributions are fairly symmetric (because the median is in the center of the box, and the boxplots are themselves fairly symmetric). However, the amount of variation in daily temperatures is much greater in Provo than in San Francisco. We can see this easily because the box is wider for Provo's temperatures.

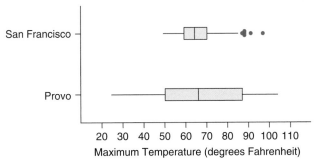

◄ **FIGURE 3.36** Boxplots of daily maximum temperatures in Provo and San Francisco emphasize the difference in variability of temperature in the two cities.

   Also note that although both cities do have days that reach about 100 degrees, these days are unusual in San Francisco—they're flagged as potential outliers—but merely fall in the upper 25% for Provo.

## Things to Watch For with Boxplots

Boxplots are best used only for unimodal distributions because they hide bimodality (or any multimodality). For example, Figure 3.37a repeats the histogram of marathon running times for two groups of women runners: amateurs and Olympians. The distribution is clearly bimodal. However, the boxplot in part (b) doesn't show us the bimodality. Boxplots can give the misleading impression that a bimodal distribution is really unimodal.

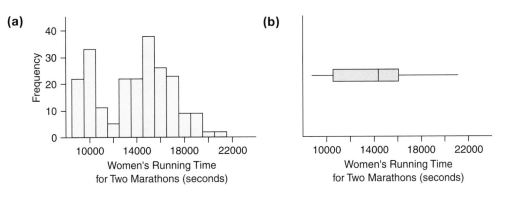

◄ **FIGURE 3.37** **(a)** Histogram of finishing times (in seconds) for two groups of women marathon runners: Olympic athletes and amateurs. The graph is bimodal because the elite athletes tend to run faster, and therefore there is a mode around 10,000 seconds (about 2.8 hours) and another mode around 15,000 seconds (about 4.2 hours). **(b)** The boxplot hides this interesting feature.

   Boxplots should *not* be used for very small data sets. It takes at least five numbers to make a boxplot, so if your data set has fewer than five observations, you can't make a boxplot.

## Finding the Five-Number Summary

Boxplots are not really pictures of a distribution in the way that histograms are. Instead, boxplots help us visualize the location of various summary statistics. The boxplot is a visualization of a numerical summary called the **five-number summary**. These five numbers are

the minimum, Q1, the median, Q3, and the maximum

For example, for daily maximum temperatures in San Francisco, the five-number summary is

49, 59, 64, 70, 97

as you can see in Figure 3.31.

Note that a boxplot is not just a picture of the five-number summary. Boxplots always show the maximum and minimum values, but sometimes they also show us potential outliers.

## SNAPSHOT  THE BOXPLOT

| | |
|---|---|
| **WHAT IS IT?** ▶ | A graphical summary. |
| **WHAT DOES IT DO?** ▶ | Provides a visual display of numerical summaries of a distribution of numerical data. |
| **HOW DOES IT DO IT?** ▶ | The box stretches from the first quartile to the third quartile, and a vertical line indicates the median. Whiskers extend to the largest and smallest values that are not potential outliers, and potential outliers are indicated with special marks. |
| **HOW IS IT USED?** ▶ | Boxplots are useful for comparing distributions of different groups of data. |

## CASE STUDY REVISITED

# Living in a Risky World

How do the men and women of this study compare when it comes to assigning risk to using a household appliance and getting an annual X-ray at the doctor's? In Chapter 2 we compared groups graphically, and this is still the first step. But in this chapter, we learned about methods for comparing groups numerically, and this will allow us to be more precise in our comparison.

Our first step is to examine the pictures of the distributions to decide which measures would be most appropriate. (We repeat Figure 3.1.)

▶ **FIGURE 3.1a (repeated)**
Histograms showing the distributions of perceived risk of using appliances. The women's data are in the left graph, and the men's data are in the right graph.

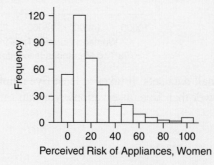
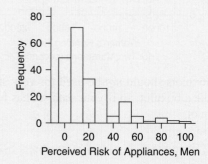

## Risk of Appliances

The histograms for the perceived risk from using appliances (Figure 3.1a) do not show large differences between men and women. We can see that both distributions are right-skewed, and both appear to have roughly the same typical value, although the women's typical value might be slightly higher than the men's. Because the distribution is right-skewed, we compute the median and IQR to compare the two groups. (Table 3.3 summarizes these comparisons.) The men assigned a median risk of 10 to using household appliances, and the women assigned a median risk of 15. We see that, first impressions aside, these women tend to feel that using appliances is a riskier activity than do these men. Also, more differences in opinion occurred among these women than among these men. The IQR was 25 for women and 20 for men. Thus the middle 50% of the women varied by as much as 25 points in how risky they saw this activity; there was less variability for the men.

## Risk of X-rays

Both distributions for the perceived risk level of X-rays (Figure 3.1b) were fairly symmetric, so it makes sense to compare the two groups using the mean and standard deviation. The mean risk level for men was 46.8 and for women 47.8. Typically, men and women feel about the same as far as the risk associated with X-rays. The standard deviations are about the same, too: men have a standard deviation of 20.0 and women 20.8. From the Empirical Rule, we know that a majority (about two-thirds) of men in this sample rated the risk level between 26.8 and 66.8. The majority of women rated it between 27.0 and 68.6.

|  | Risk of Appliances | |
|  | Median | IQR |
|---|---|---|
| Men | 10 | 20 |
| Women | 15 | 25 |

|  | Risk of X-rays | |
|  | Mean | SD |
|---|---|---|
| Men | 46.8 | 20.0 |
| Women | 47.8 | 20.8 |

▲ **TABLE 3.3** Comparison of Perceived Risks for Men and Women

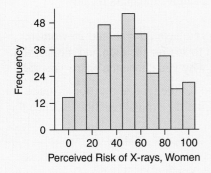

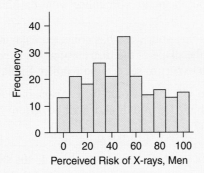

◄ **FIGURE 3.1b (repeated)** Histograms showing the distributions of perceived risk of X-rays. The women's data are shown on the left and the men's data are shown on the right.

The comparisons are summarized in Table 3.3.

# EXPLORING STATISTICS
## CLASS ACTIVITY

## Does Reaction Distance Depend on Gender?

| GOALS | MATERIALS |
|---|---|
| Apply the concepts introduced in this chapter to data you have collected in order to compare two groups. | A meter stick or stiff ruler for each group. |

**ACTIVITY**

Work in groups of two or three. One person holds the meter stick vertically, with one hand near the top of the stick, so that the 0-centimeter mark is at the bottom. The other person then positions his or her thumb and index finger about 5 cm apart (2 inches apart) on opposite sides of the meter stick at the bottom. Now the first person drops the meter stick without warning, and the other person catches it. Record the location of the middle of the thumb of the catcher. This is the distance the stick traveled and is called the reaction distance, which is related to reaction time. A student who records a small distance has a fast reaction time, and a student with a larger distance has a slower reaction time. Now switch tasks. Each person should try catching the meter stick twice, and the better (shorter) distance should be reported for each person. Then record the gender of each catcher. Your instructor will collect your data and combine the class results.

**BEFORE THE ACTIVITY**

1. Imagine that your class has collected data and you have 25 men and 25 women. Sketch the shape of the distribution you expect to see for the men and the distribution you expect to see for the women. Explain why you chose the shape you did.

2. What do you think would be a reasonable value for the typical reaction distance for the women? Do you think it will be different from the typical reaction distance for the men?

**AFTER THE ACTIVITY**

1. Now that you have actual data, how do the shapes of the distributions for men and women compare to the sketches you made before you collected data?

2. What measures of center and spread are appropriate for comparing men and women's reaction distances? Why?

3. How do the actual typical reaction distances compare to the values you predicted?

4. Using the data collected from the class, write a short paragraph (a couple of sentences) comparing the reaction distances of men and women. You should also talk about what group you could extend your findings to, and why. For example, do your findings apply to all men and women? Or do they apply only to college students?

## CHAPTER REVIEW

## Key Terms

mean, *78*
average, *80*
standard deviation, *84*
deviation, *86*
variance, *87*

Empirical Rule, *88*
standard units, *91*
z-score, *91*
median, *93*

interquartile range (IQR), *96*
first quartile (Q1), *98*
second quartile (Q2), *98*
third quartile (Q3), *98*

resistant to outliers, *101*
boxplot, *104*
potential outlier, *105*
five-number summary, *107*

## Learning Objectives

After reading this chapter and doing the assigned homework problems, you should

- Understand how measures of center and spread are used to describe characteristics of real-life samples of data.

- Understand when it is appropriate to use the mean and standard deviation and when it is better to use the median and interquartile range.

- Understand the mean as the balancing point of the distribution of a sample of data and the median as the point that has roughly 50% of the distribution below it.

- Be able to write comparisons between samples of data in context.

## Summary

The first step in any statistical investigation is to make a picture of the distribution of a numerical variable using a dotplot or histogram. Before computing any summary statistics, you must examine a graph of the distribution to determine the shape and whether or not there are outliers. As noted in Chapter 2, you should report the shape, center, and variability of every distribution.

If the shape of the distribution is symmetric and there are no outliers, you can describe the center and spread by using either the median with the interquartile range or the mean with the standard deviation, although it is customary to use the mean and standard deviation.

If the shape is skewed or there are outliers, you should use the median with the interquartile range.

If the distribution is multimodal, try to determine whether the data consist of separate groups; if so, you should analyze these groups separately. Otherwise, see whether you can justify using a single measure of center and spread.

If you are comparing two distributions and one of the distributions is skewed or has outliers, then it is usually best to compare the median and interquartile ranges for both groups.

The choices are summarized in Table 3.4.

Converting observations to z-scores changes the units to standard units, and this allows us to compare individual observations from different groups.

### Formulas

**Formula 3.1:**  Mean $= \bar{x} = \dfrac{\sum x}{n}$

The mean is the measure of center best used if the distribution is symmetric.

**Formula 3.2:**  Standard deviation $= s = \sqrt{\dfrac{\sum (x - \bar{x})^2}{n - 1}}$

The standard deviation is the measure of variability best used if the distribution is symmetric.

| Shape | Summaries of Center and Variability |
|---|---|
| If distribution is bimodal or multimodal | Try to separate groups, but if you cannot, decide whether you can justify using a single measure of center. If a single measure will not work, then report the approximate locations of the modes. |
| If any group's distribution is strongly skewed or has outliers | Use medians and interquartile ranges for all groups. |
| If all groups' distributions are roughly symmetric | Use means and standard deviations for all groups. |

▲ **TABLE 3.4** Preferred measures to report when summarizing data or comparing two or more groups.

**Formula 3.3:**  Variance $= s^2 = \dfrac{\sum (x - \bar{x})^2}{n - 1}$

The variance is another measure of variability used if the distribution is symmetric.

**Formula 3.4:**  $z = \dfrac{x - \bar{x}}{s}$

This formula converts an observation to standard units.

**Formula 3.5:**  Interquartile range $= Q3 - Q1$

The interquartile range is the measure of variability best used if the distribution is skewed.

**Formula 3.6:**  Range $=$ maximum $-$ minimum

The range is a crude measure of variability.

## Sources

Carlstrom, L., J. Woodward, P. Arthur, and G. Christina. 2000. Evaluating the simplified conjoint expected risk model: Comparing the use of objective and subjective information. *Risk Analysis* 20(3).

Baseball salaries. 2010. http://www.bizofbaseball.com/index.php?option= com_wrapper&view=wrapper&Itemid=17 (accessed June 13, 2010).

EPA. Particulate matter. http://www.epa.gov/airtrends/factbook.html

Gasoline mileage. 2010. http://www.gm.com and www.fordvehicles.com (accessed June 2010).

International math scores. 2007. http://www.edweek.org/ew/ articles/2007/05/02/35air.h26.html

Provo temperatures. 2007. http://www.pgjr.alpine.k12.ut.us/science/james/ provo2006.html (accessed August 2007).

San Francisco temperatures. 2007. http://169.237.140.1/calludt.cgi/ WXDATAREPORT (accessed August 2007).

# SECTION EXERCISES

## SECTION 3.1

**3.1 Earnings** A sociologist says, "Typically, men in the United States still earn more than women." What does this statement mean? (Pick the best choice.)

  a. All men make more than all women in the United States.

  b. All U.S. women's salaries are less varied than all men's salaries.

  c. The center of the distribution of salaries for U.S. men is greater than the center for women.

  d. The highest paid people in the United States are men.

**3.2 Houses** A real estate agent claims that all things being equal, houses with swimming pools tend to sell for less than those without swimming pools. What does this statement mean? (Pick the best choice.)

  a. There are fewer homes with swimming pools than without.

  b. The typical price for homes with pools is smaller than the typical price for homes without pools.

  c. There's more variability in the price of homes with pools than in the price of those without.

  d. The most expensive houses sold do not have pools.

TRY **3.3 Age of CEOs (Example 1)** The histogram shows the ages of 25 CEOs listed at Forbes.com. Based on the distribution, what is the approximate mean age of the CEOs in this data set? Write a sentence in context (using words in the question) interpreting the estimated mean. The typical CEO is about _____ years old.

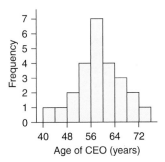

**3.4 Siblings** The histogram shows the number of siblings (brothers and sisters) for 2000 adults surveyed for the 2008 General Social Survey. Based on the distribution, what is the approximate mean number of siblings for these 2000 people? Write a sentence in context interpreting the estimated mean. You may want to refer to the model in Exercise 3.3. (Source: www.norc.uchicago.edu)

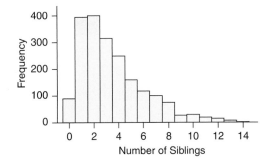

TRY **3.5 Paid Days (Example 2)** This list represents the numbers of paid vacation days required by law for different countries. (Source: *2009 World Almanac and Book of Facts*)

| United States | 0 |
|---|---|
| Australia | 20 |
| Italy | 20 |
| France | 30 |
| Germany | 24 |
| Canada | 10 |

  a. Find the mean, rounding to the nearest tenth of a day. Interpret the mean in this context: Report the mean in a sentence that includes words such as "paid vacation days."

  b. Find the standard deviation, rounding to the nearest tenth of a day. Interpret the standard deviation in context.

  c. Which number of days is farthest from the mean and therefore contributes most to the standard deviation?

**3.6 Children of First Ladies** This list represents the number of children for the first six First Ladies of the United States. (Source: *2009 World Almanac and Book of Facts*)

| Martha Washington | 0 |
|---|---|
| Abigail Adams | 5 |
| Martha Jefferson | 6 |
| Dolley Madison | 0 |
| Elizabeth Monroe | 2 |
| Louisa Adams | 4 |

  a. Find the mean number of children, rounding to the nearest tenth. Interpret the mean in this context.

  b. According to eh.net/encyclopedia, women living around 1800 tended to have between 7 and 8 children. How does the mean of these first ladies compare to that?

  c. Which of the first ladies listed here had the number of children that is farthest from the mean and therefore contributes most to the standard deviation?

  d. Find the standard deviation, rounding to the nearest tenth.

**3.7 Ages of Presidents at Inauguration** At their inauguration, the ages of the first six presidents of the United States were 57, 61, 57, 57, 58, and 57. (Source: *2009 World Almanac and Book of Facts*)

  a. Find the mean age at inauguration, rounding to the nearest tenth. The mean of the most recent six presidents (Carter to Obama) was 55.3. Did the first six presidents tend to be a bit *older* or a bit *younger* than the most recent six presidents were?

  b. Find the standard deviation of the ages at inauguration, rounding to the nearest tenth. The standard deviation of the most recent six presidents was 9.3 years. Did the first six tend to have *more* or *less* variation than the most recent presidents did?

**3.8 Ages of Chief Justices at Installation** At their installations as chief justice of the United States, the first six Chief Justices were 44, 56, 51, 45, 59, and 56 years old, respectively. (Source: *2009 World Almanac and Book of Facts*)

a. Find and interpret (report in context) the mean age at installation, rounding to the nearest tenth. The mean age at installation of the six most recent chief justices was 62. Did the first six tend to be *older* or *younger* at installation than the most recent six chief justices were?

b. Find the standard deviation of the ages, rounding to the nearest tenth. The standard deviation of the most recent six chief justices was 10.1. Did the first six tend to have *more* or *less* variation than the most recent six chief justices did?

**TRY 3.9 Areas of States (Example 3)** The Minitab output gives some numerical summaries for the areas of the 50 states and Washington D.C. (in square miles). The regions are categorized as those east (e) of the Mississippi River and those west (w) of the river. (Source: *2009 World Almanac and Book of Facts*)

**Minitab Statistics: Area square miles**

| Variable | e or w | N | Mean | StDev | Minimum |
|----------|--------|-----|--------|--------|---------|
| Area | e | 27 | 35520 | 24704 | 68 |
| | w | 24 | 117845 | 124646 | 10932 |

| Variable | e or w | Q1 | Median | Q3 | Maximum |
|----------|--------|-------|--------|--------|---------|
| Area | e | 9615 | 40411 | 53821 | 96705 |
| | w | 70103 | 84239 | 113146 | 656424 |

a. Compare the mean areas by completing this sentence: The mean of the eastern regions is _____ square miles and the mean of the western regions is _____ square miles, showing that the _____ regions tend to have smaller areas.

b. Compare the standard deviations (StDev) of the areas by completing this sentence: The standard deviations of the eastern regions is _____ square miles and the standard deviation of the western regions is _____ square miles, showing that areas of the _____ regions tend to have more variation.

c. Does the TI-83/84 output given represent the eastern or western regions?

```
1-Var Stats
x̄=35519.62963
Σx=959030
Σx²=4.99324ε10
Sx=24704.41344
σx=24242.60798
↓n=27
```

**3.10 State Populations** The StatCrunch output shows some statistics for the populations (in thousands) of the 50 states in 2010. The states are categorized by those east (e) of the Mississippi River and those west (w) of the river. (Source: www.TheGreenPapers.com)

Summary statistics for 2010 / 1000:

Group by: Mississippi

| Mississippi | n | Mean | Variance | Std. Dev. |
|-------------|-----|-----------|------------|-----------|
| e | 26 | 6871.2188 | 2.637322E7 | 5135.4863 |
| w | 24 | 5323.1333 | 6.86806E7 | 8287.376 |

a. Compare the means in a sentence or two. The means reported are in thousands of people; round to the nearest thousand. See exercise 3.9 for one possible model of a comparison.

b. Compare the standard deviations in a sentence or two. The standard deviations reported are in thousands of people; round to the nearest thousand.

**3.11 Surfing** College students and surfers Rex Robinson and Sandy Hudson collected data on the self-reported number of days surfed in a month for 30 longboard surfers and 30 shortboard surfers.

Longboard: 4, 9, 8, 4, 8, 8, 7, 9, 6, 7, 10, 12, 12, 10, 14, 12, 15, 13, 10, 11, 19, 19, 14, 11, 16, 19, 20, 22, 20, 22

Shortboard: 6, 4, 4, 6, 8, 8, 7, 9, 4, 7, 8, 5, 9, 8, 4, 15, 12, 10, 11, 12, 12, 11, 14, 10, 11, 13, 15, 10, 20, 20

a. Compare the means in a sentence or two.

b. Compare the standard deviations in a sentence or two.

**3.12 Eating Out** College student Jacqueline Loya asked a group of 50 employed students how many times they went out to eat in the last week. Half of the students had full-time jobs, and half had part-time jobs.

Full-time jobs: 5, 3, 4, 4, 4, 2, 1, 5, 6, 5, 6, 3, 3, 2, 4, 5, 2, 3, 7, 5, 5, 1, 4, 6, 7

Part-time jobs: 1, 1, 5, 1, 4, 2, 2, 3, 3, 2, 3, 2, 4, 2, 1,2, 3, 2, 1, 3, 3, 2, 4, 2, 1

a. Compare the means in a sentence or two.

b. Compare the standard deviations.

**TRY 3.13 Real Estate Prices (Example 4)** Look at the two histograms, created from 2009 real estate data taken from the *Ventura County Star*, and decide whether you think the standard deviation of home prices in Agoura, California (A), was larger or smaller than the standard deviation of home prices in Westlake, California (B). Explain.

(A) **Agoura House Prices**

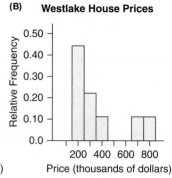

(B) **Westlake House Prices**

**3.14 Dice** The histograms contain data with a range of 1 to 6. Which group would have the larger standard deviation, group A or group B? Why?

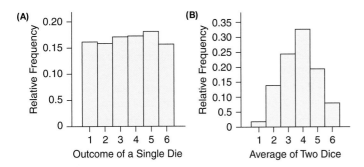

**TRY 3.15 Birth Weights (Example 5)** The mean birth weight for U.S. children born at full term (after 40 weeks) is 3462 grams (about 7.6 pounds). Suppose the standard deviation is 500 grams and the shape of the distribution is symmetric and unimodal. (Source: www.babycenter.com)

a. According to the Empirical Rule, what is the range of birth weights (in grams) of U.S.-born children from one standard deviation below the mean to one standard deviation above the mean?

b. Is a birth weight of 2800 grams (about 6.2 pounds) more than one standard deviation below the mean?

**3.16 Birth Length** The mean birth length for U.S. children born at full term (after 40 weeks) is 52.2 cm (about 20.6 inches). Suppose the standard deviation is 2.5 cm and the distributions are unimodal and symmetric. (Source: www.babycenter.com)

a. According to the Empirical Rule, what is the range of birth lengths (in centimeters) of U.S.-born children from one standard deviation below the mean to one standard deviation above the mean?

b. Is a birth length of 54 cm more than one standard deviation above the mean?

**TRY 3.17 Children's Ages (Example 6)** Mrs. Johnson's children are 2, 2, 3, and 5 years of age.

a. Calculate the standard deviation of their current ages.

b. Without doing any calculation, indicate whether the standard deviation of the ages in 20 years will be larger, smaller, or the same as the standard deviation of their current ages. Check your answer by calculating the standard deviation of the ages in 20 years. Explain how adding 20 to each number affects the standard deviation.

c. Find the mean of the children at their current ages.

d. Without doing any calculation, indicate whether the mean age in 20 years will be larger, smaller, or about the same as the mean of the current ages. Confirm your answer and describe how adding 20 to each number affects the mean.

**3.18 Pay Rate in Different Currencies** The pay rates for 3 people were the following numbers of dollars per hour: 8, 10.5, and 7. In Mexican pesos (if the exchange rate is 10 pesos per dollar), these pay rates are 80, 105, and 70 pesos per hour.

a. Compare the mean in dollars with the mean in pesos; don't forget the units. Explain what multiplying each number in a data set by 10 does to the mean.

b. Compare the standard deviations in pesos and in dollars. Explain what multiplying each number in a data set by 10 does to the standard deviation.

**3.19 Olympics** In the most recent summer Olympics, do you think the standard deviation of the running times for all men who ran the 100-meter race would be larger or smaller than the standard deviation of the running times for the men's marathon? Explain.

**3.20 Weights** Suppose you have a data set with the weights of all members of a high school soccer team and all members of a high school academic decathlon team (a team of students selected because they often answer quiz questions correctly). Which team do you think would have a larger standard deviation of weights? Explain.

## SECTION 3.2

**TRY g 3.21 Violent Crime: West (Example 7)** In 2007, the mean rate of violent crime (per 100,000 people) for the 24 states west of the Mississippi River was 406. The standard deviation was 177. Assume that the distribution of violent crime rates is approximately unimodal and symmetric. *See page 125 for guidance.* (Source: www.worldatlas.com)

a. Between what two values would you expect to find about 95% of the rates?

b. Between what two values would you expect to find about 68% of the violent crime rates?

c. If a western state had a violent crime rate of 584 crimes per 100,000 people, would you consider this unusual? Explain.

d. If a western state had a violent crime rate of 30 crimes per 100,000 people, would you consider this unusual? Explain.

**3.22 Violent Crime: East** In 2007, the mean rate of violent crime (per 100,000 people) for the 26 states east of the Mississippi River was 409; the standard deviation was 193. Assume the distribution of violent crime rates is approximately unimodal and symmetric. (Source: www.worldatlas.com)

a. Between what two values would you expect to find about 95% of the rates?

b. Between what two values would you expect to find about 68% of the rates?

c. If an eastern state had a violent crime rate of 896 crimes per 100,000 people, would you consider this unusual? Explain.

d. If an eastern state had a violent crime rate of 503 crimes per 100,000 people, would you consider this unusual? Explain.

**TRY 3.23 Property Crime (Example 8)** In 2007, the mean property crime rate (per 100,000 people) for the 24 states west of the Mississippi River was 3331; the standard deviation was 729. Assume the distribution of crime rates is unimodal and symmetric. (Source: www.worldatlas.com)

a. What percentage of western states would you expect to have property crime rates between 2602 and 4060?

b. What percentage of western states would you expect to have property crime rates between 1873 and 4789?

c. If someone guessed that the property crime rate in one state west of the Mississippi River was 9000, would you agree that that number was consistent with this data set?

**3.24 Property Crime** In 2007, the mean property crime rate (per 100,000 people) for the 26 states east of the Mississippi River was 3009; the standard deviation was 732. Assume the distribution of property crime rates is approximately unimodal and symmetric. (Source: www.worldatlas.com)

a. Approximately what percentage of eastern states would you expect to have property crime rates between 1545 and 4473?

b. Approximately what percentage of eastern states would you expect to have property crime rates between 2277 and 3741?

c. If someone guessed that the property crime rate in one state east of the Mississippi River was 500, would you agree that number was consistent with this data set?

**3.25 Heights and z-Scores** The dotplot shows heights of college women; the mean is 64 inches (5 feet 4 inches) and the standard deviation is 3 inches.

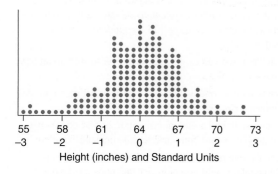

Height (inches) and Standard Units

a. What is the z-score for a height of 58 inches (4 feet 10 inches)?

b. What is the height of a woman with a z-score of 1?

**3.26 Heights** Refer to the dotplot in the previous question.

a. What is the height of a woman with a z-score of −1?

b. What is the z-score for a woman who is 70 inches tall (5 feet 10 inches)?

TRY **3.27 SAT and ACT Scores (Example 9)** Quantitative SAT scores have a mean of 500 and a standard deviation of 100, while ACT scores have a mean of 21 and a standard deviation of 5. Assuming both types of scores have distributions that are unimodal and symmetric, which is more unusual: a quantitative SAT score of 750 or an ACT score of 28? Show your work.

**3.28 Children's Heights** Mrs. Diaz has two children: a three-year-old boy 43 inches tall and a ten-year-old girl 57 inches tall. Three-year-old boys have a mean height of 38 inches and a standard deviation of 2 inches, and ten-year-old girls have a mean height of 54.5 inches and a standard deviation of 2.5 inches. Assume the distributions of boys' and girls' heights are unimodal and symmetric. Which of Mrs. Diaz's children is more unusually tall for his or her age and gender? Explain, showing any calculations you perform. (Source: www.kidsgrowth.com)

TRY **3.29 Low-Birth-Weight Babies (Example 10)** Babies born weighing 2500 grams (about 5.5 pounds) or less are called low-birth-weight babies, and this condition sometimes indicates health problems for the infant. The mean birth weight for U.S.-born children is about 3462 grams (about 7.6 pounds). The mean birth weight for babies born one month early is 2622 grams. Suppose both standard deviations are 500 grams. Also assume that the distribution of birth weights is roughly unimodal and symmetric. (Source: www.babycenter.com)

a. Find the standardized score (z-score), relative to all U.S. births, for a baby with a birth weight of 2500 grams.

b. Find the standardized score for a birth weight of 2500 grams for a child born one month early, using 2622 as the mean.

c. For which group is a birth weight of 2500 grams more common? Explain what that implies. Unusual z-scores are far from 0.

**3.30 Birth Lengths** Babies born after 40 weeks gestation have a mean length of 52.2 centimeters (about 20.6 inches). Babies born one month early have a mean length of 47.4 cm. Assume both standard deviations are 2.5 cm and the distributions are unimodal and symmetric. (Source: www.babycenter.com)

a. Find the standardized score (z-score), relative to all U.S. births, for a baby with a birth length of 45 cm.

b. Find the standardized score of a birth length of 45 cm for babies born one month early, using 47.4 as the mean.

c. For which group is a birth length of 45 cm more common? Explain what that means.

**3.31 Exam Scores** An exam has a mean of 70 and a standard deviation of 10. What exam score corresponds to a z-score of 1.5?

**3.32 Boys' Heights** Three-year-old boys in the United States have a mean height of 38 inches and a standard deviation of 2 inches. How tall is a three-year-old boy with a z-score of −1.0? (Source: www.kidsgrowth.com)

## SECTION 3.3

**3.33** Name two measures of the center of a distribution, and state the conditions under which each is preferred for describing the typical value of a single data set.

**3.34** Name two measures of the variation of a distribution, and state the conditions under which each measure is preferred for measuring the variability of a single data set.

TRY **3.35 Pixar Animated Movies (Example 11)** The ten top-grossing Pixar Animated movies for the U.S. box office up to June 2010 are shown in the table on the next page, in millions of dollars. (Source: www.pixar.com)

a. Arrange the gross income from smallest (on the left) to largest (showing the arrangement), and find the median by averaging the two middle numbers. Interpret the median in context.

b. Using the sorted data, find Q1 and Q3. Then find the interquartile range and interpret it in context.

| Movie | $ Millions |
|---|---|
| *Toy Story* | 192 |
| *A Bug's Life* | 163 |
| *Toy Story 2* | 246 |
| *Monsters, Inc.* | 256 |
| *Finding Nemo* | 340 |
| *The Incredibles* | 261 |
| *Cars* | 244 |
| *Ratatouille* | 206 |
| *WALL-E* | 224 |
| *Up* | 293 |

**3.36 DreamWorks Animated Movies** The ten top-grossing DreamWorks animated movies for the U.S. box office up to June 2010 are shown in the table, in millions of dollars. (Source: www.the-top-tens.com)

| Movie | $ Millions |
|---|---|
| *Shrek 2* | 441 |
| *Shrek the Third* | 323 |
| *Shrek* | 268 |
| *Kung Fu Panda* | 215 |
| *Monsters vs. Aliens* | 198 |
| *Madagascar* | 194 |
| *Madagascar: Escape 2 Africa* | 180 |
| *Shark Tale* | 161 |
| *Over the Hedge* | 155 |
| *Bee Movie* | 127 |

a. Find and interpret the median box office dollars for the ten top grossing DreamWorks animated movies.

b. Find and interpret the interquartile range for these movies.

**TRY 3.37 Gas Taxes (Example 12)** The gasoline taxes (in cents per gallon) are given for each state in the plains, as reported in the June 2010 issue of the *AARP Bulletin*. These are state taxes and do not include federal tax.

| State | Tax |
|---|---|
| Iowa | 22.0 |
| Kansas | 25.0 |
| Minnesota | 27.2 |
| Missouri | 17.3 |
| Nebraska | 27.7 |
| North Dakota | 23.0 |
| South Dakota | 24.0 |

a. Find the median. Show the data sorted from smallest (on the left) to largest (on the right) and indicate the location of the median.

b. Indicate the locations of the first quartile (Q1) and the third quartile (Q3). Then find the interquartile range.

**3.38 Gas Taxes** The gasoline taxes (in cents per gallon) are given for each state in the far west, as reported in the June 2010 issue of the *AARP Bulletin*. These are state taxes and do not include federal tax. (Of all 50 states, California has the highest gas tax, and Alaska has the lowest.)

| State | Tax |
|---|---|
| Alaska | 8.0 |
| California | 48.6 |
| Hawaii | 45.1 |
| Nevada | 33.1 |
| Oregon | 25.0 |
| Washington | 37.5 |

a. Find and interpret the median gas tax for these states. Between which two states is the median rate?

b. Find, Q1, Q3, and the interquartile range (IQR) for these states. Interpret the IQR.

**TRY 3.39 Surfing, Again (Example 13)** College students and surfers Rex Robinson and Sandy Hudson collected data on the self-reported numbers of days surfed in a month for 30 longboard surfers and 30 shortboard surfers.

Longboard: 4, 9, 8, 4, 8, 8, 7, 9, 6, 7, 10, 12, 12, 10, 14, 12, 15, 13, 10, 11, 19, 19, 14, 11, 16, 19, 20, 22, 20, 22

Shortboard: 6, 4, 4, 6, 8, 8, 7, 9, 4, 7, 8, 5, 9, 8, 4, 15, 12, 10, 11, 12, 12, 11, 14, 10, 11, 13, 15, 10, 20, 20

a. Compare the typical number of days surfing for these two groups by placing the correct numbers in the blanks in the following sentence: The median for the longboards was _____ days, and the median for the shortboards was _____ days, showing that those with _____ boards typically surfed more days in this month.

b. Compare the interquartile ranges by placing the correct numbers in the blanks in the following sentence: The interquartile range for the longboards was _____ days, and the interquartile range for the shortboards was _____ days, showing more variation in the days surfed this month for the _____ boards.

**3.40 Eating Out, Again** College student Jacqueline Loya asked 50 employed students how many times they went out to eat last week. Half of the students had full-time jobs and half had part-time jobs.

Full-time: 5, 3, 4, 4, 4, 2, 1, 5, 6, 5, 6, 3, 3, 2, 4, 5, 2, 3, 7, 5, 5, 1, 4, 6, 7

Part-time: 1, 1, 5, 1, 4, 2, 2, 3, 3, 2, 3, 2, 4, 2, 1, 2, 3, 2, 1, 3, 3, 2, 4, 2, 1

a. Using the median values, write a sentence comparing the typical numbers of times the two groups ate out. See exercise 3.39 for one possible model.

b. Using the interquartile ranges, write a sentence comparing the variability of these two groups.

# SECTION 3.4

**3.41 Outliers**

a. In your own words, describe to someone who has knows only a little statistics how to recognize when an observation is an outlier. What action(s) should be taken with an outlier?

b. Which measure of the center (mean or median) is more resistant to outliers, and what does "resistant to outliers" mean?

**3.42 Center and Variation** When you are comparing two sets of data, and one set is strongly skewed and the other is symmetric, which measures of the center and variation should you choose for the comparison?

**3.43 An Error** A dieter recorded the number of calories he consumed at lunch for one week. As you can see, a mistake was made on one entry. The calories are listed in increasing order:

331, 374, 387, 392, 405, 4200

When the error is corrected by removing the extra 0, will the median calories change? Will the mean? Explain without doing any calculations.

**3.44 Baseball Strike** In 1994, major league baseball players went on strike. At the time, the average salary was $1,049,589, and the median salary was $337,500. If you were representing the owners, which summary would you use to convince the public that a strike was not needed? If you were a player, which would you use? Why was there such a large discrepancy between the mean and median salaries? Explain. (Source: www.usatoday.com)

**3.45 Students' Ages** Here are the ages of some students in a statistics class: 17, 19, 35, 18, 18, 20, 27, 25, 41, 21, 19, 19, 45, and 19. The teacher's age is 66 and should be included as one of the ages when you do the calculations below. The figure shows a histogram of the data.

a. Describe the distribution of ages by giving the shape, the numerical value for an appropriate measure of the center, and the numerical value for an appropriate measure of spread, as well as mentioning any outliers.

b. Make a rough sketch (or copy) of the histogram, and mark the approximate locations of the mean and of the median. Why are they not at the same location?

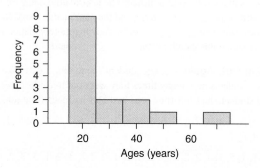

**3.46 House Prices** The figure, which is from data taken from the *Ventura County Star*, shows a histogram of house prices in Thousand Oaks, California, in 2009. The location of the mean and median are marked with letters. Which is the location of the mean, A or B? Explain why the mean and median are not the same.

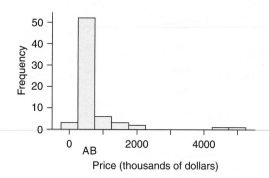

TRY **3.47 Study Hours (Example 14)** A group of 50 statistics students, 25 men and 25 women, reported the number of hours per week spent studying statistics.

a. Refer to the histograms. Which measure of the center should be compared: the means or the medians? Why?

b. Compare the distributions in context using appropriate measures. (Don't forget to mention outliers, if appropriate). Refer to the Minitab output for the summary statistics.

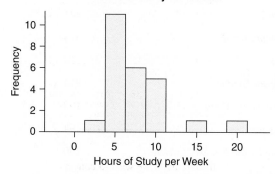

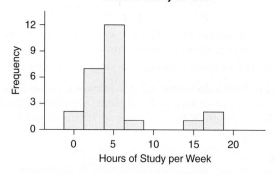

**Minitab Statistics: Men, Women**

| Variable | N | Mean | StDev | Minimum | Q1 | Median | Q3 | Maximum |
|----------|----|------|-------|---------|------|--------|------|---------|
| Men | 25 | 5.20 | 4.378 | 1.00 | 2.50 | 4.00 | 5.50 | 17.00 |
| Women | 25 | 7.52 | 3.787 | 2.00 | 5.00 | 7.00 | 9.50 | 20.00 |

**3.48 Driving Accidents** College student Sandy Hudson asked a group of college students the total number of traffic accidents they had been in as drivers. The histograms are shown, and the table displays some descriptive statistics.

**Men's Driving Accidents**

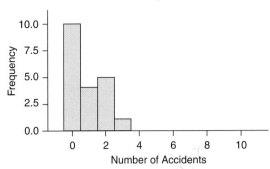

**Women's Driving Accidents**

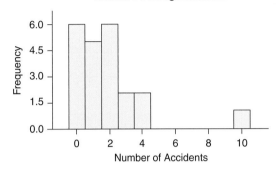

**Minitab Statistics: Men, Women**

| Variable | N | Mean | StDev | Min | Q1 | Median | Q3 | Max |
|---|---|---|---|---|---|---|---|---|
| Men | 20 | 0.850 | 0.988 | 0.00 | 0.00 | 0.505 | 2.00 | 3.00 |
| Women | 22 | 1.864 | 2.210 | 0.00 | 0.00 | 1.50 | 2.25 | 10.00 |

a. Refer to the histograms. If we wish to compare the typical numbers of accidents for these men and women, should we compare the means or the medians? Why?

b. Write a sentence or two comparing the distributions of numbers of accidents for men and women in context.

# SECTION 3.5

**3.49 Regional Rates of Millionaires** The boxplot shows the number of millionaires by state per 1000 residents for the United States, as reported by *Fortune Magazine* in March 2007. Assume all distributions are unimodal.

a. List the regions from lowest to highest in terms of the median rate of millionaires in that region.

b. Which region has the largest interquartile range?

c. Which region has the smallest interquartile range?

d. Which region has potential outliers?

*e. Why is a rate of about 24 millionaires per 1000 people in the South a potential outlier, while the rate of about 24.5 in the Northeast is not a potential outlier?

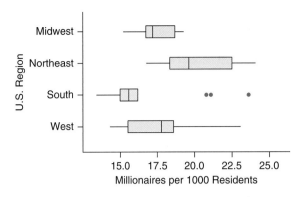

**3.50 Regional Gas Taxes** Each state in the United States can charge its own tax on gasoline. The boxplots show the gas tax (in cents per gallon) for the 50 states by region. Assume all distributions are unimodal. (Source: *AARP Bulletin,* June 2010)

a. Compare the medians. List the regions from smallest to largest in terms of the median amount of gas tax.

b. Compare the interquartile ranges. List the regions from smallest to largest interquartile range.

c. Which region had the largest range? Is it the same as the region with the largest interquartile range?

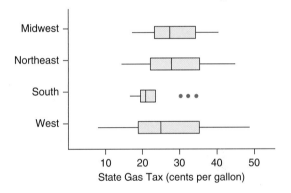

**3.51 City Temperatures** The boxplot shows temperatures for six cities. Each city's boxplot was made from 12 temperatures: the average monthly temperature over a period of years. Which city tends to be warmest? Which city has the most variation in temperatures? Compare the temperatures of the cities by interpreting the boxplots. If temperature were the only factor to consider, which city you would choose to live in, and why? (Source: *2009 World Almanac and Book of Facts*)

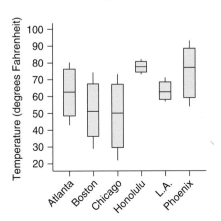

**3.52 College Tuition, Private vs. State** The figure shows side-by-side boxplots for the tuition of 71 private colleges and 105 state colleges randomly selected from the colleges listed in the *2006 World Almanac and Book of Facts*. Compare the distribution of tuitions of the two groups.

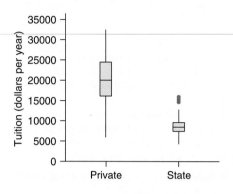

### 3.53 Matching Boxplots and Histograms

a. Match each of the boxplots (A, B, and C) with the corresponding histogram (1, 2, or 3). Explain your reasoning.

b. For each histogram, label the shape (right-skewed, left-skewed, or symmetric) and indicate whether the mean would be larger than, smaller than, or about the same as the median.

**Histogram 1**

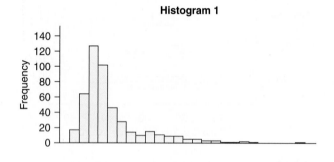

**Histogram 2**

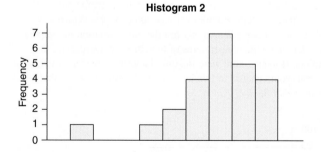

**Histogram 3**

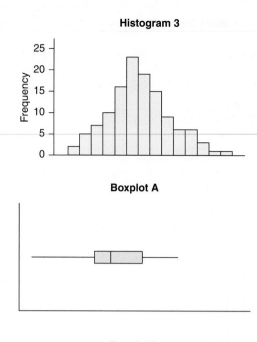

**Boxplot A**

**Boxplot B**

**Boxplot C**

**3.54 Matching Boxplots and Histograms** Match each of the histograms (X, Y, and Z) with the corresponding boxplot (C, M, or P). Explain your reasoning.

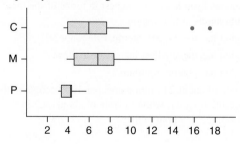

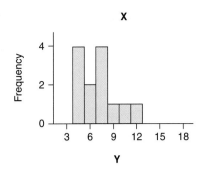

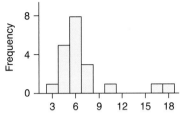

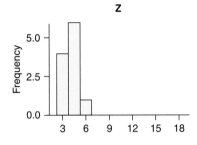

One way in which a state can have a high death rate is to have many deaths. What is the other reason why a state might have a high death rate?

```
1-Var Stats
↑n=25
 minX=16
 Q₁=17.5
 Med=20
 Q₃=23
 maxX=33
```

**3.56 Educated Residents** The U.S. Census Bureau reported the percentage of residents 25-years-old and older in each state who held a Bachelor's degree. We provide data for the states in the western U.S. in 2004. Make a boxplot of the data and describe it. If there are any potential outliers, explain what they show, including the name of the state(s).

| | |
|---|---|
| Alaska | 17.3 |
| Arizona | 17.2 |
| California | 21.2 |
| Colorado | 29.3 |
| Hawaii | 17.8 |
| Idaho | 17.4 |
| Montana | 18.3 |
| Nevada | 16.0 |
| New Mexico | 14.9 |
| Oregon | 17.6 |
| Utah | 21.2 |
| Washington | 20.2 |
| Wyoming | 16.3 |

**\*3.55 Driving Deaths** Have you ever wondered whether driving is more dangerous in some states than in others? The data on the disk give the automobile fatality rates (deaths per 100,000 residents) in 2005 for the 25 states with the highest death rates. The TI-83/84 output shows some descriptive statistics for this data set. (Source: www.nhtsa.dot.gov)

a. Make a boxplot of the data, and report the median and interquartile range. If you find potential outlier(s), identify the state(s).

b. There are two variables involved in the calculation of the death rate for a state (deaths per 100,000 residents):

$$\text{Death rate} = \frac{\text{deaths}}{\text{population}}(100{,}000)$$

# CHAPTER REVIEW EXERCISES

**3.57 Marathon Times by Gender** The figure shows side-by-side boxplots of marathon times for the 2009 Forest City Marathon. Who tended to run faster, the men or the women? Make appropriate comparisons without reporting numbers. The men's distribution and the women's distribution were both unimodal. (Remember, smaller times mean faster runners.)

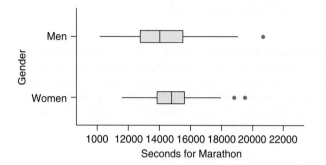

**3.58 Family Income** The boxplots show median family income in 2008 for the 50 U.S. states and Washington, D.C. by region, according to the U.S. Census Bureau. The distributions were unimodal.

a. Compare the medians and interquartile ranges without reporting numbers. List the regions from smallest (on the left) to largest median income.

b. Explain why the medians are more appropriate for comparisons than the means.

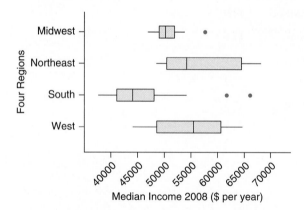

Median Income 2008 ($ per year)

**TRY** **3.59 Death Row: South (Example 15)** The table shows the numbers of capital prisoners (prisoners on death row) in 2005 in the southern U.S. states. (Source: www.statemaster.com)

a. Find the median number of prisoners and interpret (using a sentence in context).

b. Find the interquartile range (showing Q3 and Q1 in the process) to measure the variability in the number of prisoners.

c. What is the mean number of prisoners?

d. Why is the mean so much larger than the median?

e. Why is it better to report the median, instead of the mean, as a typical measure?

| State | CapPris | State | CapPris |
|-------|---------|-------|---------|
| Alabama | 193 | North Carolina | 166 |
| Arkansas | 36 | Oklahoma | 84 |
| Florida | 374 | South Carolina | 62 |
| Georgia | 105 | Tennessee | 102 |
| Kentucky | 40 | Texas | 391 |
| Louisiana | 86 | Virginia | 20 |
| Maryland | 6 | West Virginia | 0 |
| Mississippi | 69 | | |

**3.60 Death Row: West** The table shows the numbers of capital prisoners (prisoners on death row) in 2005 in the western U.S. states. (Source: www.statemaster.com)

| State | CapPris | State | CapPris |
|-------|---------|-------|---------|
| Alaska | 0 | Nevada | 82 |
| Arizona | 110 | New Mexico | 2 |
| California | 656 | Oregon | 32 |
| Colorado | 2 | Utah | 9 |
| Hawaii | 0 | Washington | 9 |
| Idaho | 18 | Wyoming | 2 |
| Montana | 2 | | |

a. Find the median.

b. Find the interquartile range (showing Q3 and Q1 in the process).

c. Find the mean number of capital prisoners.

d. Why is the mean so much larger than the median?

**TRY** **3.61 Head Circumference (Example 16)** Following are head circumferences, in centimeters, for some men and women in a statistics class.

Men: 58, 60, 62.5, 63, 59.5, 59, 60, 57, 55

Women: 63, 55, 54.5, 53.5, 53, 58.5, 56, 54.5, 55, 56, 56, 54, 56, 53, 51

Compare the circumferences of the men's and the women's heads. Start with histograms to determine shape; then compare appropriate measures of center and spread, and mention any outliers. *See page 125 for guidance.*

**3.62 Heights of Sons and Dads** The data on the disk give the heights of 18 male college students and their fathers, in inches.

a. Make histograms and describe the shapes of the two data sets from the histograms.

b. Fill in the following table to compare descriptive statistics.

|  | Mean | Median | Standard deviation | Interquartile range |
|---|---|---|---|---|
| Sons | ____ | ____ | ____ | ____ |
| Dads | ____ | ____ | ____ | ____ |

c. Compare the heights of the sons and their dads, using the means and standard deviations.

d. Compare the heights of the sons and their dads, using the medians and interquartile ranges.

e. Which pair of statistics is more appropriate for comparing these samples: the mean and standard deviation or the median and interquartile range? Explain.

**3.63 Final Exam Grades** The data that follow are final exam grades for two sections of statistics students at a community college. One class met twice a week relatively late in the day; the other class met four times a week at 11 a.m. Both classes had the same instructor and covered the same content. Is there evidence that the performances of the classes differed? Answer by making appropriate plots (including side-by-side boxplots) and reporting and comparing appropriate summary statistics. Explain why you chose the summary statistics that you used. Be sure to comment on the shape of the distributions, the center, and the spread, and be sure to mention any unusual features you observe.

11 a.m. grades: 100, 100, 93, 76, 86, 72.5, 82, 63, 59.5, 53, 79.5, 67, 48, 42.5, 39

5 p.m. grades: 100, 98, 95, 91.5, 104.5, 94, 86, 84.5, 73, 92.5, 86.5, 73.5, 87, 72.5, 82, 68.5, 64.5, 90.75, 66.5

**3.64 Speeding Tickets** College students Diane Glover and Esmeralda Olguin asked 25 men and 25 women how many speeding tickets they had received in the last three years.

Men: 14 men said they had 0 tickets, 9 said they had 1 ticket, one had 2 tickets, and 1 had 5 tickets.

Women: 18 said they had 0 tickets, 6 said they had 1 ticket, and 1 said she had 2 tickets.

Is there evidence that the men and women differed? Answer by making appropriate plots and comparing appropriate summary statistics. Be sure to comment on the shape of the distributions and to mention any unusual features you observe.

**3.65 Heights** The graph shows the heights for a large group of adults. Describe the distribution and explain what might cause this shape. (Source: www.amstat.org)

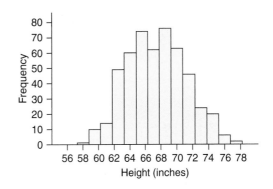

**\*3.66 Marathon Times** The histogram of marathon times includes data for men and women and also for both an Olympic marathon and an amateur marathon. Greater values indicate slower runners. (Sources: www.forestcityroadraces.com and www.runnersworld.com)

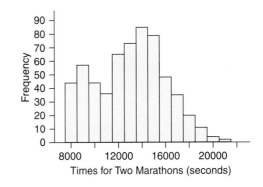

a. Describe the shape of the distribution.

b. What are two different possible reasons for the two modes?

c. Knowing that there are usually fewer women who run marathons than men and that more people ran in the amateur marathon than the Olympic marathon, look at the size of the mounds and try to decide which of the reasons stated in part b is likely to cause this. Explain.

**\*3.67 Law School GPAs** The histogram shows the minimum GPA of accepted students at 181 nationally ranked law schools in 2006. (Source: www.ilrg.com)

a. State an approximate value for the mean GPA.

b. Here is a proposed method for finding an approximation to the standard deviation based on a histogram: Find the approximate range and divide by 6. Use this method to find an approximate standard deviation for the GPAs. For comparison, the true standard deviation is 0.22.

c. Explain why dividing the range by 6 should produce a reasonable approximation of the standard deviation for symmetric, unimodal distributions.

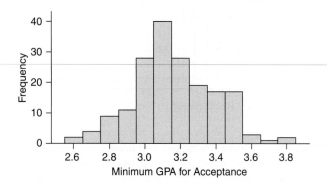

**3.68 Ideal Family** In 2008, the General Social Survey asked respondents how many children they felt would be in an "ideal" family. The histogram contains the data from 1730 people who responded to the survey.

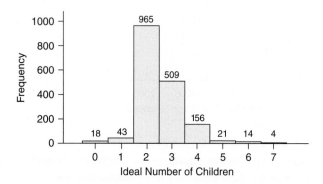

a. Approximately what is the mean ideal number of children? Explain how you chose this value.

b. What is the approximate value for the median ideal number of children? Describe how you chose this value.

c. Find the mean by completing the work that is started below:

$$\bar{x} = \frac{18(0) + 43(1) + 965(2) + \cdots}{1730}$$

d. Explain how the method in part c is related to the usual method of finding the mean, which has all the raw numbers given, without frequencies.

e. Which is more appropriate to report for these data, the mean or the median? Why?

**3.69–3.72** *Construct two sets of numbers with at least five numbers in each set (showing them as dotplots) with the following characteristics:*

**3.69** The means are the same, but the standard deviation of one of the sets is larger than that of the other. Report the mean and both standard deviations.

**3.70** The means are different, but the standard deviations are the same. Report both means and the standard deviation.

**3.71** The mean of set A is larger than that of set B, but the median of set B is larger than the median of set A. Label each dotplot with its mean and median in the correct place.

★ **3.72** The standard deviation of set A is larger, but the interquartile range of set B is larger. Report both standard deviations and interquartile ranges.

★ **3.73 Population Density** Data were recorded for each of the 50 U.S. states: the state, its population (in 2010), its area (in square miles), and region in which it is located (Northeast, Midwest, South, or West). For each state, find the population density: the number of people divided by the number of square miles. Write a few sentences comparing the distribution of population densities in the four regions. Support your description with appropriate graphs. (Sources: U.S. Census Bureau and *2009 World Almanac and Book of Facts*)

★ **3.74 Population Increase** Data were recorded for each of the 50 U.S. states: the state, its population in 2000, its population in 2010, and the region in which it is located (Northeast, Midwest, South, or West). Find the percentage population increase for each state by applying the following formula:

$$\frac{\text{pop}_{2010} - \text{pop}_{2000}}{\text{pop}_{2000}} \times 100\%$$

Write a few sentences comparing the distribution of percentage population increases for the four regions. Support your description with appropriate graphs. (Source: U.S. Census Bureau)

**3.75 Gas Barrels** Data were recorded on the number of barrels of oil per person that were used for transportation in each state (in 2008), and the region where the state is located (Northeast, Midwest, South, or West). Write a few sentences comparing the four regions. Include boxplots. (Source: www.eia.doe.gov)

**3.76 Sales Tax** Data were recorded on the sales tax as a percentage of purchase for each of the 50 states and the state's region (Northeast, Midwest, South, or West). Write a few sentences comparing the four regions. Include boxplots. (Source: www.taxadmin.org)

# gUIDED EXERCISES

g **3.21 Violent Crime: West**  In 2007, the mean rate of violent crime (per 100,000 people) for the 24 states west of the Mississippi River was 406, and the standard deviation was 177. Assume the distribution of violent crime rates is approximately unimodal and symmetric. The Empirical Rule says 95% of data should lie within two standard deviations of the mean and 68% of the data should lie within one standard deviation of the mean with unimodal and symmetric data.

## Questions
Answer these questions by following the numbered steps.

a. Between which two values would you expect to find about 95% of the violent crime rates?

b. Between which two values would you expect to find about 68% of the violent crime rates?

c. If a western state had a violent crime rate of 584 crimes per 100,000 people, would you consider this unusual? Explain.

d. If a western state had a violent crime rate of 30 crimes per 100,000 people, would you consider this unusual? Explain.

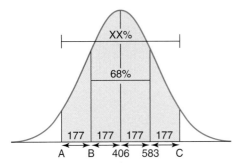

The green area is 68% of the area under the curve. The green and blue areas together shade XX% of the area. The numbers without percentage signs are crime rates, with a mean of 406 and a standard deviation of 177.

Step 1 ▶ **Percentage**
Reproduce Figure A, which is a sketch of the distribution of the crime rates. What percentage of data should occur within two standard deviations of the mean? Include this number in the figure, where it now says XX.

Step 2 ▶ **Why 583?**
How was the number 583, shown on the sketch, obtained?

Step 3 ▶ **A, B, and C**
Fill in numbers for the crime rates for areas A, B, and C.

Step 4 ▶ **Boundaries for 95%**
Read the answer from your graph:

a. Between which two values would you expect to find about 95% of the rates?

Step 5 ▶ **Boundaries for 68%**
Read the answer from your graph:

b. Between which two values would you expect to find about 68% of the violent crime rates?

Step 6 ▶ **Unusual?**

c. If a western state had a violent crime rate of 584 crimes per 100,000 people, would you consider this unusual? Many people would consider any numbers outside your boundaries of A and C to be unusual because such values occur in 5% of the states or fewer.

Step 7 ▶ **Unusual?**

d. If a western state had a violent crime rate of 30 crimes per 100,000 people, would you consider this unusual? Explain.

g **3.61 Head Circumference**  The head circumferences in centimeters for some men and women in a statistics class are given.

Men: 58, 60, 62.5, 63, 59.5, 59, 60, 57, 55

Women: 63, 55, 54.5, 53.5, 53, 58.5, 56, 54.5, 55, 56, 56, 54, 56, 53, 51

## Question
Compare the circumferences of the men's and women's heads by following the numbered steps.

Step 1 ▶ **Histograms**
Make histograms of the two sets of data separately. (You may use the same horizontal axes—if you want to—so that you can see the comparison easily.)

Step 2 ▶ **Shapes**
Report the shapes of the two data sets.

Step 3 ▶ **Measures to Compare**
If either data set is skewed or has an outlier (or more than one), you should compare medians and interquartile ranges for *both* groups. If both data sets are roughly symmetric, you should compare means and standard deviations. Which measures should be compared with these two data sets?

Step 4 ▶ **Compare Centers**
Compare the centers (means or medians) in the following sentence: The _____ (mean or median) head circumference for the men was _____ cm, and the _____ (mean or median) head circumference for the women was _____ cm. This shows that the typical head circumference was larger for the_____.

Step 5 ▶ **Compare Variations**
Compare the variations in the following sentence: The _____ (standard deviation or interquartile range) for the head circumferences for the men was _____ cm, and the _____ (standard deviation or interquartile range) for the women was _____ cm. This shows that the _____ tended to have more variation, as measured by the _____ (standard deviation or interquartile range).

Step 6 ▶ **Outliers**
Report any outliers and state which group(s) they belong to.

Step 7 ▶ **Final Comparison**
Finally, in a sentence or two, make a complete comparison of head circumferences for the men and the women.

## CHECK YOUR TECH

# Finding the Standard Deviation of Vacation Days for Several Countries

 The Minitab output shown gives the mean and standard deviation of the data set. (Source: *2009 World Almanac and Book of Facts*)

**Descriptive Statistics: Days**

```
Variable   N   Mean   StDev
Days       6   30.00  10.35
```

Minitab Output

| Country | Mean Days |
|---------|-----------|
| United States | 13 |
| Japan | 25 |
| Italy | 42 |
| France | 37 |
| Germany | 35 |
| U. K. | 28 |

The table at the left reports the mean number of vacation days per year for several countries.

$$\text{Standard deviation} = s = \sqrt{\frac{\sum (x - \bar{x})^2}{n - 1}}$$

**QUESTION**   Find the standard deviation of vacation days (by following the numbered steps) and verify that it is the same as StDev given in the Minitab output.

**SOLUTION**

**Step 1 ▶ Mean**
From the standard deviation formula above, you can see that you will need the mean, $\bar{x}$, in order to calculate the standard deviation. Check that the mean is 30 by adding the six numbers, reporting the sum, and dividing by 6.

**Step 2 ▶ Table**
Fill in the table below.

| $x$ | $x - \bar{x}$ | $(x - \bar{x})^2$ |
|-----|---------------|-------------------|
| 13 | $13 - 30 = -17$ | $(-17)^2 = 289$ |
| 25 | | |
| 42 | $42 - 30 = 12$ | $12^2 = 144$ |
| 37 | | |
| 35 | | |
| 28 | | |

**Step 3 ▶ Sum of Squares**
Add the numbers in the last column of your table to get $\sum (x - \bar{x})^2$

**Step 4 ▶ Variance**
Divide your answer to step 3 by $n - 1$, which is $6 - 1$, or 5, to get the following:

$$\frac{\sum (x - \bar{x})^2}{n - 1}$$

**Step 5 ▶ Standard Deviation**
Finally, to get the standard deviation, take the square root of your answer to step 4. Show a long version of your answer, and then round it to hundredths and compare it with the StDev in the Minitab output.

# Making a Boxplot of the Area of Western States

The area of western states is given in the table (in thousands of square miles). Verify that the boxplot given in Figure A is correct by following the numbered steps. (Source: *2009 World Almanac and Book of Facts*)

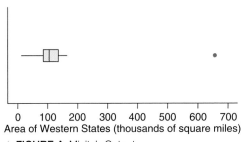

Area of Western States (thousands of square miles)

▲ **FIGURE A** Minitab Output

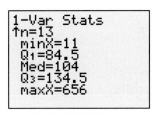

▲ **FIGURE B** TI-83/84 Output

| State | Area |
|---|---|
| Alaska | 656 |
| Arizona | 114 |
| California | 164 |
| Colorado | 104 |
| Hawaii | 11 |
| Idaho | 84 |
| Montana | 147 |
| Nevada | 111 |
| New Mexico | 122 |
| Oregon | 98 |
| Utah | 85 |
| Washington | 71 |
| Wyoming | 98 |

**QUESTION**   Make a boxplot by hand, using graph paper, a ruler, and a pencil by following the numbered steps. You may use the TI-83/84 output reported in Figure B.

**SOLUTION**

**Step 1 ▶ The Axis**
Draw a horizontal line (using a ruler or other straight edge) across the page and label it in equal intervals up to a bit above the highest area (700), and down to a bit below the lowest area (0). Below the numbers, add a label telling what the numbers are.

**Step 2 ▶ The Box**
Draw vertical lines at the positions of Q1 and Q3 and join them with horizontal lines to make a box. The height of the box is arbitrary.

**Step 3 ▶ The Median**
Put a vertical line representing the median in the correct position inside the boxplot. Which state is at the median?

**Step 4 ▶ The Interquartile Range**
Find and report the interquartile range, $Q3 - Q1$.

**Step 5 ▶ Lower Limit and Potential Lower Outliers**
Find and report the lower limit by finding

$$Q1 - 1.5 \text{ IQR}$$

If there are any points lower than that limit, make separate marks showing they are potential outliers.

**Step 6 ▶ Lower Whisker**
Draw the lower whisker (horizontal line) from the box to the lowest point that is not a potential outlier. Which state is at the left end of the lower whisker?

**Step 7 ▶ Upper Limit and Potential Upper Outliers**
Now find and report the upper limit by finding

$$Q3 + 1.5 \text{ IQR}$$

If there are any points higher than that, make separate marks showing they are potential outliers. Report the name of the state that is a potential outlier.

**Step 8 ▶ Upper Whisker**
Draw the upper whisker to the largest point that is not a potential outlier. Report this point and report the name of the state that the point represents.

**Step 9 ▶ Title**
Give your graph an informative title, different from the one given.

# TechTips

## Example

Analyze the data given by finding descriptive statistics and making boxplots. The tables give calories per ounce for sliced ham and turkey. Table 3A shows unstacked data, and Table 3B shows stacked data. We coded the meat types with numerical values (1 for ham and 2 for turkey) but you could also use descriptive terms, such as "ham" or "turkey."

| Ham | Turkey |
|-----|--------|
| 21 | 35 |
| 25 | 25 |
| 35 | 25 |
| 35 | 25 |
| 25 | 25 |
| 30 | 25 |
| 30 | 29 |
| 35 | 29 |
| 40 | 23 |
| 30 | 50 |
|    | 25 |

▲ TABLE 3A

| Cal | Meat |
|-----|------|
| 21 | 1 |
| 25 | 1 |
| 35 | 1 |
| 35 | 1 |
| 25 | 1 |
| 30 | 1 |
| 30 | 1 |
| 35 | 1 |
| 40 | 1 |
| 30 | 1 |
| 35 | 2 |
| 25 | 2 |
| 25 | 2 |
| 25 | 2 |
| 25 | 2 |
| 25 | 2 |
| 29 | 2 |
| 29 | 2 |
| 23 | 2 |
| 50 | 2 |
| 25 | 2 |

▲ TABLE 3B

## TI-83/84

Enter the unstacked data (Table 3A) into **L1** and **L2**.

### For Descriptive Comparisons of Two Groups

Follow the steps twice, first for **L1** and then for **L2**.

### Finding One-Variable Statistics

1. Press **STAT**, choose **CALC** (by using the right arrow on the keypad), and choose **1** (for **1-Var Stats**)
2. Specify **L1** (or the list containing the data) by pressing **2ND**, **1**, and **ENTER**.
3. Output: On your calculator, you will need to scroll down using the down arrow on the keypad to see all of the output.

## Making Boxplots

1. Press **2ND**, **STATPLOT**, **4 (PlotsOff)**, and **ENTER** to turn the Plots Off. This will prevent you from seeing old plots as well as the new ones.
2. Press **2ND**, **STATPLOT**, and **1**.
3. Refer to Figure 3A. Turn on **Plot1** by pressing **ENTER** when **On** is flashing. (**Off** will no longer be highlighted.)
4. Use the arrows on the keypad to locate the boxplot in the lower left corner of the six graphs (as shown highlighted in Figure 3A), and press **ENTER**. (If you accidentally choose the other boxplot—the one in the middle—there will never be any separate marks for potential outliers.)

▲ **FIGURE 3A** TI-83/84 Plot Dialogue Screen

5. Use the down arrow on the keypad to get to the **XList**. Choose **L1** by pressing **2ND** and **1**.
6. Press **GRAPH**, **ZOOM** and **9 (Zoomstat)** to make the graph.
7. Press **TRACE** and move around with the arrows on the keypad to see the numerical labels.

### Making Side-by-Side Boxplots

For side-by-side boxplots, turn on a second boxplot (**Plot2**) for data in a separate list, such as **L2**. Then, when you choose **GRAPH**, **ZOOM** and **9**, you should see both boxplots. Press **TRACE** to see numbers.

Figure 3B shows side-by-side boxplots with the boxplot from the turkey data on the bottom and the boxplot for the ham data on the top.

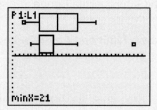

▲ **FIGURE 3B** TI-83/84 Boxplots

## MINITAB

### For Comparisons of Two Groups

Upload the data from the disk or manually enter the data given. You may use unstacked data entered into two different columns (Table 3A), or you may use stacked data and use labels such as Ham and Turkey or codes such as 1 and 2 (Table 3B).

### Finding Descriptive Statistics: One-Column Data or *Unstacked* Data in Two or More Columns

1. **Stat > Basic Statistics > Display Descriptive Statistics**.
2. Double click on the column(s) containing the data to put it (them) in the **Variables** box.
3. Ignore the **By variables** box.
4. (Optional) Click on **Statistics**; you can choose what you want to add, such as the interquartile range.
5. Click **OK**.

## Finding Descriptive Statistics: *Stacked and Coded*

1. **Stat > Basic Statistics > Display Descriptive Statistics**.
2. See Figure 3C: Double click on the column(s) containing the stack of data to put it in the **Variables** box.

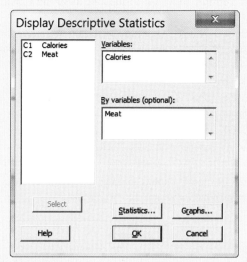

▲ **FIGURE 3C** Minitab Descriptive Statistics Dialog Box

3. When the **By variables** box is activated (by clicking in it), double click the column containing the categorical labels or code.
4. (Optional) Click on **Statistics**; you can choose what you want to add, such as the interquartile range. You can also make boxpots by clicking **Graphs**.
5. Click **OK**.

## Making Boxplots

1. **Graph > Boxplot**.
2. For a single boxplot, choose **One Y**, **Simple** and click **OK**. See Figure 3D.
3. Double click the label for the column(s) containing the data and click **OK**.
4. For side-by-side boxplots
   a. If the data are unstacked, choose **Multiple Y's Simple**, shown in Figure 3D. Then double click both labels for the columns and click **OK**.
   b. If the data are stacked, choose **One Y With Groups** (the top right in Figure 3D). Then see Figure 3E and double click on the label for the stack (such as Calories) and then click in the **Categorical variables …** box and double click the label for codes or words defining the groups such as **Meat**.
5. Labeling and transposing the boxplots. If you want to change the labeling, double click on what you want to change after the boxplot(s) are made. To change the orientation of the boxplot to horizontal, double click on the *x*-axis and select **Transpose value and category scales**.

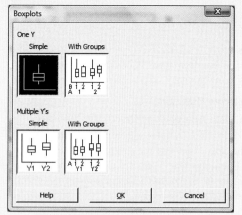

▲ **FIGURE 3D** Minitab First Boxplot Dialog Box

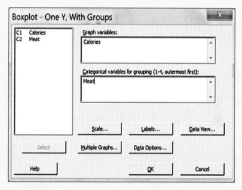

▲ **FIGURE 3E** Minitab Dialog Screen for Boxplots: One Y with Groups

Figure 3F Shows Minitab boxplots of the ham and turkey data, without transposition.

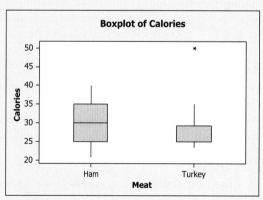

▲ **FIGURE 3F** Minitab Boxplots

### Entering Data

Input the unstacked data from the disk or enter it manually. You may have labels in the first row such as **Ham** and **Turkey**.

### Finding Descriptive Statistics

1. Click **Data, Data Analysis,** and **Descriptive Statistics**.

2. See Figure 3G: In the dialogue screen, for the **Input Range** highlight the cells containing the data (one column only) and then Click **Summary Statistics**. If you include the label in the top cell such as A1 in the **Input Range**, you need to check **Labels in First Row**. Click **OK**.

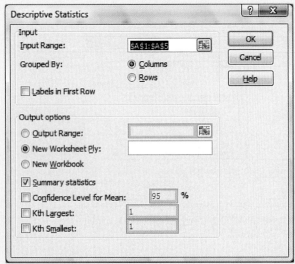

▲ **FIGURE 3G** Excel Dialog Box for Descriptive Statistics

### Comparing Two Groups

To compare two groups, use unstacked data and do the above analysis twice. (Choosing **Output Range** and selecting appropriate cells on the same sheet makes comparisons easier.)

### Boxplots (Requires XLSTAT Add-in)

1. Click **Add-ins, XLSTAT, Visualizing data,** and **Univariate plots.**

2. When the box under **Quantitative data** is activated, drag your cursor over the column containing the data including the label at the top such as **Ham.**

3. Click **Charts.**

4. Click **Box plots, Outliers** and choose **Horizontal** (or **Vertical**).

5. Click **OK** and **Continue**. See step 6 in the side-by-side instructions.

### Side-by-side Boxplots

Use unstacked data with labels in the top row.

1. Click **Add-ins, XLSTAT, Visualizing data,** and **Univariate plots**.

2. When the box under **Quantitative data** is activated, drag your cursor over all the columns containing the unstacked data including labels at the top such as **Ham** and **Turkey**.

3. Click **Charts**

4. Click **Box plots, Group plots, Outliers,** and choose **Horizontal** (or **Vertical**).

5. Click **OK** and **Continue**.

6. When you see the small labels **Turkey** and **Ham**, you may drag them to where you want them and you can increase the font size.

Figure 3H shows boxplots for the ham and turkey data. The red crosses give the location of the two means.

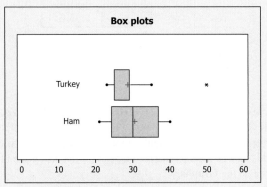

▲ **FIGURE 3H** XLSTAT Boxplots

### For Comparisons of Two Groups

Upload the data from the disk or enter the data manually. You may use stacked or unstacked data. Use the data from the example.

### Finding Summary Statistics (*stacked* or *unstacked* data)

1. **Stat > Summary Stats > Columns**

2. Refer to Figure 3I. If the data are stacked put the stack (here, Calories) in the big box on the right and put the code (here, Meat) in the rectangle labeled **Group by:**

   If the data are unstacked put both lists into the large box on the right.

3. Click **Next** for additional options such as extra statistics to be computed.

4. Click **Calculate** to compute the summary statistics.

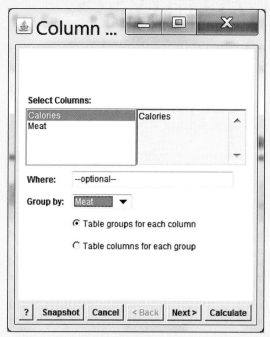

**FIGURE 3I** StatCrunch Dialog Screen for Summary Statistics

Put the one column with all the data into the big upper box and then select the column of codes or categories to put in the **Group by** small rectangle.

4. Click **Next** and check **Use fences to identify outliers** to make sure that the outliers show up as separate marks. You may also check **Draw boxes horizontally** if that is what you want.

5. Click **Create Graph!**

6. To copy your graph, click **Options** and **Copy** and paste it into a document.

Figure 3K Shows boxplots of the ham and turkey data; the top box comes from the turkey data (calories per ounce).

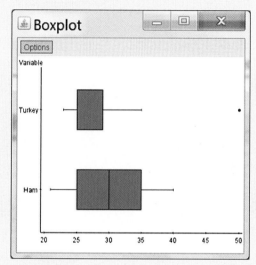

**FIGURE 3K** StatCrunch Boxplots

**▲ FIGURE 3I** StatCrunch Dialog Screen for Summary Statistics

## Making Boxplots (stacked or unstacked)

1. **Graphics > Boxplot**

2. *Unstacked* data: Select the columns to be displayed. A boxplot for each column will be included in a single graph.

3. *Stacked* and coded data: Refer to Figure 3J:

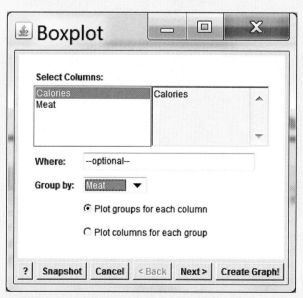

**▲ FIGURE 3J** StatCrunch Dialog Screen for Boxplots

# 4 Regression Analysis: Exploring Associations between Variables

## THEME

Relationships between two numerical variables can be described in several ways, but as always, the first step is to understand what is happening visually. When it fits the data, the linear model can help us understand how average values of one variable vary with respect to another variable, and we can use this knowledge to make predictions about one of the variables on the basis of the other.

The website www.Zillow.com can estimate the market value of any home. You need merely type in an address. Zillow estimates the value of a home even if the home has not been on the market for many years. How can this site come up with an estimate for something that is not for sale? The answer is that Zillow takes advantage of associations between the value and other easily observed variables—size of the home, selling price of nearby homes, number of bedrooms, and so on.

What role does genetics play in determining basic physical characteristics, such as height? This question fascinated nineteenth-century statistician Francis Galton (1822–1911). He examined the heights of thousands of father-son pairs to determine the nature of the relationship between these heights. If a father is 6 inches taller than average, how much taller than average will his son be? How certain can we be of the answer? Will

there be much variability? If there's a lot of variability, then perhaps factors other than the father's genetic material play a role in determining height.

Associations between variables can be used to predict as yet unseen observations. You might think that estimating the value of a piece of real estate and understanding the role of genetics in determining height are unrelated. However, both take advantage of associations between two numerical variables. They use a technique called regression, invented by Galton, to analyze these associations.

As in previous chapters, graphs play a major role in revealing patterns in data, and graphs become even more important when we have two variables, not just one. For this reason, we'll start by using graphs to visualize associations between two numerical variables, and then we'll talk about quantifying these relationships.

## CASE STUDY

## Catching Meter Thieves

Parking meters are an important source of revenue for many cities. Collecting the money from these meters is no small task, particularly in a large city. In the 1970s, New York City collected the money from its meters using several different private contractors and also some of its own employees. In 1978, city officials became suspicious that employees from one of the private contractors, Brink's Inc., were stealing some of the money. Several employees were later convicted of this theft, but the city wanted its money back, so it sued Brink's.

But how could the city tell how much money had been stolen? Fortunately, the city had collected data. For each month, it knew how much money its own employees had collected from parking meters and how much the private contractors (excluding Brink's) had collected. If there was a relationship between how much the city collected and how much the honest private contractors had collected, then that information could be used to predict about how much the city should have received from Brink's. City officials could then compare the predicted amount with the amount they actually collected from Brink's. (*Source*: De Groot et al., 1986)

At the end of this chapter, you'll see how a technique called linear regression can take advantage of patterns in the data to estimate how much money had been stolen.

# Visualizing Variability with a Scatterplot

At what age do men and women first marry? How does this vary among the 50 states in the United States? These are questions about the relationship between two variables: age at marriage for women, and age at marriage for men. The primary tool for examining two-variable relationships, when both variables are numerical, is the **scatterplot**. In a scatterplot, each point represents one observation. The location of the point depends on the values of the two variables. For example, we might expect that states where men marry later would also have women marrying at a later age. Figure 4.1 shows a scatterplot of these data, culled from the 2008 U.S. Census. Each point represents a state (and one point represents Washington, D.C.) and shows us the typical age at which men and women marry in that state. The two points in the lower left corner represent Idaho and Utah, where the typical woman first marries at about 23 years of age, the typical man at about age 25. The point in the upper right corner represents Washington, D.C.

▶ **FIGURE 4.1** A scatterplot of typical marrying ages for men and women in the United States. The points represent the 50 states and the District of Columbia.

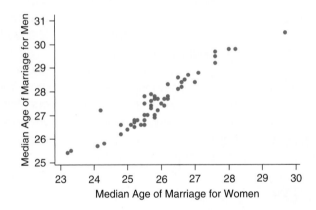

> ! **Caution**
>
> **About the Lower Left Corner**
> In scatterplots we do not require that the lower left corner be (0, 0). The reason is that we want to zoom in on the data and not show a substantial amount of empty space.

When examining histograms (or other pictures of distributions for a single variable), we look for center, spread, and shape. When studying scatterplots, we look for **trend** (which is like center), **strength** (which is like spread), and **shape** (which is like, well, shape). Let's take a closer look at these characteristics.

## Recognizing Trend

The trend of an association is the general tendency of the scatterplot as you scan from left to right. Usually trends are either increasing (uphill, /) or decreasing (downhill, \), but other possibilities exist. Increasing trends are called **positive associations** (or **positive trends**) and decreasing trends are **negative associations** (or **negative trends**).

Figure 4.2 shows examples of positive and negative trends. Figure 4.2a reveals a positive trend between the age of a used car and the miles it was driven (mileage). The positive trend matches our common sense: We expect older cars to have been driven farther, because generally, the longer a car is owned, the more miles it travels. Figure 4.2b shows a negative trend—the birthrate of a country against that country's literacy rate. The negative trend suggests that countries with higher literacy rates tend to have a lower rate of childbirth.

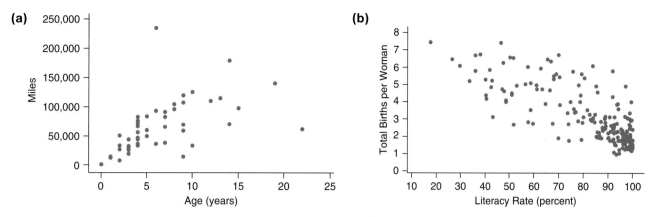

▲ **FIGURE 4.2** Scatterplots with **(a)** positive trend and **(b)** negative trend. (Source: United Nations [UN], Statistics Division, http://unstats.un.org)

Sometimes, the absence of a trend can be interesting. For example, running a marathon requires considerable training and endurance, and we might expect that the speed at which a person runs a marathon is related to his or her age. But Figure 4.3a shows no trend at all between the ages of runners in a marathon (in Forest City, Canada) and their times. The lack of a trend means that no matter what age group we examine, the runners have about the same times as any other age group, and so we conclude that at least for this group of elite runners, age is not associated with running speed in the marathon.

Figure 4.3b shows simulated data of an association between two variables that cannot be easily characterized as positive or negative—for smaller *x*-values the trend is negative (\), and for larger *x*-values it is positive (/).

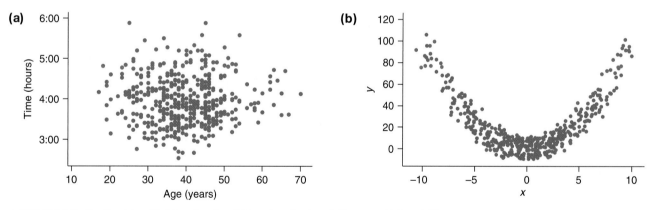

▲ **FIGURE 4.3** Scatterplots with **(a)** no trend and **(b)** a changing trend. (Sources: (a) Forest City Marathon, http://www.forestcityroadraces.com/; (b) simulated data)

## Seeing Strength of Association

Weak associations result in a large amount of scatter in the scatterplot. A large amount of scatter means that points have a great deal of spread in the vertical direction. This vertical spread makes it somewhat harder to detect a trend. Strong associations have little vertical variation.

Figure 4.4 on the next page allows us to compare the strengths of two associations. Figure 4.4a shows the association between height and weight for a sample of active adults. Figure 4.4b involves the same group of adults, but this time we examine the association between waist size and weight. Which association is stronger?

The association between waist size and weight is the stronger one (Figure 4.4b). To see this, in Figure 4.4a, consider the data for people who are 65 inches tall. Their

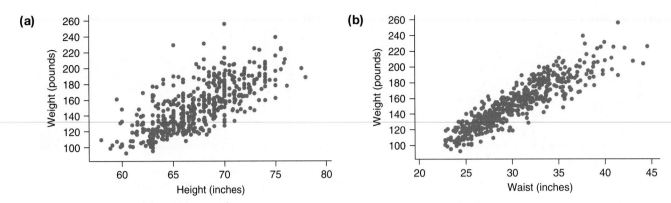

▲ **FIGURE 4.4** Part **(a)** This graph shows a relatively weaker association. **(b)** This graph shows a stronger association, because the points have less vertical spread. (Heinz et al. 2003)

weights vary anywhere from about 120 pounds to 230 pounds, a range of 110 pounds. If you were using height to predict weight, you could be off by quite a bit. Compare this with the data in Figure 4.4b for people with a waist size of 30 inches. Their weights vary from about 120 pounds to 160 pounds, only a 40-pound range. The association between waist size and weight is stronger than that between height and weight because there is less vertical spread, so tighter predictions can be made. If you had to guess someone's weight, and could ask only one question before guessing, you'd do a better job if you asked about the person's waist size than about his or her height.

Labeling a trend as strong, very strong, or weak is a subjective judgment. Different statisticians might have different opinions. Later in this section, we'll see how we can measure strength with a number.

## Identifying Shape

The simplest shape for a trend is **linear**. Fortunately, linear trends are quite common in nature. Linear trends always increase (or decrease) at the same rate. They are called linear because the trend can be summarized with a straight line. Scatterplots of linear trends often look roughly football-shaped, as shown in Figure 4.4a, particularly if there is some scatter and there are a large number of observations. Figure 4.5 shows a linear trend from data provided by Google Trends. Each point represents a week between January 2006 and December 2009. The numbers measure the frequency of Google searches for the term *vampire* or *zombie*. Figure 4.5 shows that a positive, linear association exists between the number of searches for *vampire* and the number for *zombie*. We've added a straight line to the scatterplot to highlight the linear trend.

▶ **FIGURE 4.5** A line has been inserted to emphasize the linear trend.

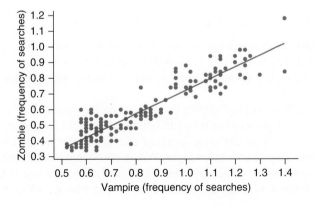

Not all trends are linear; in fact, a great variety of shapes can occur. But don't worry about that for now: All we want to do is classify trends as either linear or not linear.

Figure 4.6 shows the relationship between levels of the pollutant ozone in the air (measured in parts per million, or ppm) and air temperature (degrees Fahrenheit) in Upland, California, near Los Angeles, over the course of a year. The trend is fairly flat at first and then becomes steeper. For temperatures less than about 55 degrees, ozone levels do not vary all that much. However, higher temperatures (above 55 degrees) are associated with much greater ozone levels. The curved line superimposed on the graph shows the nonlinear trend.

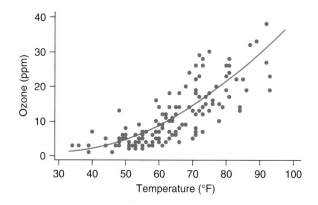

◀ **FIGURE 4.6** Ozone (ppm) is associated with temperature (degrees Fahrenheit) in a nonlinear way. (Source: Breiman, L. From the R earth package. See Faraway 2005.)

Nonlinear trends are more difficult to summarize than linear trends. This book does not cover nonlinear trends. Although our focus is on linear trends, it is very important that you first examine a scatterplot to be sure that the trend is linear. If you apply the techniques in this chapter to a nonlinear trend, you might reach disastrously incorrect conclusions!

**KEY POINT**    When examining associations, look for the trend, the strength of the trend, and the shape of the trend.

## Writing Clear Descriptions of Associations

Good communication skills are vital for success in general, and being able to clearly describe patterns in data is an important goal of this book. Here are some tips to help you describe two-variable associations.

• A written description should always include (1) trend, (2) shape, and (3) strength (not necessarily in that order) and should explain what all of these mean in (4) the context of the data. You should also mention any observations that do not fit the general trend.

Example 1 demonstrates how to write a clear, precise description of an association between numerical variables.

## EXAMPLE 1 Age and Mileage of Used Cars

Figure 4.2a on page 135 displays an association between the age and mileage of a sample of used cars.

QUESTION Describe the association.

SOLUTION The association between the age and mileage of used cars is positive and linear. This means that older cars tend to have greater mileage. The association is moderately strong; some scatter is present, but not enough to hide the shape of the relationship. There is one exceptional point: One car is only about 6 years old but has been driven many miles.

TRY THIS! Exercise 4.7

The description in Example 1 is good because it mentions trend (a "positive" association), shape ("linear"), and strength ("moderately strong") and does so in context ("older cars tend to have greater mileage").

- It is very important that your descriptions be precise. For example, it would be wrong to say that older cars have greater mileage. This statement is not true of every car in the data set. The one exceptional car (upper left corner of the plot) is relatively new (about 6 years old) but has a very high mileage (about 250,000 miles). Some older cars have relatively few miles on them. To be precise, you could say older cars *tend* to have higher mileage. The word *tend* indicates that you are describing a trend that has variability, so the trend you describe is not true of all individuals but instead is a characteristic of the entire group.

- When writing a description of a relationship, you should also mention unusual features, such as outliers, small clusters of points, or anything else that does not seem to be part of the general pattern. Figure 4.7 includes an outlier. These data are from a statistics class in which students reported their weights and heights. One student wrote the wrong height.

▶ **FIGURE 4.7** A fairly strong, positive association between height and weight for a statistics class in 2006. One student reported the wrong height. (Source: R. Gould, UCLA, Department of Statistics)

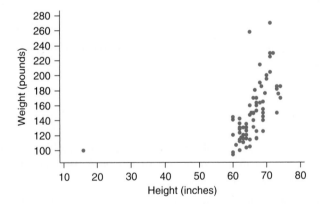

# SECTION 4.2

# Measuring Strength of Association with Correlation

The **correlation coefficient** is a number that measures the strength of the linear association between two numerical variables—for example, the relationship between people's heights and weights. We can't emphasize enough that the correlation coefficient *makes sense only if the trend is linear and if both variables are numerical.*

The correlation coefficient, represented by the letter $r$, is always a number between $-1$ and $+1$. Both the value and the sign (positive or negative) of $r$ have information we can use. If the value of $r$ is close to $-1$ or $+1$, then the association is very strong; if $r$ is close to 0, the association is weak. If the value of the correlation coefficient is positive, then the trend is positive; if the value is negative, the trend is negative.

## Visualizing the Correlation Coefficient

Figure 4.8 presents a series of scatterplots that show associations of gradually decreasing strength. The strongest linear association appears in Figure 4.8a; the points fall exactly along a line. Because the trend is positive and perfectly linear, the correlation coefficient is equal to 1.

The next scatterplot, Figure 4.8b, is slightly weaker. The points are more spread out vertically. We can see a linear trend, but the points do not fall exactly along a line. The trend is still positive, so the correlation coefficient is also positive. However, the value of the correlation coefficient is less than 1 (it is 0.98).

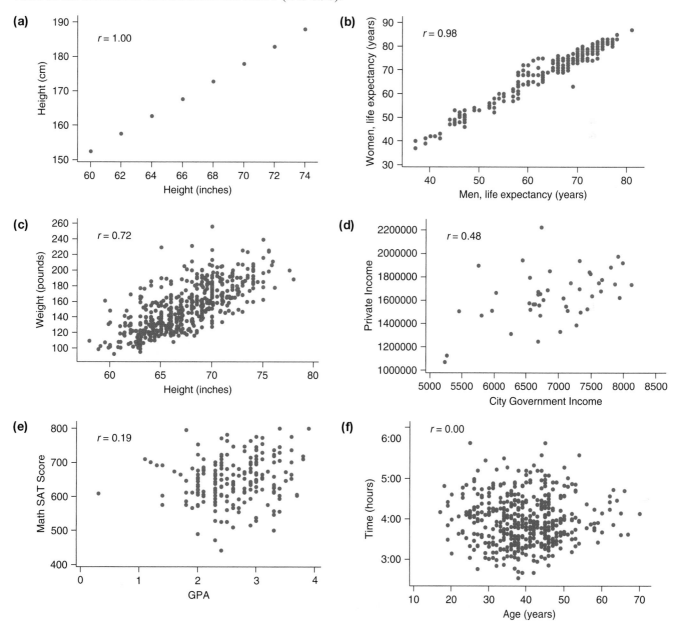

▲ **FIGURE 4.8** Scatterplots with gradually decreasing positive correlation coefficients. (Sources: (a) simulated data; (b) www.overpopulation.com; (c) Heinz, 2003; (d) Bentow and Afshartous; (e) Fathom™ Sample Document, Educational Testing Service [ETS] validation study; (f) Forest City Marathon.)

The remaining scatterplots show weaker and weaker associations, and their correlation coefficients gradually decrease. The last scatterplot, Figure 4.8f, has no association at all between the two variables, and the correlation coefficient has a value of 0.00.

The next set of scatterplots (Figure 4.9) starts with that same marathon data (having a correlation of 0.00), and the negative correlations gradually get stronger. The last figure has a correlation of −1.00.

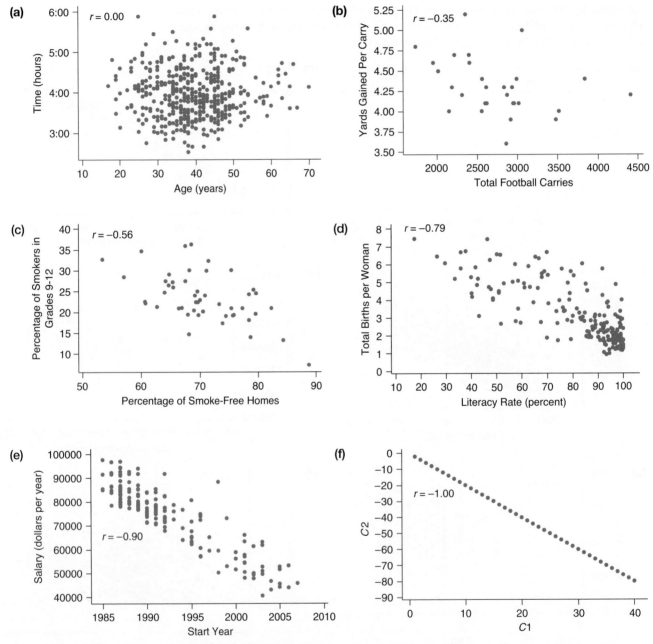

▲ **FIGURE 4.9** Scatterplots with increasingly negative correlations. (Sources: (a) Forest City Marathon; (b) http://wikipedia.org; (c) Centers for Disease Control; (d) UN Statistics; (e) Minitab Student 12 file "Salary," adjusted for inflation; (f) simulated data.)

## The Correlation Coefficient in Context

The correlation between the number of Google searches for *zombie* and *vampire* is $r = 0.924$. If we are told, or already know, that the association between these variables is linear, then we know that the trend is positive and strong. The fact that the correlation is close to 1 means that there is not much scatter in the scatterplot.

College admission offices sometimes report correlations between students' Scholastic Aptitude Test (SAT) scores and their first-year GPA. If the association is linear and the correlation is high, this justifies using the SAT to make admissions decisions, because a high correlation would indicate a strong association between SAT scores and academic performance. A positive correlation means that students who score above average on the SAT tend to get above-average grades. Conversely, those who score below average on the SAT tend to get below-average grades. Note that we're careful to say "tend to." Certainly, some students with low SAT scores do very well, and some with high SAT scores struggle to pass their classes. The correlation coefficient does not tell us about individual students; it tells us about the overall trend.

## More Context: Correlation Does Not Mean Causation!

Quite often, you'll hear someone use the correlation coefficient to support a claim of cause and effect. For example, one of the authors once read that a politician wanted to close liquor stores in a city because there was a positive correlation between the number of liquor stores in a neighborhood and the amount of crime.

As you learned in Chapter 1, we can't form cause-and-effect conclusions from observational studies. If your data came from an observational study, it doesn't matter how strong the correlation is. Even a correlation near 1 is not enough to conclude that changing one variable (closing down liquor stores) will lead to a change in the other variable (crime rate).

A positive correlation also exists between the number of blankets sold in Canada per week and the number of brush fires in Australia per week. Are brush fires in Australia caused by cold Canadians? Probably not. The correlation is likely to be the result of weather. When it is winter in Canada, people buy blankets. When winter is happening in Canada, summer is happening in Australia (which is located in the southern hemisphere), and summer is brush-fire season.

What, then, can we conclude from the fact that the number of liquor stores in a neighborhood is positively correlated with the crime rate in that neighborhood? Only that neighborhoods with a higher-than-average number of liquor stores typically (but not always) have a higher-than-average crime rate.

If you learn nothing else from this book, remember this: No matter how tempting, do *not* conclude that a cause-and-effect relationship between two variables exists just because there is a correlation, no matter how close to $+1$ or $-1$ that correlation might be!

 **KEY POINT**    Correlation does not imply causation.

## Finding the Correlation Coefficient

The correlation coefficient is best determined through the use of technology. We calculate a correlation coefficient by first converting each observation to a z-score, using the appropriate variable's mean and standard deviation. For example, to find the correlation coefficient that measures the strength of the linear relation between weight and height, we first convert each person's weight and height to z-scores. The next step is to multiply the observations' z-scores together. If both are positive or both negative—meaning that both z-scores are above average or both are below average—then the product is a positive number. In a strong positive association, most of these products are positive values. In a strong negative association, however, observations above

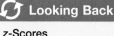

**Looking Back**

**z-Scores**
Recall that z-scores show how many standard deviations a measurement is from the mean. To find a z-score from a measurement, first subtract the mean and then divide the difference by the standard deviation.

average on one variable tend to be below average on the other variable. In this case, one z-score is negative and one positive, so the product is negative. Thus, in a strong negative association, most z-score products are negative.

To find the correlation coefficient, add the products of z-scores together and divide by $n - 1$ (where $n$ is the number of observed pairs in the sample). In mathematical terms, we get

$$\textbf{Formula 4.1: } r = \frac{\sum z_x z_y}{n - 1}$$

The following example illustrates how to use Formula 4.1 in a calculation.

 EXAMPLE **2** **Heights and Weights of Six Women**

Figure 4.10a shows the scatterplot for heights and weights of six women.

QUESTION  Using the data provided, find the correlation coefficient between heights (inches) and weights (pounds) for these six women.

| Heights | 61 | 62 | 63 | 64 | 66 | 68 |
|---------|-----|-----|-----|-----|-----|-----|
| Weights | 104 | 110 | 141 | 125 | 170 | 160 |

▶ **FIGURE 4.10a** Scatterplot showing heights and weights of six women.

SOLUTION  Before proceeding, we verify that the conditions hold. Figure 4.10a suggests that a straight line is an acceptable model; a straight line through the data might summarize the trend, although this is hard to see with so few points.

Next, we calculate the correlation coefficient. Ordinarily, we use technology to do this, and Figure 4.10b shows the output from StatCrunch, which gives us the value $r = 0.88093364$.

Tech

▶ **FIGURE 4.10b** StatCrunch, like all statistical software, lets you calculate the correlation between any two columns you choose.

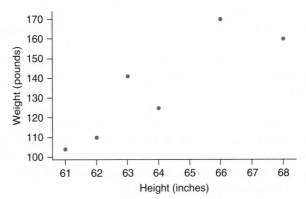

Because the sample size is small, we confirm this output using Formula 4.1. It is helpful to go through the steps of this calculation to better understand how the correlation coefficient measures linear relationships between variables.

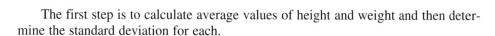

The first step is to calculate average values of height and weight and then determine the standard deviation for each.

$$\text{For the height: } \bar{x} = 64 \text{ and } s_x = 2.608$$
$$\text{For the weight: } \bar{y} = 135 \text{ and } s_y = 26.73$$

Next we convert all of the points to pairs of $z$-scores and multiply them together. For example, for the woman who is 68 inches tall and weighs 160 pounds,

$$z_x = \frac{x - \bar{x}}{s} = \frac{68 - 64}{2.608} = \frac{4}{2.608} = 1.53$$

$$z_y = \frac{y - \bar{y}}{s} = \frac{160 - 135}{26.73} = \frac{25}{26.73} = 0.94$$

The product is

$$z_x \times z_y = 1.53 \times 0.94 = 1.44$$

Note that this product is positive and shows up in the upper right quadrant in Figure 4.10c.

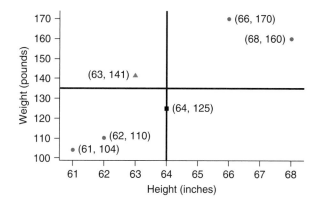

◀ **FIGURE 4.10c** The same scatterplot as in 4.10a, but with the plot divided into quadrants based on average height and weight. Points represented with blue circles contribute a positive value to the correlation coefficient (positive times positive is positive or negative times negative equals a positive). The red triangle represents an observation that contributes negatively (a negative $z$-score times a positive $z$-score is negative), and the black square contributes nothing because one of the $z$-scores is 0.

Figure 4.10c can help you visualize the rest of the process. The two blue circles in the upper right portion represent observations that are above average in both variables, so both $z$-scores are positive. The two blue circles in the lower left region represent observations that are below average in both variables; the products of the two negative $z$-scores are positive, so they add to the correlation. The red triangle has a positive $z$-score for weight (it is above average) but a negative $z$-score for height, so the product is negative. The black square is a point that makes no contribution to the correlation coefficient. This person is of average height, so her $z$-score for height is 0.

The correlation between height and weight for these six women comes out to be about 0.881.

CONCLUSION The correlation coefficient for the linear association of weights and heights of these six women is $r = 0.881$. Thus, there is a strong positive correlation between height and weight for these women. Taller women tend to weigh more.

TRY THIS! Exercise 4.19a

## Understanding the Correlation Coefficient

The correlation coefficient has a few features you should know about when interpreting a value of $r$ or deciding whether you should compute the value.

- Changing the order of the variables does not change *r*. Note that in the equation for *r* it doesn't matter which variable is called *x* and which is called *y*. In practice, this means that if the correlation between life expectancy for men and women is 0.977, then the correlation between life expectancy for women and men is also 0.977. This makes sense because the correlation measures the strength of the linear relationship between *x* and *y*, and that strength will be the same no matter which variable gets plotted on the horizontal axis and which on the vertical.

Figure 4.11a and b have the same correlation; we've just swapped axes.

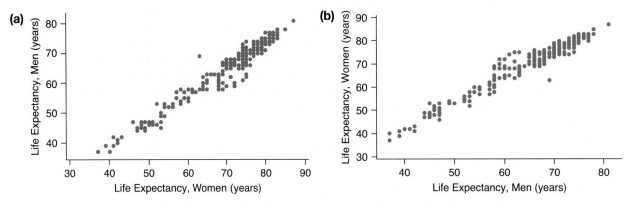

▲ **FIGURE 4.11** Scatterplots showing the relationship between men's and women's life expectancy for various countries. **(a)** Women's life expectancy is plotted on the *x*-axis. **(b)** Men's life expectancy is plotted on the *x*-axis. (Sources: http://www.overpopulation.com and Fathom™ Sample Documents)

- Adding or multiplying by a constant does not affect *r*. The correlation between the heights and weights of the six women in Example 2 was 0.881. What would happen if all six women in the sample had been asked to wear 3-inch platform heels when their heights were measured? Everyone would have been 3 inches taller. Would this have changed the value of *r*? Intuitively, you should sense that it wouldn't. Figure 4.12a shows a scatterplot of the original data, and Figure 4.12b shows the data with the women in 3-inch heels.

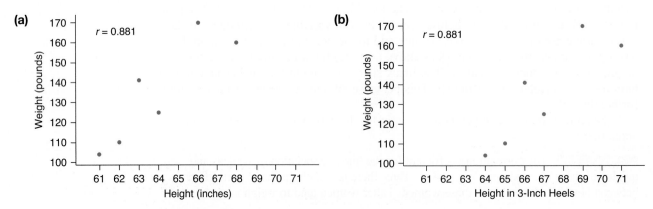

▲ **FIGURE 4.12 (a)** A repeat of the scatterplot of height and weight for six women. **(b)** The same women in 3-inch heels. The correlation remains the same.

We haven't changed the strength of the relationship. All we've done is shift the points on the scatterplot 3 inches to the right. But shifting the points doesn't change the relationship between height and weight. We can verify that the correlation is unchanged by looking at the formula. The heights will have the same *z*-scores both before and after the women put on the shoes; since everyone "grows" by the same amount, everyone is still the same number of standard deviations away from the average, which also "grows" by 3 inches. As another example, if science found a way to add 5 years to the life expectancy of men in all countries in the world, the correlation between life expectancies for men and women would still be the same.

More generally, we can add a constant (a fixed value) to all of the values of one variable, or of both variables, and not affect the correlation coefficient.

For the very same reason, we can multiply either or both variables by positive constants without changing $r$. For example, to convert the women's heights from inches to feet, we multiply their heights by 1/12. Doing this does not change how strong the association is; it merely changes the units we're using to measure height. Because the strength of the association does not change, the correlation coefficient does not change.

- The correlation coefficient is unitless. Height is measured in inches and weight in pounds, but $r$ has no units because the $z$-scores have no units. This means that we will get the same value for correlation whether we measure height in inches, meters, or fathoms.

- Linear, linear, linear. We've said it before, but we'll say it again: We're talking only about linear relationships here. The correlation can be misleading if you do not have a linear relationship. Figures 4.13a through d illustrate the fact that different nonlinear patterns can have the same correlation. All of these graphs have $r = 0.817$, but the graphs have very different shapes. The take-home message is that the correlation alone does not tell us much about the shape of a graph. We must also know that the relationship is linear to make sense of the correlation.

Remember: *Always* make a graph of your data. If the trend is nonlinear, the correlation (and, as you'll see in the next section, other statistics) can be very misleading.

> **! Caution**
>
> **Correlation Coefficient and Linearity**
> A value of $r$ close to 1 or $-1$ does *not* tell you that the relationship is linear. You must check visually; otherwise, your interpretation of the correlation coefficient might be wrong.

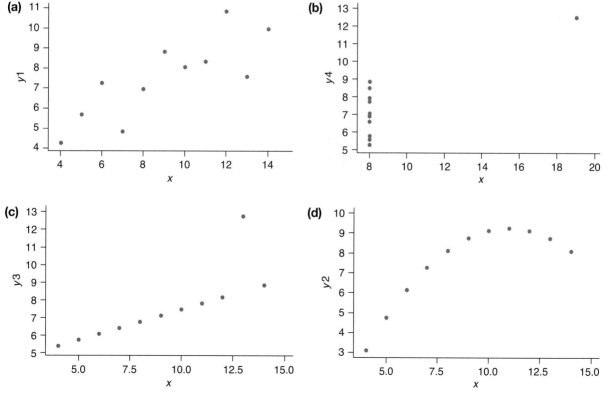

▲ **FIGURE 4.13 (a–d)** Four scatterplots with the same correlation of 0.817 have very different shapes. The correlation coefficient is meaningful only if the trend is linear. (Source: Anscombe, F. Anscombe's Quartet)

> **KEY POINT**  The correlation coefficient does not tell you whether an association is linear. However, if you already know that the association is linear, then the correlation coefficient tells you how strong the association is.

## SNAPSHOT  THE CORRELATION COEFFICIENT

| | |
|---|---|
| **WHAT IS IT?** ▶ | Correlation coefficient. |
| **WHAT DOES IT DO?** ▶ | Measures the strength of a linear association. |
| **HOW DOES IT DO IT?** ▶ | By comparing $z$-scores of the two variables. The products of the two $z$-scores for each point are averaged. |
| **HOW IS IT USED?** ▶ | The sign tells us whether the trend is positive ($+$) or negative ($-$). The value tells us the strength. If the value is close to 1 or $-1$, then the points are tightly clustered about a line; if the value is close to 0, then there is no linear association.<br><br>*Note:* The correlation coefficient can be interpreted only with linear associations. |

## SECTION 4.3

# Modeling Linear Trends

How much more do people tend to weigh for each additional inch in height? How much value do cars lose each year as they age? Are home run hitters good for their teams? Can we predict how much space a book will take on a bookshelf just by knowing how many pages are in the book? It's not enough to remark that a trend exists. To make a prediction based on data, we need to measure the trend and the strength of the trend.

To measure the trend, we're going to perform a bit of statistical sleight of hand. Rather than interpret the data themselves, we will substitute a model of the data and interpret the model. The model consists of an equation and a set of conditions that describe when the model will be appropriate. Ideally, this equation is a very concise and accurate description of the data; if so, then the model is a good fit. When the model is a good fit to the data, then any understanding we gain about the model accurately applies to our understanding of the real world. If the model is a bad fit, however, then our understanding of real situations might be seriously flawed.

## The Regression Line

The **regression line** is a tool for making predictions about future observed values. It also provides us with a useful way of summarizing a linear relationship. Recall from Chapter 3 that we could summarize a sample distribution with a mean and a standard deviation. The regression line works the same way: It reduces a linear relationship to its bare essentials and allows us to analyze a relationship without being distracted by small details.

## Review: Equation of a Line

The regression line is given by an equation for a straight line. Recall from algebra that equations for straight lines contain an **intercept** and a **slope**. The equation for a straight line is

$$y = mx + b$$

The letter $m$ represents the slope, which tells how steep the line is, and the letter $b$ represents the $y$-intercept, which is the value of $y$ when $x = 0$.

Statisticians write the equation of a line slightly differently and put the intercept first; they use the letter $a$ for the intercept and $b$ for the slope and write

$$y = a + bx$$

We often use the names of variables in place of $x$ and $y$ to emphasize that the regression line is a model about two real-world variables. We will sometimes write the word *predicted* in front of the $y$-variable to emphasize that the line consists of predictions for the $y$-variable, not actual values. A few examples should make this clear.

## Visualizing the Regression Line

Can we know how wide a book is on the basis of the number of pages in the book? A student took a random sample of books from his shelf, measured the width of the spine (in millimeters, mm), and recorded the number of pages. Figure 4.14 illustrates how the regression line captures the basic trend of a linear association between width of the book and the number of pages for this sample. The equation for this line is

$$\text{Predicted Width} = 6.22 + 0.0366 \text{ Pages}$$

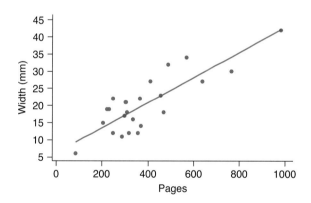

◀ **FIGURE 4.14** The regression line summarizes the relationship between the width of the book and the number of pages for a small sample of books. (Source: Onaga, E. 2005, UCLA, Department of Statistics)

In baseball, two numbers used to measure how good a batter is are the number of runs batted in (RBI) and the number of home runs. (A home run occurs when the batter rounds all three bases with one hit and scores a run. An RBI occurs when a player is already on base, and a batter hits the ball far enough for the on-base runner to score a run.) Some baseball fans believe that players who hit a lot of home runs might be exciting to watch but are not that good for the team. (Some believe that home run hitters tend to strike out more often; this takes away scoring opportunities for the team.) Figure 4.15 on the next page shows the relationship between the number of home runs and RBIs for the best home run hitters in the major leagues in the 2008 season. The association seems fairly linear, and the regression line can be used to predict how many RBIs a player will score, given the number of home runs hit. The data suggest that players who score a large number of home runs also tend to score a large number of points through RBIs.

$$\text{Predicted RBIs} = 30.46 + 2.16 \text{ Home Runs}$$

▶ **FIGURE 4.15** The regression line summarizes the relationship between RBIs and home runs for the top home run hitters in the 2008 major league baseball season. (Source: www.baseball1.com)

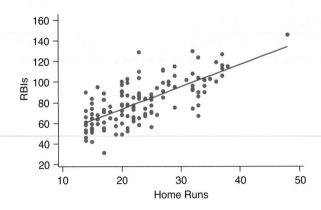

## Regression in Context

Suppose you have a 10-year-old car and want to estimate how much it is worth. One of the more important uses of the regression line is to make predictions about what *y*-values can be obtained for particular *x*-values. Figure 4.16 suggests that the relationship between age and value is linear, and the regression line that summarizes this relationship is

$$\text{Predicted Value} = 21375 - 1215\ \text{Age}$$

We can use this equation to predict approximately how much a 10-year-old car is worth:

$$\begin{aligned}
\text{Predicted Value} &= 21375 - 1215 \times 10 \\
&= 21375 - 12150 \\
&= 9225
\end{aligned}$$

▶ **FIGURE 4.16** The regression line summarizes the relationship between the value of a car, according to the Kelley Blue Book, and the car's age for a small sample of students' cars. (Source: C. Ryan 2006)

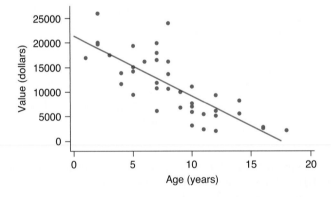

The regression line predicts that a 10-year-old car will be valued at about $9225. As we know, many factors other than age affect the value of a car, and perhaps with more information we might make a better prediction. However, if the only thing we know about a car is its age, this may be the best prediction we can get. It is also important to keep in mind that this sample comes from one particular group of students in one particular statistics class and is not representative of all used cars on the market in the United States.

Using the regression line to make predictions requires certain assumptions. We'll go into more detail about these assumptions later, but for now, just use common sense. This predicted value of $9225 is useful only if our data are valid. For instance, if all the cars in our data set are Toyotas, and our 10-year-old car is a Chevrolet, then the prediction is probably not useful.

## EXAMPLE 3 Book Width

A college instructor with far too many books on his shelf is wondering whether he has room for one more. He has about 20 mm of space left on his shelf, and he can tell from the online bookstore that the book he wants has 598 pages. The regression line is

$$\text{Predicted Width} = 6.22 + 0.0366\,\text{Pages}$$

**QUESTION** Will the book fit on his shelf?

**SOLUTION** Assuming that the data used to fit this regression line are representative of all books, we would predict the width of the book corresponding to 598 pages to be

$$\text{Predicted Width} = 6.22 + 0.0366 \times 598$$
$$= 6.22 + 21.8868$$
$$= 28.1068 \text{ mm}$$

**CONCLUSION** The book is predicted to be 28 mm wide. Even though the actual book width is likely to differ somewhat from 28 mm, it seems that the book will probably not fit on the shelf.

**TRY THIS!** Exercise 4.21

Common sense tells us that not all books with 598 pages are exactly 28 mm wide. There is a lot of variation in the width of a book for a given number of pages.

## Finding the Regression Line

In almost every case, we'll use technology to find the regression line. However, it is important to know how the technology works, and to be able to calculate the equation when we don't have access to the full data set.

To understand how technology finds the regression line, imagine trying to draw a line through the scatterplots in Figures 4.14 through 4.16 to best capture the linear trend. We could have drawn almost any line, and some of them would have looked like pretty good summaries of the trend. What makes the regression line special? How do we find the intercept and slope of the regression line?

The regression line is chosen because it is the line that comes closest to most of the points. More precisely, the square of the vertical distances between the points and the line, on average, is bigger for any other line we might draw than for the regression

> **Details**
>
> **Least Squares**
> The regression line is also called the least squares line because it is chosen so that the sum of the squares of the differences between the observed $y$-value, $y$, and the value predicted by the line, $\hat{y}$, is as small as possible. Mathematically, this means that the slope and intercept are chosen so that $\Sigma(y - \hat{y})^2$ is as small as possible.

line. Figure 4.17 shows these vertical distances with black lines. The regression line is sometimes called the "best fit" line because, in this sense, it provides the best fit to the data.

▶ **FIGURE 4.17** The "best fit" regression line, showing the vertical distance between each observation and the line. For any other line, the average of the squared distances is larger.

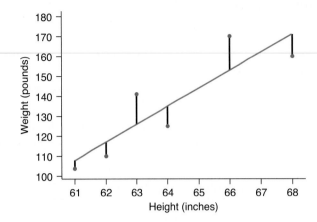

To find this best fit line, we need to find the slope and the intercept. The slope, *b*, of the regression line is the ratio of the standard deviations of the two variables, multiplied by the correlation coefficient:

**Formula 4.2a:** The slope, $b = r \dfrac{s_y}{s_x}$

Once we have found the slope, we can find the intercept. Finding the intercept, *a*, requires that we first find the means of the variables, $\bar{x}$ and $\bar{y}$:

**Formula 4.2b:** The intercept, $a = \bar{y} - b\bar{x}$

Now put these quantities into the equation of a line, and the regression line is given by

**Formula 4.2c:** The regression line, Predicted $y = a + bx$

## ✳ EXAMPLE 4 SAT Scores and GPAs

A college in the northeastern United States used SAT scores to help decide which applicants to admit. To determine whether the SAT was useful in predicting success, the college examined the relationship between the SAT scores and first-year GPAs of admitted students. Figure 4.18a shows a scatterplot of the math SAT scores and first-year GPAs for a random sample of 200 students. The scatterplot suggests a weak, positive linear association: Students with higher math SAT scores tend to get higher first-year GPAs. The average math SAT score of this sample was 649.5 with standard deviation 66.3. The average GPA was 2.63 with standard deviation 0.58. The correlation between GPA and math SAT score was 0.19.

 QUESTIONS

a. Find the equation of the regression line that best summarizes this association. Note that the *x*-variable is the math SAT score and the *y*-variable is the GPA.

b. Using the equation, find the predicted GPA score for a person in this group with a math SAT score of 650.

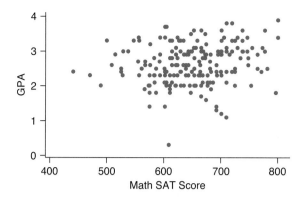

◀ **FIGURE 4.18a** Students with higher math SAT scores tend to have higher first-year GPAs, but this linear association is fairly weak. (Source: ETS validation study of an unnamed college and Fathom™ Sample Documents, 2007, Key Curriculum Press)

**SOLUTIONS**

a. With access to the full data set on the text's companion CD-ROM, we can use technology to find (and plot) the regression line. Still, when summary statistics are provided (means and standard deviations of the two variables as well as their correlation), it is not time-consuming to use Formula 4.2 to find the regression line.

Figure 4.18b shows StatCrunch output for the regression line. (StatCrunch provides quite a bit more information than we will use in this chapter.)

According to technology, the equation of the regression line is

$$\text{Predicted GPA} = 1.53 + 0.0017 \text{ Math}$$

Tech

◀ **FIGURE 4.18b** StatCrunch output for the regression line to predict GPA from math SAT score. The structure of the output is typical for many statistical software packages.

**Simple Linear Regression** Java Applet Window

Options

**Simple linear regression results:**
Dependent Variable: gpa
Independent Variable: math
gpa = 1.5264336 + 0.001699023 math
Sample size: 200
R (correlation coefficient) = 0.1942
R-sq = 0.037730377
Estimate of error standard deviation: 0.57071334

**Parameter estimates:**

| Parameter | Estimate | Std. Err. | DF | T-Stat | P-Value |
|-----------|----------|-----------|-----|--------|---------|
| Intercept | 1.5264336 | 0.39811763 | 198 | 3.8341272 | 0.0002 |
| Slope | 0.001699023 | 6.097748E-4 | 198 | 2.786312 | 0.0058 |

**Analysis of variance table for regression model:**

| Source | DF | SS | MS | F-stat | P-value |
|--------|-----|-----------|-----------|-----------|---------|
| Model | 1 | 2.5286899 | 2.5286899 | 7.7635355 | 0.0058 |
| Error | 198 | 64.49131 | 0.3257137 | | |

<- Back    Next ->

We now check this calculation by hand, using Formula 4.2.

We are given that

For math SAT scores: $\bar{x} = 649.5$, $s_x = 66.3$

For GPAs: $\bar{y} = 2.63$, $s_y = 0.580$

and $r = 0.194$

First we must find the slope:

$$b = r\frac{s_y}{s_x} = 0.194 \times \frac{0.580}{66.3} = 0.001697$$

Now we can use the slope to find the intercept:

$$a = \bar{y} - b\bar{x} = 2.63 - 0.001697 \times 649.5 = 2.63 - 1.102202 = 1.53$$

Rounding off:

Predicted GPA $= 1.53 + 0.0017$ math

b.  Predicted GPA $= 1.53 + 0.0017$ math

$$= 1.53 + 0.0017 \times 650$$
$$= 1.53 + 1.105$$
$$= 2.635$$

CONCLUSION

We would expect someone with a math SAT score of 650 to have a GPA of about 2.64.

TRY THIS! Exercise 4.23

Different software packages present the intercept and slope differently. Therefore, you need to learn how to read the output of the software you are using. Example 5 shows output from several packages.

## EXAMPLE 5 Technology Output for Regression

Figure 4.19 shows outputs from Minitab, StatCrunch, the TI-83/84, and Excel for finding the regression equation in Example 4 for GPA and math SAT scores.

QUESTION  For each software package, explain how to find the equation of the regression line from the given output.

CONCLUSION  Figure 4.19a: Minitab gives us a simple equation directly: GPA $= 1.53 + 0.00170$ Math.
   However, the more statistically correct format would be

Predicted GPA $= 1.53 + 0.00170$ Math

▶ **FIGURE 4.19a** Minitab output.

---

**Regression Analysis: GPA versus Math**
```
The regression equation is
GPA = 1.53 + 0.00170 Math
```

---

Figure 4.19b: StatCrunch gives the equation directly near the top, but it also lists the intercept and slope separately in the table near the bottom.

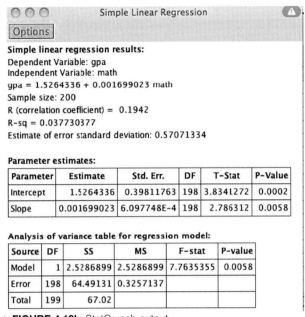

Simple Linear Regression    ⚠

**Options**

**Simple linear regression results:**
Dependent Variable: gpa
Independent Variable: math
gpa = 1.5264336 + 0.001699023 math
Sample size: 200
R (correlation coefficient) =  0.1942
R-sq = 0.037730377
Estimate of error standard deviation: 0.57071334

**Parameter estimates:**

| Parameter | Estimate | Std. Err. | DF | T-Stat | P-Value |
|-----------|----------|-----------|-----|--------|---------|
| Intercept | 1.5264336 | 0.39811763 | 198 | 3.8341272 | 0.0002 |
| Slope | 0.001699023 | 6.097748E-4 | 198 | 2.786312 | 0.0058 |

**Analysis of variance table for regression model:**

| Source | DF | SS | MS | F-stat | P-value |
|--------|-----|-----|-----|--------|---------|
| Model | 1 | 2.5286899 | 2.5286899 | 7.7635355 | 0.0058 |
| Error | 198 | 64.49131 | 0.3257137 | | |
| Total | 199 | 67.02 | | | |

▲ **FIGURE 4.19b** StatCrunch output.

```
LinReg
  y=a+bx
  a=1.526433632
  b=.0016990229
  r²=.0377303763
  r=.1942430855
```

▲ **FIGURE 4.19c** TI-83/84 output.

| | Coefficients | Standard Error | t Stat | P-valve | Lower 95% | Upper 95% | Lower 95.0% | Upper 95.0% |
|---|---|---|---|---|---|---|---|---|
| Intercept | 1.526434 | 0.398118 | 3.834127 | 0.000169 | 0.741339 | 2.311528 | 0.741339 | 2.311528 |
| Variable 1 | 0.001699 | 0.00061 | 2.786312 | 0.00585 | 0.000497 | 0.002902 | 0.000497 | 0.002902 |

▲ **FIGURE 4.19d** Excel output.

Figure 4.19c: TI-83/84 gives us the coefficients to put together. The "a" value is the intercept, and the "b" value is the slope. If the diagnostics are on (use the CATALOG button), the TI-83/84 also gives the correlation.

Figure 4.19d: Excel shows the coefficients in the column labeled "Coefficients." The intercept is in the first row, labeled "Intercept," and the slope is in the row labeled "Variable 1."

**TRY THIS!** Exercise 4.27

## Interpreting the Regression Line

An important use of the regression line is to make predictions about the value of $y$ that we will see for a given value of $x$. However, the regression line provides more information than just a predicted $y$-value. The regression line can also tell us about the rate of change of the mean value of $y$ with respect to $x$ and can help us understand the underlying theory behind cause-and-effect relationships.

### Choosing $x$ and $y$: Order Matters

In Section 4.2, you saw that the correlation coefficient is the same no matter which variable you choose for $x$ and which you choose for $y$. With regression, however, order matters.

Consider the collection of data about book widths. We used it earlier to predict the width of a book, given the number of pages. The equation of the regression line for this prediction problem (shown in Figure 4.20a) is

$$\text{Predicted Width} = 6.22 + 0.0366\,\text{Pages}$$

But what if we instead wanted to predict how many pages there are in a book on the basis of the width of the book?

To do this, we would switch the order of the variables, and use *Pages* as our *y*-variable and *Width* as our *x*-variable. Then the slope is calculated to be 19.6 (Figure 4.20b).

It is tempting to think that because we are flipping the graph over when we switch *x* and *y*, we can just flip the slope over to get the new slope. If this were true, then we could find the new slope simply by calculating 1/(old slope). However, that approach doesn't work. That would give us a slope of

$$\frac{1}{0.0366} = 27.3, \text{ which is not the same as the correct value of 19.6.}$$

How, then, do we know which variable goes where?

We use the variable plotted on the horizontal axis to make predictions about the variable plotted on the vertical axis. For this reason, the *x*-variable is called the **explanatory variable**, the **predictor variable**, or the **independent variable**. The *y*-variable is called the **response variable**, the **predicted variable**, or the **dependent variable**. These names reflect the different roles played by the *x*- and *y*-variables in regression. Which variable is which depends on what the regression line will be used to predict.

You'll see many pairs of terms used for the *x*- and *y*-variables in regression. Some are shown in Table 4.1.

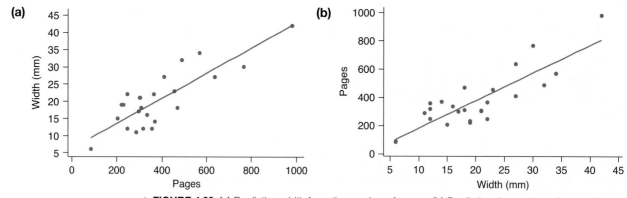

▲ **FIGURE 4.20** (a) Predicting width from the number of pages. (b) Predicting the number of pages from width.

▶ **TABLE 4.1** Terms used for the *x*- and *y*-variables.

| *x*-Variable | *y*-Variable |
| --- | --- |
| Predictor variable | Predicted variable |
| Explanatory variable | Response variable |
| Independent variable | Dependent variable |

## EXAMPLE 6 Bedridden

It is hard to measure the height of people who are bedridden, and for many medical reasons it is often important to know a bedridden patient's height. However, it is not so difficult to measure the length of the ulna (the bone that runs from the elbow to the wrist). Data collected on non-bedridden people show a strong linear association between ulnar length and height.

**QUESTION** When making a scatterplot to predict height from ulnar length, which variable should be plotted on the *x*-axis and which on the *y*-axis?

**CONCLUSION** We are measuring ulnar length to predict a person's height. Therefore, ulnar length is the predictor (independent variable) and is plotted on the *x*-axis, while height is the response (dependent variable) and is plotted on the *y*-axis.

**TRY THIS!** Exercise 4.35

## The Regression Line Is a Line of Averages

Figure 4.21 shows a histogram of the weights of a sample of 507 active people. What is the typical weight?

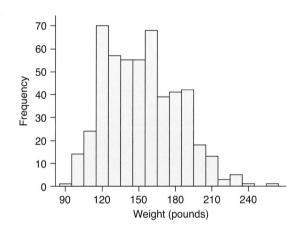

◀ **FIGURE 4.21** A histogram of the weights of 507 active people.

One way of answering this question is to calculate the mean of the sample. The distribution of weights is a little right-skewed but not terribly so, and so the mean will probably give us a good idea of what is typical. The average weight of this group is 152.5 pounds.

Now, we know that an association exists between height and weight and that shorter people tend to weigh less. If we know someone's height, then, what weight would we guess for that person? Surely not the average weight of the whole group! We can do better than that. For instance, what's the typical weight of someone 66 inches tall? To answer this, it makes sense to look only at the weights of those people in our sample who are about 66 inches tall. To make sure we have enough 66-inch-tall people in our sample, let's include everyone who is *approximately* 66 inches tall. So let's look at a slice of people whose height is between 65.5 inches and 66.5 inches. We come up with 47 people. Some of their weights are

130, 150, 142, 149, 118, 189 . . .

Figure 4.22a shows this slice, which is centered at 66 inches.

The mean of these numbers is 141.5 pounds. We put a special symbol on the plot to record this point—a triangle at the point (66, 141.5), shown in Figure 4.22b.

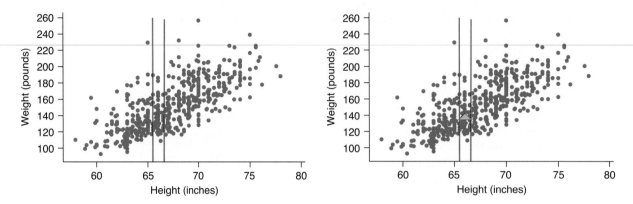

▲ **FIGURE 4.22a** Heights and weights with a slice at 66 inches.

▲ **FIGURE 4.22b** Heights and weights with the average at 66 inches, which is 141.5 pounds.

The reason for marking this point is that if we wanted to predict the weights of those who were 66 inches tall, one good answer would be 141.5 pounds.

What if we wanted to predict the weight of someone who is 70 inches tall? We could take a slice of the sample and look at those people who are between 69.5 inches and 70.5 inches tall. Typically, they're heavier than the people who are about 66 inches tall. Here are some of their weights:

$$189, 197, 151, 207, 157 \ldots$$

Their mean weight is 173.0 pounds. Let's put another special triangle symbol at (70, 173.0) to record this. We can continue in this fashion, and Figure 4.22c shows where the mean weights are for a few more heights.

▶ **FIGURE 4.22c** Heights and weights with more means marked.

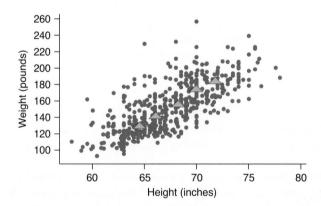

Note that the means fall (nearly) on a straight line. What could be the equation of this line? Figure 4.22d shows the regression line superimposed on the scatterplot with the means. They're nearly identical.

In theory, these means should lie exactly on the regression line. However, when working with real data, we often find that the theory doesn't always fit perfectly.

The series of graphs in Figure 4.22 illustrates a fundamental feature of the regression line: It is a line of means. You plug in a value for *x*, and the regression line "predicts" the mean *y*-value for everyone in the population with that *x*-value.

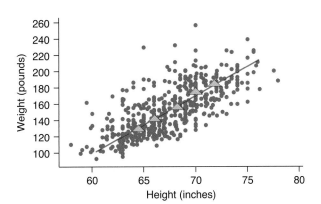

◀ **FIGURE 4.22d** Heights and weights with means and a straight line.

**KEY POINT** When the trend is linear, the regression line connects the points that represent the mean value of *y* for each value of *x*.

# EXAMPLE 7 Funny Car Race Finishing Times

A statistics student collected data on National Hot Rod Association driver John Force. The data consist of a sample of Funny Car races. The drivers race in several different trials and change speed frequently. One question of interest is whether the fastest speed (in miles per hour, mph) driven during a trial is associated with the time to finish that trial (in seconds). If a driver's fastest speed cannot be maintained and the driver is forced to slow down, then it might be best to avoid going that fast. An examination of the scatterplot Fig. 4.23 of finishing time (response variable) versus fastest speed (predictor) for a sample of trials for John Force shows a reasonably linear association.

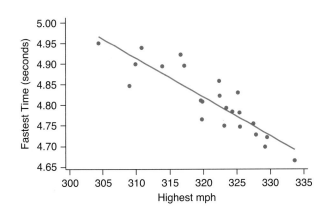

**◀ FIGURE 4.23** A scatterplot for the funny car race. (Source: Hettinga, J. 2005, UCLA, Department of Statistics)

The regression line is

Predicted Time = 7.84 − 0.0094 (Highest mph)

**QUESTION** What is the predicted mean finishing time when the driver's fastest speed is 318 mph?

**SOLUTION** We predict by estimating the mean finishing time when the speed is 318 mph. To find this, plug 318 into the regression line:

Predicted Time = 7.84 − 0.0094 × 318 = 4.85 seconds

**CONCLUSION**  We predict that the finishing time will be 4.85 seconds.

**TRY THIS!**  Exercise 4.37

## Interpreting the Slope

The slope tells us how to compare the mean *y*-values for objects that are 1 unit apart on the *x*-variable. For example, how different is the mean finishing time when the driver's fastest speed is 1 mph faster? The slope in Example 7 tells us that the means differ by a slim 0.0094 second. (Of course, races can be won or lost by such small differences.) What if the fastest time is 10 mph faster? Then the mean finishing time is $10 \times 0.0094 = 0.094$ second faster. We can see that in a typical race, if the driver wants to lose 1 full second from racing time, he or she must go over 100 mph faster, which is clearly not possible.

We should pay attention to whether the slope is 0 or very close to 0. When the slope is 0, we say that no (linear) relationship exists between *x* and *y*. The reason is that a 0 slope means that no matter what value of *x* you consider, the predicted value of *y* is always the same. The slope is 0 whenever the correlation coefficient is 0.

For example, the slope for runners' ages and their marathon times (see Figure 4.3a) is very close to 0 (0.000014 second per year of age).

## SNAPSHOT  THE SLOPE

| | |
|---|---|
| **WHAT IS IT?** ▶ | The slope of a regression line. |
| **WHAT DOES IT DO?** ▶ | Tells us how different the mean *y*-value is for observations that are 1 unit apart on the *x*-variable. |
| **HOW DOES IT DO IT?** ▶ | The regression line tells us the average *y*-values for any value of *x*. The slope tells us how these average *y*-values differ for different values of *x*. |
| **HOW IS IT USED?** ▶ | To measure the linear association between two variables. |

## EXAMPLE 8  Math and Critical Reading SAT Scores

A regression analysis to study the relationship between SAT math and SAT critical reading scores resulted in this regression model:

$$\text{Predicted Critical Reading} = 398.81 + 0.3030 \text{ Math}$$

This model was based on a sample of 200 students.

The scatterplot in Figure 4.24 shows a linear relationship with a weak positive trend.

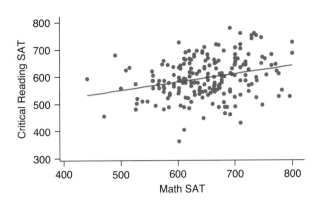

◀ **FIGURE 4.24** The regression line shows a weak positive trend between critical reading SAT score and math SAT score.

**QUESTION** Interpret the slope of this regression line.

**SOLUTION** The slope tells us that students who score 10 points higher on the math SAT had an average critical reading SAT score that was

$$10 \times 0.3030 = 3.03 \text{ points higher}$$

**CONCLUSION** Students who score 10 points higher on the math SAT score, on average, 3.03 points higher on critical reading.

**TRY THIS!** Exercise 4.55a

## Interpreting the Intercept

The intercept tells us the predicted mean *y*-value when the *x*-value is 0. Quite often, this is not terribly helpful. Sometimes it is ridiculous. For example, the regression line to predict weight, given someone's height, tells us that if a person is 0 inches tall, then his or her predicted weight is negative 231.5 pounds!

Before interpreting the intercept, ask yourself whether it makes sense to talk about the *x*-variable taking on a 0 value. For the SAT data, you might think it makes sense to talk about getting a 0 on the math SAT. However, the lowest possible score on the SAT math portion is 200, so it is not possible to get a score of 0. (One lesson statisticians learn early is that you must know something about the data you analyze—knowing only the numbers is not enough!)

## EXAMPLE 9 Predicting Funny Car Races

What can we learn about the relationship between the fastest speed of a funny car driver and the finishing time in a race by looking at the intercept and slope of the regression line? The regression model is

$$\text{Predicted Time} = 7.84 - 0.0094 \text{ (Highest mph)}$$

**QUESTION** Interpret the intercept and slope.

**CONCLUSION** The intercept is not meaningful. It tells us that if the fastest the car moves is 0 mph, the car will finish the race in 7.84 seconds. Of course, we know that if the car goes 0 mph, it will never finish the race!

The slope is meaningful. The negative sign indicates that there is a negative trend: The faster the maximum speed during the race, the shorter, on average, the driver's time. It also shows that each additional mile per hour of maximum speed shaves off, on average, 0.0094 second of time.

**TRY THIS!** Exercise 4.59

## SNAPSHOT THE INTERCEPT

| | |
|---|---|
| **WHAT IS IT?** ▶ | The intercept of a regression line. |
| **WHAT DOES IT DO?** ▶ | Tells us the average $y$-value for all observations that have a zero $x$-value. |
| **HOW DOES IT DO IT?** ▶ | The regression line is the best fit to a linear association, and the intercept is the best prediction of the $y$-value when the $x$-value is 0. |
| **HOW IS IT USED?** ▶ | It is not always useful. Often, it doesn't make sense for the $x$-variable to assume the value 0. |

## EXAMPLE 10 Age and Value of Cars

Figure 4.25 shows the relationship between age and Kelley Blue Book values for a sample of cars. These cars were owned by students in one of the author's statistics class.

▶ **FIGURE 4.25** The age of a car and its value for a chosen sample.

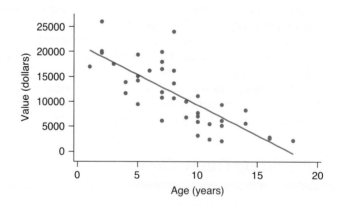

The regression line is

$$\text{Predicted Value} = 21375 - 1215 \text{ Age}$$

**QUESTION** Interpret the slope and intercept.

**CONCLUSION** The intercept estimates that the average value of a new car (0 years old) owned by students in this class is $21,375. The slope tells us that, on average, cars lost $1215 in value each year.

**TRY THIS!** Exercise 4.61

# Evaluating the Linear Model

Regression is a very powerful tool for understanding linear relationships between numerical variables. However, we need to be aware of several potential interpretation pitfalls so that we can avoid them. We will also discuss methods for determining just how well the regression model fits the data.

## Pitfalls to Avoid

You can avoid most pitfalls by simply making a graph of your data and examining it closely before you begin interpreting your linear model. This section will offer some advice for sidestepping a few subtle complications that might arise.

### Don't Fit Linear Models to Nonlinear Associations

Regression models are useful only for linear associations. If the association is not linear, a regression model can be misleading and deceiving. For this reason, before you fit a regression model, you should always make a scatterplot to verify that the association seems linear.

Figure 4.26 shows an example of a bad regression model. The association between mortality rates for several industrialized countries (deaths per 1000 people) and wine consumption (liters per person per year) is nonlinear. The regression model is

$$\text{Predicted Mortality} = 7.69 - 0.0761 \,(\text{Wine Consumption})$$

but it provides a poor fit. The regression model suggests that countries with middle values of wine consumption should have higher mortality rates than they actually do.

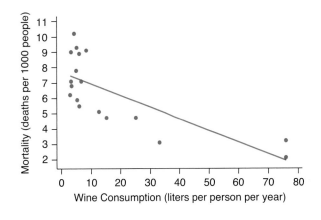

◀ **FIGURE 4.26** The straight-line regression model is a poor fit to this nonlinear relationship. (Source: Leger et al. 1979)

### Correlation Is Not Causation

One important goal of science and business is to discover relationships between variables such that if you make a change to one variable, you know the other variable will change in a reliably predictable way. This is what is meant by "*x* causes *y*": Make a change in *x,* and a change in *y* will usually follow. For example, the distance it takes to stop your car depends on how fast you were traveling when you first applied the brakes (among other things); the amount of memory an mp3 file takes on your hard drive depends on the length of the song; and the size of your phone bill depends on how many minutes you talked. In these cases, a strong causal relationship exists between two variables, and if you were to collect data and make a scatterplot, you would see an association between the variables.

In statistics, however, we are often faced with the reverse situation, in which we see an association between two variables and wonder whether there is a cause-and-effect relationship. The correlation coefficient for the association could be quite strong, but as we saw earlier, correlation does not mean cause and effect. A strong correlation or a good-fitting regression line is not evidence of a cause-and-effect relationship.

> **KEY POINT**    An association between two variables is not sufficient evidence to conclude that a cause-and-effect relationship exists between the variables, no matter how strong the correlation or how well the regression line fits the data.

Be particularly careful about drawing cause-and-effect conclusions when interpreting the slope of a regression line. For example, for the SAT data,

$$\text{Predicted Critical Reading} = 398.81 + 0.3030 \text{ Math}$$

Even if this regression line fits the association very well, it does not give us sufficient evidence to conclude that if you improve your math score by 10 points, your critical reading score will go up by 3.03 points. As you learned in Chapter 1, because these data were not collected from a controlled experiment, the presence of confounding factors could prevent you from making a causal interpretation.

Beware of the algebra trap. In algebra, you were taught to interpret the slope to mean that "as $x$ increases by 1 unit, $y$ goes up by $b$ units." However, quite often with data, the phrase "as $x$ increases" doesn't make sense. When looking at the height and weight data, where $x$ is height and $y$ is weight, to say "$x$ increases" means that people are growing taller! This is not accurate. It is much more accurate to interpret the slope as making comparisons between groups. For example, when comparing people of a certain height with those who are 1 inch taller, you can see that the taller individuals tend to weigh, on average, $b$ pounds more.

When can we conclude that an association between two variables means a cause-and-effect relationship is present? Strictly speaking, never from an observational study and only when the data were collected from a controlled experiment. (Even in a controlled experiment, care must be taken that the experiment was designed correctly.) However, for many important questions, controlled experiments are not possible. In these cases, we can sometimes make conclusions about causality after a number of observational studies have been collected and examined, and if there is a strong theoretical explanation for why the variables should be related in a cause-and-effect fashion. For instance, it took many years of observational studies to conclude that smoking causes lung cancer, including studies that compared twins—one twin who smoked and one who did not—and numerous controlled experiments on lab animals.

## Beware of Outliers

Recall that when calculating sample means, we were warned that outliers can have a big effect. Because the regression line is a line of means, you might think that outliers would have a big effect on the regression line. And you'd be right. You should always check a scatterplot before performing a regression analysis to be sure there are no outliers.

The graphs in Figure 4.27 illustrate this effect. Both graphs in Figure 4.27 show crime rates (number of reported crimes per 100,000 people) versus population density (people per square mile) in 2008. Figure 4.27a includes all 50 states and the District of Columbia. The District is an outlier and has a strong influence on the regression line. Figure 4.27b excludes the District of Columbia.

Without the District of Columbia, the slope of the regression line actually changes sign! Thus, whether you conclude that a positive or a negative association occurs between crime and population density depends on whether this one city is included.

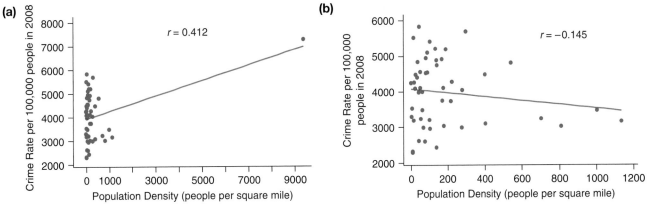

▲ **FIGURE 4.27** Crime rates and population densities in the states. Part (a) includes Washington, D.C. and part (b) does not. (Source: *The world almanac and Book of Facts 2009*)

These types of observations are called **influential points** because their presence or absence has a big effect on conclusions. When you have influential points in your data, it is good practice to try the regression and correlation with and without these points (as we did) and to comment on the difference.

## Regressions of Aggregate Data

Researchers sometimes do regression analysis based on what we call **aggregate data**. Aggregate data are those for which each plotted point is from a summary of a group of individuals. For example, in a study to examine the relationship between SAT math and critical reading scores, we might use the *mean* of each of the 50 states rather than the scores of individual students. The regression line provides a summary of linear associations between two variables. The mean provides a summary of the center of a distribution of a variable. What happens if we have a collection of means of two variables and we do a regression with these?

This is a legitimate activity, but you need to proceed with caution. For example, Figure 4.28 shows scatterplots of critical reading and math SAT scores. However, in Figure 4.28a, each point represents an individual: an SAT score for each first-year student enrolled in one northeastern university for a fall semester. The scatterplot in Figure 4.28b seems to show a much stronger relationship, because in this case each point is an aggregate. Specifically, each point represents a state in the United States: the mean SAT score for all students in the state taking SAT tests in 2005.

> **↻ Looking Back**
>
> **Outliers**
> You learned about outliers for one variable in Chapter 3. Outliers are values that lie a great distance from the bulk of the data. In two-variable associations, outliers are points that do not fit the trend or are far from the bulk of the data.

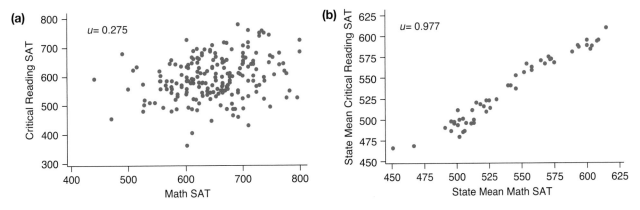

▲ **FIGURE 4.28** Critical reading and math SAT scores. **(a)** Scores for individuals. **(b)** Means for states.

We can still interpret Figure 4.28b, as long as we're careful to remember that we are talking about states, not people. We *can* say a strong correlation exists between a state's mean math SAT score and a state's mean critical reading SAT score. We *cannot* say that there is a strong correlation between individual students' math and critical reading scores.

## Don't Extrapolate!

**Extrapolation** means that we use the regression line to make predictions beyond the range of our data. This practice can be dangerous, because although the association may have a linear shape for the range we're observing, that might not be true over a larger range. This means that our predictions might be wrong outside the range of observed *x*-values.

Figure 4.29a shows a graph of height versus age for children between 2 and 9 years old from a large national study. We've superimposed the regression line on the graph (Figure 4.29a), and it looks like a pretty good fit. However, although the regression model provides a good fit for children ages 2 through 9, it fails when we use the same model to predict heights for older children.

The regression line for the data shown in Figure 4.29a is

$$\text{Predicted Height} = 31.78 + 2.45 \text{ Age}$$

However, we observed only children between the ages of 2 and 9. Can we use this line to predict the height of a 20-year-old?

The regression model predicts that the mean height of 20-year-olds is 80.78 inches:

$$\text{Predicted Height} = 31.78 + 2.45 \text{ Age} = 31.78 + 2.45 \times 20 = 80.78: \text{ nearly 7 feet!}$$

We can see from Figure 4.29b that the regression model provides a poor fit if we include people over the age of 9. Beyond that age, the trend is no longer linear, so we get bad predictions from the model.

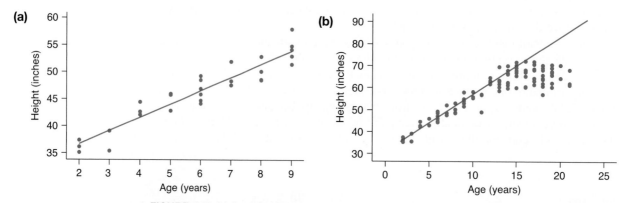

▲ **FIGURE 4.29** **(a)** Ages and heights for children between 2 and 9 years old. **(b)** Heights included for people up to age 21 years. The straight line is not valid in the upper range of ages. (Source: National Health and Nutrition Examination Survey, Centers for Disease Control)

It is often tempting to use the regression model beyond the range of data used to build the model. Don't. Unless you are very confident that the linear trend continues without change beyond the range of observed data, you must collect new data to cover the new range of values.

**KEY POINT**     Don't extrapolate!

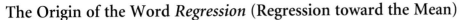

## The Origin of the Word *Regression* (Regression toward the Mean)

The first definition in the *Oxford Dictionary* says that regression is a "backward movement or return to a previous state." So why are we are using it to describe predictions of one numerical variable (such as value of a car) from another numerical variable (such as age of the car)?

The term *regression* was coined by Francis Galton, who used the regression model to study genetic relationships. He noticed that even though taller-than-average fathers tended to have taller-than-average sons, the sons were somewhat closer to average than the fathers were. Also, shorter-than-average fathers tended to have sons who were closer to the average than their fathers. He called this phenomenon regression toward mediocrity, but later it came to be known as **regression toward the mean**.

You can see how regression toward the mean works by examining the formula for the slope of the regression line:

$$b = r\frac{s_y}{s_x}$$

This formula tells us that fathers who are one standard deviation taller than average ($s_x$ inches above average) have sons who are not one standard deviation taller than average ($s_y$) but are instead $r$ times $s_y$ inches taller than average. Because $r$ is a number between $-1$ and 1, $r$ times $s_y$ is usually smaller than $s_y$. Thus the "rise" will be less than the "run" in terms of standard deviations.

The *Sports Illustrated* jinx is an example of regression toward the mean. According to the jinx, athletes who make the cover of *Sports Illustrated* end up having a really bad year after appearing. Some professional athletes have refused to appear on the cover of *Sports Illustrated*. (Once, the editors published a picture of a black cat in that place of honor, because no athlete would agree to grace the cover.) However, if an athlete's performance in the first year is several standard deviations above average, the second year is likely to be closer to average. This is an example of regression toward the mean. For a star athlete, closer to average can seem disastrous.

## The Coefficient of Determination, $r^2$, Measures Goodness of Fit

If we are convinced that the association we are examining is linear, then the regression line provides the best numerical summary of the relationship. But how good is "best"? The correlation coefficient, which measures the strength of linear relationships, can also be used to measure how well the regression line summarizes the data.

The **coefficient of determination** is simply the correlation coefficient squared: $r^2$. In fact, this statistic is often called *r-squared*. Usually, when reporting *r*-squared, we multiply by 100% to convert it to a percentage. Because $r$ is always between $-1$ and 1, *r*-squared is always between 0% and 100%. A value of 100% means the relationship is perfectly linear and the regression line perfectly predicts the observations. A value of 0% means there is no linear relationship and the regression line does a very poor job.

For example, when we predicted the width of a book from the number of pages in the book we found, the correlation between these variables to be $r = 0.9202$. So the coefficient of determination is $0.9202^2 = 0.8468$, which we report as 84.7%.

What does this value of 84.7% mean? A useful interpretation of *r*-squared is that it measures how much of the variation in the response variable is explained by the explanatory variable. For example, 84.7% of the variation in book widths was explained by the number of pages. What does this mean?

Figure 4.30 shows a scatterplot (simulated data) with a constant value for $y$ ($y = 6240$) no matter what the $x$-value is. You can see that there is no variation in $y$, so there is also nothing to explain.

▶ **FIGURE 4.30** Because $y$ has no variation, there is nothing to explain.

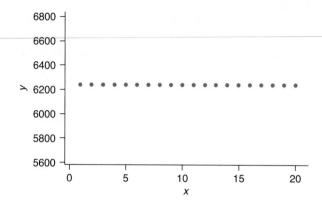

Figure 4.31 shows the height in inches and in centimeters for several people. Here, variation in the $y$-variable does occur. Height as measured in centimeters varies from about 150 cm to about 190 cm. However, the points are perfectly linear and have a correlation of 1.000. That means that if you are given an $x$-value (a person's height in inches), then you know the $y$-value (the person's height in centimeters) precisely. Thus all of the variation in $y$ is explained by the regression model. In this case, the coefficient of determination is 100%; all variation in $y$ is perfectly explained by the best-fit line.

Real data are messier. Figure 4.32 shows a plot of the age and value of some cars. The regression line has been superimposed to remind us that there is, in fact, a linear trend and that the regression line does capture it. The regression model explains some of the variation in $y$, but as we can see, it's not perfect; plugging the value of $x$ into the regression line gives us an imperfect prediction of what $y$ will be. In fact, for these data, $r = -0.778$, so we've explained $(-0.778)^2 = 0.605$, or about 60.5%, of the variation in $y$ with this regression line.

The practical implication of $r$-squared is that it helps determine which explanatory variable would be best for making predictions about the response variable. For example,

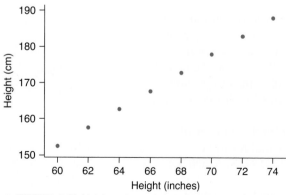

▲ **FIGURE 4.31** Heights of people in inches and in centimeters (cm) with a correlation of 1. The coefficient of determination is 100%.

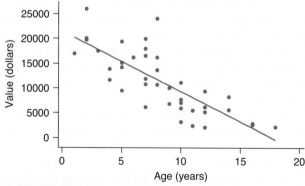

▲ **FIGURE 4.32** Age and value of cars; the correlation is $-0.778$, and the coefficient of determination is 60.5%.

is waist size or height a better predictor of weight? We can see the answer to this question from the scatterplots in Figure 4.33, which show that the linear relationship is stronger (has less scatter) for waist size.

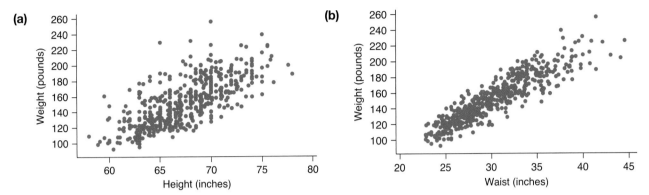

▲ **FIGURE 4.33** Scatterplots of height vs. weight **(a)** and waist size vs. weight **(b)**. Waist size has the larger coefficient of determination.

The *r*-squared for predicting weight from height is 51.4% (Figure 4.33a), and the *r*-squared for predicting weight from waist size is 81.7% (Figure 4.33b). We can explain more of the variation in these people's weights by using their waist sizes than by using their heights, and therefore, we can make better (more precise) predictions using waist size as the predictor.

**KEY POINT**     If the association is linear, the larger the coefficient of determination (*r*-squared), the smaller the variation or scatter about the regression line, and the more accurate the predictions tend to be.

**SNAPSHOT**  **R-SQUARED**

| | |
|---|---|
| **WHAT IS IT?** ▶ | *r*-squared or coefficient of determination. |
| **WHAT DOES IT DO?** ▶ | Measures how well the data fit the linear model. |
| **HOW DOES IT DO IT?** ▶ | If the points are tightly clustered around the regression line, $r^2$ has a value near 1, or 100%. If the points are more spread out, it has a value closer to 0. If there is no linear trend, and the points are a formless blob, it has a value near 0. |
| **HOW IS IT USED?** ▶ | Works only with linear associations. Large values of *r*-squared tell us that predicted *y*-values are likely to be close to actual values. The coefficient of determination shows us the percentage of variation that can be explained by the regression line. |

## CASE STUDY REVISITED

Brinks was contracted by New York City to collect money from parking meters in some months, but some Brinks employees had been stealing some of the money. The city knew how much money its own employees had collected from meters each month and how much money honest private contractors had collected. It used this relation to predict the amount that Brinks should have collected.

The first step in this analysis was to determine the relation between the amount of money collected by the city itself and that collected by the honest contractors. Figure 4.34 shows a positive (though somewhat weak) linear association. In months in which the city collected more money, the (honest) private contractors also tended to collect more money.

▶ **FIGURE 4.34** Regression plot without Brinks. The association between amounts of meter income collected from honest private contracts (vertical axis) and amounts collected by the city government. The trend is positive, linear, and somewhat weak.

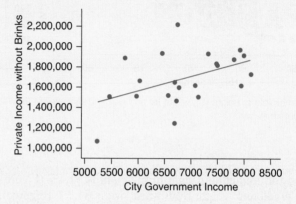

The regression line is

Predicted Contractor Collection = 688497 + 145.5 (City Income)

This model is useful for predicting how much Brinks should have collected in any given month. For example, in one particular month during which Brinks was the contractor, the city collected $7,016. The regression line predicts that if Brinks were like the other private contractors, it should have collected about

$$688497 + 145.5 \times 7016 = \$1,709,325$$

However, Brinks collected only $1,330,143—about $400,000 too little. This is just one month, but as Figure 4.35 shows, Brinks was consistently low in its collections. Figure 4.35 shows data both from Brinks and from the other private contractors. The Brinks data points are shown with red squares, and the other contractors' data points are shown with blue circles. You can see that the regression line for the Brinks collection (the dotted red line) is below the regression line for the other contractors, indicating that the mean amount collected by Brinks was consistently lower than the mean amount collected by the other contractors.

▶ **FIGURE 4.35** Scatterplot of Income from Parking Meters. The dotted red line represents the regression line for the money that Brinks collected. The solid blue line represents the money it should have collected, judging on the basis of the amount collected by the city. Notice that the predicted means for Brinks tend to be lower than those of the other contractors.

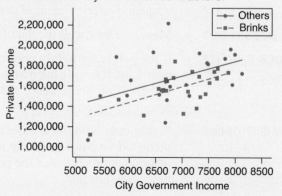

For each month that Brinks had worked for the city, the city used the regression line to compute the amount Brinks owed, by finding the difference between the amount Brinks would presumably have collected if there had been no stealing and the amount it actually collected.

# EXPLORING STATISTICS
## CLASS ACTIVITY

## Guessing the Age of Famous People

| GOALS | MATERIALS |
|---|---|
| In this activity, you will learn how to interpret the slope and intercept of a regression line, using data you collect in class. | Graph paper and a calculator or computer. |

**ACTIVITY**

Your instructor will give you a list of names of famous people. Beside each name, write your guess of the person's age, in years. Even if you don't know the age or don't know who the person is, give your best guess. If you work in a group, your group should discuss the guessed ages and record the best guess of the group. After you've finished, your instructor will give you a list of the actual ages of these people.

To examine the relationship between the actual ages and the ages you guessed, make a scatterplot with actual age on the *x*-axis and guessed age on the *y*-axis. Use technology to find the equation for the regression line and insert the line in the graph. Calculate the correlation.

**BEFORE THE ACTIVITY**

1. Suppose you guessed every age correctly. What would be the equation of the regression line? What would be the correlation?

2. What correlation do you think you will actually get? Why?

3. Suppose you consistently guess that people are older than they actually are? How will the intercept of the regression line compare with your intercept in Question 1? How about the slope?

**AFTER THE ACTIVITY**

1. How would you describe the association between the ages you guessed and the actual ages?

2. Is a regression line appropriate for your data? Explain why or why not.

3. What does it mean if a point falls above your regression line? Below your line?

4. What is the intercept of your line? What is the slope? Interpret the slope and intercept. Explain what these tell you about your ability to guess the ages of these people.

## CHAPTER REVIEW

## Key Terms

Scatterplot, *134*
Trend, *134*
Strength, *134*
Shape, *134*
Positive association (positive trend), *134*

Negative association (negative trend), *134*
Linear, *136*
Correlation coefficient, *138*
Regression line, *146*
Intercept, *147*

Slope, *147*
Explanatory variable, predictor variable, independent variable, *154*
Response variable, predicted variable, dependent variable, *154*

Influential point, *163*
Aggregate data, *163*
Extrapolation, *164*
Regression toward the mean, *165*
Coefficient of determination, $r^2$, *r*-squared, *165*

## Learning Objectives

After reading this chapter and doing the assigned homework problems, you should

• Be able to write a concise and accurate description of an association between two numerical variables based on a scatterplot.

• Understand how to use a regression line to summarize a linear association between two numerical variables.

• Interpret the intercept and slope of a regression line in context and know how to use the regression line to predict mean values of the response variable.

• Critically evaluate a regression model.

## Summary

The first step in looking at the relationship between two numerical variables is to make a scatterplot and learn as much about the association as you can. Examine trend, strength, and shape. Look for outliers.

Regression lines and correlation coefficients can be interpreted meaningfully only if the association is linear. A correlation coefficient close to 1 or −1 does *not* mean the association is linear. When the association is linear, the correlation coefficient can be calculated by Formula 4.1:

$$\textbf{Formula 4.1: } r = \frac{\sum z_x z_y}{n - 1}$$

A regression line can summarize the relationship in much the same way as a mean and standard deviation can summarize a distribution. Interpret the slope and intercept, and use the regression line to make predictions about the mean *y*-value for any given value of *x*.

The regression line is given by

$$\textbf{Formula 4.2c: } \text{Predicted } y = a + bx$$

where the slope, *b*, is

$$\textbf{Formula 4.2a: } b = r\frac{s_y}{s_x}$$

and the intercept, *a*, is

$$\textbf{Formula 4.2b: } a = \bar{y} - b\bar{x}$$

When interpreting a regression analysis, be careful:

Don't extrapolate.
Don't make cause-and-effect conclusions if the data are observational.
Beware of outliers, which may (or may not) strongly affect the regression line.
Proceed with caution when dealing with aggregated data.

## Source

Afshartous, D., Bentow, S., "Statistical Analysis of Brink's Data," www.stat.ucla.edu/cases

Faraway, J. J. 2005. *Extending the linear model with R*. Boca Raton, FL: CRC Press.

Heinz, G., L. Peterson, R. Johnson, and C. Kerk. 2003. Exploring relationships in body dimensions. *Journal of Statistics Education* 11(2). http://www.amstat.org/publications/jse/

Joyce, C. A., ed. *The 2009 World Almanac and Book of Facts*. New York: World Almanac.

Leger, A., A. Cochrane, and F. Moore. 1979. Factors associated with cardiac mortality in developed countries with particular reference to the consumption of wine. *The Lancet* 313(8124): 1017–1020.

# SECTION EXERCISES

## SECTION 4.1

**4.1 Predicting Land Value** Both figures concern the assessed value of land (with homes on the land), and both use the same data set.

a. Which do you think has a stronger relationship with value of the land—the number of acres of land or the number of rooms in the homes? Why?

b. If you were trying to predict the value of a parcel of land in this area (on which there is a home), would you be able to make a better prediction by knowing the acreage or the number of rooms in the house? Explain.

(Source: Minitab File, Student 12, "Assess.")

**(A)**

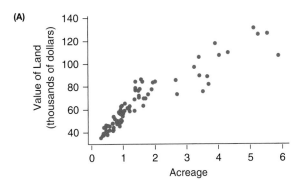

**(B)**

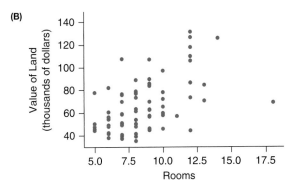

**4.2 Predicting Total Value of Property** Both figures concern the total assessed value of properties that include homes, and both use the same data set.

a. Which do you think has a stronger relationship with value of the property—the number of square feet in the home (shown in part B of the figure) or the number of fireplaces in the home (shown in part A of the figure)? Why?

b. If you were trying to predict the value of a property in this area (where there is a home), would you be able to make a better prediction by knowing the number of square feet or the number of fireplaces? Explain.

**(A)**

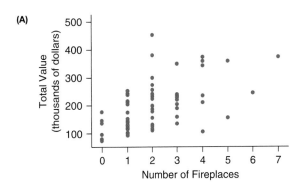

**(B)**

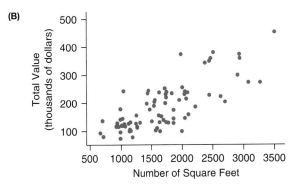

**4.3 Car Value and Age of Student** The figure shows the age of students and the value of their cars according to the Kelley Blue Book. Does it show an increasing trend, a decreasing trend, or very little trend? Explain.

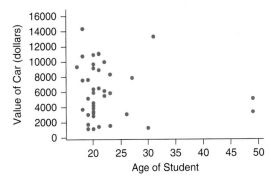

**4.4 Shoe Size and GPA** The figure shows a scatterplot of shoe size and GPA for some college students. Does it show an increasing trend, a decreasing trend, or no trend? Is there a strong relationship?

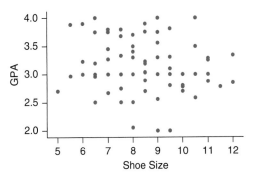

**4.5 Weight Loss** The scatterplot shows the actual weight and desired weight change of some students. Thus, if they weighed 220 and wanted to weigh 190, the desired weight change would be negative 30.

Explain what you see. In particular, what does it mean that the trend is negative?

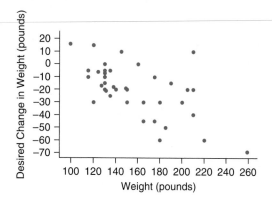

**4.6 Comparing Salaries** The figure shows the salary and year of first employment for some professors at a college. Explain, in context, what the negative trend shows. Who makes the most and who makes the least?
(Source: Minitab, Student 12, "Salary.")

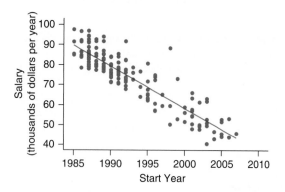

TRY **4.7 Population and Motor Vehicle Fatalities (Example 1)**
The figure shows the number of motor vehicle fatalities and the population of the 50 states and Washington, D.C., for 2006, from the U.S. Census Bureau. Explain what the trend shows.

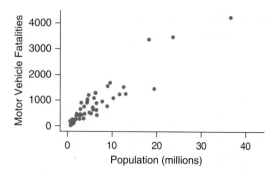

**4.8 BAs and Median Income** The scatterplot shows data from the 50 states taken from the U.S. Census—the percentage of the population (25 years or older) with a college degree or higher and the median family income. Describe and interpret the trend. (The outlier is Colorado.)

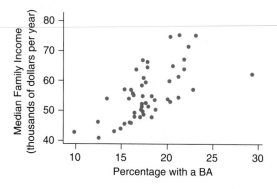

**4.9 Value of Homes and Acreage** The figure shows the total assessed value of some properties (the total value of the home and the land) and the acreage that comes with the homes. Describe what you see. Is the trend positive or negative? What does that mean?

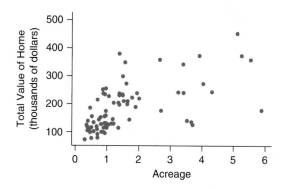

**4.10 Height and Weight for Women** The figure shows a scatterplot of the heights and weights of some women taking statistics. Describe what you see. Is the trend positive, negative, or near zero? Explain.

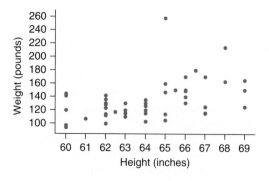

# SECTION 4.2

**4.11 Children's Ages and Heights** The figure shows information about the ages and heights of several children. Why would it not make sense to find the correlation or to perform linear regression with this data set? Explain.

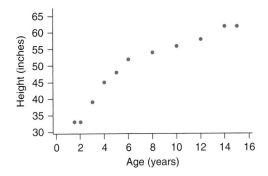

**4.12 Blackjack Tips** The figure shows the amount of money won by people playing blackjack and the amount of tips they gave to the dealer (who was a statistics student), in dollars.

Would it make sense to find a correlation for this data set? Explain.

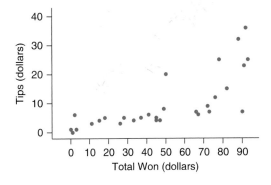

**4.13 SAT Scores and Percentage Taking SAT** The figure shows data from The College Board on the percentage of students who take the SAT and the average score on the quantitative part for the 50 states and Washington, D.C., in 2009.

a. Does the graph show an increasing or a decreasing trend? Explain what that means in the context of these data.

b. Do the points seem to follow a straight line or a curve?

c. Is it appropriate to find the correlation for these data? Why or why not?

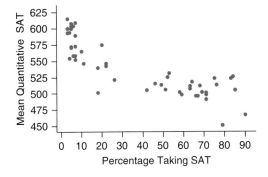

**4.14 Ages of Women Who Give Birth** The figure shows a scatterplot of birthrate (live births per 1000 women) and age of the mother in the United States. Would it make sense to find the correlation for this data set? Explain. According to this graph, at approximately what age is the highest fertility rate?
(Source: Helmut T. Wendel and Christopher S. Wendel (Editors): *Vital Statistics of the United States: Births, Life Expectancy, Deaths, and Selected Health Data.* Second Edition, 2006, Bernan Press.)

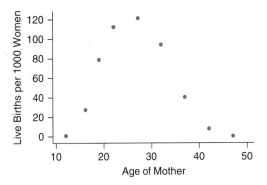

**4.15 Do Older Students Have Higher GPAs?** On the basis of the scatterplot, do you think that the correlation coefficient between age and GPA for this figure is positive, negative, or near zero?

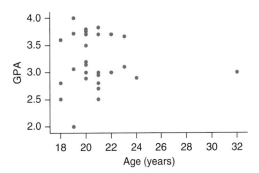

**4.16 Handspans** Refer to the figure. Is the correlation coefficient between the handspan of the dominant hand and that of the nondominant hand positive, negative, or near zero?

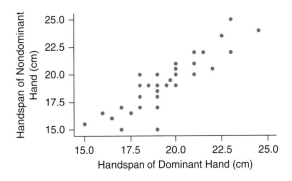

**4.17 Matching** Pick the letter of the graph that goes with each numerical value listed below for the correlation. Correlations:

0.767 _____

0.299 _____

−0.980 _____

(A)

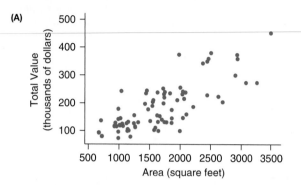

(B)

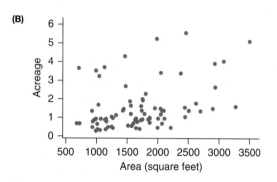

(C)

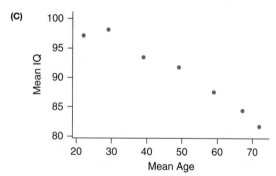

**4.18 Matching** Pick the letter of the graph that goes with each numerical value listed below for the correlation. Correlations:

−0.903 _____

0.374 _____

0.777 _____

(A)

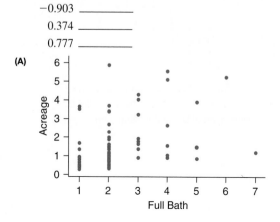

(B)

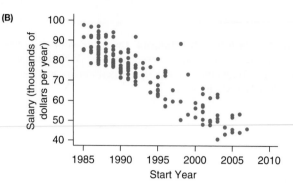

(C)

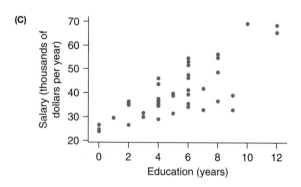

TRY **4.19 Cost of Flights (Example 2)** The table for part a shows approximate distances between selected cities and the approximate cost of flights between those cities in November 2009.

a. Calculate the correlation of the numbers shown in the part a table by using a computer or statistical calculator.

| Cost | Miles |
|------|-------|
| 180 | 960 |
| 400 | 3100 |
| 270 | 2000 |
| 100 | 430 |
| 443 | 3000 |

b. The table for part b shows the same information, except that the distance was converted to kilometers by multiplying the number of miles by 1.609. What happens to the correlation when numbers are multiplied by a constant?

| Cost | Kilometers |
|------|------------|
| 180 | 1545 |
| 400 | 4988 |
| 270 | 3218 |
| 100 | 692 |
| 443 | 4827 |

c. Suppose the Federal Aviation Administration (FAA) adds a tax to each flight. Fifty dollars is added to every flight, no matter how long it is. The table for part c shows the new data. What happens to the correlation when a constant is added to each number?

| Cost | Miles |
|------|-------|
| 230 | 960 |
| 450 | 3100 |
| 320 | 2000 |
| 150 | 430 |
| 493 | 3000 |

**4.20 Trash** The table shows the number of people living in a house and the weight of trash (in pounds) at the curb just before trash pickup.

| People | Trash (pounds) |
|--------|----------------|
| 2 | 18 |
| 3 | 33 |
| 6 | 93 |
| 1 | 23 |
| 7 | 83 |

a. Find the correlation between these numbers by using a computer or a statistical calculator.

b. Suppose some of the weight was from the container (each container weighs 3 pounds). Subtract 3 pounds from each weight and find the new correlation with the number of people. What happens to the correlation when a constant is added (we added negative 3) to each number?

c. Suppose each house contained exactly twice the number of people, but the weight of the trash was the same. What happens to the correlation when numbers are multiplied by a constant?

## SECTION 4.3

TRY **4.21 Are Men Paid More Than Women? (Example 3)** The scatterplot shows the median annual pay for college-educated men and women in the 50 states in 2005, according to the U.S. Census Bureau. The correlation is 0.834. The regression equation is above the graph.

a. Find a rough estimate (by using the scatterplot) of median pay for women in a state that has a median pay of about $60,000 for men.

Predicted Women = 8707 + 0.5932 Men

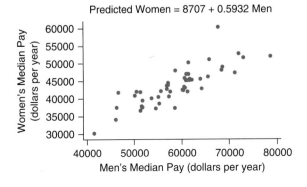

b. Use the regression equation above the graph to get a more precise estimate of the median pay for women in a state that has a median pay for men of $60,000.

**4.22 Number of Births and Population** The figure shows the number of births and the populations for the 50 states in 2005, according to the U.S. Census Bureau. The correlation is 0.99.

a. Find a rough estimate (by using the scatterplot) of number of births in a state with a population of about 10 million.

b. Use the regression equation above the graph to get a more precise estimate of the number of births for a state with a population of 10 million.

Predicted Births (K) = −4.494 + 14.66 Population (M)

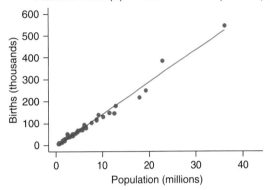

**4.23–4.26** Refer to the following table for Exercises 4.23–4.26. It provides various body measurements for some college women. The height (Ht) is self-reported in inches. All other measurements are in centimeters.

Head is the circumference of the head.
HandL is the length of the hand.
FootL is the length of the foot.
HandW is the width of the hand with the fingers spread wide.
Armspan is the armspan with the arms spread wide.

| Ht | Head | HandL | FootL | HandW | Armspan |
|------|------|-------|-------|-------|---------|
| 62.0 | 56.0 | 17.0 | 23.0 | 20.0 | 159.5 |
| 65.0 | 54.0 | 17.5 | 23.5 | 18.0 | 161.0 |
| 66.0 | 55.0 | 18.0 | 24.0 | 16.5 | 161.5 |
| 63.0 | 58.0 | 18.5 | 23.5 | 19.5 | 160.5 |
| 68.0 | 55.0 | 19.0 | 25.0 | 18.0 | 173.0 |
| 56.0 | 54.5 | 16.5 | 22.0 | 17.0 | 142.0 |
| 61.0 | 50.0 | 16.0 | 22.0 | 18.0 | 155.0 |
| 67.0 | 53.5 | 18.0 | 23.5 | 15.0 | 165.0 |
| 66.0 | 55.5 | 16.0 | 21.0 | 20.0 | 166.0 |
| 64.0 | 57.0 | 19.0 | 24.5 | 20.0 | 163.0 |
| 61.5 | 55.0 | 19.0 | 24.5 | 21.0 | 152.0 |

TRY *4.23 **Height and Armspan for Women (Example 4)** TI-83/84 output from a linear model for predicting armspan (in cm) from height (in inches) is given in the figure. Summary statistics are also provided.

```
LinReg
 y=a+bx
 a=16.80271378
 b=2.249707146
 r²=.8987441215
 r=.9480211609
```

|  | Mean | Standard Deviation |
|---|---|---|
| Height, x | 63.59 | 3.41 |
| Armspan, y | 159.86 | 8.10 |

To do parts a–c, assume that the association between armspan and height is linear.

a. Report the regression equation, using the words "Height" and "Armspan," not x and y, using the output given.

b. Verify the slope by using the formula $b = r \dfrac{s_y}{s_x}$.

c. Verify the y-intercept using $a = \bar{y} - b\bar{x}$.

d. Using the regression equation, predict the armspan (in cm) for someone 64 inches tall.

*4.24 **Hand and Foot Length for Women** Refer to the data shown in the table for Exercises 4.23–4.26. Some computer output is shown in the figure. The output is for a linear model used to predict foot length from hand length.

Assume the trend is linear. Summary statistics for these data are shown below.

|  | Mean | Standard Deviation |
|---|---|---|
| Hand, x | 17.682 | 1.168 |
| Foot, y | 23.318 | 1.230 |

```
The regression equation is
Y = 5.67 + 0.998 X
Pearson correlation of HandL and FootL = 0.948
```

a. Report the regression equation, using the words "Hand" and "Foot," not x and y.

b. Verify the slope by using the formula $b = r \dfrac{s_y}{s_x}$.

c. Verify the y-intercept by using the formula $a = \bar{y} - b\bar{x}$.

d. Using the regression equation, predict the foot length (in cm) for someone who has a hand length of 18 cm.

4.25 **Hand Width and Armspan for Women** Refer to the data shown in the table for Exercises 4.23–4.26. Use hand width as the predictor and armspan as the response.

a. Make a scatterplot of the data.

b. Explain why linear regression is probably not appropriate for these variables.

4.26 **Head Circumference and Hand Width for Women** Refer to the data shown in the table for Exercises 4.23–4.26. Use head circumference as x and hand width as y.

a. Make a scatterplot of the data.

b. Explain why linear regression is probably not appropriate for these variables.

TRY 4.27 **Height and Armspan for Men (Example 5)** Measurements were made for a sample of adult men. A regression line was fit to predict the men's armspan from their height. The output from several different statistical technologies is provided. The scatterplot confirms that the association between armspan and height is linear.

a. Report the equation for predicting armspan from height. Use words such as armspan, not just x and y.

b. Report the slope and intercept from each technology, using all the digits given.

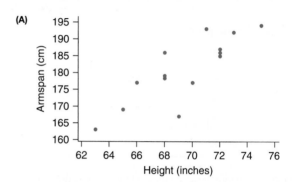

(A)

(B)
```
The regression equation is
Armspan = 6.2 + 2.51 Height
```
**From Minitab**

(C)
**Simple linear regression results:**
Dependent Variable: Armspan
Independent Variable: Height
Armspan = 6.2408333 + 2.514674 Height
Sample size: 15
R (correlation coefficient) = 0.8681
R-sq = 0.7535989
Estimate of error standard deviation: 5.409662

**From StatCrunch**

(D)

|  | Coefficients |
|---|---|
| Intercept | 6.240833 |
| X Variable | 2.514674 |

**From Excel**

**(E)**

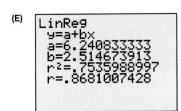

**From TI-83/84**

**4.28 Hand Length and Foot Length for Men** Measurements were made for a sample of adult men. Assume that the association between their hand length and foot length is linear. Output for predicting foot length from hand length is provided from several different statistical technologies.

a. Report the equation for predicting foot length from hand length. Use words like foot or foot length, not just $x$ and $y$.

b. Report the slope and intercept from each technology, using all the digits given.

**(A)**

```
The regression equation is
FootL = 15.8 + 0.563 HandL
```

**From Minitab**

**(B)**

**Simple linear regression results:**

Dependent Variable: FootL
Independent Variable: HandL
FootL = 15.807631 + 0.5626551 HandL
Sample size: 17
R (correlation coefficient) = 0.404
R-sq = 0.1632489
Estimate of error standard deviation: 1.6642156

**From StatCrunch**

**(C)**

|  | Coefficients |
|---|---|
| Intercept | 15.80763 |
| X Variable | 0.562655 |

**From Excel**

**(D)**

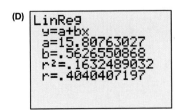

**From TI-83/84**

**4.29 Height and Head Circumference for Men** Explain what makes this scatterplot hard to interpret. What should have been done differently?

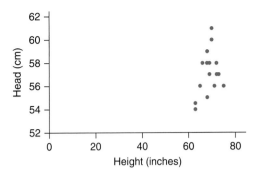

**4.30 Elbow-to-Wrist and Knee-to-Ankle Measurements** The scatterplot shows measurements (in centimeters) for the distance from elbow to wrist and from knee to ankle for some college men. Explain what makes this scatterplot difficult to interpret.

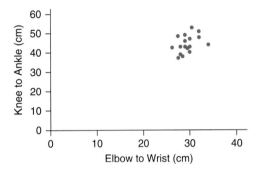

**4.31 Comparing Correlation for Armspan and Height** The correlation between height and armspan in a sample of adult women was found (in Exercise 4.23) to be $r = 0.948$. The correlation between armspan and height in a sample of adult men was found (in Exercise 4.27) to be $r = 0.868$. Which association—the association between height and armspan for women, or the association between height and armspan for men—is stronger? Explain.

* **4.32 Age and Weight for Men and Women** The scatterplot shows a solid blue line for predicting weight from age of men; the dotted red line is for predicting weight from age of women. The data were collected from a large statistics class.

a. Which line is higher and what does that mean?

b. Which line has a steeper slope and what does that mean?

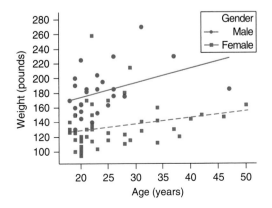

**4.33 Social Security Number and Age** The figure shows a scatterplot of the last two digits of some students' Social Security numbers and their ages.

a. If a regression line were drawn on this graph, would it have a positive slope, a negative slope, or a slope near 0?

b. Give an estimate of the numerical value of the correlation between age and Social Security number.

c. Explain what this graph tells us about the relation between Social Security number and age.

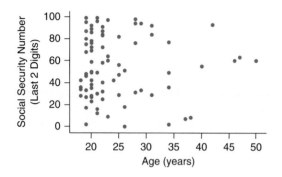

**4.34 Seesaw** The figure shows a scatterplot of the height of the left seat of a seesaw and the height of the right seat of the same seesaw. Estimate the numerical value of the correlation and explain the reason for your estimate.

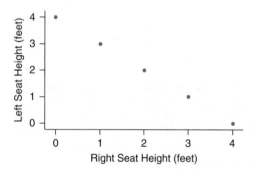

TRY **4.35 Choosing the Predictor and Response (Example 6)** Pick out which variable you think should be the predictor (x) and which variable should be the response (y). Explain your choices.

a. You collect data on the number of gallons of gas it takes to fill up the tank after driving a certain number of miles. You wish to know how many miles you've driven based on the number of gallons it took to fill up the tank.

b. Data on salaries and years of experience at a two-year college are used in a law suit to determine whether a faculty member is being paid the correct amount for her years of experience.

c. You wish to buy a belt for a friend and know only his weight. You have data on the weight and waist sizes for a large sample of adult men.

**4.36 Choosing the Predictor and Response** Pick out which variable you think should be the predictor (x) and which variable should be the response (y). Explain your choices.

a. Weights of nuggets of gold (in ounces) and their market value over the last few days are provided, and you wish to use this to estimate the value of a gold bracelet that weighs 4 ounces.

b. You have data collected on the amount of time since chlorine was added to the public swimming pool and the concentration of chlorine still in the pool. (Chlorine evaporates over time.) Chlorine was added to the pool at 8 A.M., and you wish to know what the concentration is now, at 3 P.M.

c. You have data on the circumference of oak trees (measured 12 inches from the ground) and their age (in years). An oak tree in the park has a circumference of 36 inches, and you wish to know approximately how old it is.

TRY **4.37 Percentage of Smoke-Free Homes and Percentage of High School Students Who Smoke (Example 7)** The figure shows a scatterplot with the regression line. The data are for the 50 states. The predictor is the percentage of smoke-free homes. The response is the percentage of high school students who smoke. The data came from the Centers for Disease Control and Prevention (CDC).

a. Explain what the trend shows.

b. Use the regression equation to predict the percentage of students in high school who smoke, assuming that there are 70% smoke-free homes in the state. Use 70 not 0.70.

Predicted Pct. Smokers = 56.32 − 0.464 (Pct. Smoke-free)

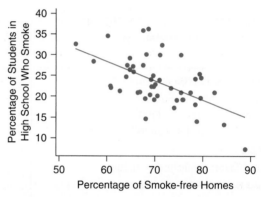

**4.38 Effect of Adult Smoking on High School Student Smoking** The figure shows a fitted line plot. The data are for the 50 states. The predictor is the percentage of adults who smoke. The response is the percentage of high school students who smoke. (The point in the lower left is Utah.)

a. Explain what the trend shows.

b. Use the regression equation to predict the percentage of high school students who smoke, assuming that 25% of adults in the state smoke. Use 25, not 0.25.

Predicted Pct. Smokers = −0.838 + 1.124 (Pct. Adult Smoke)

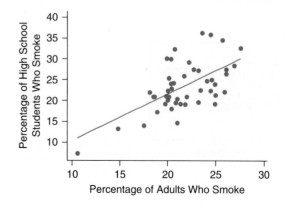

**4.39 Drivers' Deaths and Ages** The figure shows a graph of the death rate in automobile accidents and the age of the driver. The numbers came from the Insurance Information Institute.

a. Explain what the graph tells us about drivers at different ages; state which ages show the safest drivers and which show the most dangerous drivers.

b. Explain why it would not be appropriate to use these data for linear regression.

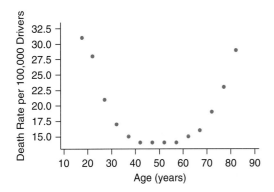

**4.40 Do Women Tend to Live Longer Than Men?** The figure shows life expectancy versus age for males and females in the United States in 2009, up to the age of 119. Females are represented by the blue circles and males by the red squares. These figures were reported by the CDC.

a. Find your own age on the graph and estimate your life expectancy from the appropriate graph.

b. Would it make sense to find the best straight line for this graph? Why or why not?

c. Is it reasonable to predict the life expectancy for a person who is 120 from the regression line for these data? Why or why not?

d. Explain what it means that nearly all of the blue circles (for women) are above the red squares (for men). (Above the age of 100, the red squares cover the blue circles because both are in the same place.)

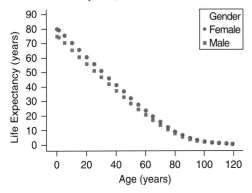

**g  4.41 Does the Cost of a Flight Depend on the Distance?**
The table gives the round-trip fare for flights between Boston and some other cities on major airlines. These were the lowest prices shown on www.Sidestep.com on January 21, 2010. The airlines varied. (The travel dates were all the same, in early February.) *See page 188 for guidance.*

**Round-trip Air Fares from Boston**

| City | Cost | Miles |
|---|---|---|
| New York | 179 | 206 |
| Chicago | 179 | 963 |
| San Francisco | 397 | 3095 |
| Denver | 274 | 1940 |
| Los Angeles | 314 | 2979 |
| Washington, D.C. | 99 | 429 |
| Atlanta | 317 | 1037 |
| Seattle | 443 | 2976 |
| Dallas | 404 | 1748 |
| St. Louis | 334 | 1141 |

How much would it cost, on average, to fly 500 miles? To answer this question, perform a complete regression analysis, including a scatterplot with a regression line.

**4.42 English in California Schools** This problem concerns the vote of the Ventura County cities on Proposition 227 in June of 1998. Proposition 227 in California stated that children should be taught primarily in English and abolished bilingual education. The statewide vote showed that 61% of the voters were in favor of this proposition. Reported in the table are the percentages of voters in each city who voted for Proposition 227 and the percentages of students in those cities with limited English, according to the *Ventura County Star*.

**Vote on Proposition 227 (English in schools)**

| City | %Yes | %Students with limited English |
|---|---|---|
| Thousand Oaks | 71.5 | 7.4 |
| Simi Valley | 71.0 | 6.0 |
| Camarillo | 68.3 | 7.8 |
| Moorpark | 67.8 | 16.8 |
| Ventura | 61.7 | 11.9 |
| Port Hueneme | 61.0 | 42.5 |
| Ojai | 57.4 | 6.5 |
| Santa Paula | 49.9 | 16.6 |
| Fillmore | 48.4 | 36.1 |
| Oxnard | 47.9 | 40.4 |

Assume that the association between the percentage of voters who favor the initiative and the percentage of students with limited English is linear enough to proceed. Find the regression equation for predicting percent Yes from percent with limited English, and report it. Interpret the sign of the slope clearly, in words that someone who had not studied statistics could understand. Pretend you are writing a conclusion for a newspaper article.

**4.43 Do States with Higher Populations Have More Millionaires?** The table gives the number of millionaires (in thousands) and the population (in hundreds of thousands) for the states in the northeastern region of the United States in 2008. The numbers of millionaires come from *Forbes Magazine* in March of 2007.

a. Without doing any calculations, predict whether the correlation and slope will be positive or negative. Explain your prediction.

b. Make a scatterplot with the population (in hundreds of thousands) on the *x*-axis and the number of millionaires (in thousands) on the *y*-axis. Was your prediction correct?

c. Find the numerical value for the correlation.

d. Find the value of the slope and explain what it means in context. Be careful with the units.

e. Explain why giving and interpreting the value for the intercept does not make sense in this situation.

| State | Millionaires | Population |
|---|---|---|
| Connecticut | 86 | 35 |
| Delaware | 18 | 8 |
| Maine | 22 | 13 |
| Massachusetts | 141 | 64 |
| New Hampshire | 26 | 13 |
| New Jersey | 207 | 87 |
| New York | 368 | 193 |
| Pennsylvania | 228 | 124 |
| Rhode Island | 20 | 11 |
| Vermont | 11 | 6 |

**\*4.44 Semesters and Units** The table shows the self-reported number of semesters completed and the number of units completed for 15 students at a community college. All units were counted, but attending summer school was not included.

a. Make a scatterplot with the number of semesters on the *x*-axis and the number of units on the *y*-axis. Does one point stand out as unusual? Explain why it is unusual. (At most colleges, full-time students take between 12 and 18 units per semester.)
Finish each part *two ways*, with and without the unusual point, and comment on the differences.

b. Find the numerical values for the correlation between semesters and units.

c. Find the two equations for the two regression lines.

d. Insert the lines. Use technology if possible.

e. Report the slopes and intercepts of the regression lines and explain what they show. If the intercepts are not appropriate to report, explain why.

| Sems | Units | Sems | Units |
|---|---|---|---|
| 2 | 21.0 | 3 | 30.0 |
| 4 | 130.0 | 4 | 60.0 |
| 5 | 50.0 | 3 | 45.0 |
| 7 | 112.0 | 5 | 70.0 |
| 3 | 45.5 | 3 | 32.0 |
| 3 | 32.0 | 8 | 70.0 |
| 8 | 140.0 | 6 | 60.0 |
| 0 | 0.0 | | |

**4.45 Motor Vehicle Fatalities and Population** The table gives the population (in hundreds of thousands of people) and the number of motor vehicle fatalities in the states in the northeastern region of the United States in 2008, according to the National Highway Traffic Safety Administration (NHTSA).

a. Do you expect the correlation to be positive or negative? Why?

b. Make a scatterplot of the data.

c. Report the correlation.

d. Report the equation of the regression line and insert the line in the scatterplot, or use technology to insert the line in your plot.

e. Report the slope of the regression line and explain what it shows.

| MVFat | Pop | MVFat | Pop |
|---|---|---|---|
| 322 | 35 | 773 | 87 |
| 47 | 8 | 1522 | 193 |
| 216 | 13 | 1614 | 124 |
| 459 | 64 | 84 | 11 |
| 127 | 13 | 78 | 6 |

**4.46 Motor Vehicle Fatalities and Population** The table gives the population (in hundreds of thousands of people) and the number of motor vehicle fatalities in the states in the western region of the United States in 2008.

a. Do you expect the correlation to be positive or negative? Why?

b. Make a scatterplot of the data. Are there any unusual points?

c. Report the correlation.

| MVFat | Pop | MVFat | Pop |
|---|---|---|---|
| 87 | 7 | 381 | 24 |
| 1117 | 50 | 449 | 19 |
| 4078 | 361 | 436 | 36 |
| 742 | 47 | 328 | 25 |
| 119 | 13 | 659 | 63 |
| 264 | 14 | 176 | 5 |
| 270 | 9 | | |

d. Report the equation of the regression line and insert the line in the scatterplot.

e. Report the slope of the regression line and explain in context what it shows.

(Source: www.encarta.msn.com and www-fars.nhtsa.dot.gov)

## SECTION 4.4

**4.47** Answer the questions using complete sentences.

a. What is an influential point?

b. It has been noted that people who go to church frequently tend to have lower blood pressure than people who don't go to church. Does this mean you can lower your blood pressure by going to church? Why or why not? Explain.

**4.48** Answer the questions, using complete sentences.

a. What is extrapolation and why is it a bad idea in regression analysis?

b. How is the coefficient of determination related to the correlation, and what does the coefficient of determination show?

∗ c. When testing the IQ of a group of adults (aged 25 to 50), an investigator noticed that the correlation between IQ and age was negative. Does this show that IQ goes down as we get older? Why or why not? Explain.

**4.49** If there is a positive correlation between number of years studying math and shoe size (for children), does that prove that larger shoes cause more studying of math, or vice versa? Can you think of a hidden variable that might be influencing both of the other variables?

**4.50** Suppose that the growth rate of children looks like a straight line if the height of a child is observed at the ages of 24 months, 28 months, 32 months, and 36 months. If you use the regression obtained from these ages and predict the height of the child at 21 years, you might find that the predicted height is 20 feet. What is wrong with the prediction and the process used?

**4.51 Coefficient of Determination** If the correlation between height and weight of a large group of people is 0.67, find the coefficient of determination (as a percent) and explain what it means. Assume that height is the predictor and weight is the response, and assume that the association between height and weight is linear.

**4.52 Coefficient of Determination** Does a correlation of −0.70 or +0.50 give a larger coefficient of determination? We say that the linear relationship having the larger coefficient of determination is more strongly correlated. Which of the values shows a stronger correlation?

∗ **4.53 Decrease in Cholesterol** A doctor is studying cholesterol readings in his patients. After reviewing the cholesterol readings, he calls the patients with the highest cholesterol readings (the top 5% of readings in his office) and asks them to come back to discuss cholesterol-lowering methods. When he tests these patients a second time, the average cholesterol readings tended to have gone down somewhat. Explain what statistical phenomenon might have been partly responsible for this lowering of the readings.

∗ **4.54 Test Scores** Suppose that students who scored much lower than the mean on their first statistics test were given special tutoring in the subject. Suppose that they tended to show some improvement on the next test. Explain what might cause the rise in grades other than the tutoring program itself.

TRY **4.55 Salary and Year of Employment (Example 8)** The equation for the regression line relating the salary and the year first employed is given above the figure.

a. Report the slope and explain what it means.

b. Either interpret the intercept (4,255,000) or explain why it is not appropriate to interpret the intercept.

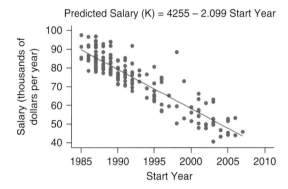

Predicted Salary (K) = 4255 − 2.099 Start Year

**4.56 MPG: Highway and City** The figure shows the relationship between the number of miles per gallon on the highway and in the city for some cars.

a. Report the slope and explain what it means.

b. Either interpret the intercept (7.792) or explain why it is not appropriate to interpret the intercept.

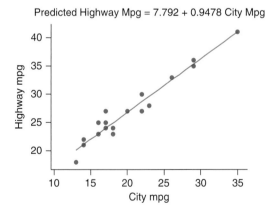

Predicted Highway Mpg = 7.792 + 0.9478 City Mpg

✱ **4.57 Cost of Turkeys** The table on the next page shows a list of the weights and prices of some turkeys at different supermarkets.

a. Make a scatterplot with weight on the x-axis and cost on the y-axis. If using computer statistical technology, include the regression line on your scatterplot.

b. Find the numerical value for the correlation between weight and price. Explain what the sign of the correlation shows.

c. Report the equation of the best straight line, using weight as the predictor (x) and cost as the response (y).

d. Insert the line on the scatterplot if you did not do it in part a.

e. Report the slope and intercept of the regression line and explain what they show. If the intercept is not appropriate to report, explain why.

f. Add a new point to your data: a 30-pound turkey that is free. Give the new value for $r$ and the new regression equation. Explain what the negative correlation implies. What happened?

| Weight (pounds) | Price |
|---|---|
| 12.3 | $17.10 |
| 18.5 | $23.87 |
| 20.1 | $26.73 |
| 16.7 | $19.87 |
| 15.6 | $23.24 |
| 10.2 | $ 9.08 |

**4.58 Iraq Casualties and Population of Hometowns** The figures show the number of Iraq casualties through October 2009 and the population of some hometowns from which the servicemen or servicewomen came, according to the *Los Angeles Times*. Comment on the difference in graphs and in the coefficient of determination between the graph that included L.A. (a) and the one that did not include L.A. (b). L.A. is the point with a population of nearly 4 million.

(A)   Predicted Casualties = 1.980 + 0.00641 Population (K)

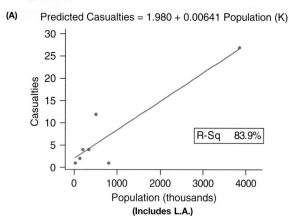

(B)   Predicted Casualties = 2.995 + 0.00309 Population (K)

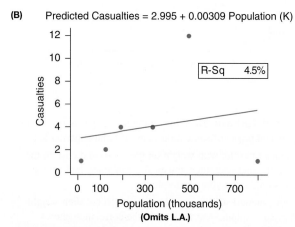

TRY **4.59 Teachers' Pay and Costs of Education (Example 9)**
The figure shows a scatterplot with a regression line for teachers' average pay and the expenditure per pupil for each state for public schooling in 2007, according to *The 2009 World Almanac and Book of Facts*.

a. From the graph, is the correlation between teachers' average pay and the expenditure per pupil positive or negative?

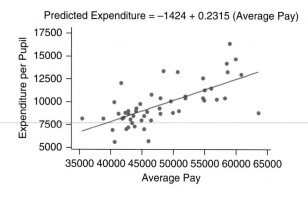

b. Interpret the slope.

c. Interpret the intercept or explain why it should not be interpreted.

**4.60 Teachers' Pay** The figure shows a scatterplot with a regression line for the average teacher's pay and the percentage of students graduating from high school for each state in 2007, according to *The 2009 World Almanac and Book of Facts*. On the basis of the graph, do you think the correlation is positive, negative, or near 0? Explain what this means.

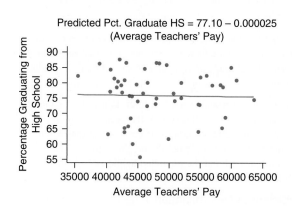

TRY **4.61 Does Having a Job Affect Students' Grades? (Example 10)** Grades on a political science test and the number of hours of paid work in the week before the test were studied. The instructor was trying to predict the grade on a test from the hours of work. The figure shows a scatterplot and the regression line for these data.

a. By looking at the plot and the line (without doing any calculations), state whether the correlation is positive or negative and explain your prediction.

b. Interpret the slope.

c. Interpret the intercept.

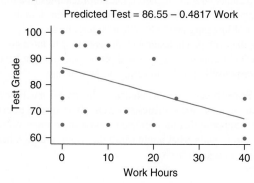

**4.62 Weight of Trash and Household Size** Data that included information on the weight of the trash (in pounds) on the street one week and the number of people who live in the house were collected. The figure shows a scatterplot with the regression line.

a. Is the trend positive or negative? What does that mean?

b. Now calculate the correlation between the weight of trash and the number of people. (Use R-Sq from the figure and take the square root of it.)

c. Report the slope. For each additional person in the house, there are, on average, how many additional pounds of trash?

d. Either report the intercept or explain why it is not appropriate to interpret it.

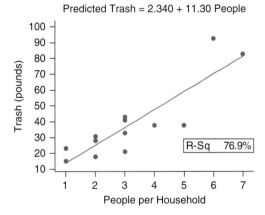

Predicted Trash = 2.340 + 11.30 People

R-Sq   76.9%

**4.63 Education of Fathers and Mothers** The data shown in the table are the number of years of formal education of the fathers and mothers of a sample of 29 statistics students at a small community college in an area with many recent immigrants. (The means are both about 8, and the standard deviations are both about 4.6.) The scatterplot (not shown) suggests a linear trend.

| Father | Mother | Father | Mother |
|--------|--------|--------|--------|
| 3 | 3 | 4 | 7 |
| 11 | 13 | 16 | 13 |
| 3 | 4 | 13 | 3 |
| 12 | 12 | 11 | 10 |
| 8 | 16 | 6 | 6 |
| 1 | 2 | 3 | 2 |
| 8 | 8 | 12 | 14 |
| 6 | 4 | 14 | 12 |
| 12 | 8 | 12 | 12 |
| 0 | 10 | 12 | 13 |
| 8 | 12 | 12 | 18 |
| 5 | 5 | 13 | 8 |
| 8 | 8 | 0 | 0 |
| 5 | 5 | 3 | 5 |
| 12 | 6 | | |

a. Find and report the regression equation for predicting the mother's years of education from the father's. Then find the predicted number of years for the mother if the father has 12 years of education, and find the predicted number of years for a mother if the father has 4 years of education.

b. Find and report the regression equation for predicting the father's years of education from the mother's. Then find the predicted number of years for the father if the mother has 12 years of education, and find the predicted number of years for the father if the mother has 4 years of education.

c. What phenomenon from the chapter does this demonstrate? Explain.

\* **4.64 Heights of Fathers and Sons** The table shows some data on a sample of heights of fathers and their sons.

You may want to use the computer or a statistics calculator to verify that

$$\bar{x}_{father} = 69.233 \quad s_{father} = 2.82$$
$$\bar{y}_{son} = 69.033 \quad s_{son} = 2.84$$
$$r = 0.762$$

The scatterplot (not shown) suggests a linear trend.

a. Find and report the regression equation for predicting the son's height from the father's height. Then predict the height of a son with a father 74 inches tall. Also predict the height of a son of a father who is 65 inches tall.

b. Find and report the regression equation for predicting the father's height from the son's height. Then predict a father's height from that of a son who is 74 inches tall and also predict a father's height from that of a son who is 65 inches tall.

c. What phenomenon does this show?

| Father's Height | Son's Height |
|-----------------|--------------|
| 75 | 74 |
| 72.5 | 71 |
| 72 | 71 |
| 71 | 73 |
| 71 | 68.5 |
| 70 | 70 |
| 69 | 69 |
| 69 | 66.5 |
| 69 | 72 |
| 68.5 | 66.5 |
| 67.5 | 65.5 |
| 67.5 | 70 |
| 67 | 67 |
| 65.5 | 64.5 |
| 64 | 67 |

g * **4.65 Test Scores** Assume that in a political science class, the teacher gives a midterm exam and a final exam. Assume that the association between midtern and final scores is linear. The summary statistics have been simplified for clarity.

Midterm:   Mean = 75,   Standard deviation = 10

Final:   Mean = 75,   Standard deviation = 10

Also, $r = 0.7$ and $n = 20$.

According to the regression equation, for a student who gets a 95 on the midterm, what is the predicted final exam grade? What phenomenon from the chapter does this demonstrate? Explain. *See page 189 for guidance.*

* **4.66 Test Scores** Assume that in a sociology class, the teacher gives a midterm exam and a final exam. Assume that the association

between midterm and final scores is linear. Here are the summary statistics:

Midterm:   Mean = 72,   Standard deviation = 8

Final:   Mean = 72,   Standard deviation = 8

Also, $r = 0.75$ and $n = 28$.

a. Find and report the equation of the regression line to predict the final exam score from the midterm score.

b. For a student who gets 55 on the midterm, predict the final exam score.

c. Your answer to part b should be higher than 55. Why?

d. For a student who gets a 100 on the midterm, without doing any calculations, state whether the predicted value would be higher, lower, or the same as 100.

## CHAPTER REVIEW EXERCISES

* **4.67 Heights and Weights of People** The table shows the height and weight of some people. The figure shows that the association is linear enough to proceed.

| Height (inches) | Weight (pounds) |
|---|---|
| 60 | 105 |
| 66 | 140 |
| 72 | 185 |
| 70 | 145 |
| 63 | 120 |

a. Calculate the correlation, and find and report the equation of the regression line, using height as the predictor and weight as the response.

b. Change the height to centimeters by multiplying each height in inches by 2.54. Find the weight in kilograms by dividing the weight in pounds by 2.205. Retain at least six digits in each number so there will be no errors due to rounding.

c. Calculate the correlation between height in centimeters and weight in kilograms, and compare it with the correlation for the heights in inches and the weights in pounds.

d. Find the equation of the regression line for predicting weight from height, using height in cm and weight in kg. Is the equation for weight (in pounds) and height (in inches) the same or different from the equation for weight (in kg) and height (in cm)?

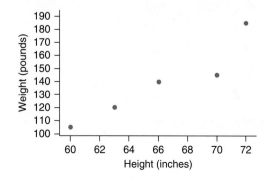

* **4.68 Heights and Weights of Men** The table shows the heights (in inches) and weights (in pounds) of 14 college men. The figure shows that the association is linear enough to proceed.

| Height (inches) | Weight (pounds) | Height (inches) | Weight (pounds) |
|---|---|---|---|
| 68 | 205 | 70 | 200 |
| 68 | 168 | 69 | 175 |
| 74 | 230 | 72 | 210 |
| 68 | 190 | 72 | 205 |
| 67 | 185 | 72 | 185 |
| 69 | 190 | 71 | 200 |
| 68 | 165 | 73 | 195 |

a. Find the equation for the regression line with weight (in pounds) as the response and height (in inches) as the predictor. Report the slope and intercept of the regression line and explain what they show. If the intercept is not appropriate to report, explain why.

b. Find the correlation between weight (in pounds) and height (in inches).

c. Find the coefficient of determination and interpret.

d. If you changed each height to centimeters by multiplying heights in inches by 2.54, what would the new correlation be? Explain.

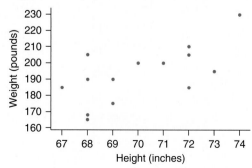

e. Find the equation with weight (in pounds) as the response and height (in cm) as the predictor, and interpret the slope.

f. Summarize what you found: Does changing units change the correlation? Does changing units change the regression equation?

**4.69 House Price and Square Feet** The figure gives a scatterplot, using data taken from *The Ventura County Star* (February, 2006), of the number of square feet in some houses and their list price for sale.

a. Choose the correct correlation from these choices: +1.00, 0, −1.00, −0.75, +0.75.

b. The equation of the regression line is given above the graph. Report the slope and intercept of the regression line and explain what they show in this context. If the intercept is not appropriate to report, explain why.

c. By looking at the graph or using the equation, predict the average cost of a house that is about 1300 square feet.

d. Find the coefficient of determination from the correlation that you chose in part a and explain what it means in the context of the problem.

Predicted Price = 223.5 + 0.174 (Square Feet)

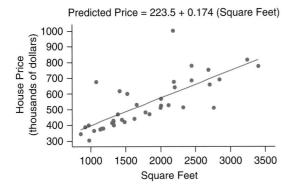

**4.70 Heights and Weights of First and Second Graders** Some statistics students recorded the heights and weights of a sample of first and second graders. Refer to the figure.

a. Choose the correct correlation coefficient from these choices: +1.00, 0, −1.00, −0.84, +0.84.

b. The equation is given above the graph. Report the slope and intercept of the regression line and explain what they show in this context. If the intercept is not appropriate to report, explain why.

c. By looking at the graph or using the equation of the regression line, predict the weight for a first or second grader who is 54 inches tall.

d. Find the coefficient of determination from the correlation that you chose in part a and explain what it means in the context of the problem.

Predicted Weight = −83.05 + 2.993 Height

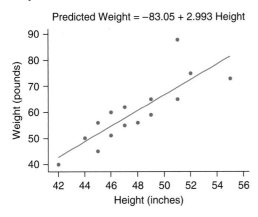

**4.71 Hours of Study and of TV Viewing** The number of hours of study and the number of hours of TV viewing per week were recorded for some community college students.

a. Make a scatterplot of the data, using hours of study as the predictor.

b. Describe the association, if any, between hours of study and hours of TV viewing.

c. Is the correlation strong and positive, strong and negative, or near zero? Do not calculate.

**4.72 Salary and Age** The age and salary (dollars per hour) were recorded for some college students with jobs.

a. Make a scatterplot of the data, using age as the predictor.

b. Find the equation of the regression line.

c. Report the slope and intercept of the regression line and explain what they show. If the intercept is not appropriate to report, explain why not.

**4.73 Age and Weight for Women** The figure shows a scatterplot and regression line for the age and weight of some women in a statistics class.

a. Find the correlation by taking the positive square root of R-Sq shown on the graph and attaching the proper sign.

b. Using the equation given, find the predicted weight for a woman 35 years old.

*c. Report the slope and explain what it means. This study did not follow the same people over time.

d. If it is appropriate, report the intercept and explain what it means. Otherwise, explain why it is not appropriate to report.

Predicted Weight = 111.4 + 0.8820 Age

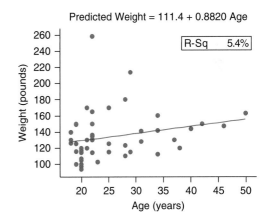

**4.74 Age and Weight for Men** The figure on the next page shows a scatterplot and regression line for the age and weight of some men in a statistics class.

a. Find the correlation by taking the positive square root of R-Sq shown on the graph and attaching the proper sign.

b. Using the equation given, find the predicted weight for a man 35 years old.

*c. Report the slope and explain what it means. (Note that this study did not follow the same people over time.)

d. Either report the intercept and explain what it shows or explain why it should not be reported.

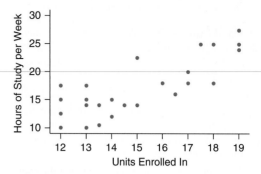

Predicted Weight = 132.9 + 2.040 Age

R-Sq    17.0%

**4.75** The figure shows a scatterplot of the educational level of twins. Describe the scatterplot. Explain the trend and mention any unusual points. (Source: www.stat.ucla.edu)

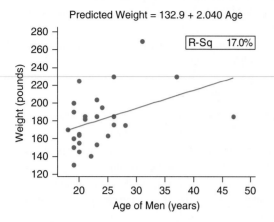

**4.76 Wages and Education** The figure shows a scatterplot of the wages and educational level of some people. Describe what you see. Explain the trend and mention any unusual points. (Source: www.stat.ucla.edu)

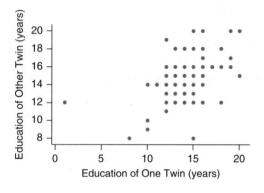

**4.77 Do Students Taking More Units Study More Hours?** The figure shows the number of units that students were enrolled in and the number of hours (per week) that they reported

studying. Do you think there is a positive trend, a negative trend, or no noticeable trend? Explain what this means about the students.

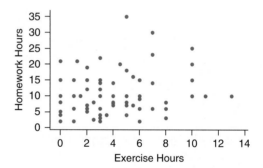

**4.78 Hours of Exercise and Hours of Homework** The scatterplot shows the number of hours of exercise per week and the number of hours of homework per week for some students. Explain what it shows.

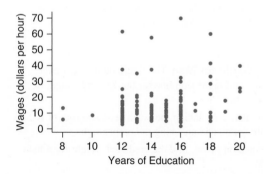

**4.79 Tree Heights** Loggers gathered information about some trees. The diameter is in inches, the height is in feet, and the volume of the wood is in cubic feet. Loggers are interested in whether they can estimate the volume of the tree given any single dimension. Which is the better predictor of volume: the diameter or the height?

**4.80 Salary and Education** Does education pay? The salary per year in dollars, the number of years employed (YrsEm), and number of years of education after high school (Educ) for the employees of a company were recorded. Determine whether number of years employed or number of years of education after high school is a better predictor of salary. Explain your thinking. (Source: Minitab File)

**4.81 Film Budgets and Grosses** Movie studios spend much effort trying to predict how much money their movies will make. One possible predictor is the amount of money spent on the production of the movie. In the table on the next page, you can see the budget and the amount of money made worldwide for the 13 movies with the highest profit. The budget and gross are in millions of dollars. Make a scatterplot and comment on what you see. If appropriate, find, report, and interpret the regression line. If it is not appropriate to do so, explain why. (Source: www.the-numbers.com)

| Movie | Budget | Gross |
|---|---|---|
| *Titanic* | 200.0 | 1848.8 |
| *The Lord of the Rings: The Return of the King* | 94.0 | 1133.0 |
| *Jurassic Park* | 63.0 | 920.1 |
| *Star Wars Ep. IV: A New Hope* | 11.0 | 797.9 |
| *Shrek 2* | 70.0 | 915.3 |
| *ET: The Extra-Terrestrial* | 10.5 | 792.9 |
| *Pirates of the Caribbean: Dead Man's Chest* | 150.0 | 1065.7 |
| *The Lord of the Rings: The Two Towers* | 94.0 | 926.3 |
| *Harry Potter and the Sorcerer's Stone* | 125.0 | 976.5 |
| *Star Wars Ep. I: The Phantom Menace* | 115.0 | 924.3 |
| *Harry Potter and the Chamber of Secrets* | 100.0 | 879.0 |
| *Finding Nemo* | 94.0 | 866.6 |
| *Avatar* | 237.0 | 2212.0 |

**4.82 Gas Mileage of Cars** The table gives the number of miles per gallon in the city and on the highway for the cars reported to have the best gasoline mileage in 2008, according to CNN. Make a scatterplot, using the city mileage as the predictor. Find the equation of the regression line for predicting the number of miles per gallon (mpg) on the highway from the number of miles per gallon in the city. Use the equation to predict the highway mileage from a city mileage of 60 mpg. Also find the coefficient of determination and explain what it means.

| | City | Highway |
|---|---|---|
| Honda Insight | 60 | 66 |
| Toyota Prius | 60 | 51 |
| Honda Civic Hybrid | 48 | 47 |
| Volkswagen Diesel | 38 | 46 |
| Honda Civic | 36 | 44 |
| Toyota Echo | 35 | 43 |
| Toyota Scion | 32 | 38 |
| Toyota Corolla | 32 | 40 |
| Toyota Celica | 29 | 36 |

For Exercises 4.83–4.86, show your points in a rough scatterplot and give the coordinates of the points.

* **4.83** Construct a small set of numbers with at least three points with a perfect positive correlation of 1.00.

* **4.84** Construct a small set of numbers with at least three points with a perfect negative correlation of −1.00.

* **4.85** Construct a set of numbers (with at least three points) with a strong negative correlation. Then add one point (an influential point) that changes the correlation to positive. Report the data and give the correlation of each set.

* **4.86** Construct a set of numbers (with at least three points) with a strong positive correlation. Then add one point (an influential point) that changes the correlation to negative. Report the data and give the correlation of each set.

# gUIDED EXERCISES

g **4.41 Does the Cost of a Flight Depend on the Distance?** The table gives the round-trip fare for flights between Boston and some other cities on major airlines. These were the lowest prices shown on www.Sidestep.com on January 21, 2010. The airlines varied. (The travel dates were all the same, in early February.)

How much would it cost, on average, to fly 500 miles? To answer this question, perform a complete regression analysis, including a scatterplot with a regression line, following the steps below.

**Round-trip Air Fares from Boston**

| City | Cost | Miles |
|------|------|-------|
| NY | 179 | 206 |
| Chicago | 179 | 963 |
| San Fran | 397 | 3095 |
| Denver | 274 | 1940 |
| LA | 314 | 2979 |
| WashDC | 99 | 429 |
| Atlanta | 317 | 1037 |
| Seattle | 443 | 2976 |
| Dallas | 404 | 1748 |
| St. Louis | 334 | 1141 |

**Step 1 ▶ Make a scatterplot**
Be sure that miles is the *x*-variable and cost is the *y*-variable. You may want to include the regression line with your scatterplot if you have that option with your technology; refer to Step 4.

**Step 2 ▶ Is the linear model appropriate?**
In this case the answer is yes, because there is a linear trend. It is hard to see with so few points, but a strong curvature is not present and the cost tends to increase as the miles increase.

**Step 3 ▶ Obtain the equation**
When finding the regression equation, be sure that you use miles as the *x*-variable and cost as the *y*-variable. For example, if you are using the TI-83/84, the predictor (which you probably put into List 2 because the second column of numbers has the miles) has to be entered first and the response (in List 1) is entered second after the comma. See the TI-83/84 figure.

**Step 4 ▶ Add the regression line to the scatterplot**
If your technology will make a plot with a line, do so. Refer to the TechTips, which begin on page 191, to see how to do this.

If your technology will not draw the line, you can choose two *x*-values (distances), find the corresponding *y*-values (cost), plot these, and then draw the line to connect them. For example,

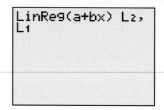

to choose the first point on the line, choose an *x*-value such as $x = 200$ and follow these additional steps:

- **First Predicted Point**
  Pick an arbitrary, fairly small distance such as 200 miles (not less than the smallest distance given) and substitute into the equation you got to find the predicted cost:

  $$\text{Predicted Cost} = 162.60 + 0.0795 \text{ Miles}$$
  $$= 162.60 + 0.0795(200)$$
  $$= 162.60 + 15.90$$
  $$= 178.5$$

  A predicted point is (200, 178.5), so a flight of about 200 miles should cost about $178.50. Put the point on the graph with a symbol you will remember.

- **Second Point**
  To get a second point, pick an arbitrary large value for distance such as 3000 (not larger than the largest distance) and substitute it into the equation to find the cost. Put the point on the graph using the same symbol you chose before.

- **The Line**
  The regression line will be a straight line between these two predicted points (use a ruler to make the line). Extend the line out to the edges of the data, to the left to about 100 miles, and to the right to about 3000 miles.

**Step 5 ▶ Interpret the slope and intercept in context**

Predicted  Y =  a  +  b  X
Predicted Cost = 162.60 + 0.0795 Miles

The slope is *b* (which is the multiplier for *Miles*), which is 0.0795, and the intercept is *a,* which is the first number, $162.60.

Fill in the blanks that follow.

For the slope: For every additional mile, on average, the price goes up by ____ dollars.
For the intercept: A trip with zero miles should cost about ____ dollars. Explain why interpreting the *y*-intercept like this is questionable.

**Step 6 ▶ Answer the question by using the regression equation**
How much would it cost to fly 500 miles?

g **4.65 Test Scores**   Assume that in a political science class, the teacher gives a midterm exam and a final exam. Assume that the association between midterm and final scores is linear. The summary statistics have been simplified for clarity.

| Midterm: | Mean = 75, | Standard deviation = 10 |
| Final: | Mean = 75, | Standard deviation = 10 |

Also, $r = 0.7$ and $n = 20$.

For a student who gets 95 on the midterm, what is the predicted final exam grade? Assume the graph is linear.

**Step 1** ▶ Find the equation of the line to predict the final exam score from the midterm score. Standard form: $y = a + bx$

    a. First find the slope: $b = r\left(\dfrac{s_{final}}{s_{midterm}}\right)$

b. Then find the $y$-intercept, $a$, from the equation

$$a = \bar{y} - b\bar{x}$$

c. Write out the following equation:

$$\text{Predicted } y = a + bx$$

However, use "Predicted Final" instead of "Predicted $y$" and "Midterm" in place of $x$.

**Step 2** ▶ Use the equation to predict the final exam score for a student who gets 95 on the midterm.

**Step 3** ▶ Your predicted final exam grade should be less than 95. Why?

## CHECK YOUR TECH

# Verifying Minitab Output for Correlation and Regression

| Husband | Wife |
|---------|------|
| 20 | 20 |
| 30 | 30 |
| 40 | 25 |

At the left, data are given for the ages at which husbands and wives married. The data were simulated to make the calculations easier.

We used Minitab to find the correlation and regression equation given in Figure A, using the husband's age ($x$) to predict his wife's age ($y$).

▶ **FIGURE A**

```
Pearson correlation of Husband and Wife = 0.500
The regression equation is
Wife = 17.5 + 0.250 Husband
```

From Figure B: for the husband, $\bar{x} = 30$ and $s_x = 10$; for the wife, $\bar{y} = 25$ and $s_y = 5$.

$$r = \frac{\sum z_x z_y}{n-1} \text{ and } z_x = \frac{x - \bar{x}}{s_x} \text{ and } z_y = \frac{y - \bar{y}}{s_y}$$

▶ **FIGURE B**

```
Variable   N    Mean   StDev
Husband    3   30.00   10.00
Wife       3   25.00    5.00
```

**QUESTION**  Using the summary statistics provided in Figure B, verify the output given in Figure A by following the steps below.

**SOLUTION**

**Step 1** ▶ Fill in the missing numbers (indicated by the blanks) in the table.

| $x$ | $x - \bar{x}$ | $z_x$ | $y$ | $y - \bar{y}$ | $z_y$ | $z_x z_y$ |
|-----|---------------|-------|-----|---------------|-------|-----------|
| 20 | $20 - 30 = -10$ | $-10/10 = -1$ | 20 | $20 - 25 = -5$ | $-5/5 = -1$ | $(-1) \times (-1) = 1$ |
| 30 | $30 - 30 = 0$ | $0/10 = \_$ | 30 | $30 - \_ = \_$ | $5/5 = \_$ | $0 \times (\_) = \_$ |
| 40 | $40 - 30 = 10$ | $\_/\_ = \_$ | 25 | $\_ - \_ = \_$ | $0/\_ = \_$ | $\_ \times (\_1) = \_$ |

**Step 2** ▶ Add the last column to get $\sum z_x z_y = \_$.

**Step 3** ▶ Find the correlation and check it with the output.

$$r = \frac{\sum z_x z_y}{n-1} = \frac{\_\_}{3-1} = \frac{\_\_}{2} = \_$$

**Step 4** ▶ Find the slope: $b = r\dfrac{s_y}{s_x} = r\dfrac{5}{10} = \_\_$.

**Step 5** ▶ Find the $y$-intercept: $a = \bar{y} - b\bar{x} = 25 - 30b = \_\_$.

**Step 6** ▶ Finally, put together the equation:

$$y = a + bx$$

Predicted Wife $= a + b$ Husband

and check the equation with the Minitab output. The equations should match.

# TechTips

## General Instructions for All Technology

Upload data from the disk or website, or enter data manually using two columns of equal length. Refer to TechTips in Chapter 2 for a review of entering data. Each row represents a single observation, and each column represents a variable. All technologies will use the example that follows.

**EXAMPLE** ►Analyze the six points in the data table with a scatterplot, correlation, and regression. Use heights (in inches) as the x-variable and weight (in pounds) as the y-variable.

| Height | Weight |
|--------|--------|
| 61 | 104 |
| 62 | 110 |
| 63 | 141 |
| 64 | 125 |
| 66 | 170 |
| 68 | 160 |

### TI-83/84

These steps assume you have entered the heights into **L1** and the weights into **L2**.

#### Making a Scatterplot

1. Press **2ND**, **STATPLOT** (which is the button above **2ND**), **4**, and **ENTER**, to turn off plots made previously.
2. Press **2ND**, **STATPLOT**, and **1** (for Plot1).
3. Refer to Figure 4A: Turn on **Plot1** by pressing **ENTER** when **On** is highlighted.

▲ **FIGURE 4A** TI-83/84 Plot1 Dialogue Screen

4. Use the arrows on the keypad to get to the scatterplot (upper left of the six plots) and press **ENTER** when the scatterplot is highlighted. Be careful with the **Xlist** and **Ylist**. To get **L1**, press **2ND** and **1**. To get **L2**, press **2ND** and **2**.
5. Press **GRAPH**, **ZOOM** and **9** (**Zoomstat**) to create the graph.
6. Press **TRACE** to see the coordinates of the points, and use the arrows on the keypad to go to other points. Your output will look like Figure 4B, but without the line.

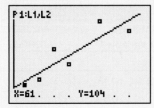

▲ **FIGURE 4B** TI-83/84 Plot with Line

7. To get the output with the line in it, shown in Figure 4B:
   **STAT**, **CALC**, 8:LinReg(a+bx), **L1 ? L2 ?** **Y1** (You get the **Y1** by pressing **VARS**, **Y-VARS**,1: **Function**, 1:**Y1**, **ENTER**.) The comma button is on the keypad above the 7 button.
8. Press **GRAPH**, **ZOOM** and **9**.
9. Press **TRACE** to see the numbers, and use the arrows on the keypad to get to other numbers.

#### Finding the Correlation and Regression Equation Coefficients

Before finding the correlation, you must turn the diagnostics on, as shown here.

Press **2ND**, **CATALOG**, and scroll down to **DiagnosticOn** and press **ENTER** twice. The diagnostics will stay on unless you **Reset** your calculator or change the batteries.

1. Press **STAT**, choose **CALC**, and **8** (for LinReg (a + bx)).
2. Press **2ND L1** (or whichever list is X, the predictor), press **?** (comma: the button above the 7), press **2ND L2** (or whichever list is Y, the response), and press **ENTER**.

Figure 4C shows the output.

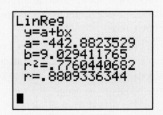

▲ **FIGURE 4C** TI-83/84 Output

## Making a Scatterplot

1. **Graph > Scatterplot**
2. Leave the default **Simple** and click **OK.**
3. Double click the column containing the weights so that it goes under the **Y Variables**. Then double click the column containing the heights so that it goes under the **X Variables**.
4. Click **OK**. After the graph is made, you can edit the labels by clicking on them.

## Finding the Correlation

1. **Stat > Basic Statistics > Correlation**
2. Double click both the predictor column and the response column (in either order).
3. Click **OK**. You will get 0.881.

## Finding the Regression Equation Coefficients

1. **Stat > Regression > Regression**
2. Put in the **Response** (y) and **Predictor** (x) columns.
3. Click **OK**. You may need to scroll up to see the regression equation. It will be easier to understand if you have put in

labels for the columns, such as "Height" and "Weight." You will get: Weight = −443 + 9.03 Height.

## To Display the Regression Line on a Scatterplot

1. **Stat > Regression > Fitted Line Plot**
2. Double click the **Response (y)** column and then double click the **Predictor (x)** column.
3. Click **OK**. Figure 4D shows the fitted line plot.

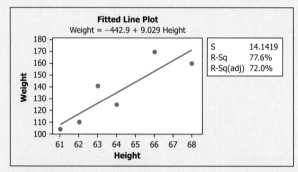

▲ **FIGURE 4D** Minitab Fitted Line Plot

## Making a Scatterplot

1. Select (highlight) the two columns containing the data, with the predictor column to the *left* of the response column. You may include the labels at the top or not include them.
2. Click **Insert**, in **Charts** click the picture of a scatterplot, and click the upper left option shown here:

3. Click **OK**.
4. Note that the lower left corner of the chart is not at the origin, (0, 0). If you want to zoom in or out on the data by changing the minimum value for an axis, right-click on the axis numbers, click **Format axis**, in **Axis Options** change the **Minimum** to **Fixed**, and put in the desired value. You may want to do this twice: once for the x-axis and once for the y-axis. Then click **Close**.
5. When the chart is active (click on it), **Chart Tools** are shown at the top of screen, right of center. Click **Layout** (not **Page Layout**), then **Axis Titles**, and **Chart Title** to add appropriate labels. After the labels are added, you can click on them to change the spelling or add words. Delete the labels on the right-hand side, such as **Series 1**, if you see any.

## Finding the Correlation

1. Click on **Data**, click on **Data Analysis**, select **Correlation**, and click **OK**.

2. For the **Input Range**, select (highlight) both columns of data (if you have highlighted the labels as well as the numbers, you must also click on the **Labels in first row**).
3. Click **OK**. You will get 0.880934.

(Alternatively, just click the $f_x$ button, for **category** choose **statistical**, select **CORREL**, click **OK**, and highlight the two columns containing the numbers, one at a time. The correlation will show up on the dialogue screen, and you do *not* have to click **OK**.)

## Finding the Coefficients of the Regression Equation

1. Click on **Data**, **Data Analysis**, **Regression**, and **OK**.
2. For the **Input Y Range**, select the column of numbers (not words) that represents the response or dependent variable. For the **Input X Range**, select the column of numbers that represents the predictor or independent variable.
3. Click **OK**.

A large summary of the model will be displayed. Look under **Coefficients** at the bottom. For the **Intercept** and the slope (next to **XVariable1**), see Figure 4E, which means the regression line is

$$y = -442.9 + 9.03x$$

|  | *Coefficients* |
|---|---|
| Intercept | -442.882 |
| X Variable 1 | 9.029412 |

▲ **FIGURE 4E** Excel Regression Output

### To Display the Regression Line on a Scatterplot

4. After making the scatterplot, under **Chart Tools** click **Design**. In the **Chart Layouts** group, click the triangle to the right of **Quick Layout**. Choose Layout 9 (the option in the lower right portion, which shows a line in it and also *fx*).

   Refer to Figure 4F.

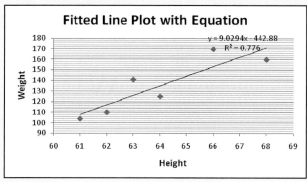

▲ **FIGURE 4F** Excel Fitted Line Plot with Equation

---

### Making a Scatterplot

1. **Graphics > Scatterplot**
2. Select an **X variable** and a **Y variable** for the plot.
3. Click **Create Graph!** to construct the plot.
4. To copy the graph, click **Options** and **Copy**.

### Finding the Correlation and Coefficients for the Equation

1. **Stat > Regression > Simple Linear**
2. Select the **X variable** and **Y variable** for the regression.
3. Click **Calculate** to view the equation and numbers, which are shown in Figure 4G.

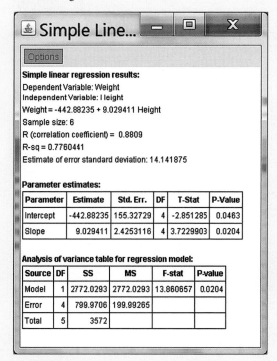

▲ **FIGURE 4G** StatCrunch Regression Output

### Plotting the Regression Line on a Scatterplot

1. **Click Stat > Regression > Simple Linear**
2. Select your columns for X and Y and click **Next**.
3. Don't select any **Options** and click **Next** again. Do this step twice.
4. Check **Plot the fitted line**.
5. Click **Calculate**. When you see the numbers in the output, enlarge it and click **Next** at the very bottom to see the Fitted line plot. Refer to Figure 4H.
6. To copy the graph, click **Options** and **Copy**.

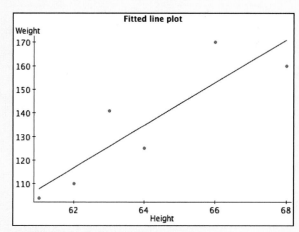

▲ **FIGURE 4H** StatCrunch Fitted Line Plot

# 5 Modeling Variation with Probability

## THEME

Probabilities are long-run relative frequencies used to describe processes where the outcome depends on chance, such as flipping a coin or rolling a die. Theoretical probabilities are based on specific assumptions (usually based on a theory) about the chance process. Empirical probabilities are based on observation of the actual process. The two types of probabilities are closely related, and we need both to analyze samples of data in the real world.

I n 1971, the United States was fighting the Vietnam War and drafting men to serve in the military. To determine who was chosen, government officials wrote the days of the year (January 1, January 2, and so on) on capsules. The capsules were placed in a large container and mixed up. They were then drawn out one at a time. The first date chosen was assigned the rank 1, the second date was assigned the rank 2, and so on. Men were drafted on the basis of their birthday. Those whose birthday had rank 1 were drafted first, then those whose birthday had rank 2, and so on until the officials had enough men.

Although the officials thought that this method was random, some fairly convincing evidence indicates that it was not (*Journal of Statistics Education*). Figure 5.1a shows boxplots with the actual ranks for each month. Figure 5.1b shows what boxplots might have looked like if the lottery had been truly random. In Figure 5.1b, each month has roughly the same rank. However, in Figure 5.1a, a few months had notably lower ranks than the other months. Bad news if you were born in December—you were more likely to be called up first.

What went wrong? The capsules, after having dates written on them, were clustered together by month as they were put into the tumbler. But the capsules weren't mixed up enough to break up these clusters. The mixing wasn't adequate to create a truly random mix.

It's not easy to generate true randomness, and humans have a hard time recognizing random events when they see them. Probability gives us a tool for understanding randomness. Probability helps us answer the question "How often does something like this happen by chance?" By answering that question, we create an important link between our data and the real world. In previous chapters, you learned how to organize, display, and summarize data to see patterns and trends. Probability is a vital tool because it gives us the ability to generalize our understanding of data to the larger world. In this chapter, we'll explore issues of randomness and probability: What is randomness? How do we measure it? And how do we use it?

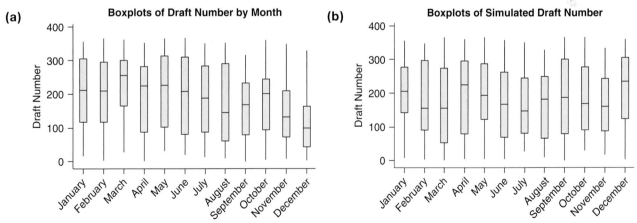

▲ **FIGURE 5.1** Boxplots of **(a)** actual Vietnam draft numbers by month, and **(b)** what might have happened if the draft had really been random.

## CASE STUDY

## SIDS or Murder?

In November 2000, Sally Clark was convicted in England of killing her two children. Her children had died several years apart, and the initial diagnosis was sudden infant death syndrome (SIDS). SIDS is the sudden, unexplained death of an infant under 1 year of age. It is estimated that in the United States, about 5500 infants die of SIDS every year. Although some risk factors have been identified (most are related to the mother's health during pregnancy), the cause or causes are as yet unknown.

Clark was convicted of murder on the basis of the expert testimony of Sir Roy Meadow, a prominent British physician who was an expert on SIDS. Dr. Meadow quoted a published statistic: The probability of an infant dying of SIDS is 1/8543. That is, in the U.K., one child in every 8543 dies of SIDS. Dr. Meadow concluded that the probability of two children in the same family dying of SIDS was therefore (1/8543) × (1/8543), or about 1 in 73 million. The jury agreed that this event—two children in the same family dying of SIDS—was so unlikely that they convicted Ms. Clark of murder, and she was sent to prison.

But then in 2003, Sally Clark's conviction was overturned and she was released from prison. Dr. Meadow was accused of professional misconduct. Why did Dr. Meadow multiply the two probabilities together? Why was the verdict overturned? We will answer these questions at the end of the chapter.

# What Is Randomness?

What exactly is randomness? The *Oxford English Dictionary* says *random* means "having no definite aim or purpose." You probably use the word to describe events that seem to happen with no purpose at all. Sometimes the word is used to describe things that have no apparent pattern. However, in statistics the word *random* has a more precise meaning, as you will see.

We can do a small experiment to compare our natural understanding of randomness with real-life randomness. We asked a student to imagine flipping a coin and to write down the resulting heads (H) and tails (T). To compare this to a real random sequence, we flipped an actual coin 20 times and recorded the results. Which of the sequences below do you think is real and which is invented by the student?

| H | H | T | H | T | H | H | T | T | T | H | T | H | H | H | H | T | H | T | T | H |
| T | T | T | H | H | H | T | T | T | T | T | H | H | H | H | T | T | H | T | H |

The first row of results is the one made up by the student, and the second row records the results of actually tossing a coin 20 times. Is it possible to tell by comparing the two sequences? Not always, but this time the student did something that most people do in this situation. Most of the time, he wrote very short "streaks." (A streak is any sequence of consecutive heads or tails. TT is a streak of two tails.) Only once did he write as many as three heads or three tails in a row. It's as if, after writing three heads in a row, he thought, "Wait! If I put a fourth, no one will believe it is random."

However, the real, truly random sequence had one streak of five tails (beginning at the seventh flip) and another streak with four heads. These long streaks are examples of the way chance creates things that look like a pattern or structure, but really are not.

## You Try It

Use a computer to flip a coin 20 times for you. Go to http://www.socr.ucla.edu.

Click on the **Experiments** tab and select **Coin Sample Experiment** from the dropdown menu that will appear. Set $n$ to be 20 (by moving the slider to the left of $n = 10$), and set $p$ to be 0.50. Press the **Play** (▶) button, and count the resulting streaks. Do this a few times, and keep track of the longest streak you get each time. How long does the longest streak have to be before you think it is unusually long?

 People are not good at identifying truly random samples or random experiments, so we need to rely on outside mechanisms such as coin flips or random number tables.

Real randomness is hard to achieve without help from a computer or some other randomizing device. If a computer is not available to generate random numbers, another useful approach is to use a random number table. (An example is provided in Appendix A of this book.) A random number table provides a sequence of digits from 0 to 9 in a random order. Here, **random** means that no predictable pattern occurs and that no digit is more likely to appear than any other. (Of course, if you use the same table often enough, it might seem predictable to you, but to an outsider, it will not seem predictable.)

For example, if we are doing a controlled experiment, we might assign each subject in our study a random number from this table as the subjects come into our office. The odd-numbered subjects would then go into the Control group, and the even-numbered would go into the Treatment group.

To use a random number table to simulate coin flipping, we assign a number to each outcome. For example, suppose we let even numbers (0, 2, 4, 6, 8) represent tails and let odd numbers (1, 3, 5, 7, 9) represent heads. Now choose any row of the table you wish and any column. For example, we arbitrarily chose column 11 and row 30 because that's the date of the day we wrote this paragraph (November 30, or 11/30). Read off the next 20 digits, but translate them into "heads" and "tails." What's the longest streak you get?

| Line | | | | | | |
|------|-------|-------|-------|-------|-------|-------|
| 28 | 3 1 4 9 8 | 8 5 3 0 4 | 2 2 3 9 3 | 2 1 6 3 4 | 3 4 5 6 0 | 7 7 4 0 4 |
| 29 | 9 3 0 7 4 | 2 7 0 8 6 | 6 2 5 5 9 | 8 6 5 9 0 | 1 8 4 2 0 | 3 3 2 9 0 |
| 30 | 9 0 5 4 9 | 5 3 0 9 4 | 7 6 2 8 2 | 5 3 1 0 5 | 4 5 5 3 1 | 9 0 0 6 1 |
| 31 | 1 1 3 7 3 | 9 6 8 7 1 | 3 8 1 5 7 | 9 8 3 6 8 | 3 9 5 3 6 | 0 8 0 7 9 |
| 32 | 5 2 0 2 2 | 5 9 0 9 3 | 3 0 6 4 7 | 3 3 2 4 1 | 1 6 0 2 7 | 7 0 3 3 6 |

◀ **TABLE 5.1** Lines 28–32 (indicated at the left side of the table) from the random number table in Appendix A. The red 7 in the 11th column and the 30th line, or row, is our starting point.

## EXAMPLE 1  Simulating Randomness

Let's play a game. You roll a six-sided die until the first 6 appears. You win one point for every roll. You pass the die to your opponent, who also rolls until the first 6 appears. Whoever has the most points, wins.

**QUESTION** Simulate rolling the die until a 6 appears. Use Table 5.1, and start at the very first entry of line 28 in Table 5.1. How many rolls did it take?

SOLUTION We'll let the digits 1 through 6 represent the outcome shown on the die. We'll ignore the digits 7, 8, 9, and 0. Starting at line 28, we get these random digits:

3, 1, 4, (ignore, ignore, ignore) 5, 3, (ignore), 4, 2, 2, 3, (ignore), 3, 2, 1, 6

CONCLUSION We rolled the die 12 times before the first 6 appeared.

TRY THIS! Exercise 5.1

Computers and calculators have random number generators that come close to true randomness. For instance, the Internet game you played (to simulate flipping the coin 20 times) uses a random number generator. Computer-generated random numbers are sometimes called pseudo-random numbers, because they are generated on the basis of a seed value—a number that starts the random sequence. If you input the same seed number, you will always see the same sequence of pseudo-random numbers. However, it is not possible to predict what number will come next in the sequence, and in this sense the generated numbers are considered random. For most practical work (and certainly for everything we cover in this book), these pseudo-random numbers are as random as we need. (You should be aware, though, that not all statistics packages produce equally convincing sequences of random numbers.)

Less technological ways exist for generating random outcomes, such as actually flipping a coin. When you play a card game, you carefully shuffle the cards to mix them up. In the games of Mah Jong and tile rummy, players create randomness by scrambling the tiles on the tabletop. In many board games, players either roll dice or spin a spinner to determine how far they will move their game pieces. In raffles, tickets are put into a basket, the basket is spun, and then a ticket is drawn by someone who does so without looking into the basket.

Such physical randomizations must be done with care. A good con artist can flip a coin so that it always comes up heads. A child learns quickly that a gentle tap of a spinner can move the spinner to the desired number. A deck of cards needs to be shuffled at least seven times in order for the result to be considered random (as mathematician and magician Persi Diaconnis proved). Many things that we think are random might not be. This is the lesson the government learned from its flawed Vietnam draft lottery, which was described in the chapter introduction. Quite often, statisticians are employed to check whether random processes, such as the way winners are selected for state lotteries, are truly random.

## Empirical and Theoretical Probabilities

**Probability** is used to measure how often random events occur. When we say, "The probability of getting heads when you flip a coin is 50%," we mean that flipping a coin results in heads about half the time (assuming that the outcome of the flips is random). This is sometimes called the "frequentist" approach to probability, because in it, probabilities are defined as relative frequencies.

We will examine two kinds of probabilities. These two kinds of probabilities are connected, as we will see later in this chapter. **Theoretical probabilities** are long-run relative frequencies. The theoretical probability is the relative frequency at which an event happens after *infinitely* many repetitions. When we say that a coin has a 50% probability of coming up heads, we mean that if it were possible to flip the coin *infinitely* many times, then exactly 50% of the flips would be heads.

Because we can't do *anything* infinitely many times, finding theoretical probabilities requires relying on theory. (That's why they're called theoretical!) As an example, a coin has two sides. My theory is that either side is equally likely to come

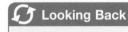

**Looking Back**

**Relative Frequencies**
Relative frequencies, introduced in Chapter 2, are the same as proportions.

up when I flip a coin. I don't know for a fact that this is true; I *assume* it is true. And so I conclude that the probability of seeing heads is 50%. However, I haven't actually done an experiment. I have simply reasoned on the basis of a theory about how coins behave.

**Empirical probabilities,** on the other hand, are short-run relative frequencies based an experiment. I toss a coin 10 times and get 6 heads. My empirical probability of getting heads is therefore $6/10 = 0.6$, or 60%.

Empirical and theoretical probabilities have some striking differences. Chief among them is that theoretical probabilities are always the same value; if we all agree on the theory, then we all agree on the value for the theoretical probabilities. Empirical probabilities, however, change with every experiment. Suppose I flip the exact same coin 10 times again, and this time I get 3 heads. Now my empirical probability of heads is 0.3. Empirical probabilities are themselves random and vary from experiment to experiment.

 **KEY POINT** Empirical probabilities tell us how often an event occurred in an actual set of experiments. Theoretical probabilities are based on theory and tell us how many times an event would occur if an experiment were repeated *infinitely* many times.

## Why Do We Need Both?

Theoretical probabilities are very abstract things. They measure what proportion of the time events would occur if an experiment were repeated *infinitely many times*. That's a very long time—forever, in fact. This means it is impossible to carry out an experiment that will provide the exact value of a theoretical probability; to do so would literally take forever. However, it turns out that we can use empirical probabilities to *estimate* and to *test* theoretical probabilities.

Why do we need to estimate a theoretical probability? Sometimes, it is just too difficult to compute a theoretical probability. Development of the mathematics of probability began in the 1600s, and we can now compute probabilities for quite complex events. But this does not mean it's easy! Sometimes, it may be adequate to get a good approximate value using an experiment that allows us to estimate how often a certain event might happen if we could repeat the experiment infinitely many times. On other occasions, the event for which we need a probability may be too complex for theory, and if we really need to know the probability, then running an experiment is the only way of finding a useful value.

Why do we need to test? We might not trust the theoretical probability value. For example, it might be based on assumptions that we are not sure are true. If done well, an experiment that produces an empirical probability can be used to verify or refute a theoretically derived value. In fact, much of the rest of this book will develop this theme.

If you do the "Let's Make a Deal" activity described at the end of this chapter, you will find empirical probabilities of winning for each of the two strategies offered to contestants: stay or switch. These empirical probabilities allow you to estimate the probability of winning, and they also allow you to test whether your own ideas about which strategy is best are correct.

**Simulations** are experiments used to produce empirical probabilities, because the investigators hope that these experiments simulate the situation they are examining. As you will see in Section 5.4, a mathematical discovery called the Law of Large Numbers tells us that if the relationship between the situation and our experiment is strong, then our empirical probability will be a good estimate for the true theoretical probability.

# Finding Theoretical Probabilities

We can use empirical probabilities as an estimate of theoretical probabilities, and if we do our experiment enough times, the estimate can be pretty good. But how good is pretty good? To understand how to use empirical probabilities to estimate and to verify theoretical probabilities, we first need to learn how to calculate theoretical probabilities.

## Facts about Theoretical Probabilities

Probabilities are always numbers between 0 and 1 (including 0 and 1). Probabilities can be expressed as fractions, decimals, or percents: 1/2, 0.50, and 50% are all used to represent the probability that a coin comes up heads.

Some values have special meanings. If the probability of an event happening is 0, then that event never happens. If you purchase a lottery ticket after all the prizes have been given out, the probability is 0 that you will win one of those prizes. If the probability of an event happening is 1, then that event always happens. If you flip a coin, the probability of a coin landing heads or tails is 1.

Another useful property to remember is that the probability that an event will *not* happen is 1 minus the probability that it will happen. If there is a 0.90 chance that it will rain, then there is a $1 - 0.9 = 0.10$ chance that it will not. If there is a 1/6 chance of rolling a "1" on a die, then there is a $1 - (1/6) = 5/6$ probability that you will not get a 1.

We call such a "not event" a **complement**. The complement of the event "it rains today" is the event "it does not rain today." The complement of the event "coin lands heads" is "coin lands tails." The complement of the event "a die lands with a 1 on top" is "a die lands with a 2, 3, 4, 5, or 6 on top"; in other words, the die lands with a number that is *not* 1 on top.

Events are usually represented by uppercase letters: A, B, C, and so on. For example, we might let A represent the event "it rains tomorrow." Then the notation P(A) means "the probability that it will rain tomorrow." In sentence form, the notation P(A) = 0.50 translates into English as "The probability that it will rain tomorrow is 0.50, or 50%."

A common misinterpretation of probability is to think that large probabilities mean that the event will certainly happen. For example, suppose your local weather reporter predicts a 90% chance of rain tomorrow. Tomorrow, however, it doesn't rain. Was the weather reporter wrong? Not necessarily. When the weather reporter says there is a 90% chance of rain, it means that on 10% of the days like tomorrow it does not rain. Thus a 90% chance of rain means that on 90% of all days just like tomorrow, it rains, but on 10% of those days it does not.

---

**Summary of Probability Rules**

**Rule 1:** A probability is always a number from 0 to 1 (or 0% to 100%) inclusive (which means 0 and 1 are allowed). It may be expressed as a *fraction,* a *decimal,* or a *percent.*

In symbols: For any event A,

$$0 \le P(A) \le 1$$

**Rule 2:** The probability that an event will not occur is 1 minus the probability that the event will occur. In symbols, for any event A,

$$P(A \text{ does } not \text{ occur}) = 1 - P(A \text{ does occur})$$

The symbol $A^c$ is used to represent the complement of A. With this notation, we can write Rule 2 as

$$P(A^c) = 1 - P(A)$$

## Finding Theoretical Probabilities with Equally Likely Outcomes

In some situations, all of the possible outcomes of a random experiment occur with the same frequency. We call these situations "equally likely outcomes." For example, when you flip a coin, heads and tails are equally likely. When you roll a die, 1, 2, 3, 4, 5, and 6 are all equally likely.

When we are dealing with equally likely outcomes, it is sometimes helpful to list all of the possible outcomes. A list that contains *all* possible (and equally likely) outcomes is called the **sample space**. We often represent the sample space with the letter S. An **event** is any collection of outcomes in the sample space. For example, the sample space S for rolling a die is the numbers 1, 2, 3, 4, 5, 6. The event "get an even number on the die" consists of the even outcomes in the sample space: 2, 4, and 6.

When the outcomes are equally likely, the probability that a particular event occurs is just the number of outcomes that make up that event, divided by the total number of equally likely outcomes in the sample space. In other words, it is the number of outcomes resulting in the event divided by the number of outcomes in the sample space.

**Summary of Probability Rules**

**Rule 3:**

$$\text{Probability of A} = P(A) = \frac{\text{Number of outcomes in A}}{\text{Number of all possible outcomes}}$$

This is true *only* for equally likely outcomes.

For example, suppose 30 people are in your class, and one person will be selected at random by a raffle to win a prize. What is the probability that you will win? The sample space is the list of the names of the 30 people. The event A is the event that contains only one outcome: your name. The probability that you win is 1/30, because there is only 1 way for you to win and there are 30 different ways that this raffle can turn out. We write this using mathematical notation as follows:

$$P(\text{you win prize}) = 1/30$$

We can be even more compact:

Let *A* represent the event that you win the raffle. Then

$$P(A) = 1/30.$$

One consequence of Rule 3 is that the probability that *something* in the sample space will occur is 1. In symbols, $P(S) = 1$. This is because

$$P(S) = \frac{\text{Number of outcomes in S}}{\text{Number of outcomes in S}} = 1$$

### EXAMPLE 2 Ten Dice in a Bowl

Reach into a bowl that contains 5 red dice, 3 green dice, and 2 white dice (Figure 5.2). But assume that, unlike what you see in Figure 5.2, the dice have been well mixed.

▶ **FIGURE 5.2** Ten dice in a bowl.

QUESTION What is the probability of picking (a) a red die? (b) a green die? (c) a white die?

SOLUTIONS The bowl contains 10 dice, so we have ten possible outcomes. All are equally likely (assuming all the dice are equal in size, they are mixed up within the bowl, and we do not peek when choosing).

a. Five dice are red, so the probability of picking a red die is 5/10, 1/2, 0.50, or 50%. That is,

$$P(\text{red die}) = 1/2, \text{ or } 50\%$$

b. Three dice are green, so the probability of picking a green die is 3/10, or 30%. That is,

$$P(\text{green die}) = 3/10, \text{ or } 30\%.$$

c. Two dice are white, so the probability of picking a white die is 2/10, 1/5, or 20%. That is,

$$P(\text{white die}) = 1/5, \text{ or } 20\%.$$

Note that the probabilities add up to 1, or 100%, as they must.

TRY THIS! Exercise 5.11

Example 3 shows that it is important to make sure the outcomes in your sample space are equally likely.

### EXAMPLE 3 Adding Two Dice

Roll two dice and add the outcomes. Assume each side of each die is equally likely to appear face up when rolled. Event A is the event that the sum of the two dice is 7.

QUESTION What is the probability of event A? In other words, find P(A).

SOLUTION This problem is harder because it takes some work to list all of the equally likely outcomes, which are shown in Table 5.2.

| Die 1 | 1 | 1 | 1 | 1 | 1 | 1 | 2 | 2 | 2 | 2 | 2 | 2 |
|-------|---|---|---|---|---|---|---|---|---|---|---|---|
| Die 2 | 1 | 2 | 3 | 4 | 5 | 6 | 1 | 2 | 3 | 4 | 5 | 6 |
| Die 1 | 3 | 3 | 3 | 3 | 3 | 3 | 4 | 4 | 4 | 4 | 4 | 4 |
| Die 2 | 1 | 2 | 3 | 4 | 5 | 6 | 1 | 2 | 3 | 4 | 5 | 6 |
| Die 1 | 5 | 5 | 5 | 5 | 5 | 5 | 6 | 6 | 6 | 6 | 6 | 6 |
| Die 2 | 1 | 2 | 3 | 4 | 5 | 6 | 1 | 2 | 3 | 4 | 5 | 6 |

◄ **TABLE 5.2** Possible outcomes for two six-sided dice.

Table 5.2 lists 36 possible equally likely outcomes. Here are the outcomes in event A:

(1, 6), (2, 5), (3, 4), (4, 3), (5, 2), (6, 1)

There are six outcomes for which the dice add to 7.

**CONCLUSION**   The probability of rolling a sum of 7 is 6/36, or 1/6.

**TRY THIS!**   Exercise 5.15

---

It is important to make sure that the outcomes in the sample space are equally likely. A common mistake when solving Example 3 is listing all the possible *sums*, instead of listing all the equally likely outcomes of the two dice. If we made that mistake here, our list of sums would look like this:

$$2, 3, 4, 5, 6, 7, 8, 9, 10, 11, 12$$

This list has 11 sums, and only 1 of them is a 7, so we would incorrectly conclude that the probability of getting a sum of 7 is 1/11.

Why didn't we get the correct answer of 1/6? The reason is that the outcomes we listed—2, 3, 4, 5, 6, 7, 8, 9, 10, 11, 12—are not equally likely. For instance, we can get a sum of 2 in only one way: roll two "aces," for 1 + 1. Similarly, we have only one way to get a 12: roll two 6's, for 6 + 6.

However, there are six ways of getting a 7: (1, 6), (2, 5), (3, 4), (4, 3), (5, 2), and (6, 1). In other words, a sum of 7 happens more often than a sum of 2 or a sum of 12. The outcomes 2, 3, 4, 5, 6, 7, 8, 9, 10, 11, 12 are not equally likely.

Usually it is not practical to list all the outcomes in a sample space—or even just those in the event you're interested in. For example, if you are dealing out 5 playing cards from a 52-card deck, the sample space has 2,598,960 possibilities. You really do not want to list all of those outcomes. Mathematicians have developed rules for counting the number of outcomes in complex situations such as these. These rules do not play an important role in introductory statistics, and we do not include them in this book.

> **! Caution**
>
> **"Equally Likely Outcomes" Assumption**
> Just wishing it were true doesn't make it true. The assumption of equally likely outcomes is not always true. And if this assumption is wrong, your theoretical probabilities will not be correct.

## Combining Events with "AND" and "OR"

As you saw in Chapter 4, we are often interested in studying more than one variable at a time. Real people, after all, often have several attributes we want to study, and we frequently want to know the relationship among these variables. The words AND and OR can be used to combine events into new, more complex events. The real people in Figure 5.3a on the next page, for example, have two attributes we decided to examine. They are either wearing a hat, or not. Also, they are either wearing glasses, or not. In the photo, the people who are wearing hats AND glasses are raising their hands.

**(a)**

Mike   Rena   Maria   Alan   John   David

**(b)**

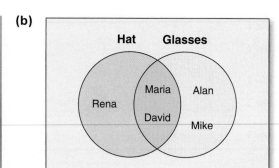

▲ **FIGURE 5.3 (a)** Raise your hand if you are wearing glasses AND a hat. **(b)** The people wearing both glasses AND a hat (Maria and David) appear in the intersection of the two circles in this Venn diagram.

 **Caution**

**Venn Diagrams**
The areas of the regions in Venn diagrams have no numerical meaning. A large area does not contain more outcomes than a small area.

Another way to visualize this situation is with a **Venn diagram**, as shown in Figure 5.3b. The rectangle represents the sample space, which consists of all possible outcomes if we were to select a person at random. The ovals represent events—for example, the event that someone is wearing glasses. The people who "belong" to *both* events are in the intersection of the two ovals.

The word **AND** creates a new event out of two other events. The probability that a randomly selected person in this photo is wearing a hat is 3/6, because three of the six people are wearing a hat. The probability that a randomly selected person wears glasses is 4/6. The probability that a person is wearing a hat AND glasses is 2/6, because only two people are in both groups. We could write this, mathematically, as

$$P(\text{wears glasses AND wears a hat}) = 2/6$$

**KEY POINT**    The word AND creates a new event out of two events A and B. The new event consists of *only* those outcomes that are in *both* event A and event B.

In most situations, you will not have a photo to rely on. A more typical situation is given in Table 5.3, which records frequencies of two attributes for the people in a random sample of a recent U.S. census (www.census.gov). The two attributes are highest educational level and current marital status. ("Single" means never married. All other categories refer to the respondent's current status. Thus a person who was divorced but then remarried is categorized as "Married.")

▶ **TABLE 5.3** Education and marital status for 665 randomly selected U.S. residents. "Less HS" means did not graduate from high school.

| Education Level | Single | Married | Divorced | Widow/Widower | Total |
|---|---|---|---|---|---|
| **Less HS** | 17 | 70 | 10 | 28 | 125 |
| **High school** | 68 | 240 | 59 | 30 | 397 |
| **College or higher** | 27 | 98 | 15 | 3 | 143 |
| **Total** | 112 | 408 | 84 | 61 | 665 |

## EXAMPLE 4 Education AND Marital Status

Suppose we select a person at random from the collection of 665 people categorized in Table 5.3.

**QUESTIONS**

a.  What is the probability that the person is married?

b.  What is the probability that the person has a college education or higher?

c.  What is the probability that the person is married AND has a college education or higher?

**SOLUTIONS**   The sample space has a total of 665 equally likely outcomes.

a.  In 408 of those outcomes, the person is married. So the probability that a randomly selected person is married is 408/665, or 61.4% (approximately).

b.  In 143 of those outcomes, the person has a college education or higher. So the probability that the selected person has a college education or higher is 143/665, or 21.5%.

c.  There are 665 possible outcomes. In 98 of them, the people are both married AND have a college degree or higher. So the probability that the selected person is married AND has a college degree is 98/665, or 14.7%.

**TRY THIS!**   Exercise 5.21

> **⚠ Caution**
>
> **AND**
> P(A AND B) will always be less than (or equal to) P(A) and also less than (or equal to) P(B). If this isn't the case, you've made a mistake!

## Using "OR" to Combine Events

The people in Figure 5.4a were asked to raise their hands if they were wearing glasses OR wearing a hat. Note that people who are wearing both also have their hands raised. If we were to select one of these people at random, the probability that this person is wearing glasses OR wearing a hat would be 5/6, because we would count people who wear glasses, people who wear hats, and people who wear both glasses AND hats.

**(a)**

**(b)**

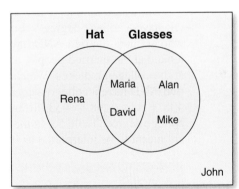

▲ **FIGURE 5.4** **(a)** Raise your hand if you are wearing a hat OR glasses. This photograph illustrates the inclusive OR. **(b)** In this Venn diagram, note the orange region for "raise your hand if you are wearing a hat OR glasses."

In a Venn diagram, OR events are represented by shading all relevant events. Here Mike, Rena, Maria, Alan, and David appear in the orange area because each is wearing glasses OR wearing a hat.

The last example illustrates a special meaning of the word OR. This word is used slightly differently in mathematics and probability than you may use it in English. In statistics and probability, we use the **inclusive OR**. For example, the people in the photo shown in Figure 5.4a were asked to raise their hands if they had a hat OR glasses. This means that the people who raise their hands have a hat only, or have glasses only, or have both hats AND glasses.

 **KEY POINT**   The word OR creates a new event out of the events A and B. The new event consists of all outcomes that are only in A, that are only in B, or that are in both.

## EXAMPLE 5 OR with Marital Status

Again, select someone at random from Table 5.3.

QUESTION What is the probability the person is single OR married?

SOLUTION The event of interest occurs if we select a person who is married, a person who is single, or a person who is both married AND single. There are 665 possible equally likely outcomes. Of these, 112 are single and 408 are married (and none are both!). Thus there are $112 + 408 = 520$ people who are single OR married.

CONCLUSION The probability that the selected person is single OR married is 520/665, or 78.2%.

TRY THIS! Exercise 5.23

## EXAMPLE 6 Education OR Marital Status

Select someone at random from Table 5.3 on page 204 (which is shown again in Table 5.4 below).

QUESTION What is the probability that the person is married OR has a college degree?

SOLUTION Table 5.3 gives us 665 possible outcomes. The event of interest occurs if we select someone who is married, or someone who has a college degree, or someone who both is married AND has a college degree. There are 408 married people and 143 people with a college degree.

But wait a minute: There are not $408 + 143$ different people who are married OR have a college degree—some of these people got counted twice! The people who are both married AND have a college degree were counted once when we looked at the married people, and they were counted a second time when we looked at the college graduates. We can see from Table 5.4 that 98 people *both* are married AND have a college degree. So we counted 98 people too many. Thus, we have $408 + 143 - 98 = 453$ different outcomes in which the person is married or has a college degree. Table 5.4 is the same as Table 5.3 except for the added ovals, which are meant to be interpreted as in a Venn diagram.

▶ **TABLE 5.4** Here we reprint Table 5.3, with ovals for Married and for College added.

| Education Level | Single | Married | Divorced | Widow/Widower | Total |
|---|---|---|---|---|---|
| Less HS | 17 | **70** | 10 | 28 | 125 |
| High school | 68 | **240** | 59 | 30 | 397 |
| **College or higher** | 27 | **98** | 15 | 3 | **143** |
| Total | 112 | **408** | 84 | 61 | 665 |

The numbers in bold type represent the people who are married OR have a college degree. This Venn-like treatment emphasizes that one group (of 98 people) is in both categories and reminds us not to count them twice.

Another way to say this is that we count the 453 distinct outcomes by adding the numbers in the ovals in the table, but not adding any of them more than once:

$$70 + 240 + 98 + 27 + 15 + 3 = 453$$

CONCLUSION The probability that the randomly selected person is married OR has a college degree is 453/665, or 68.1%.

TRY THIS! Exercise 5.25

## Mutually Exclusive Events

Did you notice that the first example of an OR (single OR married) was much easier than the second (married OR has a college degree)? In the second example, we had to be careful not to count some of the people twice. In the first example, this was not a problem. Why?

The answer is that in the first example, we were counting people who were married OR single, and no person can be in both categories. It is impossible to be simultaneously married AND single. When two events have no outcomes in common—that is, when it is impossible for both events to happen at once—they are called **mutually exclusive events**. The events "person is married" and "person is single" are mutually exclusive.

The Venn diagram in Figure 5.5 shows two mutually exclusive events. There is no intersection between these events; it is impossible for both event A AND event B to happen at once. This means that the probability that both events occur at the same time is 0.

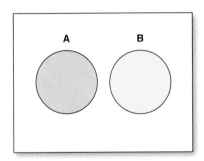

◀ **FIGURE 5.5** In a Venn diagram, two mutually exclusive events have no overlap.

## EXAMPLE 7 Mutually Exclusive Events: Education and Marital Status

Imagine selecting a person at random from those represented in Table 5.4.

**QUESTION** Name two mutually exclusive events, and name two events that are not mutually exclusive. Remember that marital status means a person's current marital status.

**SOLUTION** For mutually exclusive events, we can choose any two columns or any two rows. It is impossible for someone to be both divorced AND married (at the same time). It is impossible for someone to have less than a high school education AND to have a high school education. The events "person has a HS education" and "person has less than a HS education" are mutually exclusive, because no one can be in both categories at once. The probability that a randomly selected person has a HS education AND has less than a HS education is 0. We could have chosen other pairs of events as well.

To find two events that are not mutually exclusive, find events that have outcomes in common. There are 30 people who have a HS education AND are widows/widowers. Therefore, these events are *not* mutually exclusive. The events "person has HS education" AND "person is a widow/ widower" are not mutually exclusive.

**TRY THIS!** Exercise 5.29

**Summary of Probability Rules**

**Rule 4:** The probability that event A happens OR event B happens is

(the probability that A happens) plus (the probability that B happens) minus (the probability that both A AND B happen).

If A and B are mutually exclusive events (for example, A is the event that the selected person is single, and B is the event that the person is married), then P(A AND B) = 0. Then the rule becomes simpler:

**Rule 4a:** If A and B are mutually exclusive events, the probability that event A happens OR event B happens is the sum of the probability that A happens and the probability that B happens.

Rule 4 in symbols:

$$\text{Always: } P(A \text{ OR } B) = P(A) + P(B) - P(A \text{ AND } B)$$

Rule 4a in symbols:

$$\text{Only if A and B are mutually exclusive: } P(A \text{ OR } B) = P(A) + P(B)$$

## EXAMPLE 8 Rolling a Six-Sided Die

Roll a fair, six-sided die.

QUESTIONS

a. Find the probability that the die shows an even number OR a number greater than 4 on top.

b. Find the probability that the die shows an even number OR the number 5 on top.

SOLUTIONS

a. We could do this in two ways. First, we note that six equally likely outcomes are possible. The even numbers are (2, 4, 6) and the numbers greater than 4 are (5, 6). Thus the event "even number OR number greater than 4" has four different ways of happening: roll a 2, 4, 5, or 6. We conclude that the probability is 4/6.

The second approach is to use Rule 4. The probability of getting an even number is 3/6. The probability of getting a number greater than 4 is 2/6. The probability of getting both an even number AND a number greater than 4 is 1/6 (because the only way for this to happen is to roll a 6). So

$$P(\text{even OR greater than 4}) = P(\text{even}) + P(\text{greater than 4}) - P(\text{even AND greater than 4})$$
$$= 3/6 + 2/6 - 1/6$$
$$= 4/6$$

b. $P(\text{even OR roll 5}) = P(\text{even}) + P(\text{roll 5}) - P(\text{even AND roll 5})$

It is impossible for the die to be both even AND a 5, because 5 is an odd number. So the events "get a 5" and "get an even number" are mutually exclusive. Therefore, we get

$$P(\text{even number OR a 5}) = 3/6 + 1/6 - 0 = 4/6$$

CONCLUSIONS

a. The probability of rolling an even number OR a number greater than 4 is 4/6 (or 2/3).

b. The probability of rolling an even number OR a 5 is 4/6 (or 2/3).

TRY THIS!  Exercise 5.33

# Associations in Categorical Variables

Judging on the basis of our sample in Table 5.3, is there an association between marital status and having a college education? If so, we would expect the proportion of married people to be different for those who had a college education and those who did not have a college education. (Perhaps we would find different proportions of marital status for each category of education.)

In other words, if there is an association, we would expect the probability that a randomly selected college-educated person is married to be different from the probability that a person with less than a college education is married.

## Conditional Probabilities

Language is important here. The probability that "a college-educated person is married" is different from the probability that "a person is college-educated AND is married." In the AND case, we're looking at everyone in the sample and wondering how many have both a college degree AND are married. But when we ask for the probability that a college-educated person is married, we're taking it as given that the person is college-educated. We're *not* saying "choose someone from the whole collection." We're saying, "Just focus on the people with the college degrees. What proportion of those people are married?"

Probabilities such as these, where we focus on just one group of objects and imagine taking a random sample from that group alone, are called **conditional probabilities**.

For example, in Table 5.5 (which repeats Table 5.3), we've highlighted in red the people with college degrees. In this row, there are 143 people. If we select someone at random from among those 143 people, the probability that the person will be married is 98/143 (or about 0.685). We call this a conditional probability because we're finding the probability of being married *conditioned* on having a college education (that is, we are assuming we're selecting only from college-educated people).

| Education Level | Single | Married | Divorced | Widow/Widower | Total |
|---|---|---|---|---|---|
| **Less HS** | 17 | 70 | 10 | 28 | 125 |
| **HS** | 68 | 240 | 59 | 30 | 397 |
| College or higher | 27 | 98 | 15 | 3 | 143 |
| **Total** | 112 | 408 | 84 | 61 | 665 |

◀ **TABLE 5.5** What's the probability that a person with a college degree or higher is married? To find this, focus on the row shown in red and imagine selecting a person from this row.

## "Given That" vs. "AND"

Often, conditional probabilities are worded with the phrase *given that*, as in "Find the probability that a randomly selected person is married given that the person has a college degree." But you might also see it phrased as in the last paragraph: "Find the probability that a randomly selected person with a college degree is married." The latter phrasing is more subtle, because it implies that we're supposed to assume the selected person has a college degree: We must assume we are *given that* the person has a college degree.

Figure 5.6a on the next page shows a Venn diagram representing all of the data. The green overlap region represents the event of being married AND having completed college. By way of contrast, Figure 5.6b on the next page shows only those with college educations; it emphasizes that if we wish to find the probability of being married, given that the person has a college degree, we need to focus on only those with college degrees.

▶ **FIGURE 5.6 (a)** The probability of being married AND having a college degree; **(b)** the probability that a person with a college degree is married.

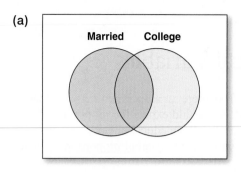

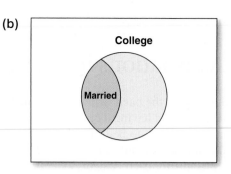

The mathematical notation for a conditional probability might seem a little unusual. We write

P(person is married | person has college degree) = 98/143

The vertical bar inside the probability notation is *not* a division sign. You should think of pronouncing it as "given that." This sentence reads, "The probability that the person is married, given that we know this person has a college degree, is 98/143." Some statisticians like to think of the vertical bar as meaning "within," and would translate this as "The probability that we randomly select a married person from within those who have a college degree." Either way you think about it is fine; use whichever makes the most sense to you.

> **KEY POINT**
>
> In the study of conditional probabilities, P(A | B) means to find the probability that event A occurs, but to restrict your consideration to those outcomes of A that occur within event B. It means "the probability of A occurring, given that event B has occurred."

## EXAMPLE 9 Teens and the Internet

Consider the following statements, which are based on a Pew Foundation report.

a. The probability that a randomly selected teenager (12–17 years old) from the United States will go online at some point during the month is about 93%. Event A: the selected person is a teenager. Event B: the selected person goes online during the month.

b. The probability that a randomly selected adult age 65 or older will go online during the month is 38%. Event A: the selected person is age 65 or older. Event B: the selected person will go online during the month.

c. The probability that a randomly selected resident of the United States is a teenager who will go on line this month is about 7%. Event A: the selected person is a teenager. Event B: the person will go online this month.

We wish to use this information to find the probability that a randomly selected resident of the United States will be 12–17 years old and will go online this month (Pew Foundation, July 2010).

**QUESTION** For each of these three statements, determine whether the events in the question are used in a conditional probability or an AND probability. Explain. Write the statement using probability notation.

**SOLUTIONS**

a. This statement is asking about a conditional probability. It says that among all teenagers, 93% go online. We are "given that" the group we're sampling from are all teenagers.

b. This statement is also a conditional probability.

c. This statement, on the other hand, is asking us to assume nothing and, instead, once the person is selected from the entire United States, to determine whether that person has these two characteristics.

Using probability notation, these statements are

a. P(person goes online | person is teenager)

b. P(person goes online | person is adult aged 65 or older)

c. P(person goes online AND person is teenager)

**TRY THIS!** Exercise 5.45

## Finding Conditional Probabilities

If you are given a table like Table 5.5, you can find conditional probabilities as we did above: by isolating the group from which you are sampling. However, a formula exists that is useful for times when you do not have such complete information.

The formula for calculating conditional probabilities is

$$P(A \mid B) = \frac{P(A \text{ AND } B)}{P(B)}$$

## EXAMPLE 10 Education and Marital Status

Suppose a person is randomly selected from those represented in Table 5.3 on page 204.

**QUESTION** Find the probability that a person with less than a high school degree (and no higher degrees) is married. Use the table, but then confirm your calculation with the formula.

**SOLUTION** We are asked to find P(married | less than high school degree)—in other words, the probability a person with less than a HS degree is married. We are told to imagine taking a random sample from only those who have less than a high school degree. There are 125 such people, of whom 70 are married.

$$P(\text{married} \mid \text{less HS}) = 70/125 = 0.560$$

The formula confirms this:

$$P(\text{married} \mid \text{less HS}) = \frac{P(\text{married AND less HS})}{P(\text{less HS})} = \frac{\frac{70}{665}}{\frac{125}{665}} = \frac{70}{125} = 0.560$$

Interestingly, the probability that a college graduate is married (0.685) is greater than the probability that someone with less than a high school education is married (0.560).

**TRY THIS!** Exercise 5.47

With a little algebra, we can discover that this formula can serve as another way of finding AND probabilities:

$$P(A \text{ AND } B) = P(A)P(B \mid A)$$

We'll make use of this formula later.

**Summary of Probability Rules**

**Rule 5a:** $P(A \mid B) = \dfrac{P(A \text{ AND } B)}{P(B)}$

**Rule 5b:** $P(A \text{ AND } B) = P(B)\,P(A \mid B)$ and also $P(A \text{ AND } B) = P(A)\,P(B \mid A)$

Both forms of Rule 5b are true, because it doesn't matter which event is called A and which is called B.

## Flipping the Condition

A common mistake with conditional probabilities is thinking that $P(A \mid B)$ is the same as $P(B \mid A)$.

$$P(B \mid A) \neq P(A \mid B)$$

A second common mistake is to confuse conditional probabilities with fractions and think that $P(B \mid A) = 1/P(A \mid B)$.

$$P(B \mid A) \neq \frac{1}{P(A \mid B)}$$

For example, using the data in Table 5.3, we earlier computed that $P(\text{married} \mid \text{college}) = 98/143 = 0.685$. What if we wanted to know the probability that a randomly selected married person is college-educated?

$$P(\text{college} \mid \text{married}) = ?$$

From Table 5.3, we can see that if we know the person is married, there are 408 possible outcomes. Of these 408 married people, 98 are college-educated, so

$$P(\text{college} \mid \text{married}) = 98/408, \text{ or about } 0.240$$

Clearly, $P(\text{college} \mid \text{married})$, which is 0.240, does not equal $P(\text{married} \mid \text{college})$, which is 0.685.

Also, it is *not* true that $P(\text{married} \mid \text{college}) = 1/P(\text{college} \mid \text{married}) = 408/98 = 4.16$, a number bigger than 1! It is impossible for a probability to be bigger than 1, so obviously,

$$P(A \mid B) \text{ does not equal } 1/P(B \mid A).$$

 **KEY POINT**   $P(B \mid A) \neq P(A \mid B)$

## Independent and Dependent Events

We saw that the probability that a person is married differs, depending on whether we know that person is a college grad or that she or he has less than a high school education. If we randomly select from different educational levels, we get a different probability of marriage. Another way of putting this is to say that marital status and education level are **associated**. We know they are associated because the conditional probabilities change depending on which educational level we condition on.

We call variables or events that are *not* associated **independent events**. Independent variables, and independent events, play a very important role in statistics. Let's first talk about independent events.

Two events are independent if knowledge that one event has happened tells you nothing about whether or not the other event has happened. In mathematical notation,

A and B are independent events means $P(A \mid B) = P(A)$.

In other words, if the event "a person is married" is independent from "a person has a college degree," then the probability that a married person has a college degree, P(college | married), is the same as the probability that any person in the sample has a college degree. We already found that

$$P(\text{college} \mid \text{married}) = 0.240 \text{ and } P(\text{college}) = 143/665 = 0.215$$

These probabilities are close—so close, in fact, that we might be willing to conclude they are "close enough." (You'll learn in Chapter 10 about how to make decisions like these.) We might conclude that the events "a randomly selected person is married" and "a randomly selected person is college-educated" are independent. However, for now we will assume that *probabilities have to be exactly the same for us to conclude independence.* We conclude that completing college and being married are *not* independent.

It doesn't matter which event you call A and which B, so events are also independent if P(B | A) = P(B).

**KEY POINT**  To say that events A and B are independent means that P(A|B) = P(A). In words: Knowledge that event B occurred does not change the probability of event A occurring.

## EXAMPLE 11 Dealing a Diamond

Figure 5.7 shows a standard deck of playing cards. When playing card games, players nearly always try to avoid showing their cards to the other players. The reason for this is that knowing the other player's cards can sometimes give you an advantage. Suppose you are wondering whether your opponent has a diamond. If you find out that one of the cards he holds is red, does this provide useful information?

**QUESTION** Suppose a deck of cards is shuffled and one card is dealt face-down on the table. Are the events "the card is a diamond" and "the card is red" independent?

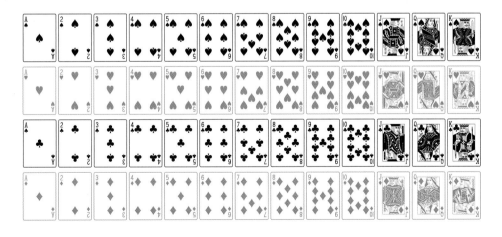

◀ **FIGURE 5.7** Fifty-two playing cards in a standard deck.

**SOLUTION** To answer this, we must apply the definition of independent events and find

$$P(\text{card is a diamond})$$

and compare it to

$$P(\text{card is a diamond} \mid \text{card is red})$$

If these probabilities are different, then the events are not independent; they are associated events.

First we find P(card is a diamond).

Out of a total of 52 cards, 13 are diamonds. Therefore,

$$P(\text{card is a diamond}) = 13/52, \text{ or } 1/4$$

Now suppose we know the card is red. What's the probability that a red card is a diamond? That is, find

$$P(\text{card is a diamond} \mid \text{card is red})$$

The number of equally likely possible outcomes is reduced from 52 to 26, because there are 26 red cards. You are now limited to the 26 cards in the middle two rows of the picture. There are still 13 diamonds. Therefore, the probability that the card is a diamond, given that it is red, is 13/26 = 1/2.

$$P(\text{card is diamond}) = 1/4$$

$$P(\text{card is diamond} \mid \text{card is red}) = 1/2$$

These probabilities are *not* equal.

**CONCLUSION** The events "select a diamond" and "select a red card" are associated, because P(select a diamond | color is red) is not the same as P(select a diamond). This means that if you learn, somehow, that the opponent's card is red, then you have gained some information that will be useful in deciding whether he has a diamond.

Note that we could also have compared P(card is red) to P(card is red | card is a diamond), and we would have reached the same conclusion.

**TRY THIS!** Exercise 5.55

---

## EXAMPLE **12** Dealing an Ace

A playing card is dealt face down. This time, you are interested in knowing whether your opponent holds an ace. You have discovered that his card is a diamond. Is this helpful information?

**QUESTION** Are the events "card is a diamond" and "card is an ace" independent?

**SOLUTION** Now we must find P(card is an ace) and compare it to P(card is an ace | card is a diamond).

P(card is an ace):
Of the 52 cards, 4 are aces.
Therefore, P(card is an ace) = 4/52 = 1/13.

P(card is an ace | card is a diamond):
There are 13 diamonds in the deck, and only one of these 13 is an ace.
Therefore, P(card is an ace | card is a diamond) = 1/13.

We find that P(card is an ace) = 1/13 = P(card is an ace | card is a diamond).

**CONCLUSION** The events "card is a diamond" and "card is an ace" are independent. This means the information that your opponent's card is a diamond will not help you determine whether it is an ace.

Note that we could also compare P(card is a diamond) to P(card is a diamond | card is an ace), and we would have reached the same conclusion.

**TRY THIS!** Exercise 5.57

## Intuition about Independence

Sometimes you can use your intuition to decide whether two events are independent. For example, flip a coin twice. You should know that P(second flip is heads) = 1/2. But what if you know that the first flip was also a head? Then you need to find

P(second flip is heads | first flip was heads)

Intuitively, we know that the coin always has a 50% chance of coming up heads. The coin doesn't know what happened earlier. Thus

P(second is heads | first is heads) = 1/2 = P(second is heads)

The two events "second flip comes up heads" and "first flip comes up heads" are independent.

Although you can sometimes feel very confident in your intuition, you should check your intuition with data whenever possible.

## EXAMPLE 13 Education and Widows

Suppose we select a person at random from the sample of people asked about education and marital status in Table 5.3. Is the event "person selected has HS education" independent from the event "person selected is widowed"? Intuitively, we would think so. After all, why should a person's educational level affect whether his or her spouse dies first?

**QUESTION** Check, using the data in Table 5.3 on page 204, whether these two events are independent.

**SOLUTION** To check independence, we need to check whether $P(A | B) = P(A)$. It doesn't matter which event we call A and which we call B, so let's check to see whether

P(person selected has HS education | person is widowed) = P(person has HS education)

From the table, we see that there are only 61 widows, so

$$P(HS | widowed) = 30/61 = 0.492$$

$$P(HS) = 397/665 = 0.597$$

The two probabilities are not equal.

**CONCLUSION** The events are associated. If you know the person is widowed, then the person is less likely to have a high school education than he or she would be if you knew nothing about his or her marital status. Our intuition was wrong, at least as far as these data are concerned. It is possible, of course, that these data are not representative of the population as a whole, or that our conclusion of association is incorrect because of chance variation in the people sampled.

**TRY THIS!** Exercise 5.59

## Sequences of Independent and Associated Events

A common challenge in probability is to find probabilities for sequences of events. By sequence, we mean events that take place in a certain order. For example, a married couple plans to have two children. What's the probability that the first will be a boy and the second a girl? When dealing with sequences, it is helpful to first determine whether the events are independent or associated.

If the two events are associated, then our knowledge of conditional probabilities is useful, and we should use Probability Rule 5b:

$$P(A \text{ AND } B) = P(A)P(B | A)$$

If the events are independent, then we know $P(B|A) = P(B)$, and this rule simplifies to $P(A \text{ AND } B) = P(A) P(B)$. This formula is often called the **multiplication rule**.

---

**Summary of Probability Rules**

**Rule 5c:** Multiplication Rule. If A and B are independent events, then

$$P(A \text{ AND } B) = P(A) P(B).$$

---

## Independent Events

When two events are independent, the multiplication rule speeds up probability calculations for events joined by AND.

For example, suppose that 51% of all babies born in the United States are boys. Then P(first child is boy) = 51%. What is the probability that a family planning to have two children will have two boys? In other words, how do we find the sequence probability

P(first child is boy AND second child is boy)?

Researchers have good reason to suspect that the genders of children in a family are independent (if you do not include identical twins). Because of this, we can apply the multiplication rule:

P(first child is boy AND second child is boy) = P(first is boy) P(second is boy)
$$= 0.51 \times 0.51 = 0.2601$$

The same logic could be applied to finding the probability that the first child is a boy and the second is a girl:

P(first child is boy AND second child is girl) = P(first is boy) P(second is girl)
$$= 0.51 \times 0.49 = 0.2499$$

## EXAMPLE 14 Three Coin Flips

Toss a fair coin three times. A fair coin is one in which each side is equally likely to land up when the coin is tossed.

**QUESTION** What is the probability that all three tosses are tails? What is the probability that the first toss is heads AND the next two are tails?

**SOLUTION** Using mathematical notation, we are trying to find P(first toss is tails AND second is tails AND third is tails). We know these events are independent (this is theoretical knowledge; we "know" this because the coin cannot change itself on the basis of its past). This means the probability is

P(first is tails) $\times$ P(second is tails) $\times$ P(third is tails) $= 1/2 \times 1/2 \times 1/2 = 1/8$

Also, P(first is heads AND second is tails AND third is tails) is

P(heads) $\times$ P(tails) $\times$ P(tails) $= 1/2 \times 1/2 \times 1/2 = 1/8$

**CONCLUSION** The probability of getting three tails is the same as that of getting first heads and then two tails: 1/8.

**TRY THIS!** Exercise 5.61

## EXAMPLE 15 Ten Coin Flips

Suppose I toss a coin 10 times and record whether each toss lands heads or tails. Assume that each side of the coin is equally likely to land up when the coin is tossed.

QUESTION Which sequence is the more likely outcome?

Sequence A: HTHTHTHTHT

Sequence B: HHTTTHTHHH

SOLUTION Because these are independent events, the probability that sequence A happens is

$$P(H)P(T)P(H)P(T)P(H)P(T)P(H)P(T)P(H)P(T) = \frac{1}{2} \times \frac{1}{2} \times \frac{1}{2} \times \frac{1}{2} \times \frac{1}{2} \times \frac{1}{2}$$

$$\times \frac{1}{2} \times \frac{1}{2} \times \frac{1}{2} \times \frac{1}{2}$$

$$= \left(\frac{1}{2}\right)^{10} = 0.0009766$$

The probability that sequence B happens is

$$P(H)P(H)P(T)P(T)P(T)P(H)P(T)P(H)P(H)P(H) = \frac{1}{2} \times \frac{1}{2} \times \frac{1}{2} \times \frac{1}{2} \times \frac{1}{2} \times \frac{1}{2}$$

$$\times \frac{1}{2} \times \frac{1}{2} \times \frac{1}{2} \times \frac{1}{2}$$

$$= \left(\frac{1}{2}\right)^{10} = 0.0009766$$

CONCLUSION Even though sequence A looks improbable, because it alternates between heads and tails, both outcomes have the same probability!

TRY THIS! Exercise 5.63

Another common probability question asks about the likelihood of "at least one" of a sequence happening a certain way.

## EXAMPLE 16 Harris Poll and "at Least One"

A Harris Poll (MetLife Survey of the American Teacher 2009) found that 67% of teachers are very satisfied with their careers. Suppose we select three teachers randomly and with replacement from the population of all teachers in the United States.

QUESTIONS

a. What is the probability that all three are very satisfied with their careers?

b. What is the probability that none of the three is satisfied?

c. What is the probability that at least one teacher is satisfied?

SOLUTIONS

a. We are asked to find P(first is satisfied AND second is satisfied AND third is satisfied). Because the teachers were selected at random from the population, these are independent events. (One teacher's answer won't affect the probability that the next one will answer one way or the other.) Because these are independent events, this is just P(first is satisfied) $\times$ P(second is satisfied) $\times$ P(third is satisfied) $= 0.67 \times 0.67 \times 0.67 = 0.3008$.

b. The probability that none is satisfied is trickier to determine. This event occurs if the first teacher is not satisfied AND the second is not AND the third is not. So we need to find P(first not satisfied AND second not satisfied AND third not satisfied) $=$ P(firstnot) $\times$ P(second not) $\times$ P(third not) $= (1 - 0.67)(1 - 0.67)(1 - 0.67) = 0.0359$.

c. The probability that at least one teacher is satisfied is the probability that one is satisfied or two are satisfied or all three are satisfied. The calculation is easier if you realize that "at least one is satisfied" is the complement of "none is satisfied" because it includes all categories except "none."

$$1 - 0.0359 = 0.9641$$

CONCLUSION  The probability that all three randomly selected teachers are satisfied with their careers is 0.3008. The probability that none is satisfied is 0.0359. The probability that at least one is satisfied is 0.9641.

TRY THIS!  Exercise 5.65

---

> ⚠ **Caution**
>
> **False Assumptions of Independence**
> If your assumption that A and B are independent events is wrong, P(A AND B) can be very wrong!

## Watch Out for Incorrect Assumptions of Independence

Do not use the multiplication rule if events are not independent. For example, suppose we wanted to find the probability that a randomly selected person is female AND has long hair (say, more than 6 inches long).

About half of the population is female, so P(selected person is female) $= 0.50$. Suppose that about 35% of everyone in the population has long hair; then P(selected person has long hair) $= 0.35$. If we use the multiplication rule, we would find that P(selected person has long hair AND is female) $= 0.35 \times 0.5 = 0.175$.

This relatively low probability of 17.5% makes it sound somewhat unusual to find a female with long hair. The reason is that we assumed that having long hair and being female are independent. This is a bad assumption: A woman is more likely than a man to have long hair. Thus "has long hair" and "is female" are associated, not independent, events. Therefore, once we know that the chosen person is female, there is a greater chance that the person has long hair.

## Associated Events with "AND"

If events are not independent, then we rely on Probability Rule 5b: P(A AND B) $=$ P(A) P(B|A). Of course, this assumes that we know the value of P(B|A).

For example, the famous cycling race the Tour de France has in recent years been plagued with accusations against cyclists for taking steroids to boost their performance. Testing for these steroids is difficult, partly because these substances occur naturally in the human body, and partly because different individuals have different levels than other individuals. Even within an individual, the level of steroid varies during the day. Also, testing for the presence of steroids is expensive and time-consuming. For this reason, racers are chosen randomly for drug tests.

Let's imagine that 2% of the racers have taken an illegal steroid. Also, let's assume that if they took the drug, there is a 99% chance that the test will return a "positive" reading. (A "positive" reading here is a negative for the athlete, who will be disqualified.)

Keep in mind that these tests are not perfect. Even if the cyclist did not take a steroid, there is the probability that the test will return a positive. Let's suppose that, given that an athlete did *not* take a steroid, there is still a 3% chance that the test will return a positive (and the athlete will be unjustly disqualified).

What is the probability that a randomly chosen cyclist will be steroid-free and will still test positive for steroids? In other words, we need to find, for a randomly chosen cyclist,

P(cyclist does not take steroids AND tests positive)

To summarize, we have these pieces of information for a randomly selected athlete at this event:

$$P(\text{took steroid}) = 0.02$$

$$P(\text{did not take steroid}) = 0.98$$

$$P(\text{tests positive} \mid \text{took steroid}) = 0.99$$

$$P(\text{tests positive} \mid \text{did not take steroid}) = 0.03$$

Note that this is a sequence of events. First, the athlete either takes or doesn't take a steroid. Then a test is given, and the results are recorded.

According to Rule 5b,

$$P(A \text{ AND } B) = P(A)\,P(B \mid A)$$

$$
\begin{aligned}
P(\text{did not take steroid AND tests positive}) &= P(\text{did not take steroid}) \\
&\quad P(\text{tests positive} \mid \text{did not take steroid}) \\
&= 0.98 \times 0.03 \\
&= 0.0294
\end{aligned}
$$

We see that roughly 2.9% of all cyclists tested will be both steroid-free AND yet test positive for steroids.

Many people find it useful to represent problems in which events occur in sequence with a tree diagram. We can represent this sequence of all possible outcomes in the tree diagram shown in Figure 5.8.

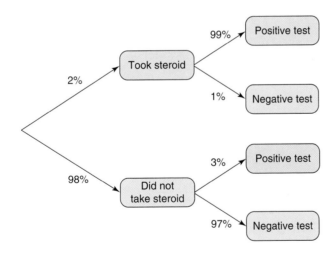

◀ **FIGURE 5.8** Tree diagram showing probabilities for the sequence of events in which an athlete either takes or does not take a drug, and then is tested for the presence of the drug.

The tree diagram shows all possible outcomes after selecting a cyclist (who either took steroids or did not) and then testing (and the test will be either positive or negative). Because the events "took steroid" and "tests positive" are associated, the probabilities of testing positive are different for the "took steroid" branch and the "did not take steroid" branch.

We want to find P(did not take steroid AND positive result), so we simply multiply along that branch of the tree that begins with "did not take steroid" and ends with "positive result":

$$
\begin{aligned}
P(\text{did not take steroid AND positive result}) &= P(\text{did not take steroid}) \times \\
&\quad P(\text{positive result} \mid \text{did not take steroid}) \\
&= 0.98 \times 0.03 \\
&= 0.0294
\end{aligned}
$$

### EXAMPLE 17 Airport Screeners

At many airports you are not allowed to take water through the security checkpoint. This means that Security screeners must check for people who accidentally pack water in their bags. Suppose that 5% of people accidentally pack a bottle of water in their bags. Also suppose that if there is a bottle of water in a bag, the Security screeners will catch it 95% of the time.

QUESTION What is the probability that a randomly chosen person with a backpack has a bottle of water in the backpack and Security finds it?

SOLUTION We are asked to find P(packed water AND Security finds the water). This is a sequence of events. First the person packs the water into the backpack. Later, Security finds (or does not find) the water.
    We are given

$$P(\text{packs water}) = 0.05$$
$$P(\text{Security finds water} \mid \text{water packed}) = 0.95$$

Therefore,

P(packs water and Security finds it) = P(packs water) ×
    P(Security finds water | water packed)
= 0.05 × 0.95
= 0.0475

The tree diagram in Figure 5.9 helps us show how to find this probability.

▶ **FIGURE 5.9** Tree diagram showing probabilities for the sequence of events in which a traveler either packs or does not pack water and then Security either finds or does not find the water.

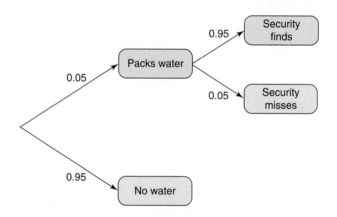

CONCLUSION There is about a 5% (4.75%) chance that a randomly selected traveler will both have packed water and will have the water found by Security.

TRY THIS! Exercise 5.67

SECTION 5.4

# The Law of Large Numbers

The **Law of Large Numbers** is a famous mathematical theorem that tells us that if our simulation is designed correctly, then the more trials we do, the closer we can expect our empirical probability to come to the true probability. The Law of Large Numbers shows that as we approach infinitely many trials, the true probability and the empirical probability approach the same value.

The Law of Large Numbers states that if an experiment with a random outcome is repeated a large number of times, the empirical probability of an event is likely to be close to the true probability. The larger the number of repetitions, the closer together these probabilities are likely to be.

The Law of Large Numbers is the reason why simulations are useful: Given enough trials, and assuming that our simulations are a good match to real life, we can get a good approximation of the true probability.

Table 5.6 shows the results of a very simple simulation. We used a computer to simulate flipping a coin, and we were interested in observing the frequency at which the "coin" comes up heads. We show the results at the end of each trial so that you can see how our approximation gets better as we perform more trials. For example, on the first trial we got a head, so our empirical probability is $1/1 = 1.00$. On the second and third trials we got heads, so the empirical probability of heads is still 1.00. On the fourth trial we got tails, and up to that point we've had 3 heads in 4 trials. Therefore, our empirical probability is $3/4 = 0.75$.

We can continue this way, but it is easier to show you the results by making a graph. Figure 5.10 shows a plot of the empirical probabilities against the number of trials.

Note that with a small number of trials (say, less than 75 or so), our empirical probability was relatively far away from the theoretical value of 0.50. Also, when the number of trials is small, the empirical probabilities can change a lot with each additional "coin flip." But eventually, things settle down, and the empirical probabilities get close to what we know to be the theoretical probability. If you were to simulate a coin flip just 20 times, you might not expect your empirical probability to be too close to the theoretical probability. But after 1000 flips, you'd expect it to be very close.

| Trial | Outcome | Empirical Probability of Heads |
|---|---|---|
| 1 | H | $1/1 = 1.00$ |
| 2 | H | $2/2 = 1.00$ |
| 3 | H | $3/3 = 1.00$ |
| 4 | T | $3/4 = 0.75$ |
| 5 | H | $4/5 = 0.80$ |
| 6 | H | $5/6 = 0.83$ |
| 7 | T | $5/7 = 0.71$ |
| 8 | H | $6/8 = 0.75$ |
| 9 | T | $6/9 = 0.67$ |
| 10 | H | $7/10 = 0.70$ |

▲ **TABLE 5.6** Simulations of heads and tails with cumulative empirical probabilities

### How Many Trials Should I Do in a Simulation?

In our simulation we did 10 trials, which is not very many. As you can tell from Figure 5.10, if we had stopped at 5 flips, we would have had an empirical probability that was pretty far from the true value.

The number of trials you need to be confident that your empirical probability is close to the theoretical value can be fairly large. In general, the more rare the event, the more trials you'll need. Fortunately, simulations are very quick when done with computers, and a variety of software tools are available to help you do fast simulations. We recommend that you do at least 100 trials in most cases.

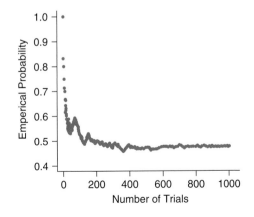

◀ **FIGURE 5.10** The Law of Large Numbers predicts that after many flips, the proportion of heads we get from flipping a real coin will get close to the true probability of getting heads. Because these empirical probabilities are "settling down" to about 0.50, this supports the theoretical probability of 0.50.

### You Try It

Toss a coin 10 times. What is your empirical probability of heads? How far from 0.50 is this? Repeat this a few times to get a sense of how far away the empirical probability can be from 0.50. What's the furthest you got?

What would happen if you were to toss a coin 20 times? Would you tend to get an empirical probability that is closer to 0.50? It's a bit time-consuming to do this with a real coin, so use a computer:

- Go to http://socr.ucla.edu/

- Click on the **Experiments** tab and select **Binomial Coin Experiment** from the dropdown menu that will appear.

- Click on the **Run** button to "toss" 10 "coins." The window below displays the number of heads and the proportion of heads.

- Move the slider beside "$n = 10$" so that "$n = 20$." The letter $n$ represents the number of tosses.

- Now click on the **Play** button a few times to get a sense of how far from 0.50 your empirical probabilities fall. What's the furthest you get? Typically, how far would you say your empirical probabilities are from 0.50? How does this compare to when you tossed the coin 10 times?

### What If My Simulation Doesn't Give the Theoretical Value I Expect?

You have no guarantee that your empirical probability will give you exactly the theoretical value. As you noticed if you did the above activity, even though it's not unusual to get 50% heads in 10 tosses, more often you get some other proportion.

When you don't get the "right" value, two explanations are possible:

1. Your theoretical value is incorrect.

2. Your empirical probability is just varying—that's what empirical probabilities do. You can make it vary less and get closer to the theoretical value by doing more trials.

How do we choose between these two alternatives? Well, that's what statistics is all about! This is one of the central questions of statistics: Are the data consistent with our expectations, or do they suggest that our expectations are wrong? We will return to this question in almost every chapter in the rest of this book.

## Some Subtleties with the Law

The Law of Large Numbers (LLN) is one law that cannot be broken. Nevertheless, many people think the law has been broken when it really hasn't, because interpreting the LLN takes some care.

The Law of Large Numbers tell us, for example, that with many flips of a coin, our empirical probability of heads will be close to 0.50. It tells us nothing about the *number* of heads we will get after some number of tosses and nothing about the order in which the heads and tails appear.

### Streaks: Tails Are Never "Due"

Many people mistakenly believe the LLN means that if they get a large number of heads in a row, then the next flip is more likely to come up tails. For example, if you just flipped five heads in a row, you might think that the sixth is more likely to be a tail than a head so that the empirical probability will work out to be 0.50. Some people might incorrectly say, "Tails are due."

This misinterpretation of the LLN has put many a gambler into debt. This is a misinterpretation for two reasons. First, the Law of Large Numbers is patient. It says that the empirical probability will equal the true probability after infinitely many trials. That's a lot of trials. Thus a streak of 10 or 20 or even 100 heads, though extremely rare, does not contradict the LLN.

## How Common Are Streaks?

Streaks are much more common than most people believe. At the beginning of the chapter, we asked you to toss a coin 20 times (or to simulate 20 tosses) and see how long your longest streak was. We now show results of a simulation to find the (empirical) probability that the longest streak is of length 6 or longer in 20 tosses of a coin.

In our 1000 trials, we saw 221 streaks that were of length 6 or greater. The longest streak we saw was of length 11. The results are shown in Figure 5.11. Our empirical probability of getting a longest streak of length 6 or more was 0.221.

There is another reason why you should not believe that heads are "due" after a long streak of tails. If this were the case, the coin would somehow have to know whether to come up heads or tails. How is the coin supposed to keep track of its past?

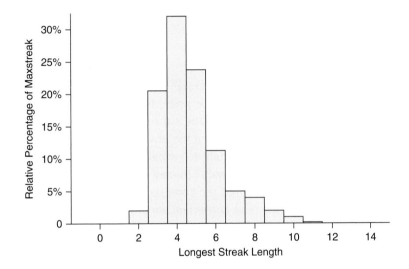

◄ **FIGURE 5.11** Longest streaks in 20 flips of a coin. In 1000 trials, 221 trials had streaks of length 6 or greater.

## CASE STUDY REVISITED

Sally Clark was convicted of murdering two children on the basis of the testimony of physician/expert Dr. Meadow. Dr. Meadow testified that the probability of two children in the same family dying of SIDS was extremely low and that, therefore, murder was a more plausible explanation. In Dr. Meadow's testimony, he assumed that the event "One baby in a family dies of SIDS" AND "a second baby in the family dies of SIDS" were independent. For this reason, he applied the multiplication rule to find the probability that two babies die of SIDS in the same family. The probability of one baby dying was 1/8543, so the probability of two independent deaths was ( 1/8543) × (1/8543), or about 1 in 73 million.

However, as noted in a press release by the Royal Statistical Society, these events may not be independent. "This approach (multiplying probabilities) . . . would only be valid if SIDS cases arose independently within families, an assumption that would need to be justified empirically. Not only was no such empirical justification provided in the case, but there are very strong *a priori* reasons for supposing that the assumption will be false. There may well be unknown genetic or environmental factors that predispose families to SIDS, so that a second case within the family becomes much more likely."

Dr. Meadow made several other errors in statistical reasoning, which are beyond the scope of this chapter but quite interesting nonetheless. There's a nice discussion that's easy to read at http://www.richardwebster.net/cotdeaths.html. For a summary and a list of supporting references, including a video by a statistician explaining the statistical errors, see http://en.wikipedia.org/wiki/Sally_Clark.

Sally Clark was released from prison but died in March 2007 of alcohol poisoning at the age of 43. Her family believes her early death was caused in part by the stress inflicted by her trial and imprisonment.

# EXPLORING STATISTICS
## CLASS ACTIVITY

## "Let's Make A Deal": Stay or Switch?

| GOALS | MATERIALS |
|---|---|
| Simulate an experiment to estimate complex (and counterintuitive) probabilities. | Three playing cards—two of one color, the third of the other color— for each group of three students. |

### ACTIVITY

| Curtain A | Curtain B | Curtain C |
|---|---|---|

▲ **FIGURE A** Three curtains, and the contestant picked curtain A.

| Curtain A |  | Curtain C |
|---|---|---|

▲ **FIGURE B** The host, who knows there is a goat behind curtain B, reveals curtain B. Should the contestant stick with curtain A or switch to curtain C?

| Curtain A |  |  |
|---|---|---|

▲ **FIGURE C** The contestant switched and won.

In the popular game show "Let's Make a Deal," contestants are shown three curtains. Behind one curtain is a high-value prize, for example a car, and behind the other two are less desirable prizes, for example, goats. The contestant picks one of the three curtains. Then the host (who knows what is behind each curtain) raises another curtain to reveal a goat. The host then asks the contestant whether he or she wants to stay or switch to the other curtain before revealing what is behind the contestant's chosen curtain.

There are three possibilities: The contestant should stay; The contestant should switch; It doesn't matter whether the contestant stays or switches.

Form groups of three and choose roles: host, contestant, and recorder. The host lays the cards face down, with the two cards of the same color representing goats and the other card representing the car. The contestant chooses a card. The host then turns over a "goat" card (but not the one just selected). The contestant now must choose a strategy: stay or switch. After that decision is made, the host reveals whether the contestant won a goat or a car. Repeat this several times, giving the contestant several opportunities to try both strategies.

For each trial, the recorder will put a tally mark in one of four categories: Stayed and Won; Stayed and Lost; Switched and Won; Switched and Lost.

After a few games, change roles so that everyone gets to try all roles. Your instructor will compile results.

| | |
|---|---|
| **BEFORE THE ACTIVITY** | Which strategy do you think will be better: staying or switching? Or does it make no difference whatsoever? Why? |
| **AFTER THE ACTIVITY** | Judging on the basis of your class data, what is the empirical probability of winning if the contestant stays? What is it if the contestant switches? Do these probabilities convince you to change your strategy? Explain. |

## CHAPTER REVIEW

## Key Terms

Random, *197*
Probability, *198*
Theoretical probabilities, *198*
Empirical probabilities, *199*
Simulation, *199*

Complement, *200*
Sample space, *201*
Event, *201*
Venn diagram, *204*
AND, *204*

Inclusive OR, *205*
Mutually exclusive events, *207*
Conditional probabilities, *209*
Associated, *212*
Independent events, *212*

Multiplication rule, *216*
Law of Large Numbers, *220*

## Learning Objectives

After reading this chapter and doing the assigned homework problems, you should

- Understand that humans can't reliably create random numbers or sequences.

- Understand that a probability is a long-term relative frequency.

- Know the difference between empirical and theoretical probabilities—and know how to calculate them.

- Be able to determine whether two events are independent or associated and understand the implications of making incorrect assumptions about independent events.

- Understand that the Law of Large Numbers allows us to use empirical probabilities to estimate and test theoretical probabilities.

- Know how to design a simulation to estimate empirical probabilities.

## Summary

Random samples or random experiments must be generated with the use of outside mechanisms such as computer algorithms or by relying on random number tables. Human intuition cannot be relied on to produce reliable "randomness."

Probability is based on the concept of long-run relative frequencies: If an action is repeated infinitely many times, how often does a particular event occur? To find theoretical probabilities, we calculate these relative frequencies on the basis of assumptions about the situation and rely on mathematical rules. In finding empirical probabilities, we actually carry out the action many times or, alternatively, rely on a simulation (using a computer or random number table) to quickly carry out the action many times. The empirical probability is the proportion of times a particular event was observed to occur. The Law of Large Numbers tells us that the empirical probability becomes closer to the true probability as the number of repetitions is increased.

### Theoretical Probability Rules

**Rule 1:** A probability is always a number from 0 to 1 (or 0% to 100%) inclusive (which means 0 and 1 are allowed). It may be expressed as a fraction, a decimal, or a percent.

$$0 \leq P(A) \leq 1$$

**Rule 2:** For any event A,

$$P(A \text{ does } not \text{ occur}) = 1 - P(A \text{ does occur})$$

$A^c$ is the complement of A:

$$P(A^c) = 1 - P(A)$$

**Rule 3:** For equally likely outcomes:

$$P(A) = \frac{\text{Number of outcomes in A}}{\text{Number of all possible outcomes}}$$

**Rule 4:** Always: $P(A \text{ OR } B) = P(A) + P(B) - P(A \text{ AND } B)$

**Rule 4a:** Only if A and B are mutually exclusive:
$$P(A \text{ OR } B) = P(A) + P(B)$$

**Rule 5a:** Conditional probabilities

Probability of A given that B occurred: $P(A \mid B) = \dfrac{P(A \text{ AND } B)}{P(B)}$

**Rule 5b:** Always: $P(A \text{ AND } B) = P(B) P(A \mid B)$

**Rule 5c:** Multiplication Rule. If A and B are independent events, then

$$P(A \text{ AND } B) = P(A) P(B)$$

This applies for any (finite) number of events. For example, $P(A \text{ AND } B \text{ AND } C \text{ AND } D) = P(A) P(B) P(C) P(D)$ if A, B, C, D are independent of each other.

## Sources

MetLife Survey of the American Teacher, 2009, http://www.harrisinteractive .com

Starr, Norton (1997), "Nonrandom Risk: The 1970 Draft Lottery," *Journal of Statistics Education,* volume 5, number 2. Vietnam-era draft data can be found, along with supporting references, at www.amstat.org/ publications/jse/datasets/draft.txt.

Pew Foundation, *Social Media and Young Adults* report, http://pewinternet .org (accessed July 2010).

## CHAPTER 5

## SECTION EXERCISES

### SECTION 5.1

TRY **5.1 Simulation (Example 1)** If we flip a coin 10 times, how often do we get 6 or more heads? A first step to answering this question would be to simulate 10 flips. Use the random number table in Appendix A to simulate flipping a coin 10 times. Let odd digits (1, 3, 5, 7, 9) represent heads, and let even digits (0, 2, 4, 6, 8) represent tails. Begin with the first digit in the third row.

a. Write the sequence of 10 random digits.

b. Write the sequence of 10 "heads" and "tails." Write H for heads and T for tails.

c. Did you get 6 or more heads? How many heads did you get?

**5.2 Simulation** Suppose you are carrying out a randomized, controlled experiment to test whether guided meditation lowers blood pressure (treatment) compared to reading in a quiet room (control). Use the random number table in Appendix A to assign the first 12 subjects to treatment or control. You want each subject to have a 50% chance of being assigned to the treatment group. Let the digits 0, 1, 2, 3, 4 represent assignment to the meditation group, and let the digits 5, 6, 7, 8, 9 represent assignment to the control group. Begin with the first digit on the fourth line of the table.

a. Write the 12 numbers, and below them write M for meditation and C for control.

b. How many subjects were assigned to the meditation group?

c. A research assistant wants to use the digits 0–5 to represent assignment to the meditation group and the digits 6–9 for assignment to the control group. Explain why this will not work.

**5.3** A Monopoly player claims that the probability of getting a 4 when rolling a six-sided die is 1/6 because the die is equally likely to land on any of the six sides. Is this an example of a theoretical probability or an empirical probability? Explain.

**5.4** A person was trying to figure out the probability of getting two heads when flipping two coins. He flipped two coins 10 times, and in 2 of these 10 times both coins landed heads. On the basis of this outcome, he claims that the probability of two heads is 2/10, or 20%. Is this an example of an empirical probability or a theoretical probability? Explain.

**5.5** A friend flips a coin 10 times and says that the probability of getting a head is 60% because he got six heads. Is the friend referring to an empirical probability or a theoretical probability? Explain.

**5.6** A magician claims that he has a fair coin—"fair" because both sides, heads and tails, are equally likely to land face up when the coin is flipped. He tells you that if you flip the coin three times, the probability of getting three tails is 1/8. Is this an empirical probability or a theoretical probability? Explain.

### SECTION 5.2

**5.7 Criminal Court Judges** Criminal cases are assigned to judges randomly. The list of the criminal judges for Memphis, Tennessee (taken from the Tennessee Court System website, www.tsc.state.tn.us), is given in the table. Assume that only Carolyn Blackett and Paula Skahan are women and the rest are men. If you were a criminal defense attorney in Memphis, you might be interested in whether the judge assigned to your case was a man or a woman.

James C. Beasley

Carolyn W. Blackett

Lee V. Coffee

Chris Craft

John T. Fowlkes

W. Otis Higgs

James M. Lammey

Paula L. Skahan

W. Mark Ward

Suppose the names are put into a pot, and a clerk pulls a name out at random.

a. List the equally likely outcomes that could occur; last names are enough.

b. Suppose the event of interest, event A, is that a judge is a woman. List the outcomes that make up event A.

c. What is the probability that one case will be assigned to a female judge?

d. List the outcomes that are in the complement of event A.

**5.8 Random Assignment of Professors** A study looked at the way students attending the Air Force Academy were randomly assigned to different professors for Calculus I, with equal numbers of students assigned to each professor. Some professors were experienced, and some were relatively inexperienced. Suppose the names of the professors are Peters, Parker, Diaz, Nguyen, and Black. Suppose Diaz and Black are inexperienced and the others are experienced. Carrell and West reported that the students who had the experienced teachers for Calculus I did better in Calculus II. (Source: Scott E. Carrell and James E. West, *Does Professor Quality Matter? Evidence from Random Assignment of Students to Professors*, 2010)

a. List the equally likely outcomes that could occur for assignment of one student to a professor.

b. Suppose the event of interest, event A, is that a teacher is experienced. List the outcomes that make up event A.

c. What is the probability that one student will be assigned to an experienced teacher?

d. List the outcomes in the complement of event A. Describe this complement in words.

e. What is the probability that one student will be assigned to an inexperienced teacher?

**5.9** Which of the following numbers could *not* be probabilities, and why?

$$0.5, 1.3, 0.001, 75\%, 1/4, 150\%, -0.5$$

**5.10** Which of the following numbers could *not* be probabilities, and why?

$$0, 2/5, 7/4, 0.002, -10\%, 0.998, 45\%$$

TRY **5.11 Playing Cards (Example 2)** There are four suits: clubs (♣), diamonds (♦), hearts (♥), and spades (♠), and the following cards appear in each suit: Ace, 2, 3, 4, 5, 6, 7, 8, 9, 10, Jack, Queen, King. The Jack, Queen, and King are called face cards because they have a drawing of a face on them. Diamonds and hearts are red, and clubs and spades are black.

If you draw 1 card randomly from a standard 52-card playing deck, what is the probability that it will be:

a. A heart?

b. A red card?

c. An ace?

d. A face card (jack, queen, or king)?

e. A three?

**5.12 Playing Cards** Refer to Exercise 5.11 for information about cards. If you draw 1 card randomly from a standard 52-card playing deck, what is the probability that it will be:

a. A black card?

b. A diamond?

c. A face card (jack, queen, or king)?

d. A nine?

e. A king or queen?

**5.13 Guessing on Tests**

a. On a true/false quiz in which you are guessing, what is the probability of guessing correctly on one question?

b. What is the probability that a guess on one true/false question will be incorrect?

**5.14 Guessing on Tests** On a multiple-choice test with a total of four possible options for each question

a. What is the probability of guessing correctly on one question? (Assume that there are three incorrect options and one correct option.)

b. What is the probability that a guess on one question will be incorrect?

TRY **5.15 Four Children (Example 3)** The sample space given here shows all possible sequences for a family with 4 children, where B stands for boy and G stands for girl.

| GGGG | GGGB | GGBB | GBBB | BBBB |
|------|------|------|------|------|
|      | GGBG | GBGB | BGBB |      |
|      | GBGG | GBBG | BBGB |      |
|      | BGGG | BGGB | BBBG |      |
|      |      | BGBG |      |      |
|      |      | BBGG |      |      |

Assume that all of the 16 outcomes are equally likely. Find the probability of having the following numbers of girls out of 4 children: (a) exactly 0 girls, (b) exactly 1 girl, (c) exactly 2 girls, (d) exactly 3 girls, (e) exactly 4 girls.

(*Hint:* The probability of having 3 girls and a boy is 4/16, or 25%, because the second column shows that there are 4 ways to have 3 girls and 1 boy.)

**5.16 Three Coins** The sample shows the possible sequences for flipping three fair coins or flipping one coin three times, where H stands for heads and T stands for tails.

| HHH | HHT | HTT | TTT |
|-----|-----|-----|-----|
|     | HTH | THT |     |
|     | THH | TTH |     |

Assume that all of the 8 outcomes are equally likely. Find the probability of having exactly the following numbers of heads out of the 3 coins: (a) exactly 0 heads, (b) exactly 1 head, (c) exactly 2 heads, (d) exactly 3 heads. What do the four probabilities add up to and why?

**5.17 Birthdays** What is the probability that a baby will be born on a Friday OR a Saturday OR a Sunday if all the days of the week are equally likely as birthdays?

**5.18 Playing Cards** If *one* card is selected from a well-shuffled deck of 52 cards, what is the probability that the card will be a club OR a diamond OR a heart? What is the probability of the complement of this event? (Refer to Exercise 5.11 for information about cards.)

**5.19 GSS: Political Party** The General Social Survey (GSS) is a survey done nearly every year at the University of Chicago. One survey, summarized in the table, asked each respondent to report her or his political party affiliation and whether she or he was liberal, moderate, or conservative. (Dem stands for Democrat, and Rep stands for Republican.)

|              | Dem | Rep | Other | Total |
|--------------|-----|-----|-------|-------|
| Liberal      | 306 | 26  | 198   | 530   |
| Moderate     | 279 | 134 | 322   | 735   |
| Conservative | 104 | 309 | 180   | 593   |
| Total        | 689 | 469 | 700   | 1858  |

a. If one person is chosen randomly from the group, what is the probability that the person is liberal?

b. If one person is chosen randomly from the group, what is the probability that the person is a Democrat?

**5.20 GSS: Political Party** Refer to the table given in Exercise 5.19.

a. If one person is chosen randomly from the group of 1858 people, what is the probability that the person is conservative?

b. If one person is chosen randomly from the group of 1858 people, what is the probability that the person is a Republican?

TRY **5.21 GSS: AND (Example 4)** Refer to the table given in Exercise 5.19. Suppose we select a person at random from this collection of 1858 people. What is the probability the person is conservative AND a Democrat? In other words, find P(person is conservative AND person is a Democrat).

**5.22 GSS: AND** Refer to the table given in Exercise 5.19. Suppose we select a person at random from this collection of 1858 people. What is the probability that the person is liberal AND a Republican? In other words, find P(person is liberal AND person is a Republican).

TRY **5.23 GSS: OR (Example 5)** Select someone at random from the 1858 people in the table given in Exercise 5.19. What is the probability that the person is a Democrat OR a Republican?

**5.24 GSS: OR** Select someone at random from the 2010 people in the table given in Exercise 5.19. What is the probability that the person is liberal OR conservative?

TRY **5.25 GSS: OR (Example 6)** Assume one person is chosen randomly from the 1858 people in the table given in Exercise 5.19. What is the probability that the person is liberal OR a Democrat? *See page 233 for guidance.*

**5.26 GSS: OR** Assume that one person is chosen randomly from the table given in Exercise 5.19. What is the probability that the person is conservative OR a Republican?

**5.27 Mutually Exclusive** Suppose a person is selected at random. Label each pair of events as *mutually exclusive* or *not mutually exclusive*.

a. The person has brown eyes; the person has blue eyes.

b. The person is 50 years old; the person is a U.S. senator.

**5.28 Mutually Exclusive** Suppose a person is selected at random. Label each pair of events as *mutually exclusive* or *not mutually exclusive*.

a. The person is a parent; the person is a toddler.

b. The person is a woman; the person is a CEO (chief executive officer).

TRY **5.29 GSS: Mutually Exclusive (Example 7)** Referring to the table given in Exercise 5.19, name a pair of mutually exclusive events that could result from selecting an individual at random from this sample.

**5.30 GSS: Mutually Exclusive** Referring to the table given in Exercise 5.19, name a pair of events that are not mutually exclusive that could result from selecting an individual at random from this sample.

**5.31 "OR"** In 2009, the Humane Society of the United States reported that 39% of households owned one or more dogs and 33% owned one or more cats. From this information, is it possible to find the percentage of households that owned a cat OR a dog? Why or why not?

**5.32 "OR"** Suppose you discovered that on your college campus, 6% of the female students were married and 4% of the female students had at least one child.

a. From this information, is it possible to determine the percentage of female students who were married OR had a child?

b. If your answer to part a is no, what additional information would you need to answer this question?

TRY **5.33 Fair Die (Example 8)** Roll a fair six-sided die.

a. What is the probability that the die shows on an odd number OR a number less than 3 on top?

b. What is the probability that the die shows on an odd number OR a number less than 2 on top?

**5.34 Roll a Die** Roll a fair six-sided die.

a. What is the probability the die shows on an odd number OR a number greater than 5 on top?

b. What is the probability the die shows on an odd number OR a number greater than 4 on top?

**5.35 Grades** Assume that the only grades possible in a history course are A, B, C, or lower than C. The probability that a randomly selected student will get an A in a certain history course is

0.18, the probability that a student will get a B in the course is 0.25, and the probability that a student will get a C in the course is 0.37.

a. What is the probability that a student will get an A OR a B?

b. What is the probability that a student will get an A OR a B OR a C?

c. What is the probability that a student will get a grade lower than a C?

**5.36 Changing Multiple-Choice Answers** One of the authors did a survey to determine the effect of students changing answers while taking a multiple-choice test on which there is only one correct answer for each question. Some students erase their initial choice and replace it with another. It turned out that 61% of the changes were from incorrect answers to correct and that 26% were from correct to incorrect. What percent of changes were from incorrect to incorrect?

**5.37 "AND" and "OR"** Consider these categories of people, assuming that we are talking about all the people in the United States:

Category 1: People who are currently married

Category 2: People who have children

Category 3: People who are currently married OR have children

Category 4: People who are currently married AND have children

a. Which of the four categories has the most people?

b. Which category has the fewest people?

**5.38 "AND" and "OR"** Assume that we are talking about all students at your college.

a. Which group is larger: students who are currently taking English AND math, or students who are currently taking English?

b. Which group is larger: students who are taking English OR math, or students who are taking English?

**5.39 Coin Flips** Let H stand for heads and let T stand for tails in an experiment where a fair coin is flipped twice. Assume that the four outcomes listed are equally likely outcomes:

HH, HT, TH, TT

What are the probabilities of getting:

a. 0 heads?

b. Exactly 1 head?

c. Exactly 2 heads?

d. At least 1 head?

e. Not more than 2 heads?

**5.40 Cubes** A hat contains a number of cubes: 15 red, 10 white, 5 blue, and 20 black. One cube is chosen at random. What is the probability that it is:

a. A red cube?

b. Not a red cube?

c. A cube that is white OR black?

d. A cube that is neither white nor black?

e. What do the answers to part a and part b add up to and why?

**5.41 Marital Status** The table gives the number of U.S. citizens (in thousands) who were 15 years of age and older and their marital status in 2009, according to the U.S Census Bureau. Each person is put in a single category on the basis of current status. Thus if a person was married, then became widowed, then remarried, and then divorced, he or she would be listed simply as divorced.

If one person 15 years of age or older is selected randomly from the people in the United States, what is the probability that:

a. The person is widowed?

b. The person is divorced?

c. The person is divorced OR widowed?

| | |
|---|---|
| Married, spouse present | 121,689 |
| Married, spouse absent | 3,348 |
| Widowed | 14,254 |
| Divorced | 23,266 |
| Separated | 5,410 |
| Never married | 72,065 |

**5.42 Marital Status** Refer to table in Exercise 5.41. If one person 15 years of age or older is selected randomly from the people in the United States, what is the probability that:

a. The person is not married (that is, the person is widowed, divorced, or never married)?

b. The person is not classified as widowed?

* **5.43 Multiple-Choice Exam** An exam consists of 12 multiple-choice questions. Each of the 12 answers is either right or wrong. Suppose the probability that a student makes fewer than 3 mistakes on the exam is 0.48 and that the probability that a student makes from 3 to 8 (inclusive) mistakes is 0.30. Find the probability that a student makes:

a. More than 8 mistakes

b. 3 or more mistakes

c. At most 8 mistakes

d. Which two of these three events are complementary and why?

* **5.44 Driving Exam** A driving exam consists of 30 multiple-choice questions. Each of the answers is either right or wrong. Suppose that the probability of making fewer than 7 mistakes is 0.23 and the probability of making from 7 to 15 mistakes is 0.41. Find the probability of making:

a. 16 or more mistakes

b. 7 or more mistakes

c. At most 15 mistakes

d. Which two of these three events are complementary? Explain.

## SECTION 5.3

TRY **5.45 Political Party, Again (Example 9)** The table summarizes results from a GSS survey that asked about political party affiliation and self-described political orientation. (Dem means Democrat, and Rep means Republican.)

| | Dem | Rep | Other | Total |
|---|---|---|---|---|
| Liberal | 306 | 26 | 198 | 530 |
| Moderate | 279 | 134 | 322 | 735 |
| Conservative | 104 | 309 | 180 | 593 |
| Total | 689 | 469 | 700 | 1858 |

A person is selected randomly from the sample summarized in the table. We want to find the probability that a conservative person is a Democrat. Which of the following statements best describes the problem? (Choose one.)

i. P(conservative | Democrat) "conservative given Democrat"

ii. P(Democrat | conservative) "Democrat given conservative"

iii. P(Democrat AND conservative)

**5.46 Political Party** Use the table in Exercise 5.45. A person is selected randomly from the sample summarized in the table. We want to determine the probability that the person is a moderate Republican. Which of the following statements best describes the problem?

i. P(moderate | Republican)

ii. P(Republican | moderate)

iii. P(moderate AND Republican)

TRY **5.47 Political Party, Again (Example 10)** Refer to the table for Exercise 5.45.

a. Find the probability that a randomly chosen respondent is a Democrat given that he or she is liberal. In other words, what percentage of the liberals are Democrats?

b. Find the probability that a randomly chosen respondent is a Democrat given he or she is conservative? In other words, what percentage of the conservatives are Democrats?

c. Which respondents are more likely to be Democrats: the liberal or the conservative respondents?

**5.48 Party, Again** Refer to the table for Exercise 5.45.

a. Find the probability that a randomly chosen respondent is conservative given that she or he is a Republican.

b. Find the probability that a randomly chosen respondent is a Republican given that he or she is conservative.

c. Find the probability that a randomly chosen respondent is both conservative AND a Republican.

**5.49 Left-handed** The percentage of left-handed people in the United States is estimated to be 10%. Men are about twice as likely to be left-handed as women. Are gender and handedness independent or associated? Explain.

**5.50 Smoking** According to the group International Network of Women Against Tobacco, 17% of adult women in the United States and 80% of adult women in Greenland smoked in 2007. Are smoking and whether the woman is from the United States or Greenland independent or associated?

**5.51 Independent?** Using your general knowledge, label the following pairs of variables as independent or associated. Explain your reasoning.

a. Hair color and age for adults 20 to 30 years old

b. Hair color and eye color for adults

**5.52 Independent?** A 2009 Gallup poll of over 14,000 people showed that 68% of weekly church attendees felt negatively about the economy. It also showed that 81% of those who seldom or never attended church felt negatively about the economy. According to the poll, are attendance at church and opinion about the economy independent? Explain.

**5.53 Same-Sex Marriage** The 2008 GSS found that 43% of women in the United States agreed that same-sex marriage should be allowed, but only 34% of men felt that way. Suppose that these figures are accurate (or nearly accurate) probabilities, and state whether gender and opinion about same-sex marriage are independent. Explain.

**5.54 Shopping** A survey published in *Vitality Magazine* suggested that 67% of women in the United States thought shopping was relaxing and enjoyable, but only 37% of men felt that way. Assuming these values are accurate, state whether or not gender and opinion about shopping are independent. Explain.

TRY **5.55 Party: Independent? (Example 11)** Refer to the table in Exercise 5.45. If a person is selected at random, is the event that he or she is a Democrat independent of the event that he or she is liberal?

* **5.56 Party** A person is selected randomly from the sample summarized in the table in Exercise 5.45. The probability of being a Democrat given that the person is conservative is 17.5%, the probability of being conservative given that the person is a Democrat is 15.1%, and the probability of being both conservative and a Democrat is 5.6%. Why is the last probability the smallest?

TRY **5.57 Hand Folding (Example 12)** When people fold their hands together with interlocking fingers, most people are more comfortable with one of two ways. In one way, the right thumb ends up on top and in the other way, the left thumb is on top. The table shows the data from one group of people. M means man, and W means woman; Right means the right thumb is on top, and Left means the left thumb is on top. Judging on the basis of this data set, are the events "right thumb on top" and "male" independent or associated? Data were collected in a class taught by one of the authors but were simplified for clarity. The conclusion remains the same as that derived from the original data. *See page 233 for guidance.*

|        | M  | W  |
|--------|----|----|
| Right  | 18 | 42 |
| Left   | 12 | 28 |

**5.58 Dice** When two dice are rolled, is the event "the first die shows a 1 on top" independent of the event "the second die shows a 1 on top"?

TRY **5.59 Happiness and Traditional Views (Example 13)** On the 2008 GSS, people were asked about their happiness and were also asked whether they agreed with the following statement: "In a marriage the husband should work, and the wife should take care of the home." The table summarizes the data collected.

|          | Agree | Don't Know | Disagree |
|----------|-------|-----------|----------|
| Happy    | 345   | 261       | 656      |
| Unhappy  | 34    | 22        | 44       |

a. Add the marginal totals and the grand total to the table. Show the complete table with marginal totals.

b. Determine whether being happy is independent of agreeing with the statement.

**5.60 Happiness** Using the table in Exercise 5.59, determine whether being unhappy is independent of disagreeing with the statement.

TRY **5.61 Coin (Example 14)** Imagine flipping three fair coins.

a. What is the theoretical probability that all three will come up heads?

b. What is the theoretical probability the first toss is tails AND the next two are heads?

**5.62 Die** Imagine rolling a fair six-sided die three times.

a. What is the theoretical probability that all three rolls of the die show a 1 on top?

b. What is the theoretical probability that the first roll of the die shows a 6 AND the next two rolls show a 1 on top?

TRY **5.63 Die Sequences (Example 15)** Roll a fair six-sided die five times and record the number of spots on top.
Which sequence is more likely? Explain.

Sequence A: 66666

Sequence B: 16643

**5.64 Babies** Assume that babies born are equally likely to be boys (B) or girls (G). Assume a woman has 6 children, none of whom are twins. Which sequence is more likely? Explain.

Sequence A: GGGGGG

Sequence B: GGGBBB

TRY **5.65 Seat Belt Use (Example 16)** According to the National Highway Traffic Safety Administration, the rate of seat belt use in the United States for 2009 was 84%. Suppose that you looked at two people, selected randomly and independently from the population, to see whether each had his or her seat belt fastened in 2009.

a. What is the probability that neither person had her or his belt fastened?

b. What is the probability that at least one person had her or his belt fastened? (*Hint:* "At least one had the belt fastened" is the complement of "neither one had the belt fastened.")

**5.66 Gender of Newborns** When a baby is born, the probability that it will be a boy is close to 51%.

a. What is the probability that a newborn will be a girl?

For the rest of the questions, assume that a woman gives birth to two babies who are not identical twins, and therefore the genders are independent.

b. What is the probability that they are both girls?

c. What is the probability of there being at least one boy?

TRY * **5.67 Cervical Cancer (Example 17)** According to a study published in *Scientific American*, about 8 women in 100,000 have cervical cancer (C), so P(C) = 0.00008. Suppose the chance that a Pap smear will detect cervical cancer when it is present is 0.84. Therefore,

$$P(\text{test pos} \mid C) = 0.84$$

What is the probability that a randomly chosen woman who has this test will both have cervical cancer AND test positive for it?

* **5.68 Cervical Cancer** About 8 women in 100,000 have cervical cancer (C), so P(C) = 0.00008 and P(no C) = 0.99992. The chance that a Pap smear will incorrectly indicate that a woman without cervical cancer has cervical cancer is 0.03. Therefore,

$$P(\text{test pos} \mid \text{no C}) = 0.03$$

What is the probability that a randomly chosen women who has this test will both be free of cervical cancer and test positive for cervical cancer (a false positive)?

## SECTION 5.4

**5.69 Law of Large Numbers** Refer to Histograms A, B, and C, which show the relative frequencies from experiments in which a fair six-sided die was rolled. One histogram shows the results from 20 rolls, one the results for 100 rolls, and another the results for 10,000 rolls. Which histogram do you think was for 10,000 rolls and why?

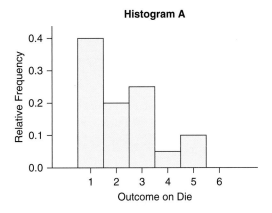

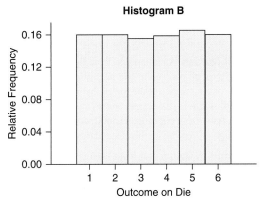

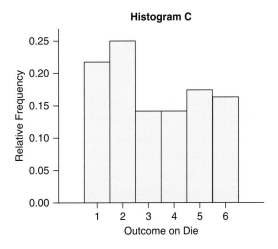

**5.70 Law of Large Numbers** The table shows the results of rolling a fair six-sided die.

| Outcome on Die | 20 Trials | 100 Trials | 1000 Trials |
|---|---|---|---|
| 1 | 8 | 20 | 167 |
| 2 | 4 | 23 | 167 |
| 3 | 5 | 13 | 161 |
| 4 | 1 | 13 | 166 |
| 5 | 2 | 16 | 172 |
| 6 | 0 | 15 | 167 |

Using the table, find the empirical probability of rolling a 1 for 20, 100, and 1000 trials. Report the theoretical probability of rolling a 1 with a fair six-sided die. Compare the empirical probabilities to the theoretical probability, and explain what they show.

**5.71 Coin Flips** Imagine flipping a fair coin many times. Explain what should happen to the proportion of heads as the number of coin flips increases.

**5.72 Coin Flips, Again** Refer to the figure.
a. After a large number of flips, the overall proportion of heads "settles down" to nearly what value?
b. Approximately how many coin flips does it take before the proportion of heads settles down?
c. What do we call the law that causes this settling down of the proportion?
d. From the graph, determine whether the first flip was heads or tails.

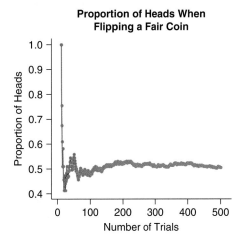

## CHAPTER REVIEW EXERCISES

**5.73 Sabotage of Airplanes** Between 1950 and 2009, there were 1300 fatal commercial airplane accidents worldwide for which a cause could be determined. Of those, 117 were due to sabotage (www.planecrashinfo.com). Using these data, find the empirical probability that sabotage will be the cause of a fatal airplane accident.

**5.74 Pilot Error** Of the 1300 fatal commercial airplane accidents between 1950 and 2009 for which a cause could be determined, 377 were due to simple pilot error (without contributing weather problems or mechanical problems). Using these data, find the empirical probability that simple pilot error will be the cause of a fatal airplane accident.

**5.75 Independent Variables** Use your general knowledge to label the following pairs of variables as independent or associated. Explain.

a. For a sample of adults, gender and shoe size

b. For a sample of football teams, win/loss record for the coin toss at the beginning of the game and number of cheerleaders for the team

**5.76 Independent Variables** Use your general knowledge to label the following pairs of variables as independent or associated. Explain.

a. The outcome on flips of two separate, fair coins

b. Breed of dog and weight of dog for dogs at a dog show

**5.77 UFOs** When two people meet, they are sometimes surprised that they have similar beliefs. A survey of 1003 random adults conducted by the Scripps Survey Research Center at Ohio University in 2008 found that 62 percent of men and 50 percent of women believe in intelligent life on other planets. Well, actually, they said it is either "very likely" or "somewhat likely" that intelligent life exists on other planets (www.reporternews.com).

a. If a man and a woman meet, what is the probability that they both believe in intelligent life on other planets?

b. If a man and a woman meet, what is the probability that neither believes in intelligent life on other planets?

c. What is the probability that the man and woman agree about life on other planets?

d. If a man and a woman meet, what is the probability that they have opposite beliefs on this issue?

**5.78 Seat Belt Use** In 2009, the National Highway Traffic Safety Administration said that 84% of drivers buckled their seat belts. Assume that this percentage is still accurate.

a. If four drivers are randomly selected, what is the probability that they are all wearing their seat belts?

b. If four drivers are randomly selected, what is the probability that at least one is not wearing his or her seat belt?

**5.79 Internet Access** A 2009 Harris poll said that 80% of adults in the United States have Internet access. Assume that this is still correct.

a. If two people are randomly selected, what is the probability that they both have Internet access?

b. If the two people chosen were a married couple living in the same residence, explain why they would not be considered independent with regard to Internet access.

**5.80 SAT Scores** The probability of a randomly selected person having a grade of 500 or above on the quantitative portion of the SAT is 0.50.

a. If two students are chosen randomly and independently, what is the probability that they both score 500 or above?

b. If two students are selected from the same high school calculus class, do you think the probability of their both scoring 500 or above is different from your answer to part a? Explain.

* **5.81 Birthdays** Suppose all the days of the week are equally likely as birthdays. Alicia and David are two randomly selected, unrelated people.

a. What is the probability that they were both born on Monday?

b. What is the probability that Alicia OR David was born on Monday? *Hint:* The answer is not 2/7. Refer to Guided Exercise 5.25 if you need help.

* **5.82 Pass Rate of Written Driver's Exam** In California, about 92% of teens who take the written driver's exam fail the first time they take it (www.teendrivingcourse.com). Suppose that Sam and Maria are randomly selected teenagers taking the test for the first time.

a. What is the probability that they both pass the test?

b. What is the probability that Sam OR Maria passes the test?

**5.83 Opinion about Nurses** A Gallup Poll from December of 2009 estimated that 83% of all people thought nurses had high or very high ethical standards, putting nurses at the top of the professions with regard to this issue. If this rate is still correct and a new poll of 5000 people were obtained, how many out of those 5000 would you expect to think nurses have high or very high ethical standards?

**5.84 Climate Change** A Gallup poll from December of 2009 asked whether U.S. residents who are aware of climate change thought their government was doing enough to reduce emissions of cars and factories. Gallup estimated that 52% of all U.S residents felt that the government was not doing enough. If another poll were taken and there were 500 participants, how many would you expect to say that the United States was not doing enough, assuming the percentage remained the same?

* **5.85 Independent** Imagine rolling a red die and a blue die. From this trial, name a pair of independent events.

* **5.86 Mutually Exclusive** Imagine rolling a red die and a blue die. From this trial, name a pair of mutually exclusive events.

* **5.87** Construct a two-way table with 60 women and 80 men in which both groups show equal percentages of right-handedness.

* **5.88** Construct a two-way table with 60 women and 80 men in which there is a higher percentage of right-handed women.

* **5.89 Law of Large Numbers** A famous study by Amos Tversky and Nobel laureate Daniel Kahneman asked people to consider two hospitals. Hospital A is small and has 15 babies born per day. Hospital B has 45 babies born each day. Over one year, each hospital recorded the number of days that it had more than 60% girls born. Assuming that 50% of all babies are girls, which hospital had the most such days? Or do you think both will have about the same number of days with more than 60% girls born? Answer,

and explain. (Source: Amos Tversky. 2004. *Preference, Belief, and Similarity: Selected Writings*, ed. Eldar Shafir. Cambridge, Mass.: MIT Press, p. 205)

* **5.90 Law of Large Numbers** A certain professional basketball player typically makes 80% of his basket attempts, which is considered to be good. Suppose you go to several games at which this player plays. Sometimes the player attempts only a few baskets, say 10. Other times, he attempts about 60. On which of those nights is the player most likely to have a "bad" night, in which he makes much fewer than 80% of his baskets?

# gUIDED EXERCISES

g **5.25 GSS: OR** Referring to the table, assume one person is chosen randomly from the 1858 people in the table. Answer the question below the table by following the numbered steps.

|  | Dem | Rep | Other | Total |
|---|---|---|---|---|
| Liberal | 306 | 26 | 198 | 530 |
| Moderate | 279 | 134 | 322 | 735 |
| Conservative | 104 | 309 | 180 | 593 |
| Total | 689 | 469 | 700 | 1858 |

QUESTION  What is the probability that the person is liberal OR a Democrat?

Step 1 ▶ What is the probability that the person is liberal?

Step 2 ▶ What is the probability that the person is a Democrat?

Step 3 ▶ If the events liberal and a Democrat were mutually exclusive, you could just add the probabilities from step 1 and step 2 to find the probability that a person is liberal OR a Democrat. Are they mutually exclusive? Why or why not?

Step 4 ▶ What is the probability that the person is liberal AND a Democrat? Use numbers from the table; do not assume independence.

Step 5 ▶ Why should you subtract the probability that the person is liberal AND a Democrat from the sum as shown in the given formula?

$$P(\text{liberal OR a Democrat}) = P(\text{liberal}) + P(\text{a Democrat})$$
$$- P(\text{liberal AND a Democrat})$$

Step 6 ▶ Do the calculation using the formula given in step 5 to find the probability that a person is liberal OR a Democrat. It is more accurate to use the numbers from the table (fractions) than to use your percentages, which are rounded off.

Step 7 ▶ Finally, report the answer in a sentence.

g **5.57 Hand Folding** When people fold their hands together with interlocking fingers, most people are more comfortable doing it in one of two ways. In one way, the right thumb ends up on top, and in the other way, the left thumb is on top. The table shows the data from one group of people.

|  | M | W |
|---|---|---|
| Right | 18 | 42 |
| Left | 12 | 28 |

M means man, W means woman, Right means the right thumb is on top, and Left means the left thumb is on top.

QUESTION  Say a person is selected from this group at random. Are the events "right thumb on top" and "male" independent or associated?

To answer, we need to determine whether the probability of having the right thumb on top given that you are a man is equal to the probability of having the right thumb on top (for the entire group). If so, the variables are independent.

Step 1 ▶ Figure out the marginal totals and put them into the table.

Step 2 ▶ Find the overall probability that the person's right thumb is on top.

Step 3 ▶ Find the probability that the right thumb is on top given that the person is a man. (What percentage of men have the right thumb on top?)

Step 4 ▶ Finally, are the variables independent? Why or why not?

# TechTips

## For All Technology

**EXAMPLE: GENERATING RANDOM INTEGERS** ▶ Generate four random integers from 1 to 6, for simulating the results of rolling a six-sided die.

---

### Seed First before the Random Integers

If you do not seed the calculator, everyone might get the same series of "random" numbers.

1. Enter the last four digits of your Social Security number or cell phone number and press **STO →** .
2. Then press **MATH**, choose **PRB**, and **ENTER ENTER** (to choose 1:rand).

You only need to seed the calculator once, unless you **Reset** the calculator. (If you want the same sequence later on, you can seed again with the same number.)

### Random Integers

1. Press **MATH**, choose **PRB**, and press **5** (to choose 5:randInt).
2. See Figure 5A. Enter **1 ❯ 6 ❯ 4)** and press **ENTER**. (The comma button is above the 7 button.)

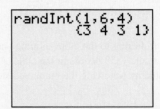

▲ **FIGURE 5A** TI-83/84

The first two digits (1 and 6 in Figure 5A) determine the smallest and largest integers, and the third digit (4 in Figure 5A) determines the number of random integers generated. To get four more random integers, press **ENTER** again.

---

### Random Integers

1. **Calc > Random Data > Integer**
2. See Figure 5B. Enter:
   **Number of rows to generate**, **4**
   **Store in column(s)**, **c1**
   **Minimum value**, **1**
   **Maximum value**, **6**
3. Click **OK**.

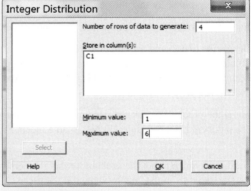

▲ **FIGURE 5B** Minitab

---

### Random Integers

1. Click **fx**, select a category **All**, and **RANDBETWEEN**.
2. See Figure 5C.
   Enter: **Bottom, 1; Top, 6**.
   Click **OK**.
   You will get one random integer in the active cell in the spreadsheet.
3. To get more random integers, put the cursor at the lower right corner of the cell containing the first integer until you see a black cross (+), and drag downward until you have as many as you need.

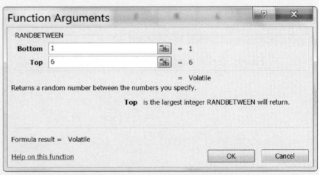

▲ **FIGURE 5C** Excel

## Random Integers

1. **Data** > **Simulate Data** > **Discrete Uniform**.

2. See Figure 5D.

   Enter **Rows**, **4**; **Columns**, **1**; **minimum**, **1**; **maximum**, **6** and leave **Split across columns** and **Use preset generator**.

3. Click **Simulate**.

   You will get four random integers (from 1 to 6) in the first empty column.

▲ **FIGURE 5D** StatCrunch

# 6 Modeling Random Events: The Normal and Binomial Models

# THEME

Probability distributions describe random outcomes in much the same way as the distributions we discussed in Chapter 2 describe collections of data. A probability distribution tells us the possible outcomes of a random trial and the probability that each of those outcomes will occur.

Random events can seem chaotic and unpredictable. If you flip a coin 10 times, there's no way you can say with absolute certainty how many heads will appear. There's no way your local weather forecaster can tell you with certainty whether it will rain tomorrow. Still, if we watch enough of these random events, patterns begin to emerge. We begin to learn how often different outcomes occur and to gain an understanding of what is common and what is unusual.

Science fiction writer Isaac Asimov is often quoted as having said, "The most exciting phrase to hear in science, the one that heralds new discoveries, is not Eureka! (I found it!) but rather 'hmm . . . that's funny . . . .'" When an outcome strikes us "funny" or unusual, it is because something unlikely occurred. This is exciting because it means a discovery has been made; the world is not the way we thought it was!

However, to know whether something unusual has happened, we first have to know how often it usually occurs. In other words, we need to know the probability. In the case study, you'll see a "homemade" example, in which one of the authors found that five ice cream cones ordered at a local McDonald's weighed more than the advertised amount. Was this a common occurrence, or did McDonald's perhaps deliberately understate the weight of the cones so that no one would ever be disappointed?

In this chapter we'll introduce a new tool to help us characterize probabilities for random events: the probability distribution. We'll examine the Normal probability distribution and the binomial probability distribution, two very useful tools for answering the question "Is this unusual?"

# CASE STUDY

## You Sometimes Get More Than You Pay For

A McDonald's restaurant near one of the authors sells ice cream cones, which, according to the "fact sheet" provided, weigh 3.18 ounces (converted from grams) and contain 150 calories. Do the ice cream cones really weigh exactly 3.18 ounces? To get 3.18 ounces for every cone would require a very fine-tuned machine or an employee with a *very* good sense of timing. Thus, we expect some natural variation in the weight of these cones. In fact, one of the authors bought five ice cream cones on different days. She found that each of the five cones weighed more than 3.18 ounces. Such an outcome, five out of five over the advertised weight, might occur just by chance. But how often? If such an outcome (five out of five) rarely happens, it's pretty surprising that it happened to us. In that case, perhaps McDonald's actually puts in more than 3.18 ounces. By the end of this chapter, after learning about the binomial and Normal probability distributions, you will be able to measure how surprising this outcome is—or is not.

SECTION 6.1

# Probability Distributions Are Models of Random Experiments

A **probability model** is a description of how a statistician thinks data are produced. We use the word *model* to remind us that our description does not really explain how the data came into existence, but we hope that it describes the actual process fairly closely. We can tell whether a model is good by noting whether the probabilities that it predicts are matched by real-life outcomes. Thus, if a model says that the probability of getting heads when we flip a coin is 0.54, but in fact we get heads 50% of the time, we suspect that the model is not a good match.

A **probability distribution**, sometimes called a **probability distribution function (pdf),** is a tool that helps us by keeping track of the outcomes of a random experiment and the probabilities associated with those outcomes. For example, suppose the playlist on your mp3 player has 10 songs: 6 are Rock, 2 are Country, 1 is Hip-hop, and 1 is Opera. Put your player on shuffle. What is the probability that the first song chosen is Rock?

The way the question is worded ("What is the probability . . . is Rock"?) means that we care about only two things: Is the song classified as Rock or is it not? We could write the probabilities as shown in Table 6.1.

Table 6.1 is a very simple probability distribution. It has two important features: It tells us all the possible outcomes of our random experiment (Rock or Not Rock), and it tells us the probability of each of these outcomes. All probability distributions have these two features, though they are not always listed so clearly.

| Outcome | Probability |
|---------|-------------|
| Rock    | 6/10        |
| Not Rock| 4/10        |

▲ **TABLE 6.1** Probability Distribution of Songs

KEY POINT — A probability distribution tells us (1) all the possible outcomes of a random experiment, and (2) the probability of each outcome.

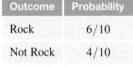

## Looking Back

**Distributions of a Sample**
Probability distributions are similar to distributions of a sample, which were introduced in Chapter 2. A distribution of a sample tells us the values in the sample and their frequency. A probability distribution tells us the possible values of the random experiment and their probability.

In Chapter 1, we classified variables as either numerical or categorical. It is now useful to break the numerical variables down into two more categories. **Discrete outcomes** (or discrete variables) are numerical values that you can list or count. An example is the number of phone numbers stored on the phones of your classmates. **Continuous outcomes** (or continuous variables) cannot be listed or counted because they occur over a range. The length of time your next phone call will last is a continuous variable. Refer to Figure 6.1 for a visual comparison of these terms.

This distinction is important because if we can list the outcomes, as we can for a discrete variable, then we have a nice way of displaying the probability distribution. However, if we are working with a continuous variable, then we can't list the outcomes, and we have to be a bit more clever in describing the probability distribution function. For this reason, we treat discrete values separately from continuous variables.

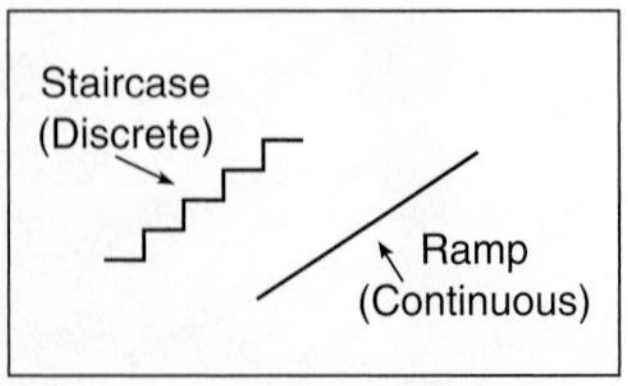

## EXAMPLE 1 Discrete or Continuous?

Consider these variables:

a. The weight of a submarine sandwich you're served at a deli.

b. The elapsed time from when you leave your house to when you arrived in class this morning.

c. The number of people in the next passing car.

d. The blood alcohol level of a driver pulled over by the police in a random sobriety check. (Blood alcohol level is measured as the percent of the blood that is alcohol.)

e. The number of eggs laid by a randomly selected salmon as observed in a fishery.

Staircase (Discrete)

Ramp (Continuous)

▲ **FIGURE 6.1** Visual representation of discrete and continuous changes in elevation. Note that for the staircase (discrete outcomes), you can count the stairs.

**QUESTION**  Identify each of these numerical variables as continuous or discrete.

**SOLUTION**  The continuous variables are variables a, b, and d. Continuous variables can take on any value in a spectrum. For example, the sandwich might weight 6 ounces or 6.1 ounces or 6.0013 ounces. The amount of time it takes you to get to class could be 5000 seconds or 5000.4 seconds or 5000.456. Blood-alcohol content can be any value between 0 and 1, including 0.0013 (or 0.13%), 0.0013333, 0.001357, and so on.

Variables c and e, on the other hand, are discrete quantities—that is, numbers that can be counted.

**TRY THIS!**  Exercise 6.1

## Discrete Probability Distributions Can Be Tables or Graphs

A statistics class at UCLA was approximately 40% male and 60% female. Let's arbitrarily code the males as 0 and the females as 1. If we select a person at random, what is the probability that the person is female?

Creating a probability distribution for this situation is as easy as listing both outcomes (0 and 1) and their probabilities (0.40 and 0.60). The easiest way to do this is in a table, as shown as Table 6.2. However, we could also do this as a graph (Figure 6.2).

| Female | Probability |
|--------|-------------|
| 0 | 0.40 |
| 1 | 0.60 |

▲ **TABLE 6.2** Probability Distribution of Gender in a Class

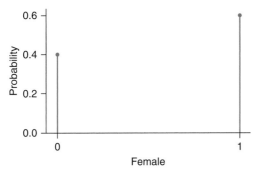

◀ **FIGURE 6.2** Probability distribution for selecting a person at random from a particular statistics class and recording whether the person is male (0) or female (1).

Amazon.com invites customers to rate books on a scale of 1 to 5 stars. At a visit to the site on July 5, 2010, *The Story of Edgar Sawtelle: A Novel*, by David Wroblewski, had received 1505 customer reviews. Suppose we randomly select a reviewer and base our decision whether to buy the book on how many stars that person gave to it. Table 6.3 and Figure 6.3 illustrate two different ways of representing the probability distribution for this random event. For example, we see that the most likely outcome is that the reviewer gave the book 1 star. The probability that the reviewer gave the book 1 star is 0.290, or about 29%.

| Number of Stars | Probability |
|-----------------|-------------|
| 1 | 0.290 |
| 2 | 0.146 |
| 3 | 0.163 |
| 4 | 0.195 |
| 5 | 0.206 |

▲ **TABLE 6.3** Distribution of Number of Stars

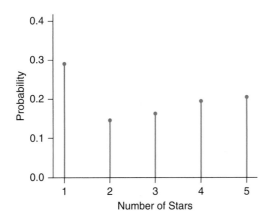

◀ **FIGURE 6.3** Probability distribution for the number of stars given, as a rating, to a particular book on amazon.com by a randomly selected customer.

Note that the probabilities in Table 6.3 add to 1. This is true of all probability distributions: When you add up the probabilities for all possible outcomes, they add to 1. They have to, because there are no other possibilities.

## Discrete Distributions Can Also Be Equations

What if we have too many outcomes to list in a table? For example, suppose a married couple decides to keep having children until they have a girl. How many children will they have, assuming that boys and girls are equally likely and that the gender of one birth doesn't depend on any of the previous births? It could very well turn out that their first child is a girl and they therefore have only one child. Or that the first is a boy but the second is a girl. Or, just possibly, they might never have a girl. The value of this experiment could be any number 1, 2, 3, . . . up to infinity. (Okay, in reality, it's impossible to have that many children. But we can imagine!)

We can't list all these values and probabilities in a table, and we can only hint at what the graph might look like. But we *can* write them in a formula:

The probability of having $x$ children before the first girl is $(1/2)^x$.

For example, the probability that they have 1 child is $(1/2)^1 = 1/2$.

The probability that they have 4 children is $(1/2)^4 = 1/16$.

The probability that they have 10 children is (thankfully) small: $(1/2)^{10} = 0.00098$.

In this book, we will give the probabilities in a table or graph whenever convenient. In fact, even if it's not especially convenient, we will often provide a graph to encourage you to visualize what the probability distribution looks like. Figure 6.4 is part of the graph of the probability distribution for the number of children a couple can have before having a girl. We see that the probability that the first child is a girl is 0.50 (50%). The probability that the couple has two children is half this: 0.25. The probabilities continue to decrease, and each probability is half the one before it.

▶ **FIGURE 6.4** Probability distribution of the number of children born until the first girl.

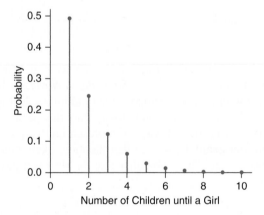

## EXAMPLE 2 Playing Dice

Roll a fair six-sided die. A fair die is one in which each side is equally likely to end up on top. You will win $4 if you roll a 5 or a 6. You will lose $5 if you roll a 1. For any other outcome, you will win or lose nothing.

**QUESTION** Give a table that shows the probability distribution for the amount of money you will win. Draw a graph of this probability distribution function.

**SOLUTION** There are three outcomes: you win $4, you win 0 dollars, you win −$5. (Winning negative five dollars is the same as losing five dollars.)

You win \$4 if you roll a 5 or 6, so the probability is $2/6 = 1/3$.

You win \$0 if you roll a 2, 3, or 4, so the probability is $3/6 = 1/2$

You "win" $-\$5$ if you roll a 1, so the probability is $1/6$.

We can put the pdf in a table (Table 6.4), or we can represent the pdf as a graph (Figure 6.5).

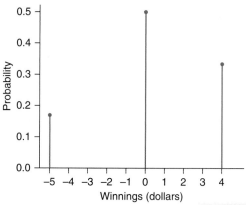

◀ **FIGURE 6.5** Probability density function of the dice game.

| Winnings | Probability |
|---|---|
| −5 | 1/6 |
| 0 | 1/2 |
| 4 | 1/3 |

▲ **TABLE 6.4** Probability density function of the dice game (Example 2).

**TRY THIS!** Exercise 6.5

## Continuous Probabilities Are Represented as Areas under Curves

Finding probabilities for continuous outcomes is more complicated, because we cannot simply list all the possible values we might see. What we can list is the *range of values* we might see.

For example, suppose you want to know the probability that you will wait in line for between 3 and 4 minutes when you go to the coffee shop. You can't list all possible outcomes that could result from your visit: 1.0 minute, 1.00032 minute, 2.00000321 minute. It would take (literally) an eternity. But you can specify a range. Suppose this particular coffee shop has done extensive research and knows that everyone gets helped in under 5 minutes. So all customers get helped within the range of 0 to 5 minutes.

If we want to find probabilities concerning a continuous-valued experiment, we also need to give a range for the outcomes. For example, the manager wants to know the probability that a customer will wait less than 2 minutes; this gives a range of 0 to 2 minutes.

The probabilities for a continuous-valued random experiment are represented as areas under curves. The curve is called a **probability density curve**. The total area under the curve is 1, because this represents the probability that the outcome will be somewhere on the *x*-axis. To find the probability of waiting between 0 and 2 minutes, we find the area under the density curve and between 0 and 2 (Figure 6.6).

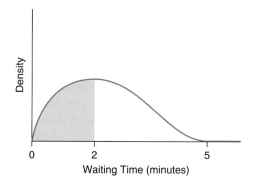

◀ **FIGURE 6.6** Probability distribution of times waiting in line at a particular coffee shop.

The *y*-axis in a continuous-valued pdf is labeled "Density." How the density is determined isn't important. What *is* important to know is that the units of density are scaled so that the area under the entire curve is 1.

You might wonder where this curve came from. How did we know the distribution was exactly this shape? In practice, it is very difficult to know which curve is correct for a real-life situation. Statisticians call these curves "probability models" because they are meant to mimic a real-life probability, but we don't know—and can never know for sure—that the curve is correct. On the other hand, we can compare our probability predictions to the actual frequencies that we see. If they are close, then our probability model is good. For example, if this probability model predicts that 45% of customers get coffee within 2 minutes, then we can compare this prediction to an actual sample of customers.

## Finding Probabilities for Continuous-Valued Outcomes

Calculating the area under a curve is not easy. If you have a formula for the probability density, then you can sometimes apply techniques from calculus to find the area. However, for many commonly used probability densities, basic calculus is not helpful, and computer-based approximations are required.

In this book, you will always find areas for continuous-valued outcomes by using a table or by using technology. In Section 6.2 we introduce a table that can be used to find areas for one type of probability density that is very common in practice: the Normal curve.

## EXAMPLE 3  Waiting for the Bus

The bus that runs near one of the authors' home arrives every 12 minutes. If the author arrives at the bus stop at a randomly chosen time, then the probability distribution for the number of minutes he must wait for the bus is shown in Figure 6.7.

▶ **FIGURE 6.7** Probability distribution function showing the number of minutes the author must wait for the bus if he arrives at a randomly determined time.

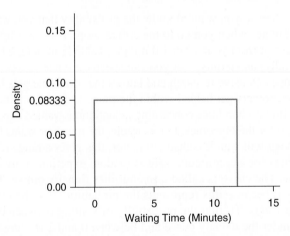

**QUESTION** Find the probability that the author will have to wait between 4 and 10 minutes for the bus. (*Hint:* Remember that the area of a rectangle is the product of the lengths of the two sides.)

**SOLUTION** The distribution shown in Figure 6.7 is called a uniform distribution. Finding areas under this curve is easy because the curve is just a rectangular shape. The area we need to find is shown in Figure 6.8. The area of a rectangle is width times height. The width is (10 − 4) = 6, and the height is 0.08333.

The probability that the author must wait between 4 and 10 minutes is 6 × 0.08333 = 0.4998, or about 0.500. Visually, we see that about half of the area in Figure 6.8 is shaded.

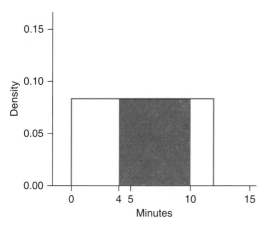

◀ **FIGURE 6.8** The shaded area represents the probability that the author will wait between 4 and 10 minutes if he arrives at the bus stop at a randomly determined time.

There is approximately a 50% chance that the author must wait between 4 and 10 minutes.

 Exercise 6.9

## SECTION 6.2

# The Normal Model

The **Normal model** is the most widely used probability model for continuous numerical variables. One reason is that many numerical variables in which researchers have historically been interested have distributions for which the Normal model provides a very close fit. Also, an important mathematical theorem called the Central Limit Theorem (to be introduced in Chapter 7) links the Normal model to several key statistical ideas, which provides good motivation for learning this model.

We begin by showing you what the Normal model looks like. Then we discuss how to find probabilities by finding areas underneath the Normal curve. We will illustrate these concepts with examples and also discuss why the Normal model is appropriate for these situations.

## Visualizing the Normal Distribution

Figure 6.9 on the next page shows several histograms of measurements taken from a sample of about 1400 adult men in the United States. All of these graphs have similar shapes: They are unimodal and symmetric. We have superimposed smooth curves over the histograms that capture this shape. You could easily imagine that if we continued to collect more and more data, the histogram would eventually fill in the curve and match the shape almost exactly.

The curve drawn on these histograms is called the **Normal curve**, or the **Normal distribution**. It is also sometimes called the Gaussian distribution, after Karl Friedrich Gauss (1777–1855), the mathematician who first derived the formula. Statisticians and scientists recognized that this curve provided a model that pretty closely described a good number of continuous-valued data distributions. Today, even though we have many other distributions to model real-life data, the Normal curve is still one of the most frequently used probability distribution functions in science.

## Center and Spread

In Chapters 2 and 3 we discussed the center and spread of a distribution of *data*. These concepts are also useful for studying distributions of *probability*. The mean of a probability distribution sits at the balancing point of the probability distribution. The

 **Looking Back**

**Unimodal and Symmetric Distributions**
Symmetric distributions have histograms whose right and left sides are roughly mirror images of each other. Unimodal distributions have histograms with one mound.

**Details**

**The Bell Curve**
The Normal or Gaussian curve is also called the bell curve.

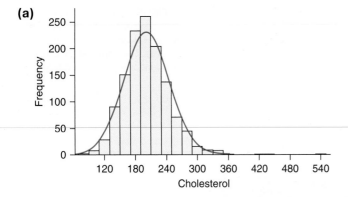

▲ **FIGURE 6.9** Measurements for a sample of men: **(a)** cholesterol levels, **(b)** diastolic blood pressure readings, and **(c)** height. For all three, a very similar shape appears: nearly symmetric and mound-shaped.

standard deviation of a probability distribution measures the spread of the distribution by telling us how far away, typically, the values are from the mean. The conceptual understanding you developed for the mean and standard deviation of a sample still apply to probability distributions.

The notation we use is slightly different, so that we can distinguish means and standard deviations of probability distributions from means and standard deviations of data. The **mean of a probability distribution** is represented by the Greek character $\mu$ (mu, pronounced "mew"), and the **standard deviation of a probability distribution** is represented by the character $\sigma$ (sigma). These Greek characters are used to avoid confusion of these concepts with their counterparts for samples of data, $\bar{x}$ and $s$.

## The Mean and Standard Deviation of a Normal Distribution

The exact shape of the Normal distribution is determined by the values of the mean and the standard deviation. Because the Normal distribution is symmetric, the mean is in the exact center of the distribution. The standard deviation determines whether the Normal curve is wide and low (large standard deviation) or narrow and tall (small standard deviation). Figure 6.10 shows two Normal curves that have the same mean but different standard deviations.

**↻ Looking Back**

**Mean and Standard Deviation**
In Chapter 3, you learned that the symbol for the mean of a *sample* of data is $\bar{x}$ and that the symbol for the standard deviation of a sample of data is $s$.

▶ **FIGURE 6.10** Two Normal curves with the same mean but different standard deviations.

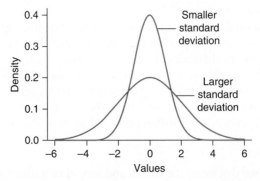

A Normal curve with a mean of 69 inches and a standard deviation of 3 inches provides a very good match to the distribution of heights of all adult men in the United States. Surprisingly, a Normal curve with the same standard deviation of 3 inches, but a smaller mean of about 64 inches, describes the distribution of adult women's heights. Figure 6.11 shows what these Normal curves look like.

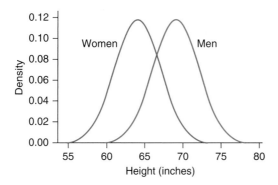

◀ **FIGURE 6.11** Two Normal curves. The blue curve represents the distribution of women's heights and has a mean of 64 inches and a standard deviation of 3 inches. The red curve represents the distribution of men's heights and has the same standard deviation, but the mean is 69 inches.

The only way to distinguish among different Normal distributions is by their means and standard deviations. We can take advantage of this fact to write a short-hand notation to represent a particular Normal distribution. The notation $N(\mu, \sigma)$ represents a Normal distribution that is centered at the value of $\mu$ (the mean of the distribution) and whose spread is measured by the value of $\sigma$ (the standard deviation of the distribution). For example, in Figure 6.11, the distribution of women's heights is $N(64, 3)$, and the distribution of men's heights is $N(69, 3)$.

**KEY POINT** The Normal distribution is symmetric and unimodal ("bell-shaped"). The notation $N(\mu, \sigma)$ tells us the mean and standard deviation of the Normal distribution.

## Finding Normal Probabilities

The Normal model $N(64, 3)$ gives a good approximation of the distribution of adult women's heights in the United States (where height is measured in inches). Suppose we were to select an adult woman from the United States at random and record her height. What is the probability that she is taller than a specified height?

Because height is a continuous numerical variable, we can answer this question by finding the appropriate area under the Normal curve. For example, Figure 6.12 shows a Normal curve that models the distribution of heights of women in the population—the same curve as in Figure 6.11. (We leave the numerical scale off the vertical axis from now on since it is not needed for doing calculations.) The area of the shaded region gives us the probability of selecting a woman taller than 62 inches. The entire area under the curve is 1.

In fact, Figure 6.12 represents both the probability of selecting a woman taller than 62 inches and also the probability of selecting a woman *62 inches tall or taller*. Because the areas for both regions (the one that is strictly greater than 62, and the other that includes 62) are the same, the probabilities are also the same. This is a convenient feature of continuous variables: We don't have to be too picky about our language when working with probabilities. This is in marked contrast with discrete variables, as you will soon see.

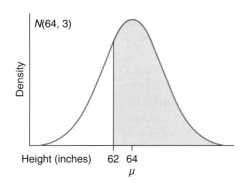

◀ **FIGURE 6.12** The area of the shaded region represents the probability of finding a woman taller than 62 inches from a $N(64, 3)$ distribution.

What if we instead wanted to know the probability that the chosen woman would be between 62 inches and 67 inches tall? That area would look like Figure 6.13.

▶ **FIGURE 6.13** The shaded area represents the probability that a randomly selected woman is between 62 and 67 inches tall. The probability distribution shown is the Normal distribution with mean 64 inches and standard deviation 3 inches.

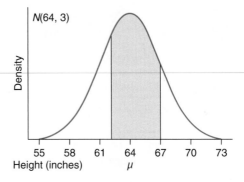

Figure 6.14 shows the area corresponding to the probability that this randomly selected woman is less than 62 inches tall.

▶ **FIGURE 6.14** The shaded area represents the probability that the randomly selected woman is less than 62 inches tall.

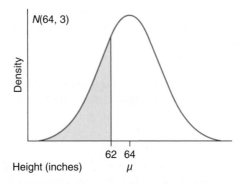

KEY
POINT ▶ When you are finding probabilities with Normal models, the first and most helpful step is to sketch the curve, label it appropriately, and shade in the region of interest.

We recommend that you always begin problems that concern Normal models by drawing a picture. One advantage to drawing a picture is sometimes that's all you need. For example, according to the McDonald's "Fact Sheet," each serving of ice cream in a cone weighs 3.18 ounces. Now, we know that in real life it is not possible to serve exactly 3.18 ounces. Probably the employees operating the machines (or the machines themselves) actually dispense a little more or a little less than 3.18 ounces. Suppose that the amount of ice cream dispensed follows a Normal distribution with a mean of 3.18 ounces. What is the probability that a hungry customer will actually get less than 3.18 ounces?

Figure 6.15a shows the situation. From it we easily see that the area to the left of 3.18 is exactly half of the total area. Thus we know that the probability of getting less than 3.18 ounces is 0.50. (We also know this because the Normal curve is symmetric, so the mean—the balancing point—must sit right in the middle. Therefore, the probability of getting a value less than the mean is 0.50.)

What if the true mean is actually larger than 3.18 ounces? How will that affect the probability of getting a cone that weighs less than 3.18 ounces? Imagine "sliding" the Normal curve to the right, which corresponds to increasing the mean. Does the area to the left of 3.18 go up or down as the curve slides to the right? Figure 6.15b shows that the area below 3.18 is now smaller than 50%. The larger the mean amount of ice cream dispensed, the less likely it is that a customer will complain about getting too little.

**(a)**

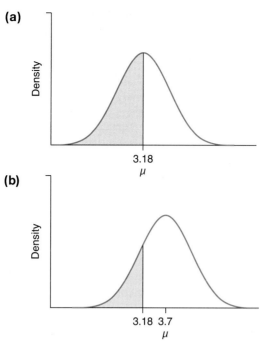

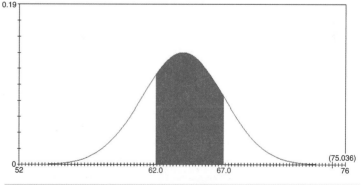

◀ **FIGURE 6.15 (a)** A N(3.18, 0.6) curve, showing that the probability of getting a cone that weighs less than 3.18 ounces is 50%. **(b)** A Normal curve with the same standard deviation (0.6) but a larger mean (3.7). The area below 3.18 is now much smaller.

**(b)**

3.18 3.7
μ

## Finding Probability with Technology

Finding the area of a Normal distribution is best done with technology. Most calculators and many software packages will show you how to do this. We illustrate one such package, available free on the Internet, that you can use. We will also show you the "old-fashioned" way, which is useful when you do not have a computer handy. The old-fashioned way is also worth learning because it helps solidify your conceptual understanding of the Normal model.

Figure 6.16a is a screenshot from the SOCR (http://socr.stat.ucla.edu) calculator. It shows the probability that a randomly selected woman is between 62 and 67 inches

**Tech**

**(a)**

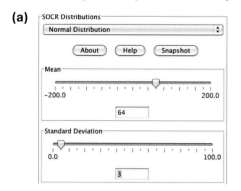

Distribution Properties
Normal (64.0, 3.0) Distribution
Mean: 64.000000
Median: 64.000000
Variance: 9.000000
Standard Deviation: 3.000000
Max Density: .132981

Probabilities

Left: .252493
Between (Red-Shaded): .588852
Right: .158655

**(b)**

Distribution Properties
Normal (64.0, 3.0) Distribution
Mean: 64.000000
Median: 64.000000
Variance: 9.000000
Standard Deviation: 3.000000
Max Density: .132981

Probabilities

Left: .252493
Between (Red-Shaded): .588852
Right: .158655

▲ **FIGURE 6.16 (a)** Screenshot showing a calculation of the area under a N(64, 3) curve between 62 and 67. **(b)** An enlarged view of the window below the graph, which shows the shaded and unshaded areas. The shaded area is given as .588852. The area to its left is given as .252493. (We will usually insert a 0 before the decimal point when we report or work with these values.)

tall if the $N(64, 3)$ model is a good description of the distribution of women's heights. To use the calculator, the user moves sliders to set the mean and standard deviation and then uses the mouse to shade in the appropriate area. The actual probability is given in a window below the graph (shown on the previous page).

This particular calculator prints the probability next to the word "Between": .588852. Thus, the probability that the woman is *between* 62 and 67 inches tall is about 59%. Note that we can also, with no extra effort, get the probability that this person will be shorter than 62 inches (.252493, or about 25%) or taller than 67 inches (about 16%.)

To use the SOCR calculator to find probabilities for the Normal model, go to http://socr.stat.ucla.edu, click on Distributions, and select Normal.

Figure 6.17 shows output from a TI-83/84 for the same calculation.

```
normalcdf(62,67,
64,3)
        .5888522734
```

▲ **FIGURE 6.17** TI-83/84 output showing that the probability that the woman is between 62 and 67 inches tall is about 59%.

## EXAMPLE 4 Baby Seals

Some research has shown that the mean length of a newborn Pacific harbor seal is 29.5 inches ($\mu = 29.5$ inches) and that the standard deviation is $\sigma = 1.2$ inches. Suppose that the lengths of these seal pups follow the Normal model.

QUESTION  Using the output in Figure 6.18, what is the probability that a randomly selected harbor seal pup is within 1 standard deviation of the mean length of 29.5 inches?

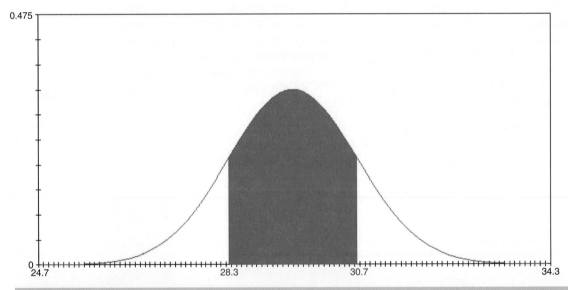

Distribution Properties
Normal (29.5, 1.2) Distribution
Mean: 29.500000
Median: 29.500000
Variance: 1.440000
Standard Deviation: 1.200000
Max Density: .332452

Probabilities

Left: .158655
Between (Red-Shaded): .682689
Right: .158655

▲ **FIGURE 6.18** The shaded region represents the area under the curve between 28.3 inches and 30.7 inches. In other words, it represents the probability that the length of a randomly selected seal pup is within 1 standard deviation of the mean.

SOLUTION  The phrase "within one standard deviation of the mean length" is one you will see often. (We used it in Chapter 3 when introducing the Empirical Rule.) It means that the pup's length will be somewhere between

mean minus 1 standard deviation

and

mean plus 1 standard deviation.

Because 1 standard deviation is 1.2 inches, this means the length must be between

$29.5 - 1.2 = 28.3$ inches

and

$29.5 + 1.2 = 30.7$ inches

From the results, we see that the probability that a randomly selected seal pup is within 1 standard deviation of mean length is about 68% (from Figure 6.18, it is 68.2689%).

**TRY THIS!** Exercise 6.15

## Without Technology: The Standard Normal

Example 4 illustrates a principle that's very useful for finding probabilities from the Normal distribution without technology. This principle is the recognition that we don't need to refer to values in our distribution in the units in which they were measured. We can also refer to them in standard units. In other words, we can ask for the probability that the man's height is between 66 inches and 72 inches (measured units), but another way to say the same thing is to ask for the probability that his height is within 1 standard deviation of the mean, or between $-1$ and $+1$ standard units.

You can still use the Normal model if you change the units to standard units, but you must also convert the mean and standard deviation to standard units. This is easy, because the mean is 0 standard deviations away from itself, and any point 1 standard deviation away from the mean is 1 standard unit. Thus, if the Normal model was a good model, then when you convert to standard units, the $N(0, 1)$ model is appropriate.

This model—the Normal model with mean 0 and standard deviation 1—has a special name: the **standard Normal model**.

**Looking Back**

**Standard Units**
Standard units (Chapter 3) tell us how many standard deviations from the mean an observation lies. Standard units are also called z-scores.

**KEY POINT**  $N(0, 1)$ is the standard Normal model: a Normal model with a mean of 0 ($\mu = 0$) and a standard deviation of 1 ($\sigma = 1$).

The standard Normal model is an important concept, because it allows us to find probabilities for any Normal model. All we need to do is first convert to standard units. We can then look up the areas in a published table that lists useful areas for the $N(0, 1)$ model. One such table is available in Appendix A.

Table 6.5 shows an excerpt from this table. The values within the table represent areas (probabilities). The numbers along the left margin, when joined to the numbers

| z | .00 | .01 | .02 | .03 | .04 | .05 | .06 | .07 | .08 | .09 |
|---|-----|-----|-----|-----|-----|-----|-----|-----|-----|-----|
| 0.9 | .8159 | .8186 | .8212 | .8238 | .8264 | .8289 | .8315 | .8340 | .8365 | .8389 |
| 1.0 | **.8413** | .8438 | .8461 | .8485 | .8508 | .8531 | .8554 | .8577 | .8599 | .8621 |
| 1.1 | .8643 | .8665 | .8686 | .8708 | .8729 | .8749 | .8770 | .8790 | .8810 | .8830 |
| 1.2 | .8849 | .8869 | .8888 | .8907 | .8925 | .8944 | .8962 | .8980 | .8997 | .9015 |

▲ **TABLE 6.5** Excerpt from the Normal table in Appendix A, which shows areas to the left of z in a standard Normal distribution. For example, the area to the left of z = 1.00 is 0.8413, and the area to the left of z = 1.01 is 0.8438.

across the top, represent *z*-scores. For instance, the boldface value in this table represents the area under the curve *and to the left* of 1.00 standard unit. This represents the probability that a randomly selected person has a height *less than* 1 standard unit. Figure 6.19 shows what this area looks like.

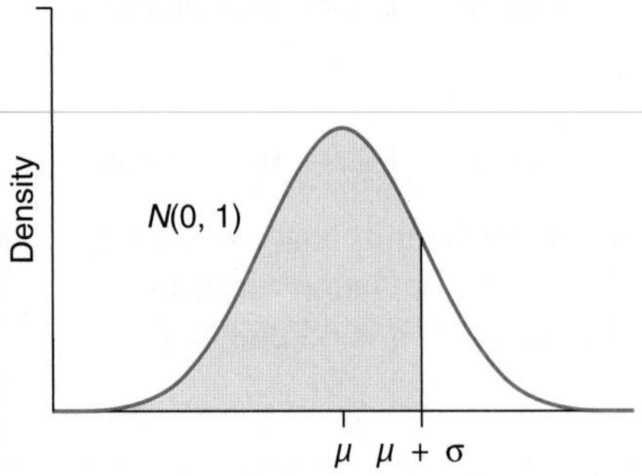

▶ **FIGURE 6.19** The area of the shaded region represents the probability that a randomly selected person (or thing) has a value less than 1.00 standard deviation above the mean, which is about 84%.

---

> **Details**
>
> ***z*-Scores**
> In Chapter 3 we gave the formula for a *z*-score in terms of the mean and standard deviation of a sample:
>
> $$z = \frac{x - \bar{x}}{s}$$
>
> The same idea works for probability distributions, but we must change the notation to indicate that we are using the mean and standard deviation of a probability distribution:
>
> $$z = \frac{x - \mu}{\sigma}$$

For example, imagine we want to find the probability that a randomly selected woman is shorter than 62 inches. This person is selected from a population of women whose heights follow a $N(64, 3)$ distribution. Our strategy is

1. Convert 62 inches to standard units. Call this number *z*.
2. Look up the area below *z* in the table for the $N(0, 1)$ distribution.

## EXAMPLE 5 Small Pups

Small newborn seal pups have a lower chance of survival. Suppose that the length of a newborn seal pup follows a Normal distribution with a mean length of 29.5 inches and a standard deviation of 1.2 inches.

**QUESTION**  What is the probability that a newborn pup selected at random is shorter than 28.0 inches?

**SOLUTION**  Begin by converting the length 28.0 inches to standard units.

$$z = \frac{28 - 29.5}{1.2} = \frac{-1.5}{1.2} = -1.25$$

Next sketch the area that represents the probability we wish to find (see Figure 6.20). We want to find the area under the Normal curve and to the left of 28 inches, or, in standard units, to the left of −1.25.

▶ **FIGURE 6.20** A standard Normal distribution, showing the shaded area that represents the probability of selecting a seal pup shorter than −1.25 standard deviations below the mean.

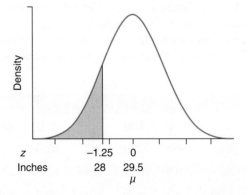

We can look this up in the standard Normal table in Appendix A. Table 6.6 on the next page shows the part we are interested in. We see that the area to the left of a *z*-score of −1.25 is 10.56%.

| z | .00 | .01 | .02 | .03 | .04 | .05 | .06 | .07 | .08 | .09 |
|---|-----|-----|-----|-----|-----|-----|-----|-----|-----|-----|
| −1.3 | .0968 | .0951 | .0934 | .0918 | .0901 | .0885 | .0869 | .0853 | .0838 | .0823 |
| −1.2 | .1151 | .1131 | .1112 | .1093 | .1075 | **.1056** | .1038 | .1020 | .1003 | .0985 |

▲ **TABLE 6.6** Part of the standard Normal table. The value printed in boldface type is the area under the standard Normal density curve to the left of −1.25.

The probability that a newborn seal pup will be shorter than 28 inches is about 11% (rounding up from 10.56%).

**TRY THIS!** Exercise 6.23

## EXAMPLE 6  A Range of Seal Pup Lengths

Again, suppose that the $N(29.5, 1.2)$ model is a good description of the distribution of seal pups' lengths.

**QUESTION** What is the probability that this randomly selected seal pup will be between 27 inches and 31 inches long?

**SOLUTION** This question is slightly tricky. The table gives us only the area *below* a given value. How do we find the area *between* two values?

We proceed in two steps. First we find the area less than 31 inches. Second, we "chop off" the area below 27 inches. The area that remains is the region between 27 and 31 inches. This process is illustrated in Figure 6.21.

To find the area less than 31 inches, we convert 31 inches to standard units:

$$z = \frac{(31 - 29.5)}{1.2} = 1.25$$

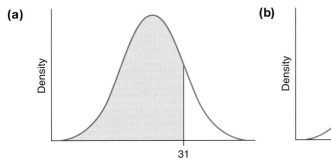

**(a)** Density, 31

**(b)** Density, 27, 31

◀ **FIGURE 6.21** Steps for finding the area between 27 and 31 inches under a $N(29.5, 1.2)$ distribution. **(a)** The area below 31. **(b)** We chop off the area below 27, and the remaining area (shaded) is what we are looking for.

Using the standard Normal table in Appendix A, we find that this probability is 0.8944. Next, we find the area below 27 inches:

$$z = \frac{(27 - 29.5)}{1.2} = -2.08$$

Again, using the standard Normal table, we find this area to be 0.0188. Finally, we subtract (or "chop off") the smaller area from the big one:

$$
\begin{array}{r}
0.8944 \\
-0.0188 \\
\hline
0.8756
\end{array}
$$

```
normalcdf(27,31,
29.5,1.2)
        .8757398007
```

▲ **FIGURE 6.22** TI-83/84 output for finding the probability that a newborn seal pup will be between 27 and 31 inches long.

---

**Details**

**Inverse Normal**
Mathematicians sometimes call finding measurements from percentiles from Normal distributions finding "inverse Normal" values.

---

**CONCLUSION** The probability that a newborn seal pup will be between 27 inches and 31 inches long is about 88%. Figure 6.22 confirms that answer.

**TRY THIS!** Exercise 6.27

---

## Finding Measurements from Percentiles for the Normal Distribution

So far we have discussed how to find the probability that you will randomly select an object with a value that is within a certain range. Thus, in Example 5 we found that if newborn seal pups' lengths follow a $N(29.5, 1.2)$ distribution, then the probability that we will randomly select a pup shorter than 28 inches is (roughly) 11%. We found this by finding the area under the Normal curve that is to the left of 28 inches.

Sometimes, though, we wish to turn this around. We are given a probability, but we want to find the value that corresponds to that probability. For instance, we might want to find the length of a seal pup such that the probability that we'll see any pups shorter than that is 11%. Such a number is called a **percentile**. The 11th percentile for seal pup lengths is 28 inches, the length that has 11% of the area under the Normal curve to its left.

Finding measurements from percentiles is simple with the right technology. The screenshot in Figure 6.23 shows how to use SOCR to find the height of a woman in the 25th percentile, assuming that women's heights follow a $N(64, 3)$ distribution. Simply move the cursor until the shaded region represents the lower 25% of the curve. (This area is given after the words "Between (Red-Shaded)" below the graph.) We see that the value that has 25.2% (as close as we could get to 0.2500) below it is 61.99. So we say that 61.99 (or 62) is the 25th percentile for this distribution.

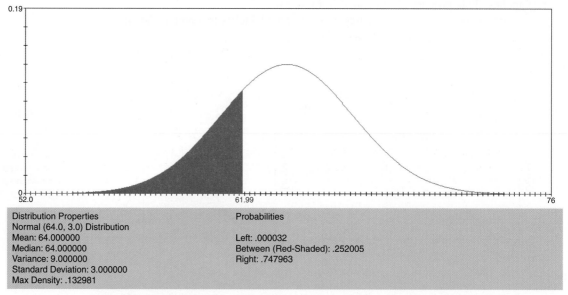

Distribution Properties
Normal (64.0, 3.0) Distribution
Mean: 64.000000
Median: 64.000000
Variance: 9.000000
Standard Deviation: 3.000000
Max Density: .132981

Probabilities
Left: .000032
Between (Red-Shaded): .252005
Right: .747963

▲ **FIGURE 6.23** Technology shows that 61.99 inches is the 25th percentile, because it has 25% of the area below it (to the left of it). Thus, if a woman is about 62 inches tall, there's a 25 percent chance that another woman will be shorter than she.

**Tech**

Things are a little trickier if you don't have technology. You first need to use the standard Normal curve, $N(0, 1)$, to find the $z$-score from the percentile. Then you must convert the $z$-score to the proper units. So without technology, finding a measurement from a percentile is a two-step process:

**Step 1.** Find the $z$-score from the percentile.

**Step 2.** Convert the $z$-score to the proper units.

Example 8 illustrates these steps.

## EXAMPLE 7 Inverse Normal or Normal?

Suppose that the amount of money people keep in their online PayPal account follows a Normal model. Consider these two situations:

a. A PayPal customer wonders how much money he would have to put into the account to be in the 90th percentile.

b. A PayPal employee wonders what the probability is that a randomly selected customer will have less than $150 in his account.

**QUESTION** For each situation, identify whether the question asks for a measurement or a Normal probability.

**SOLUTIONS**

a. This situation gives a percentile (the 90th) and asks for the measurement (in dollars) that has 90% of the other values below it. This is an inverse Normal question.

b. This situation gives a measurement ($150) and asks for a probability.

**TRY THIS!** Exercise 6.37

## EXAMPLE 8 Finding Measurements from Percentiles by Hand

Assume that women's heights follow a Normal distribution with mean 64 inches and standard deviation 3 inches: $N(64, 3)$. Earlier, we used technology to find that the 25th percentile was approximately 62 inches.

**QUESTION** Using the Normal table in Appendix A, confirm that the 25th percentile height is 62 inches.

**SOLUTION** The question asks us to find a measurement (the height) that has 25% of all women's heights below it. Use the Normal table in Appendix A. This gives probabilities and percentiles for a Normal distribution with mean 0 and standard deviation 1: a $N(0, 1)$ or standard Normal distribution.

Step 1: Find the $z$-score from the percentile. To do this, you must first find the probability within the table. For the 25th percentile, use a probability of 0.25. Usually, you will not find exactly the value you are looking for, so settle for the value that is as close as you can get. This value, 0.2514, is underlined in Table 6.7, which is an excerpt from the Normal table in Appendix A.

| z | .00 | .01 | .02 | .03 | .04 | .05 | .06 | .07 | .08 | .09 |
|------|-------|-------|-------|-------|-------|-------|-------|-------|-------|-------|
| −0.7 | .2420 | .2389 | .2358 | .2327 | .2296 | .2266 | .2236 | .2206 | .2177 | .2148 |
| −0.6 | .2743 | .2709 | .2676 | .2643 | .2611 | .2578 | .2546 | .2514 | .2483 | .2451 |

▲ **TABLE 6.7** Part of the standard Normal table.

You can now see that the $z$-score corresponding to a probability of 0.2514 is −0.67. This relation between the $z$-score and the probability is shown in Figure 6.24 on the next page.

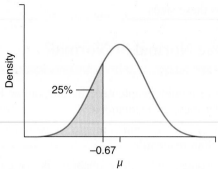

▶ **FIGURE 6.24** The percentile for 0.25 is −0.67 standard units, because 25% of the area under the standard Normal curve is below −0.67.

**Step 2: Convert the *z*-score to the proper units.**

A *z*-score of −0.67 tells us that this value is 0.67 standard deviations below the mean. We need to convert this to a height in inches.

One standard deviation is 3 inches, so 0.67 standard deviations is

$$0.67 \times 3 = 2.0 \text{ inches}$$

The height is 2.0 inches below the mean. The mean is 64.0 inches, so the 25th percentile is

$$64.0 - 2.0 = 62.0$$

**CONCLUSION** The woman's height at the 25th percentile is 62 inches, assuming that women's heights follow a $N(64, 3)$ distribution.

**TRY THIS!** Exercise 6.43

Figure 6.25 shows some percentiles and the corresponding *z*-scores to help you visualize percentiles.

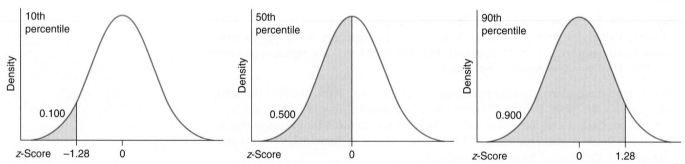

▲ **FIGURE 6.25** *z*-Scores and percentiles—the 10th, 50th, and 90th percentiles. A percentile corresponds to the percentage in the area to the left under the curve.

## The Normal Model and the Empirical Rule

In Chapter 3 we mentioned the Empirical Rule, which was not so much a rule as a guideline for helping you understand how data are distributed. The Empirical Rule is meant to be applied to any symmetric, unimodal distribution. However, the Empirical Rule is based on the Normal model. For any arbitrary unimodal, symmetric distribution, the Empirical Rule is approximate. And sometimes very approximate. But if that distribution is the Normal model, the rule is exact.

In Example 4, we found that the area between $-1$ and $+1$ in a standard Normal distribution was 68%. This is exactly what the Empirical Rule predicts. We can also find the probability that a randomly selected observation will be between $-2$ and 2 standard deviations of the mean, if that observation comes from the Normal model. Figure 6.26 shows a sketch of the $N(0, 1)$ model, with the region between $-2$ and 2 (in standard units) shaded. From the table in the appendix, the area below $+2$ is 0.9772. Also from the table, the area below $-2$ is 0.0228. The difference is the shaded area and is $0.9772 - 0.0228 = 0.9544$, or about 95%, just as the Empirical Rule predicts.

These facts from the Empirical Rule help us interpret the standard deviation in the context of the Normal distribution. For example, because the heights of women are Normally distributed and have a standard deviation of about 3 inches, we know that a majority of women (in fact, 68%) have heights within 3 inches of the mean: between 61 inches and 67 inches. Because nearly all women are within 3 standard deviations of the mean, we know not to expect too many women to be taller than $64 + 3 \times 3 = 73$ inches tall (73 inches is 6 feet and 1 inch). Such women are very rare.

**Looking Back**

**The Empirical Rule**
The Empirical Rule says that if a distribution of a sample of data is unimodal and roughly symmetric, then about 68% of the observations are within 1 standard deviation of the mean, about 95% are within 2 standard deviations of the mean, and nearly all are within 3 standard deviations of the mean.

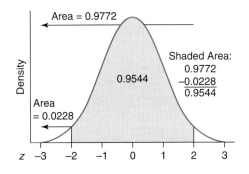

◀ **FIGURE 6.26** The area between $z$-scores of $-2.00$ and $+2.00$ on the standard Normal curve.

**SNAPSHOT** **THE NORMAL MODEL**

| | |
|---|---|
| **WHAT IS IT?** ▶ | A model of a distribution for some numerical variables. |
| **WHAT DOES IT DO?** ▶ | Provides us with a model of the distributions of probabilities for many real-life numerical variables. |
| **HOW DOES IT DO IT?** ▶ | The probabilities are represented by the area underneath the bell-shaped curve. |
| **HOW IS IT USED?** ▶ | If the Normal model is appropriate, it can be used for finding probabilities or for finding measurements associated with particular percentiles. |

## Appropriateness of the Normal Model

The Normal model does not fit all distributions of numerical variables. For example, if we are randomly selecting people who submitted tax returns to the federal government, we cannot use the Normal model to find the probability that someone's return is higher than the mean value. The reason is that incomes are right-skewed, so the Normal model will not fit.

How do we know whether the Normal model is appropriate? Unfortunately, there is no checklist. However, the Normal model is a good first-choice model if you suspect that the distribution is symmetric and has one mode. Once you collected data, you can check to see whether the Normal model closely matches the data.

In short, the Normal model is appropriate if it produces results that match what we see in real life. If the data we collect match the Normal model fairly closely, then the model is appropriate. Figure 6.27a shows a histogram of the actual heights from a sample of more than 1400 women from the National Health and Nutrition Examination Survey (NHANES, www.cdc.gov/nchs/nhanes), with the Normal model superimposed over the histogram. Note that the model, though not perfect, is a pretty good description of the shape of the distribution. Compare this to Figure 6.27b, which shows the distribution of weights for the same women. The Normal model is not a very good fit for these data. The Normal model has the peak at the wrong place; specifically, the Normal model is symmetric, whereas the actual distribution is right-skewed.

Statisticians have several ways of checking whether the Normal model is a good fit to the population, but the easiest thing for you to do is to make a histogram of your data and see whether it looks unimodal and symmetric. If so, the Normal model is likely to be a good model.

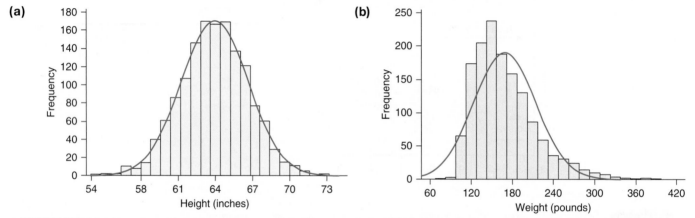

▲ **FIGURE 6.27** **(a)** A histogram of data from a large sample of adult women in the United States drawn at random from the National Health and Nutrition Examination Survey. The red curve is the Normal curve, which fits the shape of the histogram very well, indicating that the Normal model would be appropriate for these data. **(b)** A histogram of weights for the same women. Here the Normal model is a bad fit to the data.

## SECTION 6.3

# The Binomial Model (Optional)

The Normal model applies to many real-life, continuous-valued numerical variables. The **binomial probability model** is useful in many situations with discrete-valued numerical variables (typically counts, whole numbers). As with the Normal model, we will explain what the model looks like and how to calculate probabilities with it. We will also provide examples and discuss why the binomial model is appropriate to the situations.

The classic application of the binomial model is counting heads when flipping a coin. Let's stay I flip a coin 10 times. What is the probability I get 1 head? 2 heads? 10 heads? This is a situation that the binomial model fits quite well, and the probabilities of these outcomes are given by the binomial model. If we randomly select 10 people, what's the probability that exactly 5 will be Republicans? If we randomly select 100 students, what's the probability that 10 or fewer will be on the Dean's List? These are examples of situations where the binomial model applies.

How do you recognize a binomial model? The first sign that your random experiment is a candidate for the binomial model is that the outcome you are interested in is a count. If that's the case, then all four of the following characteristics must be present:

1. *A fixed number of trials.* We represent this number with the letter $n$. For example, if we flip a coin 10 times, then $n = 10$.

2. *Only two outcomes are possible at each trial.* We will call these two outcomes "success" and "failure." For example, we might consider the outcome of heads to be a success. Or we might be selecting people at random and counting the number of males; in this case, of the two outcomes "male" and "female," "male" would be considered a success.

3. *The probability of success is the same at each trial.* We represent this probability with the letter *p*. For example, the probability of getting heads after a coin flip is $p = 0.50$ and does not change from flip to flip.

4. *The trials are independent.* The outcome of one trial does not affect the outcome of any other trial.

If all four of these characteristics are present, the binomial model applies and you can easily find the probabilities by looking at the binomial probability distribution.

The binomial model provides probabilities for random experiments in which you are counting the number of successes that occur. Four characteristics must be present:

1. Fixed number of trials: *n*
2. The only two outcomes are success and failure.
3. The probability of success, *p*, is the same at each trial.
4. The trials are independent.

## EXAMPLE 9 Extrasensory Perception (Mind Reading)

Zener cards are special cards used to test whether people can read minds (telepathy). Each card in a Zener deck has one of five special designs: a star, a circle, a plus-sign, a square, or three wavy lines (Figure 6.28). In an experiment, one person, the "sender," selects a card at random, looks at it, and thinks about the symbol on the card. Another person, the "receiver," cannot see the card (and in some studies cannot even see the sender), and guesses which of the symbols was chosen. A researcher records whether the guess was correct. The card is then placed back in the deck, the deck is shuffled, and another card is drawn. Suppose this happens 10 times (10 guesses are made). The receiver gets 3 guesses correct, and the researcher wants to know the probability of this happening if the receiver is simply guessing.

◀ **FIGURE 6.28** Zener cards (ESP cards) show one of five shapes. A deck has equal numbers of each shape.

QUESTION Explain why this is a binomial experiment.

SOLUTION First, we note that we are counting something: the number of successful guesses. We need to check that the experiment meets the four characteristics of a binomial model. (1) The experiment consists of a fixed number of trials: $n = 10$. (2) The outcome of each trial is success or failure: The receiver either gets the right answer or does not. (3) The probability of a success at a trial is $p = 1/5 = 0.20$, because there are 5 cards and so the receiver has a 1-in-5 chance of getting it correct, if we assume the receiver is guessing. (4) As long as the cards are put back in the deck and reshuffled (thoroughly), the probability of a success is the same for each trial, and each trial is independent.

All four characteristics are satisfied, so the experiment fits the binomial model.

TRY THIS! Exercise 6.53

## EXAMPLE 10 Why Are They Not Binomial?

The following four experiments are almost, but not quite, binomial experiments.

a. Record the number of different eye colors in a group of 50 randomly selected people.

b. A married couple decides to have children until a girl is born, but to stop at five children if they do not have any girls. How many children will the couple have?

c. Suppose the probability that a flight will arrive on time (within 15 minutes of the scheduled arrival time) at O'Hare Airport in Chicago is 85%. How many flights arrive on time out of 300 flights scheduled to land on a day in January?

d. A student guesses on every question of a test that has 10 multiple-choice questions and 10 true-false questions. Record the number of questions the student gets right.

QUESTION For each situation, explain which of the four characteristics is not met.

SOLUTIONS

a. This is not a binomial experiment because there are more than two eye colors, so more than two outcomes may occur at each trial. However, if we reduced the eye colors to two categories by, say, recording whether the eye color was brown or not brown, then this would be a binomial experiment.

b. This is not a binomial experiment because the number of trials is not fixed before the children are born. The number of "trials" depends on when (or whether) the first girl is born. The number of trials varies depending on what happens—the word *until* tells you that.

c. This is not binomial because the flights are not independent. If the weather is bad, the chance of arriving on time for all flights is lower. Therefore, if one flight arrives late, then another flight is more likely to arrive late.

d. This is not a binomial experiment because the probability of success on each trial is not constant; the probability of success is lower on multiple-choice questions than on true/false questions. Therefore, criterion 3 is not met. However, if the test were subdivided into two sections (multiple-choice and true/false), then each separate section could be called a binomial experiment if we assumed the student was guessing.

TRY THIS! Exercise 6.55

## Visualizing the Binomial Distribution

All binomial models have the four characteristics listed above, but the list gives us flexibility in $n$ and $p$. For example, if we had flipped the coin 6 times instead of 10, it would still be a binomial experiment. Also, if the probability of a success were 0.6 instead of 0.5, we would still have a binomial experiment. For different values of $n$ and $p$, we have different binomial experiments, and the binomial distribution looks different in each case.

Figure 6.29 shows that the binomial distribution for $n = 3$ and $p = 0.5$ is symmetric. We can read from the graph that the probability of getting exactly 2 successes (2 heads in 3 flips of a coin) is almost 0.40, and the probability of getting 0 successes is the same as the probability of getting all successes.

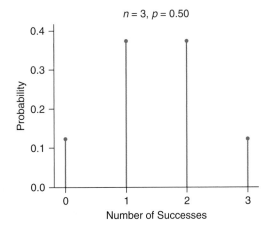

◀ **FIGURE 6.29** Binomial distribution with $n = 3$, $p = 0.50$.

If $n$ is bigger but $p$ remains fixed at 0.50, the distribution is still symmetric because the chance of a success is the same as the chance of a failure, as shown in Figure 6.30.

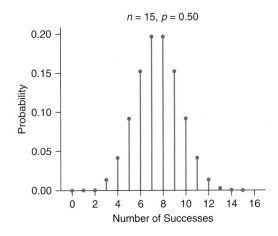

◀ **FIGURE 6.30** Binomial distribution with $n = 15$, $p = 0.50$.

If the probability of success is not 50%, the distribution might not be symmetric. Figure 6.31 on the next page shows the distribution for $p = 0.3$, which means we're less likely to get a large number of successes than a smaller number, so the probability "spikes" are taller for smaller numbers of successes. The plot is now right-skewed.

However, even if the distribution is not symmetric, if we increase the number of trials, it becomes symmetric. The shape of the distribution depends on both $n$ and $p$. If

▶ **FIGURE 6.31** Binomial distribution with $n = 15$, $p = 0.30$.

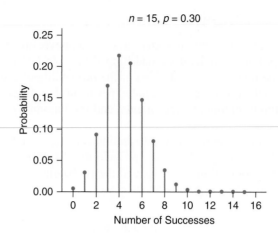

we keep $p = 0.3$ but increase $n$ to 100, we get a more symmetric shape, as shown in Figure 6.32 shown below.

▶ **FIGURE 6.32** Binomial distribution with $n = 100$, $p = 0.30$. Note that we show $x$ only for values between 15 and 45. The shape is symmetric, even though $p$ is not 0.50.

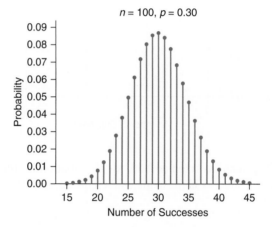

Binomial distributions have the interesting property that if the number of trials is large enough, the distributions are symmetric.

KEY POINT   The shape of a binomial distribution depends on both $n$ and $p$. Binomial distributions are symmetric when $p = 0.5$, but they are also symmetric when $n$ is large, even if $p$ is close to 0 or 1.

## Finding Binomial Probabilities

Because the binomial distribution depends only on the values of $n$ and $p$, once you have identified an experiment as binomial and have identified the values of $n$ and $p$, you can find any probability you wish. We use the notation $b(n, p, x)$ to represent the **binomial probability** of getting $x$ successes in a binomial experiment with $n$ trials and probability of success $p$. For example, imagine tossing a coin 10 times. If each side is equally likely, what is the probability of getting 4 heads? We represent this with $b(10, 0.50, 4)$.

The easiest and most accurate way to find binomial probabilities is to use technology. Statistical calculators and software have the binomial distribution built in and can easily calculate probabilities for you.

# EXAMPLE 11   Stolen Bicycles

According to the website nationalbikeregistry.com, at the campus of UC Berkeley, only 3% of stolen bicycles are returned to owners.

**QUESTIONS** Accepting for the moment that the four characteristics of a binomial experiment are satisfied, write the notation for the probability that exactly 5 bicycles will be returned if 235 are stolen in the course of a year. What is this probability?

**SOLUTION** The number of trials is 235, the probability of success is 0.03, and the number of successes is 5. Therefore, we can write the binomial probability as $b(235, 0.03, 5)$.

The command and output for finding this probability on a TI-83/84 are shown in Figure 6.33. The command binompdf stands for "binomial probability distribution function." The results show us that the probability that exactly 5 bicycles will be returned, assuming that this is in fact a binomial experiment, is about 12.6%.

**TRY THIS!** Exercise 6.59

Tech

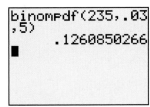

```
binompdf(235,.03
,5)
         .1260850266
```

▲ **FIGURE 6.33** TI-83/84 output for a binomial trial.

Although we applied the binomial model to the stolen bicycles example, we had to make the assumption that the trials are independent. A "trial" in this context consists of a bicycle being stolen, and a "success" occurs if the bike is returned. If one person stole several bikes on a single day from a single location, this assumption of independent trials would be wrong, because if police found one of the bikes, they would have a good chance of finding the others that the thief took. Sometimes we have no choice but to make certain assumptions to complete a problem, but we need to be sure to check our assumptions when we can, and we must be prepared to change our conclusion if the assumptions were wrong.

Another approach to finding binomial probabilities is to use a table. Published tables are available that list binomial probabilities for a variety of combinations of values for $n$ and $p$. One such table is provided in Appendix A. This table lists binomial probabilities for values of $n$ between 2 and 15 and for several different values of $p$.

# EXAMPLE 12   Recidivism in Texas

The three-year recidivism rate of parolees in Texas is 30% (www.lbb.state.tx.us). In other words, 30% of released prisoners return to prison within three years of their release. Suppose a prison in Texas released 15 prisoners.

**QUESTION** Assuming that whether one prisoner returns to prison is independent of whether any of the others returns, use Table 6.8 on the next page, which shows binomial probabilities for $n = 15$ and for various values of $p$, to find the probability that exactly 8 out of 15 will end up back in prison within three years.

**SOLUTION** Substituting the numbers, you can see that we are looking for $b(15, 0.30, 8)$. Referring to Table 6.8, on the next page, you can see—by looking in the table for $n = 15$, the row for $x = 8$, and the column for $p = 0.30$—that the probability that exactly 8 parolees will be back in prison within three years is 0.035, or about a 3.5% chance.

| x | 0.1 | 0.2 | 0.25 | 0.3 | 0.4 | 0.5 | 0.6 | 0.7 | 0.75 | 0.8 | 0.9 |
|---|-----|-----|------|-----|-----|-----|-----|-----|------|-----|-----|
| 6 | .002 | .043 | .092 | .147 | .207 | .153 | .061 | .012 | .003 | .001 | .000 |
| 7 | .000 | .014 | .039 | .081 | .177 | .196 | .118 | .035 | .013 | .003 | .000 |
| 8 | .000 | .003 | .013 | .035 | .118 | .196 | .177 | .081 | .039 | .014 | .000 |
| 9 | .000 | .001 | .003 | .012 | .061 | .153 | .207 | .147 | .092 | .043 | .002 |
| 10 | .000 | .000 | .001 | .003 | .024 | .092 | .186 | .206 | .165 | .103 | .010 |
| 11 | .000 | .000 | .000 | .001 | .007 | .042 | .127 | .219 | .225 | .188 | .043 |
| 12 | .000 | .000 | .000 | .000 | .002 | .014 | .063 | .170 | .225 | .250 | .129 |
| 13 | .000 | .000 | .000 | .000 | .000 | .003 | .022 | .092 | .156 | .231 | .267 |
| 14 | .000 | .000 | .000 | .000 | .000 | .000 | .005 | .031 | .067 | .132 | .343 |
| 15 | .000 | .000 | .000 | .000 | .000 | .000 | .000 | .005 | .013 | .035 | .206 |

▲ **TABLE 6.8** Binomial probabilities with a sample of 15 and *x*-values of 6 or higher.

```
binompdf(15,.3,8
)
          .0347700143
```

▲ **FIGURE 6.34** TI-83/84 output for *b*(15, 0.3, 8).

Using a TI-83/84 as shown in Figure 6.34, we can see another way to get the same answer.

TRY THIS! Exercise 6.61

## Finding (Slightly) More Complex Probabilities
### EXAMPLE 13 ESP with 10 Trials

For a test of psychic abilities, researchers have asked the sender to draw 10 cards at random from a large deck of Zener cards (see Example 9). Assume that the cards are replaced in the deck after each use and the deck is shuffled. Recall that this deck contains equal numbers of 5 unique shapes. The receiver guesses which card the sender has drawn.

QUESTIONS

a. What is the probability of getting *exactly 5* correct answers (out of 10 trials) if the receiver is simply guessing (and has no psychic ability)?

b. What is the probability that the receiver will get *5 or more* of the cards correct out of 10 trials?

c. What is the probability of getting *fewer than 5* correct in 10 trials with the ESP cards?

SOLUTIONS

a. In Example 9 we identified this as a binomial experiment. With that done, we must now identify *n* and *p*. The number of trials is 10, so $n = 10$. If the receiver is guessing, then the probability of a correct answer is $p = 1/5 = 0.20$. Therefore, we wish to find $b(10, 0.2, 5)$.

Figure 6.35a on the next page gives the TI-83/84 output, where you can see that the probability of getting 5 right out of 10 is only about 0.0264. Figure 6.35b shows Minitab output for the same question.

**(a)**

```
binompdf(10,.2,5
)
        .0264241152
■
```

**(b)**

```
Probability Density Function

Binomial with n = 10 and p = 0.2

x       P
5    0.0264241
```

◀ **FIGURE 6.35** Technology output for *b*(10, 0.2, 5). **(a)** Output from a TI-83/84. **(b)** Output from Minitab.

Figure 6.36a shows a graph of the pdf. The probability *b*(10, 0.2, 5) is so small that it is hard to read off the graph. The graph shows that it is unusual to get exactly 5 correct when the receiver is guessing.

b. The phrase *5 or more* means we need the probability that the receiver gets 5 correct **or** 6 correct **or** 7 **or** 8 **or** 9 **or** 10. The outcomes 5 correct, 6 correct, and so on are mutually exclusive, because if you get exactly 5 correct, you cannot possibly also get exactly 6 correct. Therefore, we can find the probability of 5 or more correct by adding the individual probabilities together:

$b(10, 0.2, 5) + b(10, 0.2, 6) + b(10, 0.2, 7) + b(10, 0.2, 8) + b(10, 0.2, 9) + b(10, 0.2, 10)$
$= 0.026 \quad + \quad 0.006 \quad + \quad 0.001 \quad + \quad 0.000 \quad + \quad 0.000 \quad + \quad 0.000$
$= 0.033$

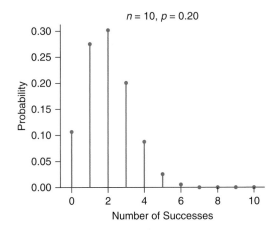

◀ **FIGURE 6.36(a)** The theoretical numbers of successes for 10 trials with the Zener deck, assuming guessing.

These probabilities are circled in Figure 6.36b.

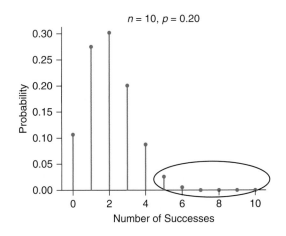

◀ **FIGURE 6.36(b)** Summing the probabilities represented by the circled bars gives us the probability of getting 5 or more correct.

c. The phrase *fewer than 5 correct* means 4, 3, 2, 1, or 0 correct. These probabilities are circled in Figure 6.36c.

▶ **FIGURE 6.36(c)** Summing the probabilities represented by the circled bars gives the probability of getting fewer than 5 correct.

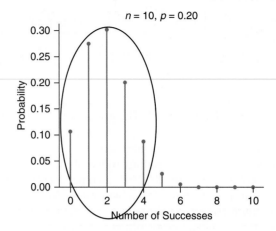

The event that we get fewer than 5 correct is the complement of the event that we get 5 or more correct, as you can see in Figure 6.37, which shows all possible numbers of successes with 10 trials. *Fewer than 5* is the same event as *4 or fewer* and is shown in the left oval in the figure.

▶ **FIGURE 6.37** The possible numbers of successes out of 10 trials with binomial data. Note that *fewer than 5* is the complement of *5 or more*.

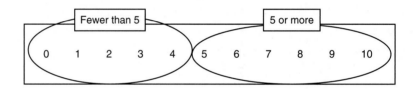

Because we know the probability of 5 or more, we can find its complement by subtracting from 1:

$$1 - 0.033 = 0.967$$

**CONCLUSIONS**

a. The probability of exactly 5 correct is 0.026.

b. The probability of 5 or more correct is 0.033.

c. The probability of fewer than 5 (that is, of 4 or fewer) is 0.967.

**TRY THIS!** Exercise 6.69

```
binomcdf(10,.2,4
)
        .9672065025
■
```

▲ **FIGURE 6.38** TI-83/84 output for $n = 10$, $p = 0.2$, and 4 or fewer successes.

Most technology also offers you the option of finding binomial probabilities for *x or fewer*. In general, probabilities of *x or fewer* are called **cumulative probabilities**. Figure 6.38 shows the cumulative binomial probabilities provided by the TI-83/84; notice the "c" in binom**c**df.

The probability of getting 5 or more correct is different from the probability of getting more than 5 correct. This very small change in the wording gives very different results. Figure 6.39 shows that *5 or more* includes the outcomes 5, 6, 7, 8, 9, and 10, whereas *more than 5* does not include the outcome of 5.

When finding Normal probabilities, we did not have to worry about such subtleties of language, because for a continuous numerical variable, the probability of getting 5 or more is the same as the probability of getting more than 5.

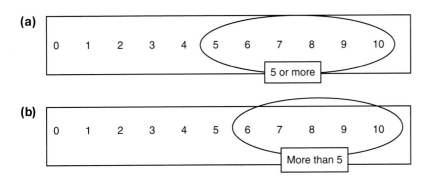

**(a)**

**(b)**

◄ FIGURE 6.39 Interpretation of words for discrete counts with *n* of 10. **(a)** Results for *5 or more*; **(b)** results for *more than 5*. Note that they are different.

## Finding Binomial Probabilities by Hand

When the number of trials is small, we can find probabilities by relying on the probability rules in Chapter 5 and listing all possible outcomes. In fact, there is a general pattern we can exploit to find a formula that will give us the probabilities for any value of *n* or *p*. Listing all of the outcomes will help us see why this formula works.

Suppose we are testing someone using the Zener cards that have five different shapes. After each card is guessed, it is returned to the deck, and the deck is shuffled before the next card is drawn. We want to count how many correct guesses a potential psychic makes in four trials. We can have five different outcomes for this binomial experiment: 0 correct answers, 1 correct, 2 correct, 3 correct, and all 4 correct.

The first step is to list all possible outcomes after four attempts. At each trial, the guesser is either right or wrong. Using R for right and W for wrong, we list all possible sequences of right or wrong results in four trials.

| 4 Right | 3 Right | 2 Right | 1 Right | 0 Right |
|---------|---------|---------|---------|---------|
| RRRR | RRRW | RRWW | RWWW | WWWW |
| | RRWR | RWRW | WRWW | |
| | RWRR | RWWR | WWRW | |
| | WRRR | WWRR | WWWR | |
| | | WRWR | | |
| | | WRRW | | |

There are 16 possible outcomes, although these are not all equally likely.

Now we find the probabilities of getting 4 right, 3 right, 2 right, 1 right, and 0 right. Because there are five shapes, and all five are equally likely to be chosen, if the receiver is simply guessing and has no psychic ability, then the probability of a right answer at each trial is $1/5 = 0.2$, and the probability of guessing wrong is 0.8.

Four right means "right AND right AND right AND right." Successive trials are independent (because we replace the card and reshuffle every time), so we can multiply the probabilities using the multiplication rule.

$$P(RRRR) = 0.2 \times 0.2 \times 0.2 \times 0.2 = 0.0016$$

This probability is just $b(4, 0.2, 4) = 1(0.0016) = 1(0.2)^4$. (We multiply by 1 because there is only one way that we can get all 4 right to happen, which is also why there is only one outcome listed in the "4 right" column.)

 **Looking Back**

**And**

The multiplication rule was Probability Rule 5c in Chapter 5 and applies only to independent events: $P(A \text{ AND } B) = P(A) P(B)$.

The probability of getting 3 right and 1 wrong includes all four options in the second group. The probability for each of these options is obtained by calculating the probability of 3 right and 1 wrong, and all of these probabilities will be the same. To get the total probability, therefore, we multiply by 4.

$$P(RRRW) = 0.2 \times 0.2 \times 0.2 \times 0.8 = 0.0064$$

$$P(\text{3 right and 1 wrong, in any order}) = 4(0.0064) = 0.0256$$

$$b(4, 0.2, 3) = 5(0.2)^3(0.8)^1 = 0.0256$$

The probability of getting 2 right and 2 wrong includes all six options in the third group. The probability for each of these options is obtained by calculating the probability of 2 right and 2 wrong; then, to get the total, we multiply by 6.

$$P(RRWW) = 0.2 \times 0.2 \times 0.8 \times 0.8 = 0.0256$$

$$P(\text{2 right and 2 wrong, in any order}) = 6(0.0256) = 0.1536$$

$$b(4, 0.2, 2) = 6(0.2)^2(0.8)^2 = 0.1536$$

The probability of getting 1 right and 3 wrong includes all four options in the fourth group. We obtain the probability for each of the options by calculating the probability of 1 right and 3 wrong; then we multiply by 4 to get the total.

$$P(RWWW) = 0.2 \times 0.8 \times 0.8 \times 0.8 = 0.1024$$

$$P(\text{1 right and 3 wrong, in any order}) = b(4, 0.2, 1) = 4(0.1024) = 0.4096$$

$$b(4, 0.2, 1) = 4(0.2)^1(0.8)^3 = 0.4096$$

Finally, the probability of getting all four wrong is

$$P(WWWW) = b(4, 0.2, 0) = 0.8 \times 0.8 \times 0.8 \times 0.8 = 0.4096$$

Because there is only one way for this to happen, we multiply 0.4096 by 1. Thus

$$b(4, 0.2, 0) = 1(0.8)^4 = 0.4096$$

Table 6.9 summarizes the results.

If you add the probabilities, you will see that they add to 1, as they should, because this list includes all possible outcomes that can happen.

Figure 6.40 shows a graph of the probability distribution. Note that the graph is right-skewed because the probability of success is less than 0.50. Also note that the probability of getting 4 out of 4 right with the Zener cards is very small.

You might compare these probabilities with those in the binomial table. You will find that they agree, if you round off the numbers we found "by hand" to three decimal places.

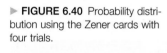

| Number Right | Probability |
|---|---|
| 4 right | 0.0016 |
| 3 right | 0.0256 |
| 2 right | 0.1536 |
| 1 right | 0.4096 |
| 0 right | 0.4096 |

▲ **TABLE 6.9** A summary of the probabilities of all possible numbers of successes in four trials with the Zener cards.

▶ **FIGURE 6.40** Probability distribution using the Zener cards with four trials.

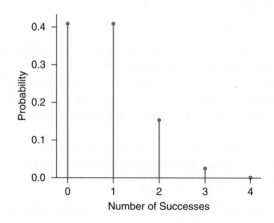

### The Formula

This approach of listing all possible outcomes is tedious when $n = 4$, and the tedium increases for larger values of $n$. (For $n = 5$ we have to list 32 possibilities, and for

$n = 6$ there are 64.) However, mathematicians have derived a formula that finds probabilities for the binomial distribution. The binomial table in this book and the results of your calculator are based on this formula:

$P(x$ successes in $n$ trials of a binomial experiment$) =$
(number of different ways of getting $x$ successes in $n$ trials) $p^x(1 - p)^{(n-x)}$

For example, for our alleged psychic who gets four guesses with the Zener cards, the probability of 3 successes ($x = 3$) in four trials ($n = 4$) is

(number of ways of getting 3 successes in 4 trials) $(0.2)^3(0.8)^1$

$= 4 (0.2)^3(0.8)^1$

$= 0.0256$

## The Shape of the Binomial Distribution: Center and Spread

Unlike the Normal distribution, the mean and standard deviation of the binomial distribution can be easily calculated. Their interpretation is the same as with all distributions: The mean tells us where the distribution balances, and the standard deviation tells us how far values are, typically, from the mean.

For example, in Figure 6.41, the binomial distribution is symmetric, so the mean sits right in the center at 7.5 successes.

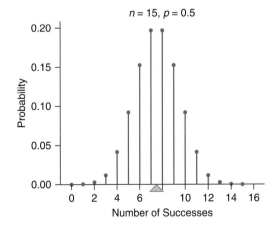

◀ **FIGURE 6.41** A binomial distribution with $n = 15$, $p = 0.50$. The mean is at 7.5 successes.

If the distribution is right-skewed, the mean will be just to the right of the peak, closer to the right tail, as shown in Figure 6.42. This is a binomial distribution with $n = 15$, $p = 0.3$, and the mean sits at the balancing point: 4.5 successes.

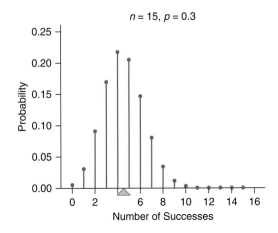

◀ **FIGURE 6.42** A binomial distribution with $n = 15$, $p = 0.3$. The mean is at the balancing point of 4.5 successes.

The mean, $\mu$, of a binomial probability distribution can be found with a simple formula:

$$\mu = np$$

In words, the mean number of successes in a binomial experiment is the number of trials times the probability of a success.

This formula should make intuitive sense. If you toss a coin 100 times, you would expect about half of the tosses to be heads. Thus the "typical" outcome is $(1/2) \times 100 = 50$, or $np = 100 \times 0.5$.

The standard deviation, $\sigma$, of a binomial probability distribution, which measures the spread, is less intuitive:

$$\sigma = \sqrt{np(1 - p)}$$

For example, Figure 6.42 shows a binomial distribution with $n = 15$, $p = 0.30$. The mean is $15 \times 0.3 = 4.5$. The standard deviation is $\sqrt{15 \times 0.3 \times 0.7} = \sqrt{3.15} = 1.775$.

> KEY POINT
> For a binomial experiment, the mean is
> $$\mu = np$$
> For a binomial experiment, the standard deviation is
> $$\sigma = \sqrt{np(1 - p)}$$

## Interpreting the Mean and Standard Deviation

The mean of any probability distribution, including the binomial, is sometimes called the **expected value**. This name gives some intuitive understanding. If you were to actually carry out a binomial experiment, you would *expect* about $\mu$ successes. If I flip a coin 100 times, I *expect* about 50 heads. If, in an ESP study, 10 trials are made and the probability of success at each trial is 0.20, we *expect* about 2 successes due to chance ($10 \times 0.2 = 2$).

Will I get exactly 50 heads? Will the ESP receiver get exactly 2 cards correct? Sometimes, yes. Usually, no. Although we expect $\mu$ successes, we usually get $\mu$ give or take some amount. That give-or-take amount is what is measured by $\sigma$.

In 100 tosses of a fair coin, we expect $\mu = 50$ heads, give-or-take $\sigma = \sqrt{100 \times 0.5 \times 0.5} = 5$ heads. We expect 50, but we will not be surprised if we get between 45 and 55 heads. In the Zener card experiment with 10 trials, we expect the receiver to guess about 2 cards correctly, but in practice we expect him or her to get 2 give or take 1.3, because

$$\sqrt{10 \times 0.2 \times 0.8)} = \sqrt{1.6} = 1.26$$

## SNAPSHOT  THE BINOMIAL DISTRIBUTION

**WHAT IS IT?** ▶ A distribution for some discrete variables.

**WHAT DOES IT DO?** ▶ Gives probabilities for the number of successes observed in a fixed number of trials in a binomial experiment.

**HOW DOES IT DO IT?** ▶ If the conditions of a binomial experiment are met, once you identify $n$ (number of trials), $p$ (the probability of success), and $x$ (the number of successes), it gives the probability.

**HOW IS IT USED?** ▶ The probabilities are generally provided in the form of a table or a formula, but if you need to calculate them, use a calculator or technology.

## EXAMPLE 14 Basketball Free-Throw Shots

LeBron James has a free-throw success percentage of 74%. Assume the free-throw shots are independent; that is, success or failure on one shot does not affect the chance of success on another shot.

**QUESTION** If he has 600 free throws in an upcoming season, how many would you expect him to sink, give or take how many?

**SOLUTION** This is a binomial experiment. (Why?) The number of trials is $n = 600$, and the probability of success at each trial is $p = 0.74$. You would expect LeBron James to sink 74% of 600, or 444, free throws:

$$\mu = np = 600\,(0.74) = 444$$

The give-or-take amount is measured by the standard deviation:

$$\sigma = \sqrt{600 \times 0.74 \times 0.26} = 10.7443$$

You should expect him to hit about 444 free throws, give or take about 10.7.

$$444 + 10.7 = 454.7$$
$$444 - 10.7 = 433.3$$

**CONCLUSION** You expect James to sink between 433 and 455 out of 600 free-throw shots.

**TRY THIS!** Exercise 6.73

## Surveys: An Application of the Binomial Model

Perhaps the most common application of the binomial model is in survey sampling. Imagine a large population of people, say the 100 million or so registered voters in the United States. Some percentage of them, call it $p$, have a certain characteristic. If we choose 10 people at random (with replacement), we can ask what the probability is that all of them, or three of them, or six of them, have that characteristic.

## EXAMPLE 15 News Survey

The Pew Research Center says that 23% of people are "news integrators": people who get their news both from traditional media (television, radio, newspapers, magazines) and from the Internet (www.people-press.org). Suppose we take a random sample of 100 people.

**QUESTIONS** If, as is claimed, 23% of the population are news integrators, then how many people in our sample should we expect to be integrators? Give or take how many? Would you be surprised if 34 people in the sample turned out to be integrators?

**SOLUTION** Assuming that all four of the characeristics of a binomial distribution are satisfied, then we would expect 23% of our sample of 100 people to be integrators—that is, 23 people. The standard deviation is $\sqrt{100 \times 0.23 \times 0.77} = 4.2$. Thus we should expect 23 people, give or take about 4.2 people, to be integrators. This means that we shouldn't be surprised if we got as many as $23 + 4.2 = 27.2$, or about 27, people. However, 34 people is quite a bit more than 1 standard deviation above what we expect. In fact, it is almost 3 standard deviations away, so it would be a surprisingly large number of people.

**TRY THIS!** Exercise 6.75

The binomial model works well for surveys when people are selected *with* replacement (which means that, once they are selected, they have a chance of being selected again). If we are counting the people who have a certain characteristic, then the four characteristics of the binomial model are usually satisfied:

1. A fixed number of people are surveyed. In Example 15, $n = 100$.
2. The outcome is success (integrator) or failure (not an integrator).
3. The probability of a success is the same at each trial: $p = 0.23$.
4. The trials are independent. (This means that if one person is found to be an integrator, no one else in the sample will change their response.)

With surveys, we usually don't report the number of people who have an interesting characteristic; instead, we report the percentage of our sample. We would not report that 23 people in our sample of 100 were integrators; we would report that 23% of our sample were integrators. But the binomial distribution still applies, because we are simply converting our counts to percentages.

In reality, surveys don't select people with replacement. The most basic surveys sample people without replacement, which means that, strictly speaking, the probability of a success is different after each trial. Imagine that the first person selected is an integrator. Now there are fewer integrators left in the population, so it is no longer true that 23% of the population are integrators; the percentage is slightly less. However, as you can imagine, this is not a problem if the population is very large. In fact, if the population size is very large relative to the sample size (at least 10 times bigger), then this difference is so slight that characteristic 3 is essentially met.

Taking a random sample, either with or without replacement, of a large and diverse population such as all U.S. voters is quite complicated. In practice, the surveys that we read about in the papers or hear about on the news use a modified approach that is slightly different from what we've discussed here. Random selection is still at the heart of these modified methods, though, and the binomial distribution often provides a good approximation for probabilities, even under these more complex schemes.

## CASE STUDY REVISITED

McDonald's claims that its ice cream cones weigh 3.18 ounces. However, one of the authors bought five cones and found that all five weighed more than that. Is this surprising?

Suppose we assume that the amount of ice cream dispensed follows a Normal distribution centered on 3.18 ounces. In other words, typically, a cone weighs 3.18 ounces, but sometimes a cone weighs a little more, and sometimes a little less. If this is the case, then, because the Normal distribution is symmetric and centered on its mean, the probability that a cone weighs more than 3.18 ounces is 0.50.

If the probability that a cone weighs more than 3.18 ounces is 0.50, what is the probability that five cones all weigh more? We can think of this as a binomial experiment, if we assume the trials are independent. (This seems like a reasonable assumption.) We can then calculate $b(5, 0.5, 5)$ to find that the probability is 0.031.

If the typical McDonald's ice cream cone really weighs 3.18 ounces, then our outcome is fairly surprising: It happens only about 3% of the time. This means it is fairly rare and raises the possibility that this particular McDonald's might deliberately dispense more than 3.18 ounces.

# EXPLORING STATISTICS
## CLASS ACTIVITY

## ESP with Coin Flipping

| GOALS | MATERIALS |
|---|---|
| Apply the binomial model to a real situation. | One coin for each pair of students. |

**ACTIVITY**

Choose a partner. One of you will play the role of the "sender" and the other will be the "receiver." The sender will flip a coin so that the receiver can't see the outcome (heads or tails). The sender will then look at the coin and concentrate, trying to mentally send a thought about whether the coin landed heads or tails. The receiver writes down (quickly) the outcome he or she believes was sent. (Just write the first thing that comes to mind.) The sender should make a tally of the number of right answers the receiver achieved in 10 trials. Now switch roles and try it again. Each of you should be prepared to report how many you got correct in your 10 trials.

**BEFORE THE ACTIVITY**

1. With 10 trials, how many would you expect to guess correctly if there is no ESP? If there are 20 people trying this, do you expect all of them to have the same results?

2. Find the standard deviation—the "give-or-take value"—of the number of correct guesses in 10 trials. Assuming that the receiver does not have ESP, what's the smallest number of correct guesses you might expect? What's the largest? How does the standard deviation help you determine this?

**AFTER THE ACTIVITY**

1. Make a histogram of the class results. Where is the distribution centered? Is this what you expected?

2. Are all of the results within the range of results predicted above?

3. Are any of the results unusually good? Does that show that the person with unusually good results has ESP? Why or why not? Explain.

## CHAPTER REVIEW

## Key Terms

Probability model, *238*
Probability distribution, *238*
Probability distribution function
(pdf), *238*
Discrete outcomes, *238*
Continuous outcomes, *238*
Probability density curve, *241*

Normal model: Notation,
$N(\mu, \sigma)$, *243*
Normal curve, *243*
Normal distribution, *243*
Mean of a probability distribu-
tion, $\mu$, *244*

Standard deviation of a probabil-
ity distribution, $\sigma$, *244*
Standard Normal model
Notation, $N(0, 1)$, *249*
Percentile, *252*
Binomial probability model
Notation, $b(n, p, x)$, *260*

$n$ is the number of trials
$p$ is the probability of success
on one trial
$x$ is the number of successes
Cumulative probabilities, *264*
Expected value, *268*

## Learning Objectives

After reading this chapter and doing the assigned homework problems, you should

- Be able to distinguish between discrete and continuous-valued
  variables.

- Know when a Normal model is appropriate and be able to
  apply the model to find probabilities.

- Know when the binomial model is appropriate and be able to
  apply the model to find probabilities.

## Summary

Probability models try to capture the essential features of real-world
experiments and phenomena that we want to study. In this chapter,
we focused on two very useful models: the Normal model and the
binomial model.

The Normal model is an example of a model of probabilities
for continuous numerical variables. The Normal probability model
is also called the Normal distribution, and the Gaussian curve. It
can be a useful model when a histogram of data collected for a variable is unimodal and symmetric. Probabilities are found by finding
the area under the appropriate region of the Normal curve. These
areas are best calculated using technology. If technology is not
available, you can also convert measures to standard units and then
use the table of areas for the standard Normal distribution, provided
in Appendix A.

The binomial model is an example of a model of probabilities
for discrete numerical outcomes. The binomial model applies to
binomial experiments, which occur when we are interested in counting the number of times some event happens. These four characteristics must be met for the binomial model to be applied:

1. There must be a fixed number of trials, $n$.
2. Each trial has exactly two possible outcomes.
3. Each of the trials must have the same probability of
   "success." This probability is represented by the letter $p$.
4. The trials must be independent of one another.

You can find binomial probabilities with technology or sometimes with a table, such as the one in Appendix A.

It is important to distinguish between continuous and discrete
numerical variables, because if the variable has discrete numerical
outcomes, then the probability of getting, say, *5 or more* (5 OR 6
OR 7 OR . . .) is different from the probability of getting *more than
5* (6 OR 7 OR . . .). This is not the case for a continuous numerical
variable.

### Formulas

For converting to standardized scores: $z = \dfrac{x - \mu}{\sigma}$

  $x$ is a measurement.
  $\mu$ is the mean of the probability distribution.
  $\sigma$ is the standard deviation of the probability distribution.

For binomial models: $\mu = np$

  $\mu$ is the binomial mean.
  $n$ is the number of trials.
  $p$ is the probability of success of one trial.

For binomial models: $\sigma = \sqrt{np(1 - p)}$

  $\sigma$ is the binomial standard deviation.
  $n$ is the number of trials.
  $p$ is the probability of success of one trial.

## Sources

Men's cholesterol levels, blood pressures, and heights throughout this
chapter: NHANES (www.cdc.gov/nchs/nhanes).

Women's heights and weights throughout this chapter: NHANES (www.cdc
.gov/nchs/nhanes).

## SECTION EXERCISES

### SECTION 6.1

**6.1–6.4 Directions** Determine whether each of the following variables would best be modeled as continuous or discrete.

TRY **6.1 (Example 1)**
a. The number of A's given to students in statistics one semester at your school
b. The distance between two cars on the freeway

**6.2** a. The weight of a newborn puppy
b. The number of newborn puppies in a litter

**6.3** a. The number of music CDs owned by a student
b. The length of a newborn baby

**6.4** a. The height of a skyscraper in New York City
b. The number of people who have climbed to the top of the skyscraper

TRY **6.5 Loaded Die (Example 2)** A magician has shaved an edge off one side of a six-sided die, and as a result, the die is no longer "fair." The figure shows a graph of the probability density function (pdf). Show the pdf in table format by listing all six possible outcomes and their probabilities.

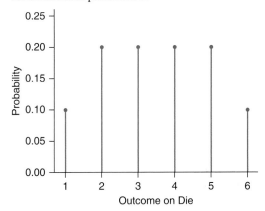

**6.6 Fair Die** Toss a fair six-sided die. The probability density function (pdf) in table form is given. Make a graph of the pdf for the die.

| Number of Spots | 1 | 2 | 3 | 4 | 5 | 6 |
|---|---|---|---|---|---|---|
| Probability | 1/6 | 1/6 | 1/6 | 1/6 | 1/6 | 1/6 |

**★6.7 Fair Die, Again** Roll a fair six-sided die. You win $3 if you roll a 1, you lose $4 if you roll a 5 or a 6, and for any other outcome you win or lose nothing.
a. Complete the table that shows the probability distribution.
b. Complete the probability distribution graph.

| Winnings | Probability |
|---|---|
| $3 | 1/6 |
| $0 | — |
| −$4 | — |

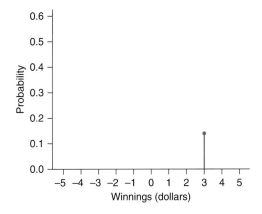

**6.8 Fair Coin** Flip a fair coin. You win $10 if it comes up heads and lose $15 if it comes up tails.
a. Create a table that shows the probability distribution of *winning*. For the loss of $15 use −$15.
b. Create a graph that shows the probability density function.
c. If someone offered to play this game with you, would it be sensible to play? Explain.

TRY **6.9 Snow Depth (Example 3)** Eric wants to go skiing tomorrow, but only if there are 3 inches or more of new snow. According to the weather report, any amount of new snow between 1 inch and 6 inches is equally likely. The probability density curve for tomorrow's new snow depth is shown. Find the probability that the new snow depth will be three inches or more tomorrow. Copy the graph, shade the appropriate area, and calculate its numerical value to find the probability. The total area is 1.

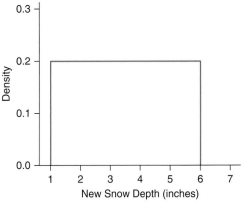

**6.10 Snow Depth** Refer to Exercise 6.9. What is the probability that the amount of new snow will be between 2 and 4 inches? Copy the graph from Exercise 6.9, shade the appropriate area, and report the numerical value of the probability.

### SECTION 6.2

**6.11 Applying the Empirical Rule with *z*-Scores** The Empirical Rule applies rough approximations to probabilities for any unimodal, symmetric distribution. But for the Normal distribution we can be more precise, as the figure shows. Use the figure and the fact that the Normal curve is symmetric to answer the questions. Do not use a Normal table or technology.

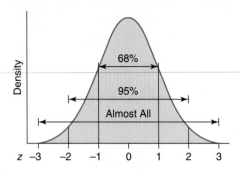

According to the Empirical Rule:

a. Roughly what percentage of z-scores are between −2 and 2?
   i. almost all    ii. 95%    iii. 68%    iv. 50%

b. Roughly what percentage of z-scores are between −3 and 3?
   i. almost all    ii. 95%    iii. 68%    iv. 50%

c. Roughly what percentage of z-scores are between −1 and 1.
   i. almost all    ii. 95%    iii. 68%    iv. 50%

d. Roughly what percentage of z-scores are more than 0?
   i. almost all    ii. 95%    iii. 68%    iv. 50%

e. Roughly what percentage of z-scores are between 1 and 2?
   i. almost all    ii. 13.5%    iii. 50%    iv. 2%

**6.12 IQs** Wechsler IQs are approximately Normally distributed with a mean of 100 and standard deviation of 15. Use the probabilities shown in the figure in Exercise 6.11 to answer the following questions. Do *not* use the Normal table or technology. You may want to label the figure with Empirical Rule probabilities to help you think about this question.

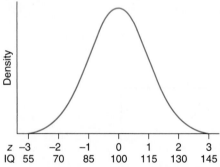

a. Roughly what percentage of people have IQs more than 100?
   i. almost all    ii. 95%    iii. 68%    iv. 50%

b. Roughly what percentage of people have IQs between 100 and 115?
   i. 34%    ii. 17%    iii. 2.5%    iv. 50%

c. Roughly what percentage of people have IQs below 55?
   i. almost all    ii. 50%    iii. 34%    iv. about 0%

d. Roughly what percentage of people have IQs between 70 and 130?
   i. almost all    ii. 95%    iii. 68%    iv. 50%

e. Roughly what percentage of people have IQs above 130?
   i. 34%    ii. 17%    iii. 2.5%    iv. 50%

f. Roughly what percentage people have IQs above 145?
   i. almost all    ii. 50%    iii. 34%    iv. about 0%

**6.13 SAT Scores** Quantitative SAT scores are approximately Normally distributed with a mean of 500 and a standard deviation of 100. On the horizontal axis of the graph, indicate the SAT scores that correspond with the provided z-scores. (See the labeling in Exercise 6.12.) Answer the questions using *only* your knowledge of the Empirical Rule and symmetry.

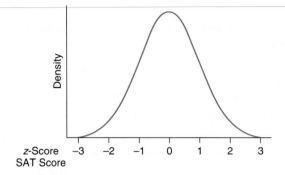

a. Roughly what percentage of students earn quantitative SAT scores more than 500?
   i. almost all    ii. 75%    iii. 50%    iv. 25%    v. about 0%

b. Roughly what percentage of students earn quantitative SAT scores between 400 and 600?
   i. almost all    ii. 95%    iii. 68%    iv. 34%    v. about 0%

c. Roughly what percentage of students earn quantitative SAT scores more than 800?
   i. almost all    ii. 95%    iii. 68%    iv. 34%    v. about 0%

d. Roughly what percentage of students earn quantitative SAT scores less than 200?
   i. almost all    ii. 95%    iii. 68%    iv. 34%    v. about 0%

e. Roughly what percentage of students earn quantitative SAT scores between 300 and 700?
   i. almost all    ii. 95%    iii. 68%    iv. 34%    v. 2.5%

f. Roughly what percentage of students earn quantitative SAT scores between 700 and 800?
   i. almost all    ii. 95%    iii. 68%    iv. 34%    v. 2.5%

**6.14 Women's Heights** Assume that college women's heights are approximately Normally distributed with a mean of 65 inches and a standard deviation of 2.5 inches. On the horizontal axis of the graph, indicate the heights that correspond to the z-scores provided. (See the labeling in Exercise 6.12.) Use only the Empirical Rule to choose your answers. Sixty inches is 5 feet and 72 inches is 6 feet.

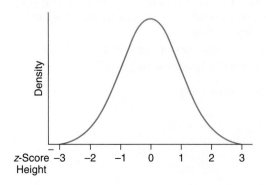

a. Roughly what percentage of women's heights are more than 72.5 inches?

   i. almost all   ii. 75%   iii. 50%   iv. 25%   v. about 0%

b. Roughly what percentage of women's heights are between 60 and 70 inches?

   i. almost all   ii. 95%   iii. 68%   iv. 34%   v. about 0%

c. Roughly what percentage of women's heights are between 65 and 67.5 inches?

   i. almost all   ii. 95%   iii. 68%   iv. 34%   v. about 0%

d. Roughly what percentage of women's heights are between 62.5 and 67.5 inches?

   i. almost all   ii. 95%   iii. 68%   iv. 34%   v. about 0%

e. Roughly what percentage of women's heights are less than 57.5 inches?

   i. almost all   ii. 95%   iii. 68%   iv. 34%   v. about 0%

f. Roughly what percentage of women's heights are between 65 and 70 inches?

   i. almost all   ii. 95%   iii. 47.5%   iv. 34%   v. 2.5%

**TRY** **6.15 IQs (Example 4)** Wechsler IQs have a population mean of 100 and a population standard deviation of 15 and are approximately Normally distributed. Use one of the StatCrunch outputs to find the probability that a randomly selected person will have an IQ of 95 or above. State which output is correct for this question, figure (A) or figure (B).

**(A)**

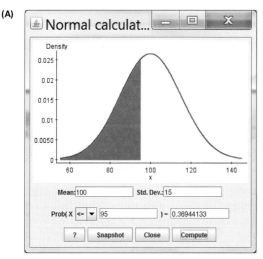

**(B)**

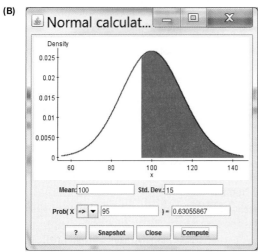

**6.16 SAT Scores** Quantitative SAT scores have a population mean of 500 and a population standard deviation of 100 and are approximately Normally distributed. Find the probability that a randomly selected person will have a quantitative SAT score of 650 or less by looking at the StatCrunch output.

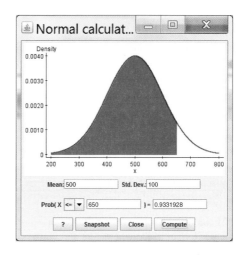

**6.17 Standard Normal** Use the table or technology to answer each question. Include an appropriately labeled sketch of the Normal curve for each part. Shade the appropriate region.

a. Find the area in a standard Normal curve to the left of 1.02 by using the Normal table. (See the excerpt provided.) Note the shaded curve.

b. Find the area in a standard Normal curve to the right of 1.02. Remember that the total area under the curve is 1.

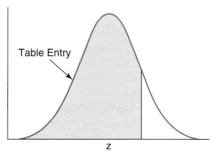

**Format of the Normal Table:** The area given is the area to the *left* of (less than) the given z-score.

| z | .00 | .01 | .02 | .03 | .04 |
|---|-----|-----|-----|-----|-----|
| **0.9** | .8159 | .8186 | .8212 | .8238 | .8264 |
| **1.0** | .8413 | .8438 | **.8461** | .8485 | .8508 |
| **1.1** | .8643 | .8665 | .8686 | .8708 | .8729 |

**6.18 Standard Normal** Use a table or technology to answer each question. Include an appropriately labeled sketch of the Normal curve for each part. Shade the appropriate region.

a. Find the area to the left of a z-score of −0.50.

b. Find the area to the right of a z-score of −0.50.

**6.19 Standard Normal** Use a table or technology to answer each question. Include an appropriately labeled sketch of the Normal curve for each part. Shade the appropriate region.

   a. Find the probability that a $z$-score will be 1.76 or less.

   b. Find the probability that a $z$-score will be 1.76 or more.

   c. Find the probability that a $z$-score will be between $-1.3$ and $-1.03$.

**6.20 Standard Normal** Use a table or technology to answer each question. Include an appropriately labeled sketch of the Normal curve for each part. Shade the appropriate region.

   a. Find the probability that a $z$-score will be $-1.00$ or less.

   b. Find the probability that a $z$-score will be more than $-1.00$.

   c. Find the probability that a $z$-score will be between 0.90 and 1.80.

**6.21 Extreme Positive z-Scores** For each question, find the area to the right of the given $z$-score in a standard Normal distribution. In this question, round your answers to the nearest 0.000. Include an appropriately labeled sketch of the $N(0, 1)$ curve

   a. $z = 4.00$

   b. $z = 10.00$ (*Hint*: Should this tail proportion be larger or smaller than the answer to part a? Draw a picture and think about it.)

   c. $z = 50.00$

   d. If you had the *exact* probability for these tail proportions, which would be the biggest and which would be the smallest?

   e. Which is equal to the area in part b: the area below (to the left of) $z = -10.00$ or the area above (to the right of) $z = -10.00$?

**6.22 Extreme Negative z-Scores** For each question, find the area to the right of the given $z$-score in a standard Normal distribution. In *this* question, round your answers to the nearest 0.000. Include an appropriately labeled sketch of the $N(0, 1)$ curve.

   a. $z = -4.00$

   b. $z = -8.00$

   c. $z = -30.00$

   d. If you had the *exact* probability for these right proportions, which would be the biggest and which would be the smallest?

   e. Which is equal to the area in part b: the area below (to the left of) $z = 8.00$ or the area above (to the right of) $z = 8.00$?

TRY g **6.23 Females' SAT Scores (Example 5)** According to data from the College Board, the mean quantitative SAT score for female college-bound high school seniors in 2009 was 500. Assume that SAT scores are approximately Normally distributed with a population standard deviation of 100. What percentage of female college-bound students had scores above 675? Please include a well-labeled Normal curve as part of your answer. *See page 281 for guidance.*

**6.24 Males' SAT Scores** According to data from the College Board, the mean quantitative SAT score for male college-bound high school seniors in 2009 was 530. Assume that SAT scores are approximately Normally distributed with a population standard deviation of 100. If a male college-bound high school senior is selected at random, what is the probability that he will score higher than 675?

**6.25 Stanford–Binet IQs** Stanford–Binet IQ scores for children are approximately Normally distributed and have $\mu = 100$ and $\sigma = 15$. What is the probability that a randomly selected child will have an IQ below 115?

**6.26 Stanford–Binet IQs** Stanford–Binet IQs for children are approximately Normally distributed and have $\mu = 100$ and $\sigma = 15$. What is the probability that a randomly selected child will have an IQ of 115 or higher?

TRY **6.27 Gestation Periods (Example 6)** The human gestation period (length of pregnancy) is 270 days with a standard deviation of 9 days and is approximately Normally distributed (assuming spontaneous labor). What is the probability that a randomly selected woman's gestation period will be between 261 and 279 days?

**6.28 Birth Weights** According to the National Vital Statistics, full-term babies' birth weights are approximately Normally distributed with a mean of 7.5 pounds and a standard deviation of 1.1 pounds. Some physicians believe that a birth weight of 5.5 pounds or less is dangerous.

   a. For a randomly selected full-term pregnancy, what is the probability that the baby's birth weight is 5.5 pounds or less?

   b. For a randomly selected full-term pregnancy, what is the probability that the baby's birth weight is between 6.4 and 8.6 pounds?

   c. If there were 600 full-term babies born at a hospital in 2010, how many of them would you expect to weigh 5.5 pounds or less?

**6.29 Total Cholesterol** According to the Centers for Disease Control, the mean total cholesterol for men between the ages of 20 and 29 is 180 milligrams per deciliter with a standard deviation of 36.2. A healthy total cholesterol level is less than 200, 200–240 is borderline, and above 240 is dangerous. Assume the distribution is approximately Normal.

   a. For a randomly selected man from this group, what is the probability that his total cholesterol level is 200 or more?

   b. For a randomly selected man from this group, what is the probability that his total cholesterol level is 240 or more?

  ⋆ c. If two randomly selected men are chosen from this group, what is the probability that both will have a total cholesterol level of 200 or more? Assume independence.

   d. If 750 randomly selected men are chosen from this group, how many (the count, not the percentage) would you expect to have a total cholesterol level of 200 or more?

**6.30 ACT Scores** In 2009, ACT reading scores had a mean of 21.4 (http://www.act.org). Suppose the population standard deviation is 5 and the distribution of ACT reading scores is approximately Normal.

   a. What percentage of people scored between 18 and 22?

   b. What percentage of people scored 25.8 or higher?

   c. If scores for 880 randomly selected people were obtained, how many would you expect to be 25.8 or higher?

**6.31 Fire Department Response Time** In New York City the average response time for calls to the fire department is about 5 minutes, according to a 2008 article published in the *New York Daily News*. Suppose the distribution of times is approximately Normal and the standard deviation is 1.5 minutes. What is the probability that the response time for a call is 5 minutes or less?

**6.32 Heights of 3-Year-Old Boys** According to the Centers for Disease Control (CDC), 3-year-old boys have a mean height of 38 inches and a standard deviation of 2 inches. The distribution is approximately Normal.

   a. What percentage of 3-year-old boys have a height of 39 inches or more?

   b. What percentage of 3-year-old boys have a height of 38 inches or more?

**6.33 Heights of 10-Year-Old Girls** According to the CDC, 10-year-old girls have a mean height of 54.5 inches and a standard deviation of 2.5 inches. The distribution of heights is approximately Normal.

   a. Find the percentage of 10-year-old girls with a height of 54.5 inches or more.

   b. Find the percentage of 10-year-old girls with a height of 52.5 inches or more.

**6.34 Boston Weather** Boston's mean minimum daily temperature in February is 23°F (http://www.city-data.com). Suppose the standard deviation of the minimum temperature is 8°F and the distribution of minimum temperatures in February is approximately Normal. What percentage of days in February have minimum temperatures below freezing (32°F)?

**6.35 New York City Weather** New York City's mean minimum daily temperature in February is 27°F (http://www.ny.com). Suppose the standard deviation of the minimum temperature is 6°F and that the distribution of minimum temperatures in February is approximately Normal. What percentage of days in February has minimum temperatures below freezing (32°F)?

**\* 6.36 Women's Heights** Assume for this question that college women's heights are approximately Normally distributed with a mean of 64.6 inches and a standard deviation of 2.6 inches. Draw a well-labeled Normal curve for each part.

   a. Find the percentage of women who should have heights of 63.5 inches or less.

   b. In a sample of 123 women, according to the probability obtained in part a, how many should have heights of 63.5 inches or less?

   c. The table shows the frequencies of heights for a sample of women, collected by statistician Brian Joiner in his statistics class. Count the women who have heights of 63.5 inches or less by looking at the table. They are in the oval.

   d. Are the answers to parts b and c the same or different? Explain.

| Height | Inches | Frequency |
|--------|--------|-----------|
|        | 59     | 2         |
| 5′     | 60     | 5         |
|        | 61     | 7         |
| 5′2″   | 62     | 10        |
|        | 63     | 16        |
| 5′4″   | 64     | 23        |
|        | 65     | 19        |
| 5′6″   | 66     | 15        |
|        | 67     | 9         |
| 5′8″   | 68     | 6         |
|        | 69     | 6         |
| 5′10″  | 70     | 3         |
|        | 71     | 1         |
| 6′     | 72     | 1         |

TRY **6.37 Probability or Measurement (Inverse)? (Example 7)** The Normal model $N(270, 9)$ describes the length of pregnancies, in days, for women who go into spontaneous labor. Which of these statements is asking for a measurement (is an inverse Normal question) and which is asking for a probability?

   a. If 60% of women have a pregnancy length below this number of days, what is the number of days?

   b. If we select, at random, a woman who goes into spontaneous labor, what is the probability that her pregnancy will be 279 days or fewer?

**6.38 Probability or Measurement (Inverse)?** The Normal model $N(100, 15)$ describes Wechsler IQs. Which of these statements is asking for a measurement (is an inverse Normal question) and which is asking for a probability?

   a. What percentage of people have IQs above 130?

   b. To get into Mensa, a high-IQ group, you must have an IQ in the upper 2%. What IQ is this?

**6.39 Inverse Normal, Standard** In a standard Normal distribution, if the area to the left of a $z$-score is about 0.6666, what is the approximate $z$-score?

   First locate inside the table, the number closest to 0.6666. Then find the $z$-score by adding 0.4 and 0.03; refer to the table. Draw a sketch of the Normal curve, showing the area and $z$-score.

| z   | .00   | .01   | .02   | .03   | .04   | .05   |
|-----|-------|-------|-------|-------|-------|-------|
| 0.4 | .6554 | .6591 | .6628 | **.6664** | .6700 | .6736 |
| 0.5 | .6915 | .6950 | .6985 | .7019 | .7054 | .7088 |
| 0.6 | .7257 | .7291 | .7324 | .7357 | .7389 | .7422 |

**6.40 Inverse Normal, Standard** In a standard Normal distribution, if the area to the left of a $z$-score is about 0.1000, what is the approximate $z$-score?

**6.41 Inverse Normal, Standard** Assume a standard Normal distribution. Draw a separate, well-labeled Normal curve for each part.

   a. Find the $z$-score that gives a left area of 0.7123.

   b. Find the $z$-score that gives a left area of 0.1587.

**6.42 Inverse Normal, Standard** Assume a standard Normal distribution. Draw a separate, well-labeled Normal curve for each part.

   a. Find an approximate $z$-score that gives a left area of 0.7000.

   b. Find an approximate $z$-score that gives a left area of 0.9500.

TRY
g **6.43 Females' SAT Scores (Example 8)** According to the College Board, the mean quantitative SAT score for female college-bound high school seniors in 2009 was 500. SAT scores are approximately Normally distributed with a population standard deviation of 100. A scholarship committee wants to give awards to college-bound women who score at the 96th percentile or above on the SAT. What score does an applicant need? Include a well-labeled Normal curve as part of your answer. *See page 282 for guidance.*

**6.44 Males' SAT Scores** According to the College Board, the mean quantitative SAT score for male college-bound high school seniors in 2009 was 530. SAT scores are approximately Normally distributed with a population standard deviation of 100. What is the SAT score at the 96th percentile for male college-bound seniors?

**6.45 Males' Body Temperatures** A study showed that males' body temperatures are approximately Normally distributed with a mean of 98.1°F and a population standard deviation of 0.70°F. What body temperature does a male have if he is at the 90th percentile? Draw a well-labeled sketch to support your answer.

**6.46 Females' Body Temperatures** A study showed that females'body temperatures are approximately Normally distributed with a mean of 98.4°F and a population standard deviation of 0.70°F. Find the female body temperature at the 90th percentile. Draw a well-labeled sketch to support your answer.

**6.47 Women's Heights** Suppose college women's heights are approximately Normally distributed with a mean of 65 inches and a population standard deviation of 2.5 inches. What height is at the 20th percentile? Include an appropriately labeled sketch of the Normal curve to support your answer.

**6.48 Men's Heights** Suppose college men's heights are approximately Normally distributed with a mean of 70.0 inches and a population standard deviation of 3 inches. What height is at the 20th percentile? Include an appropriately labeled Normal curve to support your answer.

**6.49 Wechsler IQs** Wechsler IQs have a mean of 100 and a standard deviation of 15 and are approximately Normally distributed.

  a. The Wechsler IQ at the 25th percentile is 90. What is the Wechsler IQ at the 75th percentile?

  b. The interquartile range is Q3 minus Q1. Find the interquartile range for Wechsler IQs.

  c. Is the interquartile range larger or smaller than the standard deviation, 15 IQ points?

**6.50 Gestation Period** The gestation period for humans is approximately Normally distributed with a mean of 270 days and population standard deviation of 9, assuming spontaneous labor. The gestation period at the 25th percentile is 264 days.

  a. Find the gestation period at the 75th percentile.

  b. Find the interquartile range of the gestation period.

  c. Using the distribution of gestation periods, is the standard deviation larger or smaller than the interquartile range?

**6.51 Child and Adult Heights for Males** According to the National Center for Health Statistics, 3-year-old boys have a mean height of 38 inches and a standard deviation of 2 inches. Assume the distribution is approximately Normal.

  a. Find the percentile measure for a height of 40 inches for a 3-year-old boy.

  b. If this 3-year-old boy grows up to be a man with a height at the same percentile, what will his height be? Use a population mean of 70 inches and a population standard deviation of 3 inches.

**6.52 Child and Adult Heights for Females** According to the CDC, 10-year-old girls have a mean height of 54.5 inches and a standard deviation of 2.5 inches, and the distribution of heights is approximately Normal.

  a. Find the percentile measure for a height of 57 inches for a 10-year-old girl.

  b. If this 10-year-old girl grows up to be a woman with a height at the same percentile, what will her height be? Use a population mean of 65 inches and a population standard deviation of 2.5 inches.

## SECTION 6.3

TRY **6.53 Gender of Children (Example 9)** A married couple plans to have four children, and they are wondering how many boys they should expect to have. Assume none of the children will be twins or other multiple births. Also assume the probability that a child will be a boy is 0.50. Explain why this is a binomial experiment. Check all four required conditions.

**6.54 Coin Flip** A coin will be flipped three times, and the number of heads recorded. Explain why this is a binomial experiment. Check all four required conditions.

TRY **6.55 Coin Flips (Example 10)** A teacher wants to find out whether coin flips of pennies have a 50% chance of coming up heads. In the last 5 minutes of class, he has all the students flip pennies until the end of class and then report their results to him. Which condition or conditions for use of the binomial model is or are not met?

**6.56 Twins** In Exercise 6.53 you are told to assume that none of the children will be twins or other multiple births. Why? Which of the conditions required for a binomial experiment would be violated if there were twins?

**6.57 Divorce** Suppose that the probability that a randomly selected person who has recently married for the first time will get a divorce within 5 years is 0.2. Suppose we follow 12 married couples (24 people) for 5 years and record the number of people divorced. Why is the binomial model inappropriate for finding the probability that at least 7 of these 24 people will be divorced within 5 years? List all binomial conditions that are not met.

**6.58 Divorce** Suppose that the probability that a randomly selected person who has recently married for the first time will be divorced within 5 years is 0.2, and that the probability that a randomly selected person who has recently married for the second time will be divorced within 5 years is 0.30. Take a random sample of 10 people married for the first time and 10 people married for the second time. The sample is chosen such that no one in the sample is married to anyone else in the sample. Why is the binomial model inappropriate for finding the probability that exactly 4 of the 20 people in the sample will be divorced within 5 years? List all of the binomial conditions that are not met.

TRY **6.59 Identifying n, p, and x (Example 11)** For each situation, identify the sample size $n$, the probability of success $p$, and the number of successes $x$. Give the answer in the form $b(n, p, x)$. Do *not* go on to find the probability. Assume the four conditions for a binomial experiment are satisfied.

  a. In the 2008 presidential election, 54% of the voters voted for President Obama. What is the probability that 65 out of 100 independently chosen voters voted for President Obama?

  b. The manufacturer of LoJack Stolen Vehicle Recovery System claims that the probability that a stolen vehicle using LoJack will be recovered is 90%. What is the probability that exactly 9 out of 10 independently stolen vehicles with LoJack will be recovered?

  c. A student is taking a 10-question multiple-choice test. Each question has four options: a, b, c, and d. One of these four options is correct and three of them are incorrect. What is the probability that the student correctly answers exactly 6 of the 10 questions on the test by guessing?

**6.60 Identifying n, p, and x** For each situation, identify the sample size $n$, the probability of success $p$, and the number of successes $x$.

Give the answer in the form $b(n, p, x)$. Do *not* go on to find the probability. Assume the four conditions for a binomial experiment are satisfied.

a. According to an article published on the *Daily Business Review* website, the rate of response to a jury summons is about 40% in Dade County, Florida. What is the probability that 40 out of 50 randomly selected people will respond?

b. Data from the National Comprehensive Auto Theft Research System (CARS) showed that 75% of motor vehicles stolen in Australia from 2005 to 2006 were recovered. What is the probability that in a sample of 10 stolen cars, 6 out of 10 will be recovered, assuming the same rate of recovery?

c. A student is taking a 10-question multiple-choice test. Each question has five options: a, b, c, d, and e. For each question, one of these five choices is correct and four of them are incorrect. What is the probability that the student correctly answers exactly 6 of the 10 questions on the test by guessing?

TRY **6.61 Stolen Bicycles (Example 12)** According to the *Sidney Morning Herald*, 40% of bicycles stolen in Holland are recovered. (In contrast, only 2% of bikes stolen in New York City are recovered.) Find the probability that, in a sample of 6 randomly selected cases of bicycles stolen in Holland, exactly 2 out of 6 bikes are recovered.

**6.62 Florida Recidivism Rate** The three-year recidivism rate of parolees in Florida is about 30%; that is, 30% of parolees end up back in prison within three years (http://www.floridaperforms.com). Assume that whether one parolee returns to prison is independent of whether any of the others returns.

a. Find the probability that exactly 6 out of 20 parolees will end up back in prison within three years,

b. Find the probability that 6 or fewer out of 20 parolees will end up back in prison within three years.

**6.63 MBA Job Search** Historically, about 90% of MBAs (people who have earned a master's degree in business administration) from UCLA find a job within three months of graduation, according to an article published in *Bloomberg Businessweek* magazine. Assume for the sake of simplicity that whether a graduate finds a job within three months is independent of whether any of the other graduates find jobs.

a. Find the probability that at most 7 (this means 7 or fewer) out of 10 will find a job in their chosen field.

b. Find the probability that at least 8 (this means 8 or more) out of 10 will find a job in their chosen field.

c. Find the probability that anywhere from 7 to 9 out of 10 will find a job in their chosen field. The 7 to 9 is inclusive—that is, it includes the values for 7 and 9.

**6.64 Guessing on Exams** Suppose you are taking an exam with 10 questions and you are required to get 7 or more right answers to pass.

a. With a 10-question true/false test, what is the probability of getting at least 7 answers correct by guessing?

b. With a 10-question multiple-choice test where there are three possible choices for each question, what is the probability of getting at least 7 answers correct by guessing? Only one of the choices is correct for each question.

c. With a 10-question multiple-choice test where there are five possible choices for each question, only one of which is correct, what is the probability of getting at least 7 answers correct by guessing?

d. Which test (of those described in parts a, b, and c in this exercise) would be easiest to pass by guessing, which would be hardest, and why?

**6.65 Newborn Gender** Assume that half of all children born are male and half are female. In all of the following cases, we will assume that there are no twins (or triplets or more) and that the conditions of the binomial model are satisfied.

a. If a woman plans to have two children, what is the probability that both will be girls?

b. If a woman plans to have three children, what is the probability that all will be girls?

c. If a woman plans to have three children, what is the probability that she will have at least one boy? ("At least one boy" is the complement of "all girls.")

d. Does this mean that the more children a woman has, the more likely she will be to have at least one boy? Explain.

**6.66 2010 Employment Rates** According to a 2010 Gallup poll, almost 20% of people were underemployed. That means they either were unemployed or had a job working fewer hours than they would like. (Homemakers and other people not seeking employment are not included in the data.)

a. For one randomly selected person, what is the probability that the person is fully employed (not underemployed)?

For parts b and c, assume that you randomly select five people and inquire about their employment, but you do not include those not seeking employment.

b. What is the probability that all five people are fully employed?

c. What is the probability that at least one of the five is underemployed?

**6.67 Late Flights** For commercial flights in 2009, approximately 80% arrived on time (within 15 minutes of scheduled arrival time), according to the Bureau of Transportation Statistics.

a. Assuming that this success rate still holds, if you randomly select three flights and assume they are independent, what is the probability that all three will arrive on time?

b. What is the probability that at least one of the three flights will be late?

c. If all three flights are on the same day in December and all three are flights to Boston, explain why the binomial model is not appropriate for finding the probability that at least one flight will be late.

**6.68 Crime Clearance** In 2008, the overall U.S. crime clearance rate for violent crimes was 45%, according to the FBI. The term *clearance* means that an arrest is made, a crime is charged, and a case is referred to a court. Consider 10 violent crimes that are unrelated to each other, using the 45% clearance rate.

a. What is the probability that they are all cleared?

b. What is the probability that none of them is cleared?

c. What is the probability that at least one is cleared?

d. How many of the 10 would you expect to be cleared, on average?

TRY **6.69 DWI Convictions (Example 13)** In New Mexico, about 70% of drivers who are arrested for driving while intoxicated (DWI) are convicted (http://www.drunkdrivingduilawblog.com).

a. If 15 independently selected drivers were arrested for DWI, how many of them would you expect to be convicted?

b. What is the probability that exactly 11 out of 15 independent selected drivers are convicted?

c. What is the probability that 11 or fewer are convicted?

**6.70 Internet Access** A 2010 Nielsen poll indicated that about 60% of U.S. households had access to a high-speed Internet connection.

a. Suppose 100 households were randomly selected from the United States. How many of the households would you expect to have access to a high-speed Internet connection?

b. If 10 households are selected randomly, what is the probability that exactly 6 have high-speed access?

c. If 10 households are selected randomly, what is the probability that 6 or fewer have high-speed access?

**6.71 Cell Phone Only** According to a Pew poll done in 2010, 50% of adults aged 25–29 had only a cell phone (no landline). Assume that two randomly selected adults (aged 25–29) are asked whether they have only a cell phone.

a. If a person has only a cell phone, we will record Y (for Yes). If not, we will record N. List all possible sequences of Y and N for this experiment.

b. For each sequence, find by hand the probability that it will occur, assuming each outcome is independent.

c. What is the probability that neither of the two randomly selected adults has only a cell phone?

d. What is the probability that exactly one person out of two has only a cell phone?

e. What is the probability that two out of two (both) have only a cell phone?

**6.72 Texting While Driving** According to a Pew poll in 2010, 1 in 4 teens of driving age have reported having sent or received a text message while driving. Assume that we randomly sample two teens of driving age.

a. If a teen has texted while driving, record Y; if not, record N. List all possible sequences of Y and N for this experiment.

b. For each sequence, find by hand the probability that it will occur, assuming each outcome is independent.

c. What is the probability that neither of the two randomly selected teens has texted?

d. What is the probability that exactly one out of the two teens has texted?

e. What is the probability that both have texted?

TRY **6.73 Coin Flip (Example 14)** A fair coin is flipped 50 times.

a. What is the expected number of heads?

b. Find the standard deviation for the number of heads.

c. How many heads should you expect, give or take how many? Give the range of the number of heads based on these numbers.

**6.74 Illinois Drivers** According to a 2010 *Chicago Sun Times* article, 20% of driver's license applicants fail the written test in Illinois. If 200 people take the driver's license exam, how many people should we expect to pass? Give or take how many?

TRY **6.75 Women without Children (Example 15)** According to a Pew Research poll in 2010, 20% of women in the United States have ended their childbearing years without having children. (In the 1970s, this number was 10%.)

a. If we randomly select 300 women, how many would we expect to have had no children? Give or take how many?

b. Give the range of likely values from 1 standard deviation above the mean to 1 standard deviation below the mean.

c. If you found that 62 out of 300 randomly sampled women ended their childbearing years without having children, would you be surprised? Why or why not?

**6.76 Interracial Marriage** Pew Research reported that 15% of marriages in the United States in 2008 were between people of different races. Suppose 100 random marriages were surveyed.

a. How many marriages would you expect to be between people of different races, give or take how many?

b. Give the range of likely values from 1 standard deviation above the mean to 1 standard deviation below the mean.

c. If you found that only 5 out of 100 randomly selected marriages were between people of different races, would you be surprised? Explain.

# CHAPTER REVIEW EXERCISES

**6.77 Birth Length** A study of U.S. births published on the website *Medscape from WebMD* reported that the average birth length of babies was 20.5 inches and the standard deviation was about 0.90 inch. Assume the distribution is approximately Normal. Find the percentage of babies with birth lengths of 22 inches or less.

**6.78 Birth Length** A study of U.S. births published on the website *Medscape from WebMD* reported that the average birth length of babies was 20.5 inches and the standard deviation was about 0.90 inch. Assume the distribution is approximately Normal. Find the percentage of babies who have lengths of 19 inches or less at birth.

**6.79 Males' Body Temperature** A study of human body temperatures using healthy men showed a mean of 98.1°F and a standard deviation of 0.70°F. Assume the temperatures are approximately Normally distributed.

a. Find the percentage of healthy men with temperatures below 98.6°F (that temperature was considered typical for many decades).

b. What temperature does a healthy man have if his temperature is at the 76th percentile?

**6.80 Females' Body Temperature** A study of human body temperatures using healthy women showed a mean of 98.4°F and a standard deviation of about 0.70°F. Assume the temperatures are approximately Normally distributed.

a. Find the percentage of healthy women with temperatures below 98.6°F (this temperature was considered typical for many decades).

b. What temperature does a healthy woman have if her temperature is at the 76th percentile?

**6.81 Baby's Gender** A woman with five sons and no daughters becomes pregnant. What is the likelihood that her next child will be a girl? Write a sentence or two explaining your answer.

**6.82 Coin Flip** You flip a coin and get heads four times in a row. What is the likelihood that the next flip will show heads? Write a sentence or two explaining your answer.

**6.83 Crime** According to a 2009 Gallup poll, 30% of U.S. households were the victims of a crime in the last year.

a. If there were 2000 independent households surveyed, how many would you expect to have been the victims of a crime, give or take how many?

b. Judging on the basis of your answer to part a, what range of households would you expect to have been the victims of a crime?

**6.84 Cesarean Births** According to the CDC, the rate of Cesarean births in the United States in 2007 was 32%.

a. If 200 independent births were surveyed in 2007, how many of those would you expect to be Cesarean, give or take how many?

b. Would it be unusual for a hospital with 200 births to have 110 Cesareans? Explain.

**6.85 Leukemia Survival Rate** The five-year survival rate for chronic myelogenous leukemia (CML) for those on the drug Gleevec is about 90%. Before Gleevec was introduced, the five-year survival rate was about 40%.

a. Suppose four patients have just been diagnosed with CML and will be using Gleevec. Assume that one patient's survival is independent of the survival of the others.

   i. What is the probability that the patients will all be alive in five years?

   ii. What is the probability that at least one will die within five years?

b. Suppose four patients had been diagnosed with CML 25 years ago, before Gleevec was available. Assume that one patient's survival was independent of the survival of the others.

   i. What is the probability that they all would have stayed alive for five years?

   ii. What is the probability that at least one would have been dead within five years?

c. Why is the answer to Question ii in part a smaller than the answer to Question ii in part b?

**6.86 Physician-assisted Suicide** A Field poll in 2005 showed that 70% of people polled in California answered yes to the question "Should incurably ill patients have the right to receive life-ending medication?" Assume that 70% is the correct population proportion agreeing with the statement.

a. If 300 randomly selected people in California were asked this question, how many of them would you expect to answer yes? Give or take how many?

b. Would you be surprised if 150 of the 300 answered yes to this question? Explain.

# gUIDED EXERCISES

**g 6.23 Females' SAT Scores** According to data from the College Board, the mean quantitative SAT score for female college-bound high school seniors in 2009 was 500. SAT scores are approximately Normally distributed with a population standard deviation of 100.

**QUESTION** What percentage of the female college-bound high school seniors had scores above 675? Answer this question by following the numbered steps.

Step 1 ▶ **Find the z-score**
To find the z-score for 675, subtract the mean and divide by the standard deviation. Report the z-score.

Step 2 ▶ **Explain the location of 500**
Refer to the Normal curve. Explain why the SAT score of 500 is right below the z-score of 0. The tick marks on the axis mark the location of z-scores that are integers from −3 to 3.

Step 3 ▶ **Label with SAT scores**
Carefully sketch a copy of the curve. Pencil in the SAT scores of 200, 300, 400, 600, and 700 in the correct places.

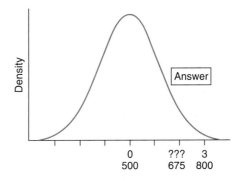

Step 4 ▶ **Add the line, z-score, and shading**
Draw a vertical line through the curve at the location of 675. Just above the 675 (indicated on the graph with "???") put in the corresponding z-score. We want to find what percentage of students had scores *above* 675. Therefore, shade the area to the *right* of this boundary, because numbers to the right are larger.

**Step 5 ▶ Use the table for the left area**

Use the following excerpt from the Normal table to find and report the area to the left of the $z$-score that was obtained from an SAT score of 675. This is the area of the unshaded region.

**Step 6 ▶ Answer**

Because you want the area to the right of the $z$-score, you will have to subtract the area you obtained in step 5 from 1. This is the area of the shaded region. Put it where the box labeled "Answer" is. Check to see that the number makes sense. For example if the shading is less than half the area, the answer should not be more than 0.5000.

**Step 7 ▶ Sentence**

Finally, write a sentence telling what you found.

| $z$ | .00 | .01 | .02 | .03 | .04 | .05 | .06 | .07 | .08 | .09 |
|-----|-----|-----|-----|-----|-----|-----|-----|-----|-----|-----|
| **1.6** | .9452 | .9463 | .9474 | .9484 | .9495 | *.9505 | .9515 | .9525 | .9535 | .9545 |
| **1.7** | .9554 | .9564 | .9573 | .9582 | .9591 | **.9599** | .9608 | .9616 | .9625 | .9633 |
| **1.8** | .9641 | .9649 | .9656 | .9664 | .9671 | .9678 | .9686 | .9693 | .9699 | .9706 |

**g 6.43 Females' SAT scores** According to the College Board, the mean quantitative SAT score for female college-bound high school seniors in 2009 was 500. SAT scores are approximately Normally distributed with a population standard deviation of 100. A scholarship committee wants to give awards to college-bound women who score at the 96th percentile or above on the SAT. What score does an applicant need? Include a well-labeled Normal curve as part of your answer.

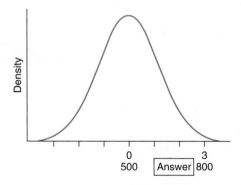

**QUESTION** What is the SAT score at the 96th percentile? Answer this question by following the numbered steps.

**Step 1 ▶ Think about it**

Will the SAT test score be above the mean or below it? Explain.

**Step 2 ▶ Label z-scores**

Label the curve with integer $z$-scores. The tick marks represent the position of integer $z$-scores from −3 to 3.

**Step 3 ▶ Use the table**

The 96th percentile has 96% of the area to the *left* because it is higher than 96% of the scores. The table above gives the areas to the *left* of $z$-scores. Therefore, we look for 0.9600 in the interior part of the table.

Use the excerpt of the Normal table given above for Exercise 6.23 to locate the area *closest* to 0.9600.

Then report the $z$-score for that area.

**Step 4 ▶ Add the z-score, line, and shading to the sketch**

Add that $z$-score to the sketch and draw a vertical line above it through the curve. Shade the left side because the area to the left is what is given.

**Step 5 ▶ Find the SAT score**

Find the SAT score that corresponds to the $z$-score. The score should be $z$ standard deviations above the mean, so

$$x = \mu + z\sigma$$

**Step 6 ▶ Add the SAT score to the sketch**

Add the SAT score on the sketch where it says "Answer."

**Step 7 ▶ Write a sentence**

Finally, write a sentence stating what you found.

# TechTips

## For All Technology

All technologies will use the two examples that follow.

**EXAMPLE A: NORMAL** ▶ Wechsler IQs have a mean of 100 and standard deviation of 15 and are Normally distributed.

  a. Find the probability that a randomly chosen person will have an IQ between 85 and 115.

  b. Find the probability that a randomly chosen person will have an IQ that is 115 or less.

  c. Find the Wechsler IQ at the 75th percentile.

*Note:* If you want to use technology to find areas from standard units ($z$-scores), use a mean of 0 and a standard deviation of 1.

**EXAMPLE B: BINOMIAL** ▶ Imagine that you are flipping a fair coin (one that comes up heads 50% of the time in the long run).

  a. Find the probability of getting 28 or fewer heads in 50 flips of a fair coin.

  b. Find the probability of getting exactly 28 heads in 50 flips of a fair coin.

---

### TI-83/84

#### NORMAL

**a. Between Two Values**

1. Press **2ND DISTR** (located below the four arrows on the keypad).
2. Select **2:normalcdf** and press **ENTER**.
3. Enter: (left boundary, right boundary, $\mu$, $\sigma$), in this example **85 ❯ 115 ❯ 100 ❯ 15**) and press **ENTER**. (The comma button is above the 7 button.)

Your screen should look like Figure 6A, which shows that the probability that a randomly selected person will have a Wechsler IQ between 85 and 115 is equal to 0.6827.

```
normalcdf(85,115
,100,15)
         .6826894809
```

▲ **FIGURE 6A** TI-83/84 normal**c**df (c stands for "cumulative")

**b. Some Value or Less**

1. Press **2ND DISTR**.
2. Select **2:normalcdf** and press **ENTER**.
3. Enter: (left boundary, right boundary, $\mu$, $\sigma$) and press **ENTER**.

The probability that a person's IQ is 115 *or less* has an *indeterminate* left boundary, for which you may use negative 1000000 or any extreme value that is clearly out of the range of data. Figure 6B shows the probability that a randomly selected person will have an IQ of 115 or less.

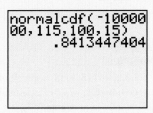

▲ **FIGURE 6B** TI-83/84 Normalcdf with indeterminate left boundary

*Caution:* The negative button (−) is to the left of the **ENTER** button and is not the same as the minus button that is above the plus button.

(If you have an indeterminate right boundary, then to find the probability that the person's IQ is 85 or more, for example, use a right boundary (such as 1000000) that is clearly above all the data.)

**c. Inverse Normal**

If you want a measurement (such as an IQ) from a proportion or percentile:

1. Press **2ND DISTR**.
2. Select **3:invNorm** and press **ENTER**.
3. Enter: (left proportion, $\mu$, $\sigma$) and press **ENTER**. Be sure to include the commas between the numbers.

Figure 6C shows the Wechsler IQ at the 75th percentile, which is 110. Note that the 75th percentile is entered as **.75**.

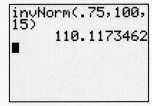

▲ **FIGURE 6C** TI-83/84 Inverse Normal

## Binomial

### a. Cumulative (or Fewer)

1. Press **2ND DISTR.**
2. Select **B:binomcdf** (you will have to scroll down to see it) and press **ENTER**. (On a TI-83, it is **A:binomcdf**.)
3. See Figure 6D. Enter: (n, p, x) and press **ENTER**. Be sure to include the commas between the numbers.

    The answer will be the probability for x or fewer. Figure 6D shows the probability of 28 or fewer heads out of 50 flips of a fair coin. (You could find the probability of 29 or more heads by subtracting your answer from 1.)

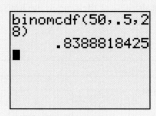

▲ **FIGURE 6D** TI-83/4
binom**c**df (cumulative)

### b. Individual (Exact)

1. Press **2ND DISTR.**
2. Select **A:binompdf** and press **ENTER**. (On a TI-83, it is **0:binompdf.**)
3. Enter: (n, p, x) and press **ENTER**. Be sure to include the commas between the numbers.

    Figure 6E shows the probability of *exactly* 28 heads out of 50 flips of a fair coin.

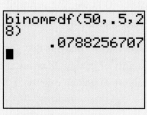

▲ **FIGURE 6E** TI-83/84
binom**p**df (individual)

---

**MINITAB**

## Normal

### a. Between Two Values

1. Enter the upper boundary, **115**, in the top cell of an empty column; here we use column C1, row 1. Enter the lower boundary, **85**, in the cell below; here column C1, row 2.
2. **Calc > Probability Distributions > Normal.**
3. See Figure 6F: Choose **Cumulative probability**. Enter: **Mean, 100; Standard deviation, 15; Input column, C1; Optional storage, C2.**
4. Click **OK**.
5. Subtract the lower probability from the larger shown in column C2.

    0.8413 − 0.1587 = 0.6836 is the probability that a Wechsler IQ is between 85 and 115.

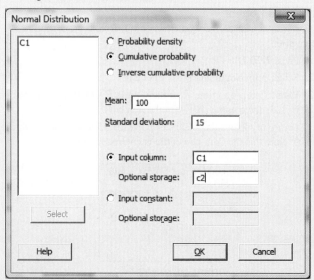

▲ **FIGURE 6F** Minitab Normal

### b. Some Value or Less

1. The probability of an IQ of 115 or less, 0.8413, is shown in column C2, row1. (In other words, do as in part a above, except do *not* enter the lower boundary, 85.)

### c. Inverse Normal

If you want a measurement (such as an IQ or height) from a proportion or percentile:

1. Enter the decimal form of the left proportion (**.75** for the 75th percentile) into a cell in an empty column in the spreadsheet; here we used column C1, row 1.
2. **Calc > Probability Distributions > Normal.**
3. See Figure 6G: Choose **Inverse cumulative probability**. Enter: **Mean, 100; Standard deviation, 15; Input column, c1;** and **Optional storage, c2** (or an empty column).
4. Click **OK**.

You will get **110**, which is the Wechsler IQ at the 75th percentile.

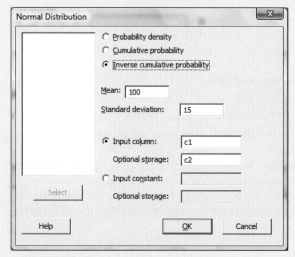

▲ **FIGURE 6G** Minitab Inverse Normal

## Binomial

### a. Cumulative (or Fewer)

1. Enter the upper bound for the number of successes in an empty column; here we used column C1, row 1. Enter **28** to get the probability of 28 or fewer heads.
2. **Calc > Probability Distributions > Binomial**.
3. See Figure 6H. Choose **Cumulative probability**.

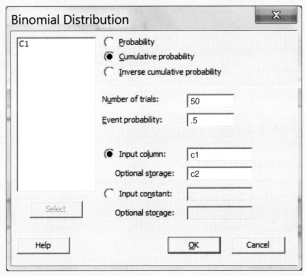

▲ **FIGURE 6H** Minitab Binomial

Enter: **Number of trials**, **50**; **Event probability**, **.5**; **Input column**, **c1**; **Optional storage**, **c2** (or an empty column).

4. Click **OK**.

Your answer will be 0.8389 for the probability of 28 *or fewer* heads.

### b. Individual (Exact)

1. Enter the number of successes at the top of column 1, **28** for 28 heads.
2. **Calc > Probability Distributions > Binomial**.
3. Choose **Probability** (at the top of Figure 6H) instead of **Cumulative Probability** and enter: **Number of trials**, **50**; **Event probability**, **.5**; **Input column**, **c1**; **Optional storage**, **c2** (or an empty column).
4. Click **OK**.

Your answer will be 0.0788 for the probability of *exactly* 28 heads.

---

### EXCEL

## Normal

Unlike the TI-83/84, Excel makes it easier to find the probability that a random person has an IQ of 115 or less) than to find than the probability that a random person has an IQ between 85 and 115). This is why, for Excel, part b appears before part a.

### b. Some Value or Less

1. Click *fx* (and **select a category All**).
2. Choose **NORMDIST**.
3. See Figure 6I. Enter: **X**, **115**; **Mean**, **100**; **Standard_dev**, **15**; **Cumulative**, **true** (for 115 *or less*). The answer is shown as 0.8413. Click **OK** to make it show up in the active cell on the spreadsheet.

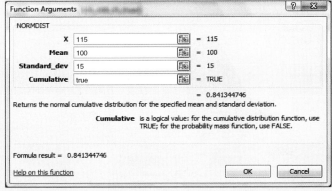

▲ **FIGURE 6I** Excel Normal

### a. Between Two Values

If you want the probability of an IQ between 85 and 115:

1. First, follow the instructions above for part b. *Do not change the active cell in the spreadsheet.*
2. You will see **=NORMDIST(115,100,15,TRUE)** in the *fx* box. Click in this box, to the right of **TRUE)** and put in a minus sign.
3. Now repeat the steps for part b starting by clicking *fx*, except enter 85 instead of 115 for X. The answer, **0.682689**, will be shown in the active cell.

    (Alternatively, just repeat steps 1−3 for part b using 85 instead of 115. Subtract the smaller probability value from the larger (0.8413 − 0.1587 = 0.6826).)

### c. Inverse Normal

If you want a measurement (such as an IQ or height) from a proportion or percentile:

1. Click *fx.*
2. Choose **NORMINV** and click **OK**.
3. See Figure 6J. Enter: **Probability**, **.75** (for the 75th percentile); **Mean**, **100**: **Standard_dev**, **15**. You may read the answer off the screen or click **OK** to see it in the active cell in the spreadsheet.

    The IQ at the 75th percentile is 110.

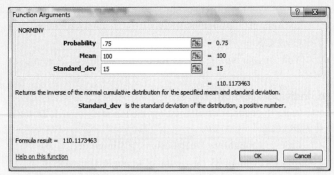

▲ FIGURE 6J Excel Inverse Normal

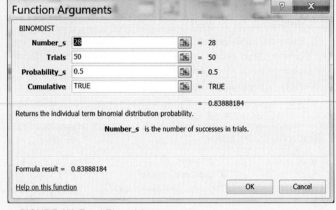

▲ FIGURE 6K Excel Binomial

## Binomial

### a. Cumulative (or Fewer)

1. Click $f_x$.
2. Choose **BINOMDIST** and click **OK**.
3. See Figure 6K. Enter: **Number_s, 28**; **Trials, 50**; **Probability_s, .5**; and **Cumulative, true** (for the probability of 28 *or fewer*).

   The answer (**0.8389**) shows up in the dialogue box and in the active cell when you click **OK**.

### b. Individual (Exact)

1. Click $f_x$.
2. Choose **BINOMDIST** and click **OK**.
3. Use the numbers in Figure 6K, but enter **False** in the Cumulative box. This will give you the probability of getting *exactly* 28 heads in 50 tosses of a fair coin.

   The answer (**0.0788**) shows up in the dialogue box and in the active cell when you click **OK**.

STATCRUNCH

## Normal

Unlike the TI-83/84, StatCrunch makes it easier to find the probability for 115 or less than to find the probability between 85 and 115. This is why part b is done before part a.

### b. Some Value or Less

1. **Stat > Calculators > Normal**
2. See Figure 6L. To find the probability of having a Wechsler IQ of 115 or less, Enter: **Mean, 100**: **Std Dev, 15**. Make sure that the arrow to the right of **Prob(X** points left (for less than). Enter the **115** in the box above **Snapshot**.
3. Click **Compute** to see the answer, **0.8413**.

### a. Between Two Values

To find the probability of having a Wechsler IQ between 85 and 115, use steps 1, 2, and 3 again, but use **85** instead of **115** in the box above **Snapshot**. When you find that probability, subtract it from the probability found in Figure 6L.

$$0.8413 - 0.1587 = 0.6826$$

### c. Inverse Normal

If you want a measurement (such as an IQ or height) from a proportion or percentile:

1. **Stat > Calculators > Normal**
2. See Figure 6M. To find the Wechsler IQ at the 75th percentile, enter: Mean, **100**; Std. Dev., **15**. Make sure that the arrow to the right of **Prob(X** points to the left, and enter **0.75** in the box above **Compute**.

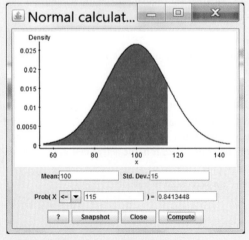

▲ FIGURE 6L StatCrunch Normal

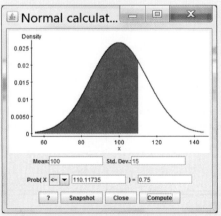

▲ FIGURE 6M StatCrunch Inverse Normal

3. Click **Compute** and the answer (**110**) is shown above **Snapshot**.

## Binomial

### a. Cumulative (or Fewer)

1. **Stat** > **Calculators** > **Binomial**
2. See Figure 6N. To find the probability of 28 or fewer heads in 50 tosses of a fair coin, enter: **n, 50**; and **p, 0.5**. The arrow after **Prob(X** should point left (for *less than*). Enter **28** in the box above **Snapshot**.
3. Click **Compute** to see the answer (**0.8389**).

### b. Individual (Exact)

1. **Stat** > **Calculators** > **Binomial**
2. To find the probability of exactly 28 heads in 50 tosses of a fair coin, use a screen similar to Figure 6N, but to the right of **Prob(X** choose the equals sign. You will get **0.0788**.

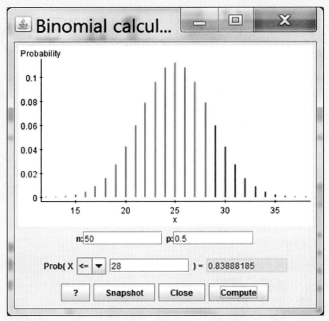

▲ **FIGURE 6N** StartCrunch Binomial

# 7 Survey Sampling and Inference

# THEME

If survey subjects are chosen randomly, then we can use their answers to infer how the entire population would answer. We can also quantify how far off our estimate is likely to be.

Somewhere in your town or city, possibly at this very moment, people are participating in a survey. Perhaps they are filling out a customer satisfaction card at a restaurant. Maybe their television is automatically transmitting information about which show is being watched so that marketers can estimate how many people are viewing their ads. They may even be text messaging in response to a television survey. Most of you will receive at least one phone call from a survey company that will ask whether you are satisfied with local government services or plan to vote for one candidate over another. The information gathered by these surveys is used to piece together, bit by bit, a picture of the larger world.

You've reached a pivotal point in the book. In this chapter, the data summary techniques you learned in Chapters 2 and 3, the probability you learned about in Chapter 5, and the Normal distribution, which you studied in Chapter 6, are all combined to allow us to generalize what we learn about a small sample to a larger group. Politicians rely on surveys of 1000 voters not because they care how those 1000 individuals will vote. Surveys are important to politicians only if they help them learn about *all* potential voters. In this and later chapters, we study ways to understand and measure just how reliable this projection from sample to the larger world is.

Whenever we draw a conclusion about a large group based on observations of some parts of that group, we are making an inference. Inferential reasoning lies at the foundation of science but is far from foolproof. As the following case study illustrates, when we make an inference we can never be absolutely certain of our conclusions. But applying the methods introduced in this chapter ensures that if we collect data carefully, we can at least measure how certain or uncertain we are.

## CASE STUDY

## Spring Break Fever: Just What the Doctors Ordered?

In 2006, the American Medical Association (AMA) issued a press release ("Sex and intoxication among women more common on spring break according to AMA poll") in which it concluded, among other things, that "eighty-three percent of the [female, college-attending] respondents agreed spring break trips involve more or heavier drinking than occurs on college campuses and 74 percent said spring break trips result in increased sexual activity." This survey made big news, particularly since the authors of the study claimed these percentages reflected the opinions not only of the 644 women who responded to the survey but of all women who participated in spring break.

The AMA's website claimed the results were based on "a nationwide random sample of 644 women who . . . currently attend college. . . . The survey has a margin of error of +/−4 percentage points at the 95 percent level of confidence." It all sounds very scientific, doesn't it?

However, some survey specialists were suspicious. After Cliff Zukin, a specialist who was president of the American Association for Public Opinion Research, corresponded with the AMA, it changed its website posting to say the results were based not on a random sample, but instead on "a nationwide sample of 644 women . . . who are part of an *online survey panel* . . . [emphasis added]." "Margin of error" is no longer mentioned.

Disagreements over how to interpret these results show just how difficult inference is. In this chapter you'll see why the method used to collect data is so important to inference, and how we use probability, under the correct conditions, to calculate a margin of error to quantify our uncertainty. At the end of the chapter, you'll see why the AMA changed its report.

# Learning about the World through Surveys

Surveys are probably the most often encountered application of statistics. Most news shows, newspapers, and magazines report on surveys or polls several times a week—and during a major election, several times a day. We can learn quite a bit through a survey if the survey is done correctly.

## Survey Terminology

A **population** is a group of objects or people we wish to study. Usually, this group is large—say, the group of all U.S. citizens, or all U.S. citizens between the ages of 13 and 18, or all senior citizens. However, it might be smaller, such as all phone calls made on your cell phone in January. We wish to know the value of a **parameter**, a numerical value that characterizes some aspect of this *population*. For example, political pollsters want to know the percentage of people who say they will vote in the next election. Drunk-driving opponents want to know the percentage of all teenagers with driver's licenses who have drunk alcohol while driving. Designers of passenger airplanes want to know the mean length of passengers' legs so that they can put the seats as close together as possible without having discomfort.

In this book we focus on two frequently used parameters: the mean of a population and the population proportion. This chapter deals with population proportions.

If the population is relatively small, we can find the exact value of the parameter by conducting a census. A **census** is a survey in which every member of the population is measured. For example, if you wish to know the percentage of people in your classroom who are left-handed, you can perform a census. The classroom is the population, and the parameter is the percentage of left-handers. We sometimes try to take a census with a large population (such as the U.S. Census), but such undertakings are too expensive for nongovernmental organizations and are filled with complications caused by trying to track down and count people who may not want to be found. (For example, the U.S. Census tends to undercount poor, urban-dwelling residents, as well as undocumented immigrants.)

In fact, most populations we find interesting are too large for a census. For this reason, we instead observe a smaller sample. A **sample** is a collection of people or objects taken from the population of interest.

Once a sample is collected, we measure the characteristic we're interested in. A **statistic** is a numerical characteristic of a sample of data. We use statistics to estimate parameters. For instance, we might be interested in knowing what proportion of all registered voters will vote in the next national election. The proportion of all registered voters who will vote in the next election is our *parameter*. Our method to estimate this parameter is to survey a small sample. The proportion of the sample who say they will vote in the next election is a *statistic*.

Statistics are sometimes called **estimators**, and the numbers that result are called **estimates**. For example, our *estimator* is the proportion of people in a sample who say they will vote in the next election. When we conduct this survey, we find, perhaps, that 0.75 of the sample say they will vote. This number, 0.75, is our *estimate*.

**KEY POINT**  A statistic is a number that is based on data and used to estimate the value of a characteristic of the population. Thus it is sometimes called an estimator.

**Statistical inference** is the art and science of drawing conclusions about a population on the basis of observing only a small subset of that population. Statistical inference always involves uncertainty, so an important component of this science is measuring our uncertainty.

## EXAMPLE 1 Pew Poll: Age and the Internet

In January 2009, the Pew Research Center surveyed 1650 adult Internet users to estimate the proportion of Internet users between 18 and 32 years old (Gen Y "millenials"). Common wisdom holds that this group of young adults is among the heavier users of the Internet. The survey found that 30% of the respondents in the sample were between ages 18 and 32.

**QUESTIONS**  Identify the population and sample. What is the parameter of interest? What is the statistic?

**CONCLUSION**  The *population* we wish to study is all American adults who are Internet users. The *sample*, which is taken from the population, consists of the 1650 Internet users surveyed. The *parameter* of interest is the percentage of all Internet users who are in Gen Y (18 to 32 years old). The *statistic*, which is the percentage of the sample who are in Gen Y, is 30%.

**TRY THIS!**  Exercise 7.1

An important difference between statistics and parameters is that statistics are knowable. Any time we collect data, we can find the value of a statistic. In Example 1, we know that 30% of the people surveyed are between 18 and 32 years old. In contrast, a parameter is typically unknown. We do not know for certain the percentage of *all* Internet users between 18 and 32. The only way to find out would be to ask all adult Internet users, and we have neither the time nor the money to do this. Table 7.1 compares the known and the unknown in this situation.

| Unknown | Known |
| --- | --- |
| *Population*<br>All adult American Internet users | *Sample*<br>A small number of adult American Internet users |
| *Parameter*<br>Percentage of all Internet users who are in Gen Y<br>Mean age of all Americans | *Statistic*<br>Percentage of sample who are in Gen Y<br>Mean age of sample |

◀ **TABLE 7.1** Some examples of unknown quantities we might wish to estimate, and their knowable counterparts.

Statisticians have developed notation for keeping track of parameters and statistics. In general, Greek characters are used to represent population parameters. For example, $\mu$ (mu, pronounced "mew," like the beginning of *music*) represents the mean of a population. Also, $\sigma$ (sigma) represents the standard deviation of a population. Statistics (estimates based on a sample) are represented by English letters: $\bar{x}$

> **↻ Looking Back**
>
> **The Sample Proportion: $\hat{p}$**
> Recall that $\hat{p}$, pronounced "p-hat," is the sample proportion, which is the number of successes divided by the number of trials.

(pronounced "*x*-bar") is the mean of a sample, and *s* is the standard deviation of a sample, for instance.

One frequently encountered exception is the use of the letter *p* to represent the proportion of a population and $\hat{p}$ (pronounced "*p*-hat") to indicate the proportion of a sample. Table 7.2 summarizes this notation. You've seen most of these symbols before, but this table organizes them in a new way that is important for statistical inference.

► **TABLE 7.2** Notation for some commonly used statistics and parameters.

| Statistics (based on data) | | Parameters (typically unknown) | |
|---|---|---|---|
| Sample mean | $\bar{x}$ (x-bar) | Population mean | $\mu$ (mu) |
| Sample standard deviation | $s$ | Population standard deviation | $\sigma$ (sigma) |
| Sample variance | $s^2$ | Population variance | $\sigma^2$ |
| Sample proportion | $\hat{p}$ (p-hat) | Population proportion | $p$ |

## What Could Possibly Go Wrong? The Problem of Bias

Unfortunately, it is far easier to conduct a bad survey than to conduct a good survey. One of the many ways in which we can reach a wrong conclusion is to use a survey method that is biased.

A method is **biased** if it has a tendency to produce an untrue value. Bias can enter a survey in three ways. The first is through **sampling bias**, which results from taking a sample that is not representative of the population. A second way is **measurement bias**, which comes from asking questions that do not produce a true answer. For example, if we ask people their income, they are likely to inflate the value. In this case, we will get a positive (or "upwards") bias: Our estimate will tend to be too high. Measurement bias occurs when measurements tend to record values larger (or smaller) than the true value.

The third way occurs because some statistics are naturally biased. For example, if you use the statistic $10\bar{x}$ to estimate the mean, you'll typically get estimates that are ten times too big. Therefore, even when no measuring or sampling bias is present, you must also take care to use an estimator that is not biased.

> ⚠ **Caution**
>
> **Bias**
> Statistical bias is different from the everyday use of the term *bias*. You might perhaps say a friend is biased if she has a strong opinion that affects her judgment. In statistics, bias is a way of measuring the performance of a method over many different applications.

### Measurement Bias

In February 2010, the *Albany Times Union* newspaper reported on two recent surveys to determine the opinions of New York State residents on taxing soda. The Quinnipiac University Polling Institute asked, "There is a proposal for an 'obesity tax' or a 'fat tax' on non-diet sugary soft drinks. Do you support or oppose such a measure?" Forty percent of respondents said they supported the tax. Another firm, Kiley and Company, asked, "Please tell me whether you feel the state should take that step in order to help balance the budget, should seriously consider it, should consider it only as a last resort, or should definitely not consider taking that step: 'Imposing a new 18 percent tax on sodas and other soft drinks containing sugar, which would also reduce childhood obesity.'" Fifty-eight percent supported the tax when asked this question. One or both of these surveys have measurement bias.

A famous example occurred in 1993, when, on the basis of the results of a Roper Organization poll, many U.S. newspapers published headlines similar to this one from the *New York Times*: "1 in 5 in New Survey Express Some Doubt About the Holocaust" (April 20, 1993). Almost a year later, the *New York Times* reported that this alarmingly high percentage of alleged Holocaust doubters could be due to measurement error. The actual question respondents were asked contained a double negative: "Does it seem possible, or does it seem impossible to you, that the Nazi extermination of the Jews never happened?" When Gallup repeated the poll but did not use a double negative, only 9% expressed doubts (*New York Times* 1994).

## Sampling Bias

Writing good survey questions to reduce measurement bias is an art and a science. This book, however, is more concerned with sampling bias, which occurs when the estimation method uses a sample that is not representative of the population. (By "not representative" we mean that the sample is fundamentally different from the population.)

Have you ever heard of Alfred Landon? Unless you're a political science student, you probably haven't. In 1936, Landon was the Republican candidate for U.S. president, running against Franklin Delano Roosevelt. The *Literary Digest*, a popular news magazine, conducted a survey with over 10 million respondents and predicted that Landon would easily win the election with 57% of the vote. The fact that you probably haven't heard of Landon suggests that he didn't win, and in fact, he lost big, setting a record at the time for the lowest number of electoral votes received by a major-party candidate. What went wrong? The *Literary Digest* had a biased sample. The journal relied largely on polling its own readers, and its readers were more well-to-do than the general public and more likely to vote for a Republican. The reputation of the *Literary Digest* was so damaged that two years later it disappeared and was absorbed into *Time* magazine.

The U.S. presidential elections of 2004 and 2008 both had candidates who claimed to have captured the youth vote, and both times, candidates claimed the polls were biased. The reason given was that the surveys used to estimate candidate support relied on landline phones, and many young voters don't own land lines, relying instead on their cell phones. Reminiscent of the 1936 *Literary Digest* poll, these surveys were potentially biased because their sample systematically excluded an important part of the population: those who do not use land lines (Cornish, 2007).

In fact, the Pew Foundation conducted a study after the 2010 congressional elections. This study found that polls that excluded cell phones had a sampling bias in favor of Republican candidates.

Today, the most commonly encountered biased-sample surveys are probably Internet polls. These can be found on many news organization websites. ("Would You Want an Asteroid Named after You?" www.cnn.com, October 3, 2007; "Who Would Wall Street Like to See in the White House in 2008: A Democrat or a Republican?" www.foxnews.com, October 7, 2007). Internet polls suffer from what is sometimes called **voluntary-response bias**. People tend to respond to such surveys only if they have strong feelings about the results; otherwise, why bother? This implies that the sample of respondents is not necessarily representative of the population. Even if the population in this case is, for example, all readers of the Foxnews.com website, the survey may not accurately reflect their views, because the voluntary nature of the survey means the sample will probably be biased. This bias might be even worse if we took the population to be all U.S. residents. Readers of Internet websites may very well not be representative of all U.S. residents, and readers of particular websites such as Fox or CNN might be even less so.

To warn readers of this fact, most Internet polls have a disclaimer: "This is not a scientific poll." What does this mean? It means we should not trust the information reported to tell us anything about anyone other than the people who responded to the poll. (And remember, we can't even trust the counts on an Internet poll, because sometimes nothing prevents people from voting many times.)

A more subtle form of bias happens when those being surveyed fail to answer a question or respond to a survey. Such **nonresponse bias** occurs when people selected to participate in the survey refuse to do so. Refusal can happen for mundane reasons (they're eating dinner when the pollster calls) or for reasons that might have a more serious effect on the results (because people are embarrassed or offended by the question). Those who choose not to respond might have different views about the survey's topics than those who do. For example, questions about income might be refused by those who are embarrassed by what they feel is too low, or too high, an income. If a large percentage declines to participate, then a biased survey could result.

When reading about a survey, it is important to know

1. what percentage of people who were asked to participate actually did so
2. whether the researchers chose people to participate in the survey or people themselves chose to participate.

If a large percentage chosen to participate refused to answer questions, nonresponse bias could result. If people themselves chose whether to participate, there will be voluntary-response bias. If a survey suffers from these problems, its conclusions are suspect.

## Simple Random Sampling Saves the Day

How do we collect a sample that has as little bias as possible and is representative of the population? Only one way works: to take a random sample.

As we explained in Chapter 5, statisticians have a precise definition of *random*. A random sample does not mean that we stand on a street corner and stop whomever we like to ask them to participate in our survey. (Statisticians call this a **convenience sample**, for obvious reasons.) A random sample must be taken in such a way that every person in our population is equally likely to be chosen.

A true random sample is difficult to achieve. (That's a big understatement!) Pollsters have invented many clever ways of pulling this off, often with great success. One basic method that's easy to understand but somewhat difficult to put into practice is **simple random sampling (SRS)**.

In SRS, we draw subjects from the population at random and without replacement. **Without replacement** means that once a subject is selected for a sample, that subject cannot be selected again. This is like dealing cards from a deck. Once a card is dealt for a hand, no one else can get the same card. A result of this method is that every sample of the same fixed size is equally likely to be chosen. As a result, we can produce unbiased estimations of the population parameters of interest and can measure the precision of our estimator.

It can't be emphasized enough that if our sample is not random, there's really nothing we can learn about the population. We can't measure a survey's precision, and we can't know how large or small the bias might be. An unscientific survey is a useless survey for the purposes of learning about a population.

In theory, we can take an SRS by assigning a number to each and every member of the population. We then use a random number table or other random number generator to select our sample, ignoring numbers that appear twice.

> **Details**
>
> Simple random sampling is not the only valid method for statistical inference. Statisticians collect representative samples using other methods, as well (for example, sampling with replacement). What these methods all have in common is that they take samples randomly.

## EXAMPLE 2  Taking a Simple Random Sample

Alberto, Justin, Michael, Audrey, Brandy, and Nicole are in a class.

QUESTION  Select an SRS of three names from these six names.

SOLUTION  First assign each person a number, as shown:

| | |
|---|---|
| Alberto | 1 |
| Justin | 2 |
| Michael | 3 |
| Audrey | 4 |
| Brandy | 5 |
| Nicole | 6 |

Next, select three of these numbers without replacement. Figure 7.1 shows how this is done in StatCrunch, and almost all statistical technologies let you do this quite easily.

Tech

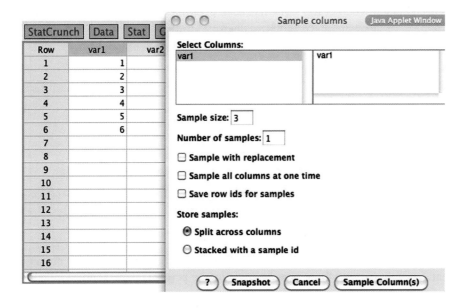

◀ **FIGURE 7.1** StatCrunch will randomly select, without replacement, three numbers from the six shown in the var1 column.

Using technology, we got these three numbers: 1, 2, and 6. These correspond to Alberto, Justin, and Nicole.

If technology is not available, a random number table, such as the one provided in Appendix A, can be used. Here are two lines from such a table:

$$77598 \quad 29511 \quad 98149 \quad 63991$$
$$31942 \quad 04684 \quad 69369 \quad 50814$$

You can start at any row or column you please. Here, we choose to start at the upper left (shown in bold face). Next, read off digits from left to right, skipping digits that are not in our population. Because no one has the number 7, skip this number, twice. The first person selected is number 5: Brandy. Then skip 9 and 8 and select number 2: Justin. Skip 9 and 5 (because you already selected Brandy) and select number 1: Alberto.

**CONCLUSION** Using technology, our sample consisted of Alberto, Justin, and Nicole. Using the random number table, we got a different sample: Brandy, Justin, and Alberto.

**TRY THIS!** Exercise 7.11

## Sampling in Practice

In practice, simple random samples are difficult to collect and often inefficient. In most situations, we can't make a list of all people in the United States and assign each of them a number. To get around this, statisticians use alternative techniques.

In addition, random sampling does not cure all ills. Nonresponse bias can still be a problem, and the possibility always exists that methods of taking the random sample are flawed (as they are if only landline telephones are used when many in the population use cell phones).

## EXAMPLE 3  Survey on Sexual Harassment

A newspaper at a large college wants to determine whether sexual harassment is a problem on campus. The paper takes a simple random sample of 1000 students and asks each person whether he or she has been a victim of sexual harassment on campus. About 35% of those surveyed refuse to answer. Of those who do answer, 2% say they have been victims of sexual harassment.

**QUESTION**  Give a reason why we should be cautious about using the 2% value as an estimate for the population percentage of those who have been victims of sexual harassment.

**CONCLUSION**  There is a large percentage of students who did not respond. Those who did not respond might be different from those who did, and if their answers had been included, the results could have been quite different. When those surveyed refuse to respond, it can create a biased sample.

**TRY THIS!**  Exercise 7.17

There are always some people who refuse to participate in a survey, but a good researcher will do everything possible to keep the percentage of nonresponders as small as possible, to reduce this source of bias.

National Public Radio reported that a relatively large number of people have dropped their land lines and use only cell phones. Cell phone users are less likely to participate in polls, since federal law forbids the use of automated dialing for cell phone numbers. Because research shows that people who use cell phones rather than land lines tend to be different from the general population, bias can occur. In the 2008 Democratic primaries, the Barack Obama campaign feared Obama was polling second to Hillary Clinton because Obama's supporters were younger and more likely not to have land lines. Therefore, Obama supporters were less likely to be polled, which led to a bias. Another problem is that cell phone users are less likely to participate in a poll, in part because they must pay for the phone call, but also because they are more likely to be doing something else— driving a car, attending a meeting—than people using land lines at home.

SECTION 7.2

# Measuring the Quality of a Survey

A frequent complaint about surveys is that a survey based on 1000 people can't possibly tell us what the entire country is thinking. This complaint raises interesting questions: How do we judge whether our estimators are working? What separates a good estimation method from a bad?

It's difficult, if not impossible, to judge whether any particular survey is good or bad. Sometimes we can find obvious sources of bias, but often we don't know whether a survey has failed unless we later learn the true parameter value. (This sometimes occurs in elections, when we learn that a survey must have had bias because it severely missed predicting the actual outcome.) Instead, statisticians evaluate the *method* used to estimate a parameter, not the outcome of a particular survey.

**KEY POINT**  Statisticians evaluate the method used for a survey, not the outcome of a single survey.

Before we talk about how to judge surveys, imagine the following scenario: We are not taking just one survey of 1000 randomly selected people. We are sending out an army of pollsters. Each pollster surveys a random sample of 1000 people, and they all use the same method for collecting the sample. Each pollster asks the same question and produces an estimate of the proportion of people in the population who would answer yes to the question. When the pollsters return home, we get to see not just a single estimate (as happens in real life) but a great many estimates. Because each estimate is based on a separate random collection of people, each one will differ slightly. We expect some of these estimates to be closer to the mark than others just because of random variation. What we really want to know is how the group did as a whole. For this reason, we talk about evaluating *estimation methods*, not estimates.

An estimation method is a lot like a golfer. To be a good golfer, we need to get the golf ball in the cup. A good golfer is both *accurate* (tends to hit the ball near the cup) and *precise* (even when she misses, she doesn't miss by very much.)

It is possible to be precise and yet be inaccurate, as shown in Figure 7.2b. Also, it is possible to aim in the right direction (be accurate) but be imprecise, as shown in Figure 7.2c. (Naturally, some of us are bad at both, as shown in Figure 7.2d.) But the best golfers can both aim in the right direction and manage to be very consistent, which Figure 7.2a shows us.

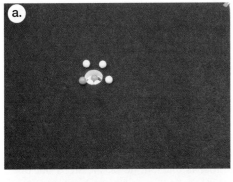

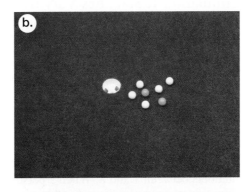

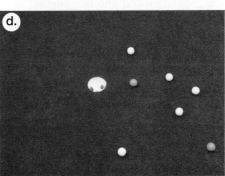

◀ **FIGURE 7.2 a.** Shots from a golfer with good aim and precision; the balls are tightly clustered and centered around the cup. **b.** Shots from a golfer with good precision but poor aim; the balls are close together but centered to the right of the cup. **c.** Shots from a golfer with good aim—the balls are centered around the cup—but bad precision. **d.** The worst-case scenario: bad precision *and* bad aim.

Think of the cup as the population parameter, and think of each golf ball as an estimate, a value of $\hat{p}$, that results from a different survey. We want an estimation method that aims in the right direction. Such a method will, on average, get the correct value of the population parameter. We also need a precise method so that if we repeated the survey, we would arrive at nearly the same estimate.

The aim of our method, which the *accuracy,* is measured in terms of the *bias.* The *precision* is measured by a number called the *standard error.* Discussion of simulation studies in the next sections will help clarify how accuracy and precision are measured. These simulation studies show how bias and standard error are used to quantify the uncertainty in our inference.

## Using Simulations to Understand the Behavior of Estimators

The three simulations that follow will help measure how well the sample proportion works as an estimator of the population proportion.

In the first simulation, imagine doing a survey of 4 people in a very small population with only 8 people. You'll see that the estimator of the population proportion is accurate (no bias) but, because of the small sample size, not terribly precise.

In the second simulation, the first simulation is repeated, using a larger population and sample. The estimator is still unbiased, and you will see a perhaps surprising change in precision. Finally, the third simulation will reveal that using a much larger sample size makes the result even more precise.

To learn how our estimation method behaves, we're going to create a very unusual, unrealistic situation: We're going to create a world in which we know the truth. In this world, there are two types of people: those who like dogs and those who like cats. No one likes both. Exactly 25% of the population are Cat People, and 75% are Dog People. We're going to take a random sample of people from this world and see what proportion of our sample are Cat People. Then we'll do it again. And again. From this repetition, we'll see some interesting patterns emerge.

### Simulation 1: Statistics Vary from Sample to Sample

To get started, let's create a very small world. This world has 8 people named 1, 2, 3, 4, 5, 6, 7, and 8. People 1 and 2 are Cat People.

▶ **FIGURE 7.3** The entire population of our simulated world; 25% are Cat People.

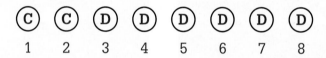

From this population, we use the random number table to generate four random numbers between 1 and 8. When a person's number is chosen, he or she steps out of the population and into our sample.

Before we tell who was selected, think for a moment about what you expect to happen. What proportion of our sample will be Cat People? Is it possible for 0% of the sample to be Cat People? For 100%?

Below is our random sample. Note that we sampled without replacement, as in a real survey. We don't want the same person to be in our sample twice.

| 6 | 8 | 4 | 5 |
|---|---|---|---|
| D | D | D | D |

None of those selected are Cat People, as Figure 7.4 indicates. The proportion of Cat People in our sample is 0%. We call this the *sample proportion* because it comes from the sample, not the population.

▶ **FIGURE 7.4** The first sample, which has 0% Cat People.

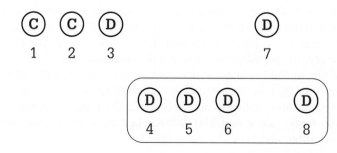

Let's take another random sample. It is possible that we will again get 0%, but it is also possible that we will get a different percentage.

| 7 | 2 | 6 | 3 |
|---|---|---|---|
| D | C | D | D |

This time, our sample proportion is 25%.
One more time:

| 2 | 8 | 6 | 5 |
|---|---|---|---|
| C | D | D | D |

Again, our sample proportion is 25%.

Table 7.3 shows what has happened so far. Even though we have done only three repetitions, we can make some interesting observations.

| Repetition | Population Parameter | Sample Statistics |
|:---:|:---:|:---:|
| 1 | $p = 25\%$ Cat People | $\hat{p} = 0\%$ Cat People |
| 2 | $p = 25\%$ Cat People | $\hat{p} = 25\%$ Cat People |
| 3 | $p = 25\%$ Cat People | $\hat{p} = 25\%$ Cat People |

◀ **TABLE 7.3** The results of three repetitions of our simulation.

First, notice that the population proportion, $p$, never changes. It can't, because in our made-up world, the population always has the same 8 people and the same 2 are Cat People. However, the sample proportion, $\hat{p}$, can be different in each sample. In fact, $\hat{p}$ is random, because it depends on a random sample.

 **KEY POINT**  No matter how many different samples we take, the value of $p$ (the population proportion) is always the same, but the value of $\hat{p}$ changes from sample to sample.

This simulation is, in fact, a random experiment and $\hat{p}$ is our outcome. Because it is random, $\hat{p}$ has a probability distribution. The probability distribution of $\hat{p}$ has a special name: **sampling distribution**. This term reminds us that $\hat{p}$ is not just any random outcome; it is a statistic we use to estimate a population parameter.

Because our world has only 8 people in it and we are taking samples of 4 people, we can write down all of the possible outcomes. There are only 70. By doing this, we can see exactly how often $\hat{p}$ will be 0%, how often 25%, and how often 50%. (Notice that it can never be more than 50%.) These probabilities are listed in Table 7.4, which presents the sampling distribution for $\hat{p}$. Figure 7.5 visually represents this sampling distribution.

| Value of | Probability of Seeing That Value |
|:---:|:---:|
| 0% | 0.21429 |
| 25% | 0.57143 |
| 50% | 0.21429 |

▲ **TABLE 7.4** The sampling distribution for $\hat{p}$, based on our random sample.

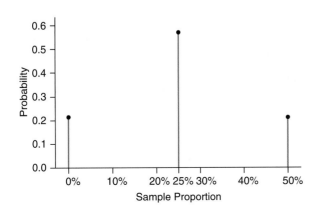

◀ **FIGURE 7.5** Graphical representation of Table 7.4, the sampling distribution for $\hat{p}$ when $p$ is 0.25.

From Table 7.4 and Figure 7.5, we learn several things:

1. Our estimator, $\hat{p}$, is not always the same as our parameter, $p$. Sometimes $\hat{p}$ turns out to be 0%, sometimes it is 50%, and sometimes it hits the target value of 25%.

2. The mean of this distribution is 25%—the same value as $p$.

3. Even though $\hat{p}$ is not always the "true" value, $p$, we are never more than 25 percentage points away from the true value.

Why are these observations important? Let's consider each one separately.

The first observation reminds us that statistics based on random samples are random. Thus, we cannot know ahead of time, with certainty, exactly what estimates our survey will produce.

The second observation tells us that our estimator has no bias—that, *on average*, it is the same as the parameter value. **Bias** is measured as the distance between the mean value of the estimator (the center of the sampling distribution) and the population parameter. In this case, the center of the sampling distribution and the population parameter are both 0.25, so the distance is 0. In other words, there is no bias.

The third observation is about precision. We know that our estimator is, on average, the same as the parameter, but the sampling distribution tells us how far away, typically, the estimator might stray from average. **Precision** is reflected in the spread of the sampling distribution and is measured by using the standard deviation of the sampling distribution. In this simulation, the standard deviation is 0.16366, or roughly 16%. The standard deviation of a sampling distribution has a special name: the **standard error (SE)**.

The standard error measures how much our estimator typically varies from sample to sample. Thus, in the above example, if we survey 4 people, we usually get 25% Cat People, but this typically varies by plus or minus 16.4% (16.4 percentage points). Looking at the graph in Figure 7.5, we might think that the variability is typically plus or minus 25 percentage points, but we must remember that the standard deviation measures how spread out observations are from the average value. Many observations are identical to the average value, so the typical, or "standard," deviation from average is only 16.4 percentage points.

 **KEY POINT**    Bias is measured using the center of the sampling distribution: It is the distance between the center and the population value.

Precision is measured using the standard deviation of the sampling distribution, which is called the standard error. When the standard error is small, we say the estimator is precise.

 **SNAPSHOT**  **SAMPLING DISTRIBUTION**

| | |
|---|---|
| **WHAT IS IT?** ▶ | A special name for the probability distribution of a statistic. |
| **WHAT DOES IT DO?** ▶ | Gives us probabilities for a statistic. |
| **WHAT IS IT USED FOR?** ▶ | It tells us how often we can expect to see particular values of our estimator, and it also gives us important characteristics of the estimator, such as bias and precision. |
| **HOW IS IT USED?** ▶ | It is used for making inferences about a population. |

# Simulation 2: The Size of the Population Does Not Affect Precision

The first simulation was very simple, because our made-up world had only 8 people. Let's consider a more realistic example with a world that has 1000 people.

In our first simulation, the bias was 0, which is good; this means we have an accurate estimator. However, the precision was fairly poor (we had a large standard error). How can we improve precision? To understand, we need a slightly more realistic simulation.

This time, we'll use the same world but make it somewhat bigger. Let's assume we have 1000 people and 25% are Cat People ($p = 0.25$). (In other words, there are 250 Cat People.) We take a random sample of 10 people and find the sample proportion, $\hat{p}$, of Cat People.

Because we've already seen how this is done, we're going to skip a few steps and show the results. This time the different outcomes are too numerous to list, so instead we just do a simulation:

1. Take a random sample, without replacement, of 10 people.

2. Calculate $\hat{p}$: the proportion of Cat People in our sample.

3. Repeat steps 1 and 2 a total of 10,000 times. Each time, calculate $\hat{p}$ and record its value.

Here are our predictions:

1. We predict that $\hat{p}$ will not be the same value every time because it is based on a random sample, so the value of $\hat{p}$ will vary randomly.

2. We predict that the mean outcome, the typical value for $\hat{p}$, will be 25%—the same as the population parameter—because our estimator is unbiased.

3. Precision: This one is left to you. Do you think the result will be more precise or less precise than in the last simulation? In the last simulation, only 4 people were sampled and the variation, as measured by the standard error, was about 16%. This time more people (10) are being sampled, but the population is much larger (1000). Will the standard error be larger (less precise) or smaller (more precise) than 16%?

Figure 7.6 shows an estimate of the sampling distribution. Figure 7.6 is not the actual sampling distribution because the histogram is based on a simulation, not a list of all possible outcomes. The relative frequencies in the figure are estimates of the probabilities, not the true probabilities. Still, with 10,000 replications, it is a very good approximation of the actual sampling distribution.

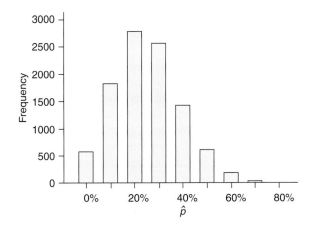

◀ **FIGURE 7.6** Simulation results for $\hat{p}$. This histogram is a simulation of the sampling distribution. The true value of $p$ is 25%. Each sample is based on 10 people, and we repeated the simulation 10,000 times.

The center of the estimated distribution is at 0.2501, which indicates that essentially no bias exists, because the population parameter is 0.25.

The outcomes ranged from $\hat{p} = 0$ to $\hat{p} = 0.8$. We can estimate the standard error by finding the standard deviation of our simulated $\hat{p}$. This turns out to be about 13.56%.

The value of the standard error tells us that if we were to take another sample of 10 people, we would expect to get about 25% Cat People, give or take 13.6 percentage points.

From Figure 7.6 we learn important information:

1. The bias of $\hat{p}$ is still 0, even though we used a larger population and a larger sample.

2. The variation of $\hat{p}$ is less; this estimator is more precise, even though the population is larger. In general, as long as the population is large relative to the sample size, the precision has *nothing* to do with the size of the *population*, but only with the size of the *sample*.

Many people are surprised to learn that precision is not affected by population size. How can the level of precision for a survey in a town of 10,000 people be the same as for one in a country of 210 *million* people?

Figure 7.7 provides an analogy. The bowls of soup represent two populations: a big one (a country, perhaps) and a small one (a city). Our goal is to taste each soup (take a sample from the population) to judge whether we like it. If both bowls are well stirred, the size of the bowl doesn't matter—using the same-size spoon, we can get the same amount of taste from either bowl.

▶ **FIGURE 7.7** The bowls of soup represent two populations and the sample size is represented by the spoons. The precision of an estimate depends only on the size of the sample, not the size of the population.

 **KEY POINT**

The precision of an estimator does not depend on the size of the population; it depends only on the sample size. An estimator based on a sample size of 10 is just as precise in a population of 1000 people as in a population of a million.

## Simulation 3: Large Samples Produce More Precise Estimators

Let's do one more simulation. This time the only change we'll make is to increase the sample size. Instead of sampling 10 people, we'll sample 100. The population size is still 1000, and Cat People still represent 25% of the population. How will the precision of the simulated sampling distribution differ from Figure 7.6, where we sampled only 10 people at a time?

Figure 7.8 shows the result. Note that the center of this estimated sampling distribution is still at 25%. Also, our estimation method remains unbiased. However, the shape looks pretty different. First, because many more outcomes are possible for $\hat{p}$, this histogram looks as though it belongs more to a continuous-valued random outcome than to a discrete value. Second, it is much more symmetric than Figure 7.6. You will see in Section 7.3 that the shape of the sampling distribution of $\hat{p}$ depends on the size of the random sample.

An important point to note is that this estimator is much more precise because it uses a larger sample size. By sampling more people, we get more information, so we can end up with a more precise estimate. The estimated standard error, which is simply the standard deviation of the data shown in Figure 7.8, is now 4.2 percentage points.

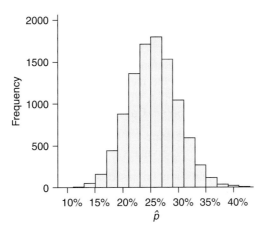

◄ **FIGURE 7.8** Simulated sampling distribution of a sample proportion of Cat People, based on a random sample of 100 people. The simulation was repeated 10,000 times.

Table 7.5 shows a summary of the three simulations.

| Simulation | Population Size | Sample Size | | Mean | Standard Error | |
|:---:|:---:|:---:|:---:|:---:|:---:|:---:|
| 1 | 8 | 4 | Increasing Sample Size | 25% | 16.4% | Increasing Precision |
| 2 | 1000 | 10 | | 25% | 13.5% | |
| 3 | 1000 | 100 | | 25% | 4.2% | |

◄ **TABLE 7.5** Increasing sample size results in increasing precision (measured as decreasing standard error).

Here is what we learned from Figure 7.8, which is based on sample sizes of 100 "people":

1. The estimator $\hat{p}$ is unbiased for all sample sizes (as long as we take random samples).

2. The precision improves as the sample size gets larger.

3. The shape of the sampling distribution is more symmetric for larger sample sizes.

**KEY POINT**    Surveys based on larger sample sizes have smaller standard error (SE) and therefore better precision. Increasing the sample size improves precision.

## Finding the Bias and the Standard Error

We've shown how to estimate bias and precision by running a simulation. But we can also do this mathematically, without running a simulation. Bias and standard error are easy to find for a sample proportion under certain conditions.

The bias of $\hat{p}$ is 0, and the standard error is

$$SE = \sqrt{\frac{p(1-p)}{n}}$$

if the following two conditions are met:

**Condition 1.** The sample must be randomly selected from the population of interest, either with or without replacement. The population parameter to be estimated is the proportion of people (or objects) with some characteristic. This proportion is denoted as $p$.

**Condition 2.** If the sampling is without replacement, the population needs to be much larger than the sample size; at least 10 times bigger is a good rule of thumb.

## EXAMPLE 4 Pet World

Suppose that in Pet World, the population is 1000 people and 25% of the population are Cat People. Cat People love cats but hate dogs. We are planning a survey in which we take a random sample of 100 people, without replacement. We calculate the proportion of people in our sample who are Cat People.

**QUESTION** What value should we expect for our sample proportion? What's the standard error? How do we interpret these values?

**SOLUTION** The sample proportion is unbiased, so we expect it to be the same as the population proportion: 25%.

$$\text{The standard error is } \sqrt{\frac{p(1 - p)}{n}} = \sqrt{\frac{0.25 \times 0.75}{100}} = \sqrt{\frac{0.1875}{100}}$$

$$= \sqrt{0.001875} = 0.04330, \text{ or about } 4.3\%.$$

This formula is appropriate because the population size is big with respect to the sample size. The population size is 1000, and the sample size is 100; $100 \times 10 = 1000$, so the population is 10 times larger than the sample size.

**CONCLUSION** We interpret the values to mean that if we were to take a survey of 100 people from Pet World, we would expect about 25% of them to be Cat People, give or take about 4.3%. The "give or take" means that if you were to draw a sample of 100 and I were to draw a sample of 100, our sample proportions would typically differ from the expected 25% by about 4.3 percentage points.

**TRY THIS!** Exercise 7.29

## Real Life: We Get Only One Chance

In simulations, we could repeat the survey many times to understand what might happen. In real life, we get just one chance. We take a sample, calculate $\hat{p}$, and then have to live with it.

It is important to realize that bias and precision are both measures of what happens if we could repeat our survey many times. Bias indicates the typical outcome of surveys repeated again and again. If the bias is 0, we will usually get the right value. If the bias is 0.10, then our estimate will characteristically be 10 percentage points too high. Precision measures how much our estimator will vary from the typical value if we do the survey again. To put it slightly differently, if someone else does the survey, precision helps determine how different her or his estimate could be from ours.

How small must the standard error be for a "good" survey? The answer varies, but the basic rule is that the precision should be small enough to be useful. A typical election poll has a sample of roughly 1000 registered voters and a standard error of about 1.5 percentage points. If the candidates are many percentage points apart, this is good precision. However, if they are neck and neck, this might not be good enough. In Section 7.4, we will discuss how to make decisions about whether the standard error is small enough.

In real life, we don't know the true value of the population proportion, $p$. This means we can't calculate the standard error. However, we can come pretty close by using the sample proportion. If $p$ is unknown, then

$$SE_{\text{est}} = \sqrt{\frac{\hat{p}(1 - \hat{p})}{n}}, \text{ where } SE_{\text{est}} \text{ is the estimated standard error,}$$

is a useful approximation to the true standard error.

SECTION 7.3

# The Central Limit Theorem for Sample Proportions

Remember that a probability tells us how often an event happens if we repeat an experiment an infinite number of times. For instance, the sampling distribution of $\hat{p}$ gives the probabilities of where our sample proportions will fall; that is, it tells us how often we would see particular values of $\hat{p}$ if we could repeat our survey infinitely many times. In the simulation, we repeated our fake survey 10,000 times. Ten thousand is a lot, but it's a far cry from infinity.

In the three simulations in Section 7.2, we saw that the shape of the sampling distribution (or our estimated version, based on simulations) changed as the sample size increased (compare Figures 7.5, 7.6, and 7.8). If we used an even larger sample size than 100 (the sample size for the last simulation), what shape would the sampling distribution have? As it turns out, we don't need a simulation to tell us. For this statistic, and for some others, a mathematical theorem called the **Central Limit Theorem (CLT)** gives us a very good approximation of the sampling distribution without our needing to do simulations.

The Central Limit Theorem is helpful because sampling distributions are important. They are important because they, along with the bias and standard error, enable us to measure the quality of our estimation methods. Sampling distributions give us the probability that an estimate falls a specified distance from the population value. For example, we don't want to know simply that 18% of our customers are likely to buy new cell phones in the next year. We also want to know the probability that the true percentage might be higher than some particular value, say, 25%.

## Meet the Central Limit Theorem for Sample Proportions

The Central Limit Theorem has several versions. The one that applies to estimating proportions in a population tells us that if some basic conditions are met, then the sampling distribution of the sample proportion is close to the Normal distribution.

More precisely, when estimating a population proportion, $p$, we must have the same conditions that were used in finding bias and precision, but with one new condition:

Condition 1. *Random and Independent.* The sample is collected randomly from the population and observations are independent at each other. The sample can be collected either with or without replacement.

Condition 2. *Large Sample.* The sample size, $n$, is large enough that the sample expects at least 10 successes (yes's) and 10 failures (no's).

Condition 3. *Big Population.* If the sample is collected without replacement, then the population size must be much (at least 10 times) bigger than the sample size.

The sampling distribution for $\hat{p}$ is then approximately Normal, with mean $p$ (the population proportion) and standard deviation the same as the standard error:

$$SE = \sqrt{\frac{p(1 - p)}{n}}$$

**Looking Back**

**Normal Notation**
Recall that the notation N(mean, standard deviation) designates a particular Normal distribution.

KEY POINT The Central Limit Theorem for Sample Proportions tells us that if we take a random sample from a population, and if the sample size is large and the population size much larger than the sample size, then the sampling distribution of $\hat{p}$ is approximately

$$N\left(p, \sqrt{\frac{p(1-p)}{n}}\right)$$

If you don't know the value of $p$, then you can substitute the value of $\hat{p}$ to calculate the estimated standard error.

Figure 7.9 illustrates the CLT for proportions. Figure 7.9a is based on simulations in which the sample size was just 10 people, which is too small for the CLT to apply. In this case, the simulated sampling distribution does not look Normal; it is right-skewed and has large gaps between values. Figure 7.9b is based on simulations of samples of 100 observations. Because the true population proportion is $p = 0.25$, a sample size of 100 is large enough for the CLT to apply, and our simulated sampling distribution looks very close to the Normal model. Figure 7.9b is actually a repeat of Figure 7.8 with the Normal curve superimposed. Now that the graphs are shown on the same scale, we can see that the sample size of 100 gives better precision than the sample size of 10—the distribution is narrower.

▶ **FIGURE 7.9 a.** Revision of Figure 7.6, a histogram of 10,000 sample proportions, each based on $n = 10$ with a population percentage $p$ equal to 25% **b.** Revision of Figure 7.8, a histogram of 10,000 sample proportions, each based on $n = 100$ percentage and with the population percentage $p$ equal to 25%.

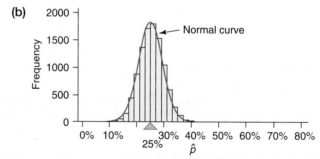

The Normal curve shown in Figure 7.9b has a mean of 0.25 because $p = 0.25$, and a standard deviation (also called the standard error) of 0.0433 because

$$\sqrt{\frac{0.25 \times 0.75}{100}} = 0.0433$$

Before illustrating how to use the CLT, we show how to check conditions to see whether the CLT applies.

## Checking Conditions for the Central Limit Theorem

The first condition states that the sample must be collected randomly. There is no way to check this just by looking at the data; you have to trust the researcher's report on how the data were collected, or, if you are the researcher, take care to use sound random sampling methods.

The second condition dictates that the sample size must be large enough. This we *can* check by looking at the data. The CLT says that the sample size needs to be sufficiently large to get at least 10 successes and 10 failures in our sample. If the probability of a success is $p$, then we would expect about $np$ successes and $n(1 - p)$ failures. One problem, though, is that we usually don't know the value of $p$. In this case, we instead check that

$$n\hat{p} \geq 10 \text{ and } n(1 - \hat{p}) \geq 10$$

For example, if our sample has 100 people and we are estimating the proportion of females in the population, and if our sample has 49% females, then we need to verify that both $100(0.49) \geq 10$ and $100(0.51) \geq 10$.

The third condition applies only to random samples done without replacement. In this case, the population must be at least 10 times bigger than the sample. In symbols, if $N$ is the number of people in the population and $n$ is the number in the sample, then

$$N \geq 10n$$

If this condition is not met, and the sample was collected without replacement, then the actual standard error will be a little smaller than what our formula says it should be.

In most real-life applications, the population size is much larger than the sample size. Over 300 million people live in the United States, so the typical survey of 1000 to 3000 easily meets this condition.

You can see how these conditions are used in the examples that follow.

## Using the Central Limit Theorem

The following examples use the CLT to find the probability that the sample proportion will be near (or far from) the population value.

## EXAMPLE 5  Pet World Revisited

Let's return to Pet World. The population is 1000 people, and the proportion of Cat People is 25%. We'll take a random sample of 100 people.

QUESTION   What is the approximate probability that the proportion in our sample will be bigger than 29%? Begin by checking conditions for the CLT.

SOLUTION   First we check conditions to see whether the Central Limit Theorem can be applied. The sample size is large enough because $np = 100(0.25) = 25$ is greater than 10 and $n(1 - p) = 100(0.75) = 75$, which is also greater than 10. Also, the population size is 10 times larger than the sample size, because $1000 = 10(100)$. Thus $N = 10(n)$; the population is just large enough. We are told that the sample was collected randomly.

According to the CLT, the sampling distribution will be approximately Normal. The mean is the same as the population proportion: $p = 0.25$. The standard deviation is the same as the standard error:

$$SE = \sqrt{\frac{p(1 - p)}{n}} = \sqrt{\frac{0.25 \times 0.75}{100}} = \sqrt{\frac{0.1875}{100}} = \sqrt{0.001875} = 0.0433$$

We can use technology to find the probability of getting a value larger than 0.29 in a N(0.25, 0.0433) distribution. Or we can standardize.

In standard units, 0.29 is

$$z = \frac{0.29 - 0.25}{0.0433} = 0.924 \text{ standard unit}$$

In a N(0,1) distribution, the probability of getting a number bigger than 0.924 is, from Table A in the appendix, about 0.18, or 18%. Figure 7.10 on the next page shows the results using technology.

▶ **FIGURE 7.10** Output from SOCR (http://socr.stat.ucla.edu): There is about an 18% chance that $\hat{p}$ will be more than 4 percentage points above 25%.

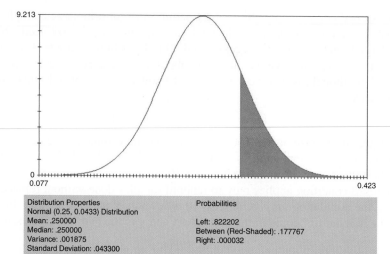

Distribution Properties
Normal (0.25, 0.0433) Distribution
Mean: .250000
Median: .250000
Variance: .001875
Standard Deviation: .043300
Max Density: 9.213448

Probabilities
Left: .822202
Between (Red-Shaded): .177767
Right: .000032

**CONCLUSION** With a sample size of 100, there is about an 18% chance that $\hat{p}$ will be more than 4 percentage points above 25%.

**TRY THIS!** Exercise 7.33

**SNAPSHOT** **THE SAMPLE PROPORTION: $\hat{p}$ (p-HAT)**

**WHAT IS IT?** ▶ The proportion of people or objects in a sample that have a particular characteristic in which we are interested.

**WHAT IT IS USED FOR?** ▶ To estimate the proportion of people or objects in a population that have that characteristic.

**WHY DO WE USE IT?** ▶ If the sample is drawn at random from the population, then the sample proportion is unbiased and has standard error $\sqrt{\dfrac{p(1-p)}{n}}$.

**HOW IS IT USED?** ▶ If, in addition to everything above, the sample size is fairly large, then we can use the Normal distribution to find probabilities concerning the sample proportion.

## EXAMPLE 6 Presidential Election Survey

In a hotly contested U.S. election, two candidates for president, a Democrat and a Republican, are running neck and neck; each candidate has 50% of the vote. Suppose a random sample of 1000 voters are asked whether they will vote for the Republican candidate.

**QUESTIONS** What percentage of the sample should be expected to express support for the Republican? What is the standard error for this sample proportion? Does the Central Limit Theorem apply? If so, what is the approximate probability that the sample proportion will fall within two standard errors of the population value of $p = 0.50$?

**SOLUTION** Because we have collected a random sample, the sample proportion has no bias (assuming there are no problems collecting the sample). Therefore, we expect that 50% of our sample supports the Republican candidate.

Because the sample size, $n = 1000$, is small relative to the population (which is over 100 million), we can calculate the standard error with

$$SE = \sqrt{\frac{(0.50)(0.50)}{1000}} = 0.0158$$

We can interpret this to mean that we expect our sample proportion to be 50%, give or take 1.58 percentage points.

Because the sample size is fairly large (the expected numbers for successes and failures are both equal to $np = 1000 \times 0.50 = 500$, which is much larger than 10), the CLT tells us we can use the Normal distribution—in particular, N(0.50, 0.0158).

We are asked to find the probability that the sample proportion will fall within two standard errors of 0.50. In other words, that it will fall somewhere between

$$0.50 - 2SE$$
and
$$0.50 + 2SE$$

Because this is a Normal distribution, we know the result will be very close to 95% (according to the Empirical Rule). But let's calculate the result anyway.

$$0.50 - 2SE = 0.50 - 2(0.0158) = 0.50 - 0.0316 = 0.4684$$

$$0.50 + 2SE = 0.50 + 0.0316 = 0.5316$$

That is, we want to find the area between 0.4684 and 0.5316 in a N(0.5, 0.0158) distribution. Figure 7.11 shows the result using technology, which tells us this probability is 0.9545.

<table>
<tr><td>

**↻ Looking Back**

**Empirical Rule**
Recall that the Empirical Rule says that roughly 68% of observations should be within one standard deviation of the mean, about 95% within two standard deviations of the mean, and nearly all within three standard deviations of the mean. In this context, the standard error is the standard deviation for the sampling distribution.

</td></tr>
</table>

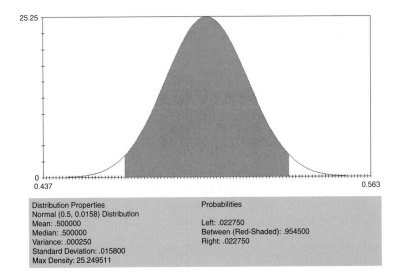

Distribution Properties
Normal (0.5, 0.0158) Distribution
Mean: .500000
Median: .500000
Variance: .000250
Standard Deviation: .015800
Max Density: 25.249511

Probabilities

Left: .022750
Between (Red-Shaded): .954500
Right: .022750

◀ **FIGURE 7.11** The probability that a sample proportion based on a random sample of 1000 people taken from a population in which p = 0.50 has about a 95% chance of falling within two standard errors of 0.50.

CONCLUSION  If each candidate truly has 50% of the vote, then we'd expect our sample proportion to be about 0.50 (or 50%). There is about a 95% chance that the sample proportion falls within two standard errors of 50%.

TRY THIS!  Exercise 7.35

The conclusion from Example 6 is useful because it implies that, in general, we can predict where $\hat{p}$ will fall, relative to $p$. It indicates that $\hat{p}$ is very likely to fall within two standard errors of the true value, as long as the sample size is large enough. If, in addition, we have a small standard error, we know that $\hat{p}$ is quite likely to fall close to $p$.

> **KEY POINT** If the conditions of a survey sample satisfy those required by the CLT, then the probability that a sample proportion will fall within two standard errors of the population value is 95%.

## EXAMPLE 7 Morse and the Proportion of E's

Samuel Morse (1791–1872), the inventor of Morse code, claimed that the letter used most frequently in the English language was E and that the proportion of E's was 0.12. We took a simple random sample with replacement from a modern-day book. Our sample consisted of 876 letters, and we found 118 E's, so $\hat{p} = 0.1347$.

QUESTION Find the probability that, if we were to take another random sample of 876 letters, the sample proportion would be greater than or equal to 0.1347. Assume that the true proportion of E's in the population is, as Morse claimed, 0.12. As a first step, check that the Central Limit Theorem can be applied in this case.

SOLUTION To check whether we can apply the Central Limit Theorem, we need to make sure the sample size is large enough. Because $p = 0.12$, we check:

$$np = 876(0.12) = 105.12, \text{ which is larger than 10,}$$

and

$$n(1 - p) = 876(0.88) = 770.88, \text{ which is also larger than 10}$$

The book contains far more than 8760 letters, so the population size is much larger than the sample size.

We can therefore use the Normal model for the distribution of sample proportions. The mean of this distribution is

$$p = 0.12$$

The standard error is

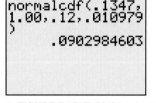

▲ FIGURE 7.12 TI-83/84 output

$$SE = \sqrt{\frac{p(1 - p)}{n}} = \sqrt{\frac{0.12(0.880)}{876}} = 0.010979$$

$$z = \frac{\hat{p} - p}{SE} = \frac{0.1347 - 0.12}{0.010979} = \frac{0.0147}{0.010979} = 1.34$$

We therefore need to determine the probability of getting a $z$-score of 1.34 or larger. We use technology (Figure 7.12) to find that it is 0.0903, or about 9.0%. This probability is represented by the shaded area in Figure 7.13. We can also find it with the Normal table; it is the area to the right of a $z$-score of 1.34.

▶ FIGURE 7.13 The shaded area represents the probability of finding a sample proportion of 0.1347 from a population with a proportion of 0.12.

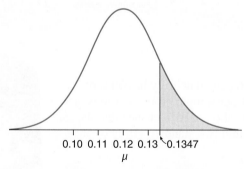

CONCLUSION If the sample is 876 letters, the probability of getting a sample proportion of 0.1347 or larger, when the true proportion of E's in the population is 0.12, is about 9%.

TRY THIS! Exercise 7.37

# Estimating the Population Proportion with Confidence Intervals

Up to this point, all of our examples have described situations in which we know the value of the population parameter, $p$. In Example 7, for instance, $p$ represented the proportion of E's found in the English language, and we could assume that $p = 0.12$. In the real world, knowing $p$ is unusual. In fact, the whole point of carrying out a survey is that we *don't* know the value of $p$.

An example of a real survey illustrates this situation. The Pew Research Center took a random sample of 2928 adults in the United States in September 2008. In this sample, 53% of the 2928 people believed that reducing the spread of acquired immune deficiency disease (AIDS) and other infectious diseases was an important policy goal for the U.S. government. However, this percentage just tells us about our sample. What percentage of the population—that is, what percentage of all adults in the United States—believes this? How much larger or smaller than 53% might it be? Can we conclude that a majority (more than 50%) of Americans share this belief?

We don't know $p$, the population parameter. We do know $\hat{p}$ for this sample; it's equal to 53%. Here's what else we know from the preceding sections:

1. Our estimator is unbiased, so while our estimate of 53% may not be exactly equal to the population parameter, it's probably just a little higher or a little lower.

2. The standard error can be estimated as

$$\sqrt{\frac{\hat{p}(1-\hat{p})}{n}} = \sqrt{\frac{(0.53)(0.47)}{2928}} = 0.0092, \text{ or about } 0.9\%.$$

   This means that the population parameter must be close by, because a standard error of 0.9 percentage point signifies a very precise estimator.

3. Because the sample size is large, we also realize that the probability distribution of $\hat{p}$ is pretty close to being Normally distributed and is centered around the true population parameter value. Thus, there's about a 68% chance that $\hat{p}$ is closer than 1 standard error away from the population proportion, and a 95% chance that it is closer than 2 standard errors away. (See Example 6.) Also, there is almost a 100% chance (99.7%, actually) that the sample proportion is closer than 3 standard errors from the population proportion. Thus we can feel very confident that the proportion of the population who support this policy goal is within 3 standard errors of 0.53. Three standard errors is $3(0.9\%) = 2.7\%$, so we can be almost certain that the value of the population parameter is within 2.7 percentage points of 53%.

In other words, we can be highly confident that the population parameter is between these two numbers:

$$53\% - 2.7\% \quad \text{to} \quad 53\% + 2.7\%, \text{ or}$$
$$50.3\% \qquad \text{to} \qquad 55.7\%$$

We have just calculated a **confidence interval**. Confidence intervals are often reported as the estimate plus or minus some amount:

$$53\% \text{ plus or minus } 2.7\%, \text{ or } 53\% \pm 2.7\%.$$

The "some amount," in this case the 2.7 percentage points, is called the **margin of error**. The margin of error tells how far from the population value our estimate can be.

A confidence interval provides two pieces of information: (1) a range of plausible values for our population parameter (50.3% to 55.7%), and (2) a **confidence level**, which expresses (no surprise here) our level of confidence in this interval. Our high confidence level of 99.7% assures us we can be very confident that a majority of Americans consider fighting AIDS and other infectious diseases an important priority, because the smallest plausible level of support in the population is 50.3%, which is (just) bigger than a majority.

An analogy can help explain confidence intervals. Imagine a city park. In this park sit a mother and her daughter, a toddler. The mother sits in the same place every day, on a bench along a walkway, while her daughter wanders here and there. Most of the time, the child stays very close to her mother, as you would expect. In fact, our studies have revealed that 68% of the days we've looked, she is within 1 yard of her mother. Sometimes she strays a little bit farther, but on 95% of the days she is still within 2 yards of her mother. Only rarely does she move much farther; she is almost always within 3 yards of her mother.

One day the unimaginable happens, and the mother and the park bench become invisible. Fortunately, the child remains visible. The problem is to figure out where the mother is sitting.

Where is the mother? On 68% of the days, the child is within 1 yard of the mother, so at these times the mother must be within 1 yard of the child. If we think the mother is within 1 yard of the child on most days—that is, 68% of the days we observe—we will be right. But this also means we will be wrong on 32% of our visits. We could be more confident of being correct if we instead guessed that the mother is within 2 yards of the child. Then we would be wrong on only 5% of the days.

In this analogy, the mother is the population proportion. Like the mother, the population proportion never moves from its spot and never changes values. And just as we cannot see the invisible mother, we don't know where the parameter sits. The toddler is like our sample proportion, $\hat{p}$; we *do* know its value, and we know that it hangs out near the population proportion and moves around from sample to sample. Thus, even though we can't know exactly what the true population proportion is, we can infer that it is near the sample proportion.

## Setting the Confidence Level

The confidence level tells us how often the estimation method is successful. Our method is to take a random sample and calculate a confidence interval to estimate the population proportion. If the method has a 100% confidence level, that method always works. If the method has a 10% confidence level, it works in 10% of surveys. We say the method works if the interval captures the true value of the population parameter. In this case, the interval works if the true population proportion is inside the interval.

Think of the confidence level as the capture rate; it tells us how often a confidence interval based on a random sample will capture the population proportion. Keep in mind that the population proportion, like the mother on the park bench, does not move—it is always the same. However, the confidence interval does change with every random sample collected. Thus, the confidence level measures the success rate of the *method*, not of any one particular interval.

KEY POINT    The confidence level measures the capture rate for our method of finding confidence intervals.

Figure 7.14 demonstrates what we mean by a 95% confidence level. Let's suppose that in the United States, 51% of all voters favor stricter laws with respect to buying and selling guns. We simulate taking a random sample of 1000 people. We calculate the percentage of the sample who favor stricter laws, and then we find the confidence interval that gives us a 95% confidence level. We do this again and keep repeating. Figure 7.14 shows 100 simulations.

Each blue point and each orange point represent a sample percentage. Note that the points are centered around the population percentage of 51%. The horizontal lines

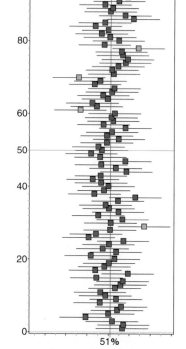

▲ **FIGURE 7.14** Results from 100 simulations in which we draw a random sample and then find and display a confidence interval with a 95% confidence level. The orange squares indicate "bad" intervals.

represent the confidence interval: the sample percentage plus or minus the margin of error. The margin of error was chosen so that the confidence level is 95%. Notice that most of the lines cross the vertical line at 51%. These are successful confidence intervals that capture the population value of 51% (in blue). However, a few sample percentages miss the mark; these are indicated by orange points. In 100 trials, our method failed 4 times and was successful 96 times. In other words, it worked in 96% of the trials. When we use a 95% confidence level, our method works in 95% of all surveys we conduct.

We can change the confidence level by changing the margin of error. The greater the margin of error, the higher our confidence level. For example, we can be 100% confident that the true percentage of Americans who favor stricter gun laws is between 0% and 100%. We're 100% confident in this interval because it can never be wrong. Of course, it will also never be useful. We really don't need to spend money on a survey to learn that the answer lies between 0% and 100%, do we?

It would be more helpful—more precise—to have a smaller margin of error than "plus or minus 50 percentage points." However, if the margin of error is too small, then we are more likely to be wrong. Think of the margin of error as a tennis racket. The bigger the racket, the more confident you are of hitting the ball. Choosing an interval that ranges from 0% to 100% is like using a racket that fills the entire court—you will definitely hit the ball, but not because you are a good tennis player. If the racket is too small, you are less confident of hitting the ball, so you don't want it too small. Somewhere between too big and too small is just right.

## Selecting a Margin of Error

We select a margin of error that will produce the desired confidence level. For instance, how can we choose a margin of error with a confidence level of 95%? We already know that if we take a large enough random sample and find the sample proportion, then the CLT tells us that 95% of the time, the sample proportion is within two standard errors of the population proportion. This is what we learned from Example 6. It stands to reason, then, that if we choose a margin of error that is two standard errors, then we'll cover the population proportion in 95% of our samples.

This means that

$$\hat{p} \pm 2SE$$

is a confidence interval with a 95% confidence level. More succinctly, we call this a 95% confidence interval.

Using the same logic, we understand that the interval

$$\hat{p} \pm 1SE$$

is a 68% confidence interval and that

$$\hat{p} \pm 3SE$$

is a 99.7% confidence interval.

Figure 7.15 shows four different margins of error for a sample in which $\hat{p} = 50\%$.

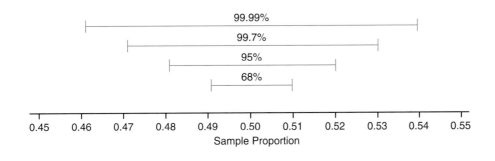

◀ **FIGURE 7.15** Four confidence intervals with confidence levels ranging from 99.99% (plus or minus 4 standard errors—top) to 68% (plus or minus 1 standard error). Notice how the interval gets wider with increasing confidence level.

This figure illustrates one reason why a 95% confidence interval is so desirable. If we increase the margin of error from 2 standard errors to 3, we gain only a small amount of confidence; the level goes from 95% to 99.7%. However, if we decrease from 2 standard errors to 1, we lose a lot of confidence; the level falls from 95% to 68%. Thus, the choice of 2 standard errors is very economical.

The margin of error has this structure:

$$\text{Margin of error} = z^*SE$$

where $z^*$ is a number that tells how many standard errors to include in the margin of error. If $z^* = 1$, the confidence level is 68%. If $z^* = 2$, the confidence level is 95%. Table 7.6 summarizes the margin of error for four commonly used confidence levels.

▶ **TABLE 7.6** We can set the confidence level to the value we wish by choosing the appropriate margin of error.

| Confidence Level | Margin of Error Is . . . |
|---|---|
| 99% | 2.58 standard errors |
| 95% | 1.96 (about 2) standard errors |
| 90% | 1.645 standard errors |
| 80% | 1.28 standard errors |

## Reality Check: Finding a Confidence Interval When $p$ Is Not Known

As we have seen, a confidence interval for a population proportion has this structure:

$$\hat{p} \pm m$$

where $m$ is the margin of error. Substituting for the margin of error, we can also write

$$\hat{p} \pm z^*SE$$

Finding the standard error requires us to know the value of $p$:

$$SE = \sqrt{\frac{p(1 - p)}{n}}$$

However, in real life, we don't know $p$. So instead, we substitute our sample proportion and use this estimated standard error:

$$SE_{\text{est}} = \sqrt{\frac{\hat{p}(1 - \hat{p})}{n}}$$

The result is a confidence interval with a confidence level close to, but not exactly equal to, the correct level. This tends to be close enough for most practical purposes.

In real life, then, Formula 7.1 is the method we use to find approximate confidence intervals for a population proportion.

**Formula 7.1:** $\hat{p} \pm m$ where $m = z^*SE_{\text{est}}$ and $SE_{\text{est}} = \sqrt{\frac{\hat{p}(1 - \hat{p})}{n}}$

where:

$m$ is the margin of error
$\hat{p}$ is the sample proportion of successes, or the proportion of people in the sample with the characteristic we are interested in
$n$ is the sample size
$z^*$ is a multiplier that is chosen to achieve the desired confidence level (Table 7.6)
$SE_{\text{est}}$ is the estimated standard error

## EXAMPLE 8 Harris Poll on Teaching

Do a majority of Americans believe the profession of teaching is prestigious? A 2007 Harris Poll took a random sample of 1011 Americans and found that 49% of the sample believed that the profession of teaching "has very great prestige."

**QUESTION** Estimate the standard error. Find an approximate 95% confidence interval for the percentage of all Americans who believe teaching has very great prestige. Is it plausible to conclude that more than half of all Americans share this opinion?

**SOLUTION** We first make sure the conditions of the Central Limit Theorem apply. We are told that the Harris Poll took a random sample. We don't know whether the pollsters sampled with or without replacement, but because the population is very large, easily 10 times larger than the sample size, we don't need to worry about the replacement issue.

Next, we need to check that the sample size is large enough for the CLT. We do not know $p$, the proportion of all Americans who believe that teaching has great prestige. We know only $\hat{p}$ (equal to 0.49), which the Harris Poll found on the basis of its sample. This means that our sample has at least 10 successes (people who believe teaching has great prestige) because $1011(0.49) = 495$, which is much larger than 10. Also, we know we have at least 10 failures (people who don't believe), because $1011(0.51)$ is even bigger than 495.

The next step is to estimate the standard error.

$$SE_{est} = \sqrt{\frac{(0.49)(0.51)}{1011}} = 0.0157$$

The problem asks for a 95% confidence level, which can be achieved by using a margin of error roughly equal to 2 standard errors.

$$\hat{p} \pm 2SE_{est} = 0.49 \pm 2(0.0157) = 0.49 \pm 0.0314,$$
$$\text{or } 49\% \pm 3.1\%$$

Expressing this as an interval, we get

$$49\% - 3.1\% = 45.9\%$$
$$49\% + 3.1\% = 52.1\%$$

The 95% confidence interval is 45.9% to 52.1%.

**CONCLUSION** The confidence interval tells us which values are plausible for the population percentage. We have to conclude that it is indeed plausible that more than half (50%) of Americans think teaching has great prestige, since the confidence interval includes values greater than 50%. It is not plausible that a large majority believes this way, though, since the largest plausible value is 52.1%.

**TRY THIS!** Exercise 7.51

## Interpreting Confidence Intervals

A confidence interval for a sample proportion gives a set of values that are plausible for the population proportion. If a value is not in the confidence interval, we conclude that it is implausible. It's not impossible that the population value is outside the interval, but it would be pretty surprising.

Suppose a candidate for political office conducts a poll and finds that a 95% confidence interval for the proportion of voters who will vote for him is 42% to 48%. He would be wise to conclude that he does *not* have 50% of the population voting for him. The reason is that the value 50% is not in the confidence interval, so it is implausible to believe that the population value is 50%.

There are many common misinterpretations of confidence intervals you must avoid. The most common mistake that students (and, indeed, many others) make is trying to turn confidence intervals into some sort of probability problem. For example, if asked to interpret a 95% confidence interval of 45.9% to 53.1%, many people would mistakenly say, "This means there is a 95% chance that the population proportion is between 45.9% and 53.1%."

What's wrong with this statement? Remember that probabilities are long-run frequencies. This sentence claims that if we were to repeat this survey many times, then in 95% of the surveys the true population percentages would be a number between 45.9% and 53.1%. This claim is wrong, because the true population percentage doesn't change. Either it is *always* between 45.9% and 53.1% or it is *never* between these two values. It can't be between these two numbers 95% of the time and somewhere else the rest of the time. In our story about the invisible mother, the mother, who represented the population proportion, *always* sat at the same place. Similarly, the population proportion (or percentage) is always the same value.

Another analogy will help make this clear. Suppose there is a skateboard factory. Say 95% of the skateboards produced by this factory are perfect, but 5% have no wheels. Once you buy a skateboard from this factory, you can't say that there is a 95% chance that it is a good board. Either it has wheels or it does not have wheels. It is not true that the board has wheels 95% of the time and, mysteriously, no wheels the other 5% of the time. A confidence interval is like one of these skateboards. Either it contains the true parameter (has wheels) or it does not. The "95% confidence" refers to the "factory" that "manufactures" confidence intervals: 95% of its products are good, 5% are bad.

Our confidence is in the process, not in the product.

> **KEY POINT**   Our confidence is in the process that produces confidence intervals, not in any particular interval. It is incorrect to say that a particular confidence interval has a 95% (or any other percent) chance of including the true population parameter. Instead, we say that the *process* that produces intervals captures the true population parameter with a 95% probability.

## EXAMPLE 9 Underwater Mortgages

A mortgage is "underwater" if the amount owed is greater than the value of the property that is mortgaged. For a 2010 Rasmussen poll, investigators interviewed a random sample of 710 homeowners in the United States and found that 35% had mortgages that were underwater. Rasmussen reports that "The margin of sampling error is plus or minus 4 percentage points with a 95% level of confidence." [Source: http://www.rasmussenreports.com, viewed August 6, 2010]

**QUESTION** State the confidence interval in interval form. How would you interpret this confidence interval? What does "95%" mean?

**CONCLUSION** The margin of error, we are told, is 4 percentage points. In interval form, then, the 95% confidence interval is

$$35\% - 4\% \quad \text{to} \quad 35\% + 4\%, \text{ or}$$

$$31\% \quad \text{to} \quad 39\%$$

We interpret this to mean that we are 95% confident that the true proportion of all homeowners in the United States whose homes are worth less than the mortgage is between 31% and 39%. The 95% indicates that if we were to conduct not just this survey, but many, then 95% of them would result in confidence intervals that include the true population proportion.

**TRY THIS!** Exercise 7.59

Example 10 demonstrates the use of confidence intervals to make decisions about population proportions.

## EXAMPLE 10 Morse and E's

Recall from Example 7 that Morse believed the proportion of E's in the English language was 0.12 and that our sample showed 118 E's out of 876 randomly chosen letters from a modern-day book.

**QUESTION** Find a 95% confidence interval for the proportion of E's in the book. Is the proportion of E's in the book consistent with Morse's 0.12?

**SOLUTION** The best approach is to use technology. Figure 7.16 shows TI-83/84 output that gives a 95% confidence interval as

$$(0.112, 0.157) \quad \text{or} \quad (11.2\%, 15.7\%)$$

If you do not have access to statistical technology, then the first step is to find the sample proportion of E's: 118/876, or 0.1347.
    The estimated standard error is

$$SE_{\text{est}} = \sqrt{\frac{\hat{p}(1-\hat{p})}{n}} = \sqrt{\frac{0.1347(0.8653)}{876}} = \sqrt{0.00012205} = 0.0115349$$

Because we want a 95% confidence level, our margin of error is plus or minus 1.96 standard errors:

$$\text{Margin of error} = 1.96 SE_{\text{est}} = 1.96(0.0115349) = 0.022608$$

The interval boundaries are

$$\hat{p} \pm 1.96 SE_{\text{est}} = 0.1347 \pm 0.0226$$

Upper end of interval: $0.1347 + 0.0226 = 0.1573$
Lower end of interval: $0.1347 - 0.0226 = 0.1121$

    This confirms the result we got through technology: A 95% confidence interval is (0.1121 to 0.1573). Note that this interval *does* include the value 0.12.

**CONCLUSION** We are 95% confident the proportion of E's in the modern book is between 0.112 and 0.157. This interval captures 0.12. This shows it is plausible that the population proportion of E's in the book is 0.12, as Morse suggested.

**TRY THIS!** Exercise 7.63

**Tech**

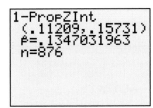

▲ **FIGURE 7.16** TI-83/84 output for a confidence interval for the proportion of E's.

## CASE STUDY REVISITED

What was wrong with the American Medical Association's spring break survey? The AMA poll was actually based on an "online survey panel," which consists of a group of people who agree to take part in several different online surveys in exchange for a small payment. Marketing companies recruit people to join panels so that the marketers can investigate trends within various slices of the public. Such a sample may or may not be representative of the population we're interested in—we have no way of knowing. And because the sample is not chosen randomly, we also have no way of knowing how our estimate will behave from sample to sample.

For such a survey, it is impossible to find a confidence interval for the true proportion of women who "agree that spring break trips involve more or heavier drinking than occurs on college campuses" because (1) our estimate might be biased, and (2) the true percentage might lie much further from our estimate than two standard errors. For this reason, the AMA ended up removing the margin of error from its website and no longer claimed that the figures were a valid inference for all college-aged women.

# EXPLORING STATISTICS
## CLASS ACTIVITY

## Simple Random Sampling Prevents Bias

| GOALS | MATERIALS |
|---|---|
| In this activity, you'll see how the sampling method affects our estimation of a population mean. | • A list of the first four amendments to the U.S. Constitution with each word numbered<br><br>• A random number table or other method of obtaining random numbers |

**ACTIVITY**

James Madison (1751–1836), who became the fourth president of the United States, wrote the first ten amendments to the U.S. Constitution, which are known as the Bill of Rights. They went into effect on December 15, 1791.

Your teacher will give you a page where the first four amendments are printed, with each word numbered. Your goal is to estimate the mean length of the words that appear in these four amendments.

You will use two estimation procedures, compare them, and decide which one works better. One method is an informal method. The other is based on random sampling. Your instructor will give you detailed instructions about each method.

The first ten words of the Bill of Rights are shown in Figure A, with each word numbered.

| 001 | 002 | 003 | 004 | 005 | 006 | 007 | 008 | 009 | 010 |
|---|---|---|---|---|---|---|---|---|---|
| Congress | shall | make | no | law | respecting | an | establishment | of | religion, |

▲ **FIGURE A**   The First Ten Words of the First Amendment

**BEFORE THE ACTIVITY**

1. What is the mean length of the ten words in the excerpt shown in Figure A?

2. Select any two words you wish from this excerpt. How different is the average length of the two words you selected from the mean length of all ten words?

**AFTER THE ACTIVITY**

Your instructor will give you data from the entire class. The data will consist of a list of estimates of mean word length based on the "informal" method, and a list based on the simple random sampling method. Make a dotplot of the "informal" estimates and a second dotplot of the estimates obtained using simple random sampling.

1. Compare the distributions from the two methods. (Comment on the shape, center, and spread of the distributions.)

2. Judging on the basis of your comparison of the dotplots, why is simple random sampling preferred over the informal method for collecting data?

This exercise is based on an activity from an INSPIRE workshop, which was based on an activity from Workshop Statistics, © 2004, Dr. Allan Rossman and Dr. Beth Chance, California Polytechnic State University.

# CHAPTER REVIEW

## Key Terms

Population, *290*
Parameter, *290*
Census, *290*
Sample, *290*
Statistic, *290*
Estimator, *290*

Estimate, *290*
Statistical inference, *291*
Biased, *292*
Sampling bias, *292*
Measurement bias, *292*
Voluntary-response bias, *293*

Nonresponse bias, *293*
Convenience sample, *294*
Simple random sample (SRS), *294*
With and without replacement, *294*
Sampling distribution, *299*
Bias (accuracy), *300*

Precision, *300*
Standard error (SE), *300*
Central Limit Theorem (CLT), *305*
Confidence interval, *311*
Margin of error, *311*
Confidence level, *311*

## Learning Objectives

After reading this chapter and doing the assigned homework problems, you should

- Be able to estimate a population proportion from a sample proportion and quantify how far off the estimate is likely to be.

- Understand that random sampling reduces bias.

- Understand when the Central Limit Theorem for sample proportions applies and know how to use it to find approximate probabilities for sample proportions.

- Understand how to find, interpret, and use confidence intervals for a single population proportion.

## Summary

The Central Limit Theorem (CLT) for sample proportions tells us that if we take a random sample from a population, and if the sample size is large and the population size much larger than the sample size, then the sampling distribution of $\hat{p}$ is approximately

$$N\left(p, \sqrt{\frac{p(1-p)}{n}}\right)$$

This result is used to infer the true value of a population proportion on the basis of the proportion in a random sample. The primary means for doing this is with a confidence interval:

**Formula 7.1:** $\hat{p} \pm m$    where    $m = z^* \times SE_{\text{est}}$

and    $SE_{\text{est}} = \sqrt{\dfrac{\hat{p}(1-\hat{p})}{n}}$

where:

$\hat{p}$ is the sample proportion of successes, the proportion of people in the sample with the characteristic we are interested in
$m$ is the margin of error
$n$ is the sample size
$z^*$ is a multiplier that is chosen to achieve the desired confidence level

An important first step is to make sure that the sample size is large enough for the CLT to work. This means that we need the sample size times the sample proportion to be at least 10 and that we need the sample size times (1 minus the sample proportion) to be at least 10.

A 95% confidence interval might or might not have the correct population value within it. However, we are confident that it does, because the method works for 95% of all samples.

## Sources

Cornish, A. Do polls miss views of the young and mobile? 2007. National Public Radio. October 1. http://www.npr.org (accessed March 29, 2010).

Crowley, C. 2010. Soda tax or flat tax? Questions can influence poll results, Cornell expert says. *Albany Times Union,* February 5. http://www.timesunion.com (accessed February 6, 2010).

Gallup. Shrunken majority now favors stricter gun laws. 2007. http://www.galluppoll.com October 11 (accessed March 29, 2010).

Jones, S., and S. Fox. 2009. Generations online in 2009. Pew Internet and American Life Project. January 28. http://pewresearch.org/pubs/1093/generations-online (accessed March 19, 2010).

Mystery Pollster. http://www.mysterypollster.com/ (accessed October 2, 2007).

*New York Times.* 1993. 1 in 5 in new survey express some doubt about the Holocaust. April 20.

*New York Times.* 1994. Pollster finds error on Holocaust doubts. May 20.

Pew Research Center for the People & the Press. 2010. The growing gap between landline and oval frame Election Polls. November 12, 2010, (accessed Dec. 20, 2010).

Pew Research Center for the People & the Press. 2008. Declining public support for global engagement. September 24. http://people-press.org (accessed March 29, 2010).

## SECTION EXERCISES

### SECTION 7.1

TRY **7.1 Parameter vs. Statistic (Example 1)** Explain the difference between a parameter and a statistic.

**7.2 Sample vs. Census** Explain the difference between a sample and a census. Every 10 years, the U.S. Census Bureau takes a census. What does that mean?

**7.3 $\bar{x}$ vs. $\mu$** Two symbols are used for the mean: $\mu$ and $\bar{x}$.
a. Which represents a parameter and which a statistic?
b. In determining the mean age of all students at your school, you survey 30 students and find the mean of their ages. Is this mean $\bar{x}$ or $\mu$?

**7.4 $\bar{x}$ vs. $\mu$** The mean GPA of all 5000 students at Uneeda College is 2.78. A sample of 50 GPAs from this school has a mean of 2.93. Which number is $\mu$ and which is $\bar{x}$?

\* **7.5 Ages of Presidents** Suppose you knew the age at inauguration of *all* the past U.S. presidents. Could you use those data to make inferences about ages of past presidents? Why or why not?

\* **7.6 Heights of Basketball Team** Suppose you find *all* the heights of the members of the men's basketball team at your school. Could you use those data to make inferences about heights of all men at your school? Why or why not?

**7.7 Sample vs. Census** You are receiving a large shipment of batteries and want to test their lifetimes. Explain why you would want to test a sample of batteries rather than the entire population.

**7.8 Sampling Strawberries** You are receiving a truckload of baskets of strawberries for your supermarket chain, and you want to check their quality. If you checked a few of the easily accessible baskets, would this necessarily be a good representative sample of the quality of all the berries? Explain.

**7.9 Sampling with and without Replacement** Explain the difference between sampling with replacement and sampling without replacement. Suppose you have the names of 10 students, each written on a 3 by 5 notecard, and want to select two names. Describe both procedures.

**7.10 Simple Random Sampling** Is simple random sampling usually done with or without replacement?

TRY **7.11 Finding a Random Sample (Example 2)** You need to select a simple random sample of four from eight friends who will participate in a survey. Assume the friends are numbered 1, 2, 3, 4, 5, 6, 7, and 8.
    Select four friends, using the two lines of numbers in the next column from a random number table.
    Read off each digit, skipping any digit not assigned to one of the friends. The sampling is without replacement, meaning that you cannot select the same person twice. Write down the numbers chosen. The first person is number 7.

```
0 7 0 3 3    7 5 2 5 0    3 4 5 4 6
7 5 2 9 8    3 3 8 9 3    6 4 4 8 7
```

Which four friends are chosen?

**7.12 Finding a Random Sample** You need to select a simple random sample of two from six friends who will participate in a survey. Assume the friends are numbered 1, 2, 3, 4, 5, and 6.
    Use technology to select your random sample. Indicate what numbers you obtained and how you interpreted them.
    If technology is not available, use the line from a random number table that corresponds to the day of the month on which you were born. For example, if you were born on the fifth day of any month, you would use line 05. Show the digits in the line and explain how you interpreted them.

**7.13 Criticize the Sampling** Marco is interested in whether Proposition P will be passed in the next election. He goes to the university library and takes a poll of 100 students. Since 58% favor Proposition P, Marco believes it will pass. Explain what is wrong with his approach.

**7.14 Criticize the Sampling** Maria opposes capital punishment and wants to find out if a majority of voters in her state support it. She goes to a church picnic and asks everyone there for their opinion. Because most of them oppose capital punishment, she concludes that a vote in her state would go against it. Explain what is wrong with Maria's approach.

**7.15 Bias** In 2003, shortly after the U.S. invasion of Iraq, www.TimeEurope.com posed a question to European readers of *Time* magazine on the Internet: "Who really poses the greatest danger to world peace? Iraq, North Korea, or the United States?" The site received 706,842 responses: 6.7% said North Korea, 6.3% Iraq, and 86.9% the United States. Identify the population, and explain why the results might not reflect true opinions in the population.

**7.16 Polling** When men and women live together, women are more likely to answer the telephone. When polling agencies call homes, they often ask to speak to the adult with the most recent birthday. Give at least one reason why polling agencies take this approach.

TRY **7.17 Questionnaire Response (Example 3)** A teacher at a community college sent out questionnaires to evaluate how well the administrators were doing their jobs. All teachers received questionnaires, but only 10% returned them. Most of the returned questionnaires contained negative comments about the administrators. Explain how an administrator could dismiss the negative findings of the report.

**7.18 Survey on Social Security** A phone survey asked whether Social Security should be continued or abandoned immediately. Only land lines (not cell phones) were called. Do you think this would introduce bias? Explain.

**7.19 Random Sampling?** If you walked around your school campus and asked people you met how many keys they were carrying, would you be obtaining a random sample? Explain.

**7.20 Biased Sample?** You want to find the mean weight of the students at your college. You calculate the mean weight of a sample of members of the football team. Is this method biased? If so, would the mean of the sample be larger or smaller than the true population mean for the whole school? Explain.

**7.21 Views on Capital Punishment** In carrying out a study on views of capital punishment, a student asked a question two ways:

1. With persuasion: "My brother has been accused of murder and he is innocent. If he is found guilty, he might suffer capital punishment. Now do you support or oppose capital punishment?"

2. Without persuasion: "Do you support or oppose capital punishment?"

Here is a breakdown of her actual data.

**Men**

|  | With persuasion | No persuasion |
|---|---|---|
| For capital punishment | 6 | 13 |
| Against capital punishment | 9 | 2 |

**Women**

|  | With persuasion | No persuasion |
|---|---|---|
| For capital punishment | 2 | 5 |
| Against capital punishment | 8 | 5 |

a. What percentage of those persuaded against it support capital punishment?

b. Find the percentage of those not persuaded who support capital punishment.

c. Compare the percentages in parts a and b. Is this what you expected? Explain.

**7.22 Views on Capital Punishment** Use the data given in Exercise 7.21.

Make the two given tables into one table by combining men for capital punishment into one group, men opposing it into another, women for it into one group, and women opposing it into another. Show your two-way table.

The student who collected the data could have made the results misleading by trying persuasion more often on one gender than on the other, but she did not do this. She used persuasion on 10 of 20 women (50%) and on 15 of 30 men (50%).

a. What percentage of the women support capital punishment? What percentage of the men support it?

b. On the basis of these results, if you were on trial for murder and did not want to suffer capital punishment, would you want men or women on your jury?

## SECTION 7.2

**7.23 Targets: Bias or Lack of Precision?**

a. If a rifleman's gunsight is adjusted incorrectly, he might shoot bullets consistently close to 2 feet left of the bull's-eye target. Draw a sketch of the target with the bullet holes. Does this show lack of precision or bias?

b. Draw a second sketch of the target if the shots are both unbiased and precise (have little variation).

The rifleman's aim is not perfect, so your sketches should show more than one bullethole.

**7.24 Targets: Bias or Lack of Precision?, Again**

a. If a rifleman's gunsight is adjusted correctly but he has shaky arms, the bullets might be scattered widely around the bull's-eye

target. Draw a sketch of the target with the bullet holes. Does this show variation (lack of precision) or bias?

b. Draw a second sketch of the target if the shots are unbiased and have precision (little variation).

The rifleman's aim is not perfect, so your sketches should show more than one bullethole.

**\* 7.25 Bias?** Suppose that, when taking a random sample of 4 from 123 women, you get a mean height of only 60 inches (5 feet). The procedure may have been biased. What else could have caused this small mean?

**\* 7.26 Bias?** Four women selected from a photo of 123 were found to have a sample mean height of 71 inches (5 feet 11 inches). The population mean for all 123 women was 64.6 inches. Is this evidence that the sampling procedure was biased? Explain.

**7.27 Proportion of Odd Digits** A large collection of one-digit random numbers should have about 50% odd and 50% even digits because five of the ten digits are odd (1, 3, 5, 7, and 9) and five are even (0, 2, 4, 6, and 8).

a. Find the proportion of odd-numbered digits in the following lines from a random number table. Count carefully.

   5 5 1 8 5        7 4 8 3 4        8 1 1 7 2
   8 9 2 8 1        4 8 1 3 4        7 1 1 8 5

b. Does the proportion found in part (a) represent $\hat{p}$ (the sample proportion) or $p$ (the population proportion)?

c. Find the error in this estimate, the difference between $\hat{p}$ and $p$ (or $\hat{p} - p$).

**7.28 Proportion of Odd Digits** 1, 3, 5, 7, and 9 are odd and 0, 2, 4, 6, and 8 are even. Consider a 30-digit line from a random number table.

a. How many of the 30 digits would you expect to be odd, on average?

b. If you actually counted, would you get exactly the number you predicted in part (a)? Explain.

**TRY 7.29 M&Ms (Example 4)** According to the Mars company, packages of milk chocolate M&Ms contain 20% orange candies. Suppose we examine 100 random candies.

a. What value should we expect for our sample percentage of orange candies?

b. What is the standard error?

c. Use your answers to fill in the blanks:
   We expect ____% orange candies, give or take ____%.

**7.30 Random Letters** Samuel Morse suggested in the nineteenth century that the letter "t" made up 9% of the English language. Assume this is still correct. A random sample of 1000 letters is taken from a randomly selected, large book and the t's are counted.

a. What value should we expect for our sample percentage of t's?

b. Calculate the standard error.

c. Use your answers to fill in the blanks:
   We expect ____% t's, give or take ____%.

**\* 7.31 What Is the Proportion of Seniors?** A population of college students is taking an advanced math class. In the class are three juniors and two seniors. Using numbers 1, 2, and 3 to represent juniors and 4 and 5 to represent seniors, sample without replacement. Draw a sample of two people four times (once in each of parts a, b, c, and d), and then fill in the following table.

a. Use the first line (reprinted here) from the random number table to select your sample of two. (The selections are underlined.)

0 2 7 7 9     7 2 6 4 5     3 2 6 9 9     8 6 0 0 9

Report the percentage of seniors in the sample. (Count the number of 4s and 5s and divide by the sample size 2.)

b. Use the next line to select your sample of two.

3 1 8 6 7     8 5 8 7 2     9 1 4 3 0     4 5 5 5 4

Report the percentage of seniors in the sample.

c. Use the next line to select your sample of two.

0 7 0 3 3     7 5 2 5 0     3 4 5 4 6     7 5 2 9 8

Report the percentage of seniors in the sample.

d. Use the last line to select your sample of two.

0 9 0 8 4     9 8 9 4 8     0 9 5 4 1     8 0 6 2 3

Report the percentage of seniors in the sample.

e. Fill in the rest of the table below, showing the results of the four samples:

| Repetition | $p$ (Population Proportion of Seniors) | $\hat{p}$ (Sample Proportion of Seniors) | Error: $\hat{p} - p$ |
|---|---|---|---|
| 1 (from part a) | 2/5 = 0.4 | 1/2 = 0.5 | 0.5 − 0.4 = 0.1 |
| 2 (from part b) | | | |
| 3 | | | |
| 4 | | | |

**7.32 Simulation** From a very large population (essentially infinite), of which half are men and half are women, you take a random sample, with replacement. Use the following random number table and assume each single digit represents selection of one person; the odd numbers represent women and the even numbers (0, 2, 4, 6, 8) men.

a. Start on the left side of the top line (with 118) and count 10 people. What percentage of the sample will be men?

1 1 8 4 8     8 0 8 0 9     2 5 8 1 8     3 8 8 5 7

2 3 8 1 1     8 0 9 0 2     8 5 7 5 7     3 3 9 6 3

9 3 0 7 6     3 9 9 5 0     2 9 6 5 8     0 7 5 3 0

b. Start in the middle of the second line (with 857) and count 20 people. What percentage of the sample will be men?

c. If you were to repeat parts a and b many times, which sample would typically come closer to 50%—the sample of 10 or the sample of 20? Why?

## SECTION 7.3

**\* 7.33 M & Ms (Example 5)** Return to Exercise 7.29 and find the approximate probability that the random sample of 100 will contain 24% or more orange candies.

**\* 7.34 Random Letters** Return to Exercise 7.30 and find the approximate probability that the random sample of 1000 letters will contain 8.1% or fewer t's.

**TRY 7.35 Jury Selection (Example 6)** Juries should have the same racial distribution as the surrounding communities. According to the U.S. Census, about 18% of residents in Minneapolis, Minnesota, are African American. Suppose a local court randomly selects 100 adult citizens of Minneapolis to participate in the jury pool.

Use the Central Limit Theorem (and the Empirical Rule) to find the approximate probability that the proportion of available African American jurors is more than two standard errors from the population value of 0.18.

The conditions for using the Central Limit Theorem are satisfied because the sample is random; the population is more than 10 times 1000; *n* times *p* is 18, and *n* times (1 minus *p*) is 82, and both are more than 10.

**\* 7.36 Mercury in Freshwater Fish** According to an article from HuffingtonPost.com, some experts believe that 20% of all freshwater fish in the United States have such high levels of mercury that they are dangerous to eat. Suppose a fish market has 250 fish we consider randomly sampled from the population of edible freshwater fish.

Use the Central Limit Theorem (and the Empirical Rule) to find the approximate probability that the market will have a proportion of fish with dangerously high levels of mercury that is more than two standard errors above 0.20.

You can use the Central Limit Theorem because the fish were randomly sampled; the population is more than 10 times 250; and *n* times *p* is 50, and *n* times (1 minus *p*) is 200, and both are more than 10.

**TRY 7.37 The Oregon Bar Exam (Example 7)** According to the Oregon Bar Association, approximately 65% of the people who take the bar exam to practice law in Oregon pass the exam. Find the approximate probability that at least 67% of 200 randomly sampled people taking the Oregon bar exam will pass. (In other words, find the probability that at least 134 out of 200 will pass.) *See page 327 for guidance.*

**\* 7.38 Feeding Vegans** A survey of eating habits showed that approximately 4% of people in Portland, Oregon, are vegans. Vegans do not eat meat, poultry, fish, seafood, eggs, or milk. A restaurant in Portland expects 300 people on opening night, and the chef is deciding on the menu. Treat the patrons as a simple random sample from Portland and the surrounding area, which has a population of about 600,000.

If 14 vegan meals are available, what is the approximate probability that there will not be enough vegan meals—that is, the probability that 15 or more vegans will come to the restaurant? Assume the vegans are independent and there are no families of vegans.

**\* 7.39 Renting Chairs for a Meeting** You have sent out 4000 invitations to hear a speaker, and you must rent chairs for the people who come. In the past, usually about 8% of the people invited have come to hear the speaker.

a. On average, what proportion of those invited should we expect to attend?

b. Suppose you assume that 8.5% of those invited will attend, and so you rent 340 chairs (because 0.085 times 4000 is 340). What is the approximate probability that more than 8.5% of those invited will show up and you will not have enough chairs? Refer to the TI-83/84 output given. Recall that this gives the Normal cumulative probability in the following format:

Normalcdf (left boundary, right boundary, mean, standard deviation)

Draw a well-labeled sketch of the Normal curve, and shade the appropriate region to represent the probability.

c. What is the approximate probability that more than 9% of the 4000 invited will show up? How many chairs would you have to rent if exactly 9% of those invited attended?

d. Why is your answer to part c smaller than your answer to part b? Draw a Normal probability distribution with a center at a proportion of 0.08 and vertical lines at about 0.085 and 0.090 so that you can visualize the size of the right-tail area.

**7.40 High School Diplomas** A U.S. government survey in 2007 said that 87% of young Americans earn a high school diploma.

a. Assume you took a simple random sample of 2000 young Americans. What proportion would you expect to earn a high school diploma? How many young Americans is this? What proportion would you expect not to earn a high school diploma? How many young Americans is this? Are both numbers greater than or equal to 10, as required by the Central Limit Theorem?

b. If you took a simple random sample of 2000 young Americans, what is the approximate probability that 88% or more of the sample will have earned their high school diploma? Refer to Example A in the TechTips for one approach to this question.

\* **7.41 Passing a Test by Guessing** A true/false test has 40 questions. A passing grade is 60% or more correct answers.

a. What is the probability that a person will guess correctly on one true/false question?

b. What is the probability that a person will guess incorrectly on one question?

c. Find the approximate probability that a person who is just guessing will pass the test.

d. If a similar test were given with multiple-choice questions with four choices for each question, would the approximate probability of passing the test by guessing be higher or lower than the approximate probability of passing the true/false test? Why?

\* **7.42 Gender: Randomly Chosen?** A large community college district has 1000 teachers, of whom 50% are men and 50% are women. In this district, administrators are promoted from among the teachers. There are currently 50 administrators, and 70% of these administrators are men.

a. If administrators are selected randomly from the faculty, what is the approximate probability that the percentage of male administrators will be 70% or more?

b. If administrators are selected randomly from the faculty, what is the approximate probability that the percentage of female administrators will be 30% or less?

c. How are part a and part b related?

d. Do your answers suggest that it is reasonable to believe the claim that the administrators have been selected randomly from the teachers? Answer yes or no, and explain your answer.

## SECTION 7.4

**7.43 Wanderlust** In a Pew Research Center Poll (Oct. 3–11, 2008), 1040 of 2260 respondents said they would prefer to live somewhere else. For example, many who lived in cities would rather not live there. Using the output given, find a 95% confidence interval for the proportion who would rather live elsewhere. Copy the statement that follows, filling in the blanks:

I am 95% confident that the population proportion of people who would like to live somewhere else is between _____ and _____.

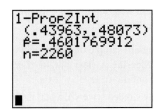

**7.44 Drug to Prevent Heart Attacks** In June of 2009, a Pew Poll asked 1005 people: Which over-the-counter drug do doctors recommend to help prevent heart attacks? Is it aspirin, cortisone, or antacids? (Options were rotated.)

Of 1005, 914 correctly chose aspirin. Report a 95% confidence interval for the population proportion who knew the correct answer. Use the output provided to find the interval. Copy the statement below, inserting the left and right endpoints of the interval.

I am 95% confident that the population proportion of people who know aspirin is used to prevent heart attacks is between _____ and _____.

**7.45 Economic Stimulus** A Gallup Poll of 1027 adults, conducted Jan. 30–Feb. 3, 2009, found that 770 of the respondents wanted Congress to pass a stimulus bill to rescue the economy.

a. What percentage of respondents wanted a stimulus bill?

b. Report a 95% confidence interval for the population percentage in favor of a stimulus bill. Refer to the Minitab output, and round the boundaries sensibly.

| Sample | X | N | Sample p | 95% CI |
|--------|-----|------|----------|---------------------|
| 1 | 770 | 1027 | 0.749757 | (0.723265, 0.776248) |

**7.46 Best President** A Gallup Poll of 1018 adults in February of 2009 asked, "If you had to choose, which of the following presidents would you regard as the greatest: Washington, Franklin Roosevelt, Lincoln, Kennedy, or Reagan?" The order was rotated to avoid biasing the response. Of the 1018 people surveyed, 224 chose Lincoln. Refer to the Minitab output.

| Sample | X | N | Sample p | 95% CI |
|--------|-----|------|----------|---------------------|
| 1 | 224 | 1018 | 0.220039 | (0.194591, 0.245488) |

a. What percentage of respondents chose Lincoln?

b. Report a 95% confidence interval for the population percentage who chose Lincoln. Refer to the figure.

**7.47 Voting** A random sample of likely voters showed that 55% planned to vote for Candidate X, with a margin of error of 2 percentage points and with 95% confidence.

a. Use a carefully worded sentence to report the 95% confidence interval for the percentage of voters who plan to vote for Candidate X.

b. Is there evidence that Candidate X could lose.

c. Suppose the survey was taken on the streets of New York City and the candidate was running for U.S. president. Explain how that would affect your conclusion.

**7.48 Voting** A random sample of likely voters showed that 49% planned to support Measure X. The margin of error is 3 percentage points with a 95% confidence level.

a. Using a carefully worded sentence, report the 95% confidence interval for the percentage of voters who plan to support Measure X.

b. Is there evidence that Measure X will fail?

c. Suppose the survey was taken on the streets of Miami and the measure was a Florida statewide measure. Explain how that would affect your conclusion.

**7.49 Estimating the Proportion of Odd Digits (with Given Standard Error)**

a. Find the proportion of odd-numbered digits in the following lines from a random number table. Count carefully; circle them!

    5 5 1 8 5   7 4 8 3 4   8 1 1 7 2
    8 9 2 8 1   4 8 1 3 4   7 1 1 8 5

b. The estimated standard error for the sample proportion of odd digits is 0.090. Find the margin of error for a 95% confidence interval for the proportion of odd digits in the entire random number table by multiplying the estimated standard error by 1.96.

c. Find a 95% confidence interval for the population proportion of odd-numbered digits in the entire random number table by adding the margin of error to the sample proportion to find the upper boundary and subtracting it to find the lower boundary.

d. The estimated standard error is 0.090. Find an 80% confidence interval for the population proportion of odd-numbered digits. For calculations by hand, multiply the standard error by 1.28 to find the margin of error because $z^*$ for 80% confidence is 1.28.

e. Find a 99% confidence interval for the population proportion of odd-numbered digits. For calculations by hand, $z^*$ for 99% is 2.58.

f. State which interval is widest and which is narrowest. Explain why.

**7.50 Proportion of Odd Digits (with Given Standard Error)**

a. Find the proportion of odd-numbered digits in the following lines from a random number table, and then find a 95% confidence interval for the population proportion of odd-numbered digits in the entire random number table. Count carefully!

    8 7 9 6 4   4 3 7 5 1   8 0 9 7 1
    5 0 6 1 3   5 1 4 4 1   3 0 5 0 5

b. The estimated standard error is 0.090. Find an 80% confidence interval for the population proportion of odd-numbered digits in the full random number table. For calculations by hand, $z^*$ for 80% is 1.28.

c. Find a 99% confidence interval for the population proportion of odd-numbered digits. For calculations by hand, $z^*$ for 99% is 2.58.

d. State which interval is wider and which is narrower. Explain why.

**TRY 7.51 High School Diplomas (Example 8)** In a simple random sample of 1500 young Americans, 87% had earned a high school diploma.

a. What is the standard error for this estimate of the percentage of all young Americans who earned a high school diploma?

b. Find the margin of error, using a 95% confidence level, for estimating the percentage of all young Americans who earned a high school diploma.

c. Report the 95% confidence interval for the percentage of all young Americans who earned a high school diploma.

d. Suppose that in the past, 80% of all young Americans earned high school diplomas. Does the confidence interval you found in part c support or refute the claim that the percentage of young Americans who earn high school diplomas has increased? Explain.

**7.52 Diabetes** In a simple random sample of 1200 Americans age 20 and over, the proportion with diabetes was found to be 0.115 (or 11.5%).

a. What is the standard error for the estimate of the proportion of all Americans age 20 and over with diabetes?

b. Find the margin of error, using a 95% confidence level, for estimating this proportion.

c. Report the 95% confidence interval for the proportion of all Americans age 20 and over with diabetes.

d. According to the Centers for Disease Control, nationally, 10.7% of all Americans age 20 or over have diabetes. Does the confidence interval you found in part c support or refute this claim? Explain.

**7.53 Dreaming in Color** In a 2003 study on dreaming, an investigator attempted to replicate an experiment done by Middleton in 1942 (Schwitzgebel, *Perceptual and Motor Skills* 2003). In the study from 2003, 92 of 113 people said they dream in color.

a. What percentage of people in 2003 said they dreamed in color?

b. Find a 95% confidence interval for the percentage who dreamed in color in 2003. The margin of error is 0.0717.

c. In the study from 1942, 29.24% of people reported dreaming in color. Does the interval for the data from 2003 capture 29.24%? Is 29.24% plausible as the 2003 population percentage? Explain.

**7.54 Seat Belt Use** In a 2008 survey, the National Highway Traffic Safety Administration (NHTSA) reported that 83% of people used seat belts. The margin of error is 3 percentage points.

a. Assuming that the confidence level is 95% and the survey was random, find a 95% confidence interval for the percentage of people who used seat belts in 2008.

b. The NHTSA said the percentage of people using seat belts in 2000 was 71%. If 71% were suggested as the percentage for 2008, would you reject that as implausible? Why or why not? Does this suggest a change in seat belt use between 2000 and 2008? Explain.

**TRY 7.55 Understanding the Meaning of Confidence Levels: 90% (Example 9)** Each student in a class of 40 was randomly assigned one line of a random number table. Each student then counted the odd-numbered digits in a 30-digit line. (Remember that 0, 2, 4, 6, and 8 are even.)

a. *On average*, in the list of 30 digits, how many odd-numbered digits would each student find?

b. If each student found a 90% confidence interval for the percentage of odd-numbered digits in the entire random number table, how many intervals (out of 40) would you expect *not* to capture the population percentage of 50%?

**7.56 Understanding the Meaning of Confidence Levels: 80%** Each student in a class of 30 was assigned one random line of a random number table. Each student then counted the even-numbered digits in a 30-digit line.

a. *On average*, in the list of 30 digits, how many even-numbered digits would each student find?

b. If each student found an 80% confidence interval for the percentage of even-numbered digits, how many intervals (out of 30) would you expect *not* to capture 50%? Explain how you arrived at your answer.

**7.57 Percentage of Female U.S. Senators in 2008** In the fall of 2008, 16 of 100 U.S. senators were female. For the year 2008, find a 95% confidence interval for the percentage of U.S. senators who were female or explain why you should *not* find a confidence interval for the percentage of U.S. senators who were female in 2008.

**7.58 Past Presidential Vote** In the 1960 presidential election, 34,226,731 people voted for Kennedy; 34,108,157 for Nixon; and 197,029 for third-party candidates (www.uselectionatlas.org).

a. What percentage of voters chose Kennedy?

b. Would it be appropriate to find a confidence interval for the proportion of voters choosing Kennedy? Why or why not?

**7.59 Newspapers Closing** A Pew Poll asked for people's opinion on whether closing local newspapers would hurt civic life; 430 of 1001 respondents said it would hurt civic life a lot.

a. Find the proportion of the respondents who said that closing local papers would hurt civic life a lot.

b. Find a 95% confidence interval for the population proportion who believed closing newspapers would hurt civic life a lot. Assume the Pew Poll used a simple random sample (SRS). (In fact, it used random sampling, but a more complex method than SRS.)

c. Find an 80% confidence interval for the population proportion who believed closing newspapers would hurt civic life a lot.

d. Which interval is wider, and why?

**7.60 Human Cloning** In a Gallup Poll, 441 of 507 adults said it was "morally wrong" to clone humans.

a. What proportion of the respondents believed it morally wrong to clone humans?

b. Find a 95% confidence interval for the population proportion who believed it is morally wrong to clone humans. Assume that Gallup used a simple random sample.

c. Find an 80% confidence interval (using a $z^*$ of 1.28 if you are calculating by hand).

d. Which interval is wider, and why?

**7.61 Desire to Lose Weight** In a Gallup Poll from November of 2008, 595 of 1009 adults surveyed said they wanted to lose weight.

a. What percentage of the sample wanted to lose weight?

b. Find a 95% confidence interval for the proportion of people in the population who wanted to lose weight.

c. Would a 99% confidence interval be wider or narrower?

**7.62 Support for Nuclear Energy** Some people fear the use of nuclear energy because of potential accidents. However, a Gallup Poll in 2009 indicated that 74% of men supported the use of nuclear energy to produce electricity. The sample size was 506.

a. In this poll, how many men supported the use of nuclear energy? Round your answer to the nearest whole number.

b. Report a 95% confidence interval for the proportion of all men in the United States who support the use of nuclear energy.

c. Would a 90% interval be wider or narrower?

 **7.63 Do People Think Astrology Is Scientific? (Example 10)** In the 2008 General Social Survey, people were asked their opinions on astrology—whether it was very scientific, somewhat scientific, or not at all scientific. Of 1438 who responded, 74 said astrology was very scientific.

a. Find the proportion of people in the survey who believe astrology is very scientific.

b. Find a 95% confidence interval for the population proportion with this belief.

c. Suppose a TV news anchor said that 5% of people in the general population think astrology is very scientific. Would you say that is plausible? Explain your answer.

**7.64 Do People Think the Sun Goes around the Earth?** In the 2008 General Social Survey, people were asked whether they thought the sun went around the earth or vice versa. Of 1381 people, 310 thought the sun went around the earth.

a. What proportion of people in the survey believed the sun went around the earth?

b. Find a 95% confidence interval for the proportion of all people with this belief.

c. Suppose a scientist said that 30% of people in the general population believe the sun goes around the earth. Using the confidence interval, would you say that was plausible? Explain your answer.

## CHAPTER REVIEW EXERCISES

**7.65 Heart Attacks and Exposure to Traffic** In 2004, *The New England Journal of Medicine* published a study of Germans who had suffered heart attacks and survived. Within the first 24 hours of the heart attack, the 691 subjects were interviewed about what they were doing before the heart attack. Of the 691 subjects, 515 reported exposure to traffic within an hour prior to the heart attack. Most of these people had been traveling in cars or trucks, although some had used public transportation and some had been riding motorcycles.

a. What percentage of people had been exposed to traffic?

b. Find a 95% confidence interval for the percentage of all heart attack victims exposed to traffic.

c. Can you conclude that traffic definitely causes heart attacks? Why or why not?

**7.66 Bariatric Surgery and Diabetes** A study published in *JAMA* in 2004 examined past results of other studies on bariatric surgery. Bariatric surgery is done to reduce the size of the stomach in various ways. It is typically used only on obese patients, and one form of surgery had a 1% mortality rate (death rate) caused by the surgery. However, in the studies reporting on the effect of this surgery on diabetes, 1417 of 1846 diabetic patients recovered completely from diabetes after surgery. These patients no longer needed diabetes medication such as insulin.

a. Find the percentage of patients who recovered from diabetes.

b. Find a 95% confidence interval for the percentage of recovery from diabetes.

c. If you had a morbidly obese relative with diabetes, what would you tell him or her about bariatric surgery? Explain.

**7.67 Opinion on Banning Smoking** A Gallup Poll of 489 adults done in July of 2005 found that 78 of respondents answered yes to the following question.

*Should smoking in this country be made totally illegal, or not?*

a. What percentage said yes?

b. Find and interpret a 95% confidence interval for the percentage of American adults in the country who would say yes.

c. Is it plausible that a majority believe smoking should be made totally illegal?

d. Suppose the sample size were 978 (twice as large) and the number were 156 which is also twice as large; would the percentage that said yes be the same or different from the percentage reported in part a? Would the interval be wider or narrower? If you don't know, try it!

**7.68 to 7.70 Opinion on Banning Smoking** A Gallup Poll (July, 2005), asked 1006 American adults to choose from the following 4 options for regulating smoking:

1. Set-aside areas,
2. Totally ban,
3. No restrictions,
4. No opinion.

The numbers replying "Totally Ban" in the various locations were:

| | |
|---|---|
| Hotels and Motels | 342 responded "Totally Ban" |
| Workplaces | 412 responded "Totally Ban" |
| Restaurants | 543 responded "Totally Ban" |
| Bars | 292 Responded "Totally Ban" |

**7.68 Banning Smoking in Workplaces**

a. What percentage said totally ban in the workplace?

b. Find and interpret a 95% confidence interval for the percentage who believe smoking should be banned in the workplace.

c. Is it plausible that a majority believe that smoking should be banned in the workplace?

d. Suppose the sample size was 503 instead of 1006 and the number that said totally ban was 206. Is this percentage the same as or different from the percentage reported in part a? Would the interval be wider or narrower?

**7.69 Banning Smoking in Bars** Answer questions a, b, and c from 7.68 but use the data from the bars.

d. If the sample size were 2012 instead of 1006 and the number that said yes was 584 (and so the percentage is the same) would the interval be wider or narrower?

**7.70 Banning Smoking in Restaurants** Answer questions a, b, and c from 7.68 but use the data from the restaurants.

d. If you wanted a narrower interval how would you change this survey?

**7.71 Terri Schiavo** In 2005, Terri Schiavo was said to have been in a "persistent vegetative state" for many years. Her husband wanted her feeding tube removed, and her parents wanted it left in place. In a Gallup Poll taken on March 18–20, 2005, 56% of adults polled answered yes to the following question. "Should Terri Schiavo's feeding tube have been removed?" The margin of error was given as 4 percentage points, assuming a 95% confidence level. Assuming the sampling was random, does this suggest that a majority of adults believe the feeding tube should have been removed? Show your work and explain your answer.

**7.72 Wording Confidence Interval Statements** From Exercise 7.71, a 95% confidence interval for the percentage who answered yes to whether Terri Schiavo's feeding tube should have been removed was (52%, 60%).

Which of the following is the best interpretation? Explain why the other wordings are incorrect.

a. We are 95% confident that the boundaries for the interval are 52% and 60%.

b. There is a 95% probability that the population percentage is between 52% and 60%.

c. There is a 95% chance that the boundaries 52% and 60% capture the population percentage.

d. Ninety-five percent of all sample percentages based on samples of the same size will be between 52% and 60%.

e. We are 95% confident that the population percentage is between 52% and 60%.

## gUIDED EXERCISES

**g * 7.37 The Oregon Bar Exam** According to the Oregon Bar Association, approximately 65% of the people who take the bar exam to practice law in Oregon pass the exam.

**QUESTION** Find the approximate probability that at least 67% of 200 randomly sampled people who take the Oregon bar exam will pass it.

Step 1 ▶ **Population proportion**
The sample proportion is 0.67. What is the population proportion?

Step 2 ▶ **Check assumptions**
Because we are asked for an approximate probability, we might be able to use the Central Limit Theorem. In order to use the Central Limit Theorem for a proportion, we must check the assumptions.

 a. Randomly sampled: Yes
 b. Sample size: If a simple random sample of 200 independent people take the bar exam, how many of them would you expect to pass, on the average? Calculate $n$ times $p$. Also calculate how many you would expect to fail, $n$ times $(1 - p)$. State whether these are both more than 10.
 c. Assume the population size is at least 10 times the sample size, which would be at least 2000.

Step 3 ▶ **Calculate**
Part of the standardization follows. Finish it, showing all the numbers.
First find the standard error:

$$SE = \sqrt{\frac{p(1 - p)}{n}} = \sqrt{\frac{0.65(0.35)}{?}} = ?$$

Then standardize:

$$z = \frac{\hat{p} - p}{SE} = \frac{0.67 - 0.65}{?} = ? \text{ standard units}$$

Find the approximate probability that at least 0.67 pass by finding the area to the right of the $z$-value of 0.59 in the Normal curve. Show a well-labeled curve, starting with what is given in Figure A.

Step 4 ▶ **Explain**
Explain why the tail area in the accompanying figure represents the correct probability.

Step 5 ▶ **Answer the question**
Find the area of the shaded region in this figure.
 For another approach to this type of question, see Example A in the TechTips.

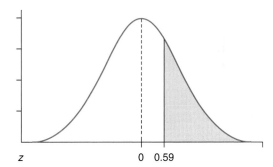

$z$      0  0.59

# TechTips

For generating random numbers, see the TechTips for Chapter 5.

**EXAMPLE A: ONE-PROPORTION PROBABILITY USING NORMAL TECHNOLOGY** ► A U.S. government survey in 2007 said that 87% of young Americans earn a high school diploma. If you took a simple random sample of 2000 young Americans, what is the approximate probability that 88% or more of the sample will have earned their high school diploma? To use the "Normal" steps from Chapter 6, we just need to evaluate the mean and standard error (the standard deviation). The population mean is 0.87. The standard error is

$$SE = \sqrt{\frac{p(1-p)}{n}} = \sqrt{\frac{0.87(1-0.87)}{2000}} = \sqrt{\frac{0.1131}{2000}} = 0.00752$$

Now you can use the "Normal" TechTips steps (for TI-83/84, Minitab, Excel, or StatCrunch) from Chapter 6 starting on page 283.

Figure 7A shows TI-83/84 output, and Figure 7B shows StatCrunch output.

Thus the probability is about 9%.

▲ **FIGURE 7A** TI-83/84 Normal Output

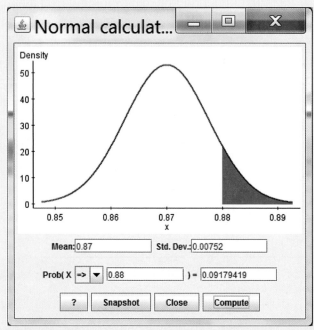

▲ **FIGURE 7B** StatCrunch Normal Output

**EXAMPLE B: FINDING A CONFIDENCE INTERVAL FOR A PROPORTION** ► Find a 95% confidence interval for a population proportion of heads when obtaining 22 heads in a sample of 50 tosses of a coin.

## TI-83/84

### Confidence Interval for One Proportion

1. Press **STAT**, choose **Tests, A: 1-PropZInt**.
2. Enter: **x, 22**; **n, 50**; **C-level, .95**.
3. Press **ENTER** when **Calculate** is highlighted.

Figure 7C shows the 95% confidence interval of (0.302, 0.578).

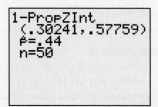

▲ **FIGURE 7C** TI-83/84 Output for a One-Proportion Z-interval

## MINITAB

### Confidence Interval for One Proportion

1. **Stat > Basic Statistics > 1 Proportion**.
2. Choose **Summarized data**.
   Enter: **Number of events, 22**; **Number of trials, 50**.
3. Click **Options** and Check **Use test and interval based on normal distribution**.
4. In the **1 Proportion – Options** box, if you want a confidence level different from 95%, change it.
5. Click **OK**: click **OK**.

The relevant part of the output is

```
         95% CI

   (0.302411,  0.577589)
```

## EXCEL

### Confidence Interval for One Proportion

1. Click **Add-Ins, XLSTAT, Parametric Test, Tests for one proportion**.
2. See Figure 7D. Enter: **Frequency, 22; Sample size, 50; Test Proportion, 0.5.** Leave the other checked options as given and click **OK**.

   (If you wanted an interval other than 95%, you would click Options and change the significance level. For a 99% interval you would use 1. For a 90% interval you would use 10.)

The relevant part of the output is

95% confidence interval on the proportion (Wald):

$$(0.302, 0.578)$$

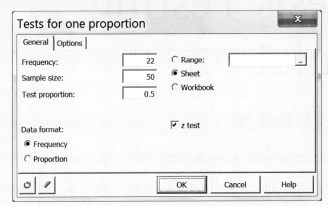

▲ **FIGURE 7D** XLSTAT Input for One Proportion

**Confidence Interval for One Proportion**

1. **Stat > Proportions > One sample > with Summary**
2. Enter: **number of successes, 22; number of observations, 50**.
3. Click **Next** and select the **Confidence Interval** option.

   Leave the default 0.95 for a 95% interval or change the **Level**. For **Method**, leave the default **Standard-Wald**.
4. Click **Calculate**.

The relevant part of the output is shown. "L. Limit" is the lower limit of the interval and "U. Limit" is the upper limit of the interval.

| L. Limit | U. Limit |
|---|---|
| 0.30241108 | 0.5775889 |

# 8 Hypothesis Testing for Population Proportions

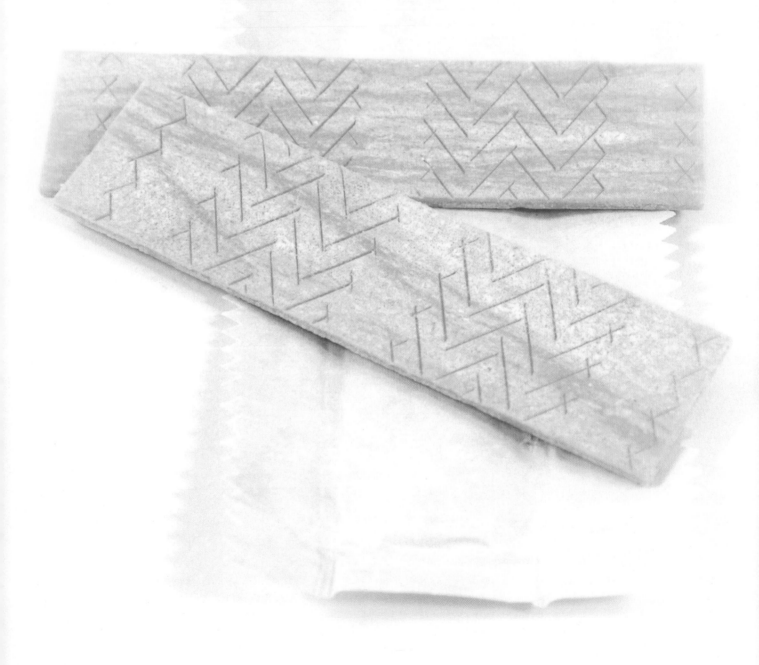

In many scientific and business contexts, decisions must be made about the values of population parameters, even though our estimates of these parameters are uncertain. Hypothesis testing provides a method for making these decisions, while controlling for the probability of making certain types of mistakes.

In science, business, and everyday life, we often have to make decisions on the basis of incomplete information. For example, a chewing gum company might need to decide what percent of customers will buy a new flavor. The company can't test the new flavor on everyone in the country—it has to base its decision on a small sample. An educational psychologist wonders whether kids who receive music training become more creative than other children. The test she uses to measure creativity does in fact show an increase, but might this increase be explained by chance alone? A sample of people who watched violent TV when they were children turn out to exhibit more violent behavior than a comparison sample made up of people who did not watch violent TV when young. Could this difference be due to chance, or is something else going on?

Hypothesis testing is a type of statistical inference. In Chapter 7, we used confidence intervals to estimate parameters and provide a margin of error for our estimates. In this chapter we make decisions on the basis of the information provided by our sample. If we knew everything about the population, we would definitely know what decision to make. But seeing only a sample from the population makes this decision harder, and mistakes are inevitable. Just as we measured our uncertainty in Chapter 7, our next task is to measure our mistake rate when testing hypotheses. In this chapter, we continue to work with population proportions. In the next chapter, we'll see how to find confidence intervals and perform hypothesis tests for means.

## CASE STUDY

### Violence on TV

Does watching violence on TV as a child lead to violent behavior as an adult? Psychologists wishing to shed some light on this question examined 329 children and divided them into two groups: those who watched a high level of violent content on television, and those who did not. Fifteen years later, when the children were in their early twenties, the researchers interviewed them again. This time they asked about acts of aggression or violence (Husemann et al. 2003).

The researchers examined many different types of violence, and one form of violence recorded was whether the now-adults had pushed, grabbed, or shoved their spouse or partner. They found that about 38% of those who reported watching violent television had pushed, grabbed, or shoved their spouse or partner, while only 22% of those who had not watched a lot of TV violence reported engaging in the same behaviors. The difference between these two groups is 16 percentage points. Is this difference due to some real behavioral differences between these two groups of people? Or is the difference due merely to chance? After all, the researchers examined only 329 adults, which is a far cry from the entire population of all adults who watched television as children. Is it possible that even if there were no real differences between these two groups of people, we might see a difference of 16 percentage points or even more? How likely is this? Should we attribute this difference to random variation, or should we conclude that something deeper is going on?

Researchers had to make a decision: Either the difference in behaviors was real or it was due to chance. In this chapter, you will see how to make this decision using a formal hypothesis-testing approach. At the end of the chapter, we will come back to this case study and see what the researchers decided and how they made their decision.

# The Main Ingredients of Hypothesis Testing

A well-known story goes something like this: Four students missed the midterm for their statistics class. They went to the professor together and said, "Please let us make up the exam. We carpool together, and on our way to the exam, we got a flat tire. That's why we missed the exam." The professor didn't believe them, but instead of arguing he said, "Sure, you can make up the exam. Be in my office tomorrow at 8." The next day, they met in his office. He sent each student to a separate room and gave them an exam. The exam consisted of only one question: "Which tire?"

We don't know the outcome of this story, but let's imagine that all four students answer, "left rear tire." The professor is surprised. He had assumed that the students were lying. "Maybe," he thinks, "they just got lucky. After all, if they just guessed, they could still all choose the same tire." But then he does a quick calculation and figures out that the probability that all four students will guess the same tire is only 1.6%. Reluctantly, he concedes that the students were not lying, and now he must give all of them an A on the exam.

The statistics professor has just performed a hypothesis test. **Hypothesis testing** is a formal procedure that enables us to choose between two hypotheses when we are uncertain about our measurements. We call hypothesis testing a formal procedure because it is based on particular terminology and a rather well-specified set of steps. However, we hope to show you that this "formal" procedure has a generous helping of common sense supporting it.

This chapter is divided into five sections. In the first section, you will learn the main ingredients and basic terminology of hypothesis testing in the context of a simple example. Next, you learn about a very useful and important concept called the p-value, and in the third section, we will further explore some details. In the fourth section, you'll learn how to alter the basic steps of this formal procedure to accommodate a more complex situation. And in the final section, we will discuss some of the subtleties behind hypothesis testing.

But before reading further, get a penny.

Football games and tennis matches begin with a coin toss to determine which team or player gets to start by playing offense. Coin tosses, in which the coin is flipped high into the air, are used because flipping a coin is believed to be "fair." The coin is equally likely to land heads or tails, so both sides have an equal chance of winning.

But what if we spin the penny (on a hard, flat surface) rather than flipping it in the air? We claim that, because the heads side of a penny bulges out, the lack of symmetry will cause the spinning coin to land on one side more often than on the other. In other words, we believe a spun penny is not fair. Some people—and they may include you—will find this claim to be outrageous and will insist it is false. So let's collect data.

## 1. Start with: A Pair of Hypotheses

We now have a research hypothesis: A spun penny is not fair. This is a good first step toward stating what we will call an alternative hypothesis. It is only a first step, however, because it is not very precise. To be more precise, we need to explain what we mean by "fair."

---

**▶ Details**

**Definition of Hypothesis**
Merriam Webster's online dictionary defines a hypothesis as "a tentative claim made in order to draw out and test its logical or empirical consequences."

When a coin is flipped or spun, it has some probability of landing heads. Let's call that probability $p$. For a fair coin, $p = 0.5$. We are claiming that for a coin that is spun, $p$ is not equal to 0.5.

We now have a pair of formal hypotheses. The *alternative hypothesis* can be written as

$$p \neq 0.50$$

The commonsense belief, which most people would probably agree with at first glance, is that a spun coin *is* fair. This is the *null hypothesis*, which we can write as

$$p = 0.50$$

In a formal hypothesis test, hypotheses are *always* statements about population parameters. In this case, our population consists of infinitely many spins of our coin, and $p$ represents the probability of getting heads.

**KEY POINT**    Hypotheses are *always* statements about population parameters; they are *never* statements about sample statistics.

Hypotheses come in pairs:

The **null hypothesis**, which we write $H_0$ (and pronounce "H-naught" or simply "the null hypothesis"), is the conservative, status-quo, business-as-usual statement about a population parameter. In the context of researching new ideas, the null hypothesis often represents "no change," "no effect," or "no difference."

The **alternative hypothesis**, $H_a$ (pronounced "H-A"), is the research hypothesis. It is usually a statement about the value of a parameter that we hope to demonstrate is true.

> **! Caution**
>
> **Pronunciation**
> Do *not* say "Ho" or "Ha," as Santa Claus might.

The most important step of a formal hypothesis test is choosing the hypotheses. In fact, there are really only two steps of a formal hypothesis test that a computer cannot do, and this is one of those steps. (The other step is checking to make sure that the conditions necessary for the probability calculations to be valid are satisfied.)

Hypothesis tests are like criminal trials. In a criminal trial, two hypotheses are placed before the jury: The defendant is not guilty, or he is guilty. These hypotheses are not given equal weight, however. The jury is told to assume the defendant is not guilty until the evidence overwhelmingly suggests this is not so. (Defendants charged with a crime in the United States must be found guilty "beyond all reasonable doubt.")

Hypothesis tests follow the same principles. The statistician plays the role of the prosecuting attorney, who hopes to show that the defendant is guilty. The hypothesis that the statistician or researcher hopes to establish plays the role of the alternative hypothesis. The null hypothesis is chosen to be a neutral, noncontroversial statement. Just as in a jury trial, where we ask the jury to believe that the defendant is not guilty unless the evidence against this belief is overwhelming, we will believe that the null hypothesis is true in the beginning. But once we examine the evidence, we may reject this belief if the evidence is overwhelmingly against it.

**KEY POINT**    The null hypothesis always gets the benefit of the doubt and is *assumed* to be true throughout the hypothesis-testing procedure. If we decide at the last step that the observed outcome is extremely unusual under this assumption, then and only then do we reject the null hypothesis.

**Looking Back**

**Statistic vs. Parameter**
In Chapter 7 you learned that $\hat{p}$, a statistic, represents the proportion of successes in a sample and that $p$, a parameter, represents the proportion of successes in the population.

Our hypotheses about spinning a coin are competing predictions about what we will see when we collect data. If the null hypothesis is right, then when we spin a coin a number of times, about half of the outcomes should be heads. If the null hypothesis is wrong, we will see either a larger or a smaller proportion. Now stop reading this book and spin a penny 20 times. Record the number of heads and calculate the sample proportion of heads, $\hat{p}$. This will only take a minute. We'll wait.

Are your results consistent with the claim that $p = 0.5$?

## EXAMPLE 1 Internet Advertising

An Internet retail business is trying to decide whether to pay a search engine company to upgrade its advertising. In the past, 15% of customers who visited the company's webpage by clicking on the advertisement bought something (this is called a "click-through"). If the business decides to purchase premium advertising, then the search engine company will make that company's ad more prominent. The search engine company offers to do an experiment: For one day, customers will see the retail business's ad in the more prominent position. The retail business can then decide whether the advertising improves the percentage of click-throughs. At the end of the experiment, the business observed that 17% of the customers now bought something. A marketing executive wrote the following hypotheses to do a hypothesis test.

$$H_0: \hat{p} = 0.15$$
$$H_a: \hat{p} = 0.17$$

where $\hat{p}$ represents the proportion of the sample that bought something from the website.

**QUESTION**  What is wrong with these hypotheses? Rewrite them so that they are correct.

**SOLUTION**  First, these hypotheses are written about the sample proportion, $\hat{p}$. We *know* that 17% of the sample bought something, so there is no need to make a hypothesis about it. What we don't know is what proportion of the entire population of people who click on the advertisement will purchase something. The hypotheses should be written in terms of $p$, the proportion of the population that will purchase something.

A second problem is with the alternative hypothesis. The research question that the company wants to answer is not whether 17% of customers will purchase something. It wants to know whether the percentage of customers who do so has increased over what has happened in the past.

The correct hypotheses are

$$H_0: p = 0.15$$
$$H_a: p > 0.15$$

where $p$ represents the proportion of all customers who click on the advertisement and purchase a product.

**TRY THIS!**  Exercise 8.5

**Details**

**$p$ Can Be Either a Proportion or a Probability**
The parameter $p$ can represent both. For example, if we were describing the voters in a city, we might say that $p = 0.54$ are Republican. Then $p$ is a proportion. But if we selected a voter at random, we might say that the probability of selecting a Republican is $p = 0.54$.

## 2. Add In: Making Mistakes

Mistakes are an inevitable part of the hypothesis-testing process. The trick is not to make them too often.

One mistake we might make is to reject the null hypothesis when it is true. For example, even a fair coin sometimes turns up heads in 20 out of 20 flips. If that happened, we might conclude that the coin was unfair when it really was fair. We can't prevent this mistake from happening, but we can try to make it happen infrequently.

The **significance level** is the name for a special probability: It is the probability of rejecting the null hypothesis when, in fact, the null hypothesis is true. It is the probability of making a very particular type of mistake. In our experiment, the significance level is the probability that we conclude that spinning a coin is *not* fair when, in fact, it really *is* fair. In a criminal justice setting, the significance level is the probability that we conclude that the suspect is guilty when he is actually innocent. The significance level is used so often that it has its own symbol, the Greek lowercase alpha: $\alpha$.

Naturally, we want a procedure with a small significance level, because we don't want to make mistakes too often. How small? Most researchers and statisticians use a significance level of 0.05. In some situations it makes sense to allow the significance level to be bigger, and some situations require a smaller significance level. But $\alpha = 0.05$ is a good place to start.

KEY POINT

The significance level, $\alpha$ (Greek lowercase alpha), represents the probability of rejecting the null hypothesis when the null hypothesis is true. For many applications, $\alpha = 0.05$ is considered acceptably small, but 0.01 and 0.10 are also sometimes used.

## EXAMPLE 2 Significance Level for Internet Advertising

In Example 1, an Internet retail business gave a pair of hypotheses about $p$, the proportion of customers who click on an advertisement and then purchase a product from the company. Recall that in the past, the proportion of customers who bought the product was 0.15, and the company hopes this proportion has increased.

$$H_0: p = 0.15$$
$$H_a: p > 0.15$$

QUESTION  Describe the significance level in context.

SOLUTION  The significance level is the probability of rejecting $H_0$ when in fact it is true. In this context, this is the probability of the company concluding that a larger proportion of customers will buy its product than in the past when, in fact, the proportion has not changed.

TRY THIS!  Exercise 8.11

## 3. Mix with: The Test Statistic

A **test statistic** compares our observed outcome to the null hypothesis. For example, in our coin-spinning example, we observed 7/20, or 0.35, heads. The null hypothesis tells us that we should expect half to be heads: 0.5. The test statistic compares the observed proportion to the expected proportion.

A useful test statistic in this situation is the **one-proportion z-test** statistic.

**Formula 8.1: The one-proportion z-test statistic**

$$z = \frac{\hat{p} - p_0}{SE}$$

$$\text{where } SE = \sqrt{\frac{p_0(1 - p_0)}{n}}$$

**Looking Back**

**Standard Error**
The standard error of the sample proportion, given in Chapter 7, is

$$SE = \sqrt{\frac{p(1-p)}{n}}$$

Formula 8.1 uses the symbol $p_0$ to remind us to use the value that the null hypothesis claims to be correct.

The symbol $p_0$ represents the value of $p$ that the null hypothesis claims is true. For example, for the coin-spinning example, $p_0$ is 0.5. Most of the test statistics you will see in this book have the same structure as Formula 8.1:

$$z = \frac{\text{observed value} - \text{null value}}{SE}$$

Because we observed 0.35 heads, the value for the test statistic for our experiment is

$$z_{\text{observed}} = \frac{0.35 - 0.5}{\sqrt{\frac{0.5(1-0.5)}{20}}} = -1.34$$

Note that the observed value of our test statistic is negative, because the proportion we observed was less than the proportion we expected.

### Why Is the *z*-Statistic Useful?

This test statistic is useful because it compares our observed sample proportion, $\hat{p}$, to the null hypothesis value, $p_0$. If the observed sample proportion is close to what the null hypothesis says we should see, then the test statistic will be close to 0. Thus, the farther the value of the test statistic is from 0 (that is, the larger its absolute value), the more suspicious we become of the null hypothesis.

 **KEY POINT** If the null hypothesis is true, then the *z*-statistic will be close to 0. Therefore, the farther the *z*-statistic is from 0, the more the null hypothesis is discredited.

 **Looking Back**

**Sampling Distribution**
A sampling distribution is the probability distribution for a statistic.

The *z*-statistic is also useful because, as you will soon see, statisticians know a good approximation to the sampling distribution when the null hypothesis is true. This is an important requirement for a test statistic, because we need to know the sampling distribution to verify that we're getting the correct significance level.

### EXAMPLE 3 Test Statistic for Spinning Heads

The authors observed 7 heads when they spun their coin 20 times, and they calculated a test statistic of $z = -1.34$. A friend of the author observed 12 heads out of 20 spins.

**QUESTION** What value of the test statistic should the friend report?

**SOLUTION** The observed sample proportion is $12/20 = 0.60$. Therefore, the friend's observed test statistic is

$$\frac{0.60 - 0.50}{\sqrt{\frac{0.50 \times 0.50}{20}}} = 0.89$$

**CONCLUSION** The observed value of the test statistic is 0.89. Note that the friend has a positive-valued test statistic, because he observed more than the expected number of heads.

**TRY THIS!** Exercise 8.13

# 4. The Final Main Ingredient: Surprise!

No, the final main ingredient in hypothesis testing is not *a* surprise. The main ingredient is surprise itself.

Surprise happens when something unexpected occurs. The null hypothesis tells us what to expect when we look at our data. If we see something unexpected—that is, if we are surprised—then we should doubt the null hypothesis and, if we are really surprised, reject it altogether.

Figure 8.1 shows all possible outcomes of our coin-spinning experiment. According to the null hypothesis, getting between about 6 and 15 heads is not that unusual; these outcomes are quite common when one is flipping a fair coin. But getting less than 6 heads (5, 4, 3, 2, 1, or 0) or more than 15 (16, 17, 18, 19, or 20) is unusual. If this happened to us, we would be surprised and would probably reject the null hypothesis.

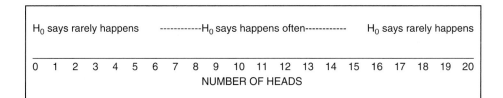

$H_0$ says rarely happens    -----------$H_0$ says happens often------------    $H_0$ says rarely happens

```
0   1   2   3   4   5   6   7   8   9   10  11  12  13  14  15  16  17  18  19  20
                                    NUMBER OF HEADS
```

 **FIGURE 8.1** Possible outcomes from spinning a coin 20 times and counting the number of heads. If the probability of heads is 0.50, then when spinning or tossing a coin 20 times, we would get between 6 and 14 heads quite often. The numbers 5 or fewer are rare, as are 15 or more.

Because we are statisticians, we have a way of measuring our surprise. The **p-value** is a number that measures our surprise by reporting the probability that if the null hypothesis is true, a test statistic will have a value as extreme as or more extreme than the value we actually observe. Small p-values (closer to 0) mean we have received a surprise. Large p-values (closer to 1) mean no surprise: The outcome happens fairly often. For instance, we can calculate that the probability of getting 4 or fewer heads when spinning a fair coin is 0.004. This tells us that getting 4 or fewer heads is highly unusual (and happens only 4 out of 1000 times). We would therefore be surprised if it were to happen.

**Details**

**"Often" and "Rarely"**
If you are wondering exactly what we mean by "often" and "rarely," don't worry. We intend these to be informal terms.

**KEY POINT**
The p-value is a probability. Assuming the null hypothesis is true, the p-value is the probability that if the experiment were repeated, you would get a test statistic as extreme or more extreme than the one you actually got. A small p-value suggests that a surprising outcome has occurred and discredits the null hypothesis.

For example, in our test to see whether a spun coin is fair, we spun a coin 20 times and saw 7 heads. The p-value associated with this outcome can be found to be about 0.18. This tells us that if spinning a coin really is fair, then the outcome we saw is not all that unusual. It happens about 1 time in 5. We would probably think that this does not discredit the hypothesis that spinning a coin is fair.

# EXAMPLE 4 Click-Throughs

The Internet business from Examples 1 and 2 wants to find out whether premium advertising will lead to a greater proportion of click-throughs—people who click on the advertisement and then actually purchase a product. Company investigators collect data and find that for one sample of customers, 18% click on the ad and then purchase a product. They perform a hypothesis test and find a p-value of 0.030. The null and alternative hypotheses are

$$H_0: p = 0.15$$
$$H_a: p > 0.15$$

where $p$ represents the proportion of all customers who click on the advertisement and purchase a product.

**QUESTION** Explain the meaning of this p-value, 0.030, in context.

**SOLUTION** The p-value of 0.030 means that if premium advertising really does nothing, then the probability of getting a proportion of purchases as high as or higher than 18% is 0.030. This tells us that if premium advertising does nothing, we have seen a fairly unusual outcome—an outcome that occurs only 3 out of 100 times.

**TRY THIS!** Exercise 8.15

---

SECTION 8.2

# Characterizing p-Values

Understanding how to interpret the p-value is crucial to understanding hypothesis testing. The computer might compute the p-value for you, but you need to understand how the computer did this calculation if you are to successfully perform a hypothesis test.

## The p-Value Is All about Extremes

The meaning of the phrase "as extreme as or more extreme than" depends on the alternative hypothesis. There are three basic pairs of hypotheses, as Table 8.1 shows.

▶ **TABLE 8.1** These three pairs of hypotheses can be used in a hypothesis test.

| Two-tailed | One-tailed (Left) | One-tailed (Right) |
|---|---|---|
| $H_0: \quad p = p_0$ | $H_0: \quad p = p_0$ | $H_0: \quad p = p_0$ |
| $H_a: \quad p \neq p_0$ | $H_a: \quad p < p_0$ | $H_a: \quad p > p_0$ |

If the alternative hypothesis is

$H_a: p \neq p_0$ (the true value of $p$ is either bigger or smaller than what the null hypothesis claims)

then "as extreme as or more extreme than" means "even farther away from 0 than the value you observed." This corresponds to finding the probability in both tails of the $N(0,1)$ distribution. This is called a **two-tailed hypothesis**.

If the alternative hypothesis is

$H_a: p < p_0$ (the true value is less than the value claimed by the null hypothesis)

then "as extreme as or more extreme than" means "less than or equal to the observed value." This corresponds to finding the probability in the left tail of $N(0, 1)$. This is an example of a **one-tailed hypothesis**. This alternative is also sometimes called a **left-tailed** hypothesis, because the p-value area is in the left tail.

Finally, if the alternative hypothesis is

$H_a: p > p_0$ (the true value is greater than the value claimed by the null hypothesis) then "as extreme as or more extreme than" means "greater than or equal to the observed value." This corresponds to finding the probability in the right tail of N(0, 1). This is another one-tailed hypothesis. This alternative hypothesis is called a **right-tailed hypothesis**.

To illustrate, suppose the observed value of the test statistic is $z = 1.56$. Figure 8.2 shows the p-value (represented by the shaded region) for each of these three possible alternative hypotheses.

**(a)**

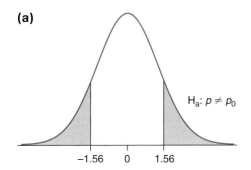

**(b)**

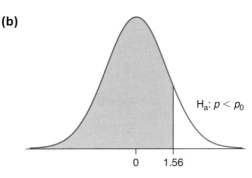

**(c)**

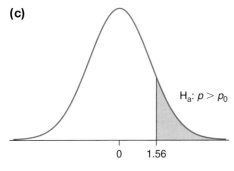

◀ **FIGURE 8.2** The shaded areas represent p-values for three different alternative hypotheses when the observed value of the test statistic is $z = 1.56$. **(a)** The p-value for a two-tailed alternative hypothesis (0.119). **(b)** The p-value for a left-tailed hypothesis (0.941). **(c)** The p-value for a right-tailed hypothesis (0.059).

Our coin-spinning experiment is an example of a two-tailed hypothesis. Figure 8.3 shows the area corresponding to the p-value for getting an outcome as extreme as or more extreme than a sample proportion of 0.35 for this two-tailed alternative hypothesis. (The units in Figure 8.3 are not "proportion of heads" but, rather, have been converted to standard units, using the test statistic above.)

As an example of a one-tailed p-value, consider Figure 8.4, which illustrates the p-value for one possible outcome in Example 4. In that example, an Internet retailer was wondering whether the new advertising campaign would result in a greater proportion of customers "clicking through" and purchasing its product. "As extreme as or more extreme than" the company planners' observed outcome means "as big as or bigger than" what the null hypothesis tells them to expect. (In Figure 8.4, the p-value shown is about 0.030. You can see that about 3% of the area under the curve is shaded.)

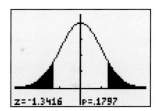

▲ **FIGURE 8.3** The shaded area represents the p-value: the probability of getting a proportion as extreme as or more extreme than 0.35 in 20 flips of a coin. Note that both the left and right tails are shaded.

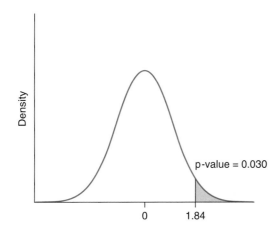

◀ **FIGURE 8.4** The shaded area represents the p-value for a z-statistic of 1.84. Because the alternative hypothesis is one-tailed ("the proportion of customers who buy a product is *larger* than it used to be"), the p-value is the area in the right tail.

## EXAMPLE 5 Identify the p-Values

Figure 8.5 shows sketches of three probability distributions. The shaded areas represent probabilities.

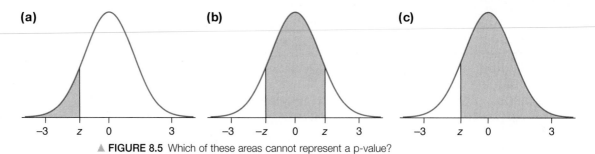

**(a)**                    **(b)**                    **(c)**

▲ **FIGURE 8.5** Which of these areas cannot represent a p-value?

**QUESTION** Which of these three sketches cannot represent a p-value? Explain your reasoning.

**SOLUTION** Figure 8.5b cannot represent a p-value because it is not a tail probability. Figure 8.5b represents the probability of getting an outcome between two values. Figure 8.5a represents the probability of getting an outcome as small or smaller than the observed outcome, and Figure 8.5c represents the probability of getting an outcome as big or bigger than the expected outcome. Both Figure 8.5a and Figure 8.5c, therefore, could represent p-values.

**TRY THIS!** Exercise 8.17

## Conditions Required for Calculating an Approximate p-Value

Note that in Figures 8.4 and 8.5 we used a Normal distribution to illustrate finding the p-value. This is no coincidence; under certain conditions, the standard Normal distribution, N(0, 1), provides a good model for the sampling distribution of the z-statistic.

These conditions are listed here and will be referred to throughout the remainder of the chapter.

1. *Random sample*: The sample is collected randomly from the population.

2. *Large enough sample size*: The sample size, $n$, is large enough that the sample has at least 10 expected successes and 10 expected failures; in other words, $np_0 \geq 10$ and $n(1 - p_0) \geq 10$.

3. *Without replacement*: If the sample is collected without replacement, then the population size is at least 10 times bigger than the sample size.

4. *Independence*: Each observation or measurement must have no influence on any others.

5. *Null hypothesis*: The null hypothesis is true.

If these conditions are satisfied, then the conditions for using the Central Limit Theorem for sample proportions (Chapter 7) have been met, and the z-statistic for the one-proportion z-test follows an approximately standard Normal distribution.

**KEY POINT** Under the appropriate conditions, the sampling distribution of the z-statistic is approximately a standard Normal distribution, N(0, 1).

## Calculating the p-Value

To find p-values, because the sampling distribution is the standard Normal, use the same techniques described in Section 6.2. As always, we recommend that you use technology to find the area under the standard Normal curve, but if you do not have technology, you can use Table 2 in Appendix A.

Figure 8.6 illustrates how the p-value depends on the observed outcome of our coin-spinning study. Each graph represents the p-value for a different outcome, with the coin spun 20 times in each case. The null hypothesis in all cases is $p = 0.5$, and the alternative is the two-tailed hypothesis that the probability of heads is *not* 0.5. Note that the closer the number of heads is to 10, the closer the z-value is to 0 and the larger the p-value is. Also note that the p-value for an outcome of 11 heads is the same as for 9 heads, and the p-value for an outcome of 12 heads is the same as for 8 heads. This happens because the alternative hypothesis is two-tailed and the Normal distribution is symmetric.

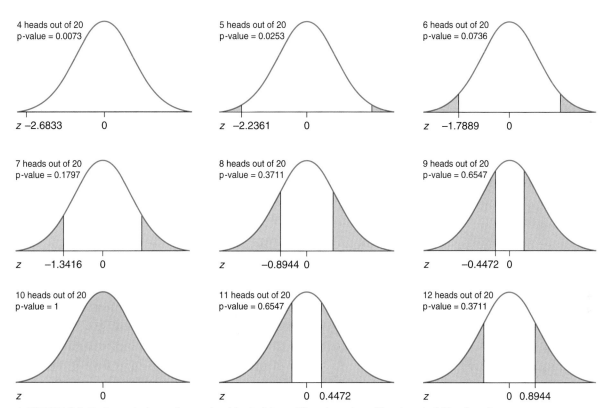

▲ **FIGURE 8.6** Each graph shows the p-value (shaded) for a different number of heads out of 20 spins of a coin, with the assumption that the coin is fair and using a two-tailed alternative hypothesis. Note that the closer the number of heads is to 10 (out of 20), the closer the z-value is to 0 and the larger the p-value. Also, as the number of heads gets farther from 10, the z-value gets farther from 0 and the p-value gets smaller.

## EXAMPLE 6 p-Values for Coin Spinning

Two different students each did the coin-spinning experiment with a two-tailed alternative. Their test statistics follow.

Study 1: $z = 1.98$

Study 2: $z = -2.02$

QUESTION Which of these test statistics has the smaller p-value and why?

> ⚠ **Caution**
>
> **So Many p's**
> $p$ is the population proportion.
> $p_0$ is the value of the population proportion according to the null hypothesis.
> $\hat{p}$ is the sample proportion.
> The p-value is the probability that if the null hypothesis is true, our test statistic will be as extreme as or more extreme than the value we actually observed.

**SOLUTION** If the null hypothesis is correct, then the test statistic should be close to 0. Values farther from 0 are more surprising and so have smaller p-values. Because –2.02 is farther from 0 than is 1.98, the area under the Normal curve in the tails is smaller for –2.02 than it is for 1.98. Thus –2.02 has the smaller p-value.

**TRY THIS!** Exercise 8.19

## SECTION 8.3

# Hypothesis Testing in Four Steps

Now that you know the main ingredients of hypothesis testing, it's time to learn the recipe.

We present hypothesis testing as a four-step procedure. In this chapter we focus on tests about population proportions. In Chapter 9 you will see tests about population means.

**Step 1: Hypothesize.**
State your hypotheses about the population parameter.

**Step 2: Prepare.**
Get ready to test: State and choose a significance level. Choose a test statistic appropriate for the hypotheses. State and check conditions required for future computations, and state any assumptions that must be made.

**Step 3: Compute to compare.**
Compute the observed value of the test statistic, and compare it to what the null hypothesis said you would get. Find the p-value to measure your level of surprise.

**Step 4: Interpret.**
Do you reject or not reject the null hypothesis? What does this mean in the context of the data?

Now we will examine in detail how these steps are applied to two different situations.

*Situation 1:* Do Political Scientists Vote at the Same Rate As the Rest of Us? In one Florida election, 46.6% of all registered voters voted. Did political scientists in Florida vote in the same proportion? The data consist of a random sample of 54 political scientists in a particular Florida election (Schwitzgebel 2010).

*Situation 2:* Do We Dream in Color?
Historically, before the age of television, color movies, and video games, 29% of the American population reported dreaming in color. A psychologist suspects that the present-day proportion might be higher, now that we are surrounded with color imagery. The data consist of a random sample of 113 people, taken in the year 2001 (Schwitzgebel 2003).

## Step 1: Hypothesize

Hypotheses, you will recall, consist of an alternative hypothesis that is based on the researchers' claim, and a null hypothesis that represents the status-quo position. The null hypothesis is assumed to be true until the very last of the four steps of our procedure.

Let's see how to create a pair of hypotheses for these two situations.

### Situation 1: Voting

Most people would probably think that political scientists are more likely to vote than the general public. However, some political scientists, a cynical lot, tend to think this group might be less likely to vote than the general public. The researcher therefore wonders whether it can be shown that political scientists do *not* vote in the same proportion as the general public. This is the alternative hypothesis. The neutral, status-quo hypothesis is that they *do* vote in the same proportion as the general public. (Recall that in the election considered, 46.6% of the registered voters voted.) So the researcher wonders whether political scientists are just as likely to vote as the general population or do not vote in the same proportion.

The null and alternative hypotheses can now be stated both in words and in symbols.

$H_0$: Political scientists vote in the same proportion as the public: 0.466.

$H_a$: Political scientists do not vote in the same proportion as the public.

$$H_0: p = 0.466$$
$$H_a: p \neq 0.466$$

This alternative hypothesis is a two-tailed hypothesis because it includes both the possibility that the population proportion might be bigger than 0.466 and the possibility that it might be smaller than 0.466.

## Situation 2: Dreaming

The researcher hopes to establish that a greater proportion of people today report that they dream in color than the proportion that reported dreaming in color historically. In the past, before color TV and movies, this proportion was 0.29. The researcher has interviewed a random sample of 113 people.

We state the null and alternative hypotheses in words and symbols.

$H_0$: The proportion of those who report dreaming in color in the population is the same as it has historically been: 0.29.

$H_a$: The proportion of those who report dreaming in color has increased.

$$H_0: p = 0.29$$
$$H_a: p > 0.29$$

This alternative hypothesis is a one-tailed hypothesis, because the hypothesis considers only the possibility that the true population proportion might be greater than—to the right of—the hypothesized value of 0.29.

>  **Details**
>
> **$H_0$ and the Equals Sign**
> In this book, the null hypothesis always has an equals sign, no matter which alternative hypothesis is used.

## EXAMPLE 7 Age Discrimination

According to the U.S. Census, about 30% of the population in Silicon Valley, a region in northern California, is between the ages of 40 and 65. A fired employee of a computer company who is in his fifties feels he is the victim of age discrimination. One piece of information he has is that only 2% of the 2100 employees at his former company are between the ages of 40 and 65 (www.workforce.com 2008). Here's what he wants to know: If the company hires at random, is this information evidence that it has a preference for younger employees?

QUESTION State the null and alternative hypotheses in words and symbols. Use the symbol $p$ to represent the proportion of all employees at this firm who are between the ages of 40 and 65.

SOLUTION

$H_0$: The proportion of employees at the company who are in this age group is 0.30, the same as that for the general population.

$H_a$: The proportion of employees at the company who are in this age group is less than that for the general population.

$$H_0: p = 0.30$$
$$H_a: p < 0.30$$

 Exercise 8.23

## Step 2: Prepare

To prepare for the actual computations, we must choose a test statistic, and we must make sure we know what the sampling distribution of this test statistic is. But even before we do that, we need to set the significance level.

### Set the Significance Level

The significance level is the probability that we will make the mistake of rejecting the null hypothesis when, in fact, the null hypothesis is true. Most scientific journals require a relatively low probability of making this mistake: 0.05. In symbols,

$$\alpha = 0.05$$

### Select a Test Statistic

Choosing the correct test statistic is not a big deal at this point, because you have seen only one test statistic to choose from—the one-proportion $z$-test. This test statistic works well, because when the null hypothesis is true we tend to get small values (close to 0), and when the null hypothesis is not true we tend to get values far from 0. The test statistic is also useful because we know a good approximation to its sampling distribution and so can compute the significance level and the p-value.

In Section 8.4 you will study a new test statistic used when comparing two population proportions. Later in the book, you will see test statistics for comparing means. Thus it is important to understand which test statistic is appropriate for the context you are working with.

### Check the Sampling Distribution Conditions

We will need to know the sampling distribution of our test statistic. At this point, you should check that the conditions listed in Section 8.1 hold. These conditions are (1) random sample, (2) large enough sample size (3) sampling without replacement, and (4) independence. Condition (5), that the null hypothesis is true, is a condition we will assume to be met for the time being. If these conditions are not met, then other approaches are needed.

> **KEY POINT**  If the null hypothesis is true, and if the four conditions for applying the Central Limit Theorem hold, then the test statistic $z$ follows a N(0, 1) distribution. Extreme values are rare in a N(0, 1) distribution, so if we see an extreme value, it is evidence that the null hypothesis is not true.

## EXAMPLE 8 Preparing for Situation 1: Voting

A researcher is wondering whether political scientists vote in the same proportion as the general public. The hypotheses for this test are

$$H_0: p = 0.466$$
$$H_a: p \neq 0.466$$

where $p$ is the proportion of all political scientists registered to vote in Florida. The data are from a random sample of 54 political scientists.

QUESTION Check that the conditions hold so that the sampling distribution of the $z$-statistic will approximately follow the standard Normal distribution.

SOLUTION We are told that the data come from a random sample, and this satisfies the first condition. We must next check that the sample size of 54 is large enough to produce at least 10 successes and 10 failures.

If the null hypothesis is true, the probability of success is $p_0 = 0.466$. Because $n = 54$,

$$np_0 = 54 \times 0.466 = 25.16 > 10$$

$$n(1 - p_0) = 54 \times (1 - 0.466) = 28.84 > 10$$

The third condition is true if the population—all political scientists registered to vote in Florida—is more than 10 times bigger than the sample size; that is,

$$\text{Population size bigger than } 10 \times 54 = 540$$

We confess that we do not know the number of political scientists in Florida. We cannot check this condition, but we *assume* that it is true.

CONCLUSION If the null hypothesis is true (as we assume when beginning these tests), then the conditions hold and the one-proportion $z$-test is appropriate to test the hypotheses.

TRY THIS! Exercise 8.29

## Step 3: Compute to Compare

At last, we are at the step where we consider actual data. Our test statistic, usually calculated with technology, will compare the value of the statistic provided by the data ($\hat{p}$) with the value that the null hypothesis says we should see ($p_0$). We then find the p-value to determine whether this value is unusual, given our assumption that the null hypothesis is true.

### Find the Observed Value of the Test Statistic

Let's see how this works for the researcher studying political scientists' voting habits. In the sample of 54 political scientists, the researcher reported that 40 of them voted. In Example 8 we found that the one-proportion $z$-test is appropriate for these data. So let's find the observed value of the $z$-statistic and interpret it. For this test, the hypotheses are

$$H_0: p = 0.466$$

$$H_a: p \neq 0.466$$

The data: The researcher reports that in his sample of 54 political scientists, 40 of them voted. The sample proportion is therefore $\hat{p} = (40/54) = 0.7407$. We wish to compare this value to 0.466, the value that the null hypothesis leads us to expect. To do this, we find the $z$-statistic (Formula 8.1), which first requires that we find the standard error.

$$SE = \sqrt{\frac{p_0(1 - p_0)}{n}} = \sqrt{\frac{0.466 \times (1 - 0.466)}{54}} = 0.067884$$

Now we substitute the standard error into our $z$-statistic:

$$z_{observed} = \frac{\hat{p} - p_0}{SE} = \frac{0.740741 - 0.466}{0.067884} = 4.05$$

## Find the p-Value

How surprised are we by this outcome, if the null hypothesis is true? Recall from Section 8.1 that the p-value, one of the main ingredients of a hypothesis test, measures our surprise, because it is the probability of getting a value as extreme as or more extreme than what we observed—in other words, as extreme as or more extreme than 4.05.

## EXAMPLE 8 (continued) Is the Vote Surprising?

In our example involving political scientists' voting behavior, our observed z-value was 4.05, and the alternative hypothesis was

$$H_a: p \neq 0.466$$

We have verified the conditions required for the sampling distribution to be well approximated by the N(0, 1) distribution.

QUESTION Find the p-value.

SOLUTION The p-value is the probability of getting a test statistic as extreme as or more extreme than 4.05. Because our alternative hypothesis is two-tailed, this translates to "the probability of getting a test statistic even farther from 0 than 4.05 is." Visually, it looks like Figure 8.7, where the shaded area corresponds to the p-value.

The p-value can be found using the methods discussed in Chapter 6 to find probabilities for the standard Normal distribution, or we can take advantage of the fact that many software packages provide the p-value automatically when asked to perform hypothesis tests. Figure 8.8 shows the p-value as calculated by the TI-83/84 one-proportion z-test.

CONCLUSION If the null hypothesis is true, then the probability of getting a z-statistic as far as or farther from 0 than 4.05 is 0.0001. This is a rather small probability, which suggests that if the null hypothesis is true, the outcome is surprising.

TRY THIS! Exercise 8.31

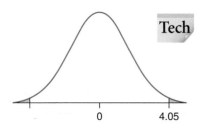

▲ **FIGURE 8.7** The p-value as a shaded area. This value is from a two-tailed test for which z comes out to be 4.05. The area has been enlarged a bit so that it can be seen readily.

Tech

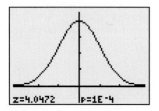

z=4.0472    p=1E-4

▲ **FIGURE 8.8** TI-83/84 output for the voting test for political scientists. The p-value is given in exponential notation, as 1 times 10 to the negative fourth power ($10^{-4}$), which is 0.0001. Note that the shaded area is so small that it is not noticeable.

## EXAMPLE 9 Dreaming in Color

In situation 2, the researchers are wondering whether a greater proportion of people now dream in color than did so before color television and movies became as prominent as they are today. The hypotheses they are checking are

$$H_0: p = 0.29$$
$$H_a: p > 0.29$$

The researchers took a random sample of 113 people. Of these 113 people, 92 reported dreaming in color.

QUESTION Find the value of the sample proportion, $\hat{p}$. Find the observed value of the test statistic, and then find the p-value associated with this observed value. In the conclusion, interpret the p-value in context. Assume that the conditions that must be met in order for us to use the N(0, 1) distribution as the sampling distribution are satisfied.

SOLUTION

The sample proportion is $\hat{p} = \dfrac{92}{113} = 0.814159$.

We can then compute the observed value of the test statistic:

$$z_{observed} = \frac{0.814159 - 0.29}{\sqrt{\frac{0.29(1 - 0.29)}{113}}} = \frac{0.524159}{0.042686} = 12.28$$

The p-value is the area under the N(0, 1) curve to the right (above) 12.28. We know, without looking, that this area must be quite small, because it is extremely rare for a value from a N(0, 1) distribution to be greater than 3, so 12.28 is a very large number. Therefore, the p-value is quite small, and we report that it is less than 0.001.

CONCLUSION If the proportion of people who now dream in color is the same as it was before color television, then the probability of getting a test statistic as large as or larger than 12.28 is extremely small.

TRY THIS! Exercise 8.33

**Details**

**Small p-Values**
When small p-values, such as 0.0001, occur, many software packages round off and report the p-value as "p < 0.001" (or use some other small value). We will follow that practice in this book.

You should always draw a sketch before you compute the p-value, even if you use technology (as we strongly recommend) to find the probability.

## EXAMPLE 10 Improving Attendance in Calculus

A very large calculus class at a university has been troubled by a high attrition rate: Over the last 5 years, 15% of the students who started the class did not finish. An instructor thinks that one reason for this is that some students need help with algebra. She randomly selects 100 students and assigns them to meet weekly with teaching assistants to study algebra. At the end of the semester, she will see what proportion of these 100 students have dropped the course. If $p$ represents the proportion of all students who drop the course and had algebra tutoring, her hypotheses can be written as

$$H_0: p = 0.15$$
$$H_a: p < 0.15$$

After the study is finished, she finds that her test statistic is $z = -1.68$. Assume that all of the required conditions hold for the one-proportion $z$-test.

QUESTION Three possible p-values are shown. One of them corresponds to the p-value for the alternative hypothesis $p > 0.15$, one to the alternative hypothesis $p < 0.15$, and one to the two-tailed alternative hypothesis $p \neq 0.15$. Without using any technology or referring to Normal tables, indicate which of the following numbers is the correct p-value. Explain.

a. p-value = 0.0930
b. p-value = 0.0465
c. p-value = 0.9535

SOLUTION Draw a sketch, as in Figure 8.9. This sketch shows that the p-value must be less than 50%, which eliminates answer c. Answer a is twice as big as answer b, so answer a must correspond to the two-tailed alternative. The correct answer is therefore answer b: p-value = 0.0465.

TRY THIS! Exercise 8.35

**Details**

**Sign of z**
$$z = \frac{\hat{p} - p}{SE}$$
The sign of $z$ is determined by the difference $\hat{p} - p$. A negative sign for $z$ means that $\hat{p}$ was smaller than $p$.

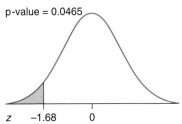

p-value = 0.0465

z   −1.68   0

▲ FIGURE 8.9 A representation of the p-value when the alternative hypothesis is left-tailed and the z-statistic is −1.68.

## Step 4: Interpret

The p-value measures our surprise (or lack of surprise) at the outcome of our test, but what should we do about this number? When is the outcome so unusual that we should reject the null hypothesis?

We apply a simple rule: Reject the null hypothesis if the p-value is smaller than (or equal to) the value chosen for the significance level, $\alpha$. If the p-value is larger than the significance level, do not reject the null hypothesis. For most applications, this means you reject the null hypothesis if the p-value is less than or equal to 0.05.

Following this rule ensures that our significance level is achieved. In other words, by following this rule and rejecting the null hypothesis when the p-value is less than or equal to 0.05, we know that there is only a 5% chance that we are mistakenly rejecting the null hypothesis (rejecting $H_0$ even though $H_0$ is true).

For the study of political scientists' voting pattern, our p-value was 0.0001, which is quite a bit smaller than 0.05. We therefore reject the null hypothesis and conclude that political scientists do not vote at the same rate as the general population.

 **KEY POINT** To achieve a significance level of $\alpha$, reject the null hypothesis if the p-value is less than (or equal to) $\alpha$. If the p-value is greater than $\alpha$, do not reject the null hypothesis.

You should always state your decision: reject the null hypothesis or do not reject the null hypothesis, and describe this decision in the context of the problem.

Now that we have illustrated all four steps, it's time to tackle an entire hypothesis test.

## EXAMPLE 11  Creative Kids and Music

Some arts educators reasoned that exposure to music instruction would help students think more creatively in a variety of educational situations than students who did not receive music instruction. To test this, elementary school children were randomly assigned, with their parents' consent, to participate either in a standard music instruction course (the treatment group) or in a recreation course (the control group). All students were given the Torrance Test of Creative Thinking, which measures several components of creativity, both before the music course began (in October) and after the course ended (in May). This test is fairly complex, and a student's score could change slightly just by chance. Of the 30 students in the treatment group, 19 increased their creativity scores by the end of the course. (The others' scores either were unchanged or decreased.) Is this evidence that the music program is effective? If the program had no effect, we would expect, due to chance, that about half of the children would exhibit an increase in creativity and that half would exhibit a decrease.

**QUESTION** Test the hypothesis that the probability that a child's creativity score will increase after participating in this music program is better than 50%. Assume the children are a random sample from a larger population of elementary school children. Use a 5% significance level.

**SOLUTION**

*Step 1: Hypothesize*
Let $p$ represent the probability that a child's creativity score will improve.

$$H_0\!: p = 0.50$$

$$H_a\!: p > 0.50$$

*Step 2: Prepare*

Use the one-proportion *z*-test. The children are assumed to be a random sample, and the measurements of their creativity scores can be assumed to be independent of each other. Thirty children is enough of a sample because $np_0 = 30(0.5) = 15$; and $n(1 - p_0) = 30(0.5) = 15$, and both are bigger than 10.

*Step 3: Compute to compare*

$$\hat{p} = \frac{19}{30} = 0.633333$$

We now compare the observed proportion to the hypothetical value of 0.50.

$$z_{\text{observed}} = \frac{0.633333 - 0.50}{\sqrt{\dfrac{(0.50)(1 - 0.50)}{30}}} = \frac{0.133333}{0.091287} = 1.46$$

The p-value is the area under a N(0, 1) curve and to the right of 1.46. This area, which is shown in Figure 8.10 and also in Figure 8.11, gives us

$$\text{p-value} = 0.072$$

*Step 4: Interpret*

Because the p-value is greater than 0.05, we cannot reject $H_0$. In other words, we do not have enough evidence to conclude that creativity scores increased. This is not the same as saying that the music program was ineffective; we are simply concluding that if it *is* effective, we do not have enough evidence to detect that effectiveness.

Most statistical software packages show you both *z* and the p-value at once. Some packages (see Figure 8.11) even show the Normal curve with the p-value shaded.

 **TRY THIS!** Exercise 8.41

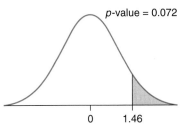

▲ **FIGURE 8.10** The shaded area represents the p-value, the probability of getting a test statistic as large as or larger than 1.46 when the true proportion is 0.50.

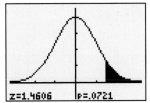

▲ **FIGURE 8.11** TI-83/84 output for 19 out of 30, one-tailed, for which $H_a$ is $p > 0.50$. (Choose Draw to get the curve.)

> ⚠ **Caution**
>
> **Do Not Accept**
>
> For reasons we'll explain in Section 8.5, it is incorrect to imply that we accept the null hypothesis. We do *not* say, "The music program was ineffective." Instead, we say, "We do not have enough evidence to show that the program is effective."

---

**SNAPSHOT** **ONE-PROPORTION *z*-TEST**

**WHAT IS IT?** ▶ A procedure for choosing between two hypotheses about the true value of a single population proportion. The test statistic is

$$z = \frac{\hat{p} - p_0}{SE} \quad \text{where } SE = \sqrt{\frac{p_0(1 - p_0)}{n}}$$

**WHAT DOES IT DO?** ▶ Because estimates of population parameters are uncertain, a hypothesis test gives us a way of making a decision while knowing the probability that we will incorrectly reject the null hypothesis.

**HOW DOES IT DO IT?** ▶ The test statistic *z* compares the sample proportion to the hypothesized population proportion. Large values of the test statistic tend to discredit the null hypothesis.

**HOW IS IT USED?** ▶ When proposing hypotheses about a single population proportion. The data must be from an independent, random sample and the sample size must be sufficiently large. The N(0, 1) distribution is used to find the p-value for the observed test statistic.

# Comparing Proportions from Two Populations

You have now seen how to carry out a hypothesis test for a single population proportion. With very few changes, this procedure can be altered to accommodate a more interesting situation: comparing proportions from two populations.

Consider as an example embryonic stem cell research, medical research that shows great promise in the treatment of several major diseases. It is controversial, however, partly because it goes against many people's religious convictions. The Pew Forum on Religion & Public Life has, over time, conducted surveys to assess Americans' support for stem cell research. These surveys raise the question of whether Americans' support has changed since 2002, when 43% of Americans in the Pew sample expressed support for stem cell research. In a more recent survey conducted in August 2009, 58% of the sample expressed support (Pew Forum 2009). Can we conclude that support has changed in the population of all Americans? Or is this difference due to chance variation from the sampling procedure?

This problem involves two populations. One population consists of all Americans in 2002, and the second population consists of all Americans in 2009. Each population has a true proportion who support stem cell research, but in each case we cannot know this true value. Instead, we have a random sample taken from both populations, and we must estimate the two proportions from these two random samples.

Here are the changes we need to make to our "ingredients" in order to compare proportions from two populations.

## Changes to Ingredients: The Hypotheses

Because we now have two population proportions to consider, we need some new notation. We'll let $p_1$ represent the proportion of Americans who supported stem cell research in 2002, and we'll let $p_2$ represent the proportion who supported it in 2009.

We are not interested in the actual numerical values of $p_1$ and $p_2$, as we were when dealing with just one population proportion. We are interested only in their relation to each other. The conservative, status-quo, not-worth-a-headline hypothesis is that these proportions are the same. In other words, there has been no change in support. We write this as

$$H_0: p_1 = p_2$$

In words, the null hypothesis says that the proportion of Americans who support stem cell research was the same in 2009 as it was in 2002.

The alternative hypothesis is that the proportion of Americans who support stem cell research has changed. If so, the two proportions are no longer equal.

$$H_a: p_1 \neq p_2$$

One-tailed hypotheses are also possible. Our research question might instead have been "Has support for stem cell research increased?" If that had been our question, then we would have used

$$H_a: p_1 < p_2$$

And if we had wished to answer the question "Has support for stem cell research decreased?" we would have used this alternative:

$$H_a: p_1 > p_2$$

These options lead to three pairs of hypotheses, as shown in Table 8.2. You choose the pair that corresponds to the research question your study hopes to answer. Note that the null hypothesis is always $p_1 = p_2$ because the neutral position is always that the two proportions are the same.

| Hypothesis | Symbols | The Alternative in Words |
|---|---|---|
| Two-tailed | $H_0: p_1 = p_2$<br>$H_a: p_1 \neq p_2$ | The proportions are different in the two populations. |
| One-tailed (left) | $H_0: p_1 = p_2$<br>$H_a: p_1 < p_2$ | The proportion in population 1 is less than the proportion in population 2. |
| One-tailed (right) | $H_0: p_1 = p_2$<br>$H_a: p_1 > p_2$ | The proportion in population 1 is greater than the proportion in population 2. |

◀ **TABLE 8.2** Possible hypotheses for a two-proportion hypothesis test.

## Changes to Ingredients: The Test Statistic

We are interested in how $p_1$ and $p_2$ differ, so our test statistic is based on the difference between our sample proportions from the two populations. The test statistic we will use has the same structure as the one-sample $z$-statistic:

$$z = \frac{\text{estimator} - \text{null value}}{SE}$$

However, the estimator for the **two-proportion $z$-test** is $\hat{p}_1 - \hat{p}_2$ because we are estimating the difference $p_1 - p_2$. Here $\hat{p}_1$ and $\hat{p}_2$ are just the sample proportions for the different samples. In our case, $\hat{p}_1$ is the sample proportion for the people surveyed in 2002 (reported as 0.43), and $\hat{p}_2$ is the sample proportion for the people surveyed in 2009 (reported as 0.58).

The null value is 0, because the null hypothesis claims these proportions are the same, so $p_1 - p_2 = 0$.

The standard error, SE, is more complicated than in the one-sample case, because the null hypothesis no longer tells us the value of the population proportion. All it tells us is that both populations have the same value. For this reason, when we estimate this single value, we pool the two samples together. Formula 8.2 shows you how to do this.

**Formula 8.2: The two-proportion $z$-test statistic**

$$z = \frac{\hat{p}_1 - \hat{p}_2 - 0}{SE}$$

where

$$SE = \sqrt{\hat{p}(1 - \hat{p})\left(\frac{1}{n_1} + \frac{1}{n_2}\right)}$$

$n_1$ = sample size in sample 1

$n_2$ = sample size in sample 2

$$\hat{p} = \frac{\text{number of successes in sample } 1 + \text{number of successes in sample } 2}{n_1 + n_2}$$

$$\hat{p}_1 = \text{proportion of successes in sample 1} = \frac{\text{number of successes in sample 1}}{n_1}$$

$$\hat{p}_2 = \text{proportion of successes in sample 2} = \frac{\text{number of successes in sample 2}}{n_2}$$

Formula 8.2 is perhaps the most elaborate formula we have shown you so far. As usual, it is much more important to be able to use technology to perform this test than to apply the formula. Still, studying the formula does help us understand why the test statistic is useful.

## EXAMPLE 12 Pew Survey on Stem Cell Research

The researchers from the Pew study interviewed two random samples. Both samples, the one taken in 2002 and the one taken in 2009, had 1500 people. In 2002, 645 people expressed support for stem cell research. In 2009, 870 expressed support. These data are summarized in Table 8.3.

▶ **TABLE 8.3** Data for the Pew Study

|  | 2002 | 2009 | Total |
|---|---|---|---|
| Support Stem Cell Research | 645 | 870 | 1515 |
| Do Not Support Stem Cell Research | 855 | 630 | 1485 |
| Total | 1500 | 1500 | 3000 |

> **Looking Back**
>
> **Two-way Tables**
> Two-way tables, such as the one used to summarize the data in Example 12, were first presented in Chapter 1.

**QUESTION** Find the observed value of the test statistic to test the hypotheses

$$H_0: p_1 = p_2$$
$$H_a: p_1 \neq p_2$$

where $p_1$ represents the proportion of Americans who supported stem cell research in 2002, and $p_2$ represents the proportion who supported this research in 2009.

**SOLUTION** We must bring all the pieces together and assemble them into the test statistic:

$$\hat{p}_1 = \frac{645}{1500} = 0.43 \text{ (a value we knew already from the original report)}$$

$$\hat{p}_2 = \frac{870}{1500} = 0.58 \text{ (another value we knew from the original report)}$$

$$\hat{p} = \frac{645 + 870}{1500 + 1500} = 0.505 \text{ (a pooled estimate of the sample proportion)}$$

$$SE = \sqrt{0.505(1 - 0.505)\left(\frac{1}{1500} + \frac{1}{1500}\right)} = 0.018257$$

> **Details**
>
> **The Sign of z**
> In a two-proportion test, whether z is positive or negative depends on which population you call "1" and which you call "2." It's important to pay attention to which proportion is subtracted from which!

Now we assemble the pieces:

$$z_{observed} = \frac{0.43 - 0.58}{0.018257} = -8.22$$

**TRY THIS!** Exercise 8.49

## Changes to Ingredients: Checking Conditions

The conditions that we need to check for a two-sample test of proportions are similar to those for a one-sample test, but with some additional things to consider.

1. *Large Samples*: Both sample sizes must be large enough. Because we don't know the value of $p_1$ or $p_2$, we must use an estimate. The null hypothesis says that these two proportions are the same, so we use $\hat{p}$, the pooled sample proportion, to check this condition. Do not use $\hat{p}_1$ or $\hat{p}_2$. This means that we need

   a. $n_1\hat{p} \geq 10$ and $n_1(1 - \hat{p}) \geq 10$

   b. $n_2\hat{p} \geq 10$ and $n_2(1 - \hat{p}) \geq 10$

2. *Random Samples*: The samples are drawn randomly from the appropriate population. In practice, this condition is often impossible to check unless we were present when the data were collected. In such cases, we have to assume that it is true.

3. *Independent Samples*: The samples are independent of each other. This condition is violated if, for example, the same individuals are in both samples that we are comparing.

4. *Independent within Samples*: The observations within each sample must be independent of one another.

5. *Null Hypothesis*: The null hypothesis is true.

If the first four conditions hold, then, assuming the null hypothesis to be true, $z$ follows a N(0, 1) distribution.

# EXAMPLE 13 Caffeine for Babies

Apnea of prematurity occurs when premature babies have shallow breathing or stop breathing for more than 20 seconds. One therapy for this condition is to give caffeine to the premature infants. Medical researchers conducted an international study in which one sample of premature infants was randomly assigned to receive caffeine therapy, and another sample received a placebo therapy. Researchers compared the rate of severely negative outcomes (death and severe disabilities) in the two groups to determine whether the caffeine therapy would lower the rate of such bad events.

The caffeine therapy group included 937 infants. Of these 937 infants, 377 suffered from death or disability. The placebo group had 932 infants, and of these, 431 suffered from death or disability (Schmidt et al., 2007). These data are summarized in Figure 8.12.

**QUESTION** Perform a four-step hypothesis test to test whether the caffeine therapy was effective (that is, whether it succeeded in lowering the death or disability proportion).

**SOLUTION** As always, hypotheses are about populations, not samples. In this example, we have two samples, consisting of 937 and 932 infants, but our hypothesis is about *all* premature infants who might receive caffeine or placebo therapy.

Let $p_1$ represent the proportion of death or disability in all infants who could receive caffeine therapy, and let $p_2$ represent the proportion of death or disability in all infants who could receive the placebo therapy.

*Step 1: Hypothesize*
The null hypothesis is neutral; it states that the therapy changes nothing. If so, then the proportions of both populations are the same.

$$H_0: p_1 = p_2$$

The alternative hypothesis is that the therapy will help, so the proportion of bad events in the caffeine therapy group should be lower than that in the placebo group.

$$H_a: p_1 < p_2$$

*Step 2: Prepare*
We use a significance level of $\alpha = 0.05$.

We are comparing proportions from two populations, so we need to make sure that we can use the two-proportion $z$-test. Here are the conditions we check:

1. Are both samples large enough?
We find that

$$\hat{p} = \frac{377 + 431}{937 + 932} = \frac{808}{1869} = 0.432317$$

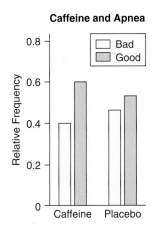

**Caffeine and Apnea**

▲ **FIGURE 8.12** Bar graph for caffeine treatment for apnea of prematurity.

First sample: $n_1 \times 0.432317 = 937 \times 0.432317 = 405$, which is greater than 10, and $937 \times (1 - 0.4323) = 532$, which is also greater than 10.

Second sample: $n_2 \times 0.432317 = 932 \times 0.4323 = 403$, which is greater than 10, and $932 \times (1 - 0.432317) = 529$, which is also greater than 10.

2. Are the samples are drawn randomly from the appropriate population?
We are not told whether the samples are random, but for the purpose of this exercise we will assume that they are.

3. Are the samples independent of each other?
The samples are independent of each other because the caffeine treatment is given to a different group of babies from those who received the placebo.

With these three conditions checked, we can proceed to step 3.

*Step 3: Compute to compare*
We must find the individual pieces of Formula 8.2. We defined $p_1$ to be the proportion of babies who could receive caffeine therapy in the population and who suffer from bad events. So let $\hat{p}_1$ be the proportion of babies in the caffeine therapy group who suffered these bad events. The sample size in this group was $n_1 = 937$, and 377 suffered bad events. Then

$$\hat{p}_1 = \frac{377}{937} = 0.4023$$

The sample proportion for the placebo group is

$$\hat{p}_2 = \frac{431}{932} = 0.4624$$

To find the standard error, we need to use $\hat{p}$, the proportion of bad events that happen in the sample if we ignore the fact that the babies belong to two different groups. Above we found that $\hat{p} = 0.4323$.

The standard error is then

$$SE = \sqrt{0.432317(1 - 0.432317)\left(\frac{1}{937} + \frac{1}{932}\right)} = 0.022918$$

Putting it all together yields

$$z_{observed} = \frac{0.402348 - 0.462446}{0.022918} = \frac{-0.060098}{0.022918} = -2.62$$

Now that we know the observed value, we must measure our surprise. The null hypothesis assumes the two population proportions are the same and that our sample proportions differed only by chance. The p-value will measure the probability of getting an outcome as extreme as or more extreme than $-2.62$, assuming that the population proportions are the same.

The p-value is calculated exactly the same way as with a one-proportion $z$-test. Our alternative hypothesis is a left-tailed hypothesis, so we need to find the probability of getting a value less than the observed value (Figure 8.13).

Figure 8.14 shows the $z$-value and the p-value from technology.

$$\text{p-value} = 0.004$$

*Step 4: Interpret*
The p-value is less than our significance level of 0.05, so we reject the null hypothesis. We conclude that the caffeine therapy does help: A lower proportion of babies will die or suffer disability with this therapy.

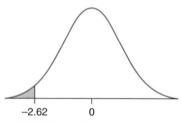

**▲ FIGURE 8.13** The shaded area represents the area to the left of a $z$-value of $-2.62$.

```
2-PropZTest
 P1<P2
 z=-2.622298815
 P=.0043669784
 P̂1=.4023479189
 P̂2=.4624463519
↓P̂=.4323167469
■
```

**▲ FIGURE 8.14** TI-83/84 output for the p-value of the test of caffeine on babies.

Note that because this is a controlled, randomized experiment (the researchers randomly assigned babies to treatment groups), we can make a cause-and-effect conclusion about the effectiveness of the treatment.

**TRY THIS!** Exercise 8.51

---

**SNAPSHOT** **TWO-PROPORTION $z$-TEST**

**WHAT IS IT?** ▶ A hypothesis test.

**WHAT DOES IT DO?** ▶ Provides a procedure for comparing two population proportions. The null hypothesis is always that the proportions are the same, and this procedure gives us a way to reject or fail to reject that hypothesis.

**HOW DOES IT DO IT?** ▶ The test statistic $z$ compares the differences between the sample proportions and the value 0 (which is what the null hypothesis says this difference should be):

$$z = \frac{\hat{p}_1 - \hat{p}_2 - 0}{SE} \text{ where } SE = \sqrt{\hat{p}(1 - \hat{p})\left(\frac{1}{n_1} + \frac{1}{n_2}\right)}$$

Values of $z$ far from 0 tend to discredit the null hypothesis.

**HOW IS IT USED?** ▶ When comparing two proportions, each from a different population. The data must come from two independent, random samples, and each sample must be sufficiently large. Then N(0, 1) can be used to compute the p-value for the observed test statistic.

---

## SECTION 8.5

# Understanding Hypothesis Testing

Now that you've seen how to do a hypothesis test, you need to know about a few subtleties before you can become an expert.

## If Conditions Fail

If the conditions fail to be met for the hypothesis test, the $z$-statistic will not follow a Normal distribution when the null hypothesis is true. This means that we cannot find the p-value using the Normal curve (using Table 2 in Appendix A or technology). However, other approaches often exist.

### Sample Sizes Too Small

The Normal distribution is only an approximation to the true distribution of the $z$-statistic. If the sample sizes are large enough, then the approximation is very good. If the sample sizes are too small, then the approximation may not be good, but other tests can be used. (One such test you might see in a future statistics class is called Fisher's Exact Test.)

## Samples Not Independent

Suppose we survey a sample of consumers and determine the proportion that will buy a new car in the next year. This same sample then watches a car commercial, and by asking again whether they plan to buy a new car in the next year, we determine a second proportion. We do this because we want to know whether the commercial changed minds. In other words, has the proportion that will buy a new car increased after the sample watched the commercial? Two populations are being studied: those who have not seen the commercial and those who have. However, the two samples we collect consist of the same people, because we interview each person both before and after the commercial. Thus the samples are *not* independent. If you find yourself in this situation, all is not lost, but you should consult a statistician to consider your options.

## Samples Not Randomly Selected

If samples are not selected randomly, then it is not possible to make inferences about the populations the samples came from. That having been said, random samples are relatively rare in medical studies. For example, medical researchers cannot take a random sample of people with the medical condition they wish to study; they must rely on recruiting patients who come into hospitals. Psychologists at universities often study students, particularly students who are willing to submit to a study for a small amount of money or the chance to win a prize in a raffle. They cannot take a random sample of all people in the population they wish to study and fly them to the university to participate in their experiment.

These convenience samples are problematic for the statistical techniques in this book—and indeed for any statistical technique. Sometimes we get around this by assuming that the samples are random, or at the very least representative of the population. However, we have no guarantee that the conclusions we make with this assumption are valid or useful.

In many situations, though, researchers are not that interested in generalizing to the larger population. For example, in the study of apnea in premature infants, the researchers were interested in knowing whether caffeine therapy worked for this group of infants or if, instead, the differences between the caffeine therapy infants and the placebo infants could be explained by chance. They concluded that the differences were too large to be due to chance, so they felt confident concluding that at least for this group, the treatment was effective. Research with other groups is still needed to see whether the results obtained are **replicable**, but the study is at least an encouraging start because researchers know the therapy works with some infants.

## Controlling Mistakes: The Role of Sample Size

In Section 8.1 we mentioned one type of mistake that we might make when carrying out a hypothesis test. This mistake occurs when we reject the null hypothesis even though it is true. The probability of making this mistake was called the significance level.

Another mistake we might make is to *fail* to reject the null hypothesis, even though it is not correct. We might make either of these two types of mistakes when doing a formal hypothesis test.

To better understand these errors, imagine sitting on a jury in a criminal trial. You are making the first type of mistake if you convict the defendant of being guilty when, in fact, he is innocent. You are making the second type if you declare the defendant not guilty when, in fact, he did commit the crime.

---

**Details**

**Replicable Results**
The results of a study are said to have been replicated when researchers using new subjects come to the same conclusion.

---

**Caution**

**Cause and Effect**
Remember that we can conclude that there is a cause-and-effect relationship between a treatment variable and a response variable only when we have a controlled experiment that uses random assignment, includes a placebo (or comparison) treatment, and is double-blind.

---

**Details**

**Mistakes (Errors)**
We are not referring here to calculation mistakes. The mistakes we refer to occur because sometimes, just by chance, we get sample data that are very unusual and come to a mistaken conclusion about the population from which the data came.

But why be so negative? Rather than focusing on what can go wrong if we mistakenly fail to reject the null hypothesis, we instead focus on its complement: correctly rejecting the null hypothesis, or, in the context of a criminal trial, correctly convicting a guilty person. The probability of rejecting the null hypothesis when the null hypothesis is wrong—the probability of doing the right thing—is called the **power**.

We can ensure that the significance level is set at 5% just by rejecting $H_0$ whenever the p-value is less than or equal to 0.05, but the power is a little more difficult to control. This is because the null hypothesis can be wrong in many different ways, so it is harder to measure the probability of correctly rejecting the null.

The power depends on three factors:

1. Just how wrong the null hypothesis truly is. If the null hypothesis value, $p_0$, is very close to the true value of $p$, then the probability of correctly identifying that they are different is small, because the two values are very similar. So the power will be low and we are unlikely to make the right decision. On the other hand, if the true value of $p$ is very different from $p_0$, then it is easy to tell them apart, and the probability of making the correct decision is high. The basic idea here is that it is easier to confuse an alligator with a crocodile than to confuse it with a chicken.

2. The sample size. The larger the sample size, the bigger the power. More is always better (although more expensive) when it comes to sample size. With a larger sample, it's easier to tell the alligators from the crocodiles.

3. The significance level. The smaller the significance level, the smaller the power. (Remember: small power, bad; small significance level, good.) If you want a small probability of mistakenly rejecting the null hypothesis when it is in fact true, then you must live with a smaller probability that you will correctly reject it when it is false. We discuss this tradeoff after the next paragraph.

Calculating the power is tricky and somewhat complex, in part because it requires that we know the true value of the population proportion. We leave this calculation to a future statistics course. For now, be aware that if you do a hypothesis test and do not reject the null, then there is always the chance that you have made a mistake because your power is too low. You simply don't have enough evidence to tell the difference between the null hypothesis and the truth.

## The Tradeoff between Significance Level and Power

We are free to choose any value we wish for the significance level. Typically, we set this probability at 0.05, but sometimes we go as low as 0.01. But why don't we make it arbitrarily small? Say 0.0000001? That way, we'd almost never make this mistake.

We can't make the significance level as small as we would like because we have a price to pay. The price is that if we make the significance level smaller, then the power gets smaller too!

To see this, think about our criminal justice example. The null hypothesis is that the defendant is innocent. The alternative is that he is guilty. A mistake occurs when we convict an innocent man (mistakenly reject the null hypothesis). Another type of mistake occurs when we free a guilty man (fail to reject the null hypothesis even though it is false). We can make the significance level, the probability of convicting an innocent man, 0 by following a simple rule: Free every defendant. If everyone goes free, then it is impossible to convict an innocent person because you are convicting no one. But now the power—the probability of correctly convicting a guilty person—is 0%.

**KEY POINT**    We cannot make the significance level arbitrarily small because doing so lowers the power—the probability that we will correctly reject the null hypothesis.

Of course, we could increase the power by changing course and convicting every defendant. Now the power is 100% because every guilty person will be convicted. But the significance level has gone to 100% as well, because every innocent person also gets convicted.

There is only one way out if you want to lower the significance level and keep the power high: Increase the sample size. This relationship between power and significance level is summarized in Table 8.4.

▶ **TABLE 8.4** Relationship between Power and Significance Level

| | Reject $H_0$ | Fail to Reject $H_0$ |
|---|---|---|
| $H_0$ True | Bad (The probability of doing this is called the significance level.) | Good |
| $H_0$ False | Good (The probability of doing this is called the power.) | Bad |

## So What? Statistical Significance vs. Practical Significance

Researchers call a result "statistically significant" when they reject the null hypothesis. This means that the difference between their data-estimated value for a parameter and the null hypothesis value is so large that it cannot be explained by chance. However, just because a difference is statistically significant does not mean it is useful or meaningful. A *practically significant* result is both statistically significant and meaningful.

For example, suppose that the proportion of people who get a certain type of cancer is 1 in ten million. However, a statistical analysis finds that those who talk on their cell phones every day have a statistically significant greater risk of getting that cancer, and that the risk is doubled. It may be true that using your cell phone is therefore more dangerous than not using it, but would you really stop talking on the phone if your risk would change from 2 in 10 million to 1 in 10 million? That's a pretty big change of habit for a pretty small change in risk. Most people would conclude that the difference in risk is statistically significant, but not practically significant.

 **KEY POINT** ▶ Statistically significant findings do not necessarily mean that the results are useful.

## Don't Change Hypotheses!

A researcher sets up a study to see whether caffeine affects our ability to concentrate. He has a large number of subjects, and he gives them a task to complete when they have not had any caffeine. The task takes some concentration to complete, and he records how long it takes them. Later, he asks them to complete the same task, only this time the subjects have had a dose of caffeine. Again he records the time, and he's interested in the proportion who take longer to complete the task with caffeine than without.

He isn't sure just what the effect of caffeine will be. It might help people concentrate, in which case only a small proportion of people will take longer. On the other hand, it might make people jittery so that a large proportion will take longer to complete the task. If caffeine has no effect, probably half will take the same amount of time or more, and half will take the same amount of time or less.

The researcher chooses a significance level of $\alpha = 0.05$ to test this pair of hypotheses:

$$H_0: p = 0.5$$

$$H_a: p \neq 0.5$$

The parameter $p$ represents the proportion of all people who would take longer to complete the task with caffeine than without. His alternative hypothesis is two-tailed because he does not know what the effect will be—that is, whether it will increase or decrease concentration.

He collects his data and gets a $z$-statistic of $-1.81$. This leads to a p-value of 0.07—and to a moral dilemma! (Figure 8.15 illustrates this p-value.) The researcher needs a p-value less than or equal to 0.05 if he is to publish this paper, because no one wants to hear about an insignificant result. Also, he needs to publish more papers if he is going to get that promotion he really wants.

However, it occurs to this researcher that if he had a different alternative hypothesis, his p-value would be different. Specifically, if he had used

$$H_a: p < 0.50$$

then the p-value would have been the area in just the lower tail. In that case, his p-value would be 0.035 and he would reject the null hypothesis.

What the researcher has just done is the sort of thing that small children (and politicians) do in a contest: They change the rules midway through so that they can win. As they say in the western movies, "You gotta dance with the gal/guy that brung ya." You can't choose your hypotheses to fit the data. By doing so, you are increasing the true significance level of your test.

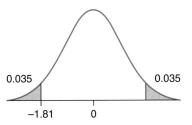

▲ **FIGURE 8.15** The shaded areas represent the p-value of 0.070 for a test statistic of $z = -1.81$ in a two-tailed hypothesis test.

## Hypothesis-Testing Logic

Statisticians and scientists are rather touchy when it comes to talk about "proving" things. They often use softer words, such as "Our data demonstrate that . . ." or "Our data are consistent with the theory that . . . ." One reason is that in mathematical and scientific circles, the word *prove* has a very precise and very definite meaning. If something is proved, then it is absolutely, positively, and without any doubt true. However, in real life, and particularly in statistics and science (which we consider to be part of real life), you can never be completely certain. For this reason we avoid saying, for example, that we have proved that more people now believe they dream in color than thought so in the past.

On a similar note, it is improper (maybe even impolite!) to say that you have "accepted" the null hypothesis when your p-value is bigger than 0.05. Instead, we say "We have failed to reject $H_0$" or "We cannot reject $H_0$." The reason for this is that several factors might make it difficult to determine whether the null hypothesis is false. It could be that our statistical power was low, and we couldn't detect that the null hypothesis was false because we didn't have enough evidence.

In Example 11, we examined a study that asked whether music instruction improved children's creativity scores. We concluded that we could not reject the null hypothesis and that there was not enough evidence to show that the scores increased. Could we make a stronger statement? Specifically, could we conclude that the music instruction was *not* effective?

No. We might have had low power: That is, the probability of detecting a difference between the students' increase in scores for an effective music instruction program was too low. With a larger sample size, we might see that, in fact, the program is effective.

**KEY POINT** Don't say you "proved" something with statistics. Say you "demonstrated" it or "showed" it. Similarly, don't say you "accept the null hypothesis"; say, rather, that you "cannot reject the null hypothesis" or that you "failed to reject the null hypothesis" or that "there is insufficient evidence to reject the null hypothesis."

## CASE STUDY REVISITED

Does watching violence on TV as a child lead to violent behavior as an adult? To find out, researchers examined two independent groups of subjects. One group had watched a lot of violent television as children; the other group had not. The researchers recorded whether the subjects enacted a form of violence with their partners or spouses, including whether they had pushed or shoved a partner or spouse, which was defined as physical abuse. About 38% of those who had watched violent television as children said yes, compared to 22% of the watchers of nonviolent TV. Can we conclude, then, that these groups are truly different? Or might the observed difference be due simply to chance?

To find out, the researchers performed a hypothesis test. We can do a similar test with what we've learned in this chapter. The data are summarized in Table 8.5 and in Figure 8.16.

▶ **TABLE 8.5** Two-way Table for the Relationship between TV Violence and Physical Abuse

|  | High TV Violence | Low TV Violence | Total |
|---|---|---|---|
| Yes, Physical Abuse | 25 | 57 | 82 |
| No, Physical Abuse | 41 | 206 | 247 |
| Total | 66 | 263 | 329 |

▶ **FIGURE 8.16** Bar Chart for the Relationship between TV Violence and Physical Abuse

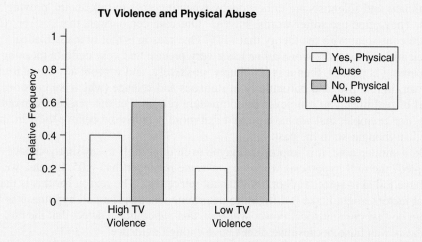

As it turns out, the differences between the two groups in the proportions of those who committed an act of violence is too large to be attributed to chance alone. Because this is an observational study, we cannot conclude that watching violent TV as a child causes adults to behave violently toward their spouses. Complicating factors could exist. For example, perhaps children raised by parents who behave violently toward each other are more likely to watch violent television *and* more likely to reenact their parents' behavior when they become adults. Still, we can rule out the possibility that these sample proportions differ only because of chance.

To come to this conclusion, we performed a two-proportion $z$-test with a significance level of $\alpha = 0.05$. The two populations consist of all the twenty-something adults who watched a lot of violent TV as children (population 1) and all the twenty-something adults who did not watch violent TV as children (population 2). The proportion of this first population that commit acts of violence directed at their partners we call $p_1$. The proportion of this second population that commit acts of violence against their partners we call $p_2$.

$H_0: p_1 = p_2$ Both groups have the same proportion of violent acts.

$H_a: p_1 > p_2$ The TV violence group has a greater proportion of violent acts.

A quick check shows that the sample sizes are large enough, and the other conditions can be assumed to hold as well. (One questionable condition is the one requiring random samples. However, we can assume the samples are random to see what the conclusion would be in this case. The fact that they are not really random means our results might not generalize to a larger group.)

The observed value of the $z$-statistic, calculated with technology, is $z = 2.72$ (see Figure 8.17). The p-value for this one-tailed hypothesis is 0.003. This small value tells us that the outcome was rather unusual if, in fact, the proportion of violent acts is the same in both populations. The fact that such an outcome is so rare leads us to reject the null hypothesis and conclude that those who watch violent TV as children do tend to commit more acts of violence toward their spouses as adults than do people who did not watch violent TV.

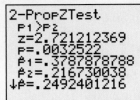

▲ **FIGURE 8.17** TI-83/84 output for a two-proportion $z$-test using a one-tailed alternative hypothesis, where the sample proportions are 25/66 and 57/263. Here $p_1$ is the proportion of those who watched violent TV who subsequently abused a partner, and $p_2$ is the proportion of those who did not watch violent TV who later abused a partner.

# EXPLORING STATISTICS
## CLASS ACTIVITY

## Identifying Flavors of Gum through Smell

| GOALS | MATERIALS |
|---|---|
| To use a hypothesis test to determine how well a person can distinguish between flavors based on smell alone. | • Gum (or candy) in two different flavors. We will call these flavor A and flavor B. Each student will need one piece of each flavor.<br>• A paper towel for each student. |

**ACTIVITY**

Pair up. One student will take the role of the sniffer, and the other will act as the researcher. Both students must know the two flavors that are being tested.

Researcher: Ask the sniffer to turn his or her back, then select a piece of gum at random and record the flavor of the gum on a sheet of paper. Place the gum on a sheet of paper towel.

Hold the gum about 2 inches below the sniffer's nose and ask him or her to identify which of the two flavors is in the hand. Sniffers may take as long as they like. Record whether the sniffer's response was correct next to the flavor that you wrote down for this trial.

Do 20 trials, then work together to determine the proportion of correct responses. If time permits, change roles.

**BEFORE THE ACTIVITY**

1. Let $p$ represent the probability that the sniffer makes a correct identification. If the sniffer is simply guessing and cannot tell the difference, what value would $p$ have?

2. If the sniffer is not guessing and really can tell the difference, would $p$ be bigger than, less than, or the same as the value you found in Question 1?

3. Suppose the sniffer cannot tell the difference between the two flavors. In 20 trials, about how many would you expect the sniffer to get right? What proportion is this? What is the greatest number of trials you'd expect the sniffer to get right if she or he were just guessing? What proportion is this?

4. How many trials would the sniffer have to get right before you would believe that he or she can tell the difference? All 20? 19? Explain.

5. Write a pair of hypotheses to test whether the sniffer is just guessing or can really tell the difference. Write the hypotheses in both words and symbols, using the parameter $p$ to represent the probability that the sniffer correctly identifies the scent.

**AFTER THE ACTIVITY**

1. Report the proportion of trials the sniffer got right.

2. Does this show that the sniffer could tell the difference in the scents? Explain.

3. If the sniffer is just guessing, what is the probability that he or she would have gotten as many right as, or more right than, the actual number recorded?

    a. Pretty likely: between 50% and 100%    b. Maybe: between 10% and 50%

    c. Fairly unlikely: between 5% and 10%    d. Very unlikely: between 0% and 5%

## CHAPTER REVIEW

### Key Terms

Hypothesis testing, *332*
Null hypothesis, $H_0$, *333*
Alternative hypothesis, $H_a$, *333*
Significance level, $\alpha$ (alpha), *335*

Test statistic, *335*
One-proportion *z*-test, *335*
*z*-statistic, *336*
p-value, *337*

Two-tailed hypothesis, *338*
One-tailed hypothesis, *338*
Left-tailed hypothesis, *338*
Right-tailed hypothesis, *339*

Two-proportion *z*-test, *351*
Replicable results, *356*
Power, *357*

### Learning Objectives

After reading this chapter and doing the assigned homework problems, you should

- Know how to test hypotheses concerning a population proportion and hypotheses concerning the comparison of two population proportions.

- Understand the meaning of p-value and how it is used.

- Understand the meaning of significance level and how it is used.

- Know the conditions required for calculating a p-value and significance level.

### Summary

Hypothesis tests are performed in the following four steps.

Step 1: Hypothesize.
Step 2: Prepare.
Step 3: Compute to compare.
Step 4: Interpret.

Step 1 is the most important, because it establishes the entire procedure. Hypotheses are *always* statements about parameters. The alternative hypothesis is the hypothesis that the researcher wishes to convince the public is true. The null hypothesis is the skeptical, neutral hypothesis. Each step of the hypothesis test is carried out assuming that the null hypothesis is true. For all tests in this book, the null hypothesis will always contain an equals ($=$) sign, whereas the alternative hypothesis can contain the symbol for "is greater than" ($>$), the symbol for "is less than" ($<$), or the symbol for "is not equal to" ($\neq$).

Step 2 requires that you decide what type of test you are doing. In this chapter, this means you are either testing the value of a proportion from a single population (one-proportion *z*-test) or comparing two proportions from different populations (two-proportion *z*-test). You must also check that the conditions necessary for using the standard Normal distribution as the sampling distribution are all met.

Step 3 is where the observed value of the statistics are compared to the null hypothesis. This step is most often handled by technology, which will compute a value of the test statistic and the p-value. These values are valid only if the conditions in step 2 are satisfied.

Step 4 requires you to compare the p-value, which measures our surprise at the outcome if the null hypothesis is true, to the significance level, which is the probability that we will mistakenly

reject the null hypothesis. If the p-value is less than (or equal to) the significance level, then you must reject the null hypothesis.

For a one-proportion *z*-test:

**Formula 8.1:** $z = \dfrac{\hat{p} - p_0}{SE}$

where

$$SE = \sqrt{\dfrac{p_0(1 - p_0)}{n}}$$

$p_0$ is the proposed population proportion
$\hat{p}$ (p-hat) is the sample proportion, $x/n$
$n$ is the sample size
For a two-proportion *z*-test:

**Formula 8.2:** $Z = \dfrac{\hat{p}_1 - \hat{p}_2 - 0}{SE}$

where

$$SE = \sqrt{\hat{p}(1 - \hat{p})\left(\dfrac{1}{n_1} + \dfrac{1}{n_2}\right)}$$

$$\hat{p} = \dfrac{\text{number of successes in both samples}}{n_1 + n_2}$$

$\hat{p}_1$ is the proportion of successes in the first sample, and $\hat{p}_2$ is the proportion in the second sample

Calculating the p-value depends on the which alternative hypothesis you are using. Figure 8.18 shows, from left to right, the p-values for a two-tailed test, a right-tailed test (one-tailed), and a left-tailed test (one-tailed).

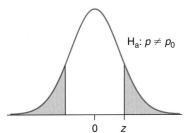
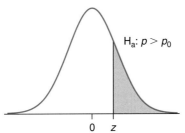
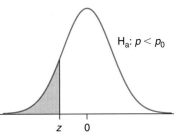

▲ **FIGURE 8.18** Representations of possible p-values for three different alternative hypotheses. The area of the shaded regions represents the p-value.

## Sources

Husemann, L. R., J. Moise-Titus, C. Podolski, and L. D. Eron. 2003. Longitudinal relations between children's exposure to TV violence and their aggressive and violent behavior in young adulthood: 1977–1992. *Developmental Psychology* 39(2), 201–221.

Pew Forum. 2009. Declining majority of Americans favor embryonic stem cell research. The Pew Forum on Religion & Public Life Issues. http://people-press.org/report/528/] http://pewforum.org/docs/?DocID=317 (accessed March 13, 2009).

Schmidt, B. et al. for the Caffeine for Apnea of Prematurity Trial Group. 2007. Long-term effects of caffeine therapy for apnea of prematurity. *New England Journal of Medicine* 357, 1893–1902.

Schwitzgebel, E., and J. Rust. 2010.Do ethicists and political philosophers vote more often than other professors? *Review of Philosophy and Psychology,* 1, 189-199.

Schwitzgebel, E. 2003. Do people still report dreaming in black and white? An attempt to replicate a questionnaire from 1942. *Perceptual & Motor Skills* 96, 25–29).

www.workforce.com. 2008. California court to hear Google age discrimination case (accessed August 26, 2009).

## SECTION EXERCISES

## SECTION 8.1

**8.1** Choose one of the answers given. The null hypothesis is always a statement about a _____ (sample statistic or population parameter).

**8.2** Choose one of the answers in each case. In statistical inference, measurements are made on a _____ (sample or population), and generalizations are made to a _____ (sample or population).

**8.3** With a two-tailed test, if the test statistic (such as $z$) is far from 0, will the p-value be large (closer to 1) or small (closer to 0)?

**8.4** With a two-tailed test, if the test statistic is 0 or close to 0, will the p-value be large (closer to 1) or small (closer to 0)?

**TRY** **8.5 Coin Flips (Example 1)** A coin is flipped 30 times and lands on heads 18 times. You want to test the hypothesis that the coin does not come up 50% heads in the long run.

Pick the correct null hypothesis for this test.

  i.  $H_0: \hat{p} = 0.60$
  ii.  $H_0: p = 0.50$
  iii.  $H_0: p = 0.60$
  iv.  $H_0: \hat{p} = 0.50$

**8.6 Die Rolls** You roll a six-sided die 30 times and land on an ace (a one) 6 times. You want to test the hypothesis that the die does not come up with an ace one-sixth of the time. Pick the correct null hypothesis.

  i. $H_0: \hat{p} = 1/5$  ii. $H_0: p = 1/5$  iii. $H_0: \hat{p} = 1/6$  iv. $H_0: p = 1/6$

**8.7** A friend claims he can predict the suit of a card drawn from a standard deck of 52 cards. There are four suits and equal numbers of cards in each suit. The parameter, $p$, is the probability of success, and the null hypothesis is that the friend is just guessing.

  a. Which is the correct null hypothesis?
   i. $p = 1/4$  ii. $p = 1/13$  iii. $p > 1/4$  iv. $p > 1/13$

  b. Which hypothesis best fits the friend's claim? (This is the alternative hypothesis.)
   i. $p = 1/4$  ii. $p = 1/13$  iii. $p > 1/4$  iv. $p > 1/13$

**8.8** A friend claims he can predict how a six-sided die will land. The parameter, $p$, is the long-run likelihood of success, and the null hypothesis is that the friend is guessing.

  a. Pick the correct null hypothesis.
   i. $p = 1/6$  ii. $p > 1/6$  iii. $p < 1/6$  iv. $p > 1/2$

  b. Which hypothesis best fits the friend's claim? (This is the alternative hypothesis.)
   i. $p = 1/6$  ii. $p > 1/6$  iii. $p < 1/6$  iv. $p > 1/2$

**8.9 Heart Attack Prevention** A new drug is being tested to see whether it can reduce the chance of heart attack in people who have previously had a heart attack. The rate of heart attack in the population of concern is 0.20. The null hypothesis is that $p$ (the population proportion using the new drug that have a heart attack) is 0.2.

Pick the correct alternative hypothesis.

  i. $p \neq 0.2$  ii. $p > 0.2$  iii. $p < 0.2$

**8.10 Stroke Survival Rate** The proportion of people who live after suffering a stroke is 0.85. Suppose there is a new treatment that is used to increase the survival rate. Use the parameter $p$ to represent the population proportion of people who survive after a stroke. For a hypothesis test of the treatment's effectiveness, researchers use a null hypothesis of $p = 0.85$. Pick the correct alternative hypothesis.

  i. $p \neq 0.85$  ii. $p > 0.85$  iii. $p < 0.85$

**TRY** **8.11 ESP (Example 2)** We are testing someone who claims to have ESP by having that person predict whether a coin will come up heads or tails. The null hypothesis is that the person is guessing and does not have ESP, and the population proportion of success is 0.50. We test the claim with a hypothesis test, using a significance level of 0.05. Select an answer and fill in the blank.

The probability of concluding that the person has ESP when in fact she or he (does/does not)_____ have ESP is _____.

**\* 8.12 Multiple-Choice Test** A teacher is giving an exam with 20 multiple-choice questions, each with four possible answers. The teacher's null hypothesis is that the student is just guessing, and the population proportion of success is 0.25. Suppose we do a test with a significance level of 0.01. Write a sentence describing the significance level in the context of the hypothesis test.

**TRY** **8.13 Votes for Independents (Example 3)** Judging on the basis of experience, a politician claims that 50% of voters in Pennsylvania have voted for an independent candidate in past elections. Suppose you surveyed 20 randomly selected people in Pennsylvania, and 12 of them reported having voted for an independent candidate. The null hypothesis is that the overall proportion of voters in Pennsylvania that have voted for an independent candidate is 50%. What value of the test statistic should you report?

**8.14 Votes for Independents** Refer to Exercise 8.13. Suppose 14 out of 20 voters in Pennsylvania report having voted for an independent candidate. The null hypothesis is that the population proportion is 0.50. What value of the test statistic should you report?

**TRY** * **8.15 Texting While Driving (Example 4)** The mother of a teenager has heard a claim that 25% of teenagers who drive and use a cell phone reported texting while driving. She thinks that this rate is too high and wants to test the hypothesis that fewer than 25% of these drivers have texted while driving. Her alternative hypothesis is that the percentage of teenagers who have texted when driving is less than 25%.

$$H_0: p = 0.25$$

$$H_a: p < 0.25$$

She polls 40 randomly selected teenagers, and 5 of them report having texted while driving, a proportion of 0.125. The p-value is 0.034. Explain the meaning of the p-value in the context of this question.

* **8.16 True/False Test** A teacher giving a true/false test wants to make sure her students do better than they would if they were simply guessing, so she forms a hypothesis to test this. Her null hypothesis is that a student will get 50% of the questions on the exam correct. The alternative hypothesis is that the student is not guessing and should get more than 50% in the long run.

$$H_0: p = 0.50$$

$$H_a: p > 0.50$$

A student gets 30 out of 50 questions, or 60%, correct. The p-value is 0.079. Explain the meaning of the p-value in the context of this question.

## SECTION 8.2

**TRY** **8.17 p-Values (Example 5)** For each graph, indicate whether the shaded area could represent a p-value. Explain why or why not. If yes, state whether the area could represent the p-value for a one-tailed or a two-tailed alternative hypothesis.

**(A)**

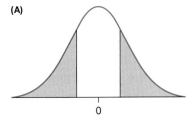

**(B)**

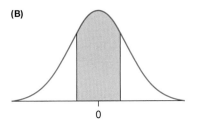

**8.18 p-Values** For each graph, state whether the shaded area could represent a p-value. Explain why or why not. If yes, state whether the area could represent the p-value for a one-tailed or a two-tailed alternative hypothesis.

**(A)**

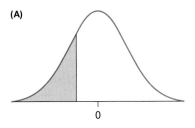

**(B)**

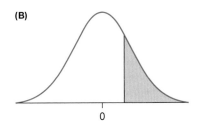

**TRY** **8.19 p-Values (Example 6)** A researcher carried out a hypothesis test using a two-tailed alternative hypothesis. Which of the following z-scores is associated with the smallest p-value? Explain.

i. $z = 0.50$   ii. $z = 1.00$   iii. $z = 2.00$   iv. $z = 3.00$

**8.20 Coin Flips** A test is conducted in which a coin is flipped 30 times to test whether the coin is unbiased. The null hypothesis is that the coin is fair. The alternative is that the coin is not fair. One of the accompanying figures represents the p-value after getting 16 heads out of 30 flips, and the other represents the p-value after getting 18 heads out of 30 flips. Which is which, and how do you know?

**(A)**

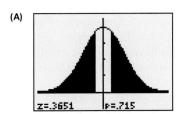

**(B)**

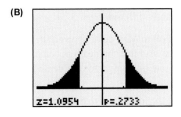

**8.21 ESP** Suppose a friend says he can predict whether a coin flip will result in heads or tails. You test him, and he gets 10 right out of 20. Do you think he can predict the coin flip (or has a way of cheating)? Or could this just be something that occurs by chance? Explain without doing any calculations.

**8.22 ESP Again** Suppose a friend says he can predict whether a coin flip will result in heads or tails. You test him, and he gets 20 right out of 20. Do you think he can predict the coin flip (or has a way of cheating)? Or could this just be something that is likely to occur by chance? Explain without performing any calculations.

## SECTION 8.3

TRY **8.23 Does Hand Washing Save Lives? (Example 7)** In the mid-1800s, Dr. Ignaz Semmelweiss decided to make the doctors wash their hands with a strong disinfectant between patients at a clinic with a death rate of 9.9%. Semmelweiss wanted to test the hypothesis that the death rate would go down after the new hand-washing procedure was used. What null and alternative hypotheses should he have used? Explain, using both words and symbols. Explain the meaning of any symbols you use.

**8.24 Healthcare Plan** Suppose you wanted to test the claim that more than half of U.S. voters support repealing the current U.S. health plan. Give the null and alternative hypotheses, and explain, using both words and symbols.

**8.25 Seatbelt Usage** The National Highway Traffic Safety Administration (NHTSA) claimed that the percentage of people who used their seatbelts in the year 2000 was 71%. The NHTSA also reported that a random sample of 3000 people showed that 2520 reported using their seatbelts in 2009.

Give the null and alternative hypotheses to test whether the percentage of people who used their seatbelts in 2009 *changed* from the 71% reported in the year 2000.

**8.26 Seatbelt Usage, Again** The NHTSA claims that the percentage of people who used their seatbelts in the year 2000 was 71%. The NHTSA recently reported that a random sample of 3000 people showed 2520 who reported using their seatbelts in 2009.

Give the null and alternative hypotheses to test whether the percentage of people who used their seatbelts in 2009 *increased* from the 71% reported in the year 2000.

g **8.27 Gun Control** Historically, the percentage of U.S. residents who support stricter gun control laws has been 52%. A recent Gallup Poll of 1011 people showed 495 in favor of stricter gun control laws. Assume the poll was given to a random sample of people. Test the claim that the proportion of those favoring stricter gun control has changed. Perform a hypothesis test, using a significance level of 0.05. *See page 373 for guidance.* Choose one of the following conclusions:

  i. The percentage is not significantly different from 52%. (A significant difference is one for which the p-value is less than or equal to 0.050.)

  ii. The percentage is significantly different from 52%.

**8.28 Death Penalty** A Gallup Poll in May 2010 showed that 669 out of 1029 randomly chosen adults said that the death penalty was "morally acceptable." In 1996 the rate of support for the death penalty was estimated to be 78%.

  a. Using the data from 2010, test the hypothesis that the rate of support has changed since 1996. Use a value of 0.05 for the significance level, $\alpha$.

  b. Choose the appropriate conclusion.

    i. The rate of support is not significantly different from 78%.

    ii. The rate of support is significantly different from 78%.

TRY **8.29 Dreaming (Example 8)** A 2003 study of dreaming found that out of a random sample of 113 people, 92 reported dreaming in color. However, the rate of reported dreaming in color that was established in the 1940s was 0.29 (Schwitzgebel 2003). Check to see whether the conditions for using a one-proportion $z$-test are met assuming the researcher wanted test to see if the proportion dreaming in color had changed since the 1940s.

**8.30 Age Discrimination** About 30% of the population in Silicon Valley, a region in California, are between the ages of 40 and 65, according to the U.S. Census. However, only 2% of the 2100 employees at a laid-off man's former Silicon Valley company are between the ages of 40 and 65. Lawyers might argue that if the company hired people regardless of their age, the distribution of ages would be the same as if they had hired people at random from the surrounding population. Check whether the conditions for using the one-proportion $z$-test are met.

TRY **8.31 Coin Spinning (Example 8 continued)** Suppose you are testing the claim that a coin comes up tails more than 50% of the time when the coin is spun on a hard surface. Steps 1 and 2 of the hypothesis test are given. Suppose that you did this experiment and got 22 tails in 30 spins. Find the value of the test statistic $z$ and the corresponding p-value.

  $p$ is the proportion of tails.

  Step 1: $H_0: p = 0.50$

          $H_a: p > 0.50$

  Step 2: Assume that the outcomes are random and the sample size is large enough because both $np$ and $n(1 - p)$ are 15.

**8.32 Coin Spinning** Repeat Exercise 8.31 assuming that you got 18 tails out of 30 spins.

TRY **8.33 Guessing on a True/False Test (Example 9)** A true/false test has 50 questions. Suppose a passing grade is 35 or more correct answers. Test the claim that a student knows more than half of the answers and is not just guessing. Assume the student gets 35 answers correct out of 50. Use a significance level of 0.05. Steps 1 and 2 of a hypothesis test procedure are given. Show steps 3 and 4, and be sure to write a clear conclusion.

  Step 1: $H_0: p = 0.50$

          $H_a: p > 0.50$

  Step 2: Choose the one-proportion $z$-test. Sample size is large enough, because $np_0$ is $50(0.5) = 25$ and $n(1 - p_0) = 50(0.50) = 25$, and both are more than 10. Assume the sample is random.

**8.34 Guessing on a Multiple-Choice Test** A multiple-choice test has 50 questions with four possible options for each question. For each question, only one of the four options is correct. A passing grade is 35 or more correct answers.

  a. What is the probability that a person will guess correctly on one multiple-choice question?

  b. Test the hypothesis that a person who got 35 right out of 50 is not just guessing, using an alpha of 0.05. Steps 1 and 2 of the hypothesis testing procedure are given. Finish the question by doing steps 3 and 4.

  Step 1: $H_0: p = 0.25$

          $H_a: p > 0.25$

Step 2: Choose the one-proportion z-test. $n$ times $p$ is 50 times 0.25, which is 12.5. This is more than 10, and 50 times 0.75 is also more than 10. Assume a random sample.

**TRY 8.35 Taste Test (Example 10)** A taste test was done in which a blindfolded subject was asked to identify the spread on English muffins. Twenty bites were taken, and they were randomly arranged such that exactly half of the bites contained butter and half contained margarine.

$$H_0: p = 0.50$$

$$H_a: p > 0.50$$

The subject got 13 out of 20 correct. The graphs show p-values (as shaded areas) corresponding to the alternative hypotheses $p \neq 0.50$, $p > 0.50$, and $p < 0.50$, *though not necessarily in that order.*

Indicate which graph matches the given alternative hypothesis, $p > 0.50$.

(A)
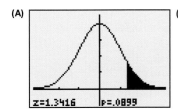
z=1.3416    P=.0899

(B)
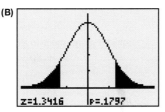
z=1.3416    P=.1797

(C)
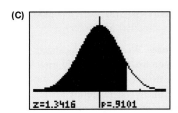
z=1.3416    P=.9101

**8.36 Penny Spinning** Suppose we spin a penny 20 times to test whether it is biased. Let $p$ represent the probability of heads.

$$H_0: p = 0.50$$

$$H_a: p \neq 0.50$$

Suppose that 8 heads were obtained out of 20 spins. The graphs show p-values (as shaded areas) corresponding to the alternative hypotheses $p \neq 0.50$, $p > 0.50$, and $p < 0.50$. Indicate which graph matches the given alternative hypothesis, $p \neq 0.50$.

(A)
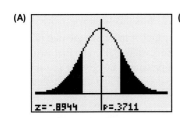
z=-.8944    P=.3711

(B)
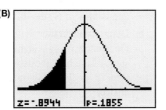
z=-.8944    P=.1855

(C)
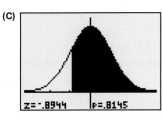
z=-.8944    P=.8145

**8.37 Banning Sugary Foods in Schools** In a Rasmussen Poll of 1000 adults in July 2010, 520 of those polled said that schools should ban sugary snacks and soft drinks.

a. Do a majority of adults (more than 50%) support a ban on sugary snacks and soft drinks? Perform a hypothesis test using a significance level of 0.05.

b. Choose the best interpretation of the results you obtained in part a:

i. The percentage of all adults who favor banning is significantly more than 50%.

ii. The percentage of all adults who favor banning is not significantly more than 50%

**8.38 Year-Round School** In July of 2010, a Rasmussen Poll of 1000 adults showed that 630 opposed students attending school year-round without the traditional summer break.

a. Test the claim that more than half of adults oppose year-round school. Use a significance level of 0.05.

b. Choose the correct interpretation:

i. The percentage opposing year-round school is significantly more than 50%.

ii. The percentage opposing year-round school is not significantly more than 50%.

**8.39 Global Warming** Historically (from about 2001 to 2005), about 58% of Americans believed that the earth's temperature was rising ("global warming"). A March 2010 Gallup Poll wanted to determine whether this proportion had changed. The poll interviewed 1014 adult Americans, and 527 said they believed that global warming was real. (Assume these 1014 adults represented a simple random sample.)

a. What percentage in the sample believed global warming was real in 2010? Is this more or less than the historical 58%?

b. Test the hypothesis that the proportion of Americans who believe global warming is real has changed. Use a significance level of 0.05.

c. Choose the correct interpretation:

i. In 2010, the percentage of Americans who believe global warming is real is not significantly different from 58%.

ii. In 2010, the percentage of Americans who believe global warming is real has changed from the historical level of 58%.

**8.40 Plane Crashes** According to one source, 50% of plane crashes are due at least in part to pilot error (http://www.planecrashinfo .com). Suppose that in a random sample of 100 separate airplane accidents, 62 of them were due to pilot error (at least in part.)

a. Test the null hypothesis that the proportion of airplane accidents due to pilot error is not 0.50. Use a significance level of 0.05.

b. Choose the correct interpretation:

i. The percentage of plane crashes due to pilot error is not significantly different from 50%.

ii. The percentage of plane crashes due to pilot error is significantly different from 50%.

**TRY 8.41 Mercury in Freshwater Fish (Example 11)** Some experts believe that 20% of all freshwater fish in the United States have such high levels of mercury that they are dangerous to eat. Suppose a fish market has 250 fish tested, and 60 of them have dangerous levels of mercury. Test the hypothesis that this sample is *not* from a population with 20% dangerous fish. Use a significance level of 0.05.

Comment on your conclusion: Are you saying that the percentage of dangerous fish is definitely 20%? Explain.

**8.42 Newspapers Closing** A 2009 Pew Poll asked people whether losing their local newspapers would hurt civic life, and 430 out of 1001 said that it would hurt civic life a lot. Another source said that 44% of people would be hurt.

a. Find the percentage from the Pew Poll that said that closing local papers would hurt civic life a lot.

b. Test the hypothesis that fewer than 44% of people say it would hurt civic life a lot, using the Pew data and a significance level of 0.05.

Comment on your conclusion: Are you saying that the percentage is definitely 44%?

**8.43 Morse's Proportion of t's** Samuel Morse determined that the percentage of t's in the English language in the 1800's was 9%. A random sample of 600 letters from a current newspaper contained 48 t's. Test the hypothesis that the proportion of t's has changed in modern times, using the 0.05 level of significance. Treat the newspaper as a random sample of all letters used.

**8.44 Morse's Proportion of a's** Samuel Morse determined that the percentage of a's in the English language in the 1800s was 8%. A random sample of 600 letters from a current newspaper contained 60 a's. Test the hypothesis that the proportion of a's has changed in modern times, using the 0.05 level of significance. Treat the newspaper as a random sample of all letters used.

## SECTION 8.4

g **8.45 Vioxx** In the fall of 2004, drug manufacturer Merck Pharmaceutical withdrew Vioxx, a drug that had been used for arthritis pain, from the market after a study revealed that its use was associated with an increase in the risk of heart attack. The experiment was placebo-controlled, randomized, and double-blind. Out of 1287 people taking Vioxx there were 45 heart attacks, and out of 1299 people taking the placebo, there were 25 heart attacks (*Los Angeles Times*, October 23, 2004). Perform a hypothesis test to test whether those who take Vioxx have a greater rate of heart attack than those who take a placebo. Use a level of significance of 0.05. *See page 373 for guidance.* Choose i or ii:

i. Those taking Vioxx did *not* have a significantly higher rate of heart attack than those taking the placebo.

ii. Those who took Vioxx had a significantly higher heart attack rate than those who took the placebo.

Can we conclude that Vioxx *causes* an increased risk of heart attack?

**8.46 Treating Traumatic Shock** Dopamine and norepinephrine are two drugs that are effective for treating traumatic shock, which results in dangerously low blood pressure and can cause death. A recent study compared the outcomes for patients who were given these drugs. After 28 days it was noted whether the patient was alive or dead. Of the 858 patients randomly assigned dopamine, 450 were alive. Of the 821 patients randomly assigned norepinephrine, 398 were alive. (Source: De Backer et al., Comparison of dopamine and norepinephrine in the treatment of shock, *New England Journal of Medicine*, vol. 362: 779–789, March 4, 2010)

a. Find and compare the sample percentages of patients alive for these groups.

b. Test the hypothesis that the rate of success (being alive) is different in the two groups, using a significance level of 0.05.

**8.47 Vaccine for Diarrhea** A vaccine to prevent severe rotavirus gastroenteritis (diarrhea) was given to African children within the first year of life as part of a drug study. The study reported that of the 3298 children randomly assigned the vaccine, 63 got the virus. Of the 1641 children randomly assigned the placebo, 80 got the virus. (Source: Madhi et al., Effect of human rotavirus vaccine on severe diarrhea in African infants, *New England Journal of Medicine*, vol. 362: 289–298, January 28, 2010)

a. Find the sample percentage of children who caught the virus in each group. Is the sample percent lower for the vaccine group, as investigators hoped?

b. Determine whether the vaccine is effective in reducing the chance of catching the virus, using a significance level of 0.05. Steps 1 and 2 of the hypothesis-testing procedure are given. Complete the question by doing steps 3 and 4.

Step 1: $H_0: p_v = p_p$ ($p_v$ is the proportion that got the virus among those who took the vaccine, and $p_p$ is the proportion that got the virus among those who took the placebo.)
$H_a: p_v < p_p$

Step 2: Although we don't have a random sample, we do have random assignment to groups.

$$\hat{p} = \frac{63 + 80}{3298 + 1641} = \frac{143}{4939} = 0.028953$$

$n_p \times \hat{p} = 3298 \times 0.028953 = 95.49$, which is more than 10

$n_p \times \hat{p} = 1641 \times 0.028953 = 47.51$, which is more than 10 (and the other two products are larger)

* **8.48 Radiation for Breast Cancer** Women with localized breast cancer are often given radiation treatment. A study tried to find out whether simplifying the treatment would lower the success rate. Each woman needing treatment was randomly assigned to receive whole-breast irradiation either at a standard dose of 50.0 Gy in 25 treatments over a period of 35 days (the control group) or at a dose of 42.5 Gy in 16 treatments over a period of 22 days (the new treatment group). The new treatment had smaller doses and also fewer treatments, so it was feared that the new treatment would be less effective than the standard treatment.

Among the 612 women randomly assigned to the standard (old) treatment, 41 had a recurrence within 10 years. Among the 622 women randomly assigned to the new treatment, 39 had a recurrence within 10 years. (Source: Whelan et al., Long-term results of hypofractionated radiation therapy for breast cancer, *New England Journal of Medicine*, vol. 362: 513–520, February 11, 2010)

a. Find the failure rate (rate of recurrence) for each treatment, and comment on the difference. Was this what was expected?

b. Determine whether the new treatment increases the rate of recurrence, using a significance level of 0.05. The first two steps of the hypothesis-testing procedure are given. You may refer to the TI-83/84 output to do steps 3 and 4 to complete the answer.

Step 1: $H_0: p_{New} = p_{Old}$ where $p_{New}$ is the probability of recurrence with the new treatment, and $p_{Old}$ is the probability of recurrence with the old treatment.

$$H_a: p_{New} > p_{Old}$$

Step 2: Although we don't have a random sample, we do have random assignment.

$$\hat{p} = \frac{41 + 39}{612 + 622} = \frac{80}{1234} = 0.064830$$

$n_{old}\hat{p} = 612 \times 0.064830 = 39.68$, which is more than 10.

$n_{new}\hat{p} = 622 \times 0.064830 = 40.32$, which is more than 10. The other two products are larger.

```
2-PropZTest
 p1<p2
 z=.3061898787
 p=.6202698963
 p̂1=.0669934641
 p̂2=.0627009646
↓p̂=.0648298217
```

In the figure, 1 refers to the old treatment and 2 refers to the new treatment.

TRY **8.49 Nicotine Gum (Example 12)** A study used nicotine gum to help people quit smoking. The study was placebo-controlled, randomized, and double-blind. Each participant was interviewed after 28 days, and success was defined as being abstinent from cigarettes for 28 days. The results showed that 174 out of 1649 people using the nicotine gum succeeded, and 66 out of 1648 using the placebo succeeded. Although the sample was not random, the assignment to groups was randomized. (Source Shiffman et al., Quitting by gradual smoking reduction using nicotine gum: A randomized controlled trial, *American Journal of Preventive Medicine*, vol. 36, issue 2, February, 2009)

a. Find the proportion of people using nicotine gum that stopped smoking and the proportion of people using the placebo that stopped smoking, and compare them. Is this what the researchers had expected?

b. Find the observed value of the test statistic, assuming that the conditions for a two-proportion z-test hold.

**8.50 Treating Prostate Cancer** A recently published article reported the results of an experiment on reducing the likelihood that men develop prostate cancer. The investigators randomly assigned 3305 men to receive the drug dutasteride (the generic name for the drug sold under the brand name Avodart) and assigned 3424 men to receive a placebo. Of those receiving the drug, 659 developed prostate cancer, and of those receiving the placebo, 858 developed prostate cancer. (Source: Andriole et al., Effect of dutasteride on the risk of prostate cancer, *New England Journal of Medicine*, vol. 362: 1192–1202, April 1, 2010)

a. Find the percentage of men that developed prostate cancer in each group. Was the sample percentage for those taking the drug lower than the sample percentage for those who took the placebo?

b. Find the observed value of the test statistic, assuming the conditions for a two-proportion z-test hold.

TRY **8.51 Smiling and Gender (Example 13)** In a 1997 study, people were observed for about 10 seconds in public places, such as malls and restaurants, to determine whether they smiled during the randomly chosen 10-second interval. The table shows the results for comparing males and females. (Source: M. S. Chappell, Frequency of public smiling over the life span, *Perceptual and Motor Skills*, vol. 45: 474, 1997)

|  | Male | Female |
|---|---|---|
| Smile | 3269 | 4471 |
| No Smile | 3806 | 4278 |

a. Find and compare the sample percentages of women who were smiling and men who were smiling.

b. Treat this as though it were a random sample, and test whether there are differences in the proportion of men and the proportion of women who smile. Use a significance level of 0.05.

c. Explain why there is such a small p-value even though there is such a small difference in sample percentages.

* **8.52 Smiling and Age** Refer to the study discussed in Exercise 8.51. The accompanying table shows the results of the study for different age groups.

| | Age Range | | | | |
|---|---|---|---|---|---|
| | **0−10** | **11−20** | **21−40** | **41−60** | **61+** |
| Smile | 1131 | 1748 | 1608 | 937 | 522 |
| No Smile | 1187 | 2020 | 3038 | 2124 | 1509 |

a. For each age group, find the percentage who were smiling.

b. Treat this as a random sample of people, and merge the groups 0–10 and 10–20 into one group (0–20) and the groups 21–40, 41–60, and 61+ into another age group (21–65+). Then determine whether these two age groups have different proportions of people who smile in the general population, using a significance level of 0.05. Please comment on the results.

## SECTION 8.5

**8.53** If we reject the null hypothesis, can we claim to have *proved* that the null hypothesis is false? Why or why not?

**8.54** If we do not reject the null hypothesis, is it valid to say that we *accept* the null hypothesis? Why or why not?

**8.55** When a person stands trial for murder, the jury is instructed to assume that the defendant is innocent. Is this claim of innocence an example of a null hypothesis, or is it an example of an alternative hypothesis?

**8.56** When, in a criminal court, a defendant is found "not guilty," is the court saying with certainty that he or she is innocent? Explain.

* **8.57 Arthritis** A magazine advertisement claims that wearing a magnetized bracelet will reduce arthritis pain in those who suffer from arthritis. A medical researcher tests this claim with 233 arthritis suffers randomly assigned to wear either a magnetized bracelet or a placebo bracelet. The researcher records the proportion of each group who report relief from arthritis pain after 6 weeks. After analyzing the data, he fails to reject the null hypothesis. Which of the following are valid interpretations of his findings? There may be more than one correct answer.

a. The magnetized bracelets are not effective at reducing arthritis pain.

b. There's insufficient evidence that the magnetized bracelets are effective at reducing arthritis pain.

c. The magnetized bracelets had exactly the same effect as the placebo at reducing arthritis pain.

d. There were no statistically significant differences between the magnetized bracelets and the placebos in reducing arthritis pain.

* **8.58 No-carb Diet** A weight-loss diet claims that it causes weight loss by eliminating carbohydrates (breads and starches)

from the diet. To test this claim, researchers randomly assign over-weight subjects to two groups. Both groups eat the same amount of calories, but one group eats almost no carbs, and the other group includes carbs in their meals. After 2 months, the researchers test the claim that the no-carb diet is better than the usual diet. They record the proportion of each group that lost more than 5% of their initial weight. They then announce that they failed to reject the null hypothesis. Which of the following are valid interpretations of the researchers' findings?

a. There were no significant differences in effectiveness between the no-carb diet and the carb diet.

b. The no-carb diet and the carb diet were equally effective.

c. The researchers did not see enough evidence to conclude that the no-carb diet was more effective.

d. The no-carb diet was less effective than the carb diet.

**8.59** When comparing two sample proportions with a two-tailed alternative hypothesis, all other factors being equal, will you get a smaller p-value if the sample proportions are close together or if they are far apart? Explain.

**8.60** When comparing two sample proportions with a two-tailed alternative hypothesis, all other factors being equal, will you get a smaller p-value with a larger sample size or a smaller sample size? Explain.

**★ 8.61 Gun Control Laws** The Gallup organization frequently conducts polls in which they ask the following question:

"In general, do you feel that the laws covering the sale of firearms should be made more strict, less strict, or kept as they are now?"

In February 1999, 60% of those surveyed said "more strict," and on April 26, 1999, shortly after the Columbine High School shootings, 66% of those surveyed said "more strict."

a. Assume that both polls used samples of 560 people. Determine the number of people in the sample that said "more strict" in February 1999, before the school shootings, and the number that said "more strict" in late April 1999, after the school shootings.

b. Do a test to see whether the proportion that said "more strict" is statistically significantly different in the two different surveys, using a *significance level of 0.01*.

c. Repeat the problem, assuming that the sample sizes were both 1120.

d. Comment on the effect of different sample sizes on the p-value and on the conclusion.

**8.62 Weight Loss in Men** Many polls have asked people whether they are trying to lose weight. A Gallup Poll in November of 2008 showed that 22% of men said they were seriously trying to lose weight. In 2006, 24% of men (with the same average weight of 194 pounds as the men polled in 2008) said they were seriously try-ing to lose weight. Assume that both samples contained 500 men.

a. Determine how many men in the sample from 2008 and how many in the sample from 2006 said they were seriously trying to lose weight.

b. Determine whether the difference in proportions is significant at the 0.05 level.

c. Repeat the problem with the same proportions but a sample size of 5000 instead of 500.

d. Comment on the different p-values and conclusions with dif-ferent sample sizes.

**★ 8.63 Effectiveness of Financial Incentives** A psychologist is interested in testing whether offering students a financial incentive improves their video-game-playing skills. She collects data and performs a hypothesis test to test whether the probability of getting to the highest level of a video game is greater with a financial incentive than without. Her null hypothesis is that the probability of getting to this level is the same with or without a financial incentive. The alternative is that this probability is greater. She gets a p-value from her hypothesis test of 0.003. Which of the following is the best interpretation of the p-value?

    i. The p-value is the probability that financial incentives are *not* effective in this context.

    ii. The p-value is the probability of getting exactly the result obtained, assuming that financial incentives are *not* effec-tive in this context.

    iii. The p-value is the probability of getting a result as extreme as or more extreme than the one obtained, assuming that financial incentives are *not* effective in this context.

    iv. The p-value is the probability of getting exactly the result obtained, assuming that financial incentives *are* effective in this context.

    v. The p-value is the probability of getting a result as extreme as or more extreme than the one obtained, assuming that financial incentives *are* effective in this context.

**8.64** Is it acceptable practice to look at your research results, note the direction of the difference, and then make the alternative hypoth-esis one-tailed in order to achieve a significant difference? Explain.

**8.65 Choosing a Test and Naming the Population** For each of the following, state whether a one-proportion z-test or a two-proportion z-test would be appropriate, and name the populations.

a. A polling agency takes a random sample to determine the pro-portion of people in Florida who support Proposition X.

b. A student asks a random sample of men and a random sample of women at her college whether they support capital punishment for some convicted murderers. She wants to determine whether the proportion of women who support capital punishment is less than the proportion of men who support it.

**8.66 Choosing a Test and Naming the Population** For each of the following, state whether a one-proportion z-test or a two-proportion z-test would be appropriate, and name the populations.

a. A student watches a random sample of men and women leaving a Milwaukee supermarket with carts to see whether they put the carts back in the designated area. She wants to compare the proportions of men and women who put the carts in the designated area.

b. The pass rate for the Oregon bar exam is 65%. A random sample of graduates from Oregon University School of Law is examined to see whether their pass rate is significantly higher than 65%.

**8.67 Choosing a Test and Giving the Hypotheses** Give the null and alternative hypotheses for each test, and state whether a one-proportion z-test or a two-proportion z-test would be appropriate.

a. You test a person to see whether he can tell tap water from bottled water. You give him 20 sips selected randomly (half from tap water and half from bottled water) and record the proportion he gets correct to test the hypothesis.

b. You test a random sample of students at your college who stand on one foot with their eyes closed and determine who can stand for at least 10 seconds, comparing athletes and nonathletes.

**8.68 Choosing a Test and Naming the Population(s)** In each case, choose whether the appropriate test is a one-proportion z-test or a two-proportion z-test. Name the population(s).

a. A researcher takes a random sample of 4-year-olds to find out whether girls or boys are more likely to know the alphabet.

b. A pollster takes a random sample of all U.S. adult voters to see whether more than 50% approve of the performance of the current U.S. president.

c. A researcher wants to know whether a new heart medicine reduces the rate of heart attacks compared to an old medicine.

d. A pollster takes a poll in Wyoming about home schooling to find out whether the approval rate for men is equal to the approval rate for women.

e. A person is studied to see whether he or she can predict the results of coin flips better than chance alone.

## CHAPTER REVIEW EXERCISES

**8.69 Cola Taste Test** A student who claims he can tell cola A from cola B is blindly tested with 20 trials. At each trial, cola A or cola B is randomly chosen and presented to the student, who must correctly identify the cola. The experiment is designed so that the student will have exactly 10 sips from each cola. He gets 6 identifications right out of 20. Can he tell cola A from cola B at the 0.05 level of significance? Explain.

**8.70 Butter Taste Test** A man is tested to determine whether he can tell butter from margarine. He is blindfolded and given small bites of English muffin to identify. At each trial, an English muffin with either butter or margarine is randomly chosen. The experiment is designed so that he will have exactly 15 bites with butter and 15 with margarine. He gets 14 right out of 30. Can he tell butter from margarine at the 0.05 level? Explain.

\* **8.71 Biased Coin?** A study is done to see whether a coin is biased. The alternative hypothesis used is two-tailed, and the obtained z-value is 2. Assuming that the sample size is sufficiently large and that the other conditions are also satisfied, use the Empirical Rule to approximate the p-value.

\* **8.72 Biased Coin?** A study is done to see whether a coin is biased. The alternative hypothesis used is two-tailed, and the obtained z-value is 1. Assuming that the sample size is sufficiently large and that the other conditions are also satisfied, use the Empirical Rule to approximate the p-value.

**8.73 ESP** A researcher studying ESP tests 200 students. Each student is asked to predict the outcome of a large number of coin flips. For each student, a hypothesis test using a 5% significance level is performed. If the p-value for the student is less than or equal to 0.05, the researcher concludes the student has ESP. Out of 200 people who do *not* have ESP, about how many would you expect the researcher to declare *do* have ESP?

**8.74 Coin Flips** Suppose you tested 50 coins by flipping each of them many times. For each coin, you perform a significance test with a significance level of 0.05 to determine whether the coin is biased. Assuming that none of the coins is biased, about how many of the 50 coins would you expect to appear biased when this procedure is applied?

**8.75 1960 Presidential Election** In the 1960 presidential election 52% of male voters voted for Kennedy, and 48% voted for Nixon. Also, 49% of female voters voted for Kennedy and 51% voted for Nixon. Would it be appropriate to do a two-proportion z-test to determine whether the proportions of men

and women voting for Kennedy were significantly different (assuming we knew the number of men and women who voted)? Explain.

**8.76 Unemployment in 2009** In February 2009, the Bureau of Labor Statistics reported that the U.S. unemployment rate was 8.1% for men and 6.7% for women. Would it be appropriate to do a two-proportion z-test to determine whether the rates for men and women were significantly different (assuming we knew the total number of men and women)? Explain.

**8.77 Do Financial Incentives Help Smokers Quit?** A controlled, randomized study compared smoking cessation for a group in which participants received $350 if they quit and a group that did not receive money. The study participants had to submit to a biochemical test to prove they had quit. Of the 442 people randomly assigned not to receive money, 52 had quit smoking after 6 months. Of the 436 people randomly assigned to receive money, 91 had quit smoking after 6 months. (Source: Volpp et al., A randomized, controlled trial of financial incentives for smoking cessation, *New England Journal of Medicine*, vol. 361: 4, 331–333, 2009.) Determine whether the proportion of success was significantly greater for those who received money than for those who did not. Use a significance level of 0.05.

**8.78 Guns: Two Polls** In 2009 a Gallup Poll reported that 40% of people said they have a gun in their home. In the same year, a Pew Poll reported that 33% of people said they have a gun in the home. Assume that each poll used a sample size of 1000.

Do these polls disagree? Test the hypothesis that the two population proportions are different. Use a significance level of 0.05.

**8.79 Aspirin for Heart Disease in Women** The *New England Journal of Medicine* in March 2005 reported on a randomized, placebo-controlled study that was done to determine whether the use of aspirin could lower the risk of heart disease in women. It was already known that aspirin reduced heart disease in men. Researchers randomly assigned 39,876 healthy women 45 years of age or older to receive either placebo or 100 milligrams of aspirin on alternate days and then monitored them for 10 years for a first major cardiovascular event (such as a nonfatal myocardial infarction, a nonfatal stroke, or death from cardiovascular causes). The table shows the results. (Source: Ridke et al., A randomized trial of low-dose aspirin in the primary prevention of cardiovascular disease in women, *New England Journal of Medicine*, vol. 352: 1293–1304, March 31, 2005)

|  | Aspirin | Placebo |
| --- | --- | --- |
| Heart Event | 477 | 522 |
| No Heart Event | 19,457 | 19,420 |
| Totals | 19,934 | 19,942 |

a. Find the percentage of those taking aspirin who had a "heart event," and compare it with the percentage of those taking a placebo who had a "heart event."

b. Use the two-proportion z-test to determine whether aspirin lowered the rate of women who suffered from cardiovascular events, using a significance level of 0.05.

**8.80 Support for Nuclear Energy** Some people fear the use of nuclear energy because of potential accidents. A Gallup Poll in 2009 showed that 74% of men and 47% of women surveyed supported the use of nuclear energy to produce electricity. The overall sample size was 1012.

a. Assume that exactly half the sample were men and half were women, and determine the numbers of men and women who said they supported the use of nuclear energy and the numbers of men and women who opposed it. Round your frequencies to whole numbers and report them in a table. Put the labels "Men" and "Women" across the top.

b. Now using the numbers from part a, determine whether women are less supportive of nuclear energy than men at the 0.05 level of significance.

* **8.81 Three-Strikes Law** California's controversial "three-strikes law" requires judges to sentence anyone convicted of three felony offenses to life in prison. Supporters say that this decreases crime both because it is a strong deterrent and because career criminals are removed from the streets. Opponents argue (among other things) that people serving life sentences have nothing to lose, so violence within the prison system increases. To test the opponents' claim, researchers examined data starting from the mid-1990s from the California Department of Corrections. "Three Strikes: Yes" means the person had committed three or more felony offenses and

was probably serving a life sentence. "Three Strikes: No" means the person had committed no more than two offenses. "Misconduct" includes serious offenses (such as assaulting an officer) and minor offenses (such as not standing for a count). "No Misconduct" means the offender had not committed any offenses in prison.

a. Compare the proportions of misconduct in these samples. Which proportion is higher, the proportion of misconduct for those who had three strikes or that for those who did not have three strikes? Explain.

b. Treat this as though it were a random sample and determine whether those with three strikes tend to have more offenses than those who do not. Use a 0.05 significance level.

|  | Three Strikes | |
| --- | --- | --- |
|  | Yes | No |
| Misconduct | 163 | 974 |
| No Misconduct | 571 | 2214 |
| Totals | 734 | 3188 |

**8.82 Sleep Medicine for Shift Workers** Shift workers, who work during the night and must sleep during the day, often become sleepy when working and have trouble sleeping during the day. In a study done at Harvard Medical School, 209 shift workers were randomly divided into two groups; one group received a new sleep medicine (modafinil or Provigil), and the other group received a placebo. During the study, 54% of the workers taking the placebo and 29% of those taking the medicine reported accidents or near accidents commuting to and from work. Assume that 104 of the people were assigned the medicine and 105 were assigned the placebo. (Source: Czeisler et al., Modafinil for excessive sleepiness associated with shift-work sleep disorder, *New England Journal of Medicine*, vol. 353: 476–486, August 4, 2005)

a. State the null and alternative hypothesis. Is the alternative hypothesis one-tailed or two-tailed? Explain your choice.

b. Perform a statistical test to determine whether the difference in proportions is significant at the 0.05 level.

## g UIDED EXERCISES

**g 8.27 Gun Control** Historically, the percentage of U.S. residents who support stricter gun control laws has been 52%. A recent Gallup Poll of 1011 people showed 495 in favor of stricter gun control laws. Assume the poll was given to a random sample of people.

**QUESTION** Test the claim that the proportion of those favoring stricter gun control has changed from 0.52. Perform a hypothesis test, using a significance level of 0.05, by following the steps.

**Step 1 ▶ Hypothesize**

$H_0$: The population proportion that supports gun control is 0.52, $p =$ _____.

$H_a$: $p$ _____.

**Step 2 ▶ Prepare**

Choose the one-proportion $z$-test.

Random sample: yes

Sample size: $np_0 = 1011(0.52) =$ about 526, which is more than 10, and $n(1 - p_0) =$ about _____, which is more than _____.

Population size is more than 10 times 1011.

**Step 3 ▶ Compute to compare**

$$\hat{p} = \underline{\quad}$$

$$SE = \sqrt{\frac{p_0(1 - p_0)}{n}} = \sqrt{\frac{0.52(\underline{\quad})}{1011}} = \underline{\quad}$$

$$z = \frac{\hat{p} - p_0}{SE} = \frac{0.4896 - \underline{\quad}}{SE} = \underline{\quad}$$

p-value = __

Please report your p-value with three decimal digits.

Check your answers with the accompanying figure. Don't worry if the last digits are a bit different (this can occur due to rounding).

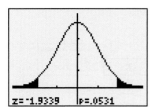

z=-1.9339    p=.0531

TI-83/84 One-Proportion $z$-Test

**Step 4 ▶ Interpret**
Reject $H_0$ (if the p-value is 0.05 or less) or do not reject $H_0$ and choose one of the following conclusions:

   i. The percentage is not significantly different from 52%. (A significant difference is one for which the p-value is less than or equal to 0.05.)

   ii. The percentage is significantly different from 52%.

---

**g 8.45 Vioxx** In the fall of 2004, drug manufacturer Merck Pharmaceutical withdrew Vioxx, a drug that had been used for arthritis pain, from the market after a study revealed that its use was associated with an increase in the risk of heart attack. The experiment was placebo-controlled, randomized, and double-blind. Out of 1287 people taking Vioxx there were 45 heart attacks, and out of 1299 people taking the placebo there were 25 heart attacks (Source: *Los Angeles Times*, October 23, 2004).

**QUESTION** Perform a hypothesis test to test whether those who take Vioxx have a greater rate of heart attack than those who take a placebo. Use a level of significance of 0.05. Follow the steps below to answer the question. In order to be able to use the typical four steps without changing the numbering, we have called the first step, step 0. Can we conclude that Vioxx *causes* an increased risk of heart attack?

**Step 0 ▶** Find the proportion of people in the sample taking Vioxx who had a heart attack and the proportion of people in the sample taking a placebo who had a heart attack. Compare these proportions.

**Step 1 ▶ Hypotheses**
Let $p_V$ be the population proportion of those taking Vioxx who had a heart attack, and let $p_p$ be the population proportion of those taking the placebo who had a heart attack.

$$H_0: p_V = p_p$$
$$H_a: \underline{\quad}$$

**Step 2 ▶ Prepare**
Choose the two-proportion $z$-test. Although we don't have a random sample, we have random assignment to groups. The pooled proportion of heart attacks is

$$\hat{p} = \frac{45 + 25}{1287 + 1299} = \frac{70}{2586} = 0.027069$$

$n_1 \times \hat{p} \geq 1287(0.027069) = 34.84$, which is more than 10
$n_1 \times (1 - \hat{p}) \geq \underline{\quad}$
$n_2 \times (1 - \hat{p}) \geq \underline{\quad}$
$n_2 \times \hat{p} \geq \underline{\quad}$

**Step 3 ▶ Compute to compare**
Refer to the accompanying figure.

$$z = \underline{\quad}$$
$$\text{p-value} = \underline{\quad}$$

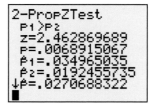

TI-83/84 Two-Proportion $z$-Test
1 is for Vioxx and 2 is for placebo.

**Step 4 ▶ Interpret**
Reject or do not reject the null hypothesis and choose i or ii:

   i. Those taking Vioxx did *not* have a significantly higher rate of heart attack than those taking the placebo.

   ii. Those who took Vioxx had a significantly higher heart attack rate than those who took the placebo.

**Causality**
Does the experiment satisfy the conditions required to conclude a cause-and-effect relationship exists between taking Vioxx and increased risk for heart attacks?

# TechTips

## General Instructions for All Technology

All technologies will use the examples that follow.

**EXAMPLE A** ▶ Do a one-proportion $z$-test to determine whether you can reject the hypothesis that a coin is a fair coin if 10 heads are obtained from 30 flips of the coin. Find $z$ and the p-value.

**EXAMPLE B** ▶ Do a two-proportion $z$-test: Find the observed value of the test statistic and the p-value that tests whether the proportion of people who support stem cell research changed from 2002 to 2007. In both years, the researchers sampled 1500 people. In 2002, 645 people expressed support. In 2007, 765 people expressed support.

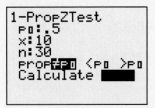

### TI-83/84

#### One-Proportion z-Test

1. Press **STAT**, choose **Tests**, and choose **5: 1-PropZTest**.
2. See Figure 8A.

Enter: $p_0$, **.5**; **x, 10**; **n, 30**.
Leave the default $\neq p_0$.
Scroll down to **Calculate** and press **ENTER**.

```
1-PropZTest
 p0:.5
 x:10
 n:30
 prop≠p0 <p0 >p0
 Calculate
```

▲ **FIGURE 8A** TI-83/84 Input for One-Proportion $z$-Test

You should get a screen like Figure 8B. If you choose **Draw** instead of **Calculate**, you can see the shading of the Normal curve.

```
1-PropZTest
 prop≠.5
 z=-1.825741858
 p=.0678890388
 p̂=.3333333333
 n=30
```

▲ **FIGURE 8B** TI-83/84 Output for One-Proportion $z$-Test

#### Two-Proportion z-Test

1. Press **STAT**, choose **Tests**, and choose **6: 2-PropZTest**.
2. See Figure 8C.

Enter: **x1, 645**; **n1, 1500**; **x2, 765**; **n2, 1500**.
Scroll down to **Calculate** (or **Draw**) and press **ENTER**.

```
2-PropZTest
 x1:645
 n1:1500
 x2:765
 n2:1500
 p1≠p2 <p2 >p2
 Calculate
```

▲ **FIGURE 8C** TI-83/84 Input for Two-Proportion $z$-Test

You should get a screen like Figure 8D. The arrow down next to $\hat{p}$ means there is more information below what is given. To see that information, scroll down, using the down arrow on your keypad.

```
2-PropZTest
 p1≠p2
 z=-4.389689024
 p=1.1360415ᴇ-5
 p̂1=.43
 p̂2=.51
↓p̂=.47
```

▲ **FIGURE 8D** TI-83/84 Output for Two-Proportion $z$-Test

*Caution!* Beware of p-values that appear at first glance to be larger than 1. In Figure 8D, the p-value is 1.1 times 10 to the negative fifth power (–5), or 0.000011.

### MINITAB

#### One-Proportion z-Test

1. **Stat** > **Basic Statistics** > **1-Proportion**.
2. Refer to Figure 8E. Click **Summarized data**. Enter: **number of events**, **10**; **Number of trials**, **30**; check **Perform hypothesis test, hypothesized proportion, .5**

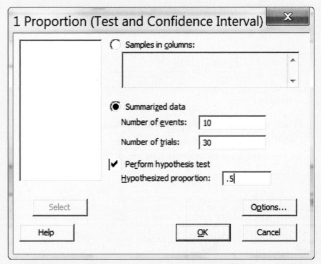

▲ **FIGURE 8E** Minitab Input for One-Proportion $z$-Test

3. Click **Options** and check the box that says **Use test and interval based on normal distribution**. (If you wanted to change the alternative hypothesis to one-tailed, you would do that also through **Options**.) Click **OK**: click **OK**.

You should get output that looks like Figure 8F. Note that you also get a 95% confidence interval (95% CI) for the proportion.

```
Test of p = 0.5 vs p not = 0.5

Sample   X   N   Sample p        95% CI        Z-Value  P-Value
1       10  30  0.333333  (0.164646, 0.502020)   -1.83    0.068

Using the normal approximation.
```

▲ **FIGURE 8F** Minitab Output for One-Proportion $z$-Test

### Two-Proportion z-Test

1. **Stat** > **Basic Statistics** > **2 Proportions**
2. See Figure 8G.

Click **Summarized data**.

| Enter | Events | Trials |
|-------|--------|--------|
| First | 645 | 1500 |
| Second | 765 | 1500 |

3. Click **OK**.

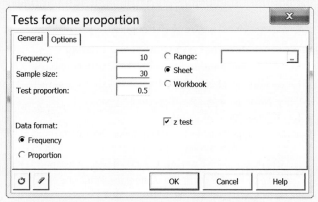

▲ **FIGURE 8G** Minitab Input for Two-Proportion z-test

Your output should look like Figure 8H. Note that it includes a 95% confidence interval for the difference between the proportions, as well as z and the p-value.

```
Test and CI for Two Proportions
Sample    X     N    Sample p
1        645   1500   0.430000
2        765   1500   0.510000


Difference = p (1) - p (2)
Estimate for difference:  -0.08
95% CI for difference:  (-0.115605, -0.0443955)
Test for difference = 0 (vs not = 0):  Z = -4.40   P-Value = 0.000
```

▲ **FIGURE 8H** Minitab Output for Two-Proportion z-Test and Interval

---

**EXCEL**

### One-Proportion z-Test

1. Click **Add-ins**, **XLSTAT**, **Parametric Tests**, **Tests for one proportion**.
2. See Figure 8I.

Enter: **Frequency**, **10**; **Sample size**, **30**; **Test proportion**, **.5**.
(If you wanted a one-tailed hypothesis, you would click **Options**.)
Click **OK**.

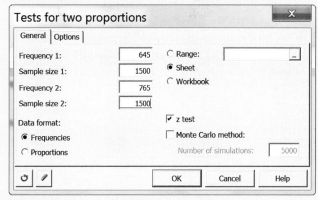

▲ **FIGURE 8I** Input for One Proportion z-Test

When the output appears, you may need to change the column width to see the answers. Click **Home**, and in the **Cells** group click **Format** and **AutoFit Column Width**. The relevant part of the output is shown here.

| | |
|---|---|
| Difference | −0.167 |
| z (Observed value) | −1.826 |
| p-value (Two-tailed) | 0.068 |

### Two-Proportion z-Test

1. Click **Add-ins**, **XLSTAT**, **Parametric Tests**, **Tests for two proportions**.
2. See Figure 8J.

Enter: **Frequency 1**, **645**; **Sample size 1**, **1500**; **Frequency 2**, **765**; **Sample size 2**, **1500**.
(If you wanted a one-tailed alternative, you would click **Options**.)
Click **OK**.

▲ **FIGURE 8J** XLSTAT Input for Two-Proportion z-Test

The relevant parts of the output are shown.

| | |
|---|---|
| Difference | −0.080 |
| z (Observed value) | −4.404 |
| p-value (Two-tailed) | <0.0001 |

## One-Proportion *z*-Test and Confidence Interval

1. **Stat** > **Proportions** > **One sample** > **with summary**
2. Enter: **number of successes**, **10**; **number of observations**, **30**.
   Click **Next** and select the **Hypothesis Test** or
   **Confidence interval** option.

   a. Leave Hypothesis test checked. Enter: **Null prop:=**, **0.5**.
      Leave the **Alternative 2-tailed**, which is the default.

   b. For a Confidence interval, leave the **Level**, **0.95** (the
      default). For **Method**, leave *Standard-Wald*, the default.

3. Click **Calculate**.
   Figure 8K shows the output for the hypothesis test.

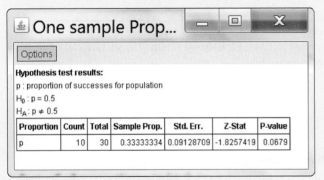

▲ **FIGURE 8K** StatCrunch Output for On-Proportion *z*-Test

## Two-Proportion *z*-Test

1. **Stat** > **Proportions** > **Two samples** > **with summary**
2. Refer to Figure 8L.

Enter:   **Sample 1: Number of successes, 645**;
         **Number of observations, 1500**:
         **Sample 2: Number of successes, 765**;
         **Number of observations, 1500**.

3. If you want to change the alternative hypothesis from the
   two-tailed default (or want a confidence interval), click **Next**.
   Otherwise, click **Calculate**.

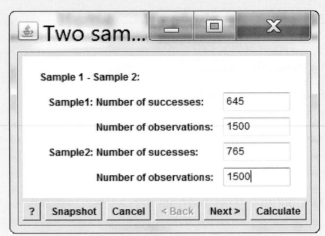

▲ **FIGURE 8L** StatCrunch Input for Two-Proportion *z*-Test

Figure 8M shows the StatCrunch output.

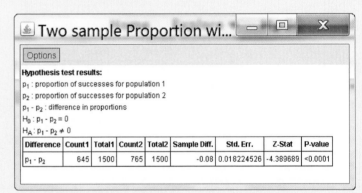

▲ **FIGURE 8M** StatCrunch Output for Two-Proportion *z*-Test

# 9 Inferring Population Means

Brewing beer is a tricky business. Beer has only four main ingredients: malted barley, hops, yeast, and water. However, these four ingredients must be mixed in precise quantities at precise temperatures. Having some experience in trying to get this mix just right, in the late 1800s the Guinness brewery in Dublin, Ireland, began to hire the best and brightest science graduates to help them perfect the brewing process. One of these, hired in 1899 at the age of 23, was William Sealy Gosset (1876–1937), who had majored in chemistry and mathematics. One of Gosset's jobs was to measure the amount of yeast in a small sample of beer. On the basis of this sample, he was to estimate the mean amount of yeast in the beers produced. If this yeast amount was too high or too low, then something was wrong and the process would have to be fixed.

Naturally, uncertainty played a role. Suppose the average yeast count in his sample was too high or too low. Did this indicate that the mean yeast count in the entire factory was off? Or was his sample different just because of chance? The statistical science of the day knew how to answer this question if the number of samples was large, but Gosset worked in a context in which large samples were just too expensive and time-consuming to collect. A decision had to be based on a small sample. Gosset solved the problem, and his approach (which is called the *t*-test) is now one of the most widely used techniques in statistics.

Estimating one or more population means is still an important part of science and public policy. How do we compare the effects of different drugs on epilepsy? How do commuting times vary between cities? How do men and women differ in their abilities to sense their surroundings? To answer these questions, we need good estimates of the population means, and these estimates must be based on reliable data from small samples.

In Chapters 7 and 8 you learned two important techniques for statistical inference: the confidence interval and the hypothesis test. In those chapters, we applied statistical inference for population proportions. In this chapter, we use the same two techniques for making inferences about means of populations. We begin with inference for one population and conclude with inferring the difference between the means of two populations.

## CASE STUDY

### Epilepsy Drugs and Children

Epilepsy is a neurological disorder that causes seizures, which can range from mild to severe. It has been estimated (wikipedia.org) that over 50 million people around the world suffer from epilepsy. It is usually treated with drugs, and four drugs that are commonly used for this purpose are carbamazepine, lamotrigine, phenytoin, and valproate. In 2009 researchers at the *New England Journal of Medicine* reported that pregnant mothers who take valproate might risk impairing the cognitive development of their children, compared to mothers who take one of the other drugs. As evidence, on the basis of a sample of pregnant women with epilepsy, they estimated the mean IQ of three-year-old children whose mothers took one of these four drugs during their pregnancies. They gave 95% confidence intervals (CI) for the mean IQ, as shown in Table 9.1 on the next page (Meador et al. 2009).

| Drug | 95% CI |
|------|--------|
| carbamazepine | (95, 101) |
| lamotrigine | (98, 104) |
| phenytoin | (94, 104) |
| valproate | (88, 97) |

▲ **TABLE 9.1** 95% confidence intervals for the average IQ of children whose mothers took various epilepsy drugs during their pregnancies.

Why did these four intervals lead the researchers to recommend that pregnant women not use valproate as a "first choice" drug for epilepsy? The researchers wrote that "Although the confidence intervals for carbamazepine and phenytoin overlap with the confidence interval for valproate, the confidence intervals for the differences between carbamazepine and valproate and between phenytoin and valproate do not include zero." What does this tell us?

In this chapter we discuss how confidence intervals can be used to estimate characteristics of a population—in this case, the population of all children of women with epilepsy who took one of these drugs during pregnancy. Confidence intervals can also be used to judge between hypotheses about the means and about differences between means. The population of pregnant women with epilepsy is large, and yet if conditions are right, we can make decisions and reach an understanding about the entire population on the basis of a small sample. At the end of this chapter, we will return to this study and see if we can better understand its conclusions.

# Sample Means of Random Samples

As you learned in Chapter 7, we estimate population parameters by collecting a random sample from that population. We use the collected data to calculate a statistic, and this statistic is used to estimate the parameter. Whether we are using the statistic $\hat{p}$ to estimate the parameter $p$ or are using $\bar{x}$ to estimate $\mu$, if we want to know how close our estimate is to the truth, we need to know how far away that statistic is, typically, from the parameter.

Just as we did in Chapter 7 with $\hat{p}$, we now examine three characteristics of the behavior of the sample mean: its accuracy, its precision, and its probability distribution. By understanding these characteristics, we'll be able to measure how well our estimate performs and thus make better decisions.

As a reminder, Table 9.2 shows the relation between some commonly used statistics and the parameters they estimate. (This table originally appeared as Table 7.2.)

▶ TABLE 9.2

| Statistic (based on data) | | Parameter (typically unknown) | |
|---------------------------|---|------------------------------|---|
| Sample mean | $\bar{x}$ | Population mean | $\mu$ (mu) |
| Sample standard deviation | $s$ | Population standard deviation | $\sigma$ (sigma) |
| Sample variance | $s^2$ | Population variance | $\sigma^2$ |
| Sample proportion | $\hat{p}$ (p-hat) | Population proportion | $p$ |

**Details**

**Mu-sings**
The mean of a population, represented by the Greek character $\mu$, is pronounced "mu" as in *music*.

This chapter uses much of the vocabulary introduced in Chapters 7 and 8, but we'll remind you of important terms as we proceed. To help you visualize how a sample mean based on randomly sampled data behaves, we'll make use of the by-now-familiar technique of simulation. Our simulation is slightly artificial, because to do a simulation, we need to know the population. However, after using a simulation to understand how the sample mean behaves in this artificial situation, we will discuss what we do in the real world when we do not know very much about the population.

## Accuracy and Precision of a Sample Mean

The reason why the sample mean is a useful estimator for the population mean is that the sample mean is accurate and, with a sufficiently large sample size, very precise. The accuracy of an estimator, you'll recall, is measured by the **bias**, and the precision is measured by the standard error. You will see in this simulation that

> **Looking Back**
>
> **Bias and Precision**
> Bias is the mean distance between the sample statistic and the parameter it is estimating. Precision is measured by the standard error, which tells us how far the statistic typically deviates from its center. Review Figure 7.2 to help visualize these two properties.

1. The sample mean is unbiased when estimating the population mean—-that is, on average, the sample mean is the same as the population mean.

2. The **precision** of the sample mean depends on the variability in the population, but the more observations we collect, the more precise the sample mean becomes.

For our simulation, we'll use the population that consists of the finishing times of everyone who ran the Cherry Blossom Ten Mile Run in 2009 (Kaplan 2009). (The finishing time is the amount of time it took to finish the race.) This race is held every spring in Washington, D.C. As in most such races, data are carefully collected on every participant. Rather than showing you the histogram of the finishing times for all 8600 runners, we're going to take advantage of the fact that the distribution closely follows a Normal model with a mean of 97 minutes and a standard deviation of 17 minutes: N(97, 17). Figure 9.1 shows the distribution of this population.

The population parameters are, in symbols,

$$\mu = 97 \text{ minutes}$$

$$\sigma = 17 \text{ minutes}$$

For our simulation, we will randomly sample 30 runners and calculate the average of their finishing times. We'll then repeat this many, many times. (The exact number of repetitions we performed isn't important for this discussion.) We are interested in two questions: (1) What is the typical value of the sample mean? If it is 97 minutes—the value of the population mean—then the sample mean is unbiased. (2) Typically, how far away is a sample mean from 97? In other words, how much spread is in the distribution of sample means? This spread helps us measure the precision of the sample mean as an estimator of the population mean.

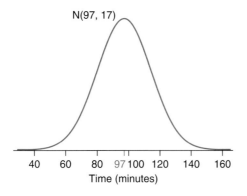

◀ **FIGURE 9.1** Finishing times in the Cherry Blossom Ten Mile Run follow a Normal model with a mean of 97 minutes and a standard deviation of 17 minutes.

For example, our very first sample of 30 runners had a mean finishing time of 95.1 minutes. We plotted this sample mean, as well as the means of the many other samples we took, in Figure 9.2 on the next page. The plot is on the same scale as Figure 9.1, the picture of the population distribution, so that you can see how much narrower this distribution of sample means is.

From this dotplot of sample means, we learn that the typical value of the sample means is the same as the population mean of 97 minutes. And we see that the sample mean is a relatively precise estimator: All of the sample means are within about 10 minutes (either above or below) the true mean value of 97 minutes.

▶ **FIGURE 9.2** Each dot represents a sample mean based on 30 runners who were randomly selected from the population whose distribution is shown in Figure 9.1. Note that the spread of this distribution is much smaller than the spread of the population, but the center looks to be at about the same place: 97 minutes.

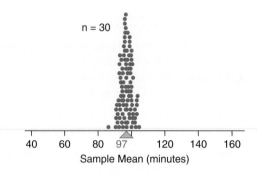

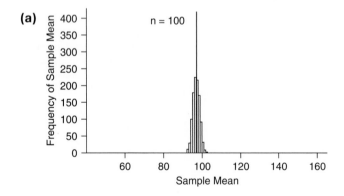

**Looking Back**

**Sampling Distribution**
In Chapter 7 we introduced the sampling distribution of sample proportions. The sampling distribution of sample means is the same concept: It is a distribution that gives us probabilities for sample means drawn from a population. The sampling distribution is the distribution of all possible sample means.

Figure 9.2 is a very approximate picture of the **sampling distribution** of the sample mean for samples of size 30. Recall that a sampling distribution is a probability distribution of a statistic; in this case, the statistic is the sample mean. You can think of the sampling distribution as the distribution of *all* possible sample means that would result from drawing repeated random samples of a certain size from the population.

When the mean of the sampling distribution is the same value as the population mean, we say that the statistic is an **unbiased estimator**. This appears to be the case here, because both the mean of the distribution of sample means in Figure 9.2 and the population mean are about 97 minutes.

The standard deviation of the sampling distribution is what we call the **standard error**. The standard error measures the precision of an estimator by telling us how much the statistic varies from sample to sample. For the sample mean, the standard error is smaller than the population standard deviation. We can see this because the spread for the sampling distribution is smaller than the spread of the population distribution. Soon you'll see how to calculate the standard error.

What happens to the center and spread of the sampling distribution if we increase the sample size? Let's start all over with the simulation. But this time, we take a random sample of 100 runners and calculate the mean. We then repeat this many hundreds of times. Figure 9.3 shows the results for this simulation and also for two new

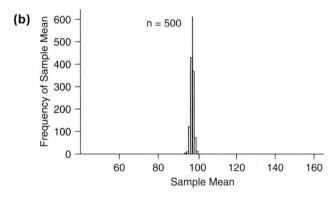

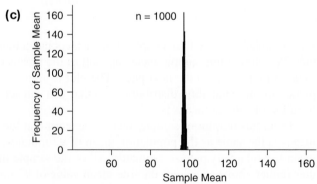

▶ **FIGURE 9.3 (a)** Histogram of a large number of sample means. Each sample mean is based on a sample of 100 randomly selected runners. **(b)** Sample means based on samples of 500 runners. **(c)** Sample means based on samples of 1000 runners. Each time, the sampling distribution gets narrower, reflecting a smaller standard error.

simulations where each sample mean is based on 500 runners and then on 1000 runners. The scale of the *x*-axis is the same as in Figure 9.1. Note that the spread of the distributions becomes quite small—so small, in fact, that we can't get a good look at the shape of the distributions.

## What Have We Demonstrated with These Simulations?

Because the sampling distributions are always centered at the population mean, we have demonstrated that the sample mean is an unbiased estimator of the population mean. We saw this for only one type of population distribution: the Normal distribution. But in fact, this is the case for any population distribution.

We have demonstrated that the standard deviation of the sampling distribution, which is called the **standard error** of the sample mean, gets smaller with larger sample size. This is true for any population distribution.

We can be more precise. If the symbol $\mu$ represents the mean of the population and if $\sigma$ represents the standard deviation of the population, then

1. The mean of the sampling distribution is also $\mu$ (which tells us that the sample mean is unbiased when estimating the population mean).

2. The standard error is $\dfrac{\sigma}{\sqrt{n}}$ (which tells us that the standard error depends on the population distribution and is smaller for larger samples).

 **Looking Back**

**Sample Proportions from Random Samples**
Compare the properties of the sample mean to those of the sample proportion, $\hat{p}$, given in Chapter 7. The sample proportion is also an unbiased estimator (for estimating the population proportion, $p$). It has standard error

$$\sqrt{\frac{p(1-p)}{n}}.$$

> **KEY POINT**  For all populations, the sample mean is unbiased when estimating the population mean. The standard error of the sample mean is $\dfrac{\sigma}{\sqrt{n}}$, so the sample mean is more precise for larger sample sizes.

## EXAMPLE 1 iTunes Library Statistics

A student's iTunes library of mp3s has a very large number of songs. The mean length of the songs is 243 seconds, and the standard deviation is 93 seconds. The distribution of song lengths is right-skewed. Using his mp3 player, this student will create a playlist that consists of 25 randomly selected songs.

QUESTIONS

a. Is the mean value of 243 minutes an example of a parameter or a statistic? Explain.

b. What should the student expect the average song length to be for his playlist?

c. What is the standard error for the mean song length of 25 randomly selected songs?

SOLUTIONS

a. The mean of 243 is an example of a parameter, because it is the mean of the population that consists of all of the songs in the student's library.

b. The sample mean length can vary, but is typically the same as the population mean: 243 seconds.

c. The standard error is $\dfrac{\sigma}{\sqrt{n}} = \dfrac{93}{\sqrt{25}} = \dfrac{93}{5} = 18.6$ seconds.

TRY THIS!  Exercise 9.9

SECTION 9.2

# The Central Limit Theorem for Sample Means

In the last simulation, all of the approximate sampling distributions (Figures 9.2 and 9.3) were Normal. This probably doesn't surprise you, because the population distribution was Normal.

What might surprise you is that the sampling distribution of the mean is always Normal (or at least approximately Normal), regardless of the shape of the population distribution. (If the sample size is small, however, the approximation can be pretty lousy.) This is the conclusion of the Central Limit Theorem, an important mathematical theorem that tells us that as long as the sample size is large, we can use the Normal distribution to perform statistical inference, regardless of the population the data are sampled from.

The **Central Limit Theorem (CLT)** assures us that no matter what the shape of the population distribution, if a sample is selected such that the following conditions are met, then the distribution of sample means follows an approximately Normal distribution. The mean of this distribution is the same as the population mean. The standard deviation (also called the standard error) of this distribution is the population standard deviation divided by the square root of the sample size. As a rule of thumb, sample sizes of 25 or more may be considered "large."

> **Looking Back**
>
> **CLT for Proportions**
> In Chapter 7, you saw that the Central Limit Theorem applies to sample proportions. Here you'll see that it also applies to sample means.

Condition 1: *Random Sample and Independence.* Each observation is collected randomly from the population, and observations are independent of each other. The sample can be collected either with or without replacement.

Condition 2: *Normal.* Either the population distribution is Normal or the sample size is large.

Condition 3: *Big Population.* If the sample is collected without replacement, then the population must be at least 10 times larger than the sample size.

**KEY POINT**

The sampling distribution of $\bar{x}$ is approximately $N\left(\mu, \dfrac{\sigma}{\sqrt{n}}\right)$, where $\mu$ is the mean of the population and $\sigma$ is the standard deviation of the population. The larger the sample size, $n$, the better the approximation. If the population is Normal to begin with, then the sampling distribution is exactly a Normal distribution.

## Visualizing Distributions of Sample Means

The histogram in Figure 9.4 shows the distribution of in-state tuition and fees for all two-year colleges in the United States for the 2008–2009 academic year (*The Chronicle of Higher Education, Facts & Figures* 2008). Note that the distribution has an unusual feature: It has a mode near 0. This is because all two-year colleges in California charged $647 for California residents. The distribution is right-skewed.

▶ **FIGURE 9.4** Distribution of annual tuitions and fees at all two-year colleges in the United States for the 2008–2009 academic year.

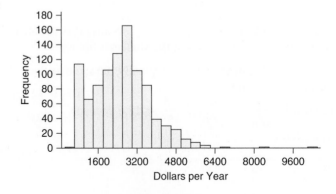

This histogram represents the distribution of a population, because it includes *all* two-year colleges. The mean of this population—the "typical" tuition of all two-year colleges—is $2535.

Using this distribution, we now show the results of a simulation that should be starting to feel familiar. First, we take a random sample of 30 colleges. The distribution of this sample is shown in Figure 9.5. We find the mean tuition of the 30 colleges in the sample and record this figure; for example, the sample mean for the sample shown in Figure 9.5 is about $2380.

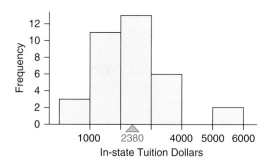

▶ **FIGURE 9.5** Distribution of a sample of 30 colleges taken from the population of all colleges. The mean of this sample, $2380, is indicated.

We repeat this activity (that is, we sample another 30 colleges from the population of all colleges and record the mean tuition of the sample) 200 times. When we are finished, we have 200 sample mean tuitions, each sample mean based on a sample of 30 colleges. Figure 9.6a shows this distribution. Figure 9.6b shows the distribution of averages when, instead of sampling 30 colleges, we double the number and sample 60 colleges. What differences do you see among the population distribution (Figures 9.4), the distribution of one of the samples (Figure 9.5), the sampling distribution when the sample size is 30 (Figure 9.6a), and the sampling distribution when the sample size is 60 (Figure 9.6b)?

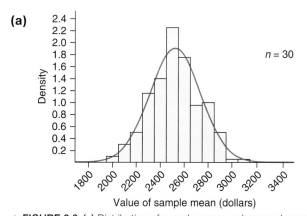

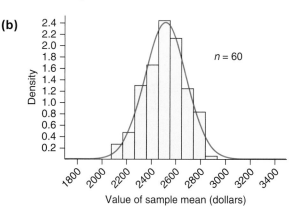

▲ **FIGURE 9.6** **(a)** Distribution of sample means, where each sample mean is based on a sample size of $n = 30$ college tuitions and is drawn from the population shown in Figure 9.4. This is (approximately) the sampling distribution of $\bar{x}$ when $n = 30$. A Normal curve is superimposed. **(b)** The approximate sampling distribution of $\bar{x}$ when $n = 60$.

Both of the sampling distributions in Figures 9.6a and 9.6b show us the values and relative frequencies for $\bar{x}$, but they are based on different sample sizes. We see that even though the *population* distribution has an unusual shape (Figure 9.4), the sampling distributions for $\bar{x}$ are fairly symmetric and unimodal. Although the Normal curve that is superimposed doesn't match the histogram very closely when $n = 30$, the match is pretty good for $n = 60$.

This is exactly what the CLT predicts. When the sample size is large enough, we can use the Normal distribution to find approximate probabilities for the values we might see for $\bar{x}$ when we take a random sample from the population.

The more observations in your sample, the better an approximation the Normal distribution provides. Generally, the CLT provides a useful approximation of the true

**↻ Looking Back**

**Distribution of a Sample vs. Sampling Distribution**
Recall that these are two different concepts. The *distribution of a sample*, from Chapter 3, is the distribution of one single sample of data (Figure 9.5). The *sampling distribution*, on the other hand, is the probability distribution of an estimator or statistic such as the sample mean (Figures 9.6a and 9.6b).

probabilities if the sample size is 25 or more. But this is just a rule of thumb. Be aware that you might need larger sample sizes in some situations. Unlike in Chapter 7, where we worked with sample proportions, we can't provide a hard-and-fast rule for sample size. For nearly all examples in this text, though not always in real life, 25 is large enough.

## Applying the Central Limit Theorem

The Central Limit Theorem helps us find probabilities for sample means when those means are based on a random sample from a population. Example 2 demonstrates how we can answer probability questions about the sample mean even if we can't answer probability questions about individual outcomes.

### EXAMPLE 2 Pulse Rates Are Not Normal

According to one very large study done in the United States, the mean resting pulse rate of adult women is about 74 beats per minute (bpm), and the standard deviation of this population is 13 bpm (NHANES 2003–2004). The distribution of resting pulse rates is known to be skewed right.

**QUESTIONS**

a. Suppose we take a random sample of 36 women from this population. What is the approximate probability that the average pulse rate of this sample will be below 71 or above 77 bpm? (In other words, what is the probability that it will be more than 3 bpm away from the population mean of 74 bpm?)

b. Can you find the probability that a single adult woman will have a resting pulse rate more than 3 bpm away from the mean value of 74?

**SOLUTION**

a. It doesn't matter that the population distribution is not Normal. Because the sample size of 36 women is relatively large, the distribution of sample means will be approximately (though not exactly) Normal.

The mean of this Normal distribution will be the same as the population mean: $\mu = 74$ bpm. The standard deviation of this distribution is the standard error:

$$SE = \frac{\sigma}{\sqrt{n}} = \frac{13}{\sqrt{36}} = \frac{13}{6} = 2.167$$

To use the Normal table to find probabilities requires that the values of 71 bpm and 77 bpm be converted to standard units:

$$z = \frac{\bar{x} - \mu}{SE} = \frac{71 - 74}{2.167} = \frac{-3}{2.167} = -1.38$$

Figure 9.7 shows that the area that corresponds to the probability that the sample mean pulse rate will be more than 1.38 standard errors away from the population mean pulse rate. This probability is calculated to be about 17%.

▶ **FIGURE 9.7** Area of the Normal curve outside of z-scores of −1.38 and 1.38.

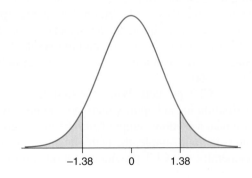

**CONCLUSIONS**

a. The approximate probability that the average pulse of 36 adult women will be more than 3 bpm away from 74 bpm is about 17%.

b. We cannot find the probability for a single woman because we do not know the probability distribution. We know only that it is "right-skewed," which is not enough information to find actual probabilities.

**TRY THIS!**  Exercise 9.11

## Many Distributions

It's natural at this point to feel that you have seen a confusingly large number of types of distributions, but it's important that you keep them straight. The *population distribution* is the distribution of values from the population. Figure 9.4 (two-year college tuitions) is an example of a population distribution because it shows the distribution of *all* two-year colleges. Figure 9.1 (runners' times for all competitors in a race) is another example of a population distribution. For some populations, we don't know precisely what this distribution is. Sometimes we assume (or know) it is Normal, sometimes we know it is skewed in one direction or the other, and sometimes we know almost nothing.

We then take a random sample of *n* observations. We can make a histogram of these data. This histogram gives us a picture of the *distribution of the sample*. If the sample size is large, and if the sample is random, then the sample will be representative of the population, and the distribution of the sample will look similar to (but not the same as!) the population distribution. Figure 9.5 is an example of the distribution of a sample of size *n* = 30 taken from the population of two-year college tuitions.

The *sampling distribution* is more abstract. If we take a random sample of data and find the sample mean (the center of the distribution of the sample), and then repeat this many, many times, we will get an idea what the sampling distribution looks like. Figures 9.6a and 9.6b are examples of approximate sampling distributions for the sample mean, based on samples from the two-year college tuition data. Note that these do not share the same shape as the population or the sample; they are both approximately Normal.

## EXAMPLE 3 Identify the Distribution

Figure 9.8 shows three distributions. One distribution is a population. The other two distributions are (approximate) sampling distributions. One sampling distribution is based on sample means of size 10, and the other is based on sample means of size 25.

**(a)**
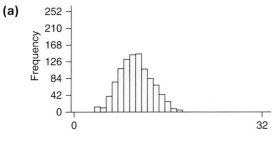

**(b)**
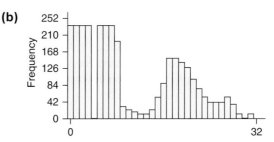

**(c)**
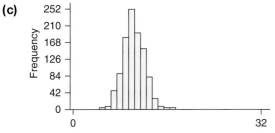

> **! Caution**
>
> **CLT Not Universal**
> The CLT does not apply to all statistics you run across. It does not apply to the sample median, for example. No matter how large the sample size, you cannot use the Normal distribution to find a probability for the median value. It also does not apply to the sample standard deviation.

◀ **FIGURE 9.8** Three distributions, all on the same scale. One is a population distribution, and the other two are sampling distributions for means sampled from the population. (Source: Rice Virtual Lab in Statistics, http://onlinestatbook.com/)

> QUESTION Which graph (a, b, or c) is the population distribution? Which shows the sampling distribution for the mean with $n = 10$? Which with $n = 25$?

> SOLUTION The Central Limit Theorem tells us that sampling distributions for means are approximately Normal. This implies that Figure 9.8b is not a sampling distribution, so it must be the population distribution from which the samples were taken. We know that the sample mean is more precise for larger samples, and because Figure 9.8a has the larger standard error (is wider), it must be the graph associated with $n = 10$. This means that Figure 9.8c is the sampling distribution of means with $n = 25$.

> TRY THIS! Exercise 9.13

## SNAPSHOT    THE SAMPLE MEAN ($\bar{x}$)

| | |
|---|---|
| **WHAT IS IT?** ▸ | The arithmetic average of a sample of data. |
| **WHAT DOES IT DO?** ▸ | Estimates the mean value of a population, $\mu$. The mean is used as a measure of what is "typical" for a population. |
| **HOW DOES IT DO IT?** ▸ | If the sample was a random sample, then the sample mean is unbiased, and we can make the precision of the estimator as good as we want by taking a large enough sample size. |
| **HOW IS IT USED?** ▸ | If the sample size is large enough (or the population is Normal), we can use the Normal distribution to find the probability that the sample mean will take on a value in any given range. This lets us know how wrong our estimate could be. |

## The *t*-Distribution

The hypothesis tests and confidence intervals that we will use for estimating and testing the mean are based on a statistic called the ***t*-statistic**:

$$t = \frac{\bar{x} - \mu}{\left(\dfrac{s}{\sqrt{n}}\right)}$$

The *t*-statistic is very similar to a *z*-score for the sample mean. In the numerator, we subtract the population mean from the sample mean. Then we divide not by the standard error but, rather, by an *estimate* of the standard error.

It would be nice if we could divide by the true standard error. But in real life, we almost never know the value of $\sigma$, the population standard deviation. So instead, we replace it with an estimate: the sample standard deviation, $s$. This gives us an estimate of the standard error:

$$SE_{\text{EST}} = \frac{s}{\sqrt{n}}$$

Compare the *t*-statistic to the *z*-statistic, and you will see that we simply replaced $\sigma$ in the *z*-statistic with $s$.

$$z = \frac{\bar{x} - \mu}{\left(\dfrac{\sigma}{\sqrt{n}}\right)}$$

The *t*-statistic does *not* follow the Normal distribution. One reason for this is that the denominator changes with every sample. For this reason, the *t*-statistic is

**Looking Back**

**Sample Standard Deviation**
In Chapter 3 we gave the formula for the sample standard deviation:

$$s = \sqrt{\frac{\Sigma(x - \bar{x})^2}{n - 1}}$$

more variable than the z-statistic (whose denominator is always the same.) Instead, the t-statistic follows a distribution called—surprise!—the **t-distribution**. This was Gosset's great discovery at the Guinness brewery. When small sample sizes were used to make inferences about the mean, even if the population was Normal, the Normal distribution just didn't fit the results that well. Gosset discovered a new distribution, which he called the t-distribution, that turned out to be a better model than the Normal for the sampling distribution of $\bar{x}$ when $\sigma$ is not known.

The t-distribution shares many characteristics with the N(0, 1) distribution. Both are symmetric, are unimodal, and might be described as "bell-shaped." However, the t-distribution has thicker tails. This means that in a t-distribution, it is more likely that we will see extreme values (values far from 0) than it is in a standard Normal distribution.

The t-distribution's shape depends on only one parameter, called the **degrees of freedom (df)**. The number of degrees of freedom is (usually) an integer: 1, 2, 3, and so on. If df is small, then the t-distribution has very thick tails. As the degrees of freedom get larger, the tails get thinner. Ultimately, when df is infinitely large, the t-distribution is exactly the same as the N(0, 1) distribution.

Figure 9.9 shows t-distributions with 1, 10, and 40 degrees of freedom. In each case, the t-distribution is shown with a N(0, 1) curve so that you can compare them. The t-distribution is the one whose tails are "higher" at the extremes. Note that by the time the degrees of freedom reaches 40 (Figure 9.9c), the t-distribution and the N(0, 1) distribution are too close to tell apart (on this scale.)

> **Details**
>
> **Degrees of Freedom**
> Degrees of freedom are related to the sample size: Generally, the larger the sample size, the larger the degrees of freedom. When estimating a single mean, as we are doing here, the number of degrees of freedom is equal to the sample size minus one.
> $$df = n - 1$$

**(a)**

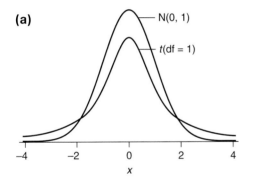

**(b)**

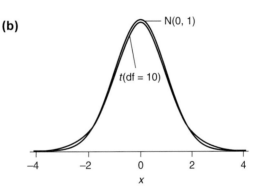

**(c)**

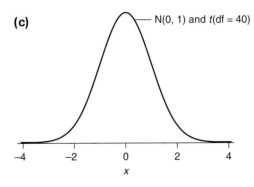

◀ **FIGURE 9.9 (a)** A t-distribution with 1 degree of freedom, along with a N(0, 1) distribution. The t-distribution has much thicker tails. **(b)** The degrees of freedom are now equal to 10, and the tails are only slightly thicker in the t-distribution. **(c)** The degrees of freedom are now 40, and the two distributions are visually indistinguishable.

SECTION 9.3

# Answering Questions about the Mean of a Population

Do you commute to work? How long does it take you to get there? Is this amount of time typical for others in your state? Which state has the greatest commuting times? This information is important not just to those of us who must fight traffic every day, but also to business leaders and politicians who make decisions about quality of living and the cost of doing business. The U.S. Census performs periodic surveys that determine, among other things, commuting times around the country. In 2007 (the last time the survey was done), the state of New York had the greatest mean commuting time, which was 31.5 minutes. Vermont was lowest at 21.2 minutes.

These means are estimates of the mean commuting time for all residents in these states who work outside of their homes. We can learn the true mean commuting time only by asking all residents, which is clearly too time-consuming to do very often. Instead, the U.S. Census takes a random sample of U.S. residents to estimate these values.

In this section we present two techniques for answering questions about the population mean. Confidence intervals are used for estimating values. Hypothesis tests are used for deciding whether the value is one thing or another. These are the same methods that were introduced in Chapter 7 (confidence intervals) and Chapter 8 (hypothesis tests) for population proportions, but here you'll see how they are modified to work with means.

## Estimation with Confidence Intervals

**Confidence intervals** are a technique for communicating an estimate of the mean, along with a measure of our uncertainty in that estimate. The job of a confidence interval is to provide us with a range of values that, according to the data, are highly plausible values for the unknown population mean. For instance, the range of values for the mean commuting time for all Vermont commuters is 21.1 to 21.3 minutes.

Not all confidence intervals do an equally good job; the "job performance" of a confidence interval is therefore measured with something called the **confidence level**. The higher this level, the better the confidence interval performs. The confidence interval for mean Vermont commuting times is 90%, which means we can be extremely confident that this interval contains the true mean.

Sometimes, you will be in a situation in which you will know only the sample mean and sample standard deviation. In these situations, you can use a calculator to find the confidence interval. However, if you have access to the actual data, you are much better off using statistical software to do all the calculations for you. We will show you how to respond to both situations.

No matter which situation you are in, you will need to judge whether a confidence interval is appropriate for the situation, and you will need to interpret the confidence interval. Therefore, we will discuss these essential skills before demonstrating the calculations.

### When Are Confidence Intervals Useful?

A confidence interval is a useful answer to the following questions: "What's the typical value for a variable in this large group of objects or people? And how far away from the truth might this estimate of the typical value be?" You should provide a confidence interval whenever you are estimating the value of a population parameter on the basis of a random sample from that population. For example, judging on the basis of a random sample of 30 adults, what's the typical body temperature of all healthy adults? On the basis of a survey of a random sample of Vermont residents, what's the typical commuting time for all Vermont residents? A confidence interval is useful for answering questions such as these because it communicates the uncertainty in our estimate and provides a range of plausible values.

A confidence interval is not appropriate if there is no uncertainty in your estimate. This would be the case if your "sample" were actually the entire population. For example, it is not necessary to find a confidence interval for the mean score on your class's statistics exam. The population is your class, and all of the scores are known. Thus the population mean is known, and there is no need to estimate it.

### Checking Conditions

In order to measure the correct confidence level, the following conditions must hold:

Condition 1: *Random Sample and Independence.* The data must be collected randomly, and each observation must be independent of the others.

Condition 2: *Normal*. Either the population must be Normally distributed or the sample size must be fairly large (at least 25).

If these conditions do not hold, then we cannot measure the job performance of the interval; the confidence level may be incorrect. This means that we may advertise a 95% confidence level when, in fact, the true performance is much worse than this.

To check the first condition, you must know how the data were collected. This is not always possible, so rather than checking these conditions, you must simply assume that they hold. If they do not, your interval will not be valid.

The requirement for independence means that measurement of one object in the sample does not affect any other. Essentially, if we know the value of any one observation, this knowledge should tell us nothing about the values of other observations. This condition might be violated if, say, we randomly sampled several schools and gave all of the students math exams. The individual math scores would not be independent, because we would expect that students within the same school might have similar scores.

The second condition is due to the Central Limit Theorem. If the population distribution is Normal (or very close to it), then we have nothing to worry about. But if it is non-Normal, then we need a large enough sample size so that the sampling distribution of sample means is approximately Normal. For many applications, a sample size of 25 is large enough, but for extremely skewed distributions, you might need an even larger sample size.

## EXAMPLE 4 Is the Cost of College Rising?

Many cities and states are finding it more difficult to offer low-cost college educations. Did the mean cost of attending two-year colleges increase in the United States from the 2007–2008 academic year to 2009–2010? In 2007–2008, the mean cost of *all* two-year colleges was $2429. A random sample of 35 two-year colleges in the United States found that the average tuition charged in 2008–2009 was $2380, with a standard deviation of $1160. Figure 9.10 shows the distribution of this sample. On the basis of this, a 90% confidence interval for the mean cost of attending two-year colleges in 2008–2009 is $2048 to $2711.

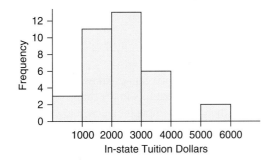

◄ **FIGURE 9.10** Distribution of in-state tuition for a random sample of 35 two-year colleges during the academic year 2008–2009.

**QUESTIONS**

a. Describe the population. Is the number $2380 an example of a parameter or a statistic?

b. Verify that the conditions for a valid confidence interval are met.

**SOLUTIONS**

a. The population consists of all two-year college tuitions (for in-state residents) in the academic year 2008–2009. (There are roughly 1000 two-year colleges in the United States.) The number $2380 is the mean of a sample of only 35 colleges. Because it is the mean of a sample, it is a statistic.

b. The first condition is that the data represent a random sample of independent observations. We are told the sample was collected randomly, so we assume this is true.

Independence also holds, because knowledge about any one school's tuition tells us nothing about other schools in the sample. The second condition requires that the population be roughly Normally distributed or the sample size be equal to or larger than 25. We do not know the distribution of the population, but because the sample size is large enough (bigger than 25), this condition is satisfied.

**TRY THIS!** Exercise 9.15

### Interpreting Confidence Intervals

To understand confidence intervals, you must know how to interpret a confidence interval and how to interpret a confidence level.

A confidence *interval* can be interpreted as a range of plausible values for the population parameter. In other words, in the case of population means, we can be confident that if we were to someday learn the true value of the population mean, it would be within the range of values given by our confidence interval. For example, the U.S. Census estimates that the mean commuting time for Vermont residents is 21.1 minutes to 21.3 minutes, with a 90% confidence level. We interpret this to mean that we can be fairly confident that the true mean commute for *all* Vermont residents is between 21.1 and 21.3 minutes. Yes, we could be wrong. The mean might be less than 21.1 minutes, or it might be more than 21.3 minutes. However, we would be rather surprised to find this was the case; we are highly confident that the mean is within this interval.

 **KEY POINT**  A confidence interval can be interpreted as a range of plausible values for the population parameter.

## EXAMPLE 5 Evidence for Changing College Costs

Based on a random sample of 35 two-year colleges, a 90% confidence interval for the mean in-state tuition at two-year colleges for the 2009–2010 academic year is $2048 to $2711. In the academic year 2007–2008, the population mean of *all* two-year colleges' in-state tuitions was $2429.

**QUESTION**  Does the confidence interval provide evidence that the mean tuition has changed?

**SOLUTION**  No, it does not. Although we cannot know the population mean of all tuitions in 2009/2010, we are highly confident that it is in the range of $2048 to $2711. This range includes the value $2429, so there is not enough evidence to conclude that the mean tuition has changed.

**TRY THIS!** Exercise 9.17

### Measuring Performance with the Confidence Level

The confidence *level*, which in the case of the interval for the mean Vermont commuting time was 90%, tells us about the method used to find the interval. A value for the level of 90% tells us that the U.S. Census used a method that works in 90% of all samples. In other words, if we were to take many same-sized samples of Vermont residents, and for each sample calculate a 90% confidence interval, then 90% of those intervals would contain the population mean.

The confidence level does *not* tell us whether the interval (21.1 to 21.3) contains or does not contain the population mean. The "90%" just tells us that the method that produced this interval is a pretty good method.

Suppose you decided to purchase an mp3 player online. You have your choice of several manufacturers, and they are rated in terms of their performance level. One manufacturer has a 90% performance level, which means that 90% of the players they produce are good ones, and 10% are defective. Some other manufacturers have lower levels: 80%, 60%, and worse. From whom do you buy? You choose to buy from the manufacturer with the 90% level, because you can be very confident that the player they send you will be good. Of course, once the player arrives at your home, the confidence level isn't too useful. Your player either works or does not work; there's no 90% about it.

Confidence levels work the same way. We prefer confidence intervals that have 90% or higher confidence levels, because then we know that the process that produced these levels is a good process, and therefore, we are confident in any decisions or conclusions we reach. But the level doesn't tell us whether this one particular interval sitting in front of us is good or bad. In fact, we shall never know that, unless we someday gain access to the entire population.

 **KEY POINT**  The confidence level is a measure of how well the method used to produce the confidence interval performs. We can interpret the confidence level to mean that if we were to take many random samples of the same size from the same population, and for each random sample calculate a confidence interval, then the confidence level is the proportion of intervals that "work"—the proportion that contain the population parameter.

Figure 9.11 illustrates this interpretation of confidence levels. From the population of 107 movies that were playing during the week of April 3, 2009 (http://www.the-numbers.com/), we took a random sample (with replacement) of 30 movies and calculated the mean revenue per theater in this sample. Because the samples were random, each sample produced a different sample mean. For each sample we also calculated a 95% confidence interval. We repeated this process 100 times, and each time we made a plot of the confidence interval. Figure 9.11a shows the results from the first 10 samples of 30 theaters. All 10 samples produced "good" intervals—intervals that contain the true population mean of $2429. Figure 9.11b shows what happened after we collected 100 95% confidence intervals. With a 95% confidence interval, we'd expect about 95% of the intervals to be good and 5% to be bad. And in fact, four intervals (shown in red) were bad.

> **⚠ Caution**
>
> **Confidence Levels Are Not Probabilities**
> A confidence level, such as 90%, is not a probability. Saying we are 90% confident the mean is between 21.1 minutes and 21.3 minutes does *not* mean that there is a 90% chance that the mean is between these two values. It either is, or isn't. There's no probability about it.

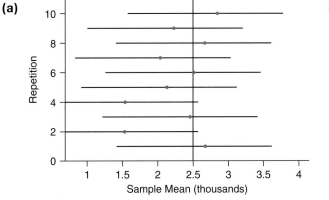

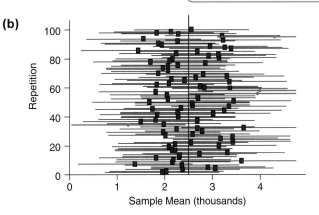

▲ **FIGURE 9.11** **(a)** Ten different 95% confidence intervals, each based on a separate random sample of 30 movie theaters. The population mean of $2500 is shown with a vertical bar. All ten intervals are good because they include this population mean. **(b)** One hundred confidence intervals, each based on a random sample of 30 theaters. Because we are using a 95% confidence level, we expect about 95% of the intervals to be good. In fact, 96 of the 100 turned out to be good, this time. The red intervals are "bad" intervals that do not contain the population mean.

## EXAMPLE 6 iPad Batteries

A consumer group wishes to test Apple's claim that the iPad has a 10-hour battery life. With a random sample of 5 iPads, and running them under identical conditions, the group finds a 95% confidence interval for the mean battery life of an iPad to be 9.5 hours to 12.5 hours. One of the following statements is a correct interpretation of the confidence level. The other is a correct interpretation of the confidence interval.

(i) We are very confident that the mean battery life of all iPads is between 9.5 and 12.5 hours.

(ii) In about 95% of all samples of 5 iPads, the resulting confidence interval will contain the mean battery life of all iPads.

**QUESTION** Which of these statements is a valid interpretation of a confidence interval? Which of these statements is a valid interpretation of a confidence level?

**SOLUTION** Statement (i) interprets a confidence *interval* (9.5, 12.5). Statement (ii) tells us the meaning of the 95% confidence *level*.

**TRY THIS!** Exercise 9.21

## Calculating the Confidence Interval

Confidence intervals for means have the same basic structure as they did for proportions:

$$\text{Estimator} \pm \text{margin of error}$$

As in Chapter 7, the margin of error has the structure

$$\text{Margin of error} = (\text{multiplier}) \times SE$$

The standard error ($SE$) is $SE = \dfrac{\sigma}{\sqrt{n}}$. Because we usually do not know the standard deviation of the population and hence the $SE$, we replace the $SE$ with its estimate. This leads to a formula similar in structure, but slightly different in details, from the one you learned for proportions.

---

**Formula 9.1: One-Sample *t*-Interval**

$$\bar{x} \pm m$$

$$\text{where} \qquad m = t^* SE_{\text{EST}} \text{ and } SE_{\text{EST}} = \frac{s}{\sqrt{n}}$$

The multiplier $t^*$ is a constant that is used to fine-tune the margin of error so that it has the level of confidence we want. This multiplier is found using a $t$-distribution with $n - 1$ degrees of freedom. $SE_{\text{EST}}$ is the estimated standard error.

To compute a confidence interval for the mean, you first need to choose the level of confidence. After that, you need either the original data or these four pieces of information:

1. The sample average, $\bar{x}$, which you calculate from the data.

2. The sample standard deviation, $s$, which you calculate from the data.

3. The sample size, $n$, which you know from looking at the data.

4. The multiplier, $t^*$, which you look up in a table (or use technology) and which is determined by your confidence level and the sample size $n$. The value of $t^*$ tells us how wide the margin of error is, in terms of standard errors. For example, if $t^*$ is 2, then our margin of error is two standard errors wide.

---

The first three steps are pretty straightforward, so let's spend a minute on finding $t^*$, the multiplier for the margin of error.

The multiplier is based on a *t*-distribution with $n - 1$ degrees of freedom. The correct values can be found in Table 4 in Appendix A, or you can use technology. Table 4 is organized such that each row represents possible values of $t^*$ for each degree of freedom. The columns contain the values of $t^*$ for a given confidence level. For example, for a 95% confidence level and a sample size of $n = 30$, we use $t^* = 2.045$. We find this in the table by looking in the row with df $= n - 1 = 30 - 1 = 29$ and using the column for a 95% confidence level. Refer to Table 9.3, which is from the table in the Appendix.

## EXAMPLE 7  Finding the Multiplier *t*\*

Suppose we collect a sample of 30 iPads and wish to calculate a 90% confidence interval for the mean battery life.

**QUESTION** Using Table 9.3, which is from the table in Appendix A, find $t^*$ for a 90% confidence interval when $n = 30$.

◀ **TABLE 9.3** Critical Values of *t*.

| DF | Confidence Level | | | |
| --- | --- | --- | --- | --- |
| | 90% | 95% | 98% | 99% |
| **28** | 1.701 | 2.048 | 2.467 | 2.763 |
| **29** | <u>1.699</u> | 2.045 | 2.462 | 2.756 |
| **30** | 1.697 | 2.042 | 2.457 | 2.750 |
| 34 | 1.691 | 2.032 | 2.441 | 2.728 |

**SOLUTION** We find the number of degrees of freedom from the sample size,

$$\text{df} = n - 1 = 30 - 1 = 29$$

And so we find, from Table 9.3, $t^* = 1.699$ (shown underlined)

**TRY THIS!** Exercise 9.23

It is best to use technology to find the multiplier, because most tables stop at 35 or 40 degrees of freedom. For a 95% confidence level, if you do not have access to technology and the sample size is bigger than 40, it is usually safe to use $t^* = 1.96$—the same multiplier that we used for confidence intervals for sample proportions (for 95% confidence). The precise value, if we used a computer, is 2.02, but this is only 0.06 unit away from 1.96, so the result is probably not going to be affected in a big way.

The wider the confidence interval, the more confident we will be that it covers the true parameter value. We can always increase our level of confidence by making the margin of error bigger. We do this by choosing larger values for $t^*$.

> **↻ Looking Back**
>
> **Why Not 100%?**
> In Chapter 7, you learned that one reason why a 95% confidence level is popular is that increasing the confidence level only a small amount beyond 95% requires a much larger margin of error.

##  EXAMPLE 8  College Tuition Costs

A random sample of 35 two-year colleges in 2008–2009 had a mean tuition (for in-state students) of $2380, with a standard deviation of $1160.

**QUESTION** Find a 90% confidence interval and a 95% confidence interval for the mean in-state tuition of all two-year colleges in 2008–2009. Interpret the intervals. Assume the necessary conditions hold.

SOLUTION We are given the desired confidence level, the standard deviation, and the sample mean, $\bar{x}$, so the next step is to calculate the estimated standard error.

$$\bar{x} = \$2380$$

$$SE_{EST} = \frac{1160}{\sqrt{35}} = 196.0758$$

$$\bar{x} \pm m = \bar{x} \pm t^*SE_{EST}$$

We find the appropriate values of $t^*$ (from Table 9.3 above):

$$t^* \text{ (for 90\%)} = 1.691$$

$$t^* \text{ (for 95\%)} = 2.032$$

For the 90% confidence interval,

$$\bar{x} \pm t^*SE_{EST} \text{ becomes } 2380 \pm (1.691 \times 196.0758) = 2380 \pm 331.5642$$

Lower limit: $2380 - 331.5642 = 2048.44$
Upper limit: $2380 + 331.5642 = 2711.56$

A 90% confidence interval for the mean tuition of all two-year colleges in the 2008–2009 academic year is ($2048, $2712). That is, we are 90% confident that the mean tuition (the typical tuition) of all two-year colleges in 2008–2009 was between $2048 and $2712.

For the 95% confidence interval,

$$\bar{x} \pm t^*SE_{EST} \text{ becomes } 2380 \pm (2.032 \times 196.0758) = 2380 \pm 398.4260$$

Lower Limit: $2380 - 398.4260 = 1981.570$
Upper Limit: $2380 + 398.4260 = 2778.4260$

CONCLUSION The 90% confidence interval is ($2048, $2712). The 95% confidence interval is ($1982, $2778), which is wider. We are 90% confident that the mean tuition of all two-year colleges is between $2048 and $2712. We are 95% confident that the mean tuition could be as low as $1982 or as high as $2778.

TRY THIS! Exercise 9.25

Tech

If you have access to the original data (and not just to the summary statistics, as we were given in Example 8), then it is always best to use a computer to find the confidence interval for you. Figure 9.12 shows StatCrunch output that indicates what we would see if we had access to the full data for Example 8 and asked the software

▶ **FIGURE 9.12** StatCrunch output showing the 95% confidence interval for the mean in-state tuition at two-year colleges in the academic year 2008–2009.

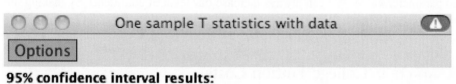

One sample T statistics with data

Options

**95% confidence interval results:**
μ : mean of Variable

| Variable | Sample Mean | Std. Err. | DF | L. Limit | U. Limit |
|---|---|---|---|---|---|
| Instate_08 | 2380.8857 | 196.01347 | 34 | 1982.5385 | 2779.233 |

to find a 95% confidence interval for the mean tuition of all two-year colleges. The output shows us the estimated mean ($2380.8857), the standard error ($196.01347), the degrees of freedom (34), the lower limit of the confidence interval ($1982.5385), and the upper limit ($2779.233). Note that these are not exactly the values we calculated in Example 8. The StatCrunch values are more accurate because there is less rounding.

## Reporting and Reading Confidence Intervals

There are two ways of reporting confidence intervals. Professional statisticians tend to report (lower boundary, upper boundary). This is what we've done so far in this chapter. Thus we reported the 95% confidence interval for the mean of two-year college tuitions in 2008–2009 as ($1982, $2778).

In the press, however, and in some scholarly publications, you'll also see confidence intervals reported as

Estimate ± margin of error

For the two-year college tuitions, we calculated the margin of error to be $398.426 for 95% confidence. Thus we could also report the confidence interval as

$2380 ± $398

This form is useful because it shows our estimate for the mean ($2380) as well as our uncertainty (the mean could plausibly be $398 lower or $398 more).

You're welcome to choose whichever you think best, although you should be familiar with both forms.

## Testing a Mean

In Chapter 8 we laid out the foundations of hypothesis testing. Here, you'll see that the same four steps can be used to test hypotheses about means of populations. These four steps are

**Step 1: Hypothesize.**
> State your hypotheses about the population parameter.

**Step 2: Prepare.**
> Get ready to test: Choose and state a significance level. Choose a test statistic appropriate for the hypotheses. State and check conditions required for the computations, and state any assumptions that must be made.

**Step 3: Compute to compare.**
> Compute the observed value of the test statistic in order to compare the null hypothesis value to our observed value. Find the p-value to measure your level of surprise.

**Step 4: Interpret.**
> Do you reject or not reject the null hypothesis? What does this mean?

As an example of testing a mean, consider this "study" one of the authors did. McDonald's advertises that its ice cream cones have a mean weight of 3.2 ounces ($\mu = 3.2$). A human server starts and stops the machine that dispenses the ice cream, so we might expect some variation in the amount. Some cones might weight slightly more, some cones slightly less. If we weighed all of the McDonald's ice cream cones at a particular store, would the mean be 3.2 ounces, as the company claims?

One of the authors collected a sample of five ice-cream cones and weighed them on a postage scale. The weights were (in ounces)

4.2, 3.6, 3.9, 3.4, and 3.3

We summarize these data as

$\bar{x} = 3.68$ ounces, $s = 0.3701$ ounce

Do these observations support the claim that the mean weight is 3.2 ounces? Or is the mean a different value? We'll apply the four steps of the hypothesis test to make a decision.

### Step 1: Hypothesize

Hypotheses come in pairs and are always statements about the population. In this case, the population consists of all ice cream cones that have been, will be, or could be dispensed from a particular McDonald's. In this chapter, our hypotheses are about the mean values of populations.

The null hypothesis is the status-quo position, which is the claim that McDonald's makes. An individual cone might weigh a little more than 3.2 ounces, or a little less, but after looking at a great many cones, we would find that McDonald's is right and the mean weight is 3.2 ounces.

We state the null hypothesis as

$$H_0: \mu = 3.2$$

Recall that the null hypothesis always contains an equals sign.

The alternative hypothesis, on the other hand, says that the mean weight is different from 3.2 ounces:

$$H_a: \mu \neq 3.2$$

This is an example of a two-tailed hypothesis. We will reject the null hypothesis if the average of our sample cones is very big (suggesting that the population mean is bigger than 3.2) or very small (suggesting the population mean is less than 3.2). It is also possible to have one-tailed hypotheses, as you will see later in this chapter.

> **KEY POINT** Hypotheses are always statements about population parameters. For the test you are about to learn, this parameter is always $\mu$, the mean of the population.

### Step 2: Prepare

The first step is to set the significance level $\alpha$ (alpha), as we discussed in Chapter 8. The significance level is a performance measure that helps us evaluate the quality of our test procedure. It is the probability of making the mistake of rejecting the null hypothesis when, in fact, the null hypothesis is true. In this case, this is the probability that we will say McDonald's cones do not weigh an average of 3.2 ounces when, in fact, they really do.

The test statistic, called the one-sample $t$-test, is very similar in structure to the test for one proportion and is based—not surprisingly, given the name of the test—on the $t$-statistic introduced in Section 9.2. The idea is simple: Compare the observed value of the sample mean, $\bar{x}$, to the value claimed by the null hypothesis, $\mu_0$.

---

**Formula 9.2: Test Statistic for the One-Sample $t$-Test**

$$t = \frac{\bar{x} - \mu_0}{SE_{EST}}, \qquad \text{where} \qquad SE_{EST} = \frac{s}{\sqrt{n}}$$

If conditions hold, the test statistic follows a $t$-distribution with df $= n - 1$

---

This test statistic works because it compares the value of the parameter that the null hypothesis says is true, $\mu_0$, to the estimate of that value that we actually observed in our data. If the estimate is close to the null hypothesis value, then the $t$-statistic is close to 0. But if the estimate is far from the null hypothesis value, then the $t$-statistic is far from 0. The farther $t$ is from 0, the worse things look for the null hypothesis.

Anyone can make a decision, but only a statistician can measure the probability that the decision is right or wrong. To do this, we need to know the sampling distribution of our test statistic.

The sampling distribution will follow the *t*-distribution under these conditions:

Condition 1: *Random Sample and Independence.* The data must be a random sample from a population, and observations must be independent of one another.

Condition 2: *Normal.* The population distribution must be Normal or the sample size must be large. For most situations, 25 is large enough.

Now let's apply this to our ice cream problem. The population for testing the mean ice cream cone weight is somewhat abstract, because a constant stream of ice cream cones is being produced by McDonald's. However, it seems logical that if some cones weigh slightly more than 3.2 ounces and some weigh slightly less, then this distribution of weights should be symmetric and not too different from a Normal distribution. Because our population distribution is Normal, the fact that we have a small sample size, $n = 5$, is not a problem here.

The ice cream cone weights are independent of each other because we were careful, when weighing, to recalibrate the scale, and each cone was obtained on a different day. The cones were not, strictly speaking, randomly sampled, although because the cones were collected on different days and at different times, we will assume that they behave as though they come from a random sample. (But if we're wrong, our conclusions could be *very* wrong!)

**Step 3: Compute to compare**

The conditions of our data tell us that our test statistic should follow a *t*-distribution with $n - 1$ degrees of freedom. Therefore, we proceed to do the calculations necessary to compare our observed sample mean to the hypothesized value of the population mean, and to measure our surprise.

To find the observed value of our *t*-statistic, we need to find the sample mean and the standard deviation of our sample. These values are given above, but you can easily calculate them from the data.

$$SE_{EST} = \frac{0.3701}{\sqrt{5}} = 0.1655$$

$$t = \frac{3.68 - 3.2}{0.1655} = \frac{0.48}{0.1655} = 2.90$$

The observed sample mean was 2.90—almost 3 standard errors above the value expected by the null hypothesis.

**KEY POINT**    The *t*-statistic measures how far away (how many standard errors) our observed mean, $\bar{x}$, lies from the hypothesized value, $\mu_0$. Values far from 0 tend to discredit the null hypothesis.

How unusual is such a value, according to the null hypothesis? The p-value tells us exactly that—the probability that we would get a *t*-statistic as extreme as or more extreme than what we observed, if in fact the mean is 3.2 ounces.

Because our alternative hypothesis says we should be on the lookout for *t*-statistic values that are much bigger or smaller than 0, we must find the probability in both tails of the *t*-distribution. The p-value is shown in the small shaded tails of Figure 9.13. Our sample size is $n = 5$, so our degrees of freedom are $n - 1 = 5 - 1 = 4$.

The p-value of 0.044 tells us that if the typical cone really weighs 3.2 ounces, our observations are somewhat unusual. We should be surprised.

**Step 4: Interpret**

The last step is to compare the p-value to the significance level and decide whether to reject the null hypothesis. If we follow a rule that says we will reject whenever the

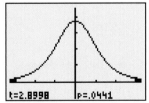

▲ FIGURE 9.13 The tail areas above 2.90 and below −2.90 are shown as the small shaded areas on both sides. The p-value is 0.0441, the probability that if $\mu = 3.2$, a test statistic will be bigger than 2.90 or smaller than −2.90. The distribution shown is a *t*-distribution with 4 degrees of freedom.

## Looking Back

**p-Values**

In Chapter 8 you learned that the p-value is the probability that when the null hypothesis is true, we will get a test statistic as extreme as or more extreme than what we actually saw. (What is meant by "extreme" depends on the alternative hypothesis.) The p-value measures our surprise at the outcome.

p-value is less than or equal to the significance level, then we know that the probability of mistakenly rejecting the null hypothesis will be the value of $\alpha$.

Our p-value (0.044) is less than the significance level we chose (0.05), so we should reject the null hypothesis and conclude that at this particular McDonald's, cones do *not* weigh, on average, 3.2 ounces.

This result makes some sense from a public relations standpoint. If the mean were 3.2 ounces, about half of the customers would be getting cones that weighed too little. By setting the mean weight a little higher than what is advertised, McDonald's can give everyone more than they thought they were getting.

## One- and Two-Tailed Alternative Hypotheses

The alternative hypothesis in the ice cream cone test was two-tailed. As you learned in Chapter 8, alternative hypotheses can also be one-tailed. The exact form of the alternative hypothesis depends on the research question. In turn, the form of the alternative hypothesis tells us how to find the p-value.

You will always use one of the following three pairs of hypotheses for the one-sample *t*-test:

| Two-Tailed | One-Tailed (Left) | One-Tailed (Right) |
|---|---|---|
| $H_0: \mu = \mu_0$ <br> $H_a: \mu \neq \mu_0$ | $H_0: \mu = \mu_0$ <br> $H_a: \mu < \mu_0$ | $H_0: \mu = \mu_0$ <br> $H_a: \mu > \mu_0$ |

You choose the pair of hypotheses on the basis of your research question. For the ice cream cone example, we asked if the mean weight is *different* from the value advertised, and so used a two-tailed alternative hypothesis. Had we instead wanted to know whether the mean weight was less than 3.2 ounces, we would have used a one-tailed (left) hypothesis.

Your choice of alternative hypothesis determines how you calculate the p-value. Figure 9.14 shows how to find the p-value for each alternative hypothesis, all using the same *t*-statistic value of $t = 2.1$ and the same sample size of $n = 30$.

▶ **FIGURE 9.14** The distributions are *t*-distributions with $n - 1 = 29$ degrees of freedom. The shaded region in each graph represents a p-value when $t = 2.1$ for **(a)** a two-tailed alternative hypothesis, **(b)** a one-tailed (left) hypothesis and **(c)** a one-tailed (right) hypothesis.

**(a)**

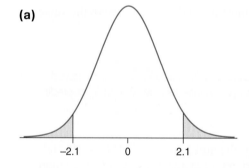

**(b)**

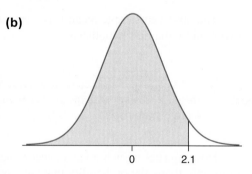

**(c)**

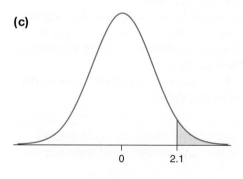

## Looking Back

**What Does "as extreme as or more extreme than" Mean?**

See Chapter 8 for a detailed discussion of how the p-value depends on the alternative hypothesis.

Note that the p-value is always an "extreme" probability; it's always the probability of the tails (even if the tail is pretty big, as it is in Figure 9.14b).

## ✳ EXAMPLE 9 Dieting

Americans who want to lose weight have many different diets among which to choose. In one study (Dansinger et al. 2005), researchers compared results from four different diets. In this example, though, we look at only a small part of these data to examine whether one of the more popular diets, the Weight Watchers diet, is effective. The researchers examined 40 subjects who were randomly assigned to this diet. Researchers recorded the change in weight after 12 months. The distribution of amount of weight lost in this sample is shown in Figure 9.15. Only 26 of the 40 subjects stayed with the diet for that long, so we have data on only these 26 people.

**QUESTION** Test the hypothesis that people on the Weight Watchers diet tend to lose weight. Summary statistics are given below. (A negative weight change means the person lost weight.)

$$\bar{x} = -4.6 \text{ kg}, \quad s = 5.4 \text{ kg}$$
(4.6 kilograms is about 10 pounds.)

**SOLUTION** From Figure 9.15, we see that although a small number of people actually gained weight, the typical experience was a loss in weight. After a year, the average change in weight of the 26 people who stayed on the diet was $-4.6$ kilograms (about 10 pounds), with a standard deviation of 5.4 kilograms.

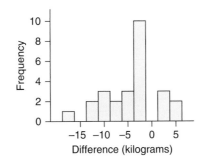

◀ **FIGURE 9.15** Change in weight for subjects after one year on a low-calorie diet (the Weight Watchers diet). Note that most values are negative, representing people who lost weight.

Our population of interest is the group of all overweight people who might go on the Weight Watchers diet and stick to the diet for one year. Could the mean weight change of this population be negative? If so, then, on average, we could say that people do lose weight on this diet.

*Step 1: Hypothesize*
Let $\mu$ represent the mean weight change of the population.

$$H_0: \mu = 0$$
$$H_a: \mu < 0$$

The null hypothesis says that the mean is 0, because the neutral position here is that no change occurs, on average. This is the same as saying that the diet is ineffective, or not different from no diet at all.

The alternative is that the mean change is negative. This differs from the ice cream cone example, where we only wanted to know whether or not the mean weight was 3.2 ounces. Here we care about the direction of the weight change: Did it go down?

*Step 2: Prepare*

We will test using a 5% significance level.

We need to check the conditions to see whether the *t*-statistic will follow the *t*-distribution (at least approximately).

Condition 1: *Random Sample and Independence.* The subjects in this study were not selected randomly from the population of all dieters. But they were selected randomly from a larger group of dieters, because one-fourth of the subjects in this study were randomly assigned to Weight Watchers and the rest to other diets. We will assume that the researchers took care so that observations are independent.

Condition 2: *Normal.* The distribution of the sample does not look Normal, which makes us suspect that the population distribution is not Normal. But because the sample size is larger than 25, this condition is satisfied.

*Step 3: Compute to compare*

$$SE = \frac{s}{\sqrt{n}} = \frac{5.4}{\sqrt{26}} = 1.0590$$

$$t = \frac{\bar{x} - \mu}{SE} = \frac{-4.6}{1.0590} = -4.34$$

This tells us that our test statistic is 4.34 standard errors below what the null hypothesis expected.

Our p-value here is the area below the observed value, because the alternative cares only whether we get values that are smaller than expected. (Remember, this is a one-tailed hypothesis.) We find the p-value, using technology, to be 0.0001 (Figure 9.16). We use a *t*-distribution with $n - 1 = 25$ degrees of freedom.

▶ **FIGURE 9.16** The p-value (shaded) for the diet data. The area has been enlarged so that it can be seen.

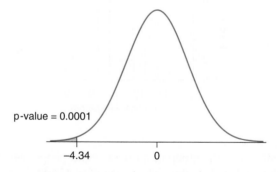

p-value = 0.0001

−4.34          0

*Step 4: Interpret*

The p-value is much smaller than 0.05, so we conclude that the mean weight change is in fact negative, meaning that people do tend to lose weight after one year on this diet.

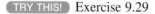 TRY THIS! Exercise 9.29

Tech

If you have access to the original data, then you should use statistical software to perform the *t*-test. Figure 9.17 shows the results from using StatCrunch to carry out the hypothesis test. We had to tell the computer only the null and alternative hypotheses, and it did the rest. Note that, because of rounding, the values are not precisely the same as those we calculated "by hand" in the example.

```
○ ○ ○          One sample T statistics with data          ⚠
```
**Options**

**Hypothesis test results:**

Where: DIET="Weight Watchers" and weightloss <>0

$\mu$ : mean of Variable

$H_0 : \mu = 0$

$H_A : \mu < 0$

| Variable | Sample Mean | Std. Err. | DF | T-Stat | P-value |
|---|---|---|---|---|---|
| weightloss | -4.5923076 | 1.0575868 | 25 | -4.3422513 | 0.0001 |

◀ **FIGURE 9.17** Output from StatCrunch, showing the test result that the mean weight loss under Weight Watchers was a negative value (Example 8). The small p-value leads us to conclude that, on average, the dieters really lost weight.

**SNAPSHOT** ONE SAMPLE *t*-TEST

**WHAT IS IT?** ▶ $t = \dfrac{\bar{x} - \mu_0}{SE_{EST}}$, where $SE_{EST} = \dfrac{s}{\sqrt{n}}$

**WHAT DOES IT DO?** ▶ It tests hypotheses about a mean of a single population.

**HOW DOES IT DO IT?** ▶ If the estimated mean differs from the hypothesized value, then the test statistic will be far from 0. Thus, values of the *t*-statistic that are unexpectedly far from 0 (in one direction or the other) discredit the null hypothesis.

**HOW IS IT USED?** ▶ When proposing values about a single population mean. The observed value of the test statistic is compared to a *t*-distribution with $n - 1$ degrees of freedom to calculate the p-value. If the p-value is small, you should be surprised by the outcome and reject the null hypothesis.

## SECTION 9.4

# Comparing Two Population Means

Does your ability to smell depend on whether you are sitting up or lying down? Do gas prices in Denver differ from those in San Francisco? These questions can be answered, in part, by comparing the means of two populations. Although we could construct separate confidence intervals to estimate each mean (for example, the mean smelling ability of people who are lying down compared to the mean smelling ability of people who are sitting), we can construct more precise estimates by focusing on the difference between the two means.

When comparing two populations, it is important to pay attention to whether the data sampled from the populations are two **independent samples** or are, in fact, one sample of related pairs (paired samples). With **paired (dependent) samples**, if you know the value that a subject has in one group, then you know something about the other group, too. In such a case, you have somewhat less information than you might have if the samples were independent. We begin with some examples to help you see which is which.

Usually, dependence occurs when the objects in your sample are all measured twice (as is common in "before and after" comparisons), or when the objects are related somehow (for example, if you are comparing twins, siblings, or spouses), or when the experimenters have deliberately matched subjects in the groups to have similar characteristics.

> ! **Caution**
>
> **Paired (Dependent) vs. Independent Samples**
> One indication that you have paired samples is that each observation in one group is coupled with *one particular observation* in the other group. In this case, the two groups will have the same sample size (assuming no observations are missing).

## EXAMPLE 10 Independent or Dependent Samples?

Here are four descriptions of research studies.

a. Subjects were tested for their sense of smell twice: once when lying down, once while sitting up. Researchers want to know whether the mean ability to detect smells differs depending on whether one is sitting up or lying down.

b. Men and women each had their sense of smell measured. Researchers want to know whether, typically, men and women differ in their ability to sense smells.

c. Researchers randomly assigned overweight people to one of two diets: Weight Watchers and Atkins. Researchers want to know whether the mean weight loss on Weight Watchers was different from that on Atkins.

d. The numbers of years of education for husbands and wives are compared to see whether the means are different.

QUESTION For each study, state whether it involves two independent samples or paired (that is, dependent) samples.

SOLUTIONS

a. This study has two populations: One population consists of people lying down, the other of people sitting up. However, the samples for the population actually consist of just one group of people. Each person has her or his sense of smell measured twice: once sitting up and once lying down. It seems reasonable to expect that a person who has a very good (or very bad) sense of smell while sitting up might also have a better (or worse) than average sense of smell while lying down. Thus, these samples are *paired* (or *dependent*); knowledge about a value in one sample could give us some information about the other value (because the same people are in both samples).

b. The two populations consist of men in one and women in the other. As long as the people are not related, knowledge about a measurement of a man could not tell us anything about any of the women. These are *independent* samples.

c. The two populations are people on the Weight Watchers diet and people on the Atkins diet. We are told that the two samples consist of different people; subjects are randomly assigned to one diet but not to the other. These are *independent* samples.

d. The populations are matched. Each husband is coupled with one particular wife, so the samples are *paired* (or *dependent*).

TRY THIS! Exercise 9.45

As you shall see, we analyze paired data differently from data that come from two independent samples. Paired data are turned into "difference" scores: We simply subtract one value in each pair from the other. We now have just a single variable, and we can analyze it using the one-sample techniques discussed in Section 9.3. The weight loss data in Example 9 was an example of this. Patients were measured before and after the diet, and these two paired values were subtracted to produce a weight loss measure.

## Estimating the Mean Difference with Confidence Intervals (Independent Samples)

The Weight Watchers diet is a very traditional, low-calorie diet. The Atkins diet, on the other hand, limits the amount of carbohydrates. Which diet is more effective? Researchers compared these two diets (as well as two others) by randomly assigning overweight subjects to the two diet groups. The boxplots in Figure 9.18 show summary statistics for the samples' weight losses after one year. The mean weight loss of the Weight Watchers dieters was 4.6 kilograms (about 10 pounds), and the mean loss of the Atkins dieters was 3.9 kilograms (Dansinger et al. 2005).

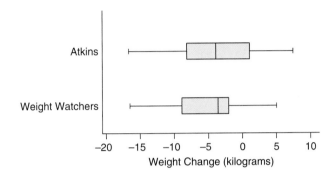

◀ **FIGURE 9.18** Weight change (kilograms) for people randomly assigned to the Atkins diet or the Weight Watchers diet. The medians are close, but because of the skew in the distributions, the sample means are slightly less close.

To guarantee a particular confidence level—for example, 95%—requires that certain conditions hold:

**Condition 1:** *Random Samples and Independence.* Both samples are randomly taken from their populations, and each observation is independent of any other.

**Condition 2:** *Independent Samples.* The two samples are independent of each other (not paired).

**Condition 3:** *Normal.* The populations are approximately Normal, or the sample size in each sample is at least 25. (In special cases, you might need even larger sample sizes.)

If these conditions hold, we can use the following procedure to find an interval with a 95% confidence level.

The formula for a confidence interval comparing two means, when the data are from independent samples, is the same structure as before:

$$(\text{Estimate}) \pm \text{ margin of error}$$

which is

$$(\text{Estimate of difference}) \pm t^*(SE_{\text{estimate of difference}})$$

We estimate the difference with

$$(\text{Mean of first sample}) - (\text{mean of second sample})$$

The standard error of this estimator depends on the sample sizes of both samples and also on the standard deviations of both samples:

$$SE_{\text{EST}} = \sqrt{\frac{s_1^2}{n_1} + \frac{s_2^2}{n_2}}$$

We can put these together into a confidence interval:

**Formula 9.3: Two-Sample *t*-Interval**

$$(\bar{x}_1 - \bar{x}_2) \pm t^* \sqrt{\frac{s_1^2}{n_1} + \frac{s_2^2}{n_2}}$$

The multiplier $t^*$ is based on an approximate *t*-distribution. If a computer is not available, you can conservatively calculate the degrees of freedom for the $t^*$ multiplier as the smaller of $n_1 - 1$ and $n_2 - 1$, but a computer provides a more accurate value.

Choosing the value of $t^*$ (the critical value of *t*) by hand to get your desired level of confidence is tricky. For reasons requiring some pretty advanced mathematics to explain, the sampling distribution is not a *t*-distribution, but only approximately a *t*-distribution. To make matters worse, to get the approximation to be good requires using a rather complex formula to find the degrees of freedom. If you must do these calculations by hand, we recommend taking a "fast and easy" (but also safe and conservative) approach instead. For $t^*$, use a *t*-distribution with degrees of freedom equal to the smaller of $n_1 - 1$ and $n_2 - 1$. That is, use the smaller of the two samples, and subtract 1. For a 95% confidence level, if both samples contain 40 or more observations, you can use 1.96 for the multiplier.

## ❋ EXAMPLE 11  Comparing Men's and Women's Senses of Smell

Researchers studying people's sense of smell devised a measure of smelling ability (Lundström et al. 2006). If you score high on this scale, you can detect smells better than others. Although it was not a goal of the study, we can use the data collected by these researchers to determine whether women and men differ in their ability to detect smells. The fact that the researchers felt it was important to record the gender of the study participants suggests that there may be some reason to think this sense might vary by gender.

For this example, we compare men and women whose sense of smell was measured while they were lying down. (The subjects' sense of smell was also measured when they were sitting up.) The summary statistics are

Men:    $\bar{x} = 10.0694$, $s = 3.3583$, $n = 18$
Women: $\bar{x} = 11.1250$, $s = 2.7295$, $n = 18$

Boxplots are shown in Figure 9.19.

▶ **FIGURE 9.19** Distribution of smelling ability for men and women. There is slight skew, and there is one potential outlier (indicated by the asterisk).

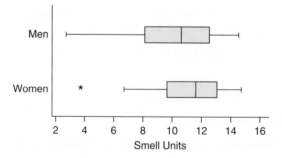

QUESTION  It looks as if women tend to have the more sensitive sense of smell. But could this difference be due to chance? Find a 95% confidence interval for the mean difference in smelling ability between men and women. Interpret the interval. Assume that the participants in this sample are random samples from the population of all adult men and women.

SOLUTION  These data consist of two independent samples: 18 men and 18 women. You might expect that the ability to smell would be Normally distributed across the population, with some people a little better than average and some people a little worse. The boxplots show some skew (and one potential outlier), but with a sample of size 18, it can be hard to tell whether a distribution is Normal on the basis of the box-plot (or histogram). Thus, even though we are not certain that the Normal condition is fulfilled, we will proceed by assuming that it is. Our assumption is based on some theoretical beliefs about how biological traits such as sense of smell are distributed.

Because the sample sizes of both groups are the same (18), our number of degrees of freedom for $t^*$ is conservatively estimated as the smaller of $18 - 1$ and $18 - 1$, which equals 17. For an approximate 95% confidence interval, we use Table 4 in Appendix A, to find $t^* = 2.110$.

Let's call the group of men group 1. (It doesn't matter which we choose for group 1 and which for group 2.)

$$\text{Estimate of difference: } 10.0694 - 11.1250 = -1.0556$$

$$m = t^*\sqrt{\frac{s_1^2}{n_1} + \frac{s_2^2}{n_1}} = t^*\sqrt{\frac{3.3583^2}{18} + \frac{2.7295^2}{18}} = t^*1.0200$$

$$m = 2.110 \times 1.0200 = 2.1522$$

Therefore, a 95% confidence interval is

$$-1.0556 \pm 2.1522, \text{ or about } (-3.2, 1.1)$$

Because the interval contains zero, we cannot rule out the possibility that the mean difference in the population is 0. This suggests that men and women may not differ in their ability to smell.

**TRY THIS!** Exercise 9.51

---

With access to the full data set, and not just to the summary statistics that were provided in Example 11, we can use statistical software to get more accurate calculations. Figure 9.20 shows StatCrunch output for the 95% confidence interval for the difference of the mean smelling ability of men and that of women. The confidence interval is (–3.1, 1.0), which is slightly different (and narrower) than what we found "by hand." The computer-produced interval is more accurate.

**Tech**

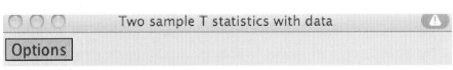

◄ **FIGURE 9.20** StatCrunch output for smelling ability confidence interval.

○ ○ ○          Two sample T statistics with data          ⚠

**Options**

**95% confidence interval results:**

$\mu_1$ : mean of Lying Down where Sex="man"

$\mu_2$ : mean of Lying Down where Sex="woman"

$\mu_1 - \mu_2$ : mean difference

(with pooled variances)

| Difference | Sample Mean | Std. Err. | DF | L. Limit | U. Limit |
|---|---|---|---|---|---|
| $\mu_1 - \mu_2$ | -1.0555556 | 1.0200313 | 34 | -3.1285086 | 1.0173975 |

## Testing Hypotheses about Mean Differences

Hypothesis tests to compare two means from independent samples follow the same structure we discussed in Chapter 8, although now we have more parameters to compare. We show this structure by revisiting the study to investigate whether men and women differ in their ability to detect smells. In Example 11 you found a confidence interval for the difference in the mean smelling ability of men and women. Here we approach the same data with a hypothesis test.

In Example 11 we used boxplots to investigate the shape of the distribution of smelling ability. Here, we examine histograms (Figure 9.21), which show a more detailed picture of the distributions. Both distributions have roughly the same amount of spread, and the histograms show only a little left skew.

▶ **FIGURE 9.21** Distributions of smelling ability for a sample of 18 men and 18 women.

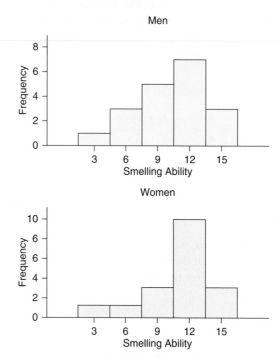

We call men "population 1" and women "population 2." Then the symbol $\mu_1$ represents the mean ability to smell for *all* men (while lying down), and $\mu_2$ represents the mean ability to smell for *all* women (while lying down.)

**Step 1: Hypothesize**

$H_0$: $\mu_1 = \mu_2$ (men and women have the same means for the sense of smell)

$H_a$: $\mu_1 \neq \mu_2$ (men and women have different means for the sense of smell)

**Step 2: Prepare**

The conditions for testing two means are not very different those for from testing one mean and are identical to those for finding confidence intervals of the difference of two means.

Condition 1: *Random Samples and Independent Observations*. Observations are taken at random from two populations, producing two samples. Observations within a sample are independent of one another, which means that knowledge of one value tells us nothing about other observed values in that sample.

Condition 2: *Independent Samples*. The samples are independent of each other. Knowledge about a value in one sample does not tell us anything about any value in the other sample.

Condition 3: *Normal*. Both populations are approximately Normal, or both sample sizes are 25 or more. (In extreme situations, larger sample sizes may be required.)

We might expect the population distributions to be Normally distributed, because measures of biological traits, such as sense of smell, are often Normally distributed.

The sample distributions show a little bit of left skew, but even Normal populations sometimes produce skewed samples when the sample sizes are small, as they are here. The skew is not so great that we would doubt that the populations are Normally distributed, so we assume that condition 2 holds. We must assume that condition 1 (random samples and independence) holds, because we don't know whether the people were randomly sampled. (In fact, they probably were not, because this is rather difficult to do for such studies.) We will assume it is true, understanding that if these people are not representative of the population, we might have substantial bias in our results. It is also safe to assume that condition 3, independent samples, holds, because these are two distinct groups of people. (If researchers had sampled married couples, for example, then this assumption would have been violated.)

Another step in our preparation is to choose a significance level. It is common to use $\alpha = 0.05$, and we will do so for this example.

**Step 3: Compute to compare**

The test statistic used to test this hypothesis is based on the difference between the sample means. Basically, the test statistic measures how far away the observed difference in sample means is from the hypothesized difference in population means. Yes, you guessed it: The distance is measured in terms of standard errors.

$$t = \frac{\text{(difference in sample means} - \text{what null hypothesis says the difference is)}}{SE_{EST}}$$

Using the test statistic is made easier by the fact that the null hypothesis almost always says that the difference is 0.

To compare the sense of smell between men and women:

$$\text{Difference in sample means} = \bar{x}_1 - \bar{x}_2 = 10.0694 - 11.125 = -1.0556$$

$$SE_{EST} = \sqrt{\frac{s_1^2}{n_1} + \frac{s_2^2}{n_2}} = \sqrt{\frac{3.3583^2}{18} + \frac{2.7295^2}{18}} = 1.0200$$

$$t = \frac{-1.0556}{1.0200} = -1.03$$

---

**Formula 9.4: Two-Sample $t$-Test**

$$t = \frac{\bar{x}_1 - \bar{x}_2 - 0}{SE_{EST}}, \quad \text{where} \quad SE_{EST} = \sqrt{\frac{s_1^2}{n_1} + \frac{s_2^2}{n_2}}$$

If conditions hold, the test statistic follows an approximate $t$-distribution, where the degrees of freedom are conservatively estimated to be the smaller of $n_1 - 1$ and $n_2 - 1$.

---

> **Details**
>
> **Null Hypotheses for Two Means**
> Mathematically, we can easily adjust our test statistic if the null hypothesis claims that the difference in means is some value other than 0. But in almost all scientific, business, and legal settings, the null hypothesis value will be 0.

This statistic tells us that the observed difference, $-1.0556$, is about one standard error below what the null hypothesis told us to expect.

We now measure how surprising this is if the null hypothesis is true. To do this, we need to know the sampling distribution of the test statistic $t$, because we measure surprise by finding the probability that if the null hypothesis is true, we would see a value as extreme as or more extreme than the value we observed. In other words, we need to find the p-value.

If the conditions listed in the Prepare step hold, then $t$ follows, approximately, a $t$-distribution with minimum $(n_1 - 1, n_2 - 1)$ degrees of freedom. This approximation can be made even better by adjusting the degrees of freedom, but this adjustment is, for most cases, too complex for a "by hand" calculation. For this reason, we recommend using technology for two-sample hypothesis tests, because you will get more accurate p-values.

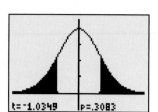

▲ **FIGURE 9.22** The shaded area represents the p-value for this test: the probability of getting a *t*-statistic more than 1.03 standard errors away from 0 when the null hypothesis is true.

Both sample sizes are 18, so $n_1 = 18$ and $n_2 = 18$. Thus we use $18 - 1 = 17$ for the degrees of freedom.

Our alternative hypothesis is two-tailed and says that the true difference might be much bigger than 0 or much smaller than 0. We therefore find the area under both tails of the *t*-distribution. Figure 9.22 shows this probability as the shaded area under the appropriate *t*-distribution.

The p-value (found with technology) is 0.3083.

**Step 4: Interpret**

Again, we compare the p-value to the significance level, $\alpha$. If the p-value is less than or equal to $\alpha$, we reject the null hypothesis. In this example, the p-value, 0.308, is bigger than 0.05. If men and women have the same mean smelling ability, then test statistics as extreme as ours occur fairly often. Our test statistic was not a surprise to the null hypothesis—it is pretty much what the null told us we would get. Therefore, we do not reject the null hypothesis.

The previous analysis was done using only the summary statistics provided. If you have the raw data, then you should use computer software to do the analysis. You will get more accurate values and save yourself lots of time. Figure 9.23 shows StatCrunch output for testing whether the mean sense of smell is different for men than for women.

▶ **FIGURE 9.23** StatCrunch output in a test of whether, on average, men's ability to smell is different from women's.

**Tech**

### Two sample T statistics with data

**Options**

**Hypothesis test results:**

$\mu_1$ : mean of Lying Down where Sex="man"

$\mu_2$ : mean of Lying Down where Sex="woman"

$\mu_1 - \mu_2$ : mean difference

$H_0 : \mu_1 - \mu_2 = 0$

$H_A : \mu_1 - \mu_2 \neq 0$

(without pooled variances)

| Difference | Sample Mean | Std. Err. | DF | T-Stat | P-value |
|---|---|---|---|---|---|
| $\mu_1 - \mu_2$ | −1.0555556 | 1.0200313 | 32.636784 | −1.0348266 | 0.3084 |

> **! Caution**
>
> **Don't Accept!**
> Remember from Chapter 8 that we do not "accept" the null hypothesis. It is possible that the sample size is too small (the test has low power) to detect the real difference that exists. Instead, we say that there is not enough evidence for us to reject the null.

> **! Caution**
>
> **Don't Pool**
> When using software to do a two-sample *t*-test, make sure it does the unpooled version. You might have to tell the software explicitly. The unpooled version is more accurate in more situations than the pooled version.

### Into the Pool

Some software packages, and some textbooks too, provide for another version of this *t*-test called the "pooled two-sample *t*-test." We have presented the unpooled version (you can see this in the StatCrunch output above the table, where it says "without pooled variances"). The unpooled version is preferred over the other version because the pooled version works only in special circumstances (when the population standard deviations are equal). The unpooled version works reasonably well in all situations, as long as the listed conditions hold.

### Test of Two Means: Dependent Samples

With **paired samples**, we turn two samples into one. We do this by finding the difference in each pair.

Recall the study to evaluate smelling ability. Earlier, you saw there were no differences in mean smelling ability between men and women. Are there differences, however, that depend on position? Researchers carried out this study to determine whether people differ in their ability to smell depending on whether they are sitting up or lying down.

## SNAPSHOT TWO SAMPLE *t*-TEST (FROM INDEPENDENT SAMPLES)

**WHAT IS IT?** ▶ A procedure for deciding whether two means, estimated from independent samples, are different. The test statistic used is

$$t = \frac{\bar{x}_1 - \bar{x}_2 - 0}{SE_{EST}}, \text{ where } SE_{EST} = \sqrt{\frac{s_1^2}{n_1} + \frac{s_2^2}{n_2}}$$

**WHAT DOES IT DO?** ▶ Provides us with a decision on whether to reject the null hypothesis that the two means are the same and lets us do so knowing the probability that we are making a mistake.

**HOW DOES IT DO IT?** ▶ Compares the observed difference in sample means to 0, the value we expect if the population means are equal.

**HOW IS IT USED?** ▶ The observed value of the test statistic can be compared to a *t*-distribution.

To test this idea, they measured each subject's sense of smell twice: once while sitting and once while lying down. This is a test of two means, because we are interested in two populations:

Population 1: All people lying down; $\mu_1$ represents the mean ability to smell while lying down.

Population 2: All people sitting up; $\mu_2$ represents the mean ability to smell while sitting up.

However, even though we have two populations, we do not have two *independent* samples. Rather, we have one sample of people who were measured twice. Thus we can change the problem slightly so that, instead of measuring the ability to smell in each position, we measure the *difference* in ability when a person goes from sitting up to lying down.

The first few lines of the data are shown in Table 9.4a.

| Subject Number | Sex | Sitting | Lying |
|:---:|:---:|:---:|:---:|
| 1 | woman | 13.5 | 13.25 |
| 2 | woman | 13.5 | 13 |
| 3 | woman | 12.75 | 11.5 |
| 4 | man | 12.5 | 12.5 |

◀ **TABLE 9.4a** Smelling ability for the first four people sitting and lying.

We create a new variable, call it *difference*, and define it to be the difference between smelling ability sitting up and smelling ability lying down. We show this new variable in Table 9.4b.

| Subject Number | Sex | Sitting | Lying | Difference |
|:---:|:---:|:---:|:---:|:---:|
| 1 | woman | 13.5 | 13.25 | 0.25 |
| 2 | woman | 13.5 | 13 | 0.50 |
| 3 | woman | 12.75 | 11.5 | 1.25 |
| 4 | man | 12.5 | 12.5 | 0 |

◀ **TABLE 9.4b** Difference between smelling ability while sitting up and smelling ability while lying down.

Our hypotheses are now about just one mean: the mean of *difference*.

$$H_0: \mu_{\text{difference}} = 0 \quad (\text{or } \mu_{\text{sitting}} = \mu_{\text{lying}})$$

$$H_a: \mu_{\text{difference}} \neq 0 \quad (\text{or } \mu_{\text{sitting}} \neq \mu_{\text{lying}})$$

Our test statistic is the same as for the one-sample *t*-test:

$$t = \frac{\bar{x}_{\text{difference}} - 0}{SE_{\text{difference}}} \quad \text{where} \quad SE_{\text{difference}} = \frac{s_{\text{difference}}}{\sqrt{n}}$$

We find $\bar{x}$ by averaging the change variable: $\bar{x} = 0.8681$.

We find $s_{\text{difference}}$ by finding the standard deviation of the difference variable: $s = 2.3946$.

There were 36 participants altogether, so

$$SE_{\text{EST}} = \frac{2.3946}{\sqrt{36}} = 0.3991$$

and then

$$t = \frac{0.8681}{0.3991} = 2.18$$

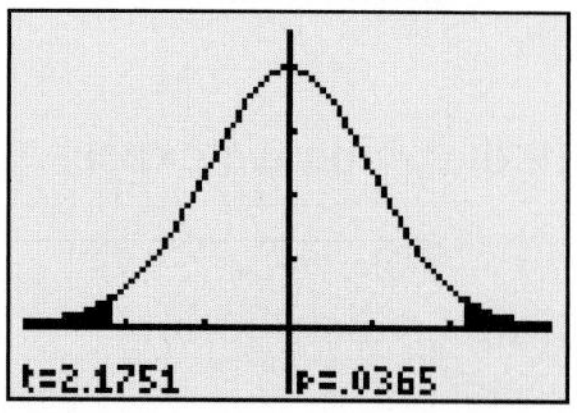

To find the p-value, we use a *t*-distribution (assuming the conditions for a one-sample *t*-test hold) with $n - 1$ degrees of freedom, where $n$ is the number of data pairs. Figure 9.24 shows a *t*-distribution with 35 degrees of freedom. The shaded areas represent the (two-sided) p-value of 0.0365.

Because the p-value is less than 0.05, we reject the null hypothesis. There is evidence that our sense of smell is affected by the position of our body.

## Paired *t*-Test vs. Two-Sample *t*-Test

If you have paired data and (incorrectly) do the two-sample *t*-test, you will generally get a p-value that is too big. Figures 9.25 and 9.26 compare the results of doing a two-sample *t*-test on paired data (Figure 9.25) and doing a paired *t*-test on the same data (Figure 9.26.) Note that the test statistic is much larger when you (correctly) use the paired *t*-test to test the paired data; as a result, the p-value is much smaller.

▲ **FIGURE 9.24** A *t*-distribution with $n - 1 = 35$ degrees of freedom. The shaded area represents the p-value for the smell study (sitting vs. lying) and illustrates that if there is no difference in our ability to smell, then our outcome was very unusual and surprising.

▶ **FIGURE 9.25** TI-83/84 output for two-sample *t*-test.

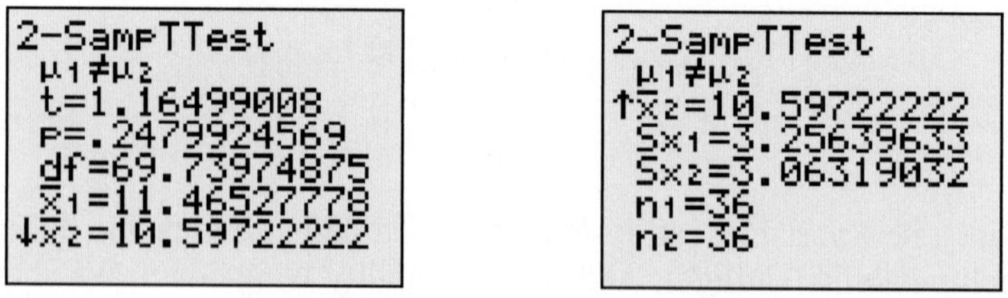

```
2-SampTTest
 μ1≠μ2
 t=1.16499008
 p=.2479924569
 df=69.73974875
 x̄1=11.46527778
↓x̄2=10.59722222
```

```
2-SampTTest
 μ1≠μ2
↑x̄2=10.59722222
 Sx1=3.25639633
 Sx2=3.06319032
 n1=36
 n2=36
```

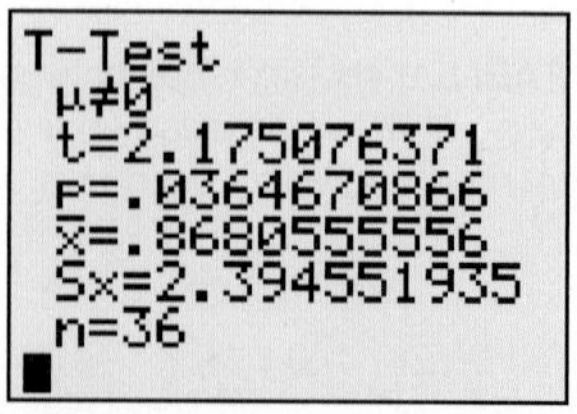

```
T-Test
 μ≠0
 t=2.175076371
 p=.0364670866
 x̄=.8680555556
 Sx=2.394551935
 n=36
```

▲ **FIGURE 9.26** TI-83/84 output for paired *t*-test.

The tests produce different values because when we convert the paired data to differences, the resulting differences have a smaller standard deviation than does either sample by itself. This smaller standard deviation leads to a smaller standard error. So even though the numerators of both *t*-statistics (the paired and the two-sample) are the same, the paired *t*-statistic is larger because its denominator is smaller.

## SNAPSHOT  PAIRED *t*-TEST (DEPENDENT SAMPLES)

**WHAT IS IT?** ▶ A procedure for deciding whether two dependent (paired) samples have different means. Each pair is converted to a difference. The test statistic is the same as for the one-sample *t*-test, except that the null-hypothesis value is 0:

$$t = \frac{\bar{x}_{\text{difference}} - 0}{SE_{\text{difference}}}$$

**WHAT DOES IT DO?** ▶ Lets us make decisions about whether the means are different, while knowing the probability that we are making a mistake.

**HOW DOES IT DO IT?** ▶ The test statistic compares the observed average difference, $\bar{x}_{\text{difference}}$, with the average difference we would expect if the means were the same: 0. Large values discredit the null hypothesis.

**HOW IS IT USED?** ▶ If the required conditions hold, the value of the observed test statistic can be compared to a *t*-distribution with $n - 1$ degrees of freedom.

## SECTION 9.5

# Overview of Analyzing Means

We hope you've been noticing a lot of repetition. The hypothesis test for two means is very similar to the test for one mean, and the hypothesis test for paired data is really a special case of the one-sample *t*-test. Also, the hypothesis tests use almost the same calculations as the confidence intervals, and they impose the same conditions, arranged slightly differently.

All the test statistics (for one proportion, for one mean, for two means, and for two proportions) have this structure:

$$\text{Test statistic} = \frac{(\text{estimated value}) - (\text{null hypothesis value})}{SE}$$

All the confidence intervals have this form:

$$\text{Estimated value} \pm (\text{multiplier})\, SE_{\text{EST}}$$

Not all confidence intervals used in statistics have this structure, but most that you will encounter do.

The method for computing a p-value is the same for all tests, although different distributions are used for different situations. The important point is to pay attention to the alternative hypothesis, which tells you whether you are finding a two-tailed or a one-tailed (and *which* tail) p-value.

## Confidence Intervals and Hypothesis Tests

In the preceding examples, we reached exactly the same conclusion about men and women and their sense of smell, whether we used a confidence interval or a hypothesis test. This is no coincidence. In fact, it has to be that way. If you have a *two-tailed* alternative hypothesis, then you actually have two choices for how to do the test. Both choices always reach the same conclusion.

| Confidence Level | Equivalent $\alpha$ (Two-Tailed) |
|---|---|
| 99% | 0.01 |
| 95% | 0.05 |
| 90% | 0.10 |

▲ **TABLE 9.5** Equivalences between confidence intervals and tests with two-tailed alternative hypotheses.

Choice 1: Perform the hypothesis test as described above with significance level $\alpha$.

Choice 2: Find a $(1 - \alpha) \times 100\%$ confidence interval (using methods given above). Reject the null hypothesis if the value does *not* appear in the interval.

**KEY POINT** A 95% confidence interval is equivalent to a test with a two-tailed alternative with a significance level of 0.05. Table 9.5 shows some other equivalences. All are true only for *two-tailed* alternative hypotheses.

## EXAMPLE **12** Calcium Levels in the Elderly

The boxplots in Figure 9.27 show the results of a study to determine whether calcium levels differ substantially between senior men and senior women (all older than 65 years). Calcium is associated with strong bones, and people with low calcium levels are believed to be more susceptible to bone fractures. The researchers carried out a hypothesis test to see whether the mean calcium levels for men and women were the same. Figure 9.28 shows the results. Calcium levels (the variable *cammol*) are measured in millimoles per liter (mmol/L).

▶ **FIGURE 9.27** Boxplots of calcium levels (mmol/L) for males and females.

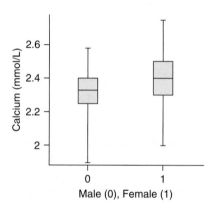

▶ **FIGURE 9.28** StatCrunch output for testing whether mean calcium levels in men and women differ. The difference estimated is mean of females minus mean of males.

**Two sample T statistics with data**

**Options**

**Hypothesis test results:**
$\mu_1$ : mean of cammol where female=1
$\mu_2$ : mean of cammol where female=0
$\mu_1 - \mu_2$ : mean difference
$H_0 : \mu_1 - \mu_2 = 0$
$H_A : \mu_1 - \mu_2 \neq 0$
(without pooled variances)

| Difference | Sample Mean | Std. Err. | DF | T-Stat | P-value |
|---|---|---|---|---|---|
| $\mu_1 - \mu_2$ | 0.07570534 | 0.019790715 | 168.45009 | 3.825296 | 0.0002 |

**QUESTIONS**

a. Assuming that all conditions necessary for carrying out *t*-tests and finding confidence intervals hold, what conclusion should the researchers make on the basis of this output? Use a significance level of 0.05.

b. Suppose the researchers calculate a confidence interval for the difference of the two means. Will this interval include the value 0? If not, will it include all negative values or all positive values? Explain.

SOLUTIONS

a. The p-value, 0.0002, is less than 0.05, so the researchers should reject the null hypothesis and conclude that men and women have different calcium levels.

b. Because we rejected the null hypothesis, we know that the confidence interval cannot include the value 0. If it did, then 0 would be a plausible difference between the means, and our hypothesis test says 0 is not plausible. The estimated difference between the two means is, from the output, 0.0757. Because this value is positive, and because the interval cannot include 0, all values in the interval must be positive.

TRY THIS!  Exercise 9.65

## Hypothesis Test or Confidence Interval?

If you can use either a confidence interval or a hypothesis test, how do you choose? First of all, remember that these two techniques produce the same results only when you have a two-tailed alternative hypothesis, so you need only make the choice when you have a two-tailed alternative.

These two approaches answer slightly different questions. The confidence interval answers the questions "What's the estimated value? And how much uncertainty do you have in this estimate?" Hypothesis tests are designed to answer the question "Is the parameter's value one thing, or another?"

For many situations, the confidence interval provides much more information than the hypothesis test. It not only tells us whether or not we should reject the null hypothesis but also gives us a plausible range for the population value. The hypothesis test, on the other hand, simply tells us whether to reject or not (although it does give us the p-value, which helps us see just how unusual our result is if the null hypothesis is true).

For example, in our test of whether people have a different sense of smell depending on whether they are sitting up or lying down, we rejected the null hypothesis and concluded that the sense of smell was affected by body position. But that is all we can say with a two-tailed test: The means are different. If we want to ask, "How much does the sense of smell change when we sit up?" then a confidence interval for the difference in these two means helps. The 95% confidence interval (smelling ability while sitting up minus smelling ability while lying down) is (0.06, 1.68). This interval tells us that the ability to smell doesn't change very much with body position.

## CASE STUDY REVISITED

The researchers reported 95% confidence intervals for the mean IQ of three-year-old children whose mothers took one of four drugs for epilepsy, as shown in Table 9.6.

| Drug | 95% CI |
|---|---|
| carbamazepine | (95, 101) |
| lamotrigine | (98, 104) |
| phenytoin | (94, 104) |
| valproate | (88, 97) |

◀ **TABLE 9.6** Confidence intervals for IQs.

The researchers could not observe all women on these drugs, so they based their observations on a random sample. If we think of these women as being sampled randomly, then the confidence intervals represent a range of plausible values for the mean IQ for the population of all three-year-olds whose mothers took these drugs.

It is helpful to display these confidence intervals graphically (as the researchers do in their paper), as shown in Figure 9.29.

▶ **FIGURE 9.29** Four confidence intervals for mean IQs of children from mothers taking different drugs for epilepsy.

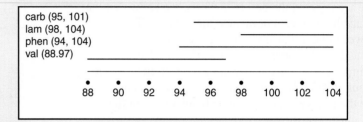

From the figure, we see that the confidence interval for valproate does not overlap with that for lamotrigine. This suggests to us visually that it is *not* plausible that the mean IQ for children whose mothers took these drugs could be the same. The confidence interval for valproate has little overlap with the others, which makes us wonder how much different the mean IQs for the children of valproate users are from those for the children of the other mothers.

For this reason, we need to focus our attention on the differences between the means, not on the individual values for the means. If we do this, as the researchers did, we will find the confidence intervals shown in Table 9.7.

▶ **TABLE 9.7** Confidence intervals for differences in mean IQs.

| Difference | 95% CI |
|---|---|
| carbamazepine – valproate | (0.6, 12.0) |
| lamotrigine – valproate | (3.1, 14.6) |
| phenytoin – valproate | (0.2, 14.0) |

The first interval tells us that the difference between the mean IQ of those under carbamazepine and that of those under valproate could be as small as 0.6 IQ point or as large as 12.0 points. None of these intervals contains 0. This tells us that if we do a hypothesis test that the means are the same, we will have to reject the null hypothesis and conclude that the means are different.

# EXPLORING STATISTICS
## CLASS ACTIVITY

## Pulse Rates

| GOALS | MATERIALS |
|---|---|
| Learn to use a confidence interval (and/or a hypothesis test) to compare the means of two populations. | • A clock on the wall with a second hand or a watch with a second hand (for the instructor).<br><br>• A computer or calculator |

**ACTIVITY**

You will take your pulse rate before and after an activity to measure the effect of the activity on your pulse rate and/or to compare pulse rates between groups.

Try to find your pulse; it is usually easiest to find in the neck on one side or the other.

After everyone has found his or her pulse, your instructor will say, "Start counting," and you will count beats until the instructor says, "Stop counting." If your instructor uses a 30-second interval, double the count to get beats per minute.

Option A: Breathe in and out ten times, taking slow and deep breaths. Now measure your pulse rate again.

Option B: Stand up and sit down five times and then measure your pulse rate again.

Your instructor will collect this data and display the values (before and after each activity) for the class.

**BEFORE THE ACTIVITY**

1. Try finding your pulse (in your neck) to see how to do it.

2. Do you think that either activity (breathing slowly or standing up and sitting down) will change your heart rate? If so, by how much, and will it raise or lower it?

3. How would you measure the typical pulse rate of the class before and after the activity? How would you measure the change in pulse rate after each activity?

4. Do you think men and women have different mean heart rates before the activities?

5. Do you think the change in pulses will be different for men and women? (*Note:* If the class includes only one gender, your instructor may ask you to compare athletes to nonathletes or the taller half to the shorter half.)

**AFTER THE ACTIVITY**

1. State a pair of hypotheses (in words) for testing whether the breathing activity changes the mean pulse rate of the class. Do the same for the standing and sitting activity.

2. State a pair of hypotheses (in words) for whether men and women have the same mean resting pulse rate.

3. Calculate a 95% confidence interval for the change in pulse rates after the activity. What does this confidence interval tell us about the effect of the activity on the mean heart rate? Suppose you did a two-sided hypothesis test. On the basis of the confidence intervals, can you tell what the conclusions of the hypothesis test will be?

## CHAPTER REVIEW

## Key Terms

You may want to review the following terms, which were introduced in Chapters 7 and 8.

> **Chapter 7:** statistic, estimator, bias, precision, sampling distribution, standard error, confidence interval, confidence level, margin of error
> **Chapter 8:** null hypothesis, alternative hypothesis, significance level, test statistic, p-value, one-tailed hypothesis, two-tailed hypothesis

| | | | |
|---|---|---|---|
| Bias, *381* | Standard error, *382* | *t*-Distribution, *389* | Independent samples, *403* |
| Precision, *381* | Central Limit Theorem | Degrees of freedom (df), *389* | Paired (dependent) samples, *403* |
| Sampling distribution, *382* | (CLT), *384* | Confidence intervals, *390* | |
| Unbiased estimator, *382* | *t*-Statistic, *388* | Confidence level, *390* | |

## Learning Objectives

After reading this chapter and doing the assigned homework problems, you should

- Understand when the Central Limit Theorem for sample means applies and know how to use it to find approximate probabilities for sample means.

- Know how to test hypotheses concerning a population mean and concerning the comparison of two population means.

- Understand how to find, interpret, and use confidence intervals for a single population mean and for the difference of two population means.

- Understand the meaning of the p-value and of significance levels.

- Understand how to use a confidence interval to carry out a two-tailed hypothesis test for a population mean or for a difference of two population means.

## Summary

The sample mean gives us an unbiased estimator of the population mean, provided that the observations are sampled randomly from a population and are independent of each other. The precision of this estimator, measured by the standard error (the standard deviation of the sampling distribution), improves as the sample size increases. If the population distribution is Normal, then the sampling distribution is Normal also. Otherwise, according to the Central Limit Theorem, the sampling distribution is approximately Normal, although for small sample sizes the approximation can be very bad. If the population distribution is not Normal, we recommend that you use a sample size of 25 or more.

Although we did not go into the (fairly complex) mathematics, our ability to measure confidence levels (which tell us how well confidence intervals perform), p-values, and significance levels ($\alpha$) depends on the Central Limit Theorem (CLT). If the conditions for applying the CLT are not satisfied, then our reported values for these performance measures may be wrong.

Confidence intervals are used to provide estimates of parameters, along with a measure of our uncertainty in that estimate. The confidence intervals in this chapter differ only a little from those for proportions (Chapter 7). All are of the form

$$\text{Estimate} \pm \text{margin of error}$$

One thing that is different is that now you must decide whether your two samples are independent or paired before performing your analysis.

Another difference between confidence intervals for means and those for proportions is that the multiplier in the margin of error is based on the *t*-distribution, not on the Normal distribution.

By this point you have learned several different hypothesis tests, including the *z*-test for one-sample proportions and for two-sample proportions; and the t-test for one-sample mean, for two means from independent samples, and for two means from dependent samples. (You may also find pooled and unpooled versions of

the two-mean independent-samples *t*-test, but you should always use the unpooled version.) It is important to learn which test to choose for the data you wish to analyze.

Hypothesis tests follow the structure described in Chapter 8, and, just as for confidence intervals, you must decide whether you have independent samples or paired samples. Tests of two means based on independent samples are based on the difference between the means. The null hypothesis is (almost) always that the difference is 0. The alternative hypothesis depends on the research question.

To find the p-value for a test of two means, use the *t*-distribution. To use the *t*-distribution, you must know the degrees of freedom (df), and this depends on whether you are doing a test for one mean (df = $n - 1$); two means from independent samples (use your computer, or, if working by hand, use the df for the smaller of $n_1 - 1$, $n_2 - 1$); or two means from paired data (number of pairs $-1$).

### Formulas

Samples are selected randomly from each population and are independent. Population distributions are Normal, or if not, sample sizes need to be 25 or bigger for each sample.

**Formula 9.1: One-Sample Confidence Interval for Mean**

$$\bar{x} \pm m$$

where $m = t^*SE_{EST}$   and   $SE_{EST} = \dfrac{s}{\sqrt{n}}$

The multiplier $t^*$ is a constant that is used to fine-tune the margin of error so that it has the level of confidence we want. It is chosen on the basis of a *t*-distribution with $n - 1$ degrees of freedom. $SE_{EST}$ is the estimated standard error.

**Paired:** $\bar{x}_{\text{difference}} \pm m$,   where $m = t^*SE_{EST}$   and

$$SE_{EST} = \dfrac{s_{\text{difference}}}{\sqrt{n}}$$

(where $\bar{x}_{\text{difference}}$ is the average difference, $s_{\text{difference}}$ is the standard deviation of the differences, and $n$ is the number of data pairs)

**Formula 9.2: The One-Sample $t$-Test for Mean**

$$t = \frac{\bar{x} - \mu}{SE_{\text{EST}}}, \text{ where } SE_{\text{EST}} = \frac{s}{\sqrt{n}} \text{ and, if conditions hold,}$$

$t$ follows a $t$-distribution with df $= n - 1$

$$\textbf{Paired: } t = \frac{\bar{x}_{\text{difference}} - 0}{SE_{\text{difference}}}, \quad \text{where } SE_{\text{difference}} = \frac{s_{\text{difference}}}{\sqrt{n}}$$

(where $\bar{x}_{\text{difference}}$ is the average difference, $s_{\text{difference}}$ is the standard deviation of the differences, and $n$ is the number of data pairs)

If conditions hold, $t$ follows a $t$-distribution with degrees of freedom df $= n - 1$ (where $n$ is the number of data pairs)

**Formula 9.3: Two-Sample Confidence Interval**

$$(\bar{x}_1 - \bar{x}_2) \pm t^* \sqrt{\frac{s_1^2}{n_1} + \frac{s_2^2}{n_2}}$$

If conditions hold, $t^*$ is based on a $t$-distribution. If no computer is available, the degrees of freedom are conservatively estimated as the smaller of $n_1 - 1$ and $n_2 - 1$.

**Formula 9.4: Two-Sample $t$-Test (Unpooled)**

$$t = \frac{\bar{x}_1 - \bar{x}_2}{SE_{\text{EST}}}, \quad \text{where } SE_{\text{EST}} = \sqrt{\frac{s_1^2}{n_1} + \frac{s_2^2}{n_2}}$$

If conditions hold, $t$ follows an approximate $t$-distribution. If no computer is available, the degrees of freedom, df, are conservatively estimated as the smaller of $n_1 - 1$ and $n_2 - 1$

(Do not use the pooled version.)

## Sources

*The Chronicle of Higher Education, Facts & Figures*. 2008. http://chronicle.com/premium/stats/tuition/2008 (accessed July 2009).

Dansinger, M., J. Gleason, J. Griffith, H. Selker, and E. Schaefer. 2005. Comparison of the Atkins, Ornish, Weight Watchers, and Zone diets for weight loss and heart disease risk reduction: A randomized trial. *Journal of the American Medical Association* 293(1), 43–53.

Kaplan, D. 2009. www.cherryblossum.org/results (accessed July 2009).

Lundström, J., Boyle, J., and Jones-Gotman, M. 2006. Sit up and smell the roses better: olfactory sensitivity to phenyl ethyl alcohol is dependent on body position", *Chemical Senses*, 31(3), March 2006: 249-252. doi: 10.1093/chemse/bjj025.

Meador, K. J., et al. 2009. Cognitive function at 3 years of age after fetal exposure to antileptic drugs, *New England Journal of Medicine* 360(16), 1597–1605.

National Health and Nutrition Examination Survey (NHANES). Centers for Disease Control and Prevention (CDC). National Center for Health Statistics (NCHS). National Health and Nutrition Examination Survey Data. Hyattsville, MD: U.S. Department of Health and Human Services, Centers for Disease Control and Prevention, 2003–2004.

# SECTION EXERCISES

## SECTION 9.1

**9.1 Library Books** An exhaustive cataloging of all hardbound books at a college library revealed that the mean age was 54.4 years, with a standard deviation of 15.5 years. Recently, the librarian commissioned a study to determine the condition of the books. A random sample of 500 books was assembled, and the mean age of this collection was found to be 50.2 years, with a standard deviation of 18.0 years.

a. Which of these numerical values are parameters?

b. Which of these numerical values are statistics?

**9.2 American Time Use** In 2010, the American Time Use Survey reported that "employed persons worked an average of 7.5 hours on the days they worked." Is the number 7.5 a statistic or a parameter? Explain. (Source: http://www.bls.gov/tus/, accessed January 16, 2011)

**9.3 Exam Scores** The distribution of the scores on a certain exam is N(70, 10), which means that the exam scores are Normally distributed with a mean of 70 and standard deviation of 10.

a. Sketch the curve and label, on the $x$-axis, the position of the mean, the mean plus or minus one standard deviation, the mean plus or minus two standard deviations, and the mean plus or minus three standard deviations.

b. Find the probability that a randomly selected score will be bigger than 80. Shade the region under the Normal curve whose area corresponds to this probability.

**9.4 Exam Scores** The distribution of the scores on a certain exam is N(70, 10), which means that the exam scores are Normally distributed with a mean of 70 and standard deviation of 10.

a. Sketch the curve and label, on the $x$-axis, the position of the mean, the mean plus or minus one standard deviation, the mean plus or minus two standard deviations, and the mean plus or minus three standard deviations.

b. Find the probability that a randomly selected score will be between 50 and 90. Shade the region under the Normal curve whose area corresponds to this probability.

**9.5 Bats** A biologist is interested in studying the effects that applying insecticide to a fruit farm has on the local bat population. She collects 23 bats from a grove of fruit trees with the insecticide and finds the mean weight of this sample to be 150.4 grams. Assuming the selected bats are a random sample, she concludes that because the sample mean is an unbiased estimator of the population mean, the mean weight of bats in the population is also 150.4 grams. Explain why this is an incorrect interpretation of what it means to have an unbiased estimator.

**9.6 Cellphone Calls** Answers.com claims that the mean length of all cell phone conversations in the United States is 3.25 minutes (3 minutes and 15 seconds). Assume that this is correct, and also assume that the standard deviation is 4.2 minutes. (Source: wiki.answers.com, accessed January 16, 2011)

* a. Describe the shape of the distribution of the length of cell phone conversations in this population. Do you expect it to be approximately Normally distributed, right-skewed, or left-skewed? Explain your reasoning.

b. Suppose that, using a phone company's records, we randomly sample 100 phone calls. We calculate the mean length from this sample and record the value. We repeat this thousands of times. What will be the (approximate) mean value of the distribution of these thousands of sample means?

c. Refer to part b. What will be the standard deviation of this distribution of thousands of sample means?

**9.7 Retirement Income** Several times during the year, the U.S. Census Bureau takes random samples from the population. One such survey is the American Community Survey. The most recent such survey, based on a large (several thousand) sample of randomly selected households, estimates the mean retirement income in the United States to be $21,201 per year. Suppose we were to make a histogram of all of the retirement incomes from this sample. Would the histogram be a display of the population distribution, the distribution of a sample, or the sampling distribution of means?

**9.8 Time Employed** A human resource manager for a large company takes a random sample of 50 employees from the company database. She calculates the mean time that they have been employed. She records this value and then repeats the process: She takes another random sample of 50 names and calculates the mean employment time. After she has done this 1000 times, she makes a histogram of the mean employment times. Is this histogram a display of the population distribution, the distribution of a sample, or the sampling distribution of means?

TRY **9.9 Women's Weights (Example 1)** Twenty-year-old women's weights have a population mean, $\mu$, of 128 pounds and a population standard deviation, $\sigma$, of 20 pounds (http://www.kidsgrowth.com). The distribution is right-skewed. Suppose we take a random sample of 100 twenty-year-old women's weights.

a. What value should we expect for the mean weight of this sample of 100 women? Why?

b. Of course, the actual sample mean will not be exactly the value you gave in part a. The amount it typically differs from this value is given by the standard error. What is the standard error for a sample mean taken from this population?

**9.10 Men's Weights** Twenty-year-old men's weights have a population mean, $\mu$, of 155 pounds and a population standard deviation, $\sigma$, of 22 pounds (http://www.kidsgrowth.com). The distribution is right-skewed. Suppose we take a random sample of 100 twenty-year-old men's weights.

a. What value should we expect for the mean weight of this sample of 100 men? Why?

b. Of course, the value that we actually see for the sample mean based on 100 men will not be exactly he same as your answer to part a. By how much, typically, will it vary from the value in part a? In other words, what is the standard error?

## SECTION 9.2

TRY **9.11 Women's Weights (Example 2)** Twenty-year-old women's weights have a population mean of 128 pounds and a population standard deviation of 20 pounds. The distribution is right-skewed. Suppose we take a random sample of 100 twenty-year-old women's weights. You will find the probability that the sample mean will be more than 4 pounds from the population mean. The Normal curve shown in figure (A) is the sampling distribution of means and therefore is not skewed.

(A)

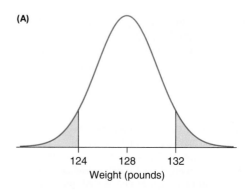

Weight (pounds)

a. The sample is random. Why is the sample size large enough for us to use the Central Limit Theorem for means? Explain.

b. What are the mean and standard deviation of the sampling distribution. (*Hint:* Refer to Exercise 9.9.)

c. Find the probability that the sample mean will be more than 132 pounds or less than 124 pounds, as shown in figure (A). Figure (B) shows TI-83/84 output for finding the probability that the mean will be *between* 124 and 132, but you want the probability that the mean will be less than 124 or more than 132.

(B)

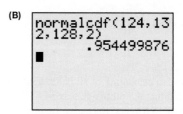

TI-83/84 Normal Curve Output

* **9.12 Men's Weights** Twenty-year-old men's weights have a population mean of 155 pounds and a population standard deviation of 22 pounds (http://www.kidsgrowth.com). The distribution is right-skewed. Suppose we take a random sample of 100 twenty-year-old men's weights.

a. Explain why the Central Limit Theorem is applicable.

b. Sketch the sampling distribution of means. Label the mean, the mean plus or minus one standard deviation, the mean plus or minus two standard deviations, and the mean plus or minus three standard deviations. (*Hint:* The standard deviation for the distribution of means—which is the standard error—was calculated in Exercise 9.10.)

c. Using your sketch of the sampling distribution, shade the area that corresponds to the probability that the sample mean will be between 152.8 pounds and 157.2 pounds.

d. What is this probability?

**TRY** **9.13 Length of Cell Phone Calls (Example 3)** One histogram shows the distribution of the length of cell phone calls for one of the authors in one month, and the other shows many sample means, in which each is the mean length for 10 randomly selected calls from the author's phone records for that month. Which is which? Explain.

**(A)**

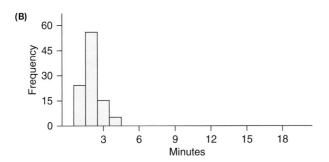

**(B)**

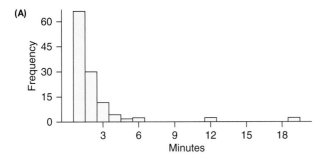

**9.14 Used Car Ages** One histogram shows the distribution of ages of all 638 used cars for sale in the *Ventura County Star* Sunday newspaper in 2008. The other three graphs show distributions of means from random samples taken from the same population of used cars. One histogram shows means based on samples of 2 cars, another shows means based on samples of 5 cars, and another shows means based on samples of 10 cars. Each graph based on means was done with many repetitions. Which distribution is which, and why?

**(A)**

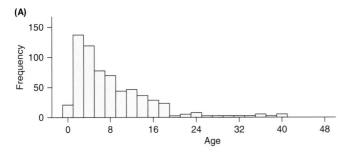

**(B)**

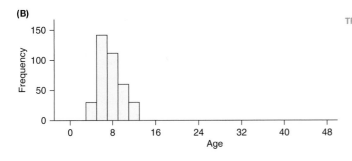

**(C)**

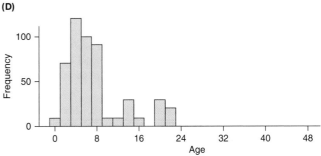

**(D)**

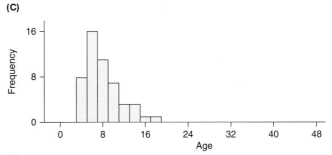

**TRY** **9.15 Used Car Ages (Example 4)** The mean age of all 638 used cars for sale in the *Ventura Country Star* one Saturday in 2008 was 7.9 years, with a standard deviation of 7.7 years. The distribution of ages is right-skewed. For a study to determine the reliability of classified ads, a reporter randomly selects 40 of these used cars and plans to visit each owner to inspect the cars. He finds that the mean age of the 40 cars he samples is 8.2 years and the standard deviation of those 40 cars is 6.0 years.

a. Which of these four numerical values are parameters and which are statistics?

b. $\mu = ?$ $\sigma = ?$ $s = ?$ $\bar{x} = ?$

c. Are the conditions for using the CLT fulfilled? What would be the shape of the approximate sampling distribution of a large number of means, each from a sample of 40 cars?

**9.16 Student Ages** The mean age of all 2550 students at a small college is 22.8 years with a standard deviation is 3.2 years, and the distribution is right-skewed. A random sample of 4 students' ages is obtained, and the mean is 23.2 with a standard deviation of 2.4 years.

a. $\mu = ?$ $\sigma = ?$ $\bar{x} = ?$ $s = ?$

b. Is $\mu$ a parameter or a statistic?

c. Are the conditions for using the CLT fulfilled? What would be the shape of the approximate sampling distribution of many means, each from a sample of 4 students? Would the shape be right-skewed, Normal, or left-skewed?

## SECTION 9.3

**TRY** **9.17 GPAs of Statistics Students (Example 5)** A random sample of 30 students taking statistics at a community college was asked their GPA. The sample mean was 3.25, and the margin of error for a 95% confidence interval was 0.18.

a. Decide whether each of the following three statements a correctly *worded* or an incorrectly worded interpretation of the confidence interval, and fill in the blanks for the correct one(s).

i. We are 95% confident that the *population* mean GPA of *statistics* students at the school is between ____ and ____.

ii. We are 95% confident that the recorded *sample* mean GPA of *statistics* students at the school is between ____ and ____.

iii. We are 95% confidence that the population mean GPA for *all* the students in the school is between ____ and ____.

b. What does the interval tell us about the population mean GPA for *statistics* students at the school? Can you reject 2.81 as the mean for statistics students? Explain.

**9.18 Gas Prices** Some students who lived on a naval base wanted to see whether the mean price of gas in the surrounding area was the same as the naval base price of $2.47 per gallon for regular unleaded gas. They took a random sample of 13 gas stations from the areas surrounding the naval base. The distribution of prices in their sample was roughly symmetric, so we will assume that the population distribution is Normal. The sample mean was 2.64, and the margin of error for a 95% confidence interval was 0.09.

a. Describe the population.

b. Decide whether each of the following three statements is a correctly worded or an incorrectly worded interpretation of the confidence interval, and fill in the blanks for any option(s) with correct wording.

   i. There is a 95% probability that all the next random samples of 13 stations will have a sample mean between _____ and ____.

   ii. We are 95% confident that the population mean is between ____ and ____.

   iii. We are 95% confident that the boundaries for the interval are ____ and ____.

c. On the basis of this confidence interval, what should the students conclude about whether gas prices in surrounding areas are different from those at the naval base? Explain.

**9.19 Oranges** A statistics instructor randomly selected four bags of oranges, each bag labeled 10 pounds, and weighed the bags. They weighed 10.2, 10.5, 10.3, and 10.3 pounds. Assume that the distribution of weights is Normal. Find a 95% confidence interval for the mean weight of all bags of oranges. Use technology for your calculations.

a. Decide whether each of the three statements is a correctly worded interpretation of the confidence interval, and fill in the blanks for the correct option(s).

   i. I am 95% confident the population mean is between ____ and ____.

   ii. There is a 95% chance that all intervals will be between ___ and ___.

   iii. I am 95% confident that the sample mean is between ___ and ___.

b. Does the interval capture 10 pounds? Is there enough evidence to reject the null hypothesis that the population mean weight is 10 pounds? Explain your answer.

**9.20 Carrots** The weights of four randomly chosen bags of horse carrots, each bag labeled 20 pounds, were 20.5, 19.8, 20.8, and 20.0 pounds. Assume that the distribution of weights is Normal. Find a 95% confidence interval for the mean weight of all bags of horse carrots. Use technology for your calculations.

a. Decide whether each of the three statements is a correctly interpretation of the confidence interval, and fill in the blanks for the correct option(s).

   i. 95% of all sample means based on samples of the same size will be between ___ and ____.

ii. I am 95% confident that the population mean is between ____ and ____.

iii. We are 95% confident that the boundaries are ____ and ____.

b. Can you reject a population mean of 20 pounds? Explain.

TRY **9.21 Human Body Temperature (Example 6)** A random sample of 10 people's body temperatures was taken in degrees Fahrenheit, and the 95% confidence interval for the mean was (97.5, 98.8). Distributions of human body temperatures are approximately Normal. Which of the following statements is the correct interpretation of the confidence *level*, and which is the correct interpretation of the confidence *interval*?

a. In about 95% of all samples of 10 temperatures, the resulting confidence interval will contain the population mean body temperature.

b. We are very confident that the mean body temperature is between 97.5 and 98.8.

**9.22 Ice Cream Cones** A random sample of 5 ice cream cones from McDonald's were weighed, and the 95% confidence interval for the mean weight in ounces was (4.2, 5.4). Assume that the distribution of weights in the population is Normal. Which of the following statements is the correct interpretation of the confidence *level*, and which is the correct interpretation of the confidence *interval*?

a. We are very confident that the mean weight of all McDonald's cones is between 4.2 and 5.4 ounces.

b. In about 95% of all samples of 5 cones, the resulting confidence intervals will contain the mean weight of all McDonald's cones.

TRY **9.23 $t^*$ (Example 7)**

a. A researcher collects a sample of 30 measurements from a population and wishes to find a 90% confidence interval for the population mean. What value should he use for $t^*$? (Recall that df $= n - 1$ for a one-sample $t$-test or interval.)

b. If he instead decides to use a 95% confidence interval, will the interval be wider, be narrower, or stay the same? Why?

| | | C-level | |
|---|---|---|---|
| **df** | **90%** | **95%** | **99%** |
| 28 | 1.701 | 2.048 | 2.763 |
| 29 | 1.699 | 2.045 | 2.756 |
| 30 | 1.697 | 2.042 | 2.750 |
| 34 | 1.691 | 2.032 | 2.728 |

**9.24 $t^*$**

a. A researcher collects a sample of 35 measurements from a population and wishes to find a 99% confidence interval for the population mean. What value should he use for $t^*$?

b. If he instead decides to use a 95% confidence interval, will the interval be wider or be narrower or stay the same? Why?

TRY ★ **9.25 Dancers' Heights (Example 8)** A random sample of 20 independent female college-aged dancers showed a sample mean height of 63.3 inches and a sample standard deviation of 2.17 inches. Assume that this distribution of heights is Normal.

| | | C-Level | |
|---|---|---|---|
| **df** | **90%** | **95%** | **99%** |
| **14** | 1.761 | 2.145 | 2.977 |
| **15** | 1.753 | 2.131 | 2.947 |
| **16** | 1.746 | 2.120 | 2.921 |
| **17** | 1.740 | 2.110 | 2.898 |
| **18** | 1.734 | 2.101 | 2.878 |
| **19** | 1.729 | 2.093 | 2.861 |

a. Find a 95% confidence interval for the population mean height of dancers. Interpret the interval. Refer to the table.

b. Find a 99% confidence interval for the population mean height. Interpret the interval.

c. Which interval is wider and why?

* **9.26 Women's Pulses** A random sample of 15 women's resting pulse rates from the National Health and Nutrition Examination Survey (NHANES) showed a mean of 73.5 beats per minute and standard deviation of 17.1. Assume that pulse rates are Normally distributed.

a. Are the mean and standard deviation that are provided statistics or parameters? Why?

b. Find a 95% confidence interval for the population mean pulse rate of women, and report it in a sentence. For $t^*$ refer to the table for Exercise 9.25.

c. Find a 90% confidence interval.

d. Which interval is wider and why?

**9.27 Confidence Interval Changes** State whether each of the following changes would make a confidence interval wider or narrower. (Assume that nothing else changes.)

a. Changing from a 90% confidence level to a 99% confidence level.

b. Changing from a sample size of 30 to a sample size of 200.

c. Changing from a standard deviation of 20 pounds to a standard deviation of 25 pounds.

**9.28 Confidence Interval Changes** State whether each of the following changes would make a confidence interval wider or narrower. (Assume that nothing else changes.)

a. Changing from a 95% level of confidence to a 90% level of confidence.

b. Changing from a sample size of 30 to a sample size of 20.

c. Changing from a standard deviation of 3 inches to a standard deviation of 2.5 inches.

TRY **9.29 Human Body Temperatures (Example 9)** A random g sample of 10 independent healthy people showed the following body temperatures (in degrees Fahrenheit):

98.5, 98.2 99.0, 96.3, 98.3, 98.7, 97.2, 99.1, 98.7, 97.2

Test the hypothesis that the population mean is not 98.6°F; use a significance level of 0.05. *See page 431 for guidance.*

**9.30 Reaction Distance** Data on the disk and website show reaction distances in centimeters for the dominant hand for a random sample of 40 independently chosen college students. Smaller distances indicate quicker reactions.

a. Make a graph of the distribution of the sample, and describe its shape.

b. Find, report, and interpret a 95% confidence interval for the population mean.

c. Suppose a professor said that the population mean should be 10 centimeters. Test the hypothesis that the population mean is not 10 cm, using the four-step procedure, with a significance level of 0.05.

**9.31 Male Height** In the United States, the population mean height for 3-year-old boys is 38 inches (http://www.kidsgrowth .com). Suppose a random sample of 15 non-U.S. 3-year-old boys showed a sample mean of 37.2 inches with a standard deviation of 3 inches. The boys were independently sampled. Assume that heights are Normally distributed in the population.

a. Determine whether the population mean for non-U.S. boys is significantly different from the U.S. population mean. Use a significance level of 0.05.

b. Now suppose the sample consists of 30 boys instead of 15, and repeat the test.

c. Explain why the $t$-values and p-values for parts a and b are different.

**9.32 Vegetarians' Weights** The mean weight of all 20-year-old women is 128 pounds (http://www.kidsgrowth.com). A random sample of 40 vegetarian women who are 20 years old showed a sample mean of 122 pounds with a standard deviation of 15 pounds. The women's measurements were independent of each other.

a. Determine whether the mean weight for 20-year old vegetarian women is significantly less than 128, using a significance level of 0.05.

b. Now suppose the sample consists of 100 vegetarian women who are 20 years old, and repeat the test.

c. Explain what causes the difference between the p-values for parts a and b.

**9.33 GPAs** Thirty GPAs from a randomly selected sample of statistics students at Oxnard College are available on the disk and website. Assume that the population distribution is approximately Normal. The technician in charge of records claimed that the population mean GPA for the whole college is 2.81.

a. What is the sample mean? Is it higher or lower than the population mean of 2.81?

b. The chair of the mathematics department claims that statistics students typically have higher GPAs than the typical college student. Use the four-step procedure and the data provided to test this claim. Use a significance level of 0.05.

**9.34 Dancers' Heights** A random sample of 20 independent female college-aged dancers was obtained and their heights (in inches) were measured. Assume the population distribution is Normal.

a. What is the sample mean? Is it above or below 64.5 inches?

b. Some people claim that the physical demands on dancers are such that dancers tend to be shorter than the typical person in the population. Use the four-step procedure to test the hypothesis that dancers have a smaller population mean height than 64.5 inches. Use a significance level of 0.05.

**9.35 Pulse and Meditation** Eleven students took their pulses while in class. Then they were asked to meditate by sitting with their eyes closed and counting their breaths for 5 minutes. After this, they took their pulse rates again. (The pulse was counted for 15 seconds and multiplied by 4 to get the number of beats per minute.) The data recorded provide differences: pulse before meditation minus pulse after meditation. So a negative value means that the pulse rates increased after meditation. Treat these students as a random sample of college students and perform a hypothesis test to determine whether the mean pulse rate decreased after meditation. In other words, determine whether the mean difference is positive. Clearly state all four steps. Assume that the conditions for using a *t*-test are met.

$$12, 4, 4, 0, -8, -8, 0, 12, 8, 0, 8$$

**9.36 Amazon Textbook Prices** Amelia Suragairin, a statistics student, was interested in the prices of textbooks and went online to the Amazon.com U.S. and Amazon.com U.K. websites to compare prices. The prices in the U.K. were in pounds and were converted to dollars using the exchange rate at that time. The data provide the differences (U.S. minus U.K.) in price between the two countries (rounded to the nearest dollar). Negative signs indicate that the U.K. price was higher. Treat these samples as random samples from their populations and perform a hypothesis test to determine whether the mean difference is significantly different from 0, using a significance level of 0.05. Clearly state all four steps. Assume that the conditions for using a *t*-test are met.

$$-4, 0, 0, -22, -21, -14, -12, 14, 3, -7, 13, -6, -12,$$
$$-11, -7, 12, -2, 2, 1, -13$$

**9.37 Carrots, Three Tests** A random and independently chosen sample of four bags of horse carrots, each bag labeled 20 pounds, had weights of 20.5, 19.8, 20.8, and 20.0 pounds. Assume that the distribution of weights in the population is Normal. In each case, report the alternative hypothesis, the *t*-value, the p-value, and the conclusion for each of the three parts.

a. Test the hypothesis that the population mean weight is not 20 pounds.

b. Test the hypothesis that the population mean is less than 20 pounds.

c. Test the hypothesis that the population mean is more than 20 pounds.

**9.38 Soft Drink Serving Size** A consumer advocate wants to find out whether the soft drinks sold at a fast-food restaurant contain less than the advertised 16 ounces. A random sample of 10 independently chosen drinks produced a mean of 16.5 ounces with a standard deviation of 0.4 ounce. Assume that the distribution of serving sizes in the population is Normally distributed.

a. State the null hypothesis in both words and symbols, using $\mu$ as the population mean weight of all the drinks labeled 16 ounces at this fast-food restaurant.

b. State the alternative hypothesis in both words and symbols.

c. Carry out the appropriate test using a 5% significance level. Explain what conclusion the consumer advocate should reach.

d. Explain why no calculations are needed to show that the population mean is not significantly less than 16 ounces.

**9.39 UCLA Textbook Prices** A 95% confidence interval for the mean price of textbooks at UCLA in the spring quarter 2010, based on a random sample taken by statistician David Diez, was ($58.30, $86.14). Assuming that nothing else changed, what would have been the effect on the confidence interval if David had used a smaller sample size? Assume that the necessary conditions for finding a confidence interval hold, and pick the best answer.

a. The confidence interval would have been wider.

b. The confidence interval would have been narrower.

c. The confidence level would have been higher.

d. The confidence level would have been lower.

e. Answers a and c are both correct.

f. Answers b and d are both correct.

**★ 9.40 UCLA Textbook Prices** A 95% confidence interval for the mean price of textbooks at UCLA in the spring quarter 2010, based on a random sample taken by David Diez (a statistician) was ($58.30, $86.14). To obtain this confidence interval, he multiplied the standard error by 2.04, which was the value of $t^*$. What would have been the effect if he had multiplied the standard error by 1.70 instead? Choose the best answer.

a. The confidence interval would have been wider.

b. The confidence interval would have been narrower.

c. The confidence level would have been higher.

d. The confidence level would have been lower.

e. Answers a and c are both correct.

f. Answers b and d are both correct.

**9.41 Student Ages** Suppose that 200 statistics students each took a random sample (with replacement) of 50 students at their college and recorded the ages of the students in their sample. Then each student used his or her data to calculate a 95% confidence interval for the mean age of all students at the college. How many of the 200 intervals would you expect to capture the true population mean age, and how many would you expect not to capture the true population mean? Explain by showing your calculation.

**9.42 Ages of Married Couples** Suppose that 300 statistics students each collected a random sample of 40 married couples from their state and recorded the ages. From the ages of these couples, they calculated the difference in ages: husband's age minus wife's age. Then each found a 90% confidence interval for the population mean age difference. How many of the 300 intervals would you expect to capture the true population mean difference, and how many would you expect to miss the true population mean difference? Explain by showing your calculation.

**9.43 Presidents' Ages at Inauguration** A 95% confidence interval for the ages of the first six presidents at their inaugurations is (56.2, 59.5). Either interpret the interval or explain why it should not be interpreted.

**9.44 Chief Justices' Ages at Installation** A 95% confidence interval for the age of the first six chief justices at their installation is (45.3, 58.4). Either interpret the interval or explain why it should not be interpreted.

## SECTION 9.4

**9.45 Independent or Paired? (Example 10)** State whether each situation has independent or paired (dependent) samples.

a. A researcher wants to know whether men and women at University X have different mean GPAs. She gathers two random samples: one of GPAs from 50 men and the other from 50 women.

b. A researcher wants to know whether husbands and wives have different mean GPAs. He collects a sample of husbands and wives who have both earned a bachelor's degree and has each person report his or her overall GPA in college.

**9.46 Independent or Paired?** State whether each situation has independent or paired (dependent) samples.

a. A researcher wants to know whether hypnosis tends to reduce pain. Research subjects place their hands in ice water to see how long they can tolerate it before the pain becomes unbearable. Then the subjects are hypnotized and again put their hands in ice water to see whether they tend to leave them in the water longer after hypnosis.

b. A researcher wants to know whether male or female professors at his college tend to have longer recorded greetings on their telephone answering machines. He calls 40 random male and 40 random female professors over the weekend and times their messages.

g **9.47 Televisions** The table shows the Minitab output for a two-sample $t$-test for the number of televisions owned in households of random samples of students at two different community colleges. Each individual was randomly chosen independently of the others; the students were not chosen as pairs or in groups. One of the schools is in a wealthy community (MC), and the other (OC) is in a less wealthy community. Test the hypothesis that the population means are not the same, using a significance level of 0.05. *See page 432 for guidance.*

```
Two-sample T for OCTV vs MCTV

        N    Mean    StDev   SE Mean
OCTV    30   3.70    1.49    0.27
MCTV    30   3.33    1.49    0.27

Difference = mu OCTV - mu MCTV
Estimate for difference: 0.367
95% CI for difference: (-0.404, 1.138)
T-Test of difference = 0 (vs not =): T-Value = 0.95  P-Value = 0.345
```

**9.48 Pulse Rates** Using data from NHANES, we looked at pulse rates of nearly 800 people to see whether men or women tended to have higher pulse rates. Refer to the Minitab output provided.

a. Report the sample means and state which group had the higher sample mean pulse rate?

b. Use the Minitab output to test the hypothesis that pulse rates for men and women are not equal, using a significance level of 0.05. The samples are large enough so that Normality is not an issue.

```
Two-sample T for Pulse

Sex     N     Mean  StDev  SE Mean
Female  384   76.3  12.8   0.65
Male    372   72.1  13.0   0.67

Difference = mu (Female) - mu (Male)
Estimate for difference: 4.248
95% CI for difference: (2.406, 6.090)
T-Test of difference = 0 (vs not =): T-Value = 4.53  P-Value = 0.000
```

**9.49 Triglycerides** Triglycerides are a form of fat found in the body. Using data from NHANES, we looked whether men have higher triglyceride levels than women.

a. Report the sample means and state which group had the higher sample mean triglyceride level. Refer to the Minitab output in figure (A).

b. Carry out a hypothesis test to determine whether men have a higher mean triglyceride level than women. Refer to the Minitab

output provided in figure (A). Output for three different alternative hypotheses is provided—see figures (B), (C), and (D)—and you must choose and state the most appropriate output.

**(A)**

```
Two-Sample T-Test and CI: Triglycerides, Gender

Two-sample T for Triglycerides

Gender   N    Mean   StDev  SE Mean
Female   44   84.4   40.2   6.1
Male     48   139.5  85.3   12

Difference = mu (Female) - mu (Male)
Estimate for difference: -55.1
95% CI for difference: (-82.5, -27.7)
```

**(B)**

```
9.49 B: T-Test of difference = 0 (vs <): T-Value = -4.02  P-Value = 0.000
```

**(C)**

```
9.49 C: T-Test of difference = 0 (vs >): T-Value = -4.02  P-Value = 1.000
```

**(D)**

```
9.49 D: T-Test of difference = 0 (vs not =): T-Value = -4.02  P-Value = 0.000
```

**9.50 Systolic Blood Pressures** When you have your blood pressure taken, the larger number is the systolic blood pressure. Using data from NHANES, we looked at whether men and women have different systolic blood pressure levels.

a. Report the two sample means and state which group had the higher sample mean systolic blood pressure. Refer to the Minitab output in figure (A).

b. Refer to the Minitab output given in figure (A) to test the hypothesis that the mean systolic blood pressures for men and women are not equal, using a significance level of 0.05. Although the distribution blood pressures in the population are right-skewed, the sample size is large enough for us to use $t$-tests. Choose from figures (B), (C), and (D) for your p-value, and explain.

**(A)**

```
Two-sample T for BPSys

gender   N     Mean   StDev  SE Mean
Female   404   116.8  22.7   1.1
Male     410   118.7  18.0   0.89

Difference = mu (Female) - mu (Male)
Estimate for difference: -1.93
95% CI for difference: (-4.75, 0.89)
```

**(B)**

```
9.50 B: T-Test of difference = 0 (vs not =): T-Value = -1.34  P-Value = 0.180
```

**(C)**

```
9.50 C: T-Test of difference = 0 (vs >): T-Value = -1.34  P-Value = 0.910
```

**(D)**

```
9.50 D: T-Test of difference = 0 (vs <): T-Value = -1.34  P-Value = 0.090
```

TRY **9.51 Triglycerides, Again (Example 11)** Report and interpret the 95% confidence interval for the difference in mean triglycerides between men and women (refer to the Minitab output in Exercise 9.49). Does this support the hypothesis that men and women differ in mean triglyceride level? Explain.

**9.52 Blood Pressures, Again** Report and interpret the 95% confidence interval for the difference in mean systolic blood pressure for men and women (refer to the Minitab output in Exercise 9.50). Does this support the hypothesis that men and women differ in mean systolic blood pressure? Explain.

**9.53 Clothes Spending** A random sample of 14 college women and a random sample of 19 college men were separately asked to estimate how much they spent on clothing in the last month. The table shows the data.

   Test the hypothesis that the population mean amounts spent on clothes are different for men and women. Use a significance level of 0.05. Assume that the distributions are Normal enough for us to use the *t*-test.

| Sex | $clothes | Sex | $clothes |
|-----|----------|-----|----------|
| m | 175 | f | 80 |
| f | 200 | m | 200 |
| m | 150 | m | 80 |
| f | 200 | m | 100 |
| f | 100 | m | 120 |
| f | 100 | m | 80 |
| f | 200 | m | 25 |
| m | 100 | f | 80 |
| m | 100 | m | 50 |
| f | 200 | m | 100 |
| m | 200 | m | 30 |
| m | 200 | f | 20 |
| m | 200 | f | 50 |
| f | 250 | m | 60 |
| f | 150 | f | 100 |
| m | 100 | f | 350 |
| m | 0 | | |

**9.54 College Athletes' Weights** A random sample of male college baseball players and a random sample of male college soccer players were obtained independently and weighed. The table shows the unstacked weights (in pounds). The distributions of both data sets suggest that the population distributions are roughly Normal. Determine whether the difference in means is significant, using a significance level of 0.05.

| Baseball | Soccer | Baseball | Soccer |
|----------|--------|----------|--------|
| 190 | 165 | 186 | 156 |
| 200 | 190 | 210 | 168 |
| 187 | 185 | 198 | 173 |
| 182 | 187 | 180 | 158 |
| 192 | 183 | 182 | 150 |
| 205 | 189 | 193 | 172 |
| 185 | 170 | 200 | 180 |
| 177 | 182 | 195 | 184 |
| 207 | 172 | 182 | 174 |
| 225 | 180 | 193 | 190 |
| 230 | 167 | 190 | 156 |
| 195 | 190 | 186 | 163 |
| 169 | 185 | | |

* **9.55 Clothes Spending** In Exercise 9.53 you could not reject the null hypothesis that the mean amount spent by men and the mean amount spent by women for clothing are the same, using a two-tailed test with a significance level of 0.05.

  a. If you found a 95% confidence interval for the difference between means, would it capture 0? Explain.

  b. If you found a 99% confidence interval, would it capture 0? Explain.

  c. Now go back to Exercise 9.53. Find a 95% confidence interval for the difference between means, and explain what it shows.

* **9.56 College Athletes' Weights** In Exercise 9.54, you could reject the null hypothesis that the mean weights of soccer and baseball players were equal using a two-tailed test with a significance level of 0.05.

  a. If you found a 95% confidence interval for the difference between means, would it capture 0? Explain.

  b. If you found a 90% interval, would it capture 0? Explain.

  c. Now go back to Exercise 9.54. Find a 95% confidence interval for the difference between means, and explain what it shows.

**9.57 Criminology** In April 2005, the *Journal of Experimental Criminology* published the results of a 1998 study in which 121 rental properties that had already been the target of drug law enforcement were randomly divided into two groups. In the experimental group, the tenants received a letter from the police describing the enforcement tactics in place. For the control group, there was no letter. The table gives summary statistics for the number of crimes reported over a 30-month interval.

| | Experimental | Control |
|---|--------------|---------|
| Mean | 3.2 | 5.1 |
| SD | 4.1 | 6.3 |
| *n* | 79 | 42 |

Determine whether the letter from the police was effective in *reducing* the number of crimes at the 0.05 level. Although the distribution of number of crimes is not Normal, assume that the sample size is large enough for the Central Limit Theorem to apply. (Source: David P. Farrington and Brandon C. Welsh, Randomized experiments in criminology: What have we learned in the last two decades?" *Journal of Experimental Criminology* (2005), 1: 9–38)

**9.58 Criminology** In April 2005, the *Journal of Experimental Criminology* reported on a study done in Washington in which 171 children were randomly divided into two groups. The experimental group went to a special cognitive preschool, and the control group went to a traditional preschool. The summary statistics for the number of self-reported criminal offenses by the age of 15 are given in the table. Although the number of criminal offenses is not Normally distributed, sample sizes are large enough that the Central Limit Theorem applies. Test the hypothesis that the mean number of self-reported offenses is not the same for each group, using a 0.05 level of significance. (Source: David P. Farrington and Brandon C. Welsh, Randomized experiments in criminology: What have we learned in the last two decades?" *Journal of Experimental Criminology* (2005), 1: 9–38)

|      | Experimental | Control |
|------|--------------|---------|
| Mean | 12.62        | 11.05   |
| SD   | 14.27        | 13.66   |
| n    | 90           | 81      |

**9.59 Females–Pulse Rates before and after a Fright** In a statistics class taught by one of the authors, students took their pulses before and after being frightened. The frightening event was having the teacher scream and run from one side of the room to the other. The pulse rates (beats per minute) of the women before and after the scream were obtained separately and are shown in the table. Treat this as though it were a random sample of female community college students. Test the hypothesis that the mean of college women's pulse rates is higher after a fright, using a significance level of 0.05. *See page 432 for guidance.*

| Women | | Women | |
|-------|-------|-------|-------|
| Pulse Before | Pulse After | Pulse Before | Pulse After |
| 64  | 68  | 84 | 88 |
| 100 | 112 | 80 | 80 |
| 80  | 84  | 68 | 92 |
| 60  | 68  | 60 | 76 |
| 92  | 104 | 68 | 72 |
| 80  | 92  | 68 | 80 |
| 68  | 72  |    |    |

**9.60 Males–Pulse Rates before and after a Fright** Follow the instructions for Exercise 9.59, but use the data for the men in the class. Test the hypothesis that the mean of college men's pulse rates is higher after a fright, using a significance level of 0.05.

| Men | |
|-------|-------|
| Pulse Before | Pulse After |
| 50 | 64  |
| 84 | 72  |
| 96 | 88  |
| 80 | 72  |
| 80 | 88  |
| 64 | 68  |
| 88 | 100 |
| 84 | 80  |
| 76 | 80  |

**9.61 Textbook Prices, UCSB vs. CSUN** The prices of a sample of books at University of California at Santa Barbara (UCSB) were obtained by statistics students Ricky Hernandez and Elizabeth Alamillo. Then the cost of books for the same subjects (at the same level) were obtained for California State University at Northridge (CSUN). Assume that the distribution of differences is Normal enough to proceed, and assume that the sampling was random.

a. First find both sample means and compare them.

b. Test the hypothesis that the population means are different, using a significance level of 0.05.

**9.62 Textbook Prices. OC vs. CSUN** The prices of a random sample of comparable (matched) textbooks from two schools were recorded. We are comparing the prices at OC (Oxnard Community College) and CSUN (California State University at Northridge). Assume that the population distribution of differences is approximately Normal. Each book was priced separately; there were no books "bundled" together.

a. Compare the sample means.

b. Determine whether the mean prices of all books are significantly different. Use a significance level of 0.05.

**9.63 Ages of Brides and Grooms** Data for the ages of grooms and their brides for a random sample of 31 couples in Ventura County, California, were obtained.

a. Compare the sample means.

b. Test the hypothesis that there is a significant difference in mean ages of brides and grooms, using a significance level of 0.05.

c. If the test had been done to determine whether the mean for the grooms was significantly larger than the mean for the brides, how would that change the alternative hypothesis and the p-value?

**9.64 Navy Commissary Prices** Amber Sanchez, a statistics student, collected data on the prices of the same items at the Navy commissary on the naval base in Ventura County, California, and a nearby Kmart. The items were matched for content, manufacturer, and size and were priced separately.

a. Report and compare the sample means.

b. Assume that they are a random sample of items, and use a significance level of 0.05 to test the hypothesis that the Navy commissary has a lower mean price. Assume that the population distribution of differences is approximately Normal.

TRY **9.65 Self-Reported Heights of Men (Example 12)** A random sample of students at Oxnard College reported what they believed to be their heights in inches. Then the students measured each others' heights in centimeters, without shoes. The data shown are for the men. Assume that conditions for *t*-tests hold.

a. Convert heights in inches to centimeters by multiplying inches by 2.54. Find a 95% confidence interval for the mean difference as measured in centimeters. Does it capture 0? What does that show?

b. Perform a *t*-test to test the hypothesis that the means are not the same. Use a significance level of 0.05, and show all four steps.

| Men | | Men | |
|---|---|---|---|
| **Centimeters** | **Inches** | **Centimeters** | **Inches** |
| 166 | 66 | 178 | 70 |
| 172 | 68 | 177 | 69 |
| 184 | 73 | 181 | 71 |
| 166 | 67 | 175 | 69 |
| 191 | 76 | 171 | 67 |
| 173 | 68 | 170 | 67 |
| 174 | 69 | 184 | 72 |
| 191 | 76 | | |

**9.66 Female Self-Reported Heights** Follow the instructions for Exercise 9.65, but use the data for the women. Assume that heights are Normally distributed in the population and the sample is random.

| Women | | Women | |
|---|---|---|---|
| **Cm** | **Inch** | **Cm** | **Inch** |
| 160 | 63 | 158 | 63 |
| 160 | 63 | 159 | 62 |
| 158 | 63 | 180 | 70 |
| 155 | 61 | 156 | 61 |
| 172 | 67 | 155 | 60 |
| 167 | 66 | 159 | 63 |
| 153 | 61 | 179 | 71 |
| 161 | 64 | 175 | 70 |
| 156 | 62 | 163 | 64 |
| 157 | 62 | 159 | 61 |
| 161 | 63 | 162 | 64 |
| 155 | 62 | 154 | 61 |
| 175 | 70 | 159 | 62 |

**9.67 Surfers** Surfers and statistics students Rex Robinson and Sandy Hudson collected data on the number of days on which surfers surfed in the last month for 30 longboard (L) users and 30 shortboard (S) users. Treat these data as though they were from two independent random samples. Test the hypothesis that the mean days surfed for all longboarders is larger than the mean days surfed for all shortboarders (because longboards can go out in many different surfing conditions). Use a level of significance of 0.05. The data can be found in Exercise 3.11 on page 114.

**9.68 Eating Out** Jacqueline Loya, a statistics student, asked students with jobs how many times they went out to eat in the last week. There were 25 students who had part-time jobs and 25 students who had full-time jobs. Carry out a hypothesis test to determine whether the mean number of meals out per week for students with full-time jobs is greater than that for those with part-time jobs. Use a significance level of 0.05. Assume that the conditions for a two-sample *t*-test hold. The data can be found in Exercise 3.12 on page 114.

# CHAPTER REVIEW EXERCISES

**9.69 Which *t*-Test?** State whether you would use a one-sample, two-sample, or paired *t*-test for each part, assuming the conditions required for *t*-tests hold.

a. The weights of a random sample of 25 people are measured before and after a diet. You want to test the hypothesis that weights tend to be lower at the end of the diet.

b. The number of felony convictions for each prisoner in a random sample of 10 prisoners at Prison A and a random sample of 10 prisoners at Prison B are obtained. You want to test the claim that the prisoners in Prison A tend to have more felony convictions than the prisoners in Prison B.

**9.70 Which *t*-Test?** State whether you would use a one-sample, two-sample, or paired *t*-test for each part, assuming the conditions required for *t*-tests hold.

a. The weights of a random sample of 10 male weight-lifters at Gold's Gym are measured. You want to test the hypothesis that the weight-lifters tend to weigh more than men in the general population, for which the population mean is 180 pounds.

b. The ages of the people in a randomly selected group of community college students are obtained. You want to find out whether the male and female students have significantly different mean ages.

c. The ages at marriage for brides and grooms are obtained from a randomly selected group of marriage licenses. You want to determine whether the mean for grooms is significantly higher than the mean for their brides.

**9.71 Marathon Times** The results for a Forest City marathon in Ontario contain times for a sample of both men and women. Below is the Minitab output for a two-sample *t*-test.

a. Test the hypothesis that men (M) are quicker, on average, than women (F), at the 0.05 level of significance. The times are in seconds. (15,455 seconds is about 4 hours and 18 minutes.)

b. Interpret the 95% CI (confidence interval).

```
Two-Sample T-Test and CI: time in secs, gender

Two-sample T for time in secs

gender    N    Mean   StDev   SE Mean
F       158   15455    1949      155
M       298   13671    1973      114


Difference = mu (F) - mu (M)
Estimate for difference: 1784
95% CI for difference: (1405, 2163)
T-Test of difference = 0 (vs not =): T-Value = 9.26   P-Value = 0.000
```

**9.72 Risks** In the Chapter 3 case study, we considered a study of the perceived risks of using appliances for both men and women. We observed that the mean perceived risk was slightly higher for women in the sample.

a. Use the Minitab output to determine whether the mean risk of using appliances for women is significantly different from the mean for men. The women are indicated with a 0 and the men with a 1.

b. Interpret the 95% CI (confidence interval).

```
Two-Sample T-Test and CI: appl, gender

gender    N    Mean   StDev   SE Mean
0       353    20.6    20.8      1.1
1       214    18.3    20.0      1.4


Difference = mu (0) - mu (1)
Estimate for difference: 2.30
95% CI for difference: (-1.16, 5.76)
T-Test of difference = 0 (vs not =): T-Value = 1.30   P-Value = 0.193
```

**9.73 Heart Rate before and after Coffee** Elena Lucin, a statistics student, collected the data in the table showing heart rate (beats per minute) for a random sample of coffee drinkers before and 15 minutes after they drank coffee. Carry out a complete analysis, using the techniques you learned in this chapter. Use a 5% significance level to test whether coffee increases heart rates. The same amount of caffeinated coffee was served to each person, and you may assume that conditions for a *t*-test hold.

| Before | After | Before | After |
|--------|-------|--------|-------|
| 90     | 92    | 74     | 78    |
| 84     | 88    | 72     | 82    |
| 102    | 102   | 72     | 76    |
| 84     | 96    | 92     | 96    |
| 74     | 96    | 86     | 88    |
| 88     | 100   | 90     | 92    |
| 80     | 84    | 80     | 74    |
| 68     | 68    |        |       |

**\* 9.74 Exam Grades** The final exam grades for a sample of daytime statistics students and evening statistics students at one college are reported. The classes had the same instructor, covered the same material, and had similar exams. Using graphical and numerical summaries, write a brief description about how grades differ for these two groups. Then carry out a hypothesis test to determine whether the mean grades are significantly different for evening and daytime students. Assume that conditions for a *t*-test hold. Select your significance level.

Daytime grades: 100, 100, 93, 76, 86, 72.5, 82, 63, 59.5, 53, 79.5, 67, 48, 42.5, 39

Evening grades: 100, 98, 95, 91.5, 104.5, 94, 86, 84.5, 73, 92.5, 86.5, 73.5, 87, 72.5, 82, 68.5, 64.5, 90.75, 66.5

**9.75 Hours of Television Viewing** The number of hours per week of television viewing for random samples of fifth grade boys and fifth grade girls were obtained. Each student logged his or her hours for one Monday-through-Friday period. Assume that the students were independent; for example, there were no pairs of siblings who watched the same shows.

Using graphical and numerical summaries, write a brief description of how the hours differed for the boys and girls. Then carry out a hypothesis test to determine whether the mean hours of television viewing are different for boys and girls. Evaluate whether the conditions for a *t*-test are met, and state any assumptions you must make in order to carry out a *t*-test.

**9.76 Reaction Distances** Reaction distances in centimeters for a random sample of 40 college students were obtained. Shorter distances indicate quicker reactions. Each student tried the experiment both with his or her dominant hand, and with his or her nondominant hand, catching the meter stick. The subjects all started with their dominant hand.

Examine summary statistics, and explain what we can learn from them. Then do an appropriate test to see whether the mean reaction distance is shorter for the dominant hand. Use a significance level of 0.05.

**9.77 Work Hours: Men and Women** Random samples of 50 men and 50 women reported how many hours they worked in the last week as part of the 2008 General Social Survey. (People who did not work were not included.)

a. Two different analyses were performed to test the claim that the population means are not the same. Choose the correct output, and carry out the hypothesis test. Both analyses were done using two-tailed alternative hypotheses.

b. Use the correct output to report and interpret an appropriate 95% confidence interval for the difference in means.

```
Two-sample T for 50 men vs 50 women

            N   Mean  StDev  SE Mean
50 men     50   49.8   16.8      2.4
50 women   50   37.8   12.2      1.7

Difference = mu (50 men) - mu (50 women)
Estimate for difference: 12.02
95% CI for difference: (6.18, 17.86)
T-Test of difference = 0 (vs not =): T-Value = 4.09  P-Value = 0.000
```

**Output A (above)**

```
Paired T for 50 men - 50 women

              N   Mean  StDev  SE Mean
50 men       50  49.80  16.81     2.38
50 women     50  37.78  12.24     1.73
Difference   50  12.02  19.91     2.82

95% CI for mean difference: (6.36, 17.68)
T-Test of mean difference = 0 (vs not = 0): T-Value = 4.27  P-Value = 0.000
```

**Output B**

**9.78 Number of Children** Random samples of 100 men and 100 women were collected from the 2008 General Social Survey, with each person reporting the number of children they had.

a. What do the sample means tell us about the differences between men and women in terms of the number of children they report having?

b. Two analyses were carried out to test the claim that the mean number of children for men is not the same as the mean number of children for women. Only one analysis is appropriate for these data. Carry out an appropriate hypothesis test, including all four steps. The sample sizes are large enough so that the shapes of the population distributions are not a concern.

c. Report and interpret the 95% confidence interval for the difference in means.

```
Two-sample T for 100 men vs 100 women

              N   Mean  StDev  SE Mean
100 men      100  1.49   1.76     0.18
100 women    100  1.85   1.57     0.16

Difference = mu (100 men) - mu (100 women)
Estimate for difference: -0.360
95% CI for difference: (-0.825, 0.105)
T-Test of difference = 0 (vs not =): T-Value = -1.53  P-Value = 0.128
```

**Output A (above)**

```
Paired T for 100 men - 100 women

               N   Mean   StDev  SE Mean
100 men       100  1.490  1.755    0.176
100 women     100  1.850  1.572    0.157
Difference    100  -0.360 2.259    0.226

95% CI for mean difference: (-0.808, 0.088)
T-Test of mean difference = 0 (vs not = 0): T-Value = -1.59  P-Value = 0.114
```

**Output B**

**9.79 Drug Testing** In 2005, the *Journal of Experimental Criminology* reported on a 1992 study in Pima County Arizona that looked at drug testing in criminal defendants. Out of 231 pretrial

defendants observed, 153 were randomly selected to have frequent pretrial drug testing (the experimental group), and 78 were merely released (the control group). The summary statistics are given in the table, and the numbers represent pretrial arrests per person. The mean of 0.04 means that on average, one pretrial defendant in 25 (4%) was arrested.

Test the hypothesis that drug testing caused the rate of arrests to go down, using a level of significance of 0.05. Assume that conditions for *t*-tests hold. Begin by choosing the appropriate output, the TI-83/84 output shown in figure (A) or that shown in figure (B), and explain your choice. In the output, 1 indicates the experimental group, and 2 indicates the control group.

|       | Experimental | Control |
|-------|--------------|---------|
| Mean  | 0.04         | 0.12    |
| SD    | 0.19         | 0.32    |
| $n$   | 153          | 78      |

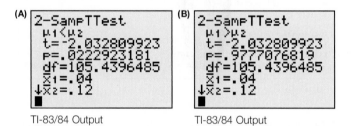

TI-83/84 Output                TI-83/84 Output

* **9.80 Drug Testing** Refer to the study description in Exercise 9.79. A similar study was performed in Maricopa County, Arizona; the data are shown in the table. Follow the instructions in Exercise 9.79 to test the hypothesis that drug testing caused the rate of arrests to go down, using a level of significance of 0.05. In the output, 1 refers to the experimental group, and 2 refers to the control group. Begin by choosing the correct TI-83/84 output—that is, the output shown in figure (A) or that shown in figure (B).

|       | Experimental | Control |
|-------|--------------|---------|
| Mean  | 0.45         | 0.37    |
| SD    | 0.50         | 0.48    |
| $n$   | 425          | 465     |

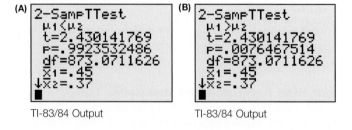

TI-83/84 Output                TI-83/84 Output

* **9.81 Why Is $n - 1$ in the Sample Standard Deviation?** Why do we calculate $s$ by dividing by $n - 1$, rather than just $n$?

$$s^2 = \frac{\sum(x - \bar{x})^2}{n - 1}$$

**(A)**

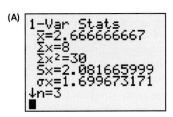

TI-83/84 Output

The reason is that if we divide by $n - 1$, then $s^2$ is an unbiased estimator of $\sigma^2$, the population variance.

We want to show that $s^2$ is an unbiased estimator of $\sigma^2$, sigma squared. The mathematical proof that this is true is beyond the scope of an introductory statistics course, but we can use an example to demonstrate that it is.

First we will use a very small population that consists only of these three numbers: 1, 2, and 5.

You can determine that the population standard deviation, $\sigma$, for this population is 1.699673 (or about 1.70), as shown in the TI-83/84 output. So the population variance, sigma squared, $\sigma^2$, is therefore 2.888889 (or about 2.89).

Now take all possible samples, with replacement, of size 2 from the population, and find the sample variance, $s^2$, for each sample. This process is started for you in the table. Average these

| Sample | $s$ | $s^2$ |
|--------|-----|-------|
| 1, 1 | 0 | 0 |
| 1, 2 | 0.7071 | 0.5 |
| 1, 5 | 2.8284 | 8.0 |
| 2, 1 | | |
| 2, 2 | | |
| 2, 5 | | |
| 5, 1 | | |
| 5, 2 | | |
| 5, 5 | | |

sample variances ($s^2$), and you should get approximately 2.88889. If you do, then you have demonstrated that $s^2$ is an unbiased estimator of $\sigma^2$, sigma squared.

Show your work by filling in the accompanying table and show the average of $s^2$.

**9.82 Is s an Unbiased Estimator of $\sigma$?** Use the data given in Exercise 9.81 to find out whether $s$ is a biased or an unbiased estimator of $\sigma$, sigma. To do this, add the standard deviations ($s$) for the nine samples, and divide by 9 to get the average. Report this average. Is it larger than, smaller than, or the same size as 1.69967? Is it a biased estimator of sigma or an unbiased estimator?

* **9.83** Construct two sets of body temperatures (in degrees Fahrenheit, such as 96.2°F), one for men and one for women, such that the sample means are different but the hypothesis test shows the population means are not different. Each set should have three numbers in it.

* **9.84** Construct heights for 3 or more sets of twins (6 or more people). Make the twins similar, but not exactly the same, in height. Put all of the shorter twins in set A and all of the taller twins in set B. Create the numbers such that a two-sample $t$-test will *not* show a significant difference in the mean heights of the shortest of each pair and the mean heights of the tallest of each pair but the paired $t$-test *does* show a significant difference. (*Hint:* Make one of the pairs really tall, one of the pairs really short, and one of the pairs in between.) Report all the numbers and the $t$- and p-values for the tests. Explain why the paired $t$-test shows a difference and the two-sample $t$-test does not show a difference. Remember that 5 feet is 60 inches and that 6 feet is 72 inches.

# gUIDED EXERCISES

g **9.29 Human Body Temperatures** A random sample of 10 independent healthy people showed body temperatures (in degrees Fahrenheit) as follows:

98.5, 98.2, 99.0, 96.3, 98.3, 98.7, 97.2, 99.1, 98.7, 97.2

The Minitab output of the results of a one-sample $t$-test is shown.

```
One-Sample T: Sample of 10
Test of mu = 98.6 vs not = 98.6

Variable   N    Mean   StDev  SE Mean     95% CI          T      P
sam10     10  98.1200  0.9187  0.2905  (97.4628, 98.7772)  -1.65  0.133
```

QUESTION  Test the hypothesis that the population mean is not 98.6°F, using a significance level of 0.05. Write out the steps given, filling in the blanks.

Step 1 ▶ **Hypothesize**

$H_0: \mu = 98.6$

$H_a:$ _____

Step 2 ▶ **Prepare**

A stemplot is shown that is not strongly skewed, suggesting that the distribution of the population is also approximately Normal.

(A histogram would also be appropriate.) Comment on the data collection, and state the test to be used. State the significance level.

| 96 | 3 |
|----|----|
| 97 | 22 |
| 98 | 23577 |
| 99 | 01 |

**Step 3 ▶ Compute to compare**

$$t = \underline{\hspace{2cm}}$$
$$\text{p-value} = \underline{\hspace{2cm}}$$

**Step 4 ▶ Interpret**
Reject or do not reject $H_0$, and choose interpretation i, ii, or iii:

i. We cannot reject 98.6 as the population mean from these data at the 0.05 level.

ii. The population mean is definitely 98.6 on the basis of these data at the 0.05 level.

iii. We can reject the null hypothesis on the basis of these data, at the 0.05 level. The population mean is not 98.6.

**g 9.47 Televisions** Minitab output is shown for a two-sample $t$-test for the number of televisions owned in households of random samples of students at two different community colleges. Each individual was randomly chosen independently of the others; the students were not chosen as pairs or in groups. One of the schools is in a wealthy community (MC), and the other (OC) is in a less wealthy community.

```
Two-Sample T-Test and CI: OCTV, MCTV

         N     Mean    StDev   SE Mean
OCTV    30     3.70    1.49    0.27
MCTV    30     3.33    1.49    0.27

Difference = mu OCTV - mu MCTV
Estimate for difference: 0.367
95% CI for difference: (-0.404, 1.138)
T-Test of difference = 0 (vs not =): T-Value = 0.95  P-Value = 0.345
```

**QUESTION** Complete the steps to test the hypothesis that the mean number of televisions per household is different in the two communities, using a significance level of 0.05.

**Step 1 ▶ Hypothesize**
Let $\mu_{oc}$ be the population mean number of televisions owned by families of students in the less wealthy community (OC), and let $\mu_{mc}$ be the population mean number of televisions owned by families of students in the wealthier community (MC).

$$H_0: \mu_{oc} = \mu_{mc}$$
$$H_a: \underline{\hspace{2cm}}$$

**Step 2 ▶ Prepare**
Choose an appropriate $t$-test. Because the sample sizes are 30, the Normality condition of the $t$-test is satisfied. State the other conditions, indicate whether they hold, and state the significance level that will be used.

**Step 3 ▶ Compute to compare**

$$t = \underline{\hspace{2cm}}$$
$$\text{p-value} = \underline{\hspace{2cm}}$$

**Step 4 ▶ Interpret**
Reject or do not reject the null hypothesis. Then choose the correct interpretation:

i. At the 5% significance level, we cannot reject the hypothesis that the mean number of televisions of all students in the wealthier community is the same as the mean number of televisions of all students in the less wealthy community.

ii. At the 5% significance level, we conclude that the mean number of televisions of all students in the wealthier community is different from the mean number of televisions of all students in the less wealthy community.

**Confidence Interval**
Report the confidence interval for the difference in means given by Minitab, and state whether it captures 0 and what that shows.

**g 9.59 Female Pulse Rates before and after a Fright** In a statistics class taught by one of the authors, students took their pulses before and after being frightened. The frightening event was having the teacher scream and run from one side of the room to the other. The pulse rates (beats per minute) of the women before and after the scream were obtained separately and are shown in the table. Treat this as though it were a random sample of female community college students.

**QUESTION** Test the hypothesis that the mean of college women's pulse rates is higher after a fright, using a significance level of 0.05, by following the steps below.

| Women | | Women | |
|-------|-------|-------|-------|
| **Pulse Before** | **Pulse After** | **Pulse Before** | **Pulse After** |
| 64 | 68 | 84 | 88 |
| 100 | 112 | 80 | 80 |
| 80 | 84 | 68 | 92 |
| 60 | 68 | 60 | 76 |
| 92 | 104 | 68 | 72 |
| 80 | 92 | 68 | 80 |
| 68 | 72 | | |

**Step 1 ▶ Hypothesize**

$\mu$ is the mean number of beats per minute.

$$H_0: \mu_{before} = \mu_{after}$$
$$H_a: \mu_{before} \underline{\hspace{2cm}} \mu_{after}$$

**Step 2 ▶ Prepare**
Choose a test: Should it be a paired $t$-test or a two-sample $t$-test? Why? Assume that the sample was random and that the distribution of differences is sufficiently Normal. Mention the level of significance.

**Step 3 ▶ Compute to compare**

$$t = \underline{\hspace{2cm}}$$
$$\text{p-value} = \underline{\hspace{2cm}}$$

**Step 4 ▶ Interpret**
Reject or do not reject $H_0$. Then write a sentence that includes "significant" or "significantly" in it. Report the sample mean pulse rate before the scream and the sample mean pulse rate after the scream.

## CHECK YOUR TECH

# Pulse Rates after Exercise: Understanding *t*

Pulse rates were observed for 35 people before and after running in place for 2 minutes. The Minitab output for a paired *t*-test is shown. We will check that the test statistic value reported by Minitab is correct by using a formula. We will not focus on the four steps.

```
Paired T-Test and CI: Pulse Before Run, Pulse After Run

Paired T for Pulse Before Run - Pulse After Run

                 N       Mean      StDev    SE Mean
Pul Bef Run     35    73.6000    11.4358     1.9330
Pul Aft Run     35    92.5143    18.9432     3.2020
Difference      35   -18.9143    15.0497     2.5439


95% CI for mean difference: (-24.0840, -13.7445)
T-Test of mean difference = 0 (vs not = 0): T-Value = -7.44   P-Value = 0.000
```

Paired *t*-test formula:

$$t = \frac{\bar{x}_{\text{diff}} - 0}{SE_{\text{diff}}}$$

where

$$SE_{\text{diff}} = \frac{s_{\text{diff}}}{\sqrt{n}}$$

$\bar{x}_{\text{diff}}$ is the average difference

$S_{\text{diff}}$ is the standard deviation of the differences

$n$ is the number of data pairs

You will verify that *t* is –7.44 by following the steps below, using the formula given.

**Step 1 ▶ Find the numerator of *t*, $\bar{x}_{\text{diff}}$**
Find the difference between the average pulse before the run and the average pulse after the run. This is the numerator. Retain four decimal digits.

**Step 2 ▶ Find the denominator of *t*, $SE_{\text{diff}}$**
To verify the standard error (*SE*) of the difference, take the standard deviation of the difference (StDev) and divide by the square root of 35, which is the sample size. (You may round to four decimal digits, which is what Minitab does.)

**Step 3 ▶ Find *t***
Divide the numerator (from step 1) by the denominator from step 2 and see whether you get –7.44 (or close to it), the *t* reported in the output.

**Step 4 ▶ Understanding What Influences *t***
In answering these questions, consider whether the variable is in the numerator or in the denominator.

  a. If the means were closer together, if all else were the same, would that cause *t* to be farther from 0 or closer to 0 than the current observed value of $-7.44$? Why? (For instance, $-8.44$ is farther from 0 than the original value of $-7.44$, and $-6.44$ is closer to 0 than is $-7.44$.)

  b. If the standard deviation ($s_{\text{diff}}$) were larger, if all else were the same, would that cause *t* to be farther from 0 or closer to 0 than the current observed value of $-7.44$? Why?

  c. If the sample size were larger, if all else were the same, would that cause *t* to be farther from 0 or closer to 0 than the current observed value of $-7.44$? Why?

# TechTips

## General Instructions for All Technology

Because of the limitations of the algorithms, precision, and rounding involved in the various technologies, there can be slight differences in the outputs. These differences can be noticeable, especially for the calculated p-values involving *t*-distributions. It is suggested that the technology that was used be reported along with the p-value, especially for two-sample *t*-tests.

**EXAMPLE A (ONE-SAMPLE *t*-TEST AND CONFIDENCE INTERVAL):** ▶ McDonald's sells ice cream cones, and the company's fact sheet says that these cones weigh 3.2 ounces. A random sample of 5 cones was obtained, and the weights were 4.2, 3.6, 3.9, 3.4, and 3.3 ounces. Test the hypothesis that the population mean is 3.2 ounces. Report the *t*- and p-values. Also find a 95% confidence interval for the population mean.

**EXAMPLE B (TWO-SAMPLE *t*-TEST AND CONFIDENCE INTERVAL):** ▶ Below are the GPAs for random samples of men and women.

Men: 3.0, 2.8, 3.5

Women: 2.2, 3.9, 3.0

Perform a two sample *t*-test to determine whether you can reject the hypothesis that the population means are equal. Find the *t*- and p-values. Also find a 95% confidence interval for the difference in means.

**EXAMPLE C (PAIRED *t*-TEST):** ▶ Here are the pulse rates (in beats per minute) before and after exercise for three randomly selected people.

| Person | Before | After |
|--------|--------|-------|
| A | 60 | 75 |
| B | 72 | 80 |
| C | 80 | 92 |

Determine whether you can reject the hypothesis that the population mean change is 0 (in other words, that the two population means are equal). Find the *t*- and p-values.

## TI-83/84

### One-Sample *t*-Test

1. Press **STAT** and choose **EDIT**, and type the data into **L1** (list one).
2. Press **STAT**, choose **TESTS**, and choose **2: T-Test**.
3. Note this but don't do it: If you did not have the data in the list and wanted to enter summary statistics such as $\bar{x}$, $s$, and $n$, you would put the cursor over **Stats** and press **ENTER**, and put in the required numbers.
4. See Figure 9A. Because you have raw data, put the cursor over **Data** and press **ENTER**.

Enter: $\mu_0$, 3.2; **List**, **L1**; **Freq: 1**; put the cursor over ≠ and press **ENTER**; scroll down to **Calculate** and press **ENTER**.

▲ **FIGURE 9A** TI-83/84 Input for One-Sample *t*-Test

Your output should look like Figure 9B.

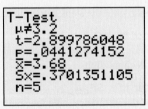

▲ **FIGURE 9B** TI-83/84 Output for One-sample *t*-Test

### One-Sample *t*-Interval

1. Press **STAT**, choose **TESTS**, and choose **8:TInterval**.
2. Choose **Data** because you have raw data. (If you had summary statistics, you would choose **Stats**.) Choose the correct **List** (to select **L1**, press **2nd** and **1**) and **C-Level**, here 0.95. Leave **Freq:1**, which is the default. Scroll down to **Calculate** and press **ENTER**.

The 95% confidence interval reported for the mean weight of the cones (ounces) is (3.2204, 4.1396).

### Two-Sample *t*-Test

1. Press **STAT**, choose **EDIT**, and put your data (GPAs) in two separate lists (unstacked). We put the men's GPAs into **L1** and the women's GPAs into **L2**.
2. Press **STAT**, choose **TESTS**, and choose **4:2-Samp-TTest**.

3. For **Inpt**, choose **Data** because we put the data into the lists. (If you had summary statistics, you would choose **Stats** and put in the required numbers.)

4. In choosing your options, be sure the lists chosen are the ones containing the data; leave the **Freq**s as 1, choose ≠ as the alternative, and choose **Pooled No** (which is the default). Scroll down to **Calculate** and press **ENTER**.

You should get the output shown in Figure 9C. The arrow down on the left-hand side means that you can scroll down to see more of the output.

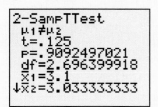

▲ **FIGURE 9C** TI-83/84 Output for Two-Sample *t*-Test

### Two-Sample *t*-Interval

1. After entering your data into two lists, press **STAT**, choose **TESTS**, and choose **0:2-SampTInt**.

2. Choose **Data** because you have raw data. (If you had summary statistics, you would choose **Stats**.) Make sure the lists chosen are the ones with your data. Leave the default for **Freq1:1**, **Freq2:1** and **Pooled No**. Be sure the **C-Level** is **.95**. Scroll down to **Calculate** and press **ENTER**.

The interval for the GPA example will be (−1.744, 1.8774) if the men's data corresponded to **L1** and the women's to **L2**.

### Paired *t*-Test

1. Enter the data given in Example 3 into **L1** and **L2** as shown in Figure 9D.

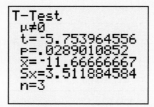

▲ **FIGURE 9D** Obtaining the List of Differences for the TI-83/84

2. See Figure 9D. Use your arrows to move the cursor to the top of **L3** so that you are in the label region. Then press **2ND L1 - 2ND L2**. For the minus sign, be sure to use the button above the plus button. Then press **ENTER**, and you should see all the differences in **L3**.

3. Press **STAT**, choose **TESTS**, and choose **2: T-Test**.

4. See Figure 9E. For **Inpt** choose **Data**. Be sure $\mu_0$ is **0** because we are testing to see whether the mean difference is 0. Also be sure to choose **L3**, if that is where the differences are. Scroll down to **Calculate** and press **ENTER**.

▲ **FIGURE 9E** TI-83/84 Input for Paired *t*-Test

Your output should look like Figure 9F.

```
T-Test
μ≠0
t=-5.753964556
p=.0289010852
x̄=-11.66666667
Sx=3.511884584
n=3
```

▲ **FIGURE 9F** TI-83/84 Output for Paired *t*-Test

MINITAB

### One-Sample *t*-Test and Confidence Interval

1. Type the weight of the cones in **C1** (column 1).

2. **Stat > Basic Statistics > 1-Sample t**

3. See Figure 9G. Click in the empty white box below **Samples in columns:**. A list of columns containing data will appear to the left. Double click **C1** to choose it. Check **Perform hypothesis test**, and put in **3.2** as the **Hypothesized mean**. (If you wanted a one-tailed test or a confidence level other than 95%, you would use **Options**.)

4. Click **OK**.

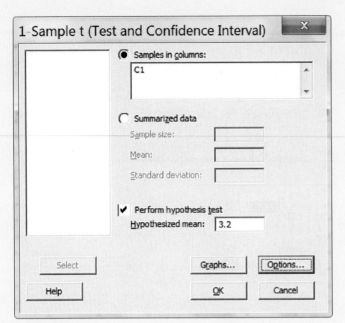

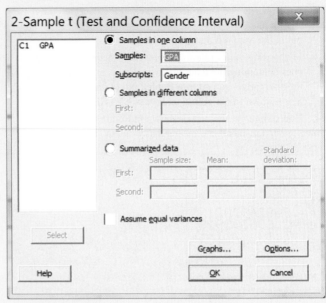

▲ **FIGURE 9G** Minitab Input Screen for One-Sample *t*-Test

▲ **FIGURE 9I** Minitab Input Screen for Two-Sample *t*-Test and Confidence Interval

The output is shown in Figure 9H.

```
One-Sample T: C1
Test of mu = 3.2 vs not = 3.2


Variable   N   Mean   StDev   SE Mean     95% CI          T      P
C1         5   3.680  0.370   0.166    (3.220, 4.140)   2.90   0.044
```

▲ **FIGURE 9H** Minitab Output for One-Sample *t*-Test and Confidence Interval

Note that the Minitab output (Figure 9H) includes the 95% confidence interval for the mean weight (3.22, 4.14) as well as the *t*-test.

### Two-Sample *t*-Test and Confidence Interval

Use stacked data with all the GPAs in one column. The second column will contain the categorical variable that designates groups: male or female.

1. Upload the data from the disk or use the following procedure: Enter the GPAs in the first column. In the second column put the corresponding **m** or **f**. (Complete words or coding are also allowed for the second column but you must decide on a system for one data set and stick to it. For example using F one time and f the other times within one data set will create problems.) Use headers for the columns: **GPA** and **Gender.**

2. **Stat > Basic Statistics > 2-Sample t**

3. Refer to Figure 9I. Choose **Samples in one column,** because we have stacked data. Click in the small box to the right of **Samples** at the top to activate the box. Then double click **GPA** and then double click **Gender** to get it into the **Subscripts** box. If you wanted to do a one-tailed test or to use a confidence level other than 95%, you would click **Options.**

   (If you had unstacked data, you would choose **Samples in different columns**, and choose both columns of data.)

4. Click **OK.**

The output is shown in Figure 9J. Note that the confidence interval is included.

```
Two-Sample T-Test and CI: GPA, Gender
Two-sample T for GPA
Gender   N    Mean   StDev   SE Mean
f        3   3.033   0.850     0.49
m        3   3.100   0.361     0.21

Difference = mu (f) - mu (m)
Estimate for difference:  -0.067
95% CI for difference:  (-2.361, 2.228)
T-Test of difference = 0 (vs not =): T-Value = -0.13 P-Value = 0.912 DF = 2
```

▲ **FIGURE 9J** Minitab Output for Two-Sample *t*-Test and Confidence Interval

### Paired *t*-Test and Confidence Interval

1. Type the numbers (pulse rates) in two columns. Label the first column "**Before**" and the second column "**After.**"

2. **Stat > Basic Statistics > Paired t**

3. Click in the small box to the right of **First sample** to activate the box. Then double click **Before** and double click **After** to get it in the **Second sample** box. (If you wanted a one-tailed test or a confidence level other than 95%, you would click **Options.**)

4. Click **OK.**

Figure 9K shows the output.

```
Paired T-Test and CI: Before, After
Paired T for Before - After

             N    Mean   StDev   SE Mean
Before       3   70.67   10.07     5.81
After        3   82.33    8.74     5.04
Difference   3  -11.67    3.51     2.03

95% CI for mean difference: (-20.39, -2.94)
T-Test of mean difference = 0 (vs not = 0): T-Value = -5.75 P-Value = 0.029
```

▲ **FIGURE 9K** Minitab Output for Paired *t*-Test and Confidence Interval

We are saving the one-sample *t*-test for last, because we have to treat it strangely.

## Two-Sample *t*-Test

1. Type the GPAs in two columns side by side.

2. Click on **Data** and **Data Analysis,** and then scroll down to **t-Test: Two-Sample Assuming Unequal Variances** and double click it.

3. See Figure 9L. For the **Variable 1 Range** select one column of numbers (don't include any labels). Then click inside the box for **Variable 2 Range**, and select the other column of numbers.
   You may leave the hypothesized mean difference empty, because the default value is 0, and that is what you want.

4. Click **OK**.

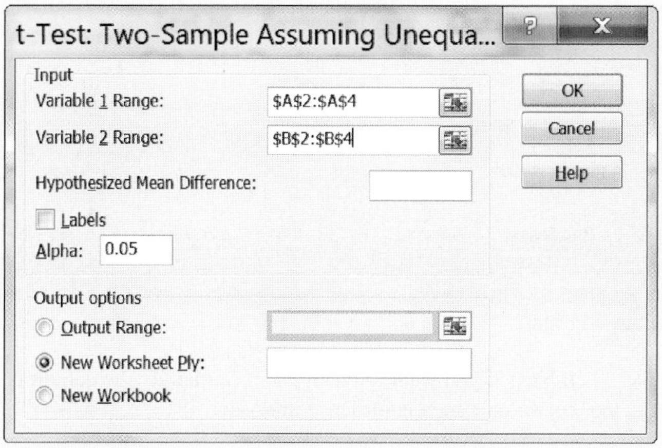

▲ **FIGURE 9L** Excel Input for Two-Sample *t*-Test

To see all of the output you may have to click **Home**, **Format** (in the **Cells** group), and **AutoFit Column Width**.
Figure 9M shows the relevant part of the output.

```
t Stat                0.125
P(T<=t)one-tail       0.454214709
P(T<-t)two-tail       0.908429419
```

▲ **FIGURE 9M** Part of the Excel Output for Two-Sample *t*-Test

For a one-tailed alternative hypothesis Excel always reports one-half the p-value for the two-tailed hypothesis. This is the correct p-value only when the observed value of the test statistic is consistent with the direction of the alternative hypothesis. (In other words, if the alternative hypothesis is ">", then the observed test statistic is positive; if "<", then observed value is negative.) If this is not the case, to find the correct p-value, calculate 1 minus the reported one-tailed p-value.

## Paired *t*-Test

1. Type the data into two columns.

2. Click on **Data**, **Data Analysis**, and **t-test: Paired Two Sample for Means**, and follow the same procedure as for the two-sample *t*-test.

## One-Sample *t*-Test

1. You need to use a trick to force Excel to do this test. Enter the weights of the cones in column A, and put zeros in column B so that the columns are equal in length, as shown in Figure 9N.

| A | B | |
|---|---|---|
| 4.2 | 0 | |
| 3.6 | 0 | |
| 3.9 | 0 | |
| 3.4 | 0 | |
| 3.3 | 0 | |

▲ **FIGURE 9N** Excel Data Entry for One-Sample *t*-Test

2. Click **Data**, **Data Analysis**, and **t-Test: Paired Two Sample for Means**.

3. See Figure 9O. After selecting the two groups of data, you need to put the hypothesized mean in the box labeled **Hypothesized Mean Difference**. For the way we set up this example, it is **3.2**. (If you had entered 3.2's in column B, you would put in 0 for the Hypothesized Mean Difference.)

4. Click **OK**.

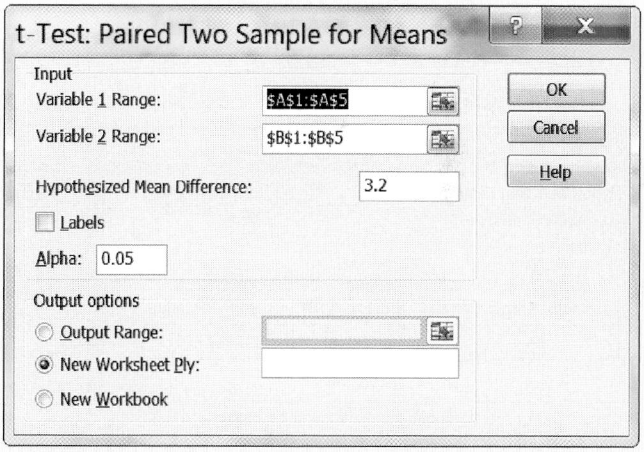

▲ **FIGURE 9O** Excel Input for One-Sample *t*-Test

Figure 9P shows the relevant part of the Excel output.

```
Hypothesized Mean Difference      3.2
Df                                4
P(T<=t) one-tail                  0.022063708
P(T<=t) two-tail                  0.044127415
```

▲ **FIGURE 9P** Relevant Part of the Excel Output for One-sample *t*-Test

Again the p-value for the one-tailed alternative hypothesis is consistent with the alternative hypothesis that the mean weight was *more* than 3.2 ounces.

### One-Sample *t*-Test

1. Type the weights of the cones into the first column.

2. **Stat > T statistics > one sample > with data**

   (If you had summary statistics, then after **one sample**, you would choose **with summary**.)

3. Click on the column containing the data, **var1**, and click **Next**.

4. See Figure 9Q. Put in the **Null mean**, which is **3.2** for the ice cream cones. Leave the default not equal for the **Alternative**, and click **Calculate**.

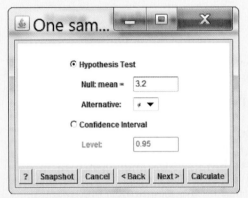

▲ **FIGURE 9Q** Part of StatCrunch Input for One-Sample *t*-Test

You will get the output shown in Figure 9R.

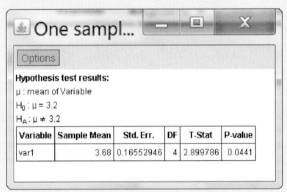

▲ **FIGURE 9R** StatCrunch Output for One-Sample *t*-Test

### One-Sample Confidence Interval

Go back and perform the same steps as for the one-sample *t*-test, but when you get to step 4, check **Confidence Interval**. You may change the confidence level from the default 0.95 if you want. See Figure 9Q.

### Two-Sample *t*-Test

Use stacked data with all the GPAs in one column. The second column will contain the categorical variable that designates groups: male or female.

1. Upload the data from the disk or follow these steps: Enter the GPAs in column 1 (**var1**). Put **m** or **f** in column 2 (**var2**) as appropriate. (Complete words or coding for column 2 are

also allowed but whatever system you use must be maintained within the data set.) Put labels at the top of the columns, change **var1** to **GPA** and **var2** to **Gender**.

2. **Stat > T statistics > Two sample > with data**

3. See Figure 9S. For **Sample 1** choose **GPA**. For the first **Where** put **Gender=m**. For **Sample 2** choose **GPA** and for the second **Where** put **Gender=f**. Click off **Pool variances**. (If you had unstacked data you would choose the two lists for the two samples and not use the **Where** boxes.)

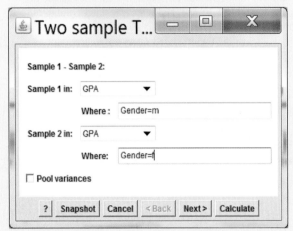

▲ **FIGURE 9S** StatCrunch Input for Two-Sample *t*-Test (Stacked Data)

4. Click **Next** if you want to change the alternative (the default is not equal) or want a confidence interval.

5. Click **Calculate**.

Figure 9T shows the output.

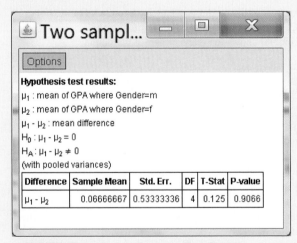

▲ **FIGURE 9T** StatCrunch Output for Two-Sample *t*-test

### Two-Sample Confidence Interval

1. Go back and do the preceding first three numbered steps.

2. When you get to step 4, click **Next**, and check **Confidence Interval**.

3. Click **Calculate**.

### Paired *t*-Test

1. Type the pulse rates Before in column 1 and the pulse rates After in column 2. The headings for the columns are not necessary.

2. **Stat > T statistics > Paired**

3. Select the two columns. Ignore the **Where** and **Group by** boxes. If you want to change your alternative hypothesis from the default two-tailed, click **Next**.

4. Click **Calculate.**

Figure 9U shows the output.

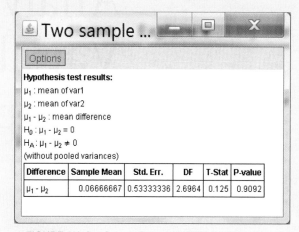

**Hypothesis test results:**

$\mu_1$ : mean of var1
$\mu_2$ : mean of var2
$\mu_1 - \mu_2$ : mean difference
$H_0 : \mu_1 - \mu_2 = 0$
$H_A : \mu_1 - \mu_2 \neq 0$
(without pooled variances)

| Difference | Sample Mean | Std. Err. | DF | T-Stat | P-value |
|---|---|---|---|---|---|
| $\mu_1 - \mu_2$ | 0.06666667 | 0.53333336 | 2.6964 | 0.125 | 0.9092 |

▲ **FIGURE 9U** StatCrunch Output for Paired *t*-Test

# 10 Analyzing Categorical Variables and Interpreting Research

## THEME

Distributions of categorical variables are often summarized in two-way tables. We can make inferences about these distributions by calculating how many observations we would expect in each cell if the null hypothesis were true, and then comparing this to the actual counts.

As we have shown in the last few chapters, statistical tests are comparisons between two different views. In one view, if two samples have different means, then it might be because the populations are truly different. In the other view, randomness rules, and this difference is due merely to chance. A sample of people who had watched violent TV as children may behave more aggressively toward their partners when adults. But this might just be due to chance. Changing the placement of an advertisement on an Internet search results page might lead to a greater number of "follow-through" visits to the company that placed the ad. Or the difference could be due to chance—that is, if we were to repeat the study, we might see a very different outcome.

In this chapter we ask the same question as before: Are the results we see due to chance, or might something

else be going on? In Chapter 8, we asked this question about one-sample and two-sample proportions. We now ask the same question with respect to categorical variables with multiple categories. For example, we might compare two categorical variables: the results of popping a batch of popcorn ("good" and "bad") for different amounts of oil ("no oil," "medium amount of oil," and "maximum oil").

This chapter goes beyond analyzing categorical variables by including suggestions for reading and interpreting published scientific studies. At this point in your studies, you know enough statistics to be able to critically evaluate scientific findings published in academic journals or reported on the nightly news, on blogs, and in newspapers. Although few of us will carry out formal experiments, we all need to know how to evaluate the strength of the conclusions of these studies, since many studies, particularly medical studies, can directly impact our lives.

## CASE STUDY

## Popping Better Popcorn

You're planning a movie night and decide to cook up some popcorn. Many factors might affect how good the popcorn tastes: the brand, for example, or how much oil you use, or how long you let the popcorn pop before stopping. Some researchers (Kukuyeva et al. 2008) investigated factors that might determine how to pop the perfect batch of popcorn. They decided that if more than half of the kernels in a bag were popped in the first 75 seconds, then the batch was a "success." If fewer than half, then it was a "failure." They popped 36 bags under three different treatments: no oil, "medium" amount of oil (1/2 tsp), and "maximum" oil (1 tsp). The bags were randomly assigned to an oil group. Each bag had 50 kernels. The outcome is

| Result | Oil Amount | | |
|---|---|---|---|
| | No Oil | Medium Oil | Maximum Oil |
| **Failure** | 23 | 22 | 33 |
| **Success** | 13 | 14 | 3 |

shown in the accompanying table. From the table it looks as though it is bad to use too much oil. But might this just be due to chance? In other words, if other investigators were to do this experiment the same way, would they also find so few successful bags with the maximum amount of oil?

This question—Is the outcome due to chance?—is one we've asked before. In this study the two variables are the amount of oil (with three values: none, medium, and maximum), and the result (success or failure). Both *Oil* and *Result* are categorical variables. In this chapter, we will see how to test the hypothesis that the amount of oil had an effect on the outcome.

# The Basic Ingredients for Testing with Categorical Variables

**Details**

**Pronunciation**
*Chi* is pronounced "kie" (rhymes with "pie"), with a silent h.

Hypothesis tests that involve categorical variables follow the same four-step recipe you studied in Chapters 8 and 9. However, the basic ingredients are slightly different. Here we introduce these ingredients: data, expected counts, the chi-square statistic (our test statistic), and the chi-square distribution.

## 1. The Data

Recall that categorical variables are those whose values are categories, as opposed to numbers. A respondent's income level can be given as a category (such as high, moderate, or low.)

When two categorical variables are analyzed, as we are doing in this chapter, they are often displayed in a **two-way table**, a summary table that displays frequencies for the outcomes. Such tables can give the impression that the data are numerical, because you are seeing numbers in the table. But it's important to keep in mind that the numbers are *summaries* of variables whose values are *categories*.

For instance, the General Social Survey (GSS) asked respondents, "There are always some people whose ideas are considered bad or dangerous by other people. For instance, somebody who is against churches and religion. If such a person wanted to make a speech in your (city/town/community) against churches and religion, should he be allowed to speak, or not?" These respondents were also asked about their income and, depending on their response, were assigned to one of four annual income levels: $0–$20K, $20–$40K, $40–$70K, and $70K and more. The first four lines of the actual data would look something like Table 10.1. A value on the boundary would be assigned to the group with the larger incomes, so an income of 40 thousand would be in the 40–60 group.

▶ **TABLE 10.1** The first four lines of the raw data (based on an actual data set) showing responses to two questions on the General Social Survey.

| Observation ID | Response | Income Level |
|:---:|:---|:---|
| 1 | allowed | 20–40 |
| 2 | allowed | 20–40 |
| 3 | not allowed | 70 and over |
| 4 | allowed | 0–20 |
| . . . | . . . | . . . |

We have chosen to display another variable, *Observation ID*, which is useful simply for keeping track of the order in which the observations are stored.

A two-way table summarizes the association between the two variables *Response* and *Income Level* as in Table 10.2.

| Response | Income | | | |
|---|---|---|---|---|
| | 0–20 | 20–40 | 40–70 | 70+ |
| **Allowed** | 2404 | 978 | 1015 | 971 |
| **Not Allowed** | 898 | 347 | 226 | 129 |

◀ **TABLE 10.2** Two-way table summarizing the association between responses to the question about whether speeches opposed to church and religion should be allowed, and income level. (Source: GSS 1972-2008, http://www .teachingwithdata.org)

## 2. Expected Counts

The **expected counts** are the numbers of observations we would see in each cell of the summary table if the null hypotheses were true.

Consider a study of the link between TV viewing and violent behavior. (We introduced these data in Chapter 8.) Researchers compared TV viewing habits for children, and then interviewed them about violent behavior many years later (Husemann et al. 2003). Table 10.3 summarizes the results.

There are two categorical variables: *TV Violence* ("High" or "Low") and *Physical Abuse* ("Yes" or "No"). (Recall that the subjects were asked whether they had pushed, grabbed, or shoved their partner.) We wish to know whether these variables are associated. The null hypothesis says that they are not—that these variables are independent. In other words, any patterns you might see are due purely to chance.

> **Details**
>
> **Expectations**
> Expected counts are actually long-run averages. When we say that we "expect" 10 observations in a cell of a table, we mean that if the null hypothesis were true and we were to repeat this data collection many, many times, then, on average, we would see 10 observations in that cell.

|  | High TV Violence | Low TV Violence | Total |
|---|---|---|---|
| **Yes, Physical Abuse** | 25 | 57 | 82 |
| **No Physical Abuse** | 41 | 206 | 247 |
| **Total** | 66 | 263 | 329 |

◀ **TABLE 10.3** A two-way summary of TV violence and later abusive behavior.

What counts should we expect if these variables are truly not related to each other? That is, if the null hypothesis is true, what would we expect the table to look like? There are two ways of answering this question, and they both lead to the same answer. Let's look at them both.

### Starting with the *Physical Abuse* Variable

We notice that out of the entire sample, 82/329 (a proportion of 0.249240, or about 24.92%) said that they had physically abused their partners. If abuse is independent of TV watching—that is, if there is no relationship between these two variables—then we should expect to find the same percentages of violent abuse in those who watched high TV violence and in those with low TV violence.

So when we consider the 66 people who watched high TV violence as children, we expect 24.92% of them to have abused their partners. This translates to $(0.249240 \times 66) = 16.4498$ people. When we consider the 263 people who watched low TV violence, we expect again that 24.92%, or $0.249240 \times 263 = 65.5501$ people to have abused their partners.

We can use similar reasoning to find the other expected counts. We know that if a proportion of 0.249240 abused their partners, then $1 - 0.249240 = 0.750760$ did not.

This proportion should be the same no matter which level of TV violence was watched. Thus, we expect the following:

In the "no physical abuse and high TV violence" group,

$$0.750760 \times 66 = 49.5502$$

In the "no physical abuse and low TV violence" group,

$$0.750760 \times 263 = 197.4499$$

(We are avoiding rounding in these intermediate steps so that our answers for the expected values will be as accurate as possible.)

We summarize these calculations in Table 10.4, which shows the expected counts in parentheses. We include the actual counts in the same table so that we can compare. Note that we rounded to two decimal digits for ease of presentation.

▶ **TABLE 10.4** TV Violence: A summary including expected values (in parentheses).

|  | High TV Violence | Low TV Violence | Total |
|---|---|---|---|
| **Yes, Physical Abuse** | 25 (16.45) | 57 (65.55) | 82 |
| **No Physical Abuse** | 41 (49.55) | 206 (197.45) | 247 |
| **Total** | 66 | 263 | 329 |

If we call the values 82 and 247 row totals (for obvious reasons, we hope!), 66 and 263 column totals, and 329 the grand total, we can generate a formula for automatically finding expected counts for each cell. This formula is rarely needed. First, you can and should always think through the calculations as we did here. Second, most software will do this automatically for you.

**Formula 10.1:** Expected count for a cell $= \dfrac{\text{(row total)} \times \text{(column total)}}{\text{grand total}}$

### Details

**Fractions of People**
Does it bother you that we have fractions of people in each category? It *is* a little strange, until you think about this in terms of an ideal model. These expected counts are like averages. We say the average family has 2.4 children, and we know very well that there is no single family with a 0.4 child. This number 2.4 is a description of the collection of all families. Our claim that we expect 16.45 people (in this group) who have seen high TV violence to be abusive is a similar idealization.

### Starting with the *TV Violence* Variable

The other way of finding the expected counts is to begin by considering the *TV Violence* variable, rather than beginning with the *Physical Abuse* variable. We see that $66/329 = 0.200608$, or 20.06%, watched high TV violence. The rest, $263/329 = 0.799392$, or 79.94%, watched low TV violence.

If these variables are not related, then when we look at the 82 people who committed physical abuse, we should expect about 20.06% of them to fall in the High TV Violence category and the rest to fall in the Low TV Violence category.

Also, when we look at the 247 who did not commit physical abuse, we should expect 20.06% of them to have viewed high TV violence.

Among the abusive, the expected number with high TV violence is $0.200608 \times 82 = 16.45$. Among the nonabusive, the expected number with high TV violence is $0.200608 \times 247 = 49.55$.

You see that you will get the same result no matter which variable you consider first.

### ❋ EXAMPLE 1 Gender and Opinion on Same-Sex Marriage

Do men and women feel differently about the issue of same-sex marriage? In 2008, the General Social Survey took a random sample of 1326 Americans, recording their gender and their level of agreement with the statement "Homosexuals should have the right to marry." The results are summarized in Table 10.5.

| Opinion | Male | Female | Total |
|---|---|---|---|
| Strongly Agree | 80 | 143 | 223 |
| Agree | 124 | 173 | 297 |
| Neutral | 79 | 91 | 170 |
| Disagree | 100 | 113 | 213 |
| Strongly Disagree | 213 | 210 | 423 |
| Total | 596 | 730 | 1326 |

◀ **TABLE 10.5** Summary of gender, and opinion that homosexuals should have the right to marry.

**QUESTION** Assuming that the two variables *Opinion* and *Gender* are *not* associated, find the expected number of males who would strongly agree and the expected number of females who would strongly agree.

**SOLUTION** We consider *Gender* first. The percentage of men in the sample is $(596/1326) \times 100\% = 44.9472\%$. The percentage of women is therefore $100\% - 44.9472\% = 55.0528\%$. If Gender is not associated with *Opinion*, then if we look at the 223 people who strongly agree, we should see that about 44.9% of those who strongly agree are male, and about 55.1% are female. In other words:

**CONCLUSION** Expected count of males who strongly agree $= 223 \times 0.449472 = 100.2323$, or about 100.23.
Expected count of females who strongly agree $= 223 \times 0.550528 = 122.7677$, or about 122.77.

**TRY THIS!** Exercise 10.7

## 3. The Chi-Square Statistic

The **chi-square statistic** is a statistic that measures the amount that our expected counts differ from our observed counts. This statistic is shown in Formula 10.2.

$$\textbf{Formula 10.2:} \quad X^2 = \sum_{\text{cells}} \frac{(O - E)^2}{E}$$

where

  $O$ is the observed count in each cell
  $E$ is the expected count in each cell
  $\Sigma$ means add the results from each cell

Why does this statistic work? The term $(O - E)$ is the difference between what we observe and what we expect under the null hypothesis. To measure the total amount of deviation between Observed and Expected, it is tempting to just add together the individual differences. But this doesn't work, because the expected counts and the observed counts always add to the same value; if we sum up the differences, they will always add to 0.

You can see that the differences between Observed and Expected add to 0 in Table 10.6 on the next page, where we've listed the data from Example 1, but we now show the expected counts as well as the differences (Observed minus Expected).

▶ **TABLE 10.6** Gender and opinion on same-sex marriage, emphasizing the Observed minus Expected values.

| Outcome | Observed Counts | Expected Counts | Observed minus Expected |
|---|---|---|---|
| Strongly Agree Male | 80 | 100.23 | −20.23 |
| Strongly Agree Female | 100.23 | 122.77 | 20.23 |
| Agree Male | 124 | 133.49 | −9.49 |
| Agree Female | 173 | 163.51 | 9.49 |
| Neutral Male | 79 | 76.41 | 2.59 |
| Neutral Female | 91 | 93.59 | −2.59 |
| Disagree Male | 100 | 95.74 | 4.26 |
| Disagree Female | 113 | 117.26 | −4.26 |
| Strongly Disagree Male | 213 | 190.13 | 22.87 |
| Strongly Disagree Female | 210 | 232.87 | −22.87 |

One reason why the chi-square statistic uses squared differences is that by squaring the differences, we always get a positive value, because even negative numbers multiplied by themselves result in positive numbers:

$$\frac{(-20.23)^2}{100.23} + \frac{20.23^2}{122.77} + \frac{(-9.49)^2}{133.49} + \frac{9.49^2}{163.51} + \cdots$$

Why divide by the expected count? The reason is that a difference between the expected and actual counts of, say, 2 is a small difference if we were expecting 1000 counts. But if we were expecting only 5 counts, then this difference of 2 is substantial. By dividing by the expected count, we're controlling for the size of the expected count. Basically, for each cell, we are finding what proportion of the expected count the squared difference is.

If we apply this formula to the data in Example 1, we get $X^2 = 14.15$. We must still decide whether this value discredits the null hypothesis that gender and the answer to the question are independent. Keep reading.

## EXAMPLE 2 Viewing Violent TV as a Child and Abusiveness as an Adult

Table 10.7 shows summary statistics from a study that asked whether there was an association between watching violent TV as a child and aggressive behavior toward one's spouse later in life. The table shows both actual counts and expected counts (in parentheses).

▶ **TABLE 10.7** A two-way summary of the effect of viewing TV violence on later abusiveness (expected values are shown in parentheses).

| | High TV Violence | Low TV Violence | Total |
|---|---|---|---|
| **Yes, Physical Abuse** | 25 (16.45) | 57 (65.55) | 82 |
| **No Physical Abuse** | 41 (49.55) | 206 (197.45) | 247 |
| **Total** | 66 | 263 | 329 |

**QUESTION** Find the chi-square statistic to measure the difference between the observed counts and expected counts for the study of the effect of violent TV on future behavior.

SOLUTION  We use Formula 10.2 with the values for $O$ and $E$ taken from Table 10.7.

$$X^2 = \sum \frac{(\text{Observed} - \text{Expected})^2}{\text{Expected}}$$

$$= \frac{(25 - 16.45)^2}{16.45} + \frac{(57 - 65.55)^2}{65.55} + \frac{(41 - 49.55)^2}{49.55} + \frac{(206 - 197.45)^2}{197.45} = 7.4047$$

CONCLUSION

$$X^2 = 7.40$$

Later we will see whether this is an unusually large value for two independent variables.

TRY THIS!  Exercise 10.9

As you might expect, for tables with many cells, these calculations can quickly become tiresome. Fortunately, technology comes to our rescue. Most statistical software will calculate the chi-square statistic for you, given data summarized in a two-way table or presented as raw data as in Table 10.1, and some software will even display the expected counts alongside the observed counts. Figure 10.1 shows the output from StatCrunch for these data.

When the null hypothesis is true, our real-life observations will usually differ slightly from the expected counts just by chance. When this happens, the chi-square statistic will be a small value.

If reality is very different from what our null hypothesis claims, then our observed counts should differ substantially from the expected counts. When that happens, the chi-square statistic is a big value.

The trick, then, is to decide what values of the chi-square statistic are "big." Big values discredit the null hypothesis. To determine whether an observed value is big, we need to know its probability distribution when the null hypothesis is true.

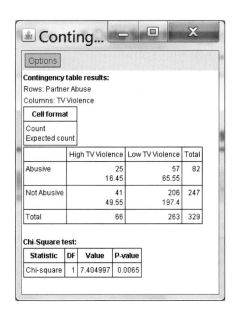

◀ **FIGURE 10.1** StatCrunch output for TV violence and abusiveness. The expected values are below the observed values.

Tech

 KEY POINT  If the data conform to the null hypothesis, then the value of the chi-square statistic will be small. For this reason, large values of the chi-square statistic make us suspicious of the null hypothesis.

# 4. The Sampling Distribution of the Chi-Square Statistic

We are wondering whether television-viewing habits as a child are associated with violent behavior as an adult. If there is *no* association, then the observed counts in Table 10.7 should be close to the expected counts, and our chi-square statistic should be small. We found that $X^2 = 7.40$. Is this small?

To help us determine whether a particular value for a chi-square statistic is big or small, we need to know the probability distribution of $X^2$. Recall that the probability distribution of a statistic, such as $X^2$, is called a sampling distribution. Using the sampling distribution for $X^2$, we can find the probability that, just through chance, a chi-square statistic would have the value of 7.40 or bigger. If this probability is large, it means that 7.40 is not an unusual value. If the probability is small, it suggests that the value might be unusually large.

If the sample size is large enough, there is a probability distribution that gives a fairly good approximation to the sampling distribution. Not surprisingly, this approximate distribution is called the **chi-square distribution**. The chi-square distribution is often represented with the Greek lowercase letter chi ($\chi$) raised to the power of 2—that is, $\chi^2$.

Unlike the normal distribution and the *t*-distribution, the $\chi^2$ distribution allows for only positive values. It also differs from the other sampling distributions you've seen in that it is (usually) not symmetric and is instead right-skewed.

Like the shape of the *t*-distribution, the shape of the chi-square distribution depends on a parameter called the **degrees of freedom**. The lower the degrees of freedom, the more skewed the shape of the chi-square distribution. Figure 10.2 shows the chi-square distribution for several different values of the degrees of freedom. Sometimes the degrees of freedom are indicated using this notation: $\chi^2_{df}$. For example, $\chi^2_6$ represents a chi-square distribution with 6 degrees of freedom, as in Figure 10.2b.

The degrees-of-freedom parameter is different for different tests, but in general it depends on the number of categories in the summary table.

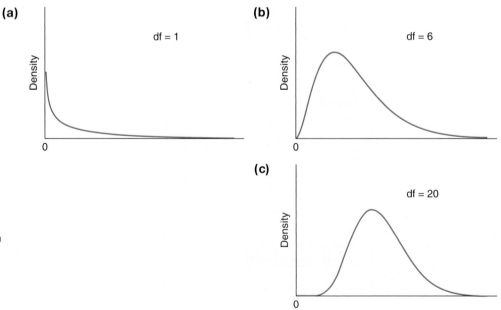

▶ **FIGURE 10.2** Three chi-square distributions for **(a)** df = 1, **(b)** df = 6, and **(c)** df = 20. Note that the shape becomes more symmetric as the degrees of freedom (df) increase. No negative values are possible with the chi-square distribution; the smallest possible value is 0.

The chi-square distribution $\chi^2$ is only an approximation of the true sampling distribution of the statistic $X^2$. The approximation is usually quite good if all of the expected counts are 5 or higher.

KEY POINT The chi-square distribution provides a good approximation to the sampling distribution of the chi-square statistic only if the sample size is large. For many applications, the sample size is large enough if each expected count is 5 or higher.

We asked whether the value $X^2 = 7.40$ was small for the problem of determining whether viewing TV violence was associated with abusiveness later in life. We'll answer this question in Section 10.2.

SECTION 10.2

# Chi-Square Tests for Associations between Categorical Variables

There are two tests to determine whether two categorical variables are associated. Which test you used depends on how the data were collected.

We often summarize two categorical variables in a two-way table, in part because it helps us see whether associations exist between these variables. When we examine a two-way table (or any data summary, for that matter), one aspect that gets hidden is the method used to collect the data. We can collect data that might appear in a two-way table in either of two ways.

The first method is to collect two or more distinct, independent samples, one from each population. Each object sampled has a categorical value that we record. For example, we could collect a random sample of men and a distinct random sample of women. We could then ask them to what extent they agree with the statement that same-sex marriage should be allowed: Strongly Agree, Agree, Neutral, Disagree, or Strongly Disagree. We now have one categorical response variable, *Opinion*. We also have another categorical variable, *Gender*, that keeps track of which population the response belongs to. Hence we have two samples, one categorical response variable, and one categorical grouping variable.

The second method is to collect just one sample. For the objects in this sample, we record two categorical response variables. For example, we might collect a large sample of people and record their marital status (single, married, divorced, or widowed) and their educational level (high school, college, graduate school). From this one sample, we get two categorical variables: *Marital Status* and *Educational Level*.

In both data collection methods, we are interested in knowing whether the two categorical variables are related or unrelated. However, because the data collection methods are different, the ways in which we test the relationship between variables differ also. That's the bad news. The good news is that this difference is all behind the scenes. The result of careful calculation shows us that no matter which method we use to collect data, we can use the same chi-square statistic and the same chi-square distribution to test the relation between variables.

These two methods have different names. If we test the association, based on two independent samples, between the grouping variable and the categorical response variable (the first method), the test is called a test of **homogeneity**. If we base the test on one sample (the second method), the test is called a test of **independence**. Two different data collection methods, two different names—but, fortunately, the same test!

> **Details**
>
> **Homogeneity**
> The word *homogeneity* is based on the word *homogeneous*, which means "of the same, or similar, kind or nature."

KEY POINT There are two tests to determine whether two categorical variables are associated. For two or more samples and one categorical response variable, we use a test of homogeneity. For one sample and two categorical response variables, we use a test of independence.

## EXAMPLE 3 Independence or Homogeneity?

The Pew Foundation conducts a regular survey to measure consumers' attitudes toward certain products. In 2006, Pew randomly sampled 2000 people and asked whether they considered a TV set a luxury or a necessity. In 2009, while the U.S. was in the middle of one of the most severe recessions in its history, Pew asked another random sample of 1003 people the same question. The researchers want to know whether consumers' attitudes about TV sets have changed since 2006, and they plan to carry out a formal hypothesis test to see whether an association exists between the variables *Year* and *Attitude*.

QUESTION  Will this be a test of independence or a test of homogeneity?

SOLUTION  The Pew Foundation took two distinct samples of people: the first sample in 2006 and the second in 2009. The variable *Year* simply tells us which group the respondents belonged to. There is one response variable: *Attitude*. So this test has two samples and one response variable, and is a test of homogeneity.

CONCLUSION  The proposed hypothesis test is a test of homogeneity

TRY THIS!  Exercise 10.11

## EXAMPLE 4 Independence or Homogeneity?

The 2008 U.S. presidential election was historic for many reasons, but one was that a woman (Hillary Clinton) and an African American (Barack Obama) both had good chances of winning the Democratic nomination. Before the Democratic National Convention, the General Social Survey polled a random sample of 1286 people eligible to vote and asked whether they would (a) vote for a woman and (b) vote for an African American. One question of interest is whether those who would take the historic leap of voting for a woman for president would also be willing to vote for an African American. (You can imagine that the Obama campaign, expecting a close battle with the Clinton campaign, might have been interested in this question.)

QUESTION  If we test whether a person's willingness to vote for a woman for president is associated with that person's willingness to vote for an African American for president, is this a test of homogeneity or of independence?

CONCLUSION  There is one sample, consisting of 1286 people. These people are asked two questions, so there are two outcome variables: "Would you vote for a woman?" and "Would you vote for an African American? This is a test of independence.

TRY THIS!  Exercise 10.13

## Tests of Independence and Homogeneity

Again, the tests follow the four-step procedure of all hypothesis tests. We'll give you an overview and then fill in the details with an example.

### Step 1: Hypothesize

The hypotheses are always the same:

$H_0$: There is *no* association between the two variables (the variables are independent).

$H_a$: There is an association between the two variables (the variables are not independent).

Although the hypotheses are always the same, you should phrase these hypotheses in the context of the problem.

**Step 2: Prepare**

Whether you are testing independence or homogeneity, the test statistic you should use to compare counts is the chi-square statistic, shown in Formula 10.2 and repeated here.

$$X^2 = \sum_{\text{cells}} \frac{(O - E)^2}{E}$$

If the conditions are right, then this statistic follows, approximately, a chi-square distribution with

$$df = (\text{number of rows} - 1)(\text{number of columns} - 1)$$

Conditions:

1. *Random Samples.* All samples were collected randomly.

2. *Independent Samples and Observations.* All samples are independent of each other. Always, the observations within a sample must be independent of each other.

3. *Large Samples.* The expected count must be 5 or more in each cell.

Note that in a test of independence, there is always only one sample. But a test of homogeneity might have several independent samples.

**Step 3: Compute to compare**

This step is best done with technology. The p-value is the probability, assuming the null hypothesis is true, of getting a value as large as or larger than the observed chi-square statistic. In other words, the p-value is the probability, if the variables really are not associated, that we would see an outcome as large or larger than the one observed. A small p-value therefore means a large test statistic, which casts doubt on the hypothesis that the variables are not associated.

**Step 4: Interpret**

If the p-value is less than or equal to the stated significance level, we reject the null hypothesis and conclude that the variables are associated.

 EXAMPLE 5 **Education and Marital Status**

Does a person's educational level affect his or her decision about marrying? With observational data, we can't know for certain, but we can see whether the data are consistent with or contradict the idea that people's educational level affects their decisions about marrying. From the U.S. Census data, we took a random sample of 665 people and measured their marital status (single, married, divorced, or widow/widower) and their educational level (less than high school, high school degree, college degree or higher). You saw these data first in Chapter 5. Figure 10.3 shows output from StatCrunch.

QUESTION Use the provided output to test whether marital status and educational level are associated. Is this a test of homogeneity or of independence?

SOLUTION Because there is one sample with two response variables, this is a test of independence.

*Step 1: Hypothesize*

$H_0$: Among all U.S. residents, marital status and educational level are independent.

$H_a$: Among all U.S. residents, marital status and educational level are associated.

*Step 2: Prepare*
We can see from the output that the expected counts are 5 or more. (The smallest expected count is 11.47.) This means that the chi-square distribution will provide a good approximation for the p-value. The data are summarized in Figure 10.3. This table displays the actual counts for a random sample of 665 people, the expected counts, and the calculated values.

▶ **FIGURE 10.3** Summary table with expected counts, chi-square statistic, and p-value to test whether marital status and educational level are associated.

**Tech**

○ ○ ○   Contingency Table with data   ⚠

Options

**Contingency table results:**

Rows: marital

Columns: education

| Cell format |
| --- |
| Count |
| Expected count |

| | College or higher | HS | Less HS | Total |
| --- | --- | --- | --- | --- |
| Divorced | 15 | 59 | 10 | 84 |
| | 18.06 | 50.15 | 15.79 | |
| Married | 98 | 240 | 70 | 408 |
| | 87.74 | 243.6 | 76.69 | |
| Single | 27 | 68 | 17 | 112 |
| | 24.08 | 66.86 | 21.05 | |
| Widow/Widower | 3 | 30 | 28 | 61 |
| | 13.12 | 36.42 | 11.47 | |
| Total | 143 | 397 | 125 | 665 |

| Statistic | DF | Value | P-value |
| --- | --- | --- | --- |
| Chi-square | 6 | 39.96996 | <0.0001 |

*Step 3: Compute to compare*
The chi-square statistic is $X^2 = 39.97$ (rounding to two decimal digits). The chi-square distribution has 6 degrees of freedom, and the p-value is reported as smaller than 0.0001.

*Step 4: Interpret*
Because the p-value is less than 0.05, we reject the null hypothesis.

CONCLUSION   Marital status and educational level are associated.

TRY THIS!   Exercise 10.17

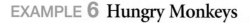

# EXAMPLE 6  Hungry Monkeys

Research in the past has suggested that mice and rats that are fed less food live longer and healthier lives. Recently, a study of Rhesus monkeys was done that involved caloric restriction (less food). It is believed that monkeys have many similarities to humans, which is what makes this study so interesting.

Seventy-six Rhesus monkeys, all young adults, were randomly divided into two groups. Half of the monkeys (38) were assigned to caloric restriction. Their food was decreased about 10% per month for three months, and as a result they were fed about 30% less food than the other 38 monkeys for the duration of the experiment.

For those on the normal diet, 14 out of 38 had died of age-related causes by the time the article was written. For those on caloric restriction, only 5 out of 38 had died of age-related causes (Colman et al. 2009).

**QUESTION** Because this is a randomized study, the hypothesis that diet is associated with aging can be stated as a cause-and-effect hypothesis. Therefore, test the hypothesis that diet causes differences in aging. Will this be a test of homogeneity or of independence? Minitab output is shown in Figure 10.4.

```
Chi-Square Test: normal diet, caloric Restriction
Expected counts are printed below observed counts
Chi-Square contributions are printed below expected counts

            Normal        Caloric
            Diet     Restriction   Total
Died         14            5        19
           9.50         9.50
           2.132        2.132

Not          24           33        57
          28.50        28.50
          0.711        0.711

Total        38           38        76

Chi-Sq = 5.684, DF = 1, P-Value = 0.017
```

◀ **FIGURE 10.4** Minitab output showing *Diet* (normal or caloric restriction) and *Aging* (died from age-related causes or not) for 76 monkeys.

**SOLUTION** There are two samples (monkeys with caloric restriction and monkeys without) and one outcome variable: whether the monkey died of age-related causes. Therefore, this is a test of homogeneity.

*Step 1: Hypothesize*

> $H_0$: For these monkeys, the amount of calories in the diet is independent of aging.
> $H_a$: For these monkeys, the amount of calories in the diet causes differences in aging. Because we do not have a random sample, our results do not generalize beyond this group of monkeys. But because we have randomized assignment to treatment groups, we are able to conclude that any differences we see in the aging process are caused by the diet.

*Step 2: Prepare*
Because we wish to use the chi-square test, we must confirm that the expected counts are all 5 or more. This is the case here, as Figure 10.4 confirms.

*Step 3: Compute to compare*
From the Minitab output, Figure 10.4, you can see that the chi-square value is 5.68 and the p-value is 0.017.

*Step 4: Interpret*
Because the p-value is less than 0.05, we can reject the null hypothesis.

**CONCLUSION** The monkeys' deaths from age-related causes were caused by differences in the number of calories in the diet.

**TRY THIS!** Exercise 10.23

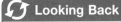
> **Looking Back**
>
> **Comparing Two Proportions**
> In Section 8.4, you learned to test two population proportions based on data from independent samples using the z-test. The test of diet in monkeys could have been done using this test. The chi-square test provides an alternative, but equivalent, approach.

The article contains other information that suggests that monkeys on a restricted diet are generally healthier than monkeys on a normal diet. Figure 10.5 shows a photo of two monkeys. The one on the right had the restricted diet and shows fewer characteristics of old age.

▶ **FIGURE 10.5** The healthier monkey (on the right) was one of those on caloric restriction.

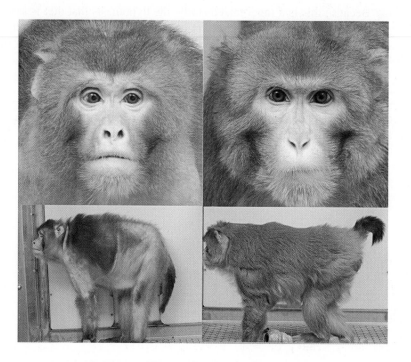

## Random Samples and Randomized Assignment

You have now seen randomization used in two different ways. Random sampling is the practice of selecting objects in our sample by choosing them at random from the population, as is done in many surveys. We can make generalizations about the population only if the sample is selected randomly, because this is the only way of ensuring that the sample is representative of the population. The General Social Survey is an example of studies based on random sampling. When we conclude, as we did in Example 5, that marital status and educational level are associated, we are stating a conclusion about the entire population—in this case, all adults in the United States. We are confident that these variables are associated in the population, because our sample was selected at random.

In Example 6, on the other hand, there was no random sample. However, the monkeys were randomly assigned to a treatment group (low-calorie diet) or the control group (normal diet). Because the monkeys were not selected randomly, we have no means of generalizing about the population as a whole, statistically speaking. (There might be a biological argument, or an assumption, that a diet that works on one group of monkeys would work on any other group, but as statisticians, we have no data to support this assumption.) However, the researchers performed this study because they were interested in a cause-and-effect relationship: Does changing the calories in a monkey's diet change the monkey's health and longevity?

Because researchers controlled which monkeys got which diet, this is a controlled experiment. And because they used randomized assignment, and because we rejected the null hypothesis that diet and health were independent, we can conclude that in fact the caloric restriction *did* affect the monkeys' health.

---

**↻ Looking Back**

**Data Collection**
Controlled experiments are those in which experimenters determine how subjects are assigned to treatment groups. In contrast, observational studies are those in which subjects place themselves into treatment groups, by behavior or innate characteristics such as gender. Causal conclusions cannot be based on a single observational study.

## SNAPSHOT  CHI-SQUARE TESTS OF INDEPENDENCE AND HOMOGENEITY

| | |
|---|---|
| **WHAT IS IT?** ▶ | A test of whether two categorical variables are associated. |
| **WHAT DOES IT DO?** ▶ | Using the chi-square statistic, we compare the observed counts in each outcome category with the counts we would expect if the variables were *not* related. If observations are too different from expectations, then the assumption that there is no association between the categorical variables looks suspicious. |
| **HOW DOES IT DO IT?** ▶ | If the sample size is large enough and basic conditions are met, then the chi-square statistic follows, approximately, a chi-square distribution with df = (number of rows − 1) × (number of columns − 1). The p-value is the probability of getting a value as large as or larger than the observed chi-square statistic, using the chi-square distribution. |
| **HOW IS IT USED?** ▶ | To compare distributions of two categorical variables. |

## Relation to Tests of Proportions

In the special case in which both categorical variables have only two categories, the test of homogeneity is identical to a $z$-test of two proportions, using a two-tailed alternative hypothesis. The following analysis illustrates this.

In a landmark study of a potential AIDS vaccine published in 2009, researchers from the U.S. Army and the Thai Ministry of Health randomly assigned about 8200 volunteers to receive a vaccine against AIDS and another 8200 to receive a placebo. (We rounded the numbers slightly to make this discussion easier.) Both groups received counseling on AIDS prevention measures and were promised life-time treatment should they contract AIDS. Of those who received the vaccine, 51 had AIDS at the end of the study (three years later). Of those that received the placebo, 74 had AIDS (http://www.hivresearch.org/, accessed September 29, 2009). We will show two ways of testing whether an association existed between receiving the vaccine and getting AIDS. The data are summarized in Table 10.8.

|  | Vaccine | No Vaccine | Total |
|---|---|---|---|
| **AIDS** | 51 | 74 | 125 |
| **No AIDS** | 8149 | 8126 | 16275 |
| **Total** | 8200 | 8200 | 16400 |

◀ **TABLE 10.8** Effect of the vaccine on AIDS.

If we use the approach of this chapter, we recognize that this is a test of homogeneity, because there are two samples (*Vaccine* and *Placebo*) and one outcome variable (*AIDS*). Although we cannot generalize to a larger population (because the volunteers were not randomly selected), we can make a cause-and-effect conclusion about whether differences in AIDS rates are due to the vaccine, because this is a controlled, randomized study.

As a first step, we calculate the expected counts, under the assumption that the two variables are not associated.

Because the proportion of those who got AIDS was 125/16400 = 0.007622, if the risk of getting AIDS had nothing to do with the vaccine, then we should see about the same proportion of those getting AIDS in both groups. If the proportion of people who got AIDS in the *Vaccine* group was 0.007622, then we would expect 8200 × 0.007622 = 62.5 people to get AIDS in the *Vaccine* group.

Both groups are the same size, so we would expect the same number of AIDS victims in the *Placebo* group. This means that in both groups, we would expect $8200 - 62.5 = 8137.5$ not to get AIDS.

The results, with expected counts in parentheses to the right of the observed counts, are shown in Table 10.9.

▶ **TABLE 10.9** Expected counts, assuming no association between variables, are shown in parentheses.

|  | Vaccine | No Vaccine | Total |
|---|---|---|---|
| **AIDS** | 51 (62.5) | 74 (62.5) | 125 |
| **No AIDS** | 8149 (8137.5) | 8126 (8137.5) | 16275 |
| **Total** | 8200 | 8200 | 16400 |

Note that all expected counts are much greater than 5. The chi-square statistic is not difficult to calculate:

$$X^2 = \sum \frac{(\text{Observed} - \text{Expected})^2}{\text{Expected}}$$

$$= \frac{(51 - 62.5)^2}{62.5} + \frac{(74 - 62.5)^2}{62.5} + \frac{(8149 - 8137.5)^2}{8137.5} + \frac{(8126 - 8137.5)^2}{8137.5}$$

$$= 4.26$$

The degree of freedom of the corresponding chi-square distribution is

$$(\text{Number of rows} - 1)(\text{number of columns} - 1) = (2 - 1)(2 - 1) = 1 \times 1 = 1$$

The p-value is illustrated in Figure 10.6. It is the area under a chi-square distribution with 1 degree of freedom and to the right of 4.26. The p-value turns out to be 0.039. We therefore reject the null hypothesis and conclude that there is an association between getting the vaccination and contracting AIDS. The difference in the numbers of AIDS victims was caused by the vaccine.

▶ **FIGURE 10.6** The area to the right of 4.26 represents the p-value to test whether there is an association between receiving the AIDS vaccine and contracting AIDS. The distribution is a chi-square distribution with 1 degree of freedom. The p-value is 0.039.

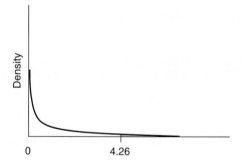

One thing that's disappointing about this conclusion is that the alternative hypothesis states only that the variables are associated. That's nice, but what we really want to know is *how* they are associated. Did the vaccine decrease the number of people who got AIDS? That's what the researchers wanted to know. They didn't want to know merely whether there was an association. They had a very specific direction in mind for this association.

One drawback with chi-square tests is that they reveal only whether two variables are associated, not how they are associated. Fortunately, when both categorical variables have only two categories, we can instead do a two-proportion z-test.

Define $p_1$ to be the probability that someone given the vaccine will get AIDS, and define $p_2$ to be the probability that someone given the placebo will get AIDS. Then proceed as follows.

**Step 1: Hypothesize**

$H_0: p_1 = p_2$

$H_a: p_1 < p_2$

In words, the null hypothesis says that the probability of getting AIDS is the same whether or not the individual gets the vaccine. The alternative hypothesis is one-tailed: If the individual gets the vaccine, the probability of contracting AIDS is less than if she or he got the placebo.

**Step 2: Prepare**

We plan to use the $z$-test for two independent proportions. Under the null hypothesis, the probability of getting AIDS is the same for both groups, and we estimate this probability with

$$\hat{p} = \frac{(51 + 74)}{16400} = 0.007622$$

(This is the same proportion we calculated above.)

We must check that both samples are large enough—both expected values must be bigger than 10. This is straightforward, because both samples are the same size:

$$8200 \times 0.007622 = 62.5$$

This means that we can use the standard Normal distribution (with mean 0 and standard deviation 1) to approximate the p-value.

**Step 3: Calculate to compare**

The observed value of the $z$-statistic is

$$z = \frac{\hat{p}_1 - \hat{p}_2}{\sqrt{\hat{p}\left(1 - \hat{p}\right)\left(\dfrac{1}{n_1} + \dfrac{1}{n_2}\right)}}$$

Note that $\hat{p}_1 = 51/8200$, $\hat{p}_2 = 74/8200$, and both $n_1$ and $n_2$ are equal to 8200. Substituting, we get

$$z = -2.065068, \text{ or about } -2.07$$

It is no coincidence that if you square this value, you get the same value as our chi-square statistic. Recall that $X^2 = 4.26$ and

$$z^2 = (-2.065068)^2 = 4.26$$

For the p-value, we find the area under a N(0, 1) curve below $-2.06$. This probability is 0.0194582, or about 0.019.

**Step 4: Interpret**

Based on the p-value, we reject the null hypothesis and conclude that people who receive the vaccine are less likely to contract AIDS than those who do not. Because this was a randomized, controlled study, we can conclude that the vaccine caused a decrease in the probability of people getting AIDS.

Note that if we had doubled the p-value—that is, if we had instead used a two-tailed alternative hypothesis—we would have got $2 \times 0.019458 = 0.039$, almost exactly what we got for the chi-square test.

Although medical researchers were very excited about this study—not too many years before, it was thought to be impossible to develop an AIDS vaccine—they caution that this vaccine lowers risk by only about 30%. Many medical professionals consider a vaccine to be useful only if it lowers risk by at least 60%. There was also some controversy over this study: Some argued that a few of the subjects who were dropped from the initial analysis (because of preexisting medical conditions) should have been included. If these subjects had been included, then the vaccine would no longer have been judged effective, statistically speaking.

> **KEY POINT** For a 2-by-2 two-way table of counts, a two-proportion z-test with a two-tailed alternative is equivalent to a test of homogeneity.

When should you use the two proportion z-test, and when should you use the test for homogeneity? If you need to use a one-tailed alternative hypothesis, then you should use the z-test. However, if you plan to use a two-tailed alternative hypothesis, then it doesn't matter which test you use.

## SECTION 10.3

# Reading Research Papers

One goal of this text is to teach you enough of the basic concepts of statistics that you can critically evaluate published research. The medical literature, in particular, records many findings that can have major consequences in our lives. When we rely on the popular media to interpret these findings, we often get contradictory messages. However, you now know enough statistics so that, with a few guiding principles, you can often make sense of research results yourself.

Before discussing ways of evaluating individual research articles, we offer a few over-arching, guiding principles.

1. *Pay attention to how randomness is used.* Random sampling is used to obtain a representative sample, so that we can make inferences about a larger population. Random assignment is used to test causal associations so that we can conclude that the treatment was truly effective (or was not) on a particular sample of subjects. Many medical studies use random assignment but do not use random sampling. This means the results are not necessarily applicable to the entire population. Table 10.10 summarizes these possibilities.

▶ **TABLE 10.10** Four different study designs and the inferences possible.

|  | Sample Selected Randomly | No Random Sample |
|---|---|---|
| **Random Assignment** | You can make a causal conclusion and conclude that the entire population would be affected similarly. | You can make a causal conclusion, but we do not know whether everyone would respond similarly. |
| **No Random Assignment** | You can assume that an association between the variables exists in the population, but you cannot conclude that it is a causal relationship. | You can conclude that an association exists within the sample but not in the entire population, and you cannot conclude that there is a causal relationship. |

2. *Don't rely solely on the conclusions of any single paper.* Research advances in small steps. A single research study, even when the conclusions are grand and ambitious, can tell us only a small part of the real story. For example, the *Los Angeles Times* reported, on the basis of a few published studies, that many people considered vitamin D to be useful for preventing cancer, cardiovascular disease, depression, and other maladies. But a panel of medical experts concluded that these beliefs were based on preliminary studies. The body of medical research, they concluded, was in fact quite mixed in its view of just how effective vitamin D really is (Healy 2011).

3. *Extraordinary claims require extraordinary evidence.* This advice was a favorite piece of wisdom of magician and professional skeptic (The Amazing) James Randi. If someone claims to have done something that had been believed impossible, don't believe it until you've seen some very compelling evidence.

4. *Be wary of conclusions based on very complex statistical or mathematical models.* Complex statistics often require complex assumptions, and one consequence of this is that the findings might be correct and true, but only in a limited set of circumstances. Also, some research studies are essentially "what if" studies: "We're studying *what* will happen *if* these assumptions hold." It is important to remember that those assumptions may not be practical—or even possible.

5. *Stick to peer-reviewed journals.* **Peer review** means that papers were read by two or three (and sometimes more) knowledgeable and experienced researchers in the same field. These reviewers can prevent papers from being published if they do not think the methods employed were sound or if they find too many mistakes. They can also demand that the authors make changes and submit the paper again. Many published papers have gone through several rounds of reviews. But be careful: Not all peer-reviewed journals are equal. The more prestigious journals have greater resources to check papers for mistakes, and they have much more careful and knowledgeable reviewers on hand to find sometimes subtle errors. Good journals have editorial boards that reflect the general practice of the community and are not restricted to people who reflect a minority point of view.

## EXAMPLE 7 Improving Tips

A sociologist wonders whether a waitress's tip is affected by whether or not she touches her customers. Below are two study designs that the sociologist might use. Read these, and then answer the questions that follow.

Design A: The sociologist chooses four large restaurants in his city and gets all waitresses at these restaurants to agree to participate in his study. On several nights during the next few weeks, the sociologist visits the restaurants and records the number of times all waitresses touch their customers on the customer's back or shoulder. He also records the total amount of tips earned by each waitress. He finds that waitresses who sometimes touched their customers earned larger tips, on average, than those who did not.

Design B: The sociologist chooses four large restaurants in his city and gets all waitresses at these restaurants to agree to participate in his study. At each restaurant, half of the waitresses are randomly assigned to the "touch" group and half assigned to the "no touch" group. The "touch" group is instructed to lightly touch each customer on the back or shoulder two or three times during the meal. The "no touch" group is instructed not to touch customers at all. At the end of the study, the sociologist finds that waitresses in the "touch" group earned larger tips, on average, than those in the "no touch" group.

QUESTION For each study design, state whether it is possible to generalize the results to a larger population and whether the researcher can make a cause-and-effect conclusion.

SOLUTION For Design A, the waitresses (or restaurants) were not randomly selected from a larger population, so we cannot generalize the results beyond the sample. This was an observational study: The waitresses themselves chose whether to be "touchers" or "no touchers." Because random assignment was not used, we cannot conclude that the difference in tips was caused by touching. A possible confounding effect is the restaurant itself, which might encourage touching and which might have a clientele that tends to tip more generously than the clientele at other restaurants.

For Design B, we also cannot generalize to a larger population. But here random assignment was used. Therefore, the environment of the restaurant is not a confounding factor, and we can conclude that the touching caused the increased amount of tips.

**TRY THIS!** Exercise 10.33

## Reading Abstracts

An **abstract** is a short paragraph at the beginning of a research article that describes the basic findings. If you look up science papers using Google Scholar, say, clicking on the link will usually take you to an abstract. Often, you will be able to read the abstract at no charge. (Reading the papers themselves often requires that you subscribe to the journal or that you read from the computers at a library that has purchased a subscription.) In addition, the websites of many journals display the abstracts to their published papers, even if they do not provide access to the papers themselves.

For example, an article recently published in the *New England Journal of Medicine* (Poordad et al. 2011) reported the findings of a study on the effectiveness of a new treatment for chronic hepatitis C virus (HCV). Currently, the standard treatment, peginterferon-ribavirin, has a relatively low success rate. A success, in treating HCV, is called a "sustained virologic response." In a sustained virologic response, the virus is not eliminated from the body, but is undetectable for a long period of time. Researchers hope that adding a new medicine, boceprevir, to the standard treatment would lead to a great proportion of patients achieving a sustained virologic response. An excerpt from the abstract follows:

> **Methods:** We conducted a double-blind study in which previously untreated adults with HCV genotype 1 infection were randomly assigned to one of three groups. In all three groups, peginterferon alfa-2b and ribavirin were administered for 4 weeks (the lead-in period). Subsequently, group 1 (the control group) received a placebo plus peginterferon-ribavirin for 44 weeks; group 2 received boceprevir plus peginterferon-ribavirin for 24 weeks, and those with a detectable HCV RNA level between weeks 8 and 24 received a placebo plus peginterferon-ribavirin for an additional 20 weeks; and group 3 received boceprevir plus peginterferon-ribavirin for 44 weeks. Nonblack patients and black patients were enrolled and analyzed separately.
>
> **Results:** A total of 938 nonblack and 159 black patients were treated. In the nonblack cohort, a sustained virologic response was achieved in 125 of the 311 patients (40%) in group 1, in 211 of the 316 patients (67%) in group 2 ($P < 0.001$), and in 213 of the 311 patients (68%) in group 3 ($P < 0.001$). In the black cohort, a sustained virologic response was achieved in 12 of the 52 patients (23%) in group 1, in 22 of the 52 patients (42%) in group 2 ($P = 0.04$), and in 29 of the 55 patients (53%) in group 3 ($P = 0.004$). In group 2, a total of 44% of patients received peginterferon-ribavirin for 28 weeks. Anemia led to dose reductions in 13% of controls and 21% of boceprevir recipients, with discontinuations in 1% and 2%, respectively.
>
> **Conclusions:** The addition of boceprevir to standard therapy with peginterferon-ribavirin, as compared with standard therapy alone, significantly increased

the rates of sustained virologic response in previously untreated adults with chronic HCV genotype 1 infection. The rates were similar with 24 weeks and 44 weeks of boceprevir. (Funded by Schering-Plough [now Merck]; SPRINT-2 ClinicalTrials.gov, number NCT00705432.)

To help you evaluate this abstract, and others like it, answer these questions:

1. What is the research question that these investigators are trying to answer?

2. What is their answer to the research question?

3. What were the methods they used to collect data?

4. Is the conclusion appropriate for the methods used to collect data?

5. To what population do the conclusions apply?

6. Have the results been replicated (that is, reproduced) in other articles? Are the results consistent with what other researchers have suggested?

The sixth question is important, although it is often difficult for a layperson to answer. Research is a difficult activity, in part because of the great variability in nature. Studies are often done with samples that do not allow generalizing to populations or are subject to bias, and sometimes researchers just make mistakes. For that reason, you should not believe in the claims of a single study unless it is consistent with currently accepted theory and supported by other research. For all of these reasons, wait until there is some accumulation of knowledge before changing your lifestyle or making a major decision on the basis of scientific research.

Let's see how we would answer these questions for this abstract.

1. *What is the research question that these investigators are trying to answer?* Does the addition of boceprevir to the standard treatment lead to a higher proportion of hepatitis C patients achieving a sustained virologic response?

2. *What is their answer to the research question?* Yes, the additional drug improved patients' responses. Researchers report an increased rate of patients achieving sustained virologic response with boceprevir plus the standard therapy compared to those receiving only standard therapy.

3. *What were the methods they used to collect data?* Patients were randomly assigned to one of three treatment groups. Assignments were double-blind. Patients were examined after 24 and 44 weeks to determine if they had achieved sustained virologic response.

4. *Is the conclusion appropriate for the methods used to collect data?* The fact that subjects were randomly assigned to treatment groups means that we can make a causal conclusion and say that the difference in virologic response rates was due to the the addition of boceprevir to the standard treatment. We also know that the observed results were too large to be explained by chance. Note that p-values are given as, for example, "$P < 0.001$". For instance, we know that group 2 (nonblack cohort) had a greater rate of sustained virologic response than the treatment group, with p-value $< 0.001$.

5. *To what population do the conclusions apply?* The sample was not randomly selected, so although we *can* conclude that the treatment was effective, we do not know if we would see the same sized effect on other samples or in the population of all hepatitis C patients. Because it is known that African Americans and non-African Americans have different responses to the standard treatment, these two groups were treated separately in the analysis.

6. *Have the results been replicated in other articles? Are the results consistent with what other researchers have suggested?* We cannot tell from the abstract.

Note that even though we don't know which statistical procedures were carried out (we could learn this from reading the article itself), we still have a good sense of the reliability of the study. We know that the reported success rates for standard treatment for hepatitis C (40% in the nonblack cohort and 23% in the black cohort) were lower than for the treatment that included boceprevir (67% for nonblack cohort after 24 weeks, 68% after three weeks; 42% for the black cohort after 24 weeks, and 53% after 44 weeks.) We know this difference cannot be plausibly assigned to chance, and we know that, because of random assignment, that boceprevir was the cause of the difference in rates.

## EXAMPLE 8 Driving While Talking on a Cell Phone

Read the abstract that follows, which concerns the hazards of driving while talking on a cell phone (Strayer et al. 2006).

**Objective:** The objective of this research was to determine the relative impairment associated with conversing on a cellular telephone while driving. **Background:** Epidemiological evidence suggests that the relative risk of being in a traffic accident while using a cell phone is similar to the hazard associated with driving with a blood alcohol level at the legal limit. The purpose of this research was to provide a direct comparison of the driving performance of a cell phone driver and a drunk driver in a controlled laboratory setting. **Method:** We used a high-fidelity driving simulator to compare the performance of cell phone drivers with drivers who were intoxicated from ethanol (i.e., blood alcohol concentration at 0.08% weight/volume). **Results:** When drivers were conversing on either a handheld or hands-free cell phone, their braking reactions were delayed and they were involved in more traffic accidents than when they were not conversing on a cell phone. By contrast, when drivers were intoxicated from ethanol, they exhibited a more aggressive driving style, following closer to the vehicle immediately in front of them and applying more force while braking. **Conclusion:** When driving conditions and time on task were controlled for, the impairments associated with using a cell phone while driving can be as profound as those associated with driving while drunk. **Application:** This research may help to provide guidance for regulation addressing driver distraction caused by cell phone conversations.

QUESTION Write a short paragraph describing this research. Use the six questions above for guidance.

SOLUTION The study asks whether talking on a cell phone impairs driving and, if so, whether the impairment is similar to the impairment caused by driving drunk. The researchers conclude that the impairment is as great when talking on a cell phone as it is when driving drunk. To reach this conclusion, the researchers had drivers use a driving simulator. Some drivers talked on their phone during the task, and others were given sufficient alcohol to make them legally drunk. However, these techniques as described in the abstract do not lead to much faith in the conclusion.

- We do not know whether drivers were randomly assigned to treatment groups.

- We do not know how these drivers were chosen.

- We do not know how many drivers were in each group.

- Although we know that all of the drivers in the alcohol group had the same blood alcohol level, we do not know whether all of the drivers in the cell phone group were equally engaged in the task. For example, perhaps some had conversations that required more concentration.

- We also do not know whether the results are due to chance, because no statistical conclusions are provided.

- Presumably, this finding is meant to apply to all cell phone drivers, but we are not given enough information about the subjects to know how they were selected, so we don't know how well we can generalize these results to a larger population.

Despite these shortcomings, the abstract states that the study's findings are consistent with a larger body of literature that finds that cell phone driving is dangerously distracting.

**TRY THIS!** Exercise 10.43

## Buyer Beware

Leo Tolstoy's *Anna Karenina* begins, "Happy families are all alike; every unhappy family is unhappy in its own way." To (very loosely) paraphrase the great Russian novelist, there are few ways that a study can be good, but there are many ways that studies can go wrong. We've given you some tips that should help you recognize a good study, but you should also be aware of some warning signs and features that indicate that a study might not be good.

### Data Dredging

Hypothesis testing is designed to test claims that result from a theory. The theory makes predictions about what we should see in the data; for example, students who write about their anxieties will do better on an exam, so the mean exam score of students who wrote about their anxieties should be greater than the mean score of those who did not. We next collect the data and then do a hypothesis test to determine whether the theory was correct. **Data dredging** is the practice of stating our hypotheses after first looking at the data. Data dredging makes it more likely that we will mistakenly reject the null hypothesis. Even when the null hypothesis is true, our data sometimes show surprising outcomes, just because of chance variation. If we first look at the test statistic to decide what the hypotheses should be, we are rigging the system in favor of the alternative hypothesis.

The situation is analogous to betting on a horse race after the race has begun. You are supposed to place a bet before the race starts, so that everyone is on equal footing. The odds on which horse will win (which determine the payoffs) are meant to estimate probabilities of a horse winning. But if you wait until the horses are running, you have a better chance than everyone else of correctly predicting which horse will win. You have "snooped" at the data to make your decision. Your probability of winning is not the same as what everyone else believed it to be.

Theories should be based on data, but the correct procedure is to use data to formulate a theory and then to collect additional data in an independent study to test that theory. If it is too costly to collect additional data, one common approach is to randomly split the data set into two (or even more) pieces. One piece of the data is then set aside—"locked away"—and the researchers are forbidden to look. The researchers can then examine the first piece as much as they want and use these data to generate hypotheses. After they have generated hypotheses, they can test these hypotheses on the second, "locked away" data set. Another possibility is to use the data for generating hypotheses and then go out and collect *more* data.

### Publication Bias

Most scientific and medical journals prefer to publish "positive" findings. A positive finding is one in which the null hypothesis is rejected (with the result that the researcher concludes that the tested treatment is effective). Some journals prefer this sort of finding, because these are generally the results that advance science. However, suppose a pharmaceutical company produces a new drug that it claims can cure depression, and suppose this drug does not work. If many researchers are interested in studying this drug and they do statistical tests with a 5% significance level, then 5% of them will conclude that the drug is effective, even though we know it is not. If a journal favors positive findings over negative findings, then we will read only about the studies that find the drug works, even though the vast majority of researchers came to the opposite conclusion.

**Publication bias** is one reason why it is important to consider several different studies of the same drug before making decisions. A new, and somewhat controversial, form of statistical analysis called meta-analysis has been developed in recent years, in part to help with problems such as these. A **meta-analysis** considers all studies done to test a particular treatment and tries to reconcile different conclusions, attempting to determine whether other factors, such as publication bias, played a role in the reported outcomes.

Psychologists conducted a meta-analysis to conclude that violent video games are not associated with violent behavior in children, despite the fact that several studies had concluded otherwise (Ferguson and Kilburn 2010). They concluded that papers that found an association between video game violence and real-life violence were more likely to be published, so researchers who found no such associations were less likely to be published. This result is not likely to settle the controversy, but it points to a potential danger that arises when journals publish only positive findings.

## Profit Motive

Much statistical research is now paid for by corporations that hope to establish that their products make life better for people. Researchers are usually required to disclose to a journal whether they are themselves making money off the drug or product that they are researching, but this does not tell us who funded the project in the first place. There is no reason to reject the conclusions of a study simply because it was paid for by a corporation or business or some other organization with a vested interest. You should always evaluate the methods of the study used and decide whether the methods are sound.

However, you should be aware that sometimes the corporation funding the research can influence whether results get published and which results get published. For example, a researcher might be funded by a pharmaceutical company to test a new drug. If he finds the drug doesn't work, the drug company might decide that it doesn't want to publicize this fact. Or perhaps the drug works on only a small subset of people. The company might then publicize that the drug is effective but fail to mention that it is effective on only a small group.

For example, a 2007 study concluded that playing "active" video games (such as the Wii) was healthier than playing "passive" games and found that children burned more calories playing the active games (Neale 2007). The study also suggested that playing real sports was much more healthful than playing any video game, but still the researchers conclude, "Nevertheless, new generation computer games stimulated positive activity behaviors. Given the current prevalence of childhood overweight and obesity, such positive behaviors should be encouraged." A press release notes that the study was funded by Cake, the "marketing arm" of Nintendo UK, which manufactures the Wii. This does not mean that the results of the study are wrong (the children playing the active games most likely did in fact burn more calories than those who played the passive games), but it may account for the positive spin despite the finding that active video games were not nearly as effective as playing sports at promoting weight loss.

## Media

The media—newspapers, magazines, television shows, and radio broadcasts—are also profit-motivated. A good journalist strives to get to the heart of the matter. Nonetheless, scientific and medical research findings are complex, and when condensing complex ideas into easily digested sound bites (and doing so in a way that entertains those who pay for the papers and magazines), the truth sometimes gets distorted.

The media often use catchy headlines, and these headlines do not always capture the true spirit of a study. The most common problem is that headlines often suggest a cause-and-effect relationship, even though the wise statistics student will quickly recognize that such a conclusion is not supported by the data. Statistician Jonathan Mueller keeps a website of such headlines. A couple of examples: "Studies say lots of candy could lead to violence." "Texting lifts crash risk by large margin." These headlines all suggest causality: Eating candy will make you violent. Texting will increase the risk of a car accident. But you now know that such a conclusion can be made only for controlled experiments, and even then we must be cautious. Ideally, you can learn what you need to know by reading the news article. But not always. The information you need to judge whether the conclusions of a study are strong is often missing from news reports.

## Clinical Significance vs. Statistical Significance

An outcome of an experiment or study that is large enough to have a real effect on people's health or lifestyle is said to have **clinical significance**. Sometimes, researchers discover that a treatment is statistically significant (meaning that the outcome is too large to be due to chance) but too small to be meaningful (so it is not clinically significant). Studies with very large sample size have large power and so are capable of detecting even very small differences between treatment groups. This does not mean the differences matter. For example, a drug might truly lower cholesterol levels, but not enough to make a real difference in someone's health. Or playing Wii might burn more calories, but maybe not enough for it to serve as a form of exercise. Treatments that cause meaningful effects are called clinically significant. Sometimes, statistically significant results are not clinically significant.

Imagine a rare disease that only 1 person in 10 million people gets. A controlled experiment finds that a new drug "significantly" reduces your risk of getting this disease; specifically, it cuts the risk in half. Here, *significantly* means that the reduction in risk is statistically significant. But given that the disease is so rare already, is it worth taking medicine to cut your risk from 1 in 10 million to 1 in 20 million? Particularly if the drug is expensive or has side effects, most people would probably decide that this treatment is not clinically significant.

## CASE STUDY REVISITED

To better understand the ideal conditions for popping corn, experimenters designed a randomized, controlled study to observe how well the popcorn popped under different conditions. Of interest here was how the amount of oil affected the outcome. A summary of results was shown in the Case Study at the beginning of the chapter, where we observed that it looked as though using the maximum amount of oil did not work well, because so few popcorn bags were successful by the criteria adopted for success.

Did the amount of oil affect the resulting quality of the popcorn? To test this using the methods of this chapter, we carry out a test of homogeneity, because we have three independent samples (no oil, medium oil, and maximum oil) and one categorical response variable (*Result*: success or failure). We have three independent samples, because bags were randomly assigned to one of these three groups. Each group was assigned 36 bags of popcorn, and each bag had 50 kernels.

The hypotheses follow.

$H_0$: The quality of popcorn and the amount of oil are independent.
$H_a$: The amount of oil affects the quality of the popcorn.

The results (and some of the raw data) are shown in Figure 10.7 in output generated by StatCrunch. The output shows that the expected counts are all greater than 5, so our sample sizes are large enough for the chi-square distribution (with 2 degrees of freedom) to produce a good approximation to the p-value.

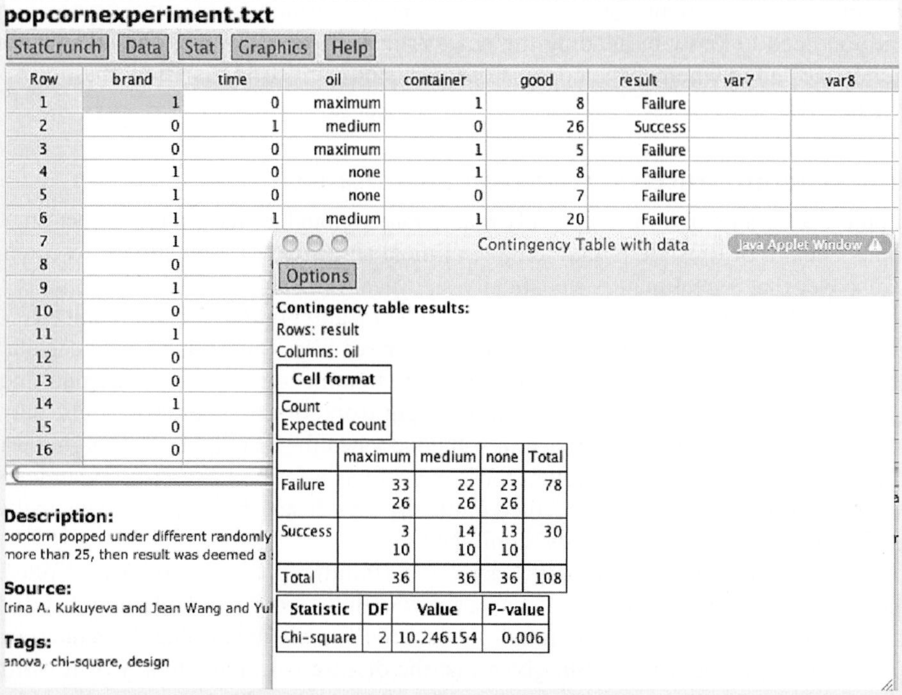

▶ **FIGURE 10.7** StatCrunch output shows the results of the analysis in the foreground and the raw data in the background. The expected values in the table are given below the observed values.

**popcornexperiment.txt**

StatCrunch | Data | Stat | Graphics | Help

| Row | brand | time | oil | container | good | result | var7 | var8 |
|---|---|---|---|---|---|---|---|---|
| 1 | 1 | 0 | maximum | 1 | 8 | Failure | | |
| 2 | 0 | 1 | medium | 0 | 26 | Success | | |
| 3 | 0 | 0 | maximum | 1 | 5 | Failure | | |
| 4 | 1 | 0 | none | 1 | 8 | Failure | | |
| 5 | 1 | 0 | none | 0 | 7 | Failure | | |
| 6 | 1 | 1 | medium | 1 | 20 | Failure | | |
| 7 | 1 | | | | | | | |
| 8 | 0 | | | | | | | |
| 9 | 1 | | | | | | | |
| 10 | 0 | | | | | | | |
| 11 | 1 | | | | | | | |
| 12 | 0 | | | | | | | |
| 13 | 0 | | | | | | | |
| 14 | 1 | | | | | | | |
| 15 | 0 | | | | | | | |
| 16 | 0 | | | | | | | |

Contingency Table with data   Java Applet Window ⚠

Options

**Contingency table results:**
Rows: result
Columns: oil

| Cell format |
|---|
| Count |
| Expected count |

| | maximum | medium | none | Total |
|---|---|---|---|---|
| Failure | 33 26 | 22 26 | 23 26 | 78 |
| Success | 3 10 | 14 10 | 13 10 | 30 |
| Total | 36 | 36 | 36 | 108 |

| Statistic | DF | Value | P-value |
|---|---|---|---|
| Chi-square | 2 | 10.246154 | 0.006 |

**Description:**
popcorn popped under different randomly
more than 25, then result was deemed a

**Source:**
Irina A. Kukuyeva and Jean Wang and Yul

**Tags:**
anova, chi-square, design

From the output, we see that the test statistic has a value of 10.25, with a p-value of 0.006. This is quite small. Certainly it is less than 0.05, so at the 5% significance level we reject the null hypothesis. We conclude that quality is affected by the amount of oil used. (At least, it is if you believe that quality is measured by the number of kernels popped after 75 seconds.)

# EXPLORING STATISTICS
## CLASS ACTIVITY

## Skittles

| GOALS | MATERIALS |
|---|---|
| Apply a chi-square test to check whether two bags of Skittles candies contain the same proportions of colors. | • One small bag of Skittles for each student<br>• Computer or TI-83/84 |

**ACTIVITY**

Open a bag of Skittles and count how many of each color are in your bag. Fill in the numbers in the table below. Then find a partner and fill in the colors from his or her bag. (If someone does not have a partner, form a group of three and use three rows.)

| | Purple | Red | Orange | Yellow | Green | Total |
|---|---|---|---|---|---|---|
| Yours | | | | | | |
| Partner's | | | | | | |

**BEFORE THE ACTIVITY**

1. Do you think that you and your partner(s) will get exactly the same number in each category?

2. Do you think you and your partner(s) will have significantly different distributions of colors? Why or why not?

**AFTER THE ACTIVITY**

1. Perform a hypothesis test to test whether the two bags have a significantly different distribution of colors, using a significance level of 0.05.

2. Throw away, save, or eat the Skittles.

## CHAPTER REVIEW

## Key Terms

## Learning Objectives

After reading this chapter and doing the assigned homework problems, you should

- Distinguish between tests of homogeneity and tests of independence.

- Understand when it is appropriate to use a chi-square statistic to test whether two categorical variables are associated; know how to perform this test and interpret the results.

- Understand how random assignment is used to allow cause-and-effect inference, and understand how random sampling is used to allow generalization to a larger population.

- Be prepared to apply knowledge about collecting and analyzing data to critically evaluate abstracts in the science literature.

## Summary

We presented two types of tests for analyzing categorical variables. Although the test of homogeneity is conceptually different from the test of independence, they are exactly the same in terms of the calculations required. Both of these tests attempt to determine whether two categorical variables are associated. The only difference is in the way the data for the study were collected. When researchers collect two or more independent samples and measure one categorical response variable, they are performing a test of homogeneity. When instead they collect one sample and measure two categorical response variables, they are performing a test of independence.

Both tests rely on the chi-square statistic. For each cell of a two-way summary table, we compare the observed count with the count we would expect if the null hypothesis were true. If the chi-square statistic is big, it means that these two counts don't agree, and it discredits the null hypothesis.

An approximate p-value is calculated by finding the area to the right of the observed chi-square statistic using a chi-square distribution. To do this, you need to know the degrees of freedom for the chi-square distribution.

Many research questions can be divided into two categories: those that ask questions about causality and those that ask about associations between variables. Questions about causality can be answered only with controlled experiments, whereas observational studies can answer questions about associations.

Interpreting conclusions in scientific studies is complex, because many things can go wrong in a study. Remember that extraordinary results must be supported by extraordinary evidence, and that you should trust studies that have been replicated (repeated, resulting in the same conclusion) over studies that have not.

## Formulas

### Expected Counts

**Formula 10.1:**   Expected count for a cell $= \dfrac{\text{(row total)} \times \text{(column total)}}{\text{grand total}}$

### Chi-Square Statistic

**Formula 10.2:**   $X^2 = \displaystyle\sum_{\text{cells}} \dfrac{(O - E)^2}{E}$

### Test of Homogeneity and Independence

#### Hypotheses

$H_0$: The variables are independent.
$H_a$: The variables are associated.

#### Conditions (Homogeneity)

1. *Random Samples.* Two or more samples, all sampled randomly.

2. *Independent Samples and Observations.* Samples are independent of each other. The observations within each sample are independent.

3. *Large Samples.* At least 5 expected counts in each cell of the summary table.

#### Conditions (Independence)

1. *Random Sample.* One sample, selected randomly.

2. *Independent Observations.* Observations are independent of each other.

3. *Large Sample.* There are at least 5 expected counts in each cell of the summary table.

#### Sampling Distribution

If conditions hold, the sampling distribution follows a chi-square distribution with degrees of freedom = (number of rows − 1) × (number of columns − 1).

# Sources

Colman, R. J., et al. 2009. Caloric restriction delays disease onset and mortality in Rhesus monkeys. *Science* 325, 201.

Husemann, L. R., J. Moise-Titus, C. Podolski, and L. D. Eron. 2003. Longitudinal relations between children's exposure to TV violence and their aggressive and violent behavior in young adulthood: 1977–1992. *Developmental Psychology* 39(2), 201–221.

I. A. Kukuyeva, J. Wang, and Y. Yaglovskaya. 2008. Popcorn popping yield: An investigation, JSM

Poordad, F., McCone, J., Bacon, B., Bruno, S., Manns, M., et al., 2011. Boceprevir for Untreated Chronis HCV Genotype 1 Infection. *New England Journal of Medicine*, v.364 (13) 1195–1206, http://www .nejm.org/doi/pdf/10.1056/NEJMoa1010494 (accessed April 26, 2011)

Strayer, D. L., F. A. Drews, and D. J. Crouch. 2006. A comparison of the cell phone driver and the drunk driver. *Journal of the Human Factors and Ergonomics Society* 48(2), 381–391.

# SECTION EXERCISES

## SECTION 10.1

### 10.1 Tests

a. In Chapter 8, you learned some tests of proportions. Are tests of proportions used for categorical or numerical data?

b. In this chapter, you are learning to use chi-square tests. Do these tests apply to categorical or numerical data?

**10.2** In Chapter 9, you learned some tests of means. Are tests of means used for numerical or categorical data?

**10.3 Crime and Gender** A statistics student conducted a study in Ventura County, California, that looked at criminals on probation who were under 15 years of age to see whether there was an association between the type of crime (violent or nonviolent) and gender. Violent crimes involve physical contact such as hitting or fighting; nonviolent crimes include vandalism, robbery, and verbal assault. The raw data are shown in Exercise 1.45 on page 31.

Create a two-way table to summarize these data. Are the two variables numerical or categorical? If you are doing this by hand, create a table with two rows and two columns. Label the columns Boy and Girl (across the top). Label the rows Violent and Nonviolent. Begin with a big table, making a tally mark in one of the four cells for each observation, and then summarize the tally marks as counts.

**10.4 Red Cars and Stop Signs** The table shows the raw data for the results of a student survey of 22 cars and whether they stopped completely at a stop sign or not. In the Color column, "Red" means the car was red and "No" means the car was not red. In the Stop column, "Stop" means the car stopped, and "No" means the car did not stop fully.

Create a two-by-two table to summarize these data. Use Red and No for the columns (across the top) and Stop and No for the rows.

(We gave you an orientation of the table so that your answers would be easy to compare.) Are the two variables categorical or numerical?

| Color | Stop | Color | Stop |
|-------|------|-------|------|
| Red | Stop | No | No |
| Red | Stop | Red | Stop |
| Red | No | Red | No |
| Red | No | Red | No |
| Red | No | No | Stop |
| No | Stop | No | Stop |
| No | Stop | No | Stop |
| No | Stop | No | Stop |
| No | No | Red | Stop |
| Red | Stop | Red | No |
| Red | No | Red | No |

**10.5** The table summarizes the outcomes of a study that students carried out to determine whether humanities students had a higher mean GPA than science students. Identify both of the variables, and state whether they are numerical or categorical. If numerical, state whether they are continuous or discrete.

|  | **Mean GPA** |
|-----------|----------|
| Science | 3.4 |
| Humanities | 3.5 |

**10.6 Finger Length** There is a theory that relative finger length depends on testosterone level. The table shows a summary of the outcomes of an observational study that one of the authors carried out to determine whether men or women were more likely to have a ring finger that appeared longer than their index finger. Identify both of the variables, and state whether they are numerical or categorical. If numerical, state whether they are continuous or discrete.

|  | Men | Women |
|---|---|---|
| Ring Finger Longer | 23 | 13 |
| Ring Finger Not Longer | 4 | 14 |

TRY **10.7 Effects of Television Violence on Men (Example 1)**
A study done by Husemann et al. and published in *Developmental Psychology* in 2003 compared men who viewed high levels of television violence as children with those who did not in order to study the differences with regard to physical abuse of their partners as adults. The men categorized as physically abusive had hit, grabbed, or shoved their partners.

|  | High TV Violence | Low TV Violence |
|---|---|---|
| Yes, Physical Abuse | 13 | 27 |
| No Physical Abuse | 18 | 95 |

a. Find the row, column, and grand totals, and prepare a table showing these values as well as the counts given.
b. Find the percentage of men overall that were abusive.
c. Find the expected number of men exposed to high levels of television violence who should say yes, if the variables are independent. Multiply the proportion overall that were abusive times the number of men exposed to high levels of television violence. Do not round off to a whole number. Round to two decimal digits.
d. Find the other expected values by knowing that the expected values must add to the row and column totals. Report them in a table with the same orientation as the one given for the data.

**10.8 Effects of Television Violence on Women** Refer to Exercise 10.7. This data table compares women who viewed high levels of television violence as children with those who did not in order to study the differences with regard to physical abuse of their partners as adults. The women categorized as physically abusive had hit, grabbed, or shoved their partners.

|  | High TV Violence | Low TV Violence |
|---|---|---|
| Yes, Physical Abuse | 12 | 30 |
| No Physical Abuse | 23 | 111 |

a. Find the row, column, and grand totals and report the table showing these values.
b. Find the percentage of all women who were abusive (who answered yes).

c. Find the expected number of women exposed to high levels of television violence who should say yes, if the variables are independent. Do not round off to a whole number. Round to two decimal digits.
d. Find the other expected values. Report them in a table with the same orientation as the one for the data.

TRY **10.9 Effects of Television Violence on Men (Example 2)**
Refer to Exercise 10.7. The data table compares men who viewed high levels of television violence as children with those who did not, in order to study the differences with regard to physical abuse of their partners as adults. Report the observed value of the chi-square statistic.

|  | High TV Violence | Low TV Violence |
|---|---|---|
| **Yes, Physical Abuse** | 13 | 27 |
| **No Physical Abuse** | 18 | 95 |

**10.10 Effects of Television Violence on Women** Refer to Exercise 10.8. The data table compares women who viewed high levels of television violence as children with those who did not, in order to study the differences with regard to physical abuse of their partners as adults. Calculate the observed value of the chi-square statistic.

|  | High TV Violence | Low TV Violence |
|---|---|---|
| **Yes, Physical Abuse** | 12 | 30 |
| **No Physical Abuse** | 23 | 111 |

# SECTION 10.2

TRY **10.11 Political Party and Right Direction? (Example 3)**
A December 2009 poll conducted by Associated Press and Roper asked a random sample of Americans whether they thought the country was heading in the right direction. Each respondent answered "Right Direction" or "Wrong Direction" and was classified as Republican, Democrat, or Independent. If we were interested in testing whether party affiliation was associated with the answer to the question, would this be a test of homogeneity or of independence? Explain.

**10.12 Antibiotic or Placebo** A large number of surgery patients get infections after surgery, which can sometimes be quite serious. Researchers randomly assigned some surgery patients to receive a simple antibiotic ointment after surgery, others to receive a placebo, and others to receive just cleansing with soap. If we wanted to test the association between treatment and whether or not patients get a, infection after surgery, would this be a test of homogeneity or of independence? Explain. (Source: Hospitals could stop infections by tackling bacteria patients bring in, studies find. *New York Times*, January 6, 2010.)

TRY **10.13 Jet Lag Drug (Example 4)** A recent study was conducted to determine whether the drug Nuvigil was effective at helping east-bound jet passengers adjust to jet lag. Subjects were randomly assigned either to one of three different doses of Nuvigil (low, medium, high) or to a placebo, flown to France in a plane in which they could not drink alcohol or coffee or take sleeping pills, and then examined in a lab where their state of wakefulness was measured and classified into categories (low, normal, alert). If we test whether treatments for jet lag are associated with wakefulness, are we doing a test of independence or of homogeneity? Explain. (Source: A drug's second act: Battling jet lag. *New York Times*, January 6, 2010.)

**10.14 Most Important Problems** A November 2009 AP–Stanford University Poll took a random sample of 1005 Americans and asked what they considered to be the most important problem facing the country today. There were 24 different categories of responses. The economy was the most popular choice. Respondents were classified by the region of the country they lived in: Northeast, Midwest, South, or West. If we wanted to test whether there was an association between the response to the question and the region of the country, would this be a test of homogeneity or of independence? Explain.

**10.15 Country of Origin** The table shows the country of origin and the percentage of foreign-born people in the United States in 2000 and 2007 for the four countries of origin with the highest percentages (*2009 World Almanac and Book of Facts*). Give two reasons why you should not do a chi-square test with these data.

|      | Mexico | China | Philippines | India |
|------|--------|-------|-------------|-------|
| **2000** | 29.5 | 4.9 | 4.4 | 3.3 |
| **2007** | 30.8 | 5.1 | 4.5 | 3.9 |

**10.16 Unemployment Rates** U.S. Unemployment rates for all residents 20 years old and older are given as a percentage. The table does not include people who are not actively seeking employment (*2009 World Almanac and Book of Facts*). Give two reasons why a chi-square test is not appropriate for this set of data.

|      | Female | Male |
|------|--------|------|
| **1990** | 4.9 | 5.0 |
| **2000** | 3.6 | 3.3 |
| **2009** | 9.5 | 10.5 |

TRY
g **10.17 Obesity and Marital Status (Example 5)** A study reported in the medical journal *Obesity* in 2009 analyzed data from the National Longitudinal Study of Adolescent Health. Obesity was defined as having a body mass index (BMI) of 30 or more. The research subjects were followed from adolescence to adulthood, and all the people in the sample were categorized in terms of whether they were obese and whether they were dating, cohabiting, or married. Test the hypothesis that relationship status and obesity are associated, using a significance level of 0.05. Can we conclude from these data that living with someone is making some people obese and that marrying is making people even more obese? Can we conclude that obesity affects relationship status? Explain. *See page 479 for guidance.*

|           | Dating | Cohabiting | Married |
|-----------|--------|------------|---------|
| **Obese** | 81 | 103 | 147 |
| **Not Obese** | 359 | 326 | 277 |

(Source: N. S. The and P. Gordon-Larsen. 2009. Entry into romantic partnership is associated with obesity. *Obesity* 17(7), 1441–1447.)

**10.18 Opinions on Household Responsibilities** In the 2008 General Social Survey, respondents were asked whether they agreed or disagreed with the following statement: "Husbands should work and wives should take care of the home." Each respondent was classified by gender. The table shows the responses.

|                            | Men | Women |
|----------------------------|-----|-------|
| **Agree**                  | 183 | 192 |
| **Neither Agree nor Disagree** | 143 | 140 |
| **Disagree**               | 288 | 412 |

```
          Men      Women    Total
Agree     183      192      375
          169.55   205.45

Neither   143      140      283
          127.95   155.05

Disagree  288      412      700
          316.49   383.51

Total     614      744      1358
Chi-Sq = 9.859, DF = 2, P-Value = 0.007
```

a. Write a sentence containing the percentages of men and women who agreed with this statement, and compare them.

b. Test the hypothesis that gender and opinion are associated, using the 0.05 level of significance. Assume the sample is random. Refer to the Minitab output, which shows expected values below the observed counts.

**10.19 Effects of Television Violence on Men** The data table compares men who viewed television violence with those who did not, in order to study the differences in physical abuse of the spouse. For the men in the table, test whether television violence and abusiveness are associated, using a significance level of 0.05. Refer to the Minitab output.

|                     | High TV Violence | Low TV Violence |
|---------------------|------------------|-----------------|
| **Yes, Physical Abuse** | 13 | 27 |
| **No Physical Abuse**   | 18 | 95 |

(Source: L. R. Husemann et al. 2003. Longitudinal relations between children's exposure to TV violence and their aggressive and violent behavior in young adulthood: 1977–1992. *Developmental Psychology* 39(2), 201–221.)

```
Chi-Square Test: High TV Violence, Low TV Violence
Expected counts are printed below observed counts
Chi-Square contributions are printed below expected counts

          High TV    Low TV
          Violence   Violence   Total
Yes, Ab      13         27       40
             8.10       31.90
             2.957      0.751

No           18         95       113
             22.90      90.10
             1.047      0.266

Total        31         122      153

Chi-Sq = 5.021, DF = 1, P-Value = 0.025
```

**10.20 Effects of Television Violence on Women** The data table compares women who viewed television violence with those who did not, in order to study the differences in physical abuse of the spouse (Husemann et al. 2003). Test whether television violence and abusiveness are associated, using a significance level of 0.05. Refer to the Minitab output.

|  | High TV Violence | Low TV Violence |
|---|---|---|
| **Yes, Physical Abuse** | 12 | 30 |
| **No Physical Abuse** | 23 | 111 |

```
Chi-Square Test: High TV Violence, Low TV Violence
Expected counts are printed below observed counts
Chi-Square contributions are printed below expected counts

          High TV    Low TV
          Violence   Violence   Total
Yes, ab      12         30        42
            8.35       33.65
            1.593       0.395

No           23        111       134
            26.65     107.35
            0.499       0.124
Total        35        141       176

Chi-Sq = 2.612, DF = 1, P-Value = 0.106
```

**10.21 Gender and Happiness of Marriage** The table shows the results of a cross tabulation of gender and whether a person is happy in his or her marriage, according to data obtained from the 2008 General Social Survey.

**Happiness of Marriage × Respondent's Gender Cross Tabulation**

| | | | Respondent's Gender | | |
|---|---|---|---|---|---|
| | | | **Male** | **Female** | **Total** |
| **Happiness of Marriage** | Very happy | Count | 297 | 299 | 596 |
| | | Expected Count | 289.1 | 306.9 | 596.0 |
| | Pretty happy | Count | 160 | 183 | 343 |
| | | Expected Count | 166.4 | 176.6 | 343.0 |
| | Not too happy | Count | 13 | 17 | 30 |
| | | Expected Count | 14.6 | 15.4 | 30.0 |
| Total | | Count | 470 | 499 | 969 |

a. If we carry out a test to determine whether these variables are associated, is this a test of independence or homogeneity?

b. Do a chi-square test with a significance level of 0.05 to determine whether gender and happiness of marriage are associated.

c. Does this suggest that women and men tend to have different levels of happiness or that their rates of happiness in marriage are about equal?

**10.22 Is Smiling Independent of Age?** Randomly chosen people were observed for about 10 seconds in several public places, such as malls and restaurants, to see whether they smiled during that time. The table shows the results for different age groups.

| | Age Group | | | | |
|---|---|---|---|---|---|
| | **0–10** | **11–20** | **21–40** | **41–60** | **61+** |
| **Smile** | 1131 | 1748 | 1608 | 937 | 522 |
| **No Smile** | 1187 | 2020 | 3038 | 2124 | 1509 |

(Source: M. S. Chapell. 1997. Frequency of public smiling over the life span. *Perceptual and Motor Skills* 85, 1326.)

a. Find the percentage of each age group that were observed smiling, and compare these percentages.

b. Treat this as a single random sample of people, and test whether smiling and age group are associated, using a significance level of 0.05. Comment on the results.

TRY **10.23 Preschool Attendance and High School Graduation Rates (Example 6)** The Perry preschool project was created in the early 1960s by David Weikart in Ypsilanti, Michigan. One hundred and twenty three African American children were randomly assigned to one of two groups: One group enrolled in the Perry preschool, and one did not enroll. Follow-up studies were done for decades to answer the research question of whether attendance at preschool had an effect on high school graduation. The table shows whether the students graduated from regular high school or not. Students who received GEDs were counted as not graduating from high school. This table includes 121 of the original 123. This is a test of homogeneity, because the students were randomized into two distinct samples.

| | Preschool | No Preschool |
|---|---|---|
| **HS Grad** | 37 | 29 |
| **No HS Grad** | 20 | 35 |

a. For those who attended preschool, the high school graduation rate was 37/57, or 64.9%. Find the high school graduation rate for those not attending preschool, and compare the two. Comment on what the rates show for *these* subjects.

b. Are attendance at preschool and high school graduation associated? Use a 0.05 level of significance.

(Source: L. J. Schweinhart et al. 2005. Lifetime effects: The High/Scope Perry Preschool Study through age 40. *Monographs of the High/Scope Educational Research Foundation,* 14. Ypsilanti, Michigan: High/Scope Press.)

**10.24 Preschool Attendance and High School Graduation Rates for Females** The Perry preschool project data presented in Exercise 10.23 (Schweinhart et al. 2005) can be divided to see whether the preschool attendance effect is different for males and females. The table shows a summary of the data for females, and the figure shows Minitab output that you may use.

```
Chi-Square Test: Preschool, No Preschool: Girls
Expected counts are printed below observed counts

                      No
         Preschool  Preschool   Total

 Grad       21          8         29
           14.50      14.50

 No Grad     4         17         21
           10.50      10.50

 Total      25         25         50

Chi-Sq = 13.875, DF = 1, P-Value = 0.000
```

Minitab Output for the Girls Only

|             | Preschool | No Preschool |
|-------------|-----------|--------------|
| **HS Grad**    | 21        | 8            |
| **HS Grad No** | 4         | 17           |

a. Find the graduation rate for those females who went to preschool, and compare it with the graduation rate for those who did not go to preschool.

b. Test the hypothesis that preschool and graduation rate are associated, using a significance level of 0.05.

**10.25 Preschool Attendance and High School Graduation Rates for Males** The Perry preschool project data presented in Exercise 10.23 can be divided to see whether there are different effects for males and females. The table shows a summary of the data for males (Schweinhart et al. 2005).

|             | Preschool | No Preschool |
|-------------|-----------|--------------|
| **HS Grad**    | 16        | 21           |
| **HS Grad No** | 16        | 18           |

a. Find the graduation rate for males who went to preschool, and compare it with the graduation rate for those who did not go to preschool.

b. Test the hypothesis that preschool and graduation are associated, using a significance level of 0.05.

c. Exercise 10.24 showed an association between preschool and graduation for just the females in this study. Write a sentence or two giving your advice to parents with preschool-eligible children about whether attending preschool is good for their children's future academic success, based on this dataset.

**10.26 Drug for Atrial Fibrillation** Atrial fibrillation is a heart problem in which the atrium (part of the heart) beats very rapidly.

Researchers compared the drug valsartan with a placebo. People in the study had had at least one bout of atrial fibrillation in the previous six months. They were randomly assigned to receive valsartan or placebo, and the counts are shown below. A "bad" outcome is when atrial fibrillation recurred again within one year, and a "good" outcome is when it did not recur during the one-year period.

|          | Valsartan | Placebo |
|----------|-----------|---------|
| **Bad**  | 371       | 375     |
| **Good** | 351       | 345     |

a. What percentage of those assigned valsartan had a bad outcome? What percentage of those assigned the placebo had a bad outcome? Compare these values.

b. Perform a hypothesis test to determine whether treatment is associated with the outcome. Use a significance level of 0.05.

c. Would you recommend valsartan for a friend with atrial fibrillation?

(Source: The GISSI-AF Investigators. 2009. Valsartan for prevention of recurrent atrial fibrillation. *New England Journal of Medicine* 360, 1606–1617.)

**10.27 Gastric Surgery and Diabetes** A study published in the *Journal of the American Medical Association* in 2008 reported on the effect of lap band surgery on diabetes. In this study, 55 obese patients with type 2 diabetes were randomly assigned to have lap band surgery or not. After two years, researchers recorded whether or not the subjects still had diabetes.

|                          | Lap Band Surgery | No Surgery |
|--------------------------|------------------|------------|
| **Free from Diabetes**     | 22               | 4          |
| **Not Free from Diabetes** | 7                | 22         |

a. What percentage of those who had the surgery were diabetes-free? What percentage of those who did not have the surgery were diabetes-free? Compare the two sample percentages.

b. Note the observed value of 4. Does this prevent us from using a chi-square test? Explain.

c. Test the hypothesis that lap bands and diabetes are associated, using a significance level of 0.05.

(Source: J. B. Dixon et al. 2008. *Adjustable gastric banding and conventional therapy for type 2 diabetes, Journal of the American Diabetes Association* 299(3), 316–323.)

**10.28 Progressive Rewards** A study was done on the effect of rewards on smokers' ability to refrain from smoking. All 60 smokers were adults who were *not* trying to quit smoking. Whether they were smoking or not was determined by a carbon monoxide meter. The smokers were studied for five days and tested three times each day. Each smoker was randomly assigned to one of three groups. The first group had "progressive rewards," in which the amount of money received per test gradually increased as they continued to refrain from smoking. The second group had "fixed rewards," in which they earned a fixed amount of money each time the test showed that they had refrained from smoking. The total dollar amounts for the first two groups were the same, assuming no smoking during the experiment. The control group did not receive any monetary reward.

The table shows whether the person smoked at all during the five days of the experiment.

|  | Progressive | Fixed | Control |
|---|---|---|---|
| **No Smoking** | 10 | 6 | 1 |
| **Some Smoking** | 10 | 14 | 19 |

a. Compare the percentage of success (no smoking) for each group.

b. Note the observed value of 1, which is less than 5. Does this mean we cannot use chi-square on this data set? Explain.

c. Determine whether the treatment (group) was associated with the outcome. Use the 0.05 level of significance.

d. Can we conclude that the type of encouragement changed smoking behaviors?

(Source: L. Shaffer and M. R. Merrens. 2001. *Research Stories for Introductory Psychology*, Allyn and Bacon, Boston. Original Source: J. M. Roll et al. 1996. An experimental comparison of three different schedules of reinforcement of drug abstinence using cigarette smoking as an exemplar. *Journal of Applied Behavioral Analysis* 29, 495–505.)

**10.29 Confederates and Compliance** A study was done to see whether participants would ignore a sign that said, "Elevator may stick between floors. Use the stairs." The people who used the stairs were said to be compliant, and those who used the elevator were noncompliant. The study was done in a university dormitory on the ground floor of a three-story building. There were three different situations, two of which involved confederates. A confederate (Conf) is a person who is secretly working with the experimenter. In the first situation, there was no other person using the stairs or elevator—that is, no confederate. In the second, there was a compliant confederate (one who used the stairs). In the third, there was a noncompliant confederate (one who used the elevator). A summary of the data is given in the table and TI-83/84 output is given.

|  | No Conf | Compliant Conf | Noncompliant Conf |
|---|---|---|---|
| **Participant Used Stairs** | 6 | 16 | 5 |
| **Participant Used Elevator** | 12 | 2 | 13 |

a. Find the percentage of participants who used the stairs in all three situations. What do these sample percentages say about the association between compliance and the existence of confederates?

b. From the figure, is the p-value 2.69? Explain. Report the actual p-value.

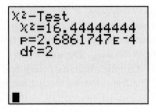

c. Determine whether there is an association between the three situations and whether the participant used the stairs (was compliant) or not. Use a significance level of 0.05.

(Source: L. Shaffer and M. R. Merrens. 2001. *Research Stories for Introductory Psychology*. Allyn and Bacon, Boston. Original Source: M. S. Wogalter, et al. 1987. Effectiveness of warnings. *Human Factors* 29, 599–612.)

**＊10.30 Effects of Intensive Glucose Control in the ICU**
Adults admitted to a hospital intensive care unit (ICU) who were expected to be there three or more days were randomly assigned to undergo either intensive glucose (blood sugar) control or conventional glucose control. Intensive glucose control means an attempt was made to make sure the patients' glucose levels were within a narrow range of values (81–108 milligrams per deciliter). Conventional control means an attempt was made to keep the glucose level below 180 mg per dl. The subjects were observed for 90 days to see whether they died from any cause.

|  | Intensive Glucose Control | Conventional Glucose Control |
|---|---|---|
| **Died** | 829 | 751 |
| **Did Not Die** | 2181 | 2261 |

a. What percentage of those who had intensive glucose control died? And what percentage of those who had conventional glucose control died? Compare these percentages, and comment on what the comparison shows.

b. Do a hypothesis test to see whether the treatment is associated with whether the patient died. Use a significance level of 0.05.

c. Judging on the basis of these data, do you think that the glucose control method affects survival?

(Source: The NICE-SUGAR Study Investigators. 2009. *Intensive versus Conventional Glucose Control in Critically Ill Patients. New England Journal of Medicine* 360(13), 1283–1297.)

**10.31 Light at Night and Tumors** A study was done on female mice to see whether the amount of light affects the risk of developing tumors. Fifty mice were randomly assigned to a regimen of 12 hours of light and 12 hours of dark (LD). Fifty similar mice were assigned to 24 hours of light (LL). The study began when the mice were 2 months old. The mice were observed for about two years. Four of the LD mice developed tumors, and 15 of the LL mice developed tumors.

a. What percentage of the LD mice in the sample developed tumors? And what percentage of the LL mice developed tumors? Compare these percentages and comment.

b. Create a two-way table showing the observed values. Label the columns (across the top) with LD and LL.

c. Test the hypothesis that the amount of light is associated with tumors (at the 0.05 level of significance).

d. Some researchers are now starting to investigate whether this phenomenon occurs in humans. They are concerned about shift workers—workers (such as nurses) who may have night shifts. Why would this be a concern?

(Source: D. Baturin et al. 2001. The effect of light regimen and melatonin on the development of spontaneous mammary tumors. *Neuroendocrinology Letters* 22, 441–447.)

**10.32 Antidepressants for Bipolar Disorder** Bipolar disorder is one in which patients cycle between mania and depression. In this study, done by G. S. Sachs et al. in 2007, the participants were randomly assigned to receive a mood stabilizer plus antidepressant or a mood stabilizer plus placebo for 26 weeks. A durable recovery means that the patients had a reasonably happy mood without being manic.

|  | **Antidepressant** | **Placebo** |
|---|---|---|
| **Durable Recovery** | 42 | 51 |
| **Not Durable Recovery** | 137 | 136 |

a. What percentage of the patients on the antidepressant had a durable recovery? What percentage of the patients on the placebo had a durable recovery? Compare these percentages, and comment.

b. Test the hypothesis that the treatment is associated with the outcome, using a significance level of 0.05.

c. If you had a friend with bipolar disorder, would you suggest that she or he ask about taking an antidepressant in addition to the mood stabilizer already prescribed? Explain why or why not.

(Source: G. S. Sachs et al. 2007. Effectiveness of adjunctive antidepressant treatment for bipolar depression. *New England Journal of Medicine* 356(17), 1711–1722.)

## SECTION 10.3

TRY **10.33 Effect of Confederates on Compliance (Example 7)** A study was done to see whether participants would ignore a sign that read, "Elevator may stick between floors. Use the stairs." The people who used the stairs were classified as compliant, those who used the elevator as noncompliant. The study was done in a university dorm on the ground floor of a building that had three floors. There were three different situations, two of which involved a person who was secretly working with the experimenter. (This person is called a confederate.) In the first situation, there was no other person using the stairs or elevator—that is, no confederate. In the second, there was a compliant confederate (one who used the stairs). In the third, there was a noncompliant confederate (one who used the elevator). Suppose that the participants (people who arrived to use the elevator at the time the experiment was going on) were randomly assigned to the three groups. There were significant differences between groups.

a. Can we generalize widely to a large group? Why or why not?

b. Can we infer causality? Why or why not?

(Source: Lary Shaffer and Matthew R. Merrens. 2001. *Research stories for introductory psychology.* Allyn and Bacon. Original Source: M. S. Wogalter, et al. 1987. Effectiveness of warnings. *Human Factors* 29, 599–612.)

**10.34 Nicotine Patch** Suppose that a new nicotine patch to help people quit smoking was developed and tested. Smokers voluntarily entered the study and were randomly assigned either the nicotine patch or a placebo patch. Suppose that a larger percentage of those using the nicotine patch were able to stop smoking.

a. Can we generalize widely to a large group? Why or why not?

b. Can we infer causality? Why or why not?

**10.35 Hospital Rooms** When patients are admitted to hospitals, they are sometimes assigned to a single room with one bed and

sometimes assigned to a double room, with a roommate. (Some insurance companies will pay only for the less-expensive, double rooms.) A researcher was interested in the effect of the type of room on the length of stay in the hospital. Assume that we are not dealing with health issues that require single rooms.

Suppose that upon admission to the hospital, the names of patients who would have been assigned a double room were put onto a list and a systematic random sample was taken; every tenth patient who would have been assigned to a double room was part of the experiment. For each participant, a coin was flipped: If it landed heads up, they got a double room, and if it landed tails up, they got a single room. Then the experimenters observed how many days the patients stayed in the hospital and compared the two groups. The experiment ran for two months. Suppose those who stayed in single rooms stayed (on average) one less day, and suppose the difference was significant.

a. Can you generalize to others from this experiment? If so, to whom can you generalize and why can you do it?

b. Can you infer causality from this study? Why or why not?

**10.36 College Tours** A random sample of 50 college first-year students (out of a total of 1000 first-years) was obtained from college records using random sampling. Half of those students had a campus tour with a sophomore student, and half had a tour with an instructor. The tour guide was determined randomly by coin flip for each student. Suppose that those with the student guide rated their experience higher than those with the instructor guide.

a. Can you generalize to other first-year students at this college? Explain.

b. Can you infer causality from this study? Explain.

**10.37 Low-Dose Flu Vaccine** In the fall of 2004, the United States experienced a shortage of flu vaccine when vaccine used in injections from one of the pharmaceutical companies was found to be contaminated. A study was done to see whether a smaller dose of the vaccine (produced by other companies) could be used successfully; if so, a small amount of vaccine could be divided into more flu shots. The usual amount of vaccine was injected into the muscle of half of the patients (intramuscular injection). The other half of the patients had only a small amount of vaccine injected under the skin (intradermal injection), not into the muscle. Assume that patients were randomly assigned to groups. The response was measured by looking at the production of antibodies. More antibodies generally result in less risk of getting the flu. The conclusion of the study is given. Tell whether the smaller dose should be used, and explain why or why not.

> *"Conclusions:* As compared with an intramuscular injection of full-dose influenza vaccine, an intradermal injection of a reduced dose resulted in similarly vigorous antibody responses among persons 18 to 60 years of age but not among those over the age of 60 years."

(Source: Robert B. Belshe et al. 2004. Serum antibody responses after intradermal vaccination against influenza. *New England Journal of Medicine* 351, 2286–2294.)

**10.38 CABG or Stent?** The following abstract is from a study to determine which is more effective for avoiding adverse health outcomes: a coronary artery bypass graft (CABG), in which arteries in one part of the body are used to replace clogged arteries, or stenting (inserting a tube into an artery to keep the artery open). The term *revascularization* means new blocking of the arteries.

a. Read the abstract, and indicate which procedure had better results, CABG or stents.

b. Does the stated conclusion make a claim for a cause-and-effect link between the treatment and response? If so, is this claim justified? If no claim of causality was made, why was it not made?

*"Methods:* We identified patients with multivessel disease who received drug-eluting stents or underwent CABG in New York State between October 1, 2003, and December 31, 2004, and we compared adverse outcomes (death, death or myocardial infarction, or repeat revascularization) through December 31, 2005, after adjustment for differences in baseline risk factors among the patients.

*Results:* In comparison with treatment with a drug-eluting stent, CABG was associated with lower 18-month rates of death and of death or myocardial infarction both for patients with three-vessel disease and for patients with two-vessel disease. . . . Patients undergoing CABG also had lower rates of repeat revascularization.

*Conclusions:* For patients with multivessel disease, CABG continues to be associated with lower mortality rates than does treatment with drug-eluting stents and is also associated with lower rates of death or myocardial infarction and repeat revascularization."

(Source: E. H. Hannan et al. 2008. Drug-eluting stents vs. coronary-artery bypass grafting in multivessel coronary disease. *New England Journal of Medicine* 358, 331–341.)

### 10.39 Endovascular or Open Repair of Aortic Aneurysms?

An aortic aneurysm is the ballooning out of the aorta. If the aorta ruptures, death usually follows. There are two methods of surgery to repair the blood vessel. Conventional repair uses a larger incision in the area. Endovascular repair uses a very small incision, and the operation is done within the small incision with special instruments.

a. Read the following abstract, and write a conclusion about which treatment you would choose and why.

b. The researchers give two confidence intervals, each associated with a different outcome. Choose one of these intervals and interpret it, as you would to someone concerned about whether she or he should have conventional or endovascular repair of the aorta.

*"Methods:* We conducted a multicenter, randomized trial comparing open repair with endovascular repair in 351 patients who had received a diagnosis of abdominal aortic aneurysm of at least 5 cm in diameter and who were considered suitable candidates for both techniques. Survival after randomization was calculated with the use of Kaplan–Meier analysis and compared with the use of the log-rank test on an intention-to-treat basis.

*Results:* Two years after randomization, the cumulative survival rates were 89.6 percent for open repair and 89.7 percent for endovascular repair (difference, −0.1 percentage point; 95 percent confidence interval, −6.8 to 6.7 percentage points). The cumulative rates of aneurysm-related death were 5.7 percent for open repair and 2.1 percent for endovascular repair (difference, 3.7 percentage points; 95 percent confidence interval, −0.5 to 7.9 percentage points).

*Conclusions:* The perioperative survival advantage with endovascular repair as compared with open repair is not sustained after the first postoperative year."

(Source: Jan D. Blankensteijn et al. 2005. Two-year outcomes after conventional or endovascular repair of abdominal aortic aneurysms. *New England Journal of Medicine* 352, 2398–2405.)

### 10.40 Antibiotics for Heart Patients

Read the following abstract, and state whether you think that coronary patients should be given azithromycin (an antibiotic) to reduce the risk of death or other bad outcomes. Explain your answer. Include the confidence interval for the difference in percentages (of bad outcomes), and explain what it shows and why.

*"Methods:* In this randomized, prospective trial, we assigned 4012 patients with documented stable coronary artery disease to receive either 600 mg of azithromycin or placebo weekly for one year. The participants were followed for a mean of 3.9 years at 28 clinical centers throughout the United States.

*Results:* The primary end point, a composite of death due to coronary heart disease, nonfatal myocardial infarction, coronary revascularization, or hospitalization for unstable angina, occurred in 446 of the participants who had been randomly assigned to receive azithromycin and 449 of those who had been randomly assigned to receive placebo. There was no significant risk reduction in the azithromycin group as compared with the placebo group with regard to the primary end point (risk reduction, 1 percent [95 percent confidence interval, −13 to 13 percent]). There were also no significant risk reductions with regard to any of the components of the primary end point, death from any cause, or stroke. The results did not differ when the participants were stratified according to sex, age, smoking status, presence or absence of diabetes mellitus, or *C. pneumoniae* serologic status at baseline.

*Conclusions:* A one-year course of weekly azithromycin did not alter the risk of cardiac events among patients with stable coronary artery disease."

(Source: J. Thomas Grayston et al. 2005. Azithromycin for the secondary prevention of coronary events. *New England Journal of Medicine* 352, 1637–1645.)

### *10.41 Alumni Donations

The alumni office wishes to determine whether students who attend a reception with alumni just before graduation are more likely to donate money within the next two years.

a. Describe a study based on a sample of students that would allow the alumni office to conclude that attending the reception *causes* future donations but that it is *not* possible to generalize this result to all students.

b. Describe a study based on a sample of students that does *not* allow fundraisers to conclude that attending receptions causes future donations but does allow them to generalize to all students.

c. Describe a study based on a sample of students that allows fundraisers to conclude that attending the reception causes future donations and also allows them to generalize to all students.

### *10.42 Recidivism Rates

The 3-year recidivism rate in Texas is about 30%, which means that 30% of released Texas prisoners return to prison within 3 years of release. There have been many attempts to reduce the recidivism rate. Suppose you want to determine whether electronic monitoring bracelets that track the location of the released prisoner reduce recidivism. Suppose that offenders released from prison are observed for three years to see whether they go back to prison and that the ones who wear electronic monitoring bracelets wear them for the first year only.

a. Describe a study based on a sample of released offenders that would allow the legal system to conclude that monitoring causes a reduction in recidivism but would not allow it to generalize this result to all released prisoners.

b. Describe a study based on a sample of released offenders that does *not allow* the legal system to conclude that monitoring causes a reduction in recidivism but does allow it to generalize to all released offenders.

c. Describe a study based on a sample of released prisoners that allows the legal system to conclude that monitoring causes a reduction in recidivism and also allows it to generalize to all released offenders.

TRY **10.43 Low-Dose Flu Vaccine (Example 8)** Refer to Exercise 10.37. If you cannot obtain the answer to any of the following questions from the excerpt of the abstract provided there, say so.

1. What is the research question that these investigators are trying to answer?

2. What is their answer to the research question?

3. What were the methods they used to collect data?

4. Is the conclusion appropriate for the methods used to collect data?

5. To what population do the conclusions apply?

6. Have the results been replicated in other articles? Are the results consistent with what other researchers have suggested?

**10.44 CABG or Stent?** Refer to the abstract in Exercise 10.38 to answer the questions listed in Exercise 10.43.

**10.45 Endovascular or Open Repair of Aortic Aneurysms?** Refer to the abstract in Exercise 10.39, and answer the questions listed in Exercise 10.43. The conclusion was "The perioperative survival advantage with endovascular repair as compared with open repair is not sustained after the first postoperative year."

*10.46 **Autism and MMR Vaccine** An article by Wakefield et al. in the British medical journal *Lancet* claimed that autism was caused by the measles, mumps, and rubella (MMR) vaccine. This vaccine is typically given to children twice, at about the age of 1 and again at about 4 years of age. The article reports a study of 12 children with autism who had all received the vaccines shortly before developing autism. The article was later retracted by *Lancet* because the conclusions were not justified by the design of the study.

Explain why *Lancet* might have felt that the conclusions were not justified by listing potential flaws in the study, as described above. (Source: A. J. Wakefield et al. 1998. Ileal–lymphoid-nodular hyperplasia, non-specific colitis, and pervasive developmental disorder in children. *Lancet* 351, 637–641.)

## CHAPTER REVIEW EXERCISES

*10.47 **Perry Preschool Arrests** The Perry preschool project discussed in Exercises 10.23–10.25 found that 8 of the 58 students who attended preschool had at least one felony arrest by age 40 and that 31 of the 65 students who did not attend preschool had at least one felony arrest (Schweinhart et al. 2005).

a. Compare the percentages descriptively. What does this comparison suggest?

b. Create a two-way table from the data and do a chi-square test on it, using a significance level of 0.05. Test the hypothesis that preschool attendance is associated with being arrested.

c. Do a two-proportion z-test. Your alternative hypothesis should be that preschool attendance lowers the chances of arrest.

d. What advantage does the two-proportion z-test have over the chi-square test?

*10.48 **Parental Training and Criminal Behavior of Children** In Montreal, Canada, an experiment was done with parents of children who were thought to have a high risk of committing crimes when they became teenagers. Some of the families were randomly assigned to receive parental training, and the others were not. Out of 43 children whose parents were randomly assigned to the parental training group, 6 had been arrested by the age of 15. Out of 123 children whose parents were not in the parental training group, 37 had been arrested by age 15.

a. Find and compare the percentages of children arrested by age 15. Is this what researchers might have hoped?

b. Create a two-way table from the data, and test whether the treatment program is associated with arrests. Use a significance level of 0.05.

c. Do a two-proportion z-test, testing whether the parental training lowers the rate of bad results. Use a significance level of 0.05.

d. Explain the difference in the results of the chi-square test and the two-proportion z-test.

e. Can you conclude that the treatment causes the better result? Why or why not?

(Source: R. E. Tremblay et al. 1996. From childhood physical aggression to adolescent maladjustment: The Montreal prevention experiment. In R.D. Peters and R. J. McMahon. *Preventing childhood disorders, substance use and delinquency*. Thousand Oaks, California: Sage, pp. 268–298.)

**10.49 Diet Drug** A randomized, placebo-controlled study of the diet drug Meridia (sibutramine) was done on overweight or obese subjects and reported in the *New England Journal of Medicine*. The patients were all 55 years old or older with a high risk of cardiovascular events. Those who had a heart event experienced a heart attack, stroke, or death from heart-related factors. The table gives the counts.

| | Drug | Placebo |
|---|---|---|
| **Heart Event** | 559 | 490 |
| **No Heart Event** | 4347 | 4408 |

a. Compare the rates of heart event for the drug group and the placebo group.

b. Test for association of the drug and a heart event, using a significance level of 0.05.

c. Can you conclude that the drug causes the difference in the rate of heart events? Why or why not?

(Source: W. P. James et al. 2010. Effect of sibutramine on cardiovascular outcomes in overweight and obese subjects. *New England Journal of Medicine* 363, 905–917.)

**10.50 Statin for Prevention of Blood Clots?** A group of 17,802 healthy people with normal cholesterol readings were randomly assigned to take the cholesterol-lowering drug Crestor (a statin) or a placebo. They were then observed for about two years to see whether they developed pulmonary embolism, or blood clots. Assume that exactly half were assigned the drug, and half were assigned the placebo.

When the results were reported in 2009, 34 of those taking Crestor and 60 of those in the placebo group had developed pulmonary embolism, or blood clots.

a. Create a two-way table showing the numbers of those who developed blood clots. Use Crestor and Placebo as labels for the columns.

b. Compare the percentages that had blood clots for these two treatments.

c. Do a test to find out whether treatment and outcome are associated, using a significance level of 0.05.

d. Can you conclude that using Crestor caused the difference? Why or why not?

(Source: R. J. Glynn et al. 2009. A randomized trial of rosuvastatin in the prevention of venous thromboembolism. *New England Journal of Medicine* 360(18), 1851–1861.)

**\* 10.51 Scared Straight** In the 1980s, the program Scared Straight was created. In this program young delinquents would go to a prison and meet prisoners who told them how difficult prison life is. The aim of the program was to scare the kids so that they would not commit crimes. A study was done in which half of the kids were randomly assigned to a Scared Straight program and half had no treatment. Then all the kids were observed for 12 months to see whether they were arrested.

Forty-three out of 53 of the Scared Straight kids were arrested, and 36 out of 53 of the group that did not see the prison were arrested.

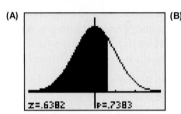

a. Find the percentage of each group that was arrested, and comment on it.

b. Perform a chi-square test of homogeneity to determine whether the treatment and outcome (Arrest or No Arrest) are independent. Use a significance level of 0.05.

c. Do a one-tailed two-proportion z-test to see whether Scared Straight decreases arrest rates. Which output shown is the correct one, output A or output B? Number 1 refers to the Scared Straight group, and number 2 refers to the control group with no treatment.

d. Compare the results of parts b and c. Include a comparison of p-values. Is there a relationship between the observed values of the z-statistic and the chi-square statistic? If so, what is it?

(Source: R. V. Lewis. 1983. Scared Straight—California style: Evaluation of the San Quentin Squires program. *Criminal Justice and Behavior* 10, 209–226.)

**\* 10.52 Boot Camp** In 1997 the California Youth Authority reported the results of a randomized experiment with young offenders. Some of the offenders were randomly assigned to a "boot camp," and some were simply confined. After they were all released, they were observed for 12 months to see whether they had been arrested. Of the 313 assigned to boot camp, 190 were arrested; of the 243 assigned to confinement, 141 were arrested.

a. Find the percentage of each group that was arrested, and comment on it.

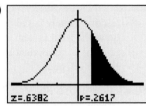

b. Perform a chi-square test of homogeneity to determine whether the variables *Treatment* and *Arrest* are associated. Use a significance level of 0.05.

c. Do a two-proportion z-test to see whether there was a significantly lower rate of arrest for the boot camp group, using a significance level of 0.05. Which of the two TI-83/84 outputs shown is correct?

d. Compare the results of parts b and c. Include a comparison of p-values. Also, is there a relationship between z and chi-square? If so, what is it?

(Source: California Youth Authority. 1997. *LEAD: A boot camp and intensive parole program: The final impact evaluation.* Sacramento, California: Department of the Youth Authority.)

**10.53 Vaccinations for Diarrhea in Mexico** Diarrhea can kill children and is often caused by rotavirus. Read the abstract below and answer the questions that follow.

"*Methods:* We obtained data on deaths from diarrhea, regardless of cause, from January 2003 through May 2009 in Mexican children under 5 years of age. We compared diarrhea-related mortality in 2008 and during the 2008 and 2009 rotavirus seasons with the mortality at baseline (2003–2006), before the introduction of the rotavirus vaccine. Vaccine coverage was estimated from administrative data.

*Results:* Diarrhea-related mortality fell from an annual median of 18.1 deaths per 100,000 children at baseline to 11.8 per 100,000 children in 2008 (rate reduction, 35%; 95% confidence interval [CI], 29 to 39; P < 0.001). . . . Mortality among unvaccinated children between the ages of 24 and 59 months was not significantly reduced. The reduction in the number of diarrhea-related deaths persisted through two full rotavirus seasons (2008 and 2009).

*Conclusions:* After the introduction of a rotavirus vaccine, a significant decline in diarrhea-related deaths among Mexican children was observed, suggesting a potential benefit from rotavirus vaccination."

a. State the death rate before vaccine and the death rate after vaccine. What was the change in deaths per 100,000 children? From the given p-value, can you reject the null hypothesis of no change in death rate?

b. Would you conclude that the vaccine was effective? Why or why not?

(Source: R. Vesta et al. 2010. Effect of rotavirus vaccination on death from childhood diarrhea in Mexico. *New England Journal of Medicine* 62, 299–305.)

**10.54 Gestational Diabetes** Comment on the following abstract, which discusses treatment of diabetes in pregnant women (gestational diabetes) and complications of childbirth.

*"[Methods:]* We conducted a randomized clinical trial to determine whether treatment of women with gestational diabetes mellitus reduced the risk of perinatal [immediately before or after childbirth] complications. We randomly assigned women between 24 and 34 weeks' gestation who had gestational diabetes to receive dietary advice, blood glucose monitoring, and insulin therapy as needed (the intervention group) or routine care. Primary outcomes included serious perinatal complications (defined as death, shoulder dystocia, bone fracture, and nerve palsy), admission to the neonatal nursery, jaundice requiring phototherapy, induction of labor, cesarean birth, and maternal anxiety, depression, and health status.

*[Results:]* The rate of serious perinatal complications was significantly lower among the infants of the 490 women in the intervention group than among the infants of the 510 women in the routine-care group (1 percent vs. 4 percent; . . . P = 0.01). . . .

*Conclusions:* Treatment of gestational diabetes reduces serious perinatal morbidity. . . ."

a. Identify the treatment and the response variables.

b. Is the stated conclusion making a claim for cause-and-effect? If so, is this claim justified? Explain.

(Source: Caroline A. Crowther et al. 2005. Effect of treatment of gestational diabetes mellitus on pregnancy outcomes. *New England Journal of Medicine* 352, 2477–2486.)

## gUIDED EXERCISES

**g  10.17 Obesity and Marital Status** A study reported in the medical journal *Obesity* in 2009 analyzed data from the National Longitudinal Study of Adolescent Health. Obesity was defined as having a body mass index (BMI) of 30 or more. The research subjects were followed from adolescence to adulthood, with all individuals in the sample categorized in terms of whether or not they were obese and whether they were dating, cohabiting, or married.

|           | Dating | Cohabiting | Married |
|-----------|--------|------------|---------|
| **Obese**     | 81     | 103        | 147     |
| **Not Obese** | 359    | 326        | 277     |

**QUESTION** Test the hypothesis that the variables *Relationship Status* and *Obesity* are associated, using a significance level of 0.05. Also consider whether the study shows causality. The steps will guide you through the process. Minitab output is provided.

**Step 1 ▶ Hypothesize**

$H_0$: Relationship status and obesity are independent.

$H_a$: ?

**Step 2 ▶ Prepare**

We choose the chi-square test of independence because the data were from *one* random sample in which the people were classified two different ways. We do not have a random sample or a random assignment, so we will test to see whether these results could easily have occurred by chance. Find the smallest expected value and report it. Is it more than 5?

**Step 3 ▶ Compute to compare**

Refer to the output given.

$X^2 =$ _____

p-value = _____

**Step 4 ▶ Interpret**

Reject or do not reject the null hypothesis, and state what that means.

**Causality**

Can we conclude from these data that living with someone is making some people obese and that marrying is making even more people obese? Can we conclude that obesity affects your relationship status? Explain why or why not.

**Percentages**

Find and compare the percentages obese in the three relationship statuses.

```
Chi-Square Test: Dating, Cohabiting, Married
Expected counts are printed below observed counts

        Dating  Cohabiting  Married  Total
Obese       81         103      147    331
        112.64      109.82   108.54

Not        359         326      277    962
        327.36      319.18   315.46

Total      440         429      424   1293

Chi-Sq = 30.829, DF = 2, P-Value = 0.000
```

# TechTips

## General Instructions for All Technology

**EXAMPLE (CHI-SQUARE TEST FOR TWO-WAY TABLES): PERRY PRESCHOOL AND GRADUATION FROM HIGH SCHOOL** ▶ In the 1960s an experiment was started in which a group of children were randomly assigned to attend preschool or not to attend preschool. They were studied for years, and whether they graduated from high school is shown in Table A.

We will show the chi-square test for two-way tables to see whether the factors are independent or not.

|  | Preschool | No Preschool |
|---|---|---|
| **Grad HS** | 37 | 29 |
| **No Grad HS** | 20 | 35 |

▲ **TABLE A** Two-way Table for Preschool and Graduation from High School

### Discussion of Data

Much of technology is set up so that you can use the table summary (such as Table A) and find the calculated results. However, it is also possible to start with a spreadsheet containing the raw data. Table B shows the beginning of the raw data, for which there would be 121 rows for the 121 children.

| Preschool | Graduate HS |
|---|---|
| Yes | No |
| Yes | Yes |
| No | Yes |
| No | Yes |
| Yes | Yes |
| No | No |

▲ **TABLE B** Some Raw Data

### TI-83/84

You will not put the data into the lists. You will use a matrix (table), and the data must be in the form of a summary such as Table A.

1. Press **2ND** and **MATRIX** (or **MATRX**).
2. Scroll over to **EDIT** and press **ENTER** when **1:** is highlighted.
3. See Figure 10A. Put in the dimensions. Because the table has two rows and two columns, press **2**, **ENTER**, **2**, **ENTER**. (The first number is the number of rows, and the second number is the number of columns.)
4. Enter each of the four numbers in the table, as shown in Figure 10A. Press **ENTER** after typing each number.
5. Press **STAT**, and scroll over to **TESTS**.

6. Scroll down (or up) to **C: $\chi^2$-Test** and press **ENTER**.
7. Leave the **Observed** as **A** and the **Expected** as **B**. Scroll down to **Calculate** and press **ENTER**.

   You should get the output shown in Figure 10B.

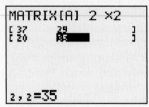

▲ **FIGURE 10A** TI-83/84 Input for Two-way Table

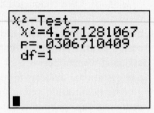

▲ **FIGURE 10B** TI-83/84 Output for a Chi-Square Test for Two-way Tables

8. To see the expected values, click **2ND**, **MATRIX**, scroll over to **EDIT**, scroll down to **2: [B]**, and press **ENTER**. You may have to scroll to the right to see some of the numbers. They will be arranged in the same order as the table of observed values. Check these numbers for the required minimum value of 5.

### MINITAB

For Minitab you may have your data as a table summary (as shown in Table A) or as raw data (as shown in Table B).

**TABLE SUMMARY**

1. Type a summary of your data into the columns.

| C1 | C2 |
|---|---|
| 37 | 29 |
| 20 | 35 |

   You may put labels above the numbers if you want to.
2. **Stat > Tables > Chi-square Test > Two-Way Table in Worksheet**
3. Select both columns (by double clicking them) and click **OK**. Figure 10C shows the output.

```
Chi-Square Test: C1, C2
Expected counts are printed below observed counts
Chi-Square contributions are printed below expected counts

        C1    C2  Total
  1     37    29     66
      31.09 34.91
      1.123 1.000

  2     20    35     55
      25.91 29.09
      1.348 1.200

Total   57    64    121

Chi-Sq = 4.671, DF = 1, P-Value = 0.031
```

▲ **FIGURE 10C** Minitab Output for Chi-Square Test for Two-way Tables

## RAW DATA

1. Make sure your raw data are in the columns. See Table B.
2. **Stat > Tables > Cross Tabulation and Chi-square**.
3. See Figure 10D: Click either C1 **For rows** and C2 **For columns** (or vice versa). Ignore **For layers**.

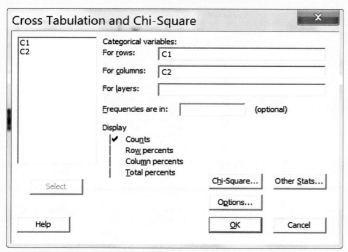

▲ **FIGURE 10D** Minitab Input for Cross Tabulation and Chi-Square

4. Click **Chi-Square**.
5. Select both **Chi-Square analysis** and **Expected cell counts**.
6. Click **OK** and click **OK**.

### EXCEL

1. Type a summary of your data into two (or more) columns, as shown in columns A and B in Figure 10E.

|   | A | B |
|---|---|---|
| 1 | 37 | 29 |
| 2 | 20 | 35 |

▲ **FIGURE 10E** Excel Input for Two-way Table

2. To get the total of 66 (from 37 + 29), click in the box to the right of 29, and then click **fx**, double click **SUM**, and click **OK**. You can do the same thing for the other sums, each time starting from the cell in which you want to put the sum. Save the grand total for last. Your table with the totals should look like columns A, B, and C in Figure 10F. Alternatively, you could simply add to get the totals.
3. To get the expected values, you will be using the formula

$$\text{Expected value} = \frac{(\text{row marginal total}) \times (\text{column marginal total})}{(\text{grand total})}$$

To get the first expected value, 31.09091, click in the empty cell where you want the expected value to be placed (here, cell E1). (An empty column, such as column D in Figure 10F, improves the clarity.) Then type = and click on the **66** in the table, type **\*** (for multiplication), click on the **57**, type **/** (for division), click on the **121**, and press **Enter**. Figure 10F shows part of the process for getting the expected value that goes

below the 31.09. For each of the expected values, you start from the cell you want filled, and you click on the row total, **\*** (for multiply), the column total, **/** (for divide), and the grand total and press **Enter**. Alternatively you could figure out the expected values by hand.

|   | A | B | C | D | E |
|---|---|---|---|---|---|
| 1 | 37 | 29 | 66 |   | 31.09091 |
| 2 | 20 | 35 | 55 |   | =A3*C2/C3 |
| 3 | 57 | 64 | 121 |   |   |

▲ **FIGURE 10F** Excel, Including Totals and One Expected Value

After you have all four expected values, be sure they are arranged in the same order as the original data:

4. Click **fx**.
5. Select a category: **Statistical** or **All**.
6. Choose **CHISQ.TEST**. For the **Actual_range**, highlight the table containing the observed counts, but do *not* include the row and column totals or the grand total. For the **Expected range**, highlight the table with the expected values.

   You will see the p-value (0.030671). Press **OK** and it will show up in the active cell in the worksheet.

   The previous steps for Excel will give you the p-value but not the value for chi-square. If you want the numerical value for chi-square, continue with the steps that follow.

7. Click in an empty cell.
8. Click **fx**.
9. Select a category: **Statistical** or **All**.
10. Choose **CHISQ.INV.RT** (for inverse, right tail).
11. For the **Probability**, click on the cell from Step 6 that shows the p-value of **0.030671**. For **Deg_freedom**, put in the degrees of freedom (df). For two-way tables,

    df = (number of rows − 1)(number of columns − 1).

    For Example A, df is 1. Click **OK**. You should get a chi-square of 4.67 for Example A.

### STATCRUNCH

#### TABLE SUMMARY

1. Enter your data summary as shown in Figure 10G. Note that you can have column labels (Preschool or No Preschool) and also row labels (GradHS or NoGrad).

| StatCrunch | Data | Stat | Graphics | Help |   |
|---|---|---|---|---|---|
| Row | var1 | Preschool | No Preschool | va |   |
| 1 | GradHS | 37 | 29 |   |   |
| 2 | NoGrad | 20 | 35 |   |   |
| 3 |   |   |   |   |   |

▲ **FIGURE 10G** StatCrunch Input for Two-way Table

2. **Stat > Tables > Contingency > with summary**
3. See Figure 10H. Select the columns that contain the summary counts, and select the column that contains the **Row labels**, here **var1**.

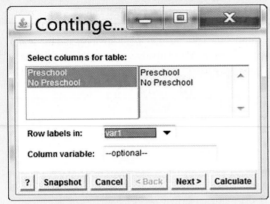

**▲ FIGURE 10H** StatCrunch Two-way-Table Options

4. Click **Next** and click **Expected Count**.
5. Click **Calculate**.
   Figure 10I shows the well-labeled output.

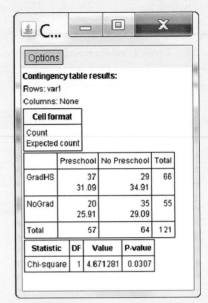

**▲ FIGURE 10I** StatCrunch Output for Two-way Table

**RAW DATA**

1. Be sure you have raw data in the columns; see Table B.
2. **Stat > Tables > Contingency > with data**
3. Select both columns.
4. Click **Next** and check **Expected Count**.
5. Click **Calculate**.

# Appendix A: Tables

## Table 1: Random Numbers

| Line | | | | | | |
|------|-------|-------|-------|-------|-------|-------|
| 01 | 21033 | 32522 | 19305 | 90633 | 80873 | 19167 |
| 02 | 17516 | 69328 | 88389 | 19770 | 33197 | 27336 |
| 03 | 26427 | 40650 | 70251 | 84413 | 30896 | 21490 |
| 04 | 45506 | 44716 | 02498 | 15327 | 79149 | 28409 |
| 05 | 55185 | 74834 | 81172 | 89281 | 48134 | 71185 |
| 06 | 87964 | 43751 | 80971 | 50613 | 81441 | 30505 |
| 07 | 09106 | 73117 | 57952 | 04393 | 93402 | 50753 |
| 08 | 88797 | 07440 | 69213 | 33593 | 42134 | 24168 |
| 09 | 34685 | 46775 | 32139 | 22787 | 28783 | 39481 |
| 10 | 07104 | 43091 | 14311 | 69671 | 01536 | 02673 |
| 11 | 27583 | 01866 | 58250 | 38103 | 35825 | 94513 |
| 12 | 60801 | 04439 | 58621 | 09840 | 35119 | 60372 |
| 13 | 62708 | 04888 | 37221 | 49537 | 96024 | 24004 |
| 14 | 21169 | 14082 | 65865 | 29690 | 00280 | 35738 |
| 15 | 13893 | 00626 | 11773 | 14897 | 37119 | 29729 |
| 16 | 19872 | 41310 | 65041 | 61105 | 31028 | 80297 |
| 17 | 29331 | 36997 | 05601 | 09785 | 18100 | 44164 |
| 18 | 76846 | 74048 | 08496 | 22599 | 29379 | 11114 |
| 19 | 11848 | 80809 | 25818 | 38857 | 23811 | 80902 |
| 20 | 85757 | 33963 | 93076 | 39950 | 29658 | 07530 |
| 21 | 71141 | 00618 | 48403 | 46083 | 40368 | 33990 |
| 22 | 47371 | 36443 | 41894 | 62134 | 86876 | 18548 |
| 23 | 46633 | 10669 | 95848 | 69055 | 49044 | 75595 |
| 24 | 79118 | 21098 | 63279 | 26834 | 43443 | 38267 |
| 25 | 91874 | 87217 | 11503 | 47925 | 13289 | 42106 |
| 26 | 85337 | 08882 | 68429 | 61767 | 18930 | 37688 |
| 27 | 88513 | 05437 | 22776 | 17562 | 03820 | 44785 |
| 28 | 31498 | 85304 | 22393 | 21634 | 34560 | 77404 |
| 29 | 93074 | 27086 | 62559 | 86590 | 18420 | 33290 |
| 30 | 90549 | 53094 | 76282 | 53105 | 45531 | 90061 |
| 31 | 11373 | 96871 | 38157 | 98368 | 39536 | 08079 |
| 32 | 52022 | 59093 | 30647 | 33241 | 16027 | 70336 |
| 33 | 14709 | 93220 | 89547 | 95320 | 39134 | 07646 |
| 34 | 57584 | 28114 | 91168 | 16320 | 81609 | 60807 |
| 35 | 31867 | 85872 | 91430 | 45554 | 21567 | 15082 |
| 36 | 07033 | 75250 | 34546 | 75298 | 33893 | 64487 |
| 37 | 02779 | 72645 | 32699 | 86009 | 73729 | 44206 |
| 38 | 24512 | 01116 | 49826 | 50882 | 44086 | 87757 |
| 39 | 52463 | 30164 | 80073 | 55917 | 60995 | 38655 |
| 40 | 82588 | 59267 | 13570 | 56434 | 66413 | 99518 |
| 41 | 20999 | 05039 | 87835 | 63010 | 82980 | 66193 |
| 42 | 09084 | 98948 | 09541 | 80623 | 15915 | 71042 |

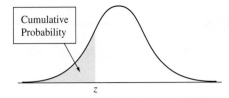

Cumulative probability for z is the area under the standard Normal curve to the left of z.

# Table 2: Standard Normal Cumulative Probabilities

| z | .00 |
|---|---|
| −5.0 | .000000287 |
| −4.5 | .00000340 |
| −4.0 | .0000317 |
| −3.5 | .000233 |

| z | .00 | .01 | .02 | .03 | .04 | .05 | .06 | .07 | .08 | .09 |
|---|---|---|---|---|---|---|---|---|---|---|
| −3.4 | .0003 | .0003 | .0003 | .0003 | .0003 | .0003 | .0003 | .0003 | .0003 | .0002 |
| −3.3 | .0005 | .0005 | .0005 | .0004 | .0004 | .0004 | .0004 | .0004 | .0004 | .0003 |
| −3.2 | .0007 | .0007 | .0006 | .0006 | .0006 | .0006 | .0006 | .0005 | .0005 | .0005 |
| −3.1 | .0010 | .0009 | .0009 | .0009 | .0008 | .0008 | .0008 | .0008 | .0007 | .0007 |
| −3.0 | .0013 | .0013 | .0013 | .0012 | .0012 | .0011 | .0011 | .0011 | .0010 | .0010 |
| −2.9 | .0019 | .0018 | .0018 | .0017 | .0016 | .0016 | .0015 | .0015 | .0014 | .0014 |
| −2.8 | .0026 | .0025 | .0024 | .0023 | .0023 | .0022 | .0021 | .0021 | .0020 | .0019 |
| −2.7 | .0035 | .0034 | .0033 | .0032 | .0031 | .0030 | .0029 | .0028 | .0027 | .0026 |
| −2.6 | .0047 | .0045 | .0044 | .0043 | .0041 | .0040 | .0039 | .0038 | .0037 | .0036 |
| −2.5 | .0062 | .0060 | .0059 | .0057 | .0055 | .0054 | .0052 | .0051 | .0049 | .0048 |
| −2.4 | .0082 | .0080 | .0078 | .0075 | .0073 | .0071 | .0069 | .0068 | .0066 | .0064 |
| −2.3 | .0107 | .0104 | .0102 | .0099 | .0096 | .0094 | .0091 | .0089 | .0087 | .0084 |
| −2.2 | .0139 | .0136 | .0132 | .0129 | .0125 | .0122 | .0119 | .0116 | .0113 | .0110 |
| −2.1 | .0179 | .0174 | .0170 | .0166 | .0162 | .0158 | .0154 | .0150 | .0146 | .0143 |
| −2.0 | .0228 | .0222 | .0217 | .0212 | .0207 | .0202 | .0197 | .0192 | .0188 | .0183 |
| −1.9 | .0287 | .0281 | .0274 | .0268 | .0262 | .0256 | .0250 | .0244 | .0239 | .0233 |
| −1.8 | .0359 | .0351 | .0344 | .0336 | .0329 | .0322 | .0314 | .0307 | .0301 | .0294 |
| −1.7 | .0446 | .0436 | .0427 | .0418 | .0409 | .0401 | .0392 | .0384 | .0375 | .0367 |
| −1.6 | .0548 | .0537 | .0526 | .0516 | .0505 | .0495 | .0485 | .0475 | .0465 | .0455 |
| −1.5 | .0668 | .0655 | .0643 | .0630 | .0618 | .0606 | .0594 | .0582 | .0571 | .0559 |
| −1.4 | .0808 | .0793 | .0778 | .0764 | .0749 | .0735 | .0721 | .0708 | .0694 | .0681 |
| −1.3 | .0968 | .0951 | .0934 | .0918 | .0901 | .0885 | .0869 | .0853 | .0838 | .0823 |
| −1.2 | .1151 | .1131 | .1112 | .1093 | .1075 | .1056 | .1038 | .1020 | .1003 | .0985 |
| −1.1 | .1357 | .1335 | .1314 | .1292 | .1271 | .1251 | .1230 | .1210 | .1190 | .1170 |
| −1.0 | .1587 | .1562 | .1539 | .1515 | .1492 | .1469 | .1446 | .1423 | .1401 | .1379 |
| −0.9 | .1841 | .1814 | .1788 | .1762 | .1736 | .1711 | .1685 | .1660 | .1635 | .1611 |
| −0.8 | .2119 | .2090 | .2061 | .2033 | .2005 | .1977 | .1949 | .1922 | .1894 | .1867 |
| −0.7 | .2420 | .2389 | .2358 | .2327 | .2296 | .2266 | .2236 | .2206 | .2177 | .2148 |
| −0.6 | .2743 | .2709 | .2676 | .2643 | .2611 | .2578 | .2546 | .2514 | .2483 | .2451 |
| −0.5 | .3085 | .3050 | .3015 | .2981 | .2946 | .2912 | .2877 | .2843 | .2810 | .2776 |
| −0.4 | .3446 | .3409 | .3372 | .3336 | .3300 | .3264 | .3228 | .3192 | .3156 | .3121 |
| −0.3 | .3821 | .3783 | .3745 | .3707 | .3669 | .3632 | .3594 | .3557 | .3520 | .3483 |
| −0.2 | .4207 | .4168 | .4129 | .4090 | .4052 | .4013 | .3974 | .3936 | .3897 | .3859 |
| −0.1 | .4602 | .4562 | .4522 | .4483 | .4443 | .4404 | .4364 | .4325 | .4286 | .4247 |
| −0.0 | .5000 | .4960 | .4920 | .4880 | .4840 | .4801 | .4761 | .4721 | .4681 | .4641 |

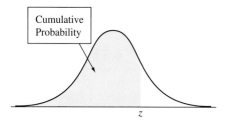

Cumulative probability for z is the area under the standard Normal curve to the left of z.

## Standard Normal Cumulative Probabilities (*continued*)

| z | .00 | .01 | .02 | .03 | .04 | .05 | .06 | .07 | .08 | .09 |
|---|---|---|---|---|---|---|---|---|---|---|
| 0.0 | .5000 | .5040 | .5080 | .5120 | .5160 | .5199 | .5239 | .5279 | .5319 | .5359 |
| 0.1 | .5398 | .5438 | .5478 | .5517 | .5557 | .5596 | .5636 | .5675 | .5714 | .5753 |
| 0.2 | .5793 | .5832 | .5871 | .5910 | .5948 | .5987 | .6026 | .6064 | .6103 | .6141 |
| 0.3 | .6179 | .6217 | .6255 | .6293 | .6331 | .6368 | .6406 | .6443 | .6480 | .6517 |
| 0.4 | .6554 | .6591 | .6628 | .6664 | .6700 | .6736 | .6772 | .6808 | .6844 | .6879 |
| 0.5 | .6915 | .6950 | .6985 | .7019 | .7054 | .7088 | .7123 | .7157 | .7190 | .7224 |
| 0.6 | .7257 | .7291 | .7324 | .7357 | .7389 | .7422 | .7454 | .7486 | .7517 | .7549 |
| 0.7 | .7580 | .7611 | .7642 | .7673 | .7704 | .7734 | .7764 | .7794 | .7823 | .7852 |
| 0.8 | .7881 | .7910 | .7939 | .7967 | .7995 | .8023 | .8051 | .8078 | .8106 | .8133 |
| 0.9 | .8159 | .8186 | .8212 | .8238 | .8264 | .8289 | .8315 | .8340 | .8365 | .8389 |
| 1.0 | .8413 | .8438 | .8461 | .8485 | .8508 | .8531 | .8554 | .8577 | .8599 | .8621 |
| 1.1 | .8643 | .8665 | .8686 | .8708 | .8729 | .8749 | .8770 | .8790 | .8810 | .8830 |
| 1.2 | .8849 | .8869 | .8888 | .8907 | .8925 | .8944 | .8962 | .8980 | .8997 | .9015 |
| 1.3 | .9032 | .9049 | .9066 | .9082 | .9099 | .9115 | .9131 | .9147 | .9162 | .9177 |
| 1.4 | .9192 | .9207 | .9222 | .9236 | .9251 | .9265 | .9279 | .9292 | .9306 | .9319 |
| 1.5 | .9332 | .9345 | .9357 | .9370 | .9382 | .9394 | .9406 | .9418 | .9429 | .9441 |
| 1.6 | .9452 | .9463 | .9474 | .9484 | .9495 | .9505 | .9515 | .9525 | .9535 | .9545 |
| 1.7 | .9554 | .9564 | .9573 | .9582 | .9591 | .9599 | .9608 | .9616 | .9625 | .9633 |
| 1.8 | .9641 | .9649 | .9656 | .9664 | .9671 | .9678 | .9686 | .9693 | .9699 | .9706 |
| 1.9 | .9713 | .9719 | .9726 | .9732 | .9738 | .9744 | .9750 | .9756 | .9761 | .9767 |
| 2.0 | .9772 | .9778 | .9783 | .9788 | .9793 | .9798 | .9803 | .9808 | .9812 | .9817 |
| 2.1 | .9821 | .9826 | .9830 | .9834 | .9838 | .9842 | .9846 | .9850 | .9854 | .9857 |
| 2.2 | .9861 | .9864 | .9868 | .9871 | .9875 | .9878 | .9881 | .9884 | .9887 | .9890 |
| 2.3 | .9893 | .9896 | .9898 | .9901 | .9904 | .9906 | .9909 | .9911 | .9913 | .9916 |
| 2.4 | .9918 | .9920 | .9922 | .9925 | .9927 | .9929 | .9931 | .9932 | .9934 | .9936 |
| 2.5 | .9938 | .9940 | .9941 | .9943 | .9945 | .9946 | .9948 | .9949 | .9951 | .9952 |
| 2.6 | .9953 | .9955 | .9956 | .9957 | .9959 | .9960 | .9961 | .9962 | .9963 | .9964 |
| 2.7 | .9965 | .9966 | .9967 | .9968 | .9969 | .9970 | .9971 | .9972 | .9973 | .9974 |
| 2.8 | .9974 | .9975 | .9976 | .9977 | .9977 | .9978 | .9979 | .9979 | .9980 | .9981 |
| 2.9 | .9981 | .9982 | .9982 | .9983 | .9984 | .9984 | .9985 | .9985 | .9986 | .9986 |
| 3.0 | .9987 | .9987 | .9987 | .9988 | .9988 | .9989 | .9989 | .9989 | .9990 | .9990 |
| 3.1 | .9990 | .9991 | .9991 | .9991 | .9992 | .9992 | .9992 | .9992 | .9993 | .9993 |
| 3.2 | .9993 | .9993 | .9994 | .9994 | .9994 | .9994 | .9994 | .9995 | .9995 | .9995 |
| 3.3 | .9995 | .9995 | .9995 | .9996 | .9996 | .9996 | .9996 | .9996 | .9996 | .9997 |
| 3.4 | .9997 | .9997 | .9997 | .9997 | .9997 | .9997 | .9997 | .9997 | .9997 | .9998 |

| z | .00 |
|---|---|
| 3.5 | .999767 |
| 4.0 | .9999683 |
| 4.5 | .9999966 |
| 5.0 | .999999713 |

# Table 3: Binomial Probabilities

| | | | | | | | | $p$ | | | | | | | |
|---|---|---|---|---|---|---|---|---|---|---|---|---|---|---|---|
| $n$ | $x$ | .01 | .05 | .10 | .20 | .30 | .40 | .50 | .60 | .70 | .80 | .90 | .95 | .99 | $x$ |
| 2 | 0 | .980 | .902 | .810 | .640 | .490 | .360 | .250 | .160 | .090 | .040 | .010 | .002 | 0+ | 0 |
| | 1 | .020 | .095 | .180 | .320 | .420 | .480 | .500 | .480 | .420 | .320 | .180 | .095 | .020 | 1. |
| | 2 | 0+ | .002 | .010 | .040 | .090 | .160 | .250 | .360 | .490 | .640 | .810 | .902 | .980 | 2 |
| 3 | 0 | .970 | .857 | .729 | .512 | .343 | .216 | .125 | .064 | .027 | .008 | .001 | 0+ | 0+ | 0 |
| | 1 | .029 | .135 | .243 | .384 | .441 | .432 | .375 | .288 | .189 | .096 | .027 | .007 | 0+ | 1 |
| | 2 | 0+ | .007 | .027 | .096 | .189 | .288 | .375 | .432 | .441 | .384 | .243 | .135 | .029 | 2 |
| | 3 | 0+ | 0+ | .001 | .008 | .027 | .064 | .125 | .216 | .343 | .512 | .729 | .857 | .970 | 3 |
| 4 | 0 | .961 | .815 | .656 | .410 | .240 | .130 | .062 | .026 | .008 | .002 | 0+ | 0+ | 0+ | 0 |
| | 1 | .039 | .171 | .292 | .410 | .412 | .346 | .250 | .154 | .076 | .026 | .004 | 0+ | 0+ | 1 |
| | 2 | .001 | .014 | .049 | .154 | .265 | .346 | .375 | .346 | .265 | .154 | .049 | .014 | .001 | 2 |
| | 3 | 0+ | 0+ | .004 | .026 | .076 | .154 | .250 | .346 | .412 | .410 | .292 | .171 | .039 | 3 |
| | 4 | 0+ | 0+ | 0+ | .002 | .008 | .026 | .062 | .130 | .240 | .410 | .656 | .815 | .961 | 4 |
| 5 | 0 | .951 | .774 | .590 | .328 | .168 | .078 | .031 | .010 | .002 | 0+ | 0+ | 0+ | 0+ | 0 |
| | 1 | .048 | .204 | .328 | .410 | .360 | .259 | .156 | .077 | .028 | .006 | 0+ | 0+ | 0+ | 1 |
| | 2 | .001 | .021 | .073 | .205 | .309 | .346 | .312 | .230 | .132 | .051 | .008 | .001 | 0+ | 2 |
| | 3 | 0+ | .001 | .008 | .051 | .132 | .230 | .312 | .346 | .309 | .205 | .073 | .021 | .001 | 3 |
| | 4 | 0+ | 0+ | 0+ | .006 | .028 | .077 | .156 | .259 | .360 | .410 | .328 | .204 | .048 | 4 |
| | 5 | 0+ | 0+ | 0+ | 0+ | .002 | .010 | .031 | .078 | .168 | .328 | .590 | .774 | .951 | 5 |
| 6 | 0 | .941 | .735 | .531 | .262 | .118 | .047 | .016 | .004 | .001 | 0+ | 0+ | 0+ | 0+ | 0 |
| | 1 | .057 | .232 | .354 | .393 | .303 | .187 | .094 | .037 | .010 | .002 | 0+ | 0+ | 0+ | 1 |
| | 2 | .001 | .031 | .098 | .246 | .324 | .311 | .234 | .138 | .060 | .015 | .001 | 0+ | 0+ | 2 |
| | 3 | 0+ | .002 | .015 | .082 | .185 | .276 | .312 | .276 | .185 | .082 | .015 | .002 | 0+ | 3 |
| | 4 | 0+ | 0+ | .001 | .015 | .060 | .138 | .234 | .311 | .324 | .246 | .098 | .031 | .001 | 4 |
| | 5 | 0+ | 0+ | 0+ | .002 | .010 | .037 | .094 | .187 | .303 | .393 | .354 | .232 | .057 | 5 |
| | 6 | 0+ | 0+ | 0+ | 0+ | .001 | .004 | .016 | .047 | .118 | .262 | .531 | .735 | .941 | 6 |
| 7 | 0 | .932 | .698 | .478 | .210 | .082 | .028 | .008 | .002 | 0+ | 0+ | 0+ | 0+ | 0+ | 0 |
| | 1 | .066 | .257 | .372 | .367 | .247 | .131 | .055 | .017 | .004 | 0+ | 0+ | 0+ | 0+ | 1 |
| | 2 | .002 | .041 | .124 | .275 | .318 | .261 | .164 | .077 | .025 | .004 | 0+ | 0+ | 0+ | 2 |
| | 3 | 0+ | .004 | .023 | .115 | .227 | .290 | .273 | .194 | .097 | .029 | .003 | 0+ | 0+ | 3 |
| | 4 | 0+ | 0+ | .003 | .029 | .097 | .194 | .273 | .290 | .227 | .115 | .023 | .004 | 0+ | 4 |
| | 5 | 0+ | 0+ | 0+ | .004 | .025 | .077 | .164 | .261 | .318 | .275 | .124 | .041 | .002 | 5 |
| | 6 | 0+ | 0+ | 0+ | 0+ | .004 | .017 | .055 | .131 | .247 | .367 | .372 | .257 | .066 | 6 |
| | 7 | 0+ | 0+ | 0+ | 0+ | 0+ | .002 | .008 | .028 | .082 | .210 | .478 | .698 | .932 | 7 |
| 8 | 0 | .923 | .663 | .430 | .168 | .058 | .017 | .004 | .001 | 0+ | 0+ | 0+ | 0+ | 0+ | 0 |
| | 1 | .075 | .279 | .383 | .336 | .198 | .090 | .031 | .008 | .001 | 0+ | 0+ | 0+ | 0+ | 1 |
| | 2 | .003 | .051 | .149 | .294 | .296 | .209 | .109 | .041 | .010 | .001 | 0+ | 0+ | 0+ | 2 |
| | 3 | 0+ | .005 | .033 | .147 | .254 | .279 | .219 | .124 | .047 | .009 | 0+ | 0+ | 0+ | 3 |
| | 4 | 0+ | 0+ | .005 | .046 | .136 | .232 | .273 | .232 | .136 | .046 | .005 | 0+ | 0+ | 4 |
| | 5 | 0+ | 0+ | 0+ | .009 | .047 | .124 | .219 | .279 | .254 | .147 | .033 | .005 | 0+ | 5 |
| | 6 | 0+ | 0+ | 0+ | .001 | .010 | .041 | .109 | .209 | .296 | .294 | .149 | .051 | .003 | 6 |
| | 7 | 0+ | 0+ | 0+ | 0+ | .001 | .008 | .031 | .090 | .198 | .336 | .383 | .279 | .075 | 7 |
| | 8 | 0+ | 0+ | 0+ | 0+ | 0+ | .001 | .004 | .017 | .058 | .168 | .430 | .663 | .923 | 8 |

NOTE: 0+ represents a positive probability less than 0.0005.

(*continued*)

# Binomial Probabilities (*continued*)

| n | x | .01 | .05 | .10 | .20 | .30 | .40 | .50 | .60 | .70 | .80 | .90 | .95 | .99 | x |
|---|---|-----|-----|-----|-----|-----|-----|-----|-----|-----|-----|-----|-----|-----|---|
| 9 | 0 | .914 | .630 | .387 | .134 | .040 | .010 | .002 | 0+ | 0+ | 0+ | 0+ | 0+ | 0+ | 0 |
|   | 1 | .083 | .299 | .387 | .302 | .156 | .060 | .018 | .004 | 0+ | 0+ | 0+ | 0+ | 0+ | 1 |
|   | 2 | .003 | .063 | .172 | .302 | .267 | .161 | .070 | .021 | .004 | 0+ | 0+ | 0+ | 0+ | 2 |
|   | 3 | 0+ | .008 | .045 | .176 | .267 | .251 | .164 | .074 | .021 | .003 | 0+ | 0+ | 0+ | 3 |
|   | 4 | 0+ | .001 | .007 | .066 | .172 | .251 | .246 | .167 | .074 | .017 | .001 | 0+ | 0+ | 4 |
|   | 5 | 0+ | 0+ | .001 | .017 | .074 | .167 | .246 | .251 | .172 | .066 | .007 | .001 | 0+ | 5 |
|   | 6 | 0+ | 0+ | 0+ | .003 | .021 | .074 | .164 | .251 | .267 | .176 | .045 | .008 | 0+ | 6 |
|   | 7 | 0+ | 0+ | 0+ | 0+ | .004 | .021 | .070 | .161 | .267 | .302 | .172 | .063 | .003 | 7 |
|   | 8 | 0+ | 0+ | 0+ | 0+ | 0+ | .004 | .018 | .060 | .156 | .302 | .387 | .299 | .083 | 8 |
|   | 9 | 0+ | 0+ | 0+ | 0+ | 0+ | 0+ | .002 | .010 | .040 | .134 | .387 | .630 | .914 | 9 |
| 10 | 0 | .904 | .599 | .349 | .107 | .028 | .006 | .001 | 0+ | 0+ | 0+ | 0+ | 0+ | 0+ | 0 |
|   | 1 | .091 | .315 | .387 | .268 | .121 | .040 | .010 | .002 | 0+ | 0+ | 0+ | 0+ | 0+ | 1 |
|   | 2 | .004 | .075 | .194 | .302 | .233 | .121 | .044 | .011 | .001 | 0+ | 0+ | 0+ | 0+ | 2 |
|   | 3 | 0+ | .010 | .057 | .201 | .267 | .215 | .117 | .042 | .009 | .001 | 0+ | 0+ | 0+ | 3 |
|   | 4 | 0+ | .001 | .011 | .088 | .200 | .251 | .205 | .111 | .037 | .006 | 0+ | 0+ | 0+ | 4 |
|   | 5 | 0+ | 0+ | .001 | .026 | .103 | .201 | .246 | .201 | .103 | .026 | .001 | 0+ | 0+ | 5 |
|   | 6 | 0+ | 0+ | 0+ | .006 | .037 | .111 | .205 | .251 | .200 | .088 | .011 | .001 | 0+ | 6 |
|   | 7 | 0+ | 0+ | 0+ | .001 | .009 | .042 | .117 | .215 | .267 | .201 | .057 | .010 | 0+ | 7 |
|   | 8 | 0+ | 0+ | 0+ | 0+ | .001 | .011 | .044 | .121 | .233 | .302 | .194 | .075 | .004 | 8 |
|   | 9 | 0+ | 0+ | 0+ | 0+ | 0+ | .002 | .010 | .040 | .121 | .268 | .387 | .315 | .091 | 9 |
|   | 10 | 0+ | 0+ | 0+ | 0+ | 0+ | 0+ | .001 | .006 | .028 | .107 | .349 | .599 | .904 | 10 |
| 11 | 0 | .895 | .569 | .314 | .086 | .020 | .004 | 0+ | 0+ | 0+ | 0+ | 0+ | 0+ | 0+ | 0 |
|   | 1 | .099 | .329 | .384 | .236 | .093 | .027 | .005 | .001 | 0+ | 0+ | 0+ | 0+ | 0+ | 1 |
|   | 2 | .005 | .087 | .213 | .295 | .200 | .089 | .027 | .005 | .001 | 0+ | 0+ | 0+ | 0+ | 2 |
|   | 3 | 0+ | .014 | .071 | .221 | .257 | .177 | .081 | .023 | .004 | 0+ | 0+ | 0+ | 0+ | 3 |
|   | 4 | 0+ | .001 | .016 | .111 | .220 | .236 | .161 | .070 | .017 | .002 | 0+ | 0+ | 0+ | 4 |
|   | 5 | 0+ | 0+ | .002 | .039 | .132 | .221 | .226 | .147 | .057 | .010 | 0+ | 0+ | 0+ | 5 |
|   | 6 | 0+ | 0+ | 0+ | .010 | .057 | .147 | .226 | .221 | .132 | .039 | .002 | 0+ | 0+ | 6 |
|   | 7 | 0+ | 0+ | 0+ | .002 | .017 | .070 | .161 | .236 | .220 | .111 | .016 | .001 | 0+ | 7 |
|   | 8 | 0+ | 0+ | 0+ | 0+ | .004 | .023 | .081 | .177 | .257 | .221 | .071 | .014 | 0+ | 8 |
|   | 9 | 0+ | 0+ | 0+ | 0+ | .001 | .005 | .027 | .089 | .200 | .295 | .213 | .087 | .005 | 9 |
|   | 10 | 0+ | 0+ | 0+ | 0+ | 0+ | .001 | .005 | .027 | .093 | .236 | .384 | .329 | .099 | 10 |
|   | 11 | 0+ | 0+ | 0+ | 0+ | 0+ | 0+ | 0+ | .004 | .020 | .086 | .314 | .569 | .895 | 11 |
| 12 | 0 | .886 | .540 | .282 | .069 | .014 | .002 | 0+ | 0+ | 0+ | 0+ | 0+ | 0+ | 0+ | 0 |
|   | 1 | .107 | .341 | .377 | .206 | .071 | .017 | .003 | 0+ | 0+ | 0+ | 0+ | 0+ | 0+ | 1 |
|   | 2 | .006 | .099 | .230 | .283 | .168 | .064 | .016 | .002 | 0+ | 0+ | 0+ | 0+ | 0+ | 2 |
|   | 3 | 0+ | .017 | .085 | .236 | .240 | .142 | .054 | .012 | .001 | 0+ | 0+ | 0+ | 0+ | 3 |
|   | 4 | 0+ | .002 | .021 | .133 | .231 | .213 | .121 | .042 | .008 | .001 | 0+ | 0+ | 0+ | 4 |
|   | 5 | 0+ | 0+ | .004 | .053 | .158 | .227 | .193 | .101 | .029 | .003 | 0+ | 0+ | 0+ | 5 |
|   | 6 | 0+ | 0+ | 0+ | .016 | .079 | .177 | .226 | .177 | .079 | .016 | 0+ | 0+ | 0+ | 6 |
|   | 7 | 0+ | 0+ | 0+ | .003 | .029 | .101 | .193 | .227 | .158 | .053 | .004 | 0+ | 0+ | 7 |
|   | 8 | 0+ | 0+ | 0+ | .001 | .008 | .042 | .121 | .213 | .231 | .133 | .021 | .002 | 0+ | 8 |
|   | 9 | 0+ | 0+ | 0+ | 0+ | .001 | .012 | .054 | .142 | .240 | .236 | .085 | .017 | 0+ | 9 |
|   | 10 | 0+ | 0+ | 0+ | 0+ | 0+ | .002 | .016 | .064 | .168 | .283 | .230 | .099 | .006 | 10 |
|   | 11 | 0+ | 0+ | 0+ | 0+ | 0+ | 0+ | .003 | .017 | .071 | .206 | .377 | .341 | .107 | 11 |
|   | 12 | 0+ | 0+ | 0+ | 0+ | 0+ | 0+ | 0+ | .002 | .014 | .069 | .282 | .540 | .886 | 12 |

*NOTE:* 0+ represents a positive probability less than 0.0005.

(*continued*)

# Binomial Probabilities (*continued*)

| | | | | | | | | $p$ | | | | | | | |
|---|---|---|---|---|---|---|---|---|---|---|---|---|---|---|---|
| $n$ | $x$ | .01 | .05 | .10 | .20 | .30 | .40 | .50 | .60 | .70 | .80 | .90 | .95 | .99 | $x$ |
| 13 | 0 | .878 | .513 | .254 | .055 | .010 | .001 | 0+ | 0+ | 0+ | 0+ | 0+ | 0+ | 0+ | 0 |
| | 1 | .115 | .351 | .367 | .179 | .054 | .011 | .002 | 0+ | 0+ | 0+ | 0+ | 0+ | 0+ | 1 |
| | 2 | .007 | .111 | .245 | .268 | .139 | .045 | .010 | .001 | 0+ | 0+ | 0+ | 0+ | 0+ | 2 |
| | 3 | 0+ | .021 | .100 | .246 | .218 | .111 | .035 | .006 | .001 | 0+ | 0+ | 0+ | 0+ | 3 |
| | 4 | 0+ | .003 | .028 | .154 | .234 | .184 | .087 | .024 | .003 | 0+ | 0+ | 0+ | 0+ | 4 |
| | 5 | 0+ | 0+ | .006 | .069 | .180 | .221 | .157 | .066 | .014 | .001 | 0+ | 0+ | 0+ | 5 |
| | 6 | 0+ | 0+ | .001 | .023 | .103 | .197 | .209 | .131 | .044 | .006 | 0+ | 0+ | 0+ | 6 |
| | 7 | 0+ | 0+ | 0+ | .006 | .044 | .131 | .209 | .197 | .103 | .023 | .001 | 0+ | 0+ | 7 |
| | 8 | 0+ | 0+ | 0+ | .001 | .014 | .066 | .157 | .221 | .180 | .069 | .006 | 0+ | 0+ | 8 |
| | 9 | 0+ | 0+ | 0+ | 0+ | .003 | .024 | .087 | .184 | .234 | .154 | .028 | .003 | 0+ | 9 |
| | 10 | 0+ | 0+ | 0+ | 0+ | .001 | .006 | .035 | .111 | .218 | .246 | .100 | .021 | 0+ | 10 |
| | 11 | 0+ | 0+ | 0+ | 0+ | 0+ | .001 | .010 | .045 | .139 | .268 | .245 | .111 | .007 | 11 |
| | 12 | 0+ | 0+ | 0+ | 0+ | 0+ | 0+ | .002 | .011 | .054 | .179 | .367 | .351 | .115 | 12 |
| | 13 | 0+ | 0+ | 0+ | 0+ | 0+ | 0+ | 0+ | .001 | .010 | .055 | .254 | .513 | .878 | 13 |
| 14 | 0 | .869 | .488 | .229 | .044 | .007 | .001 | 0+ | 0+ | 0+ | 0+ | 0+ | 0+ | 0+ | 0 |
| | 1 | .123 | .359 | .356 | .154 | .041 | .007 | .001 | 0+ | 0+ | 0+ | 0+ | 0+ | 0+ | 1 |
| | 2 | .008 | .123 | .257 | .250 | .113 | .032 | .006 | .001 | 0+ | 0+ | 0+ | 0+ | 0+ | 2 |
| | 3 | 0+ | .026 | .114 | .250 | .194 | .085 | .022 | .003 | 0+ | 0+ | 0+ | 0+ | 0+ | 3 |
| | 4 | 0+ | .004 | .035 | .172 | .229 | .155 | .061 | .014 | .001 | 0+ | 0+ | 0+ | 0+ | 4 |
| | 5 | 0+ | 0+ | .008 | .086 | .196 | .207 | .122 | .041 | .007 | 0+ | 0+ | 0+ | 0+ | 5 |
| | 6 | 0+ | 0+ | .001 | .032 | .126 | .207 | .183 | .092 | .023 | .002 | 0+ | 0+ | 0+ | 6 |
| | 7 | 0+ | 0+ | 0+ | .009 | .062 | .157 | .209 | .157 | .062 | .009 | 0+ | 0+ | 0+ | 7 |
| | 8 | 0+ | 0+ | 0+ | .002 | .023 | .092 | .183 | .207 | .126 | .032 | .001 | 0+ | 0+ | 8 |
| | 9 | 0+ | 0+ | 0+ | 0+ | .007 | .041 | .122 | .207 | .196 | .086 | .008 | 0+ | 0+ | 9 |
| | 10 | 0+ | 0+ | 0+ | 0+ | .001 | .014 | .061 | .155 | .229 | .172 | .035 | .004 | 0+ | 10 |
| | 11 | 0+ | 0+ | 0+ | 0+ | 0+ | .003 | .022 | .085 | .194 | .250 | .114 | .026 | 0+ | 11 |
| | 12 | 0+ | 0+ | 0+ | 0+ | 0+ | .001 | .006 | .032 | .113 | .250 | .257 | .123 | .008 | 12 |
| | 13 | 0+ | 0+ | 0+ | 0+ | 0+ | 0+ | .001 | .007 | .041 | .154 | .356 | .359 | .123 | 13 |
| | 14 | 0+ | 0+ | 0+ | 0+ | 0+ | 0+ | 0+ | .001 | .007 | .044 | .229 | .488 | .869 | 14 |
| 15 | 0 | .860 | .463 | .206 | .035 | .005 | 0+ | 0+ | 0+ | 0+ | 0+ | 0+ | 0+ | 0+ | 0 |
| | 1 | .130 | .366 | .343 | .132 | .031 | .005 | 0+ | 0+ | 0+ | 0+ | 0+ | 0+ | 0+ | 1 |
| | 2 | .009 | .135 | .267 | .231 | .092 | .022 | .003 | 0+ | 0+ | 0+ | 0+ | 0+ | 0+ | 2 |
| | 3 | 0+ | .031 | .129 | .250 | .170 | .063 | .014 | .002 | 0+ | 0+ | 0+ | 0+ | 0+ | 3 |
| | 4 | 0+ | .005 | .043 | .188 | .219 | .127 | .042 | .007 | .001 | 0+ | 0+ | 0+ | 0+ | 4 |
| | 5 | 0+ | .001 | .010 | .103 | .206 | .186 | .092 | .024 | .003 | 0+ | 0+ | 0+ | 0+ | 5 |
| | 6 | 0+ | 0+ | .002 | .043 | .147 | .207 | .153 | .061 | .012 | .001 | 0+ | 0+ | 0+ | 6 |
| | 7 | 0+ | 0+ | 0+ | .014 | .081 | .177 | .196 | .118 | .035 | .003 | 0+ | 0+ | 0+ | 7 |
| | 8 | 0+ | 0+ | 0+ | .003 | .035 | .118 | .196 | .177 | .081 | .014 | 0+ | 0+ | 0+ | 8 |
| | 9 | 0+ | 0+ | 0+ | .001 | .012 | .061 | .153 | .207 | .147 | .043 | .002 | 0+ | 0+ | 9 |
| | 10 | 0+ | 0+ | 0+ | 0+ | .003 | .024 | .092 | .186 | .206 | .103 | .010 | .001 | 0+ | 10 |
| | 11 | 0+ | 0+ | 0+ | 0+ | .001 | .007 | .042 | .127 | .219 | .188 | .043 | .005 | 0+ | 11 |
| | 12 | 0+ | 0+ | 0+ | 0+ | 0+ | .002 | .014 | .063 | .170 | .250 | .129 | .031 | 0+ | 12 |
| | 13 | 0+ | 0+ | 0+ | 0+ | 0+ | 0+ | .003 | .022 | .092 | .231 | .267 | .135 | .009 | 13 |
| | 14 | 0+ | 0+ | 0+ | 0+ | 0+ | 0+ | 0+ | .005 | .031 | .132 | .343 | .366 | .130 | 14 |
| | 15 | 0+ | 0+ | 0+ | 0+ | 0+ | 0+ | 0+ | 0+ | .005 | .035 | .206 | .463 | .860 | 15 |

*NOTE:* 0+ represents a positive probability less than 0.0005.

From Frederick C. Mosteller, Robert E. K. Rourke, and George B. Thomas, Jr., *Probability with Statistical Applications*, 2nd ed., copyright © 1970 Pearson Education. Reprinted with permission of the publisher.

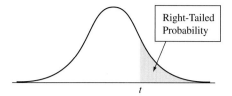

Right-Tailed Probability

$t$

# Table 4: $t$-Distribution Critical Values

| | Confidence Level | | | | | |
|---|---|---|---|---|---|---|
| | 80% | 90% | 95% | 98% | 99% | 99.8% |
| | Right-Tailed Probability | | | | | |
| df | $t_{.100}$ | $t_{.050}$ | $t_{.025}$ | $t_{.010}$ | $t_{.005}$ | $t_{.001}$ |
| 1 | 3.078 | 6.314 | 12.706 | 31.821 | 63.656 | 318.289 |
| 2 | 1.886 | 2.920 | 4.303 | 6.965 | 9.925 | 22.328 |
| 3 | 1.638 | 2.353 | 3.182 | 4.541 | 5.841 | 10.214 |
| 4 | 1.533 | 2.132 | 2.776 | 3.747 | 4.604 | 7.173 |
| 5 | 1.476 | 2.015 | 2.571 | 3.365 | 4.032 | 5.894 |
| 6 | 1.440 | 1.943 | 2.447 | 3.143 | 3.707 | 5.208 |
| 7 | 1.415 | 1.895 | 2.365 | 2.998 | 3.499 | 4.785 |
| 8 | 1.397 | 1.860 | 2.306 | 2.896 | 3.355 | 4.501 |
| 9 | 1.383 | 1.833 | 2.262 | 2.821 | 3.250 | 4.297 |
| 10 | 1.372 | 1.812 | 2.228 | 2.764 | 3.169 | 4.144 |
| 11 | 1.363 | 1.796 | 2.201 | 2.718 | 3.106 | 4.025 |
| 12 | 1.356 | 1.782 | 2.179 | 2.681 | 3.055 | 3.930 |
| 13 | 1.350 | 1.771 | 2.160 | 2.650 | 3.012 | 3.852 |
| 14 | 1.345 | 1.761 | 2.145 | 2.624 | 2.977 | 3.787 |
| 15 | 1.341 | 1.753 | 2.131 | 2.602 | 2.947 | 3.733 |
| 16 | 1.337 | 1.746 | 2.120 | 2.583 | 2.921 | 3.686 |
| 17 | 1.333 | 1.740 | 2.110 | 2.567 | 2.898 | 3.646 |
| 18 | 1.330 | 1.734 | 2.101 | 2.552 | 2.878 | 3.611 |
| 19 | 1.328 | 1.729 | 2.093 | 2.539 | 2.861 | 3.579 |
| 20 | 1.325 | 1.725 | 2.086 | 2.528 | 2.845 | 3.552 |
| 21 | 1.323 | 1.721 | 2.080 | 2.518 | 2.831 | 3.527 |
| 22 | 1.321 | 1.717 | 2.074 | 2.508 | 2.819 | 3.505 |
| 23 | 1.319 | 1.714 | 2.069 | 2.500 | 2.807 | 3.485 |
| 24 | 1.318 | 1.711 | 2.064 | 2.492 | 2.797 | 3.467 |
| 25 | 1.316 | 1.708 | 2.060 | 2.485 | 2.787 | 3.450 |
| 26 | 1.315 | 1.706 | 2.056 | 2.479 | 2.779 | 3.435 |
| 27 | 1.314 | 1.703 | 2.052 | 2.473 | 2.771 | 3.421 |
| 28 | 1.313 | 1.701 | 2.048 | 2.467 | 2.763 | 3.408 |
| 29 | 1.311 | 1.699 | 2.045 | 2.462 | 2.756 | 3.396 |
| 30 | 1.310 | 1.697 | 2.042 | 2.457 | 2.750 | 3.385 |
| 40 | 1.303 | 1.684 | 2.021 | 2.423 | 2.704 | 3.307 |
| 50 | 1.299 | 1.676 | 2.009 | 2.403 | 2.678 | 3.261 |
| 60 | 1.296 | 1.671 | 2.000 | 2.390 | 2.660 | 3.232 |
| 80 | 1.292 | 1.664 | 1.990 | 2.374 | 2.639 | 3.195 |
| 100 | 1.290 | 1.660 | 1.984 | 2.364 | 2.626 | 3.174 |
| $\infty$ | 1.282 | 1.645 | 1.960 | 2.326 | 2.576 | 3.091 |

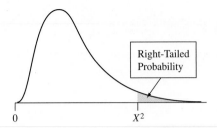

0     $X^2$

## Table 5: Chi-Square Distribution for Values of Various Right-Tailed Probabilities

| df | Right-Tailed Probability | | | | | | |
|---|---|---|---|---|---|---|---|
| | 0.250 | 0.100 | 0.050 | 0.025 | 0.010 | 0.005 | 0.001 |
| 1 | 1.32 | 2.71 | 3.84 | 5.02 | 6.63 | 7.88 | 10.83 |
| 2 | 2.77 | 4.61 | 5.99 | 7.38 | 9.21 | 10.60 | 13.82 |
| 3 | 4.11 | 6.25 | 7.81 | 9.35 | 11.34 | 12.84 | 16.27 |
| 4 | 5.39 | 7.78 | 9.49 | 11.14 | 13.28 | 14.86 | 18.47 |
| 5 | 6.63 | 9.24 | 11.07 | 12.83 | 15.09 | 16.75 | 20.52 |
| 6 | 7.84 | 10.64 | 12.59 | 14.45 | 16.81 | 18.55 | 22.46 |
| 7 | 9.04 | 12.02 | 14.07 | 16.01 | 18.48 | 20.28 | 24.32 |
| 8 | 10.22 | 13.36 | 15.51 | 17.53 | 20.09 | 21.96 | 26.12 |
| 9 | 11.39 | 14.68 | 16.92 | 19.02 | 21.67 | 23.59 | 27.88 |
| 10 | 12.55 | 15.99 | 18.31 | 20.48 | 23.21 | 25.19 | 29.59 |
| 11 | 13.70 | 17.28 | 19.68 | 21.92 | 24.72 | 26.76 | 31.26 |
| 12 | 14.85 | 18.55 | 21.03 | 23.34 | 26.22 | 28.30 | 32.91 |
| 13 | 15.98 | 19.81 | 22.36 | 24.74 | 27.69 | 29.82 | 34.53 |
| 14 | 17.12 | 21.06 | 23.68 | 26.12 | 29.14 | 31.32 | 36.12 |
| 15 | 18.25 | 22.31 | 25.00 | 27.49 | 30.58 | 32.80 | 37.70 |
| 16 | 19.37 | 23.54 | 26.30 | 28.85 | 32.00 | 34.27 | 39.25 |
| 17 | 20.49 | 24.77 | 27.59 | 30.19 | 33.41 | 35.72 | 40.79 |
| 18 | 21.60 | 25.99 | 28.87 | 31.53 | 34.81 | 37.16 | 42.31 |
| 19 | 22.72 | 27.20 | 30.14 | 32.85 | 36.19 | 38.58 | 43.82 |
| 20 | 23.83 | 28.41 | 31.41 | 34.17 | 37.57 | 40.00 | 45.32 |
| 25 | 29.34 | 34.38 | 37.65 | 40.65 | 44.31 | 46.93 | 52.62 |
| 30 | 34.80 | 40.26 | 43.77 | 46.98 | 50.89 | 53.67 | 59.70 |
| 40 | 45.62 | 51.80 | 55.76 | 59.34 | 63.69 | 66.77 | 73.40 |
| 50 | 56.33 | 63.17 | 67.50 | 71.42 | 76.15 | 79.49 | 86.66 |
| 60 | 66.98 | 74.40 | 79.08 | 83.30 | 88.38 | 91.95 | 99.61 |
| 70 | 77.58 | 85.53 | 90.53 | 95.02 | 100.43 | 104.21 | 112.32 |
| 80 | 88.13 | 96.58 | 101.88 | 106.63 | 112.33 | 116.32 | 124.84 |
| 90 | 98.65 | 107.57 | 113.15 | 118.14 | 124.12 | 128.30 | 137.21 |
| 100 | 109.14 | 118.50 | 124.34 | 129.56 | 135.81 | 140.17 | 149.45 |

# Appendix B: Answers to Check Your Tech

## CHAPTER 3

### Finding the Standard Deviation of Vacation Days for Several Countries

**1:** $180/6 = 30$

**2:**

| $x$ | $x - \bar{x}$ | $(x - \bar{x})^2$ |
|-----|---------------|--------------------|
| 13 | $13 - 30 = -17$ | $(-17)^2 = 289$ |
| 25 | $25 - 30 = -5$ | $(-5)^2 = 25$ |
| 42 | $42 - 30 = 12$ | $12^2 = 144$ |
| 37 | $37 - 30 = 7$ | $7^2 = 49$ |
| 35 | $35 - 30 = 5$ | $5^2 = 25$ |
| 28 | $28 - 30 = -2$ | $(-2)^2 = 4$ |

**3:** $289 + 25 + 144 + 49 + 25 + 4 = 536$

**4:** $536/5 = 107.2$

**5:** $\sqrt{107.2} = 10.3537$, which rounds to 10.35. This is the same value shown in Figure A.

### Making a Boxplot of the Area of Western States

**1:** The axis is shown in the graph.
**2:** The box goes from 84.5 on the left to 134.5 on the right.
**3:** The median line is at 104, which is Colorado.
**4:** IQR $= 134.5 - 84.5 = 50$
**5:** Lower limit $= Q1 - (1.5 \times IQR) = 84.5 - (1.5 \times 50) = 84.5 - 75 = 9.5$
The smallest area is 11, which is not below 9.5, so there are no low-end potential outliers.
**6:** The low whisker goes down to 11 (Hawaii).
**7:** Upper limit $= Q3 + (1.5 \times IQR) = 134.5 + (1.5 \times 50) = 134.5 + 75 = 209.5$
The only area larger than 209.5 is 656 (Alaska), which is an outlier and should have a separate mark.
**8:** The upper whisker goes to 164 (California), which is the largest area that is not a potential outlier.
**9:** One title is shown in the graph.

## CHAPTER 4

**1:**

| $x$ | $x - \bar{x}$ | $z_x$ | $y$ | $y - \bar{y}$ | $z_y$ | $z_x z_y$ |
|-----|---------------|-------|-----|---------------|-------|-----------|
| 20 | $20 - 30 = -10$ | $-10/10 = -1$ | 20 | $20 - 25 = -5$ | $-5/5 = -1$ | $(-1) \times (-1) = 1$ |
| 30 | $30 - 30 = 0$ | $0/10 = 0$ | 30 | $30 - 25 = 5$ | $5/5 = 1$ | $0 \times (1) = 0$ |
| 40 | $40 - 30 = 10$ | $10/10 = 1$ | 25 | $25 - 25 = 0$ | $0/5 = 0$ | $1 \times 0 = 0$ |

**2:** Add the last column to get $\sum z_x z_y = 1 + 0 + 0 = 1$.

**3:** Find the correlation and check it with the output.

$$r = \frac{\sum z_x z_y}{n - 1} = \frac{1}{3 - 1} = \frac{1}{2} = 0.5$$

This is the same correlation as the output.

**4:** Find the slope.

$$b = r \frac{s_y}{s_x} = r \times \frac{5}{10} = 0.50 \times (0.50) = 0.25$$

**5:** Find the $y$-intercept.

$$a = \bar{y} - b\bar{x} = 25 - 0.25 \times 30 = 25 - 7.5 = 17.5$$

**6:** Finally, put together the equation:

$$y = a + bx$$

$$\text{Predicted Wife} = a + b \text{ Husband}$$

$$\text{Predicted Wife} = 17.5 + 0.25 \text{ Husband}$$

The equation is the same as the Minitab output.

## CHAPTER 9

**1:** $73.6 - 92.5143 = -18.9143$ (which is also shown in the Minitab Output)

**2:** $SE_{\text{diff}} = \dfrac{S_{\text{diff}}}{\sqrt{n}} = \dfrac{15.0497}{\sqrt{35}} = \dfrac{15.0497}{5.9161} = 2.5439$

**3:** $t = \dfrac{\bar{x}_{\text{diff}} - 0}{SE_{\text{diff}}} = \dfrac{-18.9143}{2.5439} = -7.44$

**4:** a. If the means were farther apart, that would cause $t$ to be farther from 0. Because the difference between means is in the numerator, a bigger difference in the numerator means a value farther from 0. b. If the standard deviation ($S_{\text{diff}}$) were larger, that would cause $t$ to be closer to 0, because the standard deviation is in the denominator, and a larger denominator results in a value closer to 0. c. If the sample size were larger, that would cause $t$ to be farther from 0. The larger sample size would cause $SE_{\text{diff}}$ to be smaller (because the sample size is in the denominator of $SE_{\text{diff}}$), and the smaller $SE_{\text{diff}}$ would cause $t$ to be farther from 0 (because $SE_{\text{diff}}$ is in the denominator of $t$).

# Appendix C:
# Answers to Odd-Numbered Exercises

## CHAPTER 1

## Section 1.2

**1.1 a.** Handedness: categorical **b.** Age: numerical

**1.3** Male is categorical with two categories. The 1's represent males, and the 0's represent females. If you added the numbers, you would get the number of males, so it makes sense here.

**1.5 a.** Stacked **b.** 1 means male, and 0 means female.

**c.**

| Female | Male |
|--------|------|
| 9.5 | 9.4 |
| 9.5 | 9.5 |
| 9.9 | 9.5 |
| | 9.7 |

## Section 1.3

**1.7 a.** $189/29{,}617 = 0.64\%$ changes **b.** $124/14{,}513 = 0.85\%$ changes

**1.9 a.** $15/38 = 39.5\%$ of the class were male **b.** $0.641(234) = 149.994$, or about 150, men in the class

**c.** $0.40(x) = 20$
$20/0.4 = 50$ people in the class

**1.11** The frequency of women is 7, the proportion is $7/11$, and the percentage is 63.6%.

**1.13** $0.202x = 88{,}547{,}000$
$x = 438{,}351{,}485$ or a rounded version of this

**1.15** Steps 1–3 are shown in the accompanying table.

| State | AIDS | Rank Cases | Population | Population (thousands) | AIDS per 1000 | Rank Rate |
|-------|------|-----------|-----------|----------------------|---------------|-----------|
| New York | 75,253 | 1 | 19,297,729 | 19,298 | 3.90 | 2 |
| California | 65,582 | 2 | 36,553,215 | 36,553 | 1.79 | 4 |
| Florida | 48,059 | 3 | 18,251,243 | 18,251 | 2.63 | 3 |
| Texas | 34,940 | 4 | 23,904,380 | 23,904 | 1.46 | 6 |
| Pennsylvania | 19,236 | 5 | 12,432,792 | 12,433 | 1.55 | 5 |
| District of Columbia | 8,895 | 6 | 588,292 | 588 | 15.13 | 1 |

4: No, the ranks are not the same. The District of Columbia had the highest rate and had the lowest number of cases. (Also, the rate for Florida puts its rank above California, and the rate for Pennsylvania puts it above Texas in ranking.)

5: The District of Columbia is the place (among these six regions) where you would be most likely to meet a person living with AIDS, and Texas is the place (among these six regions) where you would be least likely to do so.

**1.17** 1990: 58.7%, 1997: 56.4%, 2000: 56.2%, 2007: 55.1% The percentage of married people is decreasing over time (at least with these dates).

**1.19** We don't know the percentage of female students in the two classes. The larger number of women at 8 a.m. may just result from a larger number of students at 8 a.m., which may be because the class can accommodate more students because perhaps it is in a large lecture hall.

## Section 1.4

**1.21** Observational study

**1.23** Controlled experiment

**1.25** Controlled experiment

**1.27** Observational study

**1.29** This was an observational study, and from it you cannot conclude that the tutoring raises the grades. Possible confounders (answers may vary): 1. It may be the more highly motivated who attend the tutoring, and this motivation is what causes the grades to go up. 2. It could be that those with more time attend the tutoring, and it is the increased time studying that causes the grades to go up.

**1.31** It was an observational study. There was no random assignment. They simply looked at records. We cannot say that CABG causes better results, because there may have been confounding variables.

**1.33** The students should have been randomly assigned to the treatment. Half of the students should have been given a placebo. Also, ideally the person to whom the subjects report their results should not know whether they are taking vitamin C or a placebo, making the study double-blind.

**1.35** Because this was not a randomized experiment (no one assigned people to be happy), we should not infer causation. It is possible that healthier people tend to be happier, and it was the health of those people that caused them not to catch a cold. If we cannot infer causation, then you should not believe that a change in happiness would change the likelihood of catching a cold.

**1.37** Ask whether the patients were randomly assigned the full or the half dose. Without randomization there could be bias, and we cannot infer causation. With randomization we can infer causation.

**1.39** This was an observational study: vitamin C and breast milk. We cannot conclude cause and effect from observational studies.

**1.41 a.** LD: 8% tumors; LL: 28% tumors **b.** A controlled experiment. You can tell by the random assignment. **c.** Yes, we can conclude cause and effect because it was a controlled experiment, and random assignment will balance out potential confounding variables.

### Chapter Review Exercises

**1.43 a.** Dating: $81/440 = 18.4\%$ **b.** Cohabiting: $103/429 = 24.0\%$ **c.** Married: $147/424 = 34.7\%$ **d.** No, this was an observational study. Confounding variables may vary. Perhaps married people are likely to be older, and older people are more likely to be obese.

**1.45 a.** The two-way table follows.

| | Boy | Girl |
|-----------|-----|------|
| Violent | 10 | 11 |
| Nonviolent | 19 | 4 |

**b.** For the boys, $10/29$, or 34.5%, were on probation for violent crime. For the girls, $11/15$, or 73.3%, were on probation for violent crime. **c.** The girls were more likely to be on probation for violent crime.

**1.47** Writing: Answers will vary but should include randomization, placebo, control, and blinds.

**1.49** This was an observational study, not a controlled experiment. From observational studies you cannot conclude cause and effect. Therefore, offering to participate in clinical trials would probably have very little effect on one's chance of surviving. Possible confounders will vary. However, the people who offer to be in clinical trials may tend to follow the advice of doctors more, and it may be the practice of following advice that raises their chances of survival.

**1.51** No, we cannot conclude causation. There was no control group for comparison, and the sample size was very small.

## CHAPTER 2

### Sections 2.1 and 2.2

**2.1 a.** 11 are morbidly obese. **b.** 11/134 is about 8%, which is much more than 3%.

**2.3** New vertical axis labels: 0.04, 0.08, 0.12, 0.16, 0.20, 0.24, 0.28. Note that 0.04 comes from 1/25

**2.5 a.** 1 (or 2) have no TVs **b.** 9 TVs **c.** Between 25 and 30 **d.** Around 6 **e.** 6/90, or 0.0667

**2.7 a.** NYC **b.** Las Vegas **c.** Las Vegas

**2.9** It should be right-skewed.

**2.11** It would be bimodal because the men and women tend to have different heights and therefore different armspans.

**2.13** About 58 years (between 56 and 60)

**2.15** Riding the bus shows a larger typical value and also more variation.

**2.17 a.** Multimodal with modes at 12 years (high school), 14 years (junior college), 16 years (bachelor's degree), and 18 years (possible master's degree). It is also left-skewed with numbers as low as 0. **b.** Estimate: 300 + 50 + 100 + 40 + 50, or about 500 to 600, had 16 or more years. **c.** This is between 25% (from 500/2018) and 30% (from 600/2018) have a bachelor's degree or higher. This is very similar to the 27% given.

**2.19** Both graphs go from about 0 to about 20, but the data for years of formal education for the respondents (compared to their mothers) include more with education above 12 years. For example, the bar at 16 (college bachelor's degree) is higher for the respondents than for the mothers, which shows that the respondents tend to have a bit more education than their mothers. Also, the bar at 12 is taller for the mothers, showing that the mothers were more likely to get only a high school diploma. Furthermore, the bar graph for the mothers includes more people (taller bars) at lower numbers of years, such as 0 and 3 and 6.

**2.21 1.** B **2.** C **3.** A

**2.23 1.** See the dotplots. Histograms would also be good for visualizing the distributions. Stemplots would not work with these data sets because all the observed values have only one digit.

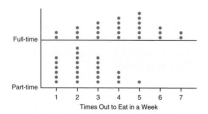

**2.** Full-time is a bit left-skewed, and part-time is a bit right-skewed.

**3.** Those with full-time jobs tend to go out to eat more than those with part-time jobs.

**4.** The full-time workers have a distribution that is more spread out; full-time goes from 1 to 7, whereas part-time goes only from 1 to 5.

**5.** There are no outliers—that is, no dots detached from the main group with an empty space between.

**6.** For the full-time workers the distribution is a little left-skewed, and for the part-time workers it is a little right-skewed. The full-time workers tend to go out to eat more, and their distribution is more spread out.

**2.25** See histogram.

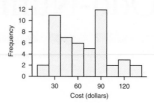

The histogram is bimodal with modes at about $30 and about $90.

**2.27** See histogram. The histogram is right-skewed. The typical value is around 12 (between 10 and 15) years, and there are three outliers: Asian elephant (40), African elephant (35), and hippo (41). Humans (75 years) would be way off to the right; they live much longer than other mammals.

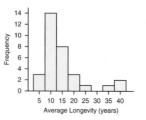

**2.29 1.** B **2.** A **3.** C

### Sections 2.3 and 2.4

**2.31** No, the largest category is Wrong to Right, which suggests that changes tend to make the answers more likely to be right.

**2.33 a.** Baidu is the mode (the most frequently used search engine) for both time periods. It becomes even more popular in 2010. **b.** There was more variability in 2009, because in 2010 the predominance of Baidu was more noticeable.

**2.35 a.** Democrat (not strong). It is easier to compare using the heights of the bars in the bar chart. **b.** Republican (not strong)

**2.37 a.** The percentage of old people is increasing, the percentage of those 25–64 is decreasing, and the percentage of those 24 and below is relatively constant. **b.** The money for Social Security normally comes from those in a working age range (which includes those 25–64), and that group is decreasing in percentage. Also, the group receiving Social Security (those 65 and older) is becoming larger. This suggests that in the future, Social Security might not get enough money from the workers to support the old people.

**2.39** A Pareto chart or pie chart would also be appropriate.

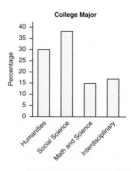

Note that the mode is Social Science and that there is substantial variation. (Of course, individual majors such as chemistry were grouped into Math and Science.)

## Chapter Review Exercises

**2.41** TV: Histograms: One for the males and one for the females would be appropriate. Dotplots or stemplots would also work for this numerical data set.

**2.43 a.** The diseases with higher rates for HRT were heart attack, stroke, pulmonary embolism, and breast cancer. The diseases with lower rates for HRT were endometrial cancer, colorectal cancer, and hip fracture. **b.** Comparing the rates makes more sense than comparing just the numbers, in case there were more women in one group than in the other.

**2.45** The vertical axis does not start at zero and exaggerates the differences. Make a graph for which the vertical axis starts at zero.

**2.47 a.** Pie chart **b.** Histogram

**2.49** The shapes are roughly Normal; the later period is warmer, but the spread is similar. This is consistent with theories on global warming. The difference is $57.9 - 56.7 = 1.2$, so the difference is only a bit more than 1 degree Fahrenheit.

**2.51** The created 10-point dotplots will vary. The dotplot should have skew.

**2.53** Graphs will vary. Histograms, dotplots, or stemplots are all appropriate. The prices in West Los Angeles tend to be higher and more varied than the prices in Midtown.

**2.55** The data set should be right-skewed with some unusually high numbers.

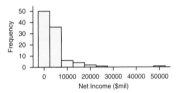

Exxon Mobil is an outlier.

## CHAPTER 3

### Section 3.1

**3.1** c

**3.3** The typical age of the CEOs is between about 56 and 60 (or any number from 56 to 60). (The distribution is symmetric, so the mean should be about in the middle.)

**3.5 a.** The typical number of vacation days is about 17.3, the mean. One could also say that number of vacation days tends to be about 17.3, on average. **b.** The standard deviation of paid vacation days for the six countries is 10.7 **c.** The United States, at 0, is farthest from the mean and contributes most to the standard deviation.

**3.7 a.** 57.8, Older: 57.8 (the mean age for the first six presidents) is more than 55.3 for the most recent six presidents. **b.** Less: 1.6 (the standard deviation of ages for the first six presidents) tends to be much less varied than the 9.3 for the most recent six presidents.

**3.9 a.** 35,520; 117,845, eastern **b.** 24,704; 124,646, western **c.** Eastern

**3.11 a.** The mean for longboards is 12.4 days which is more than the mean for shortboards, which is 9.8 days. So longboarders tend to get more surfing days. **b.** The standard deviation of 5.2 days for the longboarders was larger than the standard deviation of 4.2 days for the shortboarders. So the longboarders have more variation in days.

**3.13** The prices of the houses in Westlake (Figure B) have a larger standard deviation than the prices in Agoura, because the data from Agoura show a lot of prices near the center of the graph and the prices in Westlake show a lot of prices far from the center.

**3.15 a.** Top: $3462 + 500 = 3962$
Bottom: $3462 - 500 = 2962$
**b.** Yes, a birth weight of 2800 grams is more than one SD below the mean because it is less than 2962.

**3.17 a.** The standard deviation is 1.4 years. **b.** The same. The standard deviation in 20 years is still 1.4. Adding 20 to each number does not affect the standard deviation. Standard deviation does not depend on the size of the numbers, only on how far apart they are. **c.** The mean is 3 years. **d.** Larger: The mean is 23 years of age. When 20 is added to each number, the mean increases by 20.

**3.19** SD for the 100-meter event would be less. All the runners come to the finish line within a few seconds of each other. In the marathon, the runners can be quite widely spread after running that long distance.

### Section 3.2

**3.21** Answers correspond to the guided steps.
1: 95% (See the accompanying curve.)
2: 583 is from $406 + 177$, because it is one standard deviation above the mean.
3: As shown on the curve, A is 52, B is 229, and C is 760.
4: Answer a. About 95% between 52 and 760.
5: Answer b: About 68% between 229 and 583.
6: Answer c: Most would not consider 584 unusual because it is between 52 and 760.
7: Answer d: 30 is unusually small, because it is less than 52, which means it is more than two standard deviations below the mean, and so less than 2.5% of the population have values lower than this.

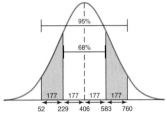

**3.23** $z = \dfrac{4060 - 3331}{729} = \dfrac{729}{729} = 1$

**a.** About 68% according to the Empirical Rule, because the $z$-scores are 1 and −1. **b.** About 95% according to the Empirical Rule **c.** Nearly all the data should be within three standard deviations of the mean. Three standard deviations above the mean is

$$3331 + 3(729) = 5518$$

Because 9000 is quite a bit above 5518, you might think it unlikely that one state's rate would be 9000.

**3.25 a.** −2 **b.** 67 inches (or 5 feet 7 inches)

**3.27** The $z$-score for the SAT of 750 is 2.5, and the $z$-score for the ACT of 28 is 1.4. The score of 750 is more unusual, because its $z$-score is farther from 0.

**3.29 a.** $z = \dfrac{2500 - 3462}{500} = \dfrac{-962}{500} = -1.92$

**b.** $z = \dfrac{2500 - 2622}{500} = \dfrac{-122}{500} = -0.24$

**c.** A birth rate of 2500 grams is more common (the $z$-score is closer to 0) for babies born one month early. In other words, there are a higher percentages of babies with low birth weight among those born one month early. This makes sense because babies gain weight during gestation, and babies born one month early have had less time to gain weight.

**3.31** $70 + 1.5(10) = 85$

### Section 3.3

**3.33** Two measures of the center of data are the mean and the median. The median is preferred for data that are strongly skewed or have outliers. If the data are relatively symmetric, the mean is preferred but the median is also OK.

**3.35**

| 163 | 192 | 206 | 224 | 244 | | 246 | 256 | | 261 | 293 | 340 |
|-----|-----|-----|-----|-----|-----|-----|-----|-----|-----|-----|-----|
| | | | Q1 | | | Med | | | Q3 | | |

**a.** The median of 245 million is the typical income for the top ten grossing Pixar animated movies. **b.** IQR = 261 − 206 = 55 million. This is the range of the middle 50% of the sorted incomes in the top ten grossing Pixar animated movies.

**3.37**

| 17.3 | 22.0 | 23.0 | 24.0 | 25.0 | 27.2 | 27.7 |
|------|------|------|------|------|------|------|
|      | Q1   |      | Med  |      | Q3   |      |

**a.** 24.0 cents per gallon **b.** IQR = Q3 − Q1 = 27.2 − 22.0 = 5.2 cents per gallon.

**3.39 a.** 11.5, 9.5, longboards **b.** 8, 5, longboards (Answers for interquartile range may vary with different technology.)

## Section 3.4

**3.41 a.** Outliers are observed values that are far from the main group of data. In a histogram they are separated from the others by space. If they are mistakes, they should be removed. If they are not mistakes, do the analysis twice: once with and once without outliers. **b.** The median is more resistant, which implies that it changes less than the mean (when comparing the data with and without outliers).

**3.43** The corrected value will give a different mean but not a different median. Medians are not as affected by the size of extreme scores, but the mean is affected.

**3.45 a.** The shape is right-skewed, the median is 20, and the interquartile range is 35 − 19 = 16. There is an outlier at 66. **b.** The mean is 27.3 and the median is 20. The mean is much larger because of the outlier, 66. The mean and median should be marked on the histogram of the data.

**3.47 a.** Both data sets are right-skewed and have outliers that represent large numbers of hours of study, so the medians (and interquartile ranges) should be compared. **b.** The median of 7 was larger for the women; the men's median was only 4. The interquartile range was 4.5 for the women and 3 for the men, so the IQR was larger for the women. Both data sets are right-skewed with outliers at around 15 or 20 hours. Summary: The women tended to study more and had more variation as measured by the interquartile range.

## Section 3.5

**3.49 a.** S, MW, W, NE (lowest to highest medians) **b.** The Northeast has the largest IQR. **c.** The South has the smallest IQR. **d.** The South has potential outliers. **e.** The box for the South is very small and far to the left, so the numbers in the 20's are very far from the box. The box for the Northeast is large and far to the right, so 24.5 is not a potential outlier.

**3.51** Answers will vary: Phoenix tends to be hottest, since median temperatures are high, although Honolulu and Phoenix have approximately the same medians. The most varied temperatures are in Chicago. Both Honolulu and L.A. have relatively little variation, as determined by the interquartile range.

    The choice of favorite city will vary.

**3.53 a.** Histogram 1 goes with boxplot C.
    Histogram 2 goes with boxplot B.
    Histogram 3 goes with boxplot A.
    Reasoning: A boxplot with a whisker (or potential outliers) on the far right corresponds to a histogram that is right-skewed. A boxplot that is centered approximately symmetrically corresponds to a histogram that is not very skewed.
    **b.** Histogram 1 is strongly right-skewed, histogram 2 is left-skewed, and histogram 3 is relatively symmetric.
    Histogram 1 would have a mean larger than the median.
    Histogram 2 would have a mean smaller than the median.
    Histogram 3 would have a mean and median that were very similar.

**3.55 a.** The median is 20. Q1 is 17.5 and Q3 is 23, so the IQR is 5.5. There is a high-end outlier of about 33, which is Wyoming.

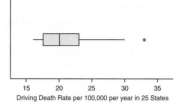

Driving Death Rate per 100,000 per year in 25 States

**b.** A low population relative to the number of deaths is another way of seeing a high death rate. (Wyoming has a low population.)

## Chapter Review Exercises

**3.57** Both data sets are slightly right-skewed because of the potential outliers. The men tend to be faster than the women, as shown by the lower median line for the men. The interquartile range is larger for the men, as shown by the larger box.

**3.59 a.** The median number of death row prisoners in the states in the South is 84 **b.** IQR = Q3 − Q1 = 166 − 36 = 130
    The range of the middle 50% of the state data on death row prisoners in the South is 130.
**c.** The mean number of death row prisoners in the states in the South is 115.6 **d.** The mean is pulled up by the really large numbers, such as the numbers from Texas and Florida. The median is not affected by the size of these large numbers. **e.** The median is more stable with outliers.

**3.61** The answers given follow the steps in the Guided Exercises.
1:

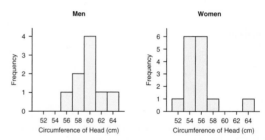

2: The men's histogram is roughly symmetric (bell-shaped). The women's histogram is bell-shaped except for the high-end outlier, so it may be called right-skewed.
3: Compare the medians and interquartile ranges because of the outlier for the women.
4: median, 59.5, median, 55, men
5: Men's IQR = 3.75. Women's IQR = 2. IQR may vary with different technology.
6: The measurement of 63 for the women was an outlier.
7: Both data sets are unimodal and roughly symmetric except for one large outlier for the women. The men tended to have larger heads with more variation in size than the women. However, there was a large outlier (63 cm) for the women.

**3.63** Summary statistics are shown below. The 5 p.m. class did better, typically; both the mean and the median are higher**.** Also, the spread (as reflected in both the standard deviation and the IQR) is larger for the 11 a.m. class, so the 5 p.m. class has less variation.
    The visual comparison is shown by the boxplots. Both distributions are slightly left-skewed. Therefore, you can compare the means and standard deviations *or* the medians and IQRs.

```
Minitab Statistics
Variable   N      Mean    Median   StDev   Min    Max     Q1   Q3
11am      15     70.73    72.5     19.84   39     100     53   86
5pm       19     84.78    86.5     11.95   64.5   104.5   73   94
```

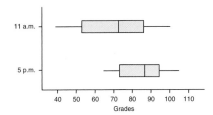

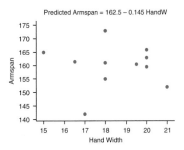
Predicted Armspan = 162.5 − 0.145 HandW

**3.65** The graph is bimodal, with modes around 65 inches (5 feet 5 inches) and around 69 inches (5 feet 9 inches). There are two modes because men tend to be taller than women.

**3.67 a.** The approximate mean GPA is around 3.1 or 3.2 (GPA) **b.** The approximate standard deviation of the GPAs is about 0.2, from $(3.8 − 2.6)/6$ **c.** The data set is unimodal and roughly symmetric and so all (or nearly all) of the data should be within three standard deviations of the mean. Thus one boundary is up 3SD and one is down 3SD, which is a difference of 6SD.

**3.69** and **3.71** Constructed numbers will vary.

**3.73** and **3.75** Answers will vary but should include graphs and comparison of centers and variation.

## CHAPTER 4

### Section 4.1

**4.1 a.** Acres: The number of acres has a stronger relationship with the value of the land, as shown by the fact that the points are less scattered in a vertical direction. **b.** Acreage. The association is stronger between the value of land and acreage than with the number of fireplaces because the vertical spread is less.

**4.3** Very little trend.

**4.5** The more people weigh, the more weight they tend to want to lose.

**4.7** The states with higher populations tend to have more motor vehicle fatalities.

**4.9** The trend is positive, but the trend tends to level off at higher acreages. Generally, homes on larger acreages have larger assessed values.

### Section 4.2

**4.11** Linear regression is not appropriate because the trend is not linear, it is curved.

**4.13 a.** Decreasing: States with a higher percentage of students who take the SAT generally have lower mean math SATs. **b.** Somewhat curved. **c.** No. The trend is not linear.

**4.15** The correlation between age and GPA would be near zero.

**4.17**  0.767 A
     0.299 B
    −0.980 C

**4.19 a.** 0.987 **b.** The correlation remains the same ($r = 0.987$) if you multiply a constant. **c.** The correlation remains the same ($r = 0.987$) if you add a constant.

### Section 4.3

**4.21 a.** Between 40,000 and 50,000 **b.** 44,299

**4.23 a.** Predicted Armspan = 16.8 + 2.25 height
**b.** $b = 0.948(8.10/3.41) = 2.25$ **c.** $a = 159.86 − 2.25(63.59) = 16.8$
**d.** Armspan = 16.8 + 2.25(64) = 160.8, or about 161 cm

**4.25 a.** Refer to the scatterplot. **b.** The trend is not linear.

**4.27 a.** Predicted Armspan = 6.24 + 2.515 Height (Rounding may vary.)
**b.** Minitab: slope = 2.51, intercept = 6.2
StatCrunch: slope = 2.514674, intercept = 6.2408333
Excel: slope = 2.514674, intercept = 6.240833
TI-83/84: slope = 2.514673913, intercept = 6.240833333

**4.29** The horizontal axis starts at 0. To zoom in on the data, the horizontal axis should have been started at about 60. This would have rescaled the graph, making it easier to interpret.

**4.31** The correlation for the women is stronger because the $r$-squared value is closer to 1.

**4.33 a.** The slope would be near 0. **b.** $r$ is about 0 **c.** Last two digits of Social Security Number is not associated with age.

**4.35** Explanations will vary.

|      | $x$    | $y$       |
|------|--------|-----------|
| **a.** | gas    | miles     |
| **b.** | years  | salary    |
| **c.** | weight | belt size |

**4.37 a.** The higher the percentage of smoke-free homes in a state, the lower the percentage of high school students who smoke tends to be.
**b.** $56.32 − 0.464(70) = 23.84$, or about 24%

**4.39 a.** The graph shows that young drivers and old drivers have more fatalities and that the safest drivers are between about 40 and 60 years of age. **b.** It would not be appropriate for linear regression because the trend is not linear.

**4.41.** The answers are given in the order shown in the Guidance section.

1: The regression line is shown with the scatterplot.

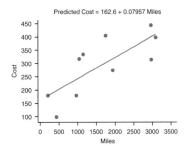
Predicted Cost = 162.6 + 0.07957 Miles

2: Is the linear model appropriate?

Given: In this case the answer is yes, because there is a linear trend. It is hard to see with so few points, but a strong curvature is not present and the cost tends to increase as the miles increase.

3: Predicted Cost = 162.60 + 0.07957 Miles

4: Put in the correct regression line (shown in Step 1).

5: Interpret the slope and intercept in context.

For the slope: For every additional mile, on average, the price goes up by 0.07957 dollars (or about 8 cents).

For the intercept: A trip of 0 miles should cost about 162.60 dollars on average. Beware of extrapolation. The linear trend might not continue near 0 miles and so this prediction could be very different from truth.

Predicted cost = 162.60 + 0.07957 miles

6: Final answer: For the hypothetical flight of 500 miles, you would pay

$$162.60 + 0.07957 \times 500 = 202.385, \text{ or about 202 dollars}$$

**4.43 a.** Positive: The more population, the more millionaires there tend to be. **b.** See scatterplot. **c.** $r = 0.992$ **d.** For each additional hundred thousand in the population, there is an additional 1.9 thousand millionaires.

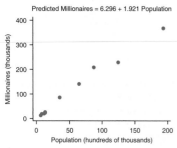

Predicted Millionaires = 6.296 + 1.921 Population

**e.** Do not focus on the intercept, because it does not make sense to look for millionaires in states with no people.

**4.45 a.** The correlation should be positive. The more people living in the state, the more drivers, and the more motor vehicle fatalities there should be. **b.** See figure. **c.** $r = 0.949$ **d.** Equation and line are shown on plot. **e.** For each additional hundred thousand people in a state, there are about 9 more fatalities on average.

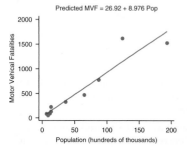

Predicted MVF = 26.92 + 8.976 Pop

## Section 4.4

**4.47 a.** An influential point is a point that changes the regression equation by a large amount. **b.** Going to church may not cause lower blood pressure; just because two variables are related does not show that one caused the other. It could be that healthy people are more likely to go to church, or there could be other confounding factors.

**4.49** Older children have larger shoes and have studied math longer. Large shoes do not cause higher grades. Both are affected by age.

**4.51** Square 0.67 and you get 0.4489, so the coefficient of determination is about 45%. Therefore, 45% of the variation in weight can be explained by the regression line.

**4.53** The cholesterol going down might be partly caused by regression toward the mean.

**4.55 a.** The salary is $2.099 thousand less for each year later that the person was hired, or $2.099 thousand more for each year earlier. **b.** The intercept ($4,255,000) would be the salary for a person who started in the year 0, which is inappropriate (and ridiculous).

**4.57 a.** See graph.

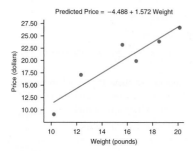

Predicted Price = −4.488 + 1.572 Weight

**b.** $r = 0.933$
A positive correlation suggests that larger turkeys tend to have a higher prices. **c.** Predicted Price $= -4.49 + 1.57$ Weight **d.** The line is on the graph. **e.** The slope: For each additional pound, the price goes up by $1.57. The interpretation of the intercept is inappropriate, because it is not possible to have a turkey that weighs 0 pounds.
**f.** The 30-pound free turkey changes the correlation to $-0.375$ and changes the equation to

$$\text{Predicted Cost} = 26.87 - 0.553 \text{ Weight}$$

This implies that the bigger the turkey, the less it costs! The 30-pound free turkey was an influential point, which really changed the results.

**4.59 a.** Positive correlation **b.** For each additional dollar spent on teachers' pay, the expenditure per pupil goes up about 23 cents on the average.
**c.** The intercept might represent the mean cost of education excluding the cost of paying teachers. However, doing so requires extrapolation which may not be valid.

**4.61 a.** Correlation is negative. **b.** For each additional hour of work, the score tended to go down by 0.48 point. **c.** A student who did not work would expect to get about 87 on average.

**4.63 a.** $Predicted\ Mother's\ Education = 3.12 + 0.637(12) = 3.12 = 7.64 = 10.76\ (12, 10.76)$

$Predicted\ Mother's\ Education = 3.12 + 0.637\ (4.0) = 3.12 + 2.548 = 5.67\ (4, 5.67)$
**b.** The new equation is given below.

$Predicted\ Father's\ Education = 2.78 + 0.637\ (Mother's\ Education)$

$Predicted\ Father's\ Education = 2.78 + 0.637\ (12) = 10.42\ (12, 10.42)$

$Predicted\ Father's\ Education = 2.78 + 0.637\ (4) = 5.33\ (4, 5.33)$
**c.** Regression toward the mean: Values for the predictor variable that are far from the mean lead to responses that are closer to the mean.

**4.65** 1: i. Slope: $b = r\dfrac{s_y}{s_x} = 0.7\dfrac{10}{10} = 0.7$

ii. Intercept: $a = \bar{y} - b\bar{x}$

$$a = 75 - 0.7(75) = 22.5$$

iii. Equation

$$\text{Predicted final} = 22.5 + 0.7 \text{ Midterm}$$

2: Predicted final $= 22.5 + 0.7$ Midterm
$$= 22.5 + 0.7(95)$$
$$= 89$$

3: The score of 89 is lower than 95 because of regression toward the mean.

## Chapter Review Exercises

**4.67 a.** $r = 0.941$

Predicted Weight $= -245 + 5.80$ Height

**b.**

| Height | Weight |
|---|---|
| 60(2.54) = 152.4 | 105/2.205 = 47.6190 kg |
| 66(2.54) = 167.64 | 140/2.205 = 63.4921 kg |
| 72(2.54) = 182.88 | 185/2.205 = 83.9002 kg |
| 70(2.54) = 177.8 | 145/2.205 = 65.7596 kg |
| 63(2.54) = 160.02 | 120/2.205 = 54.4218 kg |

**c.** The correlation between height and weight is 0.941. It does not matter whether you use inches and pounds or centimeters and kilograms. A change of units does not affect the correlation because it has no units.
**d.** The equations are different.

Predicted Weight Pounds $= -245 + 5.80$ Height (in inches)

Predicted Weight Kilograms $= -111 + 1.03$ Height (in centimeters)

**4.69 a.** Choose $r = +0.75$ **b.** Price $= 223.5 + 0.174$ square feet

The slope is about 0.174 thousand dollars per square foot. On average, for every additional square foot, the price is $174 more. The intercept is not appropriate to interpret because there are no houses with 0 square feet. **c.** 1300 sq ft: about $450 thousand **d.** $0.75^2 = 56.25\%$ About 56% of the variation in price can be explained by the regression line.

**4.71 a.** See figure.

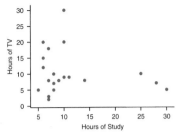

**b.** The trend is not linear. **c.** The correlation is near zero ($-0.176$).

**4.73 a.** $r$ is the square root of 0.054, which is 0.23. It is positive because the slope is positive. **b.** Predicted weight $= 111.4 + 0.882(35) = 111.4 + 30.87 = 142.27$, or about 142 pounds **c.** For each additional year of age, a woman weighs 0.88 pound more, on average, a little less than 1 pound. **d.** The intercept is not appropriate to interpret because no woman is near 0 years of age.

**4.75** The trend is positive. In general, if one twin has a higher-than-average level of education, so does the other twin. The point that shows one twin with 1 year of education and the other twin with 12 years is an outlier. (Another point showing one twin with 15 years and the other with 8 years is a bit unusual, as well.)

**4.77** There appears to be a positive trend. It appears that the number of hours of homework tends to increase slightly with enrollment in more units.

**4.79** Corelation table:

|  | Diameter | Height |
|---|---|---|
| Height | 0.519 | |
| Volume | 0.967 | 0.598 |

The diameter is a better predictor of volume than the height because there is a larger correlation between diameter and volume than between height and volume.

**4.81** You can see the tendency for big-budget films to gross more. One point in the upper right corner jumps out at you. *Avatar* had the largest budget (of these) and also the largest gross. *Titanic* is the other outlier. Linear regression is not appropriate because the trend is non-linear.

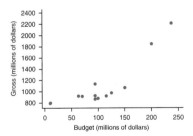

**4.83** and **4.85** Answers will vary.

## CHAPTER 5

### Section 5.1

**5.1 a.** 2 6 4 2 7   4 0 6 5 0
**b.** TTTTH   TTTHT
**c.** No. We got 2 heads.

**5.3** Theoretical probability: because it is not based on an experiment.

**5.5** Empirical probability: because it is based on an experiment.

### Section 5.2

**5.7 a.** The nine equally likely outcomes are Beasley, Blackett, Coffee, Craft, Fowlkes, Higgs, Lammey, Skahan, Ward. **b.** The outcomes that make up event A are Blackett and Skahan. **c.** 2/9, or about 22.2% **d.** Beasley, Coffee, Craft, Fowlkes, Higgs, Lammey, Ward.

**5.9** 1.3 could not be a probability because it is larger than 1.
150% could not be a probability because it is larger than 100%.
$-0.50$ could not be a probability because it is negative (less than 0).

**5.11 a.** A heart: 13/52 or 1/4 **b.** A red card: 26/52 or 1/2 **c.** An ace: 4/52 or 1/13 **d.** A face card: 12/52 or 3/13 **e.** A three: 4/52 or 1/13 Answers may also be in decimal or percentage form.

**5.13 a.** P(guessing correctly) $= 1/2$ **b.** P(guessing incorrectly) $= 1/2$

**5.15**

|  | Number of Girls | Probability |
|---|---|---|
| **a.** | 0 | 1/16 |
| **b.** | 1 | 4/16 = 1/4 |
| **c.** | 2 | 6/16 = 3/8 |
| **d.** | 3 | 4/16 = 1/4 |
| **e.** | 4 | 1/16 |

**5.17** The probability of being born on a Friday OR Saturday OR Sunday is 3/7, or 42.86%.

**5.19 a.** 530/1858 = 28.5% **b.** 689/1858 = 37.1%

**5.21** 104/1858 = 5.6%

**5.23** (689 + 469)/1858 = 62.3%

**5.25** The answers follow the guided steps.
  **1.** P(liberal) = 530/1858 = 28.5%
  **2.** P(Dem) = 689/1858 = 37.1%
  **3.** They are not mutually exclusive because there are people who are both liberal AND Democrats.
  **4.** P(liberal AND Dem) = 306/1858 = 16.5%
  **5.** If you don't subtract the probability of being liberal AND a Democrat you will be counting those people twice, once when you count the liberals and once when you count the Democrats.
  **6.** $\dfrac{530}{1858} + \dfrac{689}{1858} - \dfrac{306}{1858} = \dfrac{913}{1858} = 49.1\%$
  **7.** The probability of being liberal OR a Democrat is 49.1%.

**5.27 a.** Mutually exclusive **b.** Not mutually exclusive

**5.29** Any two column headings or any two row headings are mutually exclusive. For example, you cannot be both a Republican AND a Democrat, so those are mutually exclusive. Likewise, you cannot be both liberal AND moderate, so those are mutually exclusive.

**5.31** You don't know what percentage of households have at least one cat AND at least one dog. You cannot simply add the percentages, because the events are not mutually exclusive and you would count households with at least one cat AND at least one dog twice.

**5.33 a.** 4/6, or 66.7% **b.** 3/6 = 1/2, or 50%

**5.35 a.** A OR B: 0.18 + 0.25 = 0.43 **b.** A OR B OR C: 0.18 + 0.25 + 0.37 = 0.80 **c.** Lower than a C: 1 − 0.80 = 0.20

**5.37 a.** Most: Category 3: married OR have children **b.** Fewest: Category 4: married AND have children

**5.39** Two coin flips
**a.** 0 heads: ¼ (Decimals and percentages are also acceptable.) **b.** 1 head: ½ (from 2/4) **c.** 2 heads: ¼ **d.** At least 1 head: ¾ **e.** Not more than 2 heads (which means 2 or fewer heads): 1 (from 4/4)

**5.41** The total is 240,032 thousand.
**a.** Widowed: 14,254/240,032, or 5.9% **b.** Divorced: 23,266/240,032, or 9.7% **c.** Divorced OR widowed: (23,266 + 14,254)/240,032, or 15.6% (You can also get this from adding the percentages.)

**5.43 a.** More than 8 mistakes: $1 - 0.48 - 0.30 = 0.22$ **b.** 3 or more mistakes: $0.30 + 0.22 = 0.52$ **c.** At most 8 mistakes: $0.48 + 0.30 = 0.78$ **d.** The events in parts a and c are complementary because "at most 8 mistakes" means from 0 mistakes up to 8 mistakes. "More than 8 mistakes" means 9, 10, up to 12 mistakes. Together, these mutually exclusive events include the entire sample space

### Section 5.3

**5.45 ii.** D | C

**5.47 a.** $306/530 = 57.7\%$ **b.** $104/593 = 17.5\%$ **c.** Liberals

**5.49** Because men are more likely to be left-handed than women, gender and handedness are associated.

**5.51 a.** Hair color and age are independent because hair color does not change in this age range. (If you included all adults, you would have to say the variables are not independent, because old people are more likely to have gray hair.) **b.** Hair color and eye color are not independent, because people with dark eyes are more likely to have dark hair.

**5.53** Gender and opinion on same-sex marriage are not independent; they are associated. The females were more likely to support same sex marriage.

**5.55** They are not independent because liberals are more likely than conservatives to be Democrats. We see that P(Dem), which is about 37.1%, does not equal P(Dem | liberal), which is about 57.7%.

**5.57** The answers follow the format for the Guided Exercises.

|       | M  | W  |     |
|-------|----|----|-----|
| Right | 18 | 42 | 60  |
| Left  | 12 | 28 | 40  |
|       | 30 | 70 | 100 |

Step 1: See table.

Step 2: 60/100, or 60%

Step 3: 18/30 = 60%

Step 4: The variables are independent because the probability of having the right thumb on top given a person is a man is equal to the probability that a person has the right thumb on top (for the whole data set).

**5.59**

|         | Agree | Don't Know | Disagree |      |
|---------|-------|------------|----------|------|
| Happy   | 345   | 261        | 656      | 1262 |
| Unhappy | 34    | 22         | 44       | 100  |
|         | 379   | 283        | 700      | 1362 |

P(happy | agree) = 345/379 = 91.0%

P(happy) = 1262/1362 = 92.7%

These percentages are very close, so you might think the factors are independent. But we said in the chapter that for this chapter, independence requires precisely equal probabilities, so they are *not independent* from that point of view.

**5.61 a.** 1/8 **b.** 1/8

**5.63** They are the same. Both probabilities are $\left(\frac{1}{6}\right)^5$ or $\frac{1}{7776}$

**5.65 a.** Neither fastened: $0.16(0.16) = 0.0256$ **b.** At least one fastened: $1.0000 - 0.0256 = 0.9744$

**5.67** P(have C AND test positive) = P(have C) P(test positive | have C)
$$= 0.00008(0.84)$$
$$= 0.000067$$

### Section 5.4

**5.69** Histogram B was for 10,000 rolls because it has nearly a flat top. In theory, there should be the same number of each outcome, and the Law of

Large Numbers says that the one with the largest sample should be closest to the theory.

**5.71** The proportion should get closer to 0.5 as the number of flips increases.

### Chapter Review Exercises

**5.73** $117/1300 = 9.0\%$

**5.75 a.** Gender and shoe size are associated, because men tend to wear larger shoe sizes than women. **b.** Win/loss record is independent of the number of cheerleaders. The coin does not "know" how many cheerleaders there are or have an effect on the number of cheerleaders.

**5.77 a.** Both believe: $(0.62)(0.50) = 0.31$ **b.** Neither believes: $(0.38)(0.50) = 0.19$ **c.** Same beliefs: $0.31 + 0.19 = 0.50$ **d.** Different beliefs: $1 - 0.50 = 0.50$

**5.79 a.** $0.80(0.80) = 0.64$ **b.** The married couple might have Internet access at home, in which case, if one of them has Internet access, then the other also does (unless one of them prohibits the other's use).

**5.81 a.** Both born Monday: $\frac{1}{7}\left(\frac{1}{7}\right) = \frac{1}{49}$

**b.** Alicia OR David were born on Monday:
P(A OR B) = P(A) + P(B) − P(A AND B)
$$= \frac{1}{7} + \frac{1}{7} - \frac{1}{7}\left(\frac{1}{7}\right)$$
$$= \frac{7}{49} + \frac{7}{49} - \frac{1}{49}$$
$$= \frac{13}{49}$$

**5.83** $0.83(5000) = 4150$

**5.85** Answers will vary. Red die is 1, blue die is 1.

**5.87** Answers will vary.

**5.89** The smaller hospital will have more than 60% girls born more often, because, according to the Law of Large Numbers, there's more variability in proportions for small sample sizes. For the larger sample size ($n = 45$), the proportion will be more "settled" and will vary less from day to day. Over half of the subjects in Tversky and Kahneman's study said that "both hospitals will be the same." But *you* didn't, did you?

## CHAPTER 6

Answers may vary slightly due to rounding or type of technology used.

### Section 6.1

**6.1 a.** Discrete **b.** Continuous

**6.3 a.** Discrete **b.** Continuous

**6.5**

| Number of Spots | 1   | 2   | 3   | 4   | 5   | 6   |
|-----------------|-----|-----|-----|-----|-----|-----|
| Probability     | 0.1 | 0.2 | 0.2 | 0.2 | 0.2 | 0.1 |

**The table may have a different orientation.**

**6.7 a.** See table.

| Win | Prob |
|-----|------|
| 3   | 1/6  |
| 0   | 1/2  |
| −4  | 1/3  |

**4.37 a.** The higher the percentage of smoke-free homes in a state, the lower the percentage of high school students who smoke tends to be.
**b.** $56.32 - 0.464(70) = 23.84$, or about 24%

**4.38 a.** Trend: The higher the percentage of adults who smoke, the higher the percentage of smoking high school students tends to be.
**b.** $-0.838 + 1.124(25) = 27.262$, or about 27%

**4.39 a.** The graph shows that young drivers and old drivers have more fatalities and that the safest drivers are between about 40 and 60 years of age. **b.** It would not be appropriate for linear regression because the trend is not linear.

**4.40 a.** Answers will vary. However, a woman of 20 has a life expectancy of about 60 years, and a man of 20 has a life expectancy of about 55 years. **b.** No. Linear regression would not be appropriate, because the trend is not linear. **c.** Don't extrapolate. Do not make predictions for a person 120 years old, because that would be extrapolation. **d.** Women tend to live longer than men.

**4.41.** The answers are given in the order shown in the Guidance section.

1: The regression line is shown with the scatterplot.

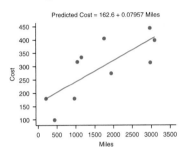

Predicted Cost = 162.6 + 0.07957 Miles

2: Is the linear model appropriate?

Given: In this case the answer is yes, because there is a linear trend. It is hard to see with so few points, but a strong curvature is not present and the cost tends to increase as the miles increase.

3: Predicted Cost $= 162.60 + 0.07957$ Miles

4: Put in the correct regression line (shown in Step 1).

5: Interpret the slope and intercept in context.

For the slope: For every additional mile, on average, the price goes up by 0.07957 dollars (or about 8 cents).

For the intercept: A trip of 0 miles should cost about 162.60 dollars on average. Beware of extrapolation. The linear trend might not continue near 0 miles and so this prediction could be very different from truth.

$$\text{Predicted cost} = 162.60 + 0.07957 \text{ miles}$$

6: Final answer: For the hypothetical flight of 500 miles, you would pay

$$162.60 + 0.07957 \times 500 = 202.385, \text{ or about 202 dollars}$$

**4.42** Predicted Yes% $= 68.1 - 0.399$ LimEng%
The negative sign of the slope means that the larger the percentage of people with limited English background, the smaller the percentage of voters who were for Proposition 227. (In other words, the English speakers tended to vote for it, and the people with limited English tended to vote against it.)

**4.43 a.** Positive: The more population, the more millionaires there tend to be. **b.** See scatterplot. **c.** $r = 0.992$ **d.** For each additional hundred thousand in the population, there is an additional 1.9 thousand millionaires.

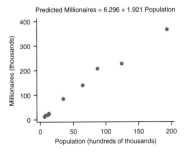

Predicted Millionaires = 6.296 + 1.921 Population

**e.** Do not focus on the intercept, because it does not make sense to look for millionaires in states with no people.

**4.44 a.** See Figure A for all points and Figure B for the graph without one point. The unusual point in Figure A is the person with four semesters who has more than 120 units. **b.** $r = 0.755$ for all points $r = 0.883$ without one point

The correlation increases without the omitted point, because this point is far from the line.

For all: Predicted Units $= 2.7 + 13.4$ Semesters

Without one point: Predicted Units $= -3.96 + 13.69$ Semesters

**d.** The line is shown in Figure A with all points and in Figure B without one point. **e.** The slope is 13.4 units per semester. For each additional semester, an average student has 13.4 additional units.

When we leave out one point, the slope is 13.69. For each additional semester, a student has another 13.69 units on the average. The two slopes are not very different, because the omitted point is near the middle of the horizontal axis.

Intercept: It is meaningful to expect people to report 0 semesters reported, so the intercept is potentially meaningful. One such report was recorded, and so we are not extrapolating to see what the line predicts.

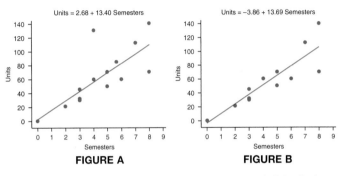

**FIGURE A**       **FIGURE B**

**4.45 a.** The correlation should be positive. The more people living in the state, the more drivers, and the more motor vehicle fatalities there should be. **b.** See figure. **c.** $r = 0.949$ **d.** Equation and line are shown on plot. **e.** For each additional hundred thousand people in a state, there are about 9 more fatalities on average.

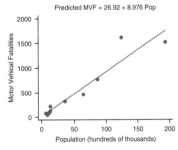

Predicted MVF = 26.92 + 8.976 Pop

**4.46 a.** The correlation should be positive. The more people living in the state, the more drivers, and the more motor vehicle fatalities there tend to be. **b.** See the figure. There are no unusual points.
**c.** $r = 0.989$.
**d.** The equation and the line are shown on the plot.

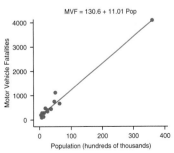

MVF = 130.6 + 11.01 Pop

**e.** For each additional hundred thousand people in a state, there are an additional 11 fatalities, on average.

**c.** The flights might not be independent, because the weather could be bad and that might cause most of the flights to be late. Thus if one flight is late, the others are more likely to be late.

**6.69 a.** $0.7(15) = 10.5$; expect 10 or 11 **b.** $b(15, 0.70, 11) = 0.219$
**c.** $b(15, 0.70, 11 \text{ or fewer}) = 0.703$

**6.71 a.** YY   YN   NY   NN
   **b.** YY ¼
       YN ¼
       NY ¼
       NN ¼
**c.** Neither has only cell: ¼ **d.** Exactly one has only cell (from ¼ + ¼): ½
**e.** Both have only cell: ¼

**6.73 a.** In 50 flips, expect 25 heads, because $np$ is 50 times ½.
   **b.** $\sigma = \sqrt{np(1-p)} = \sqrt{50(0.5)(1-0.5)} = \sqrt{12.5} = 3.54$, or about 4. **c.** Therefore, you should expect between 21 and 29 heads.

**6.75 a.** Expect 60 give or take 7, because
   $\sigma = \sqrt{np(1-p)} = \sqrt{300(0.2)(0.8)} = \sqrt{48} = 6.93$, or about 7
   **b.** 53 to 67 **c.** No, because 62 is close to the mean (within 1 SD).

**Chapter Review Exercises**

**6.77** 95.2%, or about 95%

**6.79 a.** 0.7625, or about 76% **b.** 98.6°F

**6.81** Answers will vary. However, the probability that her next child will be a girl should be about 50%, because genders of children is independent unless there are identical twins (or selection of gender with in vitro fertilization).

**6.83 a.** $\mu = np = 2000(0.3) = 600$ give or take about 20, because
$\sigma = \sqrt{np(1-p)} = \sqrt{2000(0.3)(1-0.3)} = \sqrt{420} = 20.49$, or about 20
**b.** 580 to about 620

**6.85 a. i.** $0.9^4 = 0.656$;
       **ii.** $1 - 0.656 = 0.344$
   **b. i.** $0.4^4 = 0.026$;
       **ii.** $1 - 0.026 = 0.974$
**c.** They are less likely to die because of the Gleevec, so part (a) has a smaller probability of at least one dead.

## CHAPTER 7

Answers may vary slightly due to rounding or type of technology used.

**Section 7.1**

**7.1** A parameter is a measure of the population, and a statistic is a measure of a sample.

**7.3 a.** $\bar{x}$ is a statistic, and $\mu$ is a parameter. **b.** $\bar{x}$

**7.5** If you know all the ages at inauguration, you should not make inferences because you have the population, not a sample from the population.

**7.7** You want to test a sample of batteries. If you tested them all until they burned out, no usable batteries would be left.

**7.9** First, all 10 cards are put in a bowl. Then one is drawn out and noted.
   "With replacement": The card that is selected is replaced in the bowl, and a second draw is done. It is possible that the same student could be picked twice.
   "Without replacement": After the first card is drawn out, it is not replaced, and the second draw must be a different card.

**7.11** Chosen: 7, 3, 5, 2

**7.13** Marco took a convenience sample. The students may not be representative of the voting population, so the proposition may not pass.

**7.15** The population is all readers of Time Magazine on the Internet. The *Time* Internet poll used a voluntary-response sample. It could be that only

people who were really angry with the United States took time to submit their answers. The poll may have shown bias because of that fact.

**7.17** The administrator might dismiss the negative findings by saying the results could be biased because the small percentage who chose to return the survey might be very different from the majority who did not return the survey.

**7.19** No, the people you met would not be a random sample but a convenience sample.

**7.21**

|  | With Persuasion | Without Persuasion |
|---|---|---|
| Support Cap | 6 + 2 | 13 + 5 |
| Oppose | 9 + 8 | 2 + 5 |

Support capital punishment
**a.** With persuasion $8/25 = 32\%$
**b.** Without persuasion $18/25 = 72\%$
**c.** Yes, she spoke against it, and fewer who heard her statements against it (32%) supported capital punishment, compared with those who did not hear her persuasion (72%).

**Section 7.2**

**7.23 a.** The sketch should show bullet holes consistently to the left of the target and close to each other. If the bullets go consistently to the left, then there is bias, not lack of precision. **b.** The sketch should show bullet holes that are all near the center of the target.

**7.25** The small mean might have occurred by chance.

**7.27 a.** $17/30$, or about 56.7%, odd digits **b.** $17/30$ is $\hat{p}$ ($p$-hat), the sample proportion. **c.** The error is $17/30$ minus $15/30$ (or $2/30$) or about 6.7%

**7.29 a.** We should expect 0.20, or 20%, orange candies.
   **b.** $\sqrt{\dfrac{p(1-p)}{n}} = \sqrt{\dfrac{0.2 \times 0.8}{100}} = \sqrt{\dfrac{0.16}{100}} = \sqrt{0.0016} = 0.04$, or 4%
**c.** We expect about 20% orange candies, give or take 4%.

**Table for 7.31**

| Repetition | $p$ (pop prop of seniors) | $\hat{p}$ (sample prop of seniors) | Error: $\hat{p} - p$ |
|---|---|---|---|
| 1 (from part a) | $2/5 = 0.4$ | $1/2 = 0.5$ | $0.5 - 0.4 = 0.1$ |
| 2 (from part b) | $2/5 = 0.4$ | $0/2 = 0.0$ | $0.0 - 0.4 = -0.4$ |
| 3 | $2/5 = 0.4$ | $1/2 = 0.5$ | $0.5 - 0.4 = 0.1$ |
| 4 | $2/5 = 0.4$ | $2/2 = 1.0$ | $1.0 - 0.4 = 0.6$ |

**7.31 a.** 50% seniors **b.** 0% seniors **c.** 50% seniors **d.** 100% seniors

**Section 7.3**

**7.33** $z = \dfrac{0.24 - 0.20}{0.04} = \dfrac{0.04}{0.04} = 1.00$

   The area to the right of a $z$-score of 1.00 is 0.1587.
   The probability that the percentage of orange candies will be 24% or more is about 16%.

**7.35** Because the sampling distribution for the sample proportion is approximately Normal, we know the probability of falling within two standard errors is about 0.95. Therefore, the probability of falling more than two standard errors away from the mean is about 0.05.

**7.37** Answers are given in the order shown in the Guided Exercises.

Step 1: $p = 0.65$

Step 2: $np = 200(0.65) = 130$ which is more than 10

$n(1-p) = 200(0.35) = 70$ which is more than 10

The other assumptions were given.

Step 3:

$$SE = \sqrt{\frac{p(1-p)}{n}} = \sqrt{\frac{0.65(0.35)}{200}} = \sqrt{\frac{0.2275}{200}} = \sqrt{0.0011375} = 0.03373$$

$$z = \frac{\hat{p} - p}{SE} = \frac{0.67 - 0.65}{0.03373} = 0.59$$

The area to the right of a $z$-value of 0.59 is 0.2776 (or 0.2766 with technology and no rounding of intermediate steps).

Step 4: The probability is represented by the area to the right of 0.59, because the question asks for the probability that the sample proportion will be "at least" 0.67 (134 out of 200), which translates to a $z$-score of 0.59. This means we are asked to find the probability that the $z$-score will be 0.59 or greater.

Step 5: The probability that at least 67% of 200 pass the exam is about 28% because the area to the left of a $z$-score of 0.59 is about 72%, and so the area to the right is about 28%.

**7.39 a.** Expect 0.08. **b.** The approximately probability is 0.122. It should be shown in a Normal curve. **c.** 0.0099 or about 0.010 and 360 chairs **d.** The answer to part c is smaller because 9% is farther out in the right tail than 8.5%, and it is the tail area that gives the probability we want.

**7.41 a.** 0.5 **b.** 0.5 **c.** 0.103 **d.** Lower, because the probability of guessing correctly on each question is lower when there are four options.

### Section 7.4

**7.43** (0.440, 0.481)

**7.45 a.** 770/1027, or 75.0%, want a stimulus bill. **b.** (72.3%, 77.6%)

**7.47 a.** I am 95% confident that the population percentage of voters supporting Candidate X is between 53% and 57%. **b.** There is no evidence that he could lose, because the interval is entirely above 50%. **c.** A sample from New York City would not be representative of the entire country and would be worthless in this context.

**7.49 a.** 17/30, or 0.5667 (or round it off to 0.567) **b.** $SE_{est}$ given as 0.090 $m = 1.96 \times 0.090 = 0.1764$ **c.** 95% CI (0.390, 0.743) **d.** $m = 1.28 \times 0.090 = 0.115$ 80% CI (0.452, 0.682) **e.** 99% CI (0.334, 0.799) **f.** The 99% confidence interval is widest, and the 80% interval is narrowest. A greater confidence level requires a larger $z^*$.

**7.51 a.** Sqrt [(0.87 × 0.13)/1500] = 0.0087 or 0.87% **b.** 1.70% **c.** (85.3%, 88.7%) **d.** The interval supports this claim. Because 80% is not in the interval, and all values are above 80%, we are confident that the current percentage is higher than 80%.

**7.53 a.** 92/113, or 81.4%, reported dreaming in color. **b.** (0.742, 0.886) **c.** The interval does not capture 29.24%. This shows that the results in 2003 were quite different from the results in 1942.

**7.55 a.** He or she should find 15, because 15 is half of 30. **b.** You would expect about 4 out of 40 not to capture 50%, because with a 90% confidence interval about 10% should not capture, and 10% of 40 is 4.

**7.57** The proportion 16/100 is the population proportion, not a sample proportion. You should not find a confidence interval unless you have a sample and are making statements about the population from which the sample has been drawn.

**7.59 a.** 430/1001 = 43.0% **b.** 95% CI (0.399, 0.460) **c.** 80% CI (0.410, 0.450) **d.** The 95% interval is wider and the 80% interval is narrower. To get a higher degree of certainty, we need to widen the interval.

**7.61 a.** 595/1009 = 59.0% **b.** 95% CI (0.559, 0.620) **c.** A 99% CI would be wider.

**7.63 a.** 74/1438 = 5.15% **b.** 95% CI (0.040, 0.063) **c.** 5% is plausible because it is inside the interval.

### Chapter Review Exercises

**7.65 a.** 515/691, or 74.5%, had been exposed to traffic. **b.** 95% CI (0.713, 0.778) **c.** No. Without random assignment, you cannot draw conclusions about cause and effect.

**7.67 a.** 78/489 is 16.0% **b.** I am 95% confident the percentage of American adults who would say yes to making smoking totally illegal is between 12.7% and 19.2%. **c.** No, the entire interval is below 50% and it is not plausible to think a majority would say smoking should be made totally illegal. **d.** The percentage is the same. If the sample size were larger the interval would be narrower.

**7.69 a.** 292/1006 is 29.0% **b.** I am 95% confident that the percentage of American adults who would say yes to banning smoking in bars is between 26.2% and 31.8%. **c.** No, the entire interval is below 50% **d.** If the sample size were larger the interval would be narrower.

**7.71** 95% confidence interval (52%, 60%) or (0.52, 0.60) This suggests that a majority of adults believe the feeding tube should have been removed, because the interval is entirely above 50%.

## CHAPTER 8

Answers may vary slightly due to type of technology or rounding.

### Section 8.1

**8.1** population parameter

**8.3** Small.

**8.5** ii. $p = 0.50$

**8.7 a.** i. **b.** iii

**8.9** iii.

**8.11** does not, 0.05

**8.13** $z = 0.89$.

**8.15** The p-value tells us that if the true proportion of those who text while driving is 0.25, then there is only a 0.034 probability that one would get a sample proportion of 0.125 or smaller with a sample size of 40.

### Section 8.2

**8.17** In Figure (A), the shaded area could be a p-value because it includes tail areas only; it would be for a two-tailed alternative because both tails are shaded. In Figure (B) the shaded area would not be a p-value because it is the area between two z-values.

**8.19** $z = 3.00$. It is farthest from 0 and therefore has the smallest tail area.

**8.21** He has not demonstrated ESP; 10 right out of 20 is only 50% right, which you should expect from guessing.

### Section 8.3

**8.23** $H_0$: The death rate after starting hand washing is still 9.9%, or $p = 0.099$ ($p$ is the proportion of all deaths at the clinic.)
$H_a$: The death rate after starting hand washing is less than 9.9%, or $p < 0.099$.

**8.25** $H_0$: The rate of using seat belts has not changed and is 0.71, or $p = 0.71$.
$H_a$: The rate of using seat belts has changed and is not 0.71, or $p \neq 0.71$.

**8.27** *Step 1:* $H_0$: $p = 0.52$, $H_a$: $p \neq 0.52$ where $p$ is the population proportion favoring stricter gun control. *Step 2:* One-proportion z-test (given), sample size large (1011 times 0.48 = about 485 which is more than 10), sampling random, and population large. *Step 3:* $\hat{p} = 0.4896$, $SE = 0.01571$, $z = -1.93$, p-value = 0.053. *Step 4:* Do not reject $H_0$. Choose conclusion i.

**8.29** Random sample was mentioned.
Large sample: $np_0 = 113(0.29) = 32.77 > 10$ and $n(1 - p_0) = 113(0.71) = 80.23 > 10$.
Large population: There are more than 1130 people in the population of dreamers.
So the conditions are met.

**8.31** $z = 2.56$, p-value = 0.005.

**8.33** *Step 3:* $z = 2.83$, p-value = 0.002. *Step 4:* Reject $H_0$. The probability of doing this well by chance alone is so small that we must conclude that the student is not guessing.

**8.35** Pick the graph shown in Figure (A). The observed value of 13 right out of 20 is more than half, so the $z$-statistic will be bigger than 0. The right tail area corresponds to the p-value.

**8.37 a.** *Step 1:* $H_a$: $p > 0.50$ where $p$ is the population proportion of people who believe that schools should ban sugary foods. *Step 2:* One-proportion $z$-test, the sampling is random, sample size large, and population large. *Step 3:* $z = 1.26$, $p = 0.103$ (one-tailed). *Step 4:* Do not reject the null hypothesis. **b.** Pick ii.

**8.39 a.** $527/1014 = 52.0\%$. This is less than the 58%. **b.** *Step 1:* $H_a$: $p \neq 0.58$ where $p$ is the population proportion of people who believe there is global warming. *Step 2:* One-proportion $z$-test, sample size large, sampling random, population more than 10 times sample. *Step 3:* $z = -3.89$, p-value $< 0.001$. *Step 4:* Reject $H_0$. **c.** Choose ii.

**8.41** *Step 1:* $H_a$: $p \neq 0.20$ where $p$ is the population proportion of dangerous fish. *Step 2:* One-proportion $z$-test, sample size large enough, population large, assume a random sample. *Step 3:* $z = 1.58$, p-value $= 0.114$. *Step 4:* Do not reject $H_0$.

We are not saying the percentage is 20%. We are only saying that we cannot reject 20%. (We might have been able to reject the value of 20% if we had had a larger sample.)

**8.43** *Step 1:* $H_a$: $p \neq 0.09$ where $p$ is the population proportion of t's in the English language. *Step 2:* One-proportion $z$-test, the sample is random and large enough, and population large. *Step 3:* $z = -0.86$, p-value $= 0.392$. *Step 4:* Do not reject $H_0$. We cannot reject 9% as the current proportion of t's.

## Section 8.4

**8.45**

*0:* Vioxx percentage heart attack $= 3.5\%$

Placebo percentage heart attack $= 1.9\%$

So the risk of heart attack in the sample was greater for those taking Vioxx.

*1:* $H_0$: $p_V = p_p$ where $p_V$ is the population proportion of heart attacks for those who could use Vioxx and $p_p$ is the population proportion of heart attacks for those who could use the placebo.

$H_a$: $p_V > p_p$

*2:* We don't have a random sample but we do have random assignment.

$n_1 \times \hat{p} = 1287(0.027069) = 34.84$, which is more than 10

$n_1 \times (1 - \hat{p}) = 1287(0.972931) = 1252.16$, which is more than 10

$n_2 \times \hat{p} = 1299(0.027069) = 35.16$, which is more than 10

$n_2 \times (1 - \hat{p}) \geq 1299 (0.972931) = 1263.84$, which is more than 10

*3:* $z = 2.46$ (or $-2.46$)

p-value $= 0.007$.

*4:* Reject $H_0$ and choose ii. Do not generalize because you do not have a random sample.

Causality: Yes, you can conclude cause and effect for this group of patients because of the format of the study, including random assignment to groups. Vioxx caused a significantly increased rate of heart attacks. (Vioxx was taken off the market because of this study.)

**8.47 a.** Vaccine: 1.9% got the virus. Placebo: 4.9% got the virus. So the vaccine group got the virus at a lower rate, which is what was hoped for. **b.** *Step 3:* $z = 5.85$ (or $-5.85$), p-value $< 0.001$. *Step 4:* Reject $H_0$. The vaccine causes a significantly lower chance of getting the virus, but don't generalize.

**8.49 a.** For nicotine gum, the proportion quitting was 0.106. For the placebo, it was 0.040. This was what was hoped for—that the drug was helpful compared to the placebo. **b.** $z = 7.23$ (or $-7.23$).

**8.51 a.** Men: 46.2% smiling. Women 51.1% smiling. **b.** *Step 1:* $H_a$: $p_{men} \neq p_{women}$. (where $p$ is the proportion smiling). *Step 2:* Two-proportion $z$-test, large sample size, assume random sample. *Step 3:* $z = 6.13$ (or $-6.13$), p-value $< 0.001$. *Step 4:* Reject $H_0$. There is a significant difference in rate of smiling for men and women. **c.** The difference is significant because of the large sample size.

## Section 8.5

**8.53** No; we don't use "prove" because we cannot be 100% sure of conclusions based on chance processes.

**8.55** It is a null hypothesis.

**8.57** Interpretations b and d are valid. Interpretations a and c are both "accepting" the null hypothesis claim, which is an incorrect way of expressing the outcome.

**8.59** Far apart. Assuming the standard errors are the same, the farther apart the two proportions are, the larger the absolute value of the numerator of $z$, and therefore the larger the absolute value of $z$ and the smaller the p-value.

**8.61 a.** 336 (from 0.60 times 560) said more strict in February 1999, and 370 said more strict in late April 1999. **b.** *Step 1:* $H_a$: $p_{more\ strict\ Feb} \neq p_{more\ strict\ April}$. *Step 2:* Two-proportion $z$-test, samples are random, sample size large, population large. *Step 3:* $z = 2.10$ (or $-2.10$), p-value $= 0.035$. *Step 4:* Do not reject $H_0$, because alpha is 0.01. **c.** 672 out of 1120 in February vs. 739 out of 1120 in late April, 1999 *Step 1:* $H_a$: $p_{more\ strict\ Feb} \neq p_{more\ strict\ April}$. *Step 2:* Two-proportion $z$-test, samples are random, sample size large, population large. *Step 3:* $z = 2.93$ (or $-2.93$), p-value $= 0.003$. *Step 4:* Reject $H_0$. The results of the polls were significantly different from each other. **d.** With a larger sample size (more evidence), we got a smaller p-value and were able to reject $H_0$.

**8.63** Interpretation iii.

**8.65 a.** One-proportion $z$-test. The population is all Florida voters. **b.** Two-proportion $z$-test. One population is all men at the college, and the other population is all women at that college.

**8.67 a.** $p = $ the population proportion of correct answers.

$H_0$: $p = 0.50$ (he is just guessing), $H_a$: $p > 0.50$ (he is not just guessing).

One-proportion $z$-test. **b.** $p_a$ is the population proportion of athletes at your college who can balance for at least 10 seconds. $p_n$ is the population proportion of nonathletes at your college who can balance for at least 10 seconds.

$H_0$: $p_a = p_n$

$H_a$: $p_a \neq p_n$

Two-proportion $z$-test.

## Chapter Review Exercises

**8.69** 6 right out of 20 is less than half, so he cannot tell the difference. (Or: $H_a$: $p > 0.50$, $z = -1.79$, p-value $= 0.963$, do not reject $H_0$.)

**8.71** 0.05 (because $1 - 0.95 = 0.05$).

**8.73** 5% of 200, or 10.

**8.75** It would not be appropriate to do such a test, because the data were the entire population of people who voted. This was not a sample, so inference is not reasonable.

**8.77** *Step 1:* $H_a$: $p_{money} > p_{no\ money}$ where $p$ is the population proportion that successfully would quit smoking. *Step 2:* Two-proportion $z$-test, the samples large enough, not a random sample. *Step 3:* $z = 3.65$ (or $-3.65$), p-value $< 0.001$. *Step 4:* Reject $H_0$. The proportion of success was significantly higher for those who received the money. We can conclude the money caused the difference but cannot generalize to a larger group of smokers.

**8.79 a.** Aspirin: 2.39%, Placebo 2.62%. There was a slightly lower heart event rate for those using aspirin. **b.** *Step 1:* $H_a$: $p_{aspirin} < p_{placebo}$ (where $p$ is the proportion of those who had heart events). *Step 2:* Two-proportion $z$-test, sample large, random assignment but not a random sample. *Step 3:* $z = -1.44$ (or 1.44), p-value $= 0.076$. *Step 4:* You cannot reject the null hypothesis. There is not enough evidence to conclude that aspirin prevents heart events in women.

**8.81 a.** The misconduct rate was higher for those in the sample who did *not* have three strikes (30.6%) than for those in the sample who had three strikes (22.2%). This was not what was expected. **b.** *Step 1:* $H_a$: $p_{three-strikers} >$

$p_{\text{others}}$ (where $p$ is the proportion of those with misconduct). *Step 2:* Two-proportion $z$-test, samples large enough, assume random samples. *Step 3:* $z = -4.49$ (or 4.49), p-value > 0.999. *Step 4:* Do not reject $H_0$. The three-strikers do not have a greater rate of misconduct than the other prisoners.

(If a two-tailed alternative had been used, the p-value would have been < 0.001, and we would have rejected the null hypothesis because the three-strikers had *less* misconduct.)

## CHAPTER 9

Answers may vary due to rounding or type of technology used.

### Section 9.1

**9.1 a.** Parameters: 54.4 years, 15.5 years **b.** Statistics: 50.2 years, 18.0 years

**9.3 a.** See the accompanying figure.

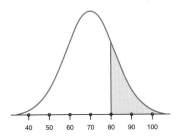

**b.** 16% (By the Empirical Rule, 68% of the observed values are between 60 and 80, which leaves 32% outside of those boundaries. But we want only the right half.)

**9.5** The sample mean is based on a random sample, so different samples will have different means. Having an "unbiased" estimator does not mean that the sample mean will be equal to the population mean. Instead, it means that if we took all possible samples of bats, the mean of all of the sample means would be the same as the population mean.

**9.7** The distribution of *a* sample (*one* sample).

**9.9 a.** About 128 pounds, because the sample mean is unbiased.

**b.** $\dfrac{20}{\sqrt{100}} = 2.0$ pounds

### Section 9.2

**9.11 a.** It is large enough because the sample size of 100 is larger than 25. **b.** The mean is 128 pounds.

$$SE = \frac{\sigma}{\sqrt{n}} = \frac{20}{\sqrt{100}} = \frac{20}{10} = 2$$

**c.** $1 - 0.9545 = 0.0455$, or about 5%

**9.13** A: Calls B: Means: smaller standard deviation (narrower) and more Normal (less right-skewed)

**9.15 a.** 7.9 and 7.7 are parameters; 8.2 and 6.0 are statistics. **b.** $\mu = 7.9$, $\sigma = 7.7$, $\bar{x} = 8.2$, $s = 6.0$ **c.** Random: OK. Sample size: 40 > 25 OK. The shape would be approximately Normal.

### Section 9.3

**9.17 a.** i. is correct: (3.07, 3.43); ii. and iii. are both incorrect. **b.** It tells us a range of plausible values for the population mean GPA, where the population is all statistics students at the school. Yes, reject 2.81, because it is not in the interval. We are confident that the statistics students have a higher population mean than 2.81.

**9.19 a.** i. is correct: (10.125, 10.525). Both ii. and iii. are incorrect. **b.** No, it does not capture 10. Reject the claim of 10 pounds because 10 is not in the interval.

**9.21 a.** Confidence level; **b.** Confidence interval

**9.23 a.** 1.699 **b.** Wider because $t^*$ is bigger (2.045), which makes the margin of error bigger:

$$\text{Margin of error} = t^* \frac{s}{\sqrt{n}}$$

**9.25 a.** $m = t^* \dfrac{s}{\sqrt{n}} = 2.093 \dfrac{2.17}{\sqrt{20}} = 1.016$, or about 1.0

$63.3 \pm 1.0$ 95% CI (62.3 to 64.3). I am 95% confident the population mean height is between 62.3 and 64.3 inches.

**b.** $m = t^* \dfrac{s}{\sqrt{n}} = 2.861 \dfrac{2.17}{\sqrt{20}} = 1.388$, or about 1.4

$63.3 \pm 1.4$ 99% CI (61.9 to 64.7). I am 99% confident the population mean height is between 61.9 and 64.7 inches.

**c.** The 99% interval is wider because it has a greater confidence level, and therefore we use a bigger value of $t^*$, which creates a wider interval.

**9.27 a.** Wider **b.** Narrower **c.** Wider

**9.29** *Step 1:* $H_a$: $\mu \neq 98.6$. *Step 2:* One-sample $t$-test, random sample, not strongly skewed, use 0.05. *Step 3:* $t = -1.65$, p-value = 0.133. *Step 4:* Do not reject $H_0$. Choose i.

**9.31 a.** *Step 1:* $H_a$: $\mu \neq 38$. *Step 2:* One-sample $t$-test, Normal and random. *Step 3:* $t = -1.03$, p-value = 0.319. *Step 4:* Do not reject $H_0$. The mean for non-U.S. boys is not significantly different from 38. **b.** *Step 3:* $t = -1.46$, p-value = 0.155. *Step 4:* Do not reject $H_0$. The mean for non-U.S. boys is not significantly different from 38. **c.** Larger $n$, smaller standard error (narrower sampling distribution) with less area in the tails, as shown by the smaller p-value.

**9.33 a.** 3.25, higher **b.** *Step 1:* $H_a$: $\mu > 2.81$. *Step 2:* One-sample $t$-test, random and 40 > 25. *Step 3:* $t = 4.91$, p-value < 0.001. *Step 4:* Reject $H_0$. The mean GPA for statistics students is significantly higher than 2.81.

**9.35** *Step 1:* $H_0$: $\mu_{\text{before}} = \mu_{\text{after}}$ [or $\mu_{\text{difference}} = 0$]; $H_a$: $\mu_{\text{before}} > \mu_{\text{after}}$ [or $\mu_{\text{difference}} > 0$]. *Step 2:* One-sample $t$-test, assume conditions are met (given). *Step 3:* $t = 1.39$, p-value = 0.098. *Step 4:* Do not reject $H_0$. Meditation does not lower pulse rates significantly.

**9.37** $t = 1.20$ for all three.

**a.** $H_a$: $\mu \neq 20$, p-value = 0.315, do not reject $H_0$.

**b.** $H_a$: $\mu < 20$, p-value = 0.842, do not reject $H_0$.

**c.** $H_a$: $\mu > 20$, p-value = 0.158, do not reject $H_0$.

**9.39** a

**9.41** Expect $0.95(200) = 190$ to capture and 10 to miss.

**9.43** It should not be interpreted. The data are not a random sample, and so inference based on a confidence interval is not possible.

### Section 9.4

**9.45 a.** Independent; **b.** Paired (or dependent)

**9.47** *Step 1:* $H_a$: $\mu_{\text{oc}} \neq \mu_{\text{mc}}$ where $\mu$ is the population mean number of TVs. *Step 2:* Two-sample $t$-test, samples large ($n = 30$), independent, and random, use 0.05. *Step 3:* $t = 0.95$, p-value = 0.345. *Step 4:* Do not reject $H_0$. Choose i. Confidence Interval: $(-0.404, 1.138)$. Because the interval for the difference captures 0, we cannot reject the hypothesis that the mean difference in number of TVs is 0.

**9.49 a.** The men's sample mean triglyceride level of 139.5 was higher than the women's sample mean of 84.4. **b.** *Step 1:* $H_a$: $\mu_{\text{men}} > \mu_{\text{women}}$ where $\mu$ is the population mean trigliceride level. *Step 2:* Two-sample $t$-test, assume the conditions are met. *Step 3:* $t = 4.02$ or $-4.02$, p-value < 0.001. *Step 4:* Reject $H_0$. The mean triglyceride level is significantly higher for men than for women. Choose output B: Difference = $\mu_{\text{female}} - \mu_{\text{male}}$, which tests whether this difference is less than 0, which is the one-tailed hypothesis that we want.

**9.51** $(-82.5, -27.7)$; because the difference of 0 is not captured, it shows there is a significant difference. Also, the difference $\mu_{\text{female}} - \mu_{\text{male}}$ is negative, which shows that the men's mean (triglyceride level) is significantly higher than the women's mean.

**9.53** *Step 1:* $H_a$: $\mu_{men} \neq \mu_{women}$ where $\mu$ is the population mean clothing expense for one month. *Step 2:* Two-sample *t*-test, assume Normal and random. *Step 3:* $t = 1.42$ or $-1.42$, p-value $= 0.171$. *Step 4:* Do not reject $H_0$. The mean clothing expense is not significantly different for men and women.

**9.55 a.** The 95% interval would capture 0, because we could not reject the hypothesis that the mean amounts spent on clothing are the same. **b.** A 99% interval would also capture 0, because it is wider than the 95% interval and centered at the same place. **c.** $(-18.4, 97.7)$; because the interval captures 0, we cannot reject the hypothesis that the mean difference in spending on clothing is 0, which shows we cannot reject the hypothesis that the means are the same.

**9.57** *Step 1:* $H_a$: $\mu_{letter} < \mu_{no\ letter}$ where $\mu$ is the population mean rate of crime. *Step 2:* Two-sample *t*-test, large sample sizes, not random samples, so don't generalize to a larger population; random assignment allows us to conclude causality. *Step 3:* $t = 1.77$ or $-1.77$, p-value $= 0.041$ (one-tailed). *Step 4:* Reject $H_0$. The police letter reduced the number of crimes reported for these properties, but we cannot generalize to other properties.

**9.59** *Step 1:* $H_a$: $\mu_{before} < \mu_{after}$ where $\mu$ is the population mean pulse rate. *Step 2:* Paired *t*-test, each woman is measured twice (repeated measures), so a measurement in the first column is coupled with a measurement of the same person in the second column, assume random and Normal, use 0.05. *Step 3:* $t = 4.90$ or $-4.90$, p-value $< 0.001$. *Step 4:* Reject $H_0$. The sample mean before was 74.8 and the sample mean after was 83.7. The pulse rates of women go up significantly after a scream.

**9.61 a.** $\bar{x}_{UCSB} = \$61.01$ and $\bar{x}_{CSUN} = \$75.55$, so the sample mean at CSUN was larger. **b.** *Step 1:* $H_a$: $\mu_{UCSB} \neq \mu_{CSUN}$ where $\mu$ is the population mean book price. *Step 2:* Paired *t*-test, matched pairs, assume random and Normal (given). *Step 3:* $t = -3.21$ or 3.21, p-value $= 0.004$. *Step 4:* You can reject $H_0$. The means are significantly different.

**9.63 a.** $\bar{x}_{groom} = 27.3$ was larger ($\bar{x}_{bride} = 25.9$).

**b.** *Step 1:* $H_a$: $\mu_{bride} \neq \mu_{groom}$ where $\mu$ is the population mean age at marriage. *Step 2:* Paired *t*-test, random, and large samples. *Step 3:* $t = 2.24$ or $-2.24$, p-value $= 0.033$. *Step 4:* Reject $H_0$. The mean ages of brides and grooms are significantly different. **c.** $H_a$: $\mu_{bride} < \mu_{groom}$. The new p-value would be half of 0.033, or about 0.017.

**9.65 a.** 95% CI $(-1.44, 0.25)$ captures 0, so the hypothesis that the means are equal cannot be rejected. **b.** *Step 1:* $H_a$: $\mu_{measured} \neq \mu_{reported(converted\ to\ cm)}$ where $\mu$ is population mean height of men. *Step 2:* Paired *t*-test, each person is the source of two numbers, assume conditions for *t*-tests hold (given). *Step 3:* $t = 1.50$ or $-1.50$, p-value $= 0.155$. *Step 4:* Do not reject $H_0$. The mean measured and reported heights are not significantly different for men or there's not enough evidence to support the claim that the typical self-reported height differs from the typical measured height for men.

**9.67** *Step 1:* $H_a$: $\mu_{longboard} > \mu_{shortboard}$ where $\mu$ is the population mean number of surfing days last month. *Step 2:* Two-sample *t*-test, assume random, large samples. *Step 3:* $t = 2.11$, p-value $= 0.020$. *Step 4:* Reject $H_0$. Longboard users have a significantly larger mean number of days than shortboard users.

## Chapter Review Exercises

**9.69 a.** Paired *t* **b.** Two-sample *t*

**9.71 a.** *Step 1:* $H_a$: $\mu_{men} < \mu_{women}$ where $\mu$ is the population mean number of seconds for marathon runners. *Step 2:* Two-sample *t*-test, large sample, assume random. *Step 3:* $t = 9.26$, p-value $< 0.001$. *Step 4:* Reject $H_0$. The mean time for men is significantly less than the mean time for women or the typical man is quicker than the typical woman. **b.** 95% CI for the mean difference: (1405, 2163). Because the interval does not capture 0, it shows

there is a significant difference. I am 95% confident that the population mean time for men is between 1405 and 2163 seconds less than the population mean time for women.

**9.73** *Step 1:* $H_a$: $\mu_{before} < \mu_{after}$ where $\mu$ is the population mean pulse rate (before and after coffee). *Step 2:* Paired *t*-test (repeated measures), assume condition hold (given). *Step 3:* $t = -2.96$, p-value $= 0.005$. *Step 4:* Reject $H_0$. Heart rates increase significantly after coffee. (The average rate before coffee was 82.4, and the average rate after coffee was 87.5.)

**9.75** The typical number of hours was a little higher for the boys and the variation was almost the same. $\bar{x}_{girls} = 9.8$ $\bar{x}_{boys} = 10.3$. $s_{girls} = 5.4$ and $s_{boys} = 5.5$. See the histograms.

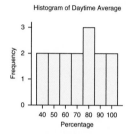

Histogram of Daytime Average

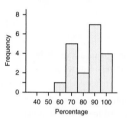

Histogram of Evening Average

*Step 1:* $H_a$: $\mu_{girls} \neq \mu_{boys}$ where $\mu$ is the population mean number of TV viewing hours. *Step 2:* Two-sample *t*-test, random samples, assume the sample sizes of 32 girls and 22 boys is large enough that slight non-Normality is not a problem. *Step 3:* $t = -0.38$ or 0.38, p-value $= 0.706$. *Step 4:* You cannot reject the null hypothesis. There is not enough evidence to conclude that boys and girls differ in the typical hours of TV watched.

**9.77 a.** Choose A, the two-sample t-test; the data are not paired. *Step 1:* $H_a$: $\mu_{men} \neq \mu_{women}$ where $\mu$ is the population mean number of work hours. *Step 2:* Two sample *t*-test, random, $n = 50 > 25$ *Step 3:* $t = 4.09$, p-value $< 0.001$. *Step 4:* Reject $H_0$. The mean number of work hours for men and women are significantly different. 95% CI for the mean difference in work hours, $\mu_{men} - \mu_{women}$: (6.18, 17.86). Because it does not capture 0, you can reject the hypothesis of no difference. Because the interval contains only positive values, we know that the mean number of work hours is greater for men than for women.

**9.79** Choose A because it is testing $\mu_1 < \mu_2$, which is what we want to test, assuming that 1 is the experimental group. *Step 1:* $H_a$: $\mu_{drug\ tested} < \mu_{not\ drug\ tested}$ where $\mu$ is the population mean arrest rate. *Step 2:* Two-sample *t*-test, assume conditions hold (given) but not random sample, random assignment. *Step 3:* $t = 2.03$ or $-2.03$, p-value $= 0.022$. *Step 4:* Reject $H_0$. The drug testing lowered the rate of arrest significantly but do not generalize.

**9.81** The table shows the results. The average of $s^2$ in the table is 2.8889 (or about 2.89), and if you take the square root, you get about 1.6997 (or about 1.70), which is the value for sigma ($\sigma$) given in the TI-83/84 output shown in the exercise. This demonstrates that $s^2$ is an unbiased estimator of $\sigma^2$, sigma squared.

| Sample | $s$ | $s^2$ |
|---|---|---|
| 1, 1 | 0 | 0 |
| 1, 2 | 0.7071 | 0.5 |
| 1, 5 | 2.8284 | 8.0 |
| 2, 1 | 0.7071 | 0.5 |
| 2, 2 | 0 | 0 |
| 2, 5 | 2.1213 | 4.5 |
| 5, 1 | 2.8284 | 8.0 |
| 5, 2 | 2.1213 | 4.5 |
| 5, 5 | 0 | 0 |
| | | Sum 26.0 |

$$26/9 = 2.8889$$

**9.83** Answers will vary.

## CHAPTER 10

Answers may vary slightly due to type of technology or rounding.

### Section 10.1

**10.1 a.** Proportions are used for categorical data. **b.** Chi-square tests are used for categorical data.

**10.3**

| | Boy | Girl |
|---|---|---|
| Violent | 10 | 11 |
| Nonviolent | 19 | 4 |

The table may have a different orientation. Both *Gender* and *Type of Crime* are categorical variables, as you can see from the raw data.

**10.5** *Mean GPA*: numerical and continuous. *Field of Study*: categorical.

**10.7 a.**

| | High TV Violence | Low TV Violence | Total |
|---|---|---|---|
| Yes, Physical Abuse | 13 | 27 | 40 |
| No Physical Abuse | 18 | 95 | 113 |
| Total | 31 | 122 | 153 |

**b.** $40/153 = 26.1\%$ **c.** $0.261438(31) = 8.10$
**d.** Expected values are shown in the table.

| | High TV Violence | Low TV Violence | Total |
|---|---|---|---|
| Yes, Physical Abuse | 8.10 | 31.90 | 40 |
| No Physical Abuse | 22.90 | 90.10 | 113 |
| Total | 31 | 122 | 153 |

**10.9**

$$X^2 = \frac{(13 - 8.10)^2}{8.1} + \frac{(27 - 31.90)^2}{31.9} + \frac{(18 - 22.90)^2}{22.9} + \frac{(95 - 90.10)^2}{90.1}$$

$$= 2.964 + 0.753 + 1.048 + 0.266 = 5.03$$

From technology to avoid rounding: Chi-square = 5.02

### Section 10.2

**10.11** Independence: one sample.

**10.13** Homogeneity: random assignment (to four groups).

**10.15** The data are of the entire population (not a sample), and therefore there is no need for inference. The data are given as rates (percentages), not frequencies (counts), and there is not enough information for us to convert these percentages to counts.

**10.17**

*1:* $H_a$: The variables *Relationship Status* and *Obesity* are not independent (are associated).

*2:* Chi-square test of independence (given). The smallest expected value is 108.54, which is much more than 5.

*3:* $X^2 = 30.83$, p-value $< 0.001$.

*4:* Reject $H_0$. There is a connection between obesity and marital status; they are not independent. However, we should not generalize.
Causality? No, it is an observational study.
Percentage Obese: Dating: $81/440 = 18.4\%$ Cohabiting: $103/429 = 24.0\%$
Married: $147/424 = 34.7\%$

**10.19** *Step 1:* $H_a$: For men, watching violent TV is not independent of abusiveness. *Step 2:* Chi-square test of independence (one sample), expected values all more than 5, sample not random. *Step 3:* Chi-square = 5.02, p-value = 0.025. *Step 4:* Reject $H_0$. High TV violence as a child is associated with abusiveness as an adult in men, but don't generalize to all males and don't conclude causality.

**10.21 a.** Independence: one sample with two variables **b.** *Step 1:* $H_a$: Gender and happiness of marriage are associated (not independent). *Step 2:* Chi-square test of independence (one sample), random sample, minimum expected count is $14.6 > 5$. *Step 3:* Chi-square = 1.22, p-value = 0.545. *Step 4:* You cannot reject $H_0$. Gender and happiness have not been shown to be associated. **c.** The rate of happiness in marriage has not been found to be significantly different for men and women.

**10.23 a.** HS Grad rate for no preschool: 29/64, or 45.3%. The preschool kids had a higher graduation rate. **b.** *Step 1:* $H_a$: Graduation and preschool are not independent (they are associated). *Step 2:* Chi-square test of homogeneity, random assignment, not a random sample, the smallest expected value is $25.91 > 5$. *Step 3:* $X^2 = 4.67$, p-value = 0.031. *Step 4:* Reject $H_0$. Graduation and preschool are associated: causality, yes, generalization, no.

**10.25 a.** For preschool, 50% graduated, and for no preschool, $21/39 = 53.8\%$ graduated. It is surprising to see that the boys who did not go to preschool had a bit higher graduation rate. **b.** *Step 1:* $H_a$: For the boys, graduation and preschool are associated. *Step 2:* Chi-square test for homogeneity, random assignment, not a random sample, the smallest expected value $15.32 > 5$. *Step 3:* $X^2 = 0.10$, p-value = 0.747. *Step 4:* Do not reject $H_0$. For the boys, there is no evidence that attending preschool is associated with graduating from high school. c. The results do not generalize to other groups of boys and girls, but what evidence we have suggests that although preschool might be effective for girls, it may not be for boys, at least with regard to graduation from high school.

**10.27 a.** Surgery: $22/29 = 75.9\%$ free from diabetes. No surgery: $4/26 = 15.4\%$ free from diabetes. The surgery patients were much more likely to be free from diabetes. **b.** Although there is an observed value that is less than 5, there is no expected value less than 5, so you can use chi-square.

**c.** *Step 1:* $H_a$: Surgery and recovery from diabetes are not independent. *Step 2:* Chi-square test of homogeneity, random assignment, not a random sample, there are no expected values less than 5. *Step 3:* $X^2 = 20.12$, p-value $< 0.001$. *Step 4:* Reject $H_0$. The surgery significantly affects the chance of recovery from diabetes but don't generalize.

**10.29 a.** With no confederate, 6/18 (33.3%) followed the directions and took the stairs. With a compliant confederate, 16/18 (88.9%) followed directions. With a noncompliant confederate, 5/18 (27.8%) followed directions. Thus the subject was influenced to tend to do the same thing as the confederate. **b.** A p-value can never be larger than 1. The p-value is about 2.7 times 10 to the negative fourth power or 0.00027, which is less than 0.001. **c.** *Step 1:* $H_a$: Treatment and compliance are not independent. *Step 2:* Chi-square test of homogeneity, random assignment, not a random sample, expected values are all 9 > 5. *Step 3:* $X^2 = 16.44$, p-value < 0.001. *Step 4:* Reject $H_0$. There is a significant effect: causality, yes; generalization, no.

**10.31 a.** LD: 4/50 = 8% tumor; LL 15/50 = 30% tumor. Thus there is a higher rate of tumors in the mice that were exposed to light 24 hours a day.

**b.**

|  | LD | LL |
|---|---|---|
| Mice with Tumor(s) | 4 | 15 |
| Mice with No Tumors | 46 | 35 |

**c.** *Step 1:* $H_a$: The rate of tumors is associated with the lighting conditions. *Step 2:* Chi-square test for homogeneity, random assignment, none of the *expected values* is less than 5. *Step 3:* $X^2 = 7.86$, p-value = 0.005. *Step 4:* Reject $H_0$. Light affects tumor development in mice. **d.** Some shift workers are exposed to light at night and also during the day, because they may sleep where there is light. Perhaps they will be more likely to develop tumors.

### Section 10.3

**10.33 a.** No, we cannot generalize, because this was not a random sample. **b.** Yes, we can infer causality because of random assignment.

**10.35 a.** You can generalize to other people admitted to this hospital who would have been assigned a double room because of the random sampling from that group. **b.** Yes, you can infer causality because of the random assignment.

**10.37** Because a random sample was not used (or we do not know whether one was used), we cannot generalize to the larger public. Because patients were randomly assigned to treatment groups (full-dose or reduced dose), we can conclude cause and effect. However, it might appear that both full-dose and reduced-dose have the same effect only because researchers did not have a large enough sample to detect a difference. We *can* say that among persons 60 years of age or older, the intradermal injection causes a less vigorous antibody response.

**10.39 a.** It does not matter which treatment is chosen; the treatments are not significantly different. **b.** Report one interval: The first interval given, (−6.8% to 6.7%) represents the difference in survival rates for patients receiving the two treatments. The fact that the interval includes 0 means we cannot rule out the possibility that the survival rate is the same for both treatments. The other interval (−0.5, 7.90) also does not show a significant difference because it captures 0.

**10.41 a.** Take a nonrandom sample of students and randomly assign some to the reception and some to attend a "control group" meeting where they do something else (such as learn the history of the college). **b.** Take a random sample of students and offer them the choice of attending the reception or attending a "control group" meeting where they do something else (such as learn the history of the college). **c.** Take a random sample of students. Then randomly assign some of the students in this sample to the reception and some to the "control group" meeting.

**10.43**

1. They want to know whether low-dose vaccine is as effective as the regular dose.

2. It appears to be effective in people aged 18 to 60 but not for those above 60.

3. You cannot tell whether random assignment was used, because it was not mentioned in this excerpt.

4. Because we do not know the methods, we do not know whether the conclusion was appropriate.

5. Because there was no random sampling and we cannot tell whether there is random assignment, we cannot generalize beyond these subjects.

6. No other studies are mentioned, so this might be the first study of its kind.

**10.45**

1. Which is the better procedure for aneurysms, open repair or endovascular repair?

2. There is no difference between them in terms of the outcomes observed.

3. They used a randomized design.

4. They used random assignment and found no difference, so the conclusion was appropriate. However, we cannot eliminate the possibility that the effect is subtle and the sample size wasn't large enough to detect it.

5. Because there was not a random sample, we cannot generalize. But the random assignment would have allowed us to make cause-and-effect conclusions, had we rejected the null hypothesis.

6. We cannot tell whether previous research was done or not from the excerpt.

### Chapter Review Exercises

**10.47 a.** 31/65, or 47.7%, of those in the control group were arrested, and 8/58, or 13.8%, of those who attended preschool were arrested. So there was a lower rate of arrest for those who went to preschool.

**b.**

|  | Preschool | No Preschool |
|---|---|---|
| Arrest | 8 | 31 |
| No Arrest | 50 | 34 |

*Step 1:* $H_a$: The treatment and arrest rate are associated. *Step 2:* Chi-square for homogeneity, random assignment, not a random sample, all the expected values are more than 5. *Step 3:* Chi-square = 16.27, p-value = 0.000055 (or < 0.001). *Step 4:* We can reject the hypothesis of no association at the 0.05 level. Don't generalize. We conclude that preschool attendance affects the arrest rate. **c.** Two-proportion z-test. *Step 1:* $p_{pre} < p_{nopre}$ (p is the rate of arrest). *Step 2:* Two-proportion z-test, all expected values are large. *Step 3:* $z = 4.03$, p-value = 0.000028 (or < 0.001). *Step 4:* Reject the null hypothesis; preschool lowers the rate of arrest, but we cannot generalize. **d.** The z-test allows us to test the alternative hypothesis that preschool attendance *lowers* the risk of later arrest. The chi-square test allows only for testing for some sort of association, but we can't specify whether it is a positive or a negative association. Note that the p-value for the one-tailed hypothesis with the z-test is half the p-value for the two-tailed hypothesis with the chi-square test.

**10.49 a.** 11.4% of those on the drug had a heart event, and 10.0% of those on the placebo had a heart event. So those on the drug were more likely to have a heart event. **b.** *Step 1:* $H_a$: There is an association between use of the drug and heart events. *Step 2:* Chi-square test of homogeneity, random assignment, not a random sample, all expected values are large. *Step 3:* Chi-square = 4.96, p-value = 0.026. *Step 4:* Reject $H_0$. The drug affects the rate of heart events. **c.** Yes, because of the random assignment to groups and the statistically significant result.

**10.51 a.** In the experimental group the percentage arrested was 43/53, or 81.1%. In the control group the percentage was 36/53, or 67.9%. It seems that Scared Straight had a bad influence, because the control group was less likely to be arrested, but we don't yet know whether this difference was significant. **b.** *Step 1:* $H_a$: Treatment and arrests are not independent. *Step 2:* Chi-square test of homogeneity, random assignment, not random samples, no expected values less than 5. *Step 3:* Chi-square = 2.44, p-value = 0.119. *Step 4:* We cannot reject the hypothesis of independence at the 0.05 level. **c.** *Step 1:* $H_a$: $p_1 < p_2$ where $p_1$ is the population proportion of arrest for Scared Straight and $p_2$ is the population proportion of arrest for no treatment. *Step 2:* Two-proportion z-test, all expected values are large enough, random assignment, not a random sample. *Step 3:* $z = 1.56$, p-value = 0.941 (use output A). *Step 4:* Don't reject the null hypothesis; we cannot conclude that the Scared Straight program lowered the arrest rate. **d.** To get the p-value for the one-tailed alternative, divide 0.119 by 2 and subtract that quotient from 1.000. The value of the chi-square test statistic is the square of the z-statistic.

**10.53 a.** The death rate before the vaccine was 18.1 deaths per 100,000 children. After the vaccine, the death rate fell to 11.8 deaths per 100,000 children. The difference is 6.3 fewer deaths per 100,000 children after the vaccine was introduced. The small p-value (less than 0.001) means we can reject the null hypothesis that the death rate was unchanged and conclude that the death rate decreased. **b.** Although there are many indications that the vaccine is successful, this was not a randomized study. We cannot rule out the possibility that a confounding variable, not the vaccine, caused the decrease in death rates. (For example, because the comparison was done using different years, there might have been a difference in weather that contributed to the difference in disease rates.)

# Credits

# Index